Handbook of Research on Globalized Agricultural Trade and New Challenges for Food Security

Vasilii Erokhin
Harbin Engineering University, China

Tianming Gao
Harbin Engineering University, China

A volume in the Advances in Environmental Engineering and Green Technologies (AEEGT) Book Series

Published in the United States of America by
IGI Global
Engineering Science Reference (an imprint of IGI Global)
701 E. Chocolate Avenue
Hershey PA, USA 17033
Tel: 717-533-8845
Fax: 717-533-8661
E-mail: cust@igi-global.com
Web site: http://www.igi-global.com

Copyright © 2020 by IGI Global. All rights reserved. No part of this publication may be reproduced, stored or distributed in any form or by any means, electronic or mechanical, including photocopying, without written permission from the publisher. Product or company names used in this set are for identification purposes only. Inclusion of the names of the products or companies does not indicate a claim of ownership by IGI Global of the trademark or registered trademark.

Library of Congress Cataloging-in-Publication Data

Names: Erokhin, Vasily, 1980- editor. | Gao, Tianming, 1974- editor.
Title: Handbook of research on globalized agricultural trade and new
   challenges for food security / Vasilii Erokhin, Tianming Gao, editors.
Description: Hershey, PA : Engineering Science Reference, [2020] | Includes
   bibliographical references and index. | Summary: "This book addresses
   the contemporary issues of agricultural trade, including major
   commodities and food products traded between major countries, the import
   and export of trade, and trends in trade"-- Provided by publisher.
Identifiers: LCCN 2019023696 (print) | LCCN 2019023697 (ebook) | ISBN
   9781799810421 (hardcover) | ISBN 9781799810438 (ebook)
Subjects: LCSH: Produce trade. | Food security. | Agriculture--Economic
   aspects.
Classification: LCC HD9000.5 .H3625 2020 (print) | LCC HD9000.5 (ebook) |
   DDC 382/.41--dc23
LC record available at https://lccn.loc.gov/2019023696
LC ebook record available at https://lccn.loc.gov/2019023697

This book is published in the IGI Global book series Advances in Environmental Engineering and Green Technologies (AEEGT) (ISSN: 2326-9162; eISSN: 2326-9170)

British Cataloguing in Publication Data
A Cataloguing in Publication record for this book is available from the British Library.

All work contributed to this book is new, previously-unpublished material. The views expressed in this book are those of the authors, but not necessarily of the publisher.

For electronic access to this publication, please contact: eresources@igi-global.com.

Advances in Environmental Engineering and Green Technologies (AEEGT) Book Series

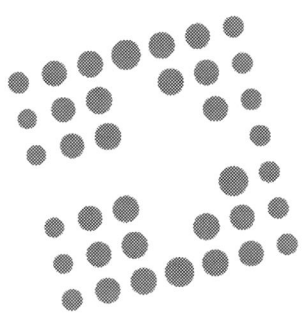

Sang-Bing Tsai
University of Electronic Science and Technology of China Zhongshan Institute, China
Ming-Lang Tseng
Lunghwa University of Science and Technology, Taiwan
Yuchi Wang
University of Electronic Science and Technology of China Zhongshan Institute, China

ISSN:2326-9162
EISSN:2326-9170

Mission

Growing awareness and an increased focus on environmental issues such as climate change, energy use, and loss of non-renewable resources have brought about a greater need for research that provides potential solutions to these problems. Research in environmental science and engineering continues to play a vital role in uncovering new opportunities for a "green" future.

The **Advances in Environmental Engineering and Green Technologies (AEEGT)** book series is a mouthpiece for research in all aspects of environmental science, earth science, and green initiatives. This series supports the ongoing research in this field through publishing books that discuss topics within environmental engineering or that deal with the interdisciplinary field of green technologies.

Coverage

- Renewable Energy
- Industrial Waste Management and Minimization
- Air Quality
- Biofilters and Biofiltration
- Policies Involving Green Technologies and Environmental Engineering
- Sustainable Communities
- Pollution Management
- Contaminated Site Remediation
- Radioactive Waste Treatment
- Electric Vehicles

IGI Global is currently accepting manuscripts for publication within this series. To submit a proposal for a volume in this series, please contact our Acquisition Editors at Acquisitions@igi-global.com or visit: http://www.igi-global.com/publish/.

The Advances in Environmental Engineering and Green Technologies (AEEGT) Book Series (ISSN 2326-9162) is published by IGI Global, 701 E. Chocolate Avenue, Hershey, PA 17033-1240, USA, www.igi-global.com. This series is composed of titles available for purchase individually; each title is edited to be contextually exclusive from any other title within the series. For pricing and ordering information please visit http://www.igi-global.com/book-series/advances-environmental-engineering-green-technologies/73679. Postmaster: Send all address changes to above address. Copyright © 2020 IGI Global. All rights, including translation in other languages reserved by the publisher. No part of this series may be reproduced or used in any form or by any means – graphics, electronic, or mechanical, including photocopying, recording, taping, or information and retrieval systems – without written permission from the publisher, except for non commercial, educational use, including classroom teaching purposes. The views expressed in this series are those of the authors, but not necessarily of IGI Global.

Titles in this Series

For a list of additional titles in this series, please visit: www.igi-global.com/book-series

Modern Techniques for Agricultural Disease Management and Crop Yield Prediction
N. Pradeep (Bapuji Institute of Engineering and Technology, India) Sandeep Kautish (Lord Buddha Education Foundation, Nepal & Asia Pacific University, Malaysia) C.R. Nirmala (Bapuji Institute of Engineering and Technology, India) Vishal Goyal (Punjabi University, Patiala, India) and Sonia Abdellatif (Seoul National University, South Korea)
Engineering Science Reference • copyright 2020 • 291pp • H/C (ISBN: 9781522596325) • US $195.00 (our price)

Novel Advancements in Electrical Power Planning and Performance
Smita Shandilya (Sagar Institute of Research, Technology and Science, India) Shishir Kumar Shandilya (Vellore Institute of Technology, India) Tripta Thakur (Maulana Azad National Institute of Technology, India) and Atulya K. Nagar (Liverpool Hope University, UK)
Engineering Science Reference • copyright 2020 • 388pp • H/C (ISBN: 9781522585510) • US $225.00 (our price)

Global Food Politics and Approaches to Sustainable Consumption Emerging Research and Opportunities
Luke Amadi (University of Port Harcourt, Nigeria) and Fidelis Allen (University of Port Harcourt, Nigeria)
Engineering Science Reference • copyright 2020 • 242pp • H/C (ISBN: 9781799801252) • US $185.00 (our price)

Fuzzy Expert Systems and Applications in Agricultural Diagnosis
A.V. Senthil Kumar (Hindusthan College of Arts and Science, India) and M. Kalpana (Tamil Nadu Agricultural University, India)
Engineering Science Reference • copyright 2020 • 335pp • H/C (ISBN: 9781522591757) • US $245.00 (our price)

Handbook of Research on Environmental and Human Health Impacts of Plastic Pollution
Khursheed Ahmad Wani (Government Degree College Bijbehara, India) Lutfah Ariana (Indonesian Institute of Sciences, Indonesia) and S.M. Zuber (Government Degree College Bijbehara, India)
Engineering Science Reference • copyright 2020 • 600pp • H/C (ISBN: 9781522594529) • US $285.00 (our price)

Decision Support Methods for Assessing Flood Risk and Vulnerability
Ahmed Karmaoui (Southern Center for Culture and Sciences (SCCS), Morocco)
Engineering Science Reference • copyright 2020 • 332pp • H/C (ISBN: 9781522597711) • US $215.00 (our price)

Handbook of Research on Energy-Saving Technologies for Environmentally-Friendly Agricultural Development
Valeriy Kharchenko (Federal Scientific Agroengineering Center VIM, Russia) and Pandian Vasant (Universiti Teknologi Petronas, Malaysia)
Engineering Science Reference • copyright 2020 • 554pp • H/C (ISBN: 9781522594208) • US $295.00 (our price)

701 East Chocolate Avenue, Hershey, PA 17033, USA
Tel: 717-533-8845 x100 • Fax: 717-533-8661
E-Mail: cust@igi-global.com • www.igi-global.com

Editorial Advisory Board

Ilan Alon, *University of Agder, Norway*
Jean-Vasile Andrei, *Petroleum-Gas University of Ploiesti, Romania*
Alexander Arskiy, *Russian Academy of Personnel Support for the Agroindustrial Complex, Russia*
Jędrzej Gorski, *City University of Hong Kong, Hong Kong SAR, China*
Julien Chaisse, *Chinese University of Hong Kong, Hong Kong SAR, China*
Anna Ivolga, *Stavropol State Agrarian University, Russia*
Alexander Voronenko, *Khabarovsk State University of Economics and Law, Russia*
Kathleen Zabelin, *Key West University, USA*

List of Reviewers

Sinmi Abosede, *Pan Atlantic University, Nigeria*
Albina Assenova, *Harbin Engineering University, China*
Sukalpa Chakrabarti, *Symbiosis International (Deemed University), India*
Drago Cvijanović, *University of Kragujevac, Republic of Serbia*
Teodoro Gallucci, *University of Bari Aldo Moro, Italy*
Ishita Ghosh, *Symbiosis International (Deemed University), India*
Ishita Ghoshal, *Fergusson College (Autonomous), India*
Mykhailo Guz, *National University of Life and Environmental Sciences of Ukraine, Ukraine*
Zhang Hongzhou, *Nanyang Technological University, Singapore*
Svetlana Ignjatijević, *University Business Academy in Novi Sad, Republic of Serbia*
Nina Kazydub, *Omsk State Agrarian University, Russia*
Mikail Khudzhatov, *Peoples' Friendship University of Russia, Russia*
Nikolay Kotlyarov, *Financial University under the Government of the Russian Federation, Russia*
Oleksander Labenko, *National University of Life and Environmental Sciences of Ukraine, Ukraine*
Henrietta Nagy, *Szent Istvan University, Hungary*
György Iván Neszmélyi, *Budapest Business School – University of Applied Sciences, Hungary*
Muhamad Rusliyadi, *Polytechnic of Agricultural Development Yogyakarta-Magelang, Indonesia*
Olga Storozhenko, *Bauman Moscow State Technical University, Russia*
Natalia Staurskaya, *Omsk State Technical University, Russia*
Ismail Taaricht, *Cadi Ayyad University, Morocco*
Bistra Vassileva, *University of Economics-Varna, Bulgaria*
Anna Veber, *Omsk State Agrarian University, Russia*
Oleg Zabelin, *Moscow University of Finance and Law, Russia*

List of Contributors

Abdulkadr, Ahmed Abduletif / *Szent Istvan University, Hungary* 425
Abosede, Sinmi / *Pan Atlantic University, Nigeria* 384
Aleshkov, Aleksey / *Khabarovsk State University of Economics and Law, Russia* 337
Amicarelli, Vera / *University of Bari Aldo Moro, Italy* 147
Arskiy, Alexander / *Russian Academy of Personnel Support for the Agroindustrial Complex, Russia* 105
Bogdanov, Dmitry / *Kutafin Moscow State Law University, Russia* 170
Bogdanova, Svetlana / *Bauman Moscow State Technical University, Russia* 170
Bux, Christian / *University of Bari Aldo Moro, Italy* 147
Chakrabarti, Sukalpa / *Symbiosis International University (Deemed), India* 400
Chaunina, Elena / *Omsk State Agrarian University, Russia* 242
Cvijanović, Drago / *University of Kragujevac, Serbia* 443
Elfimova, Yulia / *Stavropol State Agrarian University, Russia* 129
Erokhin, Vasilii / *Harbin Engineering University, China* 319
Esmurzaeva, Zhanbota / *Omsk State Agrarian University, Russia* 252
Fedorova, Maria / *Omsk State Technical University, Russia* 465
Gallucci, Teodoro / *University of Bari Aldo Moro, Italy* 147
Ghosh, Ishita / *Symbiosis International University (Deemed), India* 400
Ghoshal, Ishita / *Fergusson College (Autonomous), India* 400
Gorodilov, Alexey / *Tomsk Polytechnic University, Russia* 189
Greizik, Sergei / *Khabarovsk State University of Economics and Law, Russia* 361
Ignjatijević, Svetlana / *University Business Academy in Novi Sad, Serbia* 443
Ivolga, Anna / *Stavropol State Agrarian University, Russia* 88, 129
Jamil, Azaharaini Bin Hj. Mohd. / *University Brunei Darussalam, Brunei* 481
Khudzhatov, Mikail / *Peoples' Friendship University of Russia (RUDN University), Russia* 105
Korsheva, Inna / *Omsk State Agrarian University, Russia* 242
Lagioia, Giovanni / *University of Bari Aldo Moro, Italy* 147
Leonova, Svetlana / *Bashkir State Agrarian University, Russia* 252
Lipski, Stanislav / *State University of Land Use Planning, Russia* 215
Meleshkina, Elena / *All-Russian Research Institute of Grain and Grain Products, Russia* 252
Nagy, Henrietta / *Szent Istvan University, Hungary* 425
Neszmélyi, György Iván / *Budapest Business School, University of Applied Sciences, Hungary* 425
Nikiforova, Tamara / *Orenburg State University, Russia* 252
Pasko, Olga / *Tomsk Polytechnic University, Russia* 43, 189
Perkel, Roman / *Peter the Great Saint Petersburg Polytechnic University, Russia* 277

Pertsevyi, Fedor / *Sumy National Agrarian University, Ukraine* 296
Rusliyadi, Muhamad / *Polytechnic of Agricultural Development Yogyakarta-Magelang,
   Indonesia* 481
Ryazantsev, Ivan / *Stavropol State Agrarian University, Russia* 129
Ryumkin, Sergey / *Novosibirsk State Agrarian University, Russia* 63
Ryumkina, Inga / *Irkutsk State Agrarian University, Russia* 63
Sapunov, Valentin / *Saint Petersburg State Agrarian University, Russia* 1
Simakova, Inna / *Saratov State Vavilov Agrarian University, Russia* 277, 296
Staurskaya, Natalia / *Omsk State Technical University, Russia* 43, 189
Storozhenko, Olga / *Bauman Moscow State Technical University, Russia* 215
Strizhevskaya, Victoria / *Saratov State Vavilov Agrarian University, Russia* 296
Taaricht, Ismail / *Cadi Ayyad University, Morocco* 465
Tomilov, Mikhail / *Economic Research Institute, Far Eastern Branch, Russian Academy of
   Science, Russia* 361
Trukhachev, Alexander / *Ministry of Tourism and Recreation of Stavropol Region, Russia* 88
Vassileva, Bistra / *University of Economics-Varna, Bulgaria* 13
Veber, Anna / *Omsk State Agrarian University, Russia* 252
Voronenko, Alexander / *Khabarovsk State University of Economics and Law, Russia* 361
Vorotnikov, Igor / *Saratov State Vavilov Agrarian University, Russia* 296
Zakharchenko, Alexander / *Institute of the Problems of Northern Development, Russian
   Academy of Sciences, Russia* 189
Zemliak, Kirill / *Khabarovsk State University of Economics and Law, Russia* 337
Zhebo, Anna / *Khabarovsk State University of Economics and Law, Russia* 337

Table of Contents

Foreword xxi

Preface xxii

Acknowledgment xxx

Section 1
Trade-Related Aspects of Food Security

Chapter 1
Real Need of the World in Food: A Future of Agricultural Production 1
Valentin Sapunov, Saint Petersburg State Agrarian University, Russia

Chapter 2
Current Tendencies in Agricultural Trade and the Roles of Major Actors 13
Bistra Vassileva, University of Economics-Varna, Bulgaria

Chapter 3
Food Security-Related Issues and Solutions 43
Olga Pasko, Tomsk Polytechnic University, Russia
Natalia Staurskaya, Omsk State Technical University, Russia

Chapter 4
World Market of Organic Food Products 63
Inga Ryumkina, Irkutsk State Agrarian University, Russia
Sergey Ryumkin, Novosibirsk State Agrarian University, Russia

Section 2
Establishing Food Security: Policies, Practices, and Innovations

Chapter 5
Innovative Approaches to Regulation of Agricultural Production and Trade: Protectionism-Related Effects on Food Security 88
Anna Ivolga, Stavropol State Agrarian University, Russia
Alexander Trukhachev, Ministry of Tourism and Recreation of Stavropol Region, Russia

Chapter 6
Technical Equipment of Agricultural Production: The Effects for Food Security 105
   Mikail Khudzhatov, Peoples' Friendship University of Russia (RUDN University), Russia
   Alexander Arskiy, Russian Academy of Personnel Support for the Agroindustrial Complex, Russia

Chapter 7
Food Security Through Rational Land Management: Innovative Aspects and Practices 129
   Yulia Elfimova, Stavropol State Agrarian University, Russia
   Anna Ivolga, Stavropol State Agrarian University, Russia
   Ivan Ryazantsev, Stavropol State Agrarian University, Russia

Chapter 8
Food Waste Reduction Towards Food Sector Sustainability 147
   Giovanni Lagioia, University of Bari Aldo Moro, Italy
   Vera Amicarelli, University of Bari Aldo Moro, Italy
   Teodoro Gallucci, University of Bari Aldo Moro, Italy
   Christian Bux, University of Bari Aldo Moro, Italy

Chapter 9
Comparative Analysis of Food Security and Tort Liability Doctrines in Russia and China: Harm Caused by Poor-Quality Agricultural Products 170
   Dmitry Bogdanov, Kutafin Moscow State Law University, Russia
   Svetlana Bogdanova, Bauman Moscow State Technical University, Russia

Chapter 10
Agrarian Reforms of the 20th Century in Russia: Impacts on Agricultural Sector and Food Security 189
   Olga Pasko, Tomsk Polytechnic University, Russia
   Natalia Staurskaya, Omsk State Technical University, Russia
   Alexey Gorodilov, Tomsk Polytechnic University, Russia
   Alexander Zakharchenko, Institute of the Problems of Northern Development, Russian Academy of Sciences, Russia

Chapter 11
Economic Aspects of Agricultural Public Policy as a Key Factor of Establishing Food Security: Retrospectives of Post-Soviet Russia 215
   Stanislav Lipski, State University of Land Use Planning, Russia
   Olga Storozhenko, Bauman Moscow State Technical University, Russia

Section 3
Agricultural Trade and Quality of Nutrition

Chapter 12
Agricultural Trade and Quality of Nutrition: Impacts on Undernourishment and Dietary Diversity 242
   Elena Chaunina, Omsk State Agrarian University, Russia
   Inna Korsheva, Omsk State Agrarian University, Russia

Chapter 13
Agricultural Trade and Undernourishment, Nutrition, and Dietary Diversity: The Use of Elite Selection Cultivars of Legumes 252
    Anna Veber, Omsk State Agrarian University, Russia
    Svetlana Leonova, Bashkir State Agrarian University, Russia
    Elena Meleshkina, All-Russian Research Institute of Grain and Grain Products, Russia
    Zhanbota Esmurzaeva, Omsk State Agrarian University, Russia
    Tamara Nikiforova, Orenburg State University, Russia

Chapter 14
Imports and Use of Palm Oil as a Way to Increase Safety of Food Fats 277
    Inna Simakova, Saratov State Vavilov Agrarian University, Russia
    Roman Perkel, Peter the Great Saint Petersburg Polytechnic University, Russia

Chapter 15
Resource-Saving Technology of Dehydration of Fruit and Vegetable Raw Materials: Scientific Rationale and Cost Efficiency 296
    Inna Simakova, Saratov State Vavilov Agrarian University, Russia
    Victoria Strizhevskaya, Saratov State Vavilov Agrarian University, Russia
    Igor Vorotnikov, Saratov State Vavilov Agrarian University, Russia
    Fedor Pertsevyi, Sumy National Agrarian University, Ukraine

Section 4
Regional Aspects of Food Security and Trade in Agricultural Products

Chapter 16
Emerging Trade-Related Threats to Food Security: Evidence From China 319
    Vasilii Erokhin, Harbin Engineering University, China

Chapter 17
Food Security and Self-Sufficiency as a Basis for National Security and Sovereignty: Evidence From Russia 337
    Kirill Zemliak, Khabarovsk State University of Economics and Law, Russia
    Anna Zhebo, Khabarovsk State University of Economics and Law, Russia
    Aleksey Aleshkov, Khabarovsk State University of Economics and Law, Russia

Chapter 18
Integrative Associations and Food Security: Case of China-Russia Interregional Cooperation 361
    Alexander Voronenko, Khabarovsk State University of Economics and Law, Russia
    Sergei Greizik, Khabarovsk State University of Economics and Law, Russia
    Mikhail Tomilov, Economic Research Institute, Far Eastern Branch, Russian Academy of
       Science, Russia

Chapter 19
Water, Food Security, and Trade in Sub-Saharan Africa 384
    Sinmi Abosede, Pan Atlantic University, Nigeria

Chapter 20
Prospects, Challenges, and Policy Directions for Food Security in India-Africa Agricultural
Trade 400
    Ishita Ghosh, Symbiosis International University (Deemed), India
    Sukalpa Chakrabarti, Symbiosis International University (Deemed), India
    Ishita Ghoshal, Fergusson College (Autonomous), India

Chapter 21
The Role of Agricultural Production and Trade Integration in Sustainable Rural Development:
Evidence From Ethiopia 425
    Henrietta Nagy, Szent Istvan University, Hungary
    György Iván Neszmélyi, Budapest Business School, University of Applied Sciences, Hungary
    Ahmed Abduletif Abdulkadr, Szent Istvan University, Hungary

Chapter 22
International Competitiveness of Niche Agricultural Products: Case of Honey Production in
Serbia 443
    Drago Cvijanović, University of Kragujevac, Serbia
    Svetlana Ignjatijević, University Business Academy in Novi Sad, Serbia

Chapter 23
Agricultural Cooperatives for Sustainable Development of Rural Territories and Food Security:
Morocco's Experience 465
    Maria Fedorova, Omsk State Technical University, Russia
    Ismail Taaricht, Cadi Ayyad University, Morocco

Chapter 24
Participatory Poverty Assessment Effort in Food Security and Extension Policy: Evidence From
Indonesia 481
    Muhamad Rusliyadi, Polytechnic of Agricultural Development Yogyakarta-Magelang,
        Indonesia
    Azaharaini Bin Hj. Mohd. Jamil, University Brunei Darussalam, Brunei

Compilation of References 500

About the Contributors 559

Index 571

Detailed Table of Contents

Foreword xxi

Preface xxii

Acknowledgment xxx

Section 1
Trade-Related Aspects of Food Security

Chapter 1
Real Need of the World in Food: A Future of Agricultural Production 1
    Valentin Sapunov, Saint Petersburg State Agrarian University, Russia

Mankind has minimal areas of agricultural land that produces more food than required to feed the world's population. When allocating forces and assets within the framework of the global policy of investing in agriculture, it can be safely reduced. What is food policy in the 21st century? First of all, it is advisable to increase investments in the study of food opportunities, the development of technology for the collection and processing of aborigine animals and plants in particular territories with a further increase in investments in the methods of biological technology. It is advisable to increase the investments in industrial methods for obtaining food products from animals, plants, microorganisms, in the future – in the course of chemical industrial synthesis. Vernadsky predicted that in the future, mankind will switch to autotrophic nutrition, i.e. artificial synthesis of food from inorganic materials. Biotechnology will gradually reduce the volume of traditional agricultural production.

Chapter 2
Current Tendencies in Agricultural Trade and the Roles of Major Actors 13
    Bistra Vassileva, University of Economics-Varna, Bulgaria

Globalization was signified as a quintessence of the last decade of the century. Its effect on international trade is outstanding because of the interconnectedness of the international markets. The current wave of globalization is driven by major technological breakthroughs in transportation and communication. Major market actors are witnessing lower trade barriers and reduced transaction costs due to numerous technological innovations. The chapter starts with a literature review on global market tendencies and their implications on agricultural trade. The first section begins with a comparative analysis of the agricultural trade flows by globalization waves. The second section describes the major actors in agricultural trade and their global market positions. In the third section, network analysis is used to visualize the connections between the major actors. Their behavioral pattern is presented as well as the key interactions are revealed. The chapter ends with recommendations aimed at the future development of agricultural trade with a focus on global intervention policies.

Chapter 3
Food Security-Related Issues and Solutions... 43
    Olga Pasko, Tomsk Polytechnic University, Russia
    Natalia Staurskaya, Omsk State Technical University, Russia

The food problem has been and has remained relevant throughout the history of mankind. At the end of 20th and the beginning of the 21st century, in the lives of many nations and countries, there have been significant changes. Health status and level of education of the population, such as, for example, food security, is the priority in many countries since, in the absence of sufficient food reserves, there is an economic and political dependence of some countries on others. Having not yet received the required amount of food, the world is faced with the problem of ensuring security in its quality. Anthropogenic pollution of the environment complicates the problem with the quality of food and the exception of harmful chemicals in food. There is a problem of using environmentally friendly agrotechnical means, ensuring the production of high yields of environmentally safe products with a desirable reduction in their cost, and shortening the time required for their production.

Chapter 4
World Market of Organic Food Products... 63
    Inga Ryumkina, Irkutsk State Agrarian University, Russia
    Sergey Ryumkin, Novosibirsk State Agrarian University, Russia

In the past two decades, the role of international relations in various spheres has increased significantly. The world market for agricultural products is not an exception. Agricultural production is influenced by many factors, including climate, development strategies, and financing of agricultural research centers, among others. The factor of organic production should form both domestic and global markets of agricultural products and food since the health of people and the environment depends on the quality of food products. Therefore, the agrarian policy should primarily focus on the development of markets of organic food. In this chapter, the authors attempt to identify major actors in the world market of organic food products.

Section 2
Establishing Food Security: Policies, Practices, and Innovations

Chapter 5
Innovative Approaches to Regulation of Agricultural Production and Trade: Protectionism-
Related Effects on Food Security.. 88
    Anna Ivolga, Stavropol State Agrarian University, Russia
    Alexander Trukhachev, Ministry of Tourism and Recreation of Stavropol Region, Russia

The chapter studies contemporary innovative approaches to and practices of state support of agricultural production and trade in food and agricultural products. The authors attempt to discover how protectionist policies in the sphere of production and trade affect the level of food security in the conditions of expanding globalization. The chapter focuses on the investigation of advanced innovative practices of state support in the case of selected OECD countries. The authors reveal that the introduction of innovations into the system of state regulation is one of the key determinants of achieving food security in the conditions of the volatile market. Both the volume and priority directions of innovations in agricultural protectionism policies are discussed and evaluated.

Chapter 6
Technical Equipment of Agricultural Production: The Effects for Food Security 105
    Mikail Khudzhatov, Peoples' Friendship University of Russia (RUDN University), Russia
    Alexander Arskiy, Russian Academy of Personnel Support for the Agroindustrial Complex,
      Russia

The guarantee of a sufficient food supply is one of the challenges in both international and national economic security. Development of world agriculture is impossible without using advanced technologies. Their application in agriculture depends on the security of agricultural producers with highly effective agricultural machinery for which agricultural engineering serves. Agricultural engineering is an important element of the agro-industrial complex of any state providing it with necessary machinery and equipment. One of the important directions of the establishment of food security is the development of agricultural engineering. In this regard, the chapter analyses the current state of the world market of agricultural machinery; develops the methodology of assessment of the competitiveness of agricultural machinery in the domestic market; and elaborates the definition of effective methods of management of logistics costs at the operation of agricultural machinery.

Chapter 7
Food Security Through Rational Land Management: Innovative Aspects and Practices 129
    Yulia Elfimova, Stavropol State Agrarian University, Russia
    Anna Ivolga, Stavropol State Agrarian University, Russia
    Ivan Ryazantsev, Stavropol State Agrarian University, Russia

Food security is challenged by the growth of the world population, intensification of agricultural production, depletion of scarce agricultural and environmental resources, and the consequent introduction of innovations to farming processes, one of which is land management. In this chapter, the authors discuss the role of innovative aspects and practices of land management in the establishment of food security and the provision of rational, sound, and effective administration of scarce land resources. In the case of Russia and other countries, the authors justify the necessity of rational land management focused on the innovative way of development of agricultural production.

Chapter 8
Food Waste Reduction Towards Food Sector Sustainability 147
    Giovanni Lagioia, University of Bari Aldo Moro, Italy
    Vera Amicarelli, University of Bari Aldo Moro, Italy
    Teodoro Gallucci, University of Bari Aldo Moro, Italy
    Christian Bux, University of Bari Aldo Moro, Italy

FAO estimates on average more than 1.3 billion tons of food loss and waste (FLW) along the whole food supply chain (equivalent to one-third of total food production) of which more than 670 million tons in developed countries and approximately 630 million tons in developing ones, showing wide differences between countries. In particular, EU data estimates an amount of more than 85 million tons of FLW, equal to approximately 20% of total food production. This research presents two main goals. First, to review the magnitude of FLW at a global and European level and its environmental, social and economic implications. Second, use Material Flow Analysis (MFA) to support and improve FLW management and its application in an Italian potato industry case study. According to the case study presented, MFA has demonstrated the advantages of tracking input and output to prevent FLW and how they provide economic, social, and environmental opportunities.

Chapter 9
Comparative Analysis of Food Security and Tort Liability Doctrines in Russia and China: Harm Caused by Poor-Quality Agricultural Products... 170
    Dmitry Bogdanov, Kutafin Moscow State Law University, Russia
    Svetlana Bogdanova, Bauman Moscow State Technical University, Russia

Provision of the population with environmentally friendly and safe agricultural products is an important challenge in the developed states. This chapter analyzes the issues of food safety and quality. The indemnification caused by low-quality products stimulates producers to ensure the quality and safety of food resources. The institute of indemnification caused by low-quality agricultural products is analyzed in the chapter. Special attention is paid to the issues of consumer protection in the legislation of Russia and China.

Chapter 10
Agrarian Reforms of the 20th Century in Russia: Impacts on Agricultural Sector and Food Security 189
    Olga Pasko, Tomsk Polytechnic University, Russia
    Natalia Staurskaya, Omsk State Technical University, Russia
    Alexey Gorodilov, Tomsk Polytechnic University, Russia
    Alexander Zakharchenko, Institute of the Problems of Northern Development, Russian
        Academy of Sciences, Russia

Current political and economic reforms, as well as the development of market relations and private property rights, need a retrospect to the experience of the past. An ambitious reform implemented by Russian public entities in the early 20th century was a result of a compromise between the government, society, and individuals. The goals of the reforms offered by Pyotr Stolypin were similar to those of the contemporary ones. Stolypin's reforms aimed at the substitution of group type of land use by public property. The reforms were not evolutional but were motivated by the explosive political and social-economic situation. Another agrarian reform took place in the early 1990s in the Soviet bloc, including the USSR. It aimed at state land property and a centrally planned agrarian economy, the domination of big manufacturers like collective and communal farms, and state pricing control. Despite similar basic principles, the states chose different strategies for the implementation of agrarian reforms.

Chapter 11
Economic Aspects of Agricultural Public Policy as a Key Factor of Establishing Food Security: Retrospectives of Post-Soviet Russia 215
    Stanislav Lipski, State University of Land Use Planning, Russia
    Olga Storozhenko, Bauman Moscow State Technical University, Russia

In Russia, food security is ensured by sustainable development of domestic agriculture and related industries. Arable lands, the key agricultural resource in Russia, account for about 9% of the world's total. This study investigates changes in public policy related to agricultural lands in post-Soviet period, namely, arguments for land redistribution; privatization that covered over 60% of agricultural lands and resulted in appearance of land shares owned by about 12 million rural citizens barely understanding what to do with their land shares; post-privatization issues and problems concerned with the involvement of agricultural and other lands in economic activity; implementation of public economic policy measures aimed to resolve the above-mentioned issues (transfer of unclaimed land shares to municipalities); current transformation of ownership structure of agricultural lands; specifics of demarcation of un-privatized lands between federal, regional, and local authorities.

Section 3
Agricultural Trade and Quality of Nutrition

Chapter 12
Agricultural Trade and Quality of Nutrition: Impacts on Undernourishment and Dietary Diversity.. 242
    Elena Chaunina, Omsk State Agrarian University, Russia
    Inna Korsheva, Omsk State Agrarian University, Russia

Proper nutrition is not only a biological but also a social, economic, and political issue. Insufficient intake of essential elements may result in the occurrence of hidden hunger and metabolic disorders. Some regions of the world are characterized by a lack of certain nutrients in the environment which leads to their lack in plant and animal products. The most common problem is a deficiency of iodine and selenium. To solve this problem, the government takes various measures, such as direct inclusion of necessary additives in food products, as well as the modernization of technological process of crop and livestock production. In this chapter, the authors analyze the provision of the population in various countries and regions with limiting nutrients. The study specifically aims at exploring the issues of production and trade in fortified (modified) food products that can directly fill in the lack of essential elements in particular territories.

Chapter 13
Agricultural Trade and Undernourishment, Nutrition, and Dietary Diversity: The Use of Elite Selection Cultivars of Legumes 252
    Anna Veber, Omsk State Agrarian University, Russia
    Svetlana Leonova, Bashkir State Agrarian University, Russia
    Elena Meleshkina, All-Russian Research Institute of Grain and Grain Products, Russia
    Zhanbota Esmurzaeva, Omsk State Agrarian University, Russia
    Tamara Nikiforova, Orenburg State University, Russia

The results described in this chapter are of the investigation based on the collaborative research of scientists from the three Russian universities (Omsk State Agrarian University, Bashkir State Agrarian University, and Orenburg State University) which started in 2014. The authors assess various indicators of food safety. The study includes physical and chemical properties, technological characteristics, and chemical composition of new elite selection cultivars of pea ("Pisum arvense", the harvest of 2018, Bashkir Scientific and Research Institute of Agriculture) and haricot bean (harvest of 2018, Omsk State Agrarian University). Most of the samples have increased phytochemical capacity and high protein concentration (21.15-22.49% in haricot bean; 19.38-23.75% in pea). The authors demonstrate that these cultivars can be used for the enrichment of foodstuff and the creation of new functional foods.

Chapter 14
Imports and Use of Palm Oil as a Way to Increase Safety of Food Fats 277
    Inna Simakova, Saratov State Vavilov Agrarian University, Russia
    Roman Perkel, Peter the Great Saint Petersburg Polytechnic University, Russia

The authors compare the biological value and safety of hydrogenated fat containing trans-isomers of oleic acid and palm oil-based fat. The chapter assesses the potential of replacing hydrogenated fats by palm oil in the production of special fat products. Hematological and histological studies are carried out in a form of biological experiment on animals (white rats). The study reveals the explicit negative effect of trans-isomers even with a relatively low concentration of trans-isomers in a diet. Pathological changes are not observed in animals when palm-based fat is introduced into their ration. The findings suggest that palm oil along with its fractions may be considered as an alternative to hydrogenated fats in the production of margarine, cooking, baking, and deep-frying fats. The use of palm oil in the production of special fats of increased hardness (spreads. confectionery, waffles and fillings, and chocolate coating) requires the application of modern methods for modifying triglyceride composition of fats – biocatalytic interesterification and fractionation.

Chapter 15
Resource-Saving Technology of Dehydration of Fruit and Vegetable Raw Materials: Scientific
Rationale and Cost Efficiency 296
Inna Simakova, Saratov State Vavilov Agrarian University, Russia
Victoria Strizhevskaya, Saratov State Vavilov Agrarian University, Russia
Igor Vorotnikov, Saratov State Vavilov Agrarian University, Russia
Fedor Pertsevyi, Sumy National Agrarian University, Ukraine

Thousands of tons of fruit and vegetables are lost annually during harvesting, transportation, and storage. Meanwhile, there is a problem of insufficient consumption of fruit and vegetables in the diet of modern people which results in an increase in the occurrence of alimentary-dependent diseases. One of the possible solutions to these two interrelated problems is the development of a technology of processing of substandard raw materials directly at the harvesting site. This study aims at the development of the technology of dehydration of fruit and vegetables applicable in a field. The economic effect of the proposed solution is contingent on the reduction of losses at the stage of cleaning and saving water resources and saving transportation and storage costs.

Section 4
Regional Aspects of Food Security and Trade in Agricultural Products

Chapter 16
Emerging Trade-Related Threats to Food Security: Evidence From China 319
Vasilii Erokhin, Harbin Engineering University, China

It is generally believed that free trade plays a vital role in stabilizing food supplies and food prices since abundant foods stocks in some countries coexist with shortages in some others. Contemporary global trade system, however, is becoming increasingly distorted by unfair and inefficient policies in many countries, creating both winners and losers among not only small developing economies, but also largest producers of food and agricultural products. One of the recent examples of such distortion is US-China trade tensions and potential tariff escalations where the agricultural sector is the most vulnerable. By raising import tariffs on food and agricultural products in response to protectionist policies, the countries may face a situation of rising prices for consumers, limited market access for producers, and increasing pressures on food security. In this chapter, the author develops the theme of the effects of globalized agricultural trade on food security with a critical focus on the importance of balancing trade liberalization and protectionism.

Chapter 17
Food Security and Self-Sufficiency as a Basis for National Security and Sovereignty: Evidence
From Russia 337
    Kirill Zemliak, Khabarovsk State University of Economics and Law, Russia
    Anna Zhebo, Khabarovsk State University of Economics and Law, Russia
    Aleksey Aleshkov, Khabarovsk State University of Economics and Law, Russia

The study discusses one of the global problems of mankind—ensuring food security for the population. The historical context of the food problem, the formation of the concept of food security, the approaches of the world community and individual countries to its provision and evaluation are considered. The case of Russia reveals the role of food security in ensuring economic, social, and political security and sovereignty of a state. Special attention is paid to the state of agriculture in Russia as a source of raw materials for ensuring food security, problems of its development, and ways to solve them. The place of Russia in ensuring the food security of the world is shown.

Chapter 18
Integrative Associations and Food Security: Case of China-Russia Interregional Cooperation 361
    Alexander Voronenko, Khabarovsk State University of Economics and Law, Russia
    Sergei Greizik, Khabarovsk State University of Economics and Law, Russia
    Mikhail Tomilov, Economic Research Institute, Far Eastern Branch, Russian Academy of
       Science, Russia

This chapter presents the analysis of agricultural regulations in the frames of the World Trade Organization (WTO) along with an overview of control measures in the sphere of food security in bilateral and multilateral trade unions. The main attention is given to the associations in the Asia-Pacific Region (APR). The chapter concludes with an overview of interregional cooperation between Russia and China in the sphere of agriculture and analysis of its impact on food security of the region. Recommendations for improving and establishing food security are made and future research directions are discussed.

Chapter 19
Water, Food Security, and Trade in Sub-Saharan Africa 384
    Sinmi Abosede, Pan Atlantic University, Nigeria

Water is essential for food production and it plays an important role in helping countries achieve food security. The effect of climate change poses significant threats to agricultural productivity in Sub-Saharan Africa, where 95% of agriculture is rain-fed. Changes in weather patterns in the form of prolonged drought and severe flooding, in addition to poor water and land agricultural management practices, has resulted in a significant decline in crop and pasture production in several African countries. The agricultural sector in the region faces the challenge of using the existing scarce water resources in a more efficient way. Most of the countries have failed to achieve food self-sufficiency and rely on imports to meet the demand for food. Agricultural trade can play a significant role in helping countries in Africa achieve food security by increasing availability and access to food in countries that are experiencing food insecurity.

Chapter 20
Prospects, Challenges, and Policy Directions for Food Security in India-Africa Agricultural
Trade 400
    Ishita Ghosh, Symbiosis International University (Deemed), India
    Sukalpa Chakrabarti, Symbiosis International University (Deemed), India
    Ishita Ghoshal, Fergusson College (Autonomous), India

This chapter focusses on the agricultural and investment potential between Sub-Saharan Africa and India, in order to combat food insecurity. There is much scope for meaningful collaboration with governments and public-private partnerships, which could be instrumental in reducing hunger and poverty and managing the adverse effects of climate change. Increasing inclusivity, devising sustainable land-holding policies, incentivizing exporters, knowledge sharing in terms of technology and expertise will also boost employability, production and trade potential. Moreover, effective financial and technical cooperation between India and the African countries may be the key to achieving the desired synergies that will bring about positive changes towards ensuring food security.

Chapter 21
The Role of Agricultural Production and Trade Integration in Sustainable Rural Development: Evidence From Ethiopia 425

    Henrietta Nagy, Szent Istvan University, Hungary
    György Iván Neszmélyi, Budapest Business School, University of Applied Sciences, Hungary
    Ahmed Abduletif Abdulkadr, Szent Istvan University, Hungary

Ethiopia is the second-most populous country in Africa with rainfed agriculture as a backbone of its economy. Most of the population, 79.3%, are rural residents. Sustainable rural development can be achieved if great attention is given to the labor-intensive sector of the country, agriculture, by improving the level of productivity through research-based information and technologies, increasing the supply of industrial and export crops, and ensuring the rehabilitation and conservation of natural resource bases with special consideration packages. The improvement in agricultural productivity alone cannot bring sustainable development unless supported by appropriate domestic and international trade. The main objective of this study is to identify and examine key determinants that influence agricultural productivity to assure food security, as well as to analyze domestic and foreign trade in agricultural products in Ethiopia.

Chapter 22
International Competitiveness of Niche Agricultural Products: Case of Honey Production in Serbia 443

    Drago Cvijanović, University of Kragujevac, Serbia
    Svetlana Ignjatijević, University Business Academy in Novi Sad, Serbia

The subject of the study is the analysis of honey production in Fruska Gora. Specifically, the authors determine the possibilities of honey production in the monasteries of Fruska Gora and the possibilities of increasing production and its impact on rural development. The chapter introduces the readers to the sector of honey production and sale in the monasteries of Fruska Gora, as well as to the problems related to the procurement of production material, state of marketing, and engagement of human resources. Since the monasteries represent sacred places, honey produced in such an environment can be considered a unique and special product. The authors reveal the factors which have a restrictive effect on the development of the honey sector in a specified geographical area, as well as explore the significance and role of production growth in the economic development of both the sector and the area.

Chapter 23
Agricultural Cooperatives for Sustainable Development of Rural Territories and Food Security: Morocco's Experience 465

    Maria Fedorova, Omsk State Technical University, Russia
    Ismail Taaricht, Cadi Ayyad University, Morocco

This chapter deals with the elaboration of a conceptual framework for agricultural cooperatives in Morocco: sustainable development of rural territories. The farming cooperative associations form an effective means for the advancement of the agricultural sector, being one of the elements of agricultural policy, which play an important role in the development of agricultural production, both plant and animal, as well as in the development process in Morocco, especially for rural development, and through it, rural income of the farmers and their social statuses. In this chapter, the authors have taken the Moroccan agriculture cooperatives as a case of cooperative longevity and survival in order to observe the evolution and processes of adaptation to the distinct economic, social, and environmental demands of a broad range of member-owners. The demands of the farming community, members, and society have resulted in social and environmental factors being as much a priority as economic aspects.

Chapter 24
Participatory Poverty Assessment Effort in Food Security and Extension Policy: Evidence From Indonesia 481
    Muhamad Rusliyadi, Polytechnic of Agricultural Development Yogyakarta-Magelang,
       Indonesia
    Azaharaini Bin Hj. Mohd. Jamil, University Brunei Darussalam, Brunei

The impact study assessment aims to evaluate policies and monitor the achievement of targets and the results of a development program such as DMP. The output obtained is information that is an evaluation of how the policy was planned, initiated, and implemented. Participatory monitoring and evaluation analyze the outcome and impact of the DMP Program. PPA seeks to answer the question of whether or not the policy or program is working properly. A participatory approach may improve the outcomes in the form of a new policy model for the future. The output of the PPA process from this study is the agricultural policy formulated in terms of practical ways of approaching poverty problems from a local perspective. The success of alternative policy options applied by local government such as physical, human resources, and institution development at the grassroots level should be adopted at the national level. It should represent the best example of a case of successful program implementation at the grassroots level which can then be used in formulating national policies and strategies.

Compilation of References 500

About the Contributors 559

Index 571

Foreword

In a globalized market, trade plays a vital role in stabilizing food supplies and food prices since abundant foods stocks in some countries coexist with shortages in others. The objective of the book is to take a close look at food security and food sovereignty in domestic markets in a globalized economy. The authors argue that the current global trade systems are becoming increasingly distorted by unfair and inefficient policies in many countries, creating both winners and losers. This publication describes and analyzes the effects of the globalized agricultural trade on food security. Balancing trade liberalization versus protectionism is the critical issue here

This publication edited by Dr. Erokhin and Dr. Gao contains chapters that discuss trade-related aspects of agriculture and food. It includes, among others, policies, practices, and innovations in the areas of food security, agricultural trade, and quality of nutrition. The book also focuses on the protection and domestic support of agriculture, the effects of climate change on crop productivity, and advanced farming technologies aimed at increasing output. Further, it pays attention to the development of value-added chains in agriculture and the potential of traditional foods, local plants, and novel foods for ensuring food security of vulnerable populations in various parts of the world.

Written by an international interdisciplinary team of experts representing over 30 universities and research institutions from many countries, the publication offers a wealth of information and critical analyses. It identifies key knowledge gaps, provides salient recommendations for prioritizing future research, and contains a strong practical component which makes the book appropriate reading for government officials, policymakers, and those engaged in agribusiness worldwide.

I highly recommend this book to those interested in contemporary developments in global food markets and food security in various parts of the planet. The editors should be complimented for bringing together a large number of contributors that are experts in the described fields. The variety of talents, expertise, and experience, along the wide scope of topics, make this book a unique collection of knowledge, and therefore a most valuable resource for anyone studying food security and globalized agricultural trade.

Wim Heijman
Wageningen University and Research, The Netherlands

Preface

Both global and national dimensions of food security of a particular country are developed under the influence of two major types of economic policy, which are often countered with one another. On the one side, there is a global trend towards liberalization of economic and trade relations, which is facilitated in the framework of international organizations and various alliances of countries. On the other side, countries tend to protect domestic producers, support the purchasing power of consumers, especially those with low income, and promote domestic products abroad.

Although free trade is a key factor in promoting economic growth through international competition and the efficient allocation of resources, the food market is very specific. Trade plays a vital role in stabilizing food supplies and food prices since abundant foods stocks in some countries coexist with shortages in some others. Unforeseen price surges can push millions of people in developing countries into poverty, aggravating income inequalities and threatening social cohesion. Malnutrition can result in maternal mortality and stunted growth of children, reducing their learning ability and lowering their productivity in adulthood.

Governments are pursuing various policy options for ensuring food security: expanding investment in agriculture; encouraging climate-friendly technology; restoring degraded farmland; improving postharvest storage and supply chains; and even indulging in the promotion of niche products. They face special challenges as their growing middle-class shifts from traditional staples to more nutritious products such as meat, fish, and dairy whose higher resource intensity requires expansion of domestic agricultural capacity or greater reliance on imports. But importing countries also worry about unreliability of world markets in times of need and promote self-sufficiency in food along with trade policy restrictions.

Current global trade systems are becoming increasingly distorted by unfair and inefficient policies in many countries, creating both winners and losers. One of the recent examples of such distortion is US-China trade tensions and potential tariff escalations where agricultural sector is the most vulnerable. By raising import tariffs on food and agricultural products in response to protectionist policies, the countries may face a situation of rising prices for consumers, limited market access for producers, and increasing pressures on food security.

This publication develops the theme of the effects of globalized agricultural trade on food security with a critical focus on the importance of balancing trade liberalization and protectionism. It probes into many of the choices that link national, regional and global policies extensively with the provision of food security for all. The publication addresses the contemporary issues of agricultural trade, including major commodities and food products traded between major countries, directions of trade, and trends (liberalization, protectionism, and recent trade tensions). The retrospective of economic and trade policies in the sphere of agricultural production and trade in America, Asia, Europe, Latin America, and Africa is

Preface

provided. The authors discuss the current tendencies in agricultural trade and the roles of major producers, exporters, and importers of agricultural commodities, raw materials, and food products. The special focus is made on the challenges of globalized agricultural trade to food security, food independence, and food sovereignty. The publication addresses the effects of tariff escalations, administrative restrictions, and other forms of trade protectionism on food security. The book is concluded with a discussion of the role of multinational trade alliances and other forms of collaboration on global food security and sustainable development.

The book is structured in four sections and 24 chapters, which address various trade-related aspects of food security; policies, practices, and innovations in establishing food security; effects of agricultural trade on quality of nutrition; and regional aspects of food security and trade in agricultural products. The book has been contributed by the respected experts representing eleven countries of Europe and Asia and 34 research institutions, universities, and organizations.

ORGANIZATION OF THE BOOK

Part 1: Trade-Related Aspects of Food Security

The publication opens with an overview of the current needs of the mankind in food. In Chapter 1, Prof. Valentin Sapunov from Saint Petersburg State Agrarian University (Russia) states that the mankind having minimal areas of agricultural land produces more food than required to feed the world's population. When allocating forces and assets within the framework of the global policy of investing in agriculture, it can be safely reduced. Prof. Sapunov studies food policy in the XXI century and substantiates the ways to improve food security by increasing investments in the study of food opportunities, development of technology for the collection and processing of aborigine animals and plants in particular territories with a further increase in investments in the methods of biological technology. He also recommends increasing investments in industrial methods for obtaining food products from animals, plants, microorganisms, in the future – in the course of chemical industrial synthesis.

In Chapter 2, Dr. Bistra Vassileva from University of Economics-Varna (Bulgaria) emphasizes the effects of globalization on international trade as a quintessence of the last decade of the 20th century. The current wave of globalization is driven by major technological breakthroughs in transportation and communication. Major market actors are witnessing lower trade barriers and reduced transaction costs due to numerous technological innovations. Dr. Vassileva reviews the literature on global market tendencies and their implications on agricultural trade, conducts a comparative analysis of agricultural trade flows by globalization waves, describes the major actors in agricultural trade and their global market positions, visualizes the connections between the major actors, presents their behavioral pattern, and reveals the key interactions. The chapter ends with recommendations aimed at the future development of agricultural trade with a focus on global intervention policies.

Prof. Olga Pasko from Tomsk Polytechnic University (Russia) and Dr. Natalia Staurskaya from Omsk State Technical University (Russia) introduce the readers to global problems of food insecurity. In Chapter 3, they state that health status and level food security are among the priorities in many countries since, in the absence of sufficient food reserves, there is an economic and political dependence of some countries on the others. Having not yet received the required amount of food, the world is faced with the problem of ensuring security in its quality. Anthropogenic pollution of the environment complicates the problem with the quality of food and the exception of harmful chemicals in food. Prof. Pasko and Dr. Staurskaya emphasize a problem of using environmentally friendly agrotechnical means, ensuring the production of high yields of environmentally safe products with a desirable reduction in their cost, and shortening the time required for their production.

Part 1 is concluded by an overview of the world market of organic food products prepared by Dr. Inga Ryumkina from Irkutsk State Agrarian University (Russia) and Dr. Sergey Ryumkin from Novosibirsk State Agrarian University (Russia). In Chapter 4, the authors attempt to identify major actors in the global production of and trade in organic food products. Agricultural production is influenced by many factors, including climate, development strategies, and financing of agricultural research centers, among others. According to the authors, the factor of organic production should form both domestic and global markets of agricultural products and food since the health of people and the environment depends on the quality of food products. Therefore, the agrarian policy should primarily focus on the development of markets of organic food.

Part 2: Establishing Food Security: Policies, Practices, and Innovations

Part 2 focuses on the analysis of policies, practices, and innovations applied in various countries with an aim of ensuring food security. In Chapter 5, Dr. Anna Ivolga from Stavropol State Agrarian University (Russia) and Prof. Alexander Trukhachev from the Ministry of Tourism and Recreation of Stavropol Region (Russia) study the contemporary innovative approaches to and practices of state support of agricultural production and trade in food and agricultural products. The authors attempt to discover how protectionist policies in the sphere of production and trade affect the level of food security in the conditions of expanding globalization. The chapter focuses on the investigation of advanced innovative practices of state support in the case of selected OECD countries. Dr. Ivolga and Prof. Trukhachev reveal that the introduction of innovations into the system of state regulation is one of the key determinants of achieving food security in the conditions of the volatile market. In the chapter, they discuss and evaluate the volume and priority directions of innovations in agricultural protectionism policies.

Development of world agriculture is impossible without using advanced technologies. Their application in agriculture depends on the security of agricultural producers with highly effective agricultural machinery for which agricultural engineering serves. In Chapter 6, Dr. Mikail Khudzhatov from Peoples' Friendship University of Russia (RUDN University) (Russia) and Dr. Alexander Arskiy from Russian Academy of Personnel Support for the Agroindustrial Complex (Russia) state that agricultural engineering is an important element of the agro-industrial complex of any state providing it with necessary machinery and equipment. One of the important directions of the establishment of food security is the development of agricultural engineering. In this regard, Dr. Khudzhatov and Dr. Arskiy analyze the current state of the world market of agricultural machinery; develop the methodology of assessment of the competitiveness of agricultural machinery in the domestic market; and elaborate the definition of effective methods of management of logistics costs at the operation of agricultural machinery.

Food security is challenged by the growth of the world population, intensification of agricultural production, depletion of scarce agricultural and environmental resources, and the consequent introduction of innovations to farming processes, one of which is land management. In Chapter 7, Dr. Yulia Elfimova, Dr. Anna Ivolga, and Dr. Ivan Ryazantsev (all three from Stavropol State Agrarian University, Russia) discuss the role of innovative aspects and practices of land management in the establishment of food security and the provision of rational, sound, and effective administration of scarce land resources.

Preface

In the case of Russia and other countries, the authors justify the necessity of rational land management focused on the innovative way of development of agricultural production.

One of the most relevant concerns of food security is the reduction of food waste. Food and Agriculture Organization of the United Nations (FAO) estimates on average more than 1.3 billion tons of food loss and waste (FLW) along the whole food supply chain (equivalent to one-third of total food production) of which more than 670 million tons in developed countries and approximately 630 million tons in developing ones, showing wide differences between countries. In particular, EU data estimates an amount of more than 85 million tons of FLW, equal to approximately 20% of total food production. In Chapter 8, Prof. Giovanni Lagioia, Prof. Vera Amicarelli, Dr. Teodoro Gallucci, and Dr. Christian Bux (all four from the University of Bari Aldo Moro, Italy) review the magnitude of FLW at a global and European level and study its environmental, social and economic implications. The authors use Material Flow Analysis (MFA) to support and improve FLW management and its application in an Italian potato industry case study. According to the case study presented in Chapter 8, MFA has demonstrated the advantages of tracking input and output to prevent FLW and how they provide economic, social, and environmental opportunities.

Provision of the population with environmentally friendly and safe agricultural products is an important challenge in the developed states. Prof. Dmitry Bogdanov from Kutafin Moscow State Law University (Russia) and Dr. Svetlana Bogdanova from Bauman Moscow State Technical University (Russia) analyze the issues of food safety and quality in Chapter 9. They analyze the institute of indemnification caused by low-quality agricultural product since indemnification stimulates producers to ensure the quality and safety of food resources. Special attention is paid to the issues of consumer protection in the legislation of Russia and China.

Current political and economic reforms, as well as the development of market relations and private property rights, need a retrospect to the experience of the past. In Chapter 10, Prof. Olga Pasko and Alexey Gorodilov (both from Tomsk Polytechnic University, Russia), Dr. Natalia Staurskaya from Omsk State Technical University (Russia), and Prof. Alexander Zakharchenko from Institute of the Problems of Northern Development, Russian Academy of Sciences (Russia) review the agrarian reform implemented by Russian public entities in the early XX century. According to the authors, the goals of the reforms offered by Pyotr Stolypin were similar to those of the contemporary ones. Stolypin's reforms aimed at the substitution of group type of land use by public property. The reforms were not evolutional but were motivated by the explosive political and social-economic situation. Another agrarian reform took place in the early 1990s in the Soviet bloc, including the USSR. It aimed state land property and centrally planned agrarian economy, the domination of big manufacturers like collective and communal farms, and state pricing control. Despite similar basic principles, the states chose different strategies for the implementation of agrarian reforms.

Part 2 is concluded with an analysis of economic aspects of public policy in post-Soviet Russia as a key factor of establishing food security conducted by Prof. Stanislav Lipski from State University of Land Use Planning (Russia) and Dr. Olga Storozhenko from Bauman Moscow State Technical University (Russia). In Chapter 11, the authors investigate the arguments for land redistribution; privatization that resulted in appearance of land shares owned by about 12 million rural citizens barely understanding what to do with their land shares; post-privatization issues and problems concerned with the involvement of agricultural and other lands in economic activity; implementation of public economic policy measures aimed to resolve the above-mentioned issues (transfer of unclaimed land shares to municipalities); current transformation of ownership structure of agricultural lands; specifics of demarcation of un-privatized lands between federal, regional, and local authorities.

Part 3: Agricultural Trade and Quality of Nutrition

Agricultural trade affects not only availability of food in various parts of the planet, but also a quality of nutrition. Proper nutrition is a complex issue of food security which involves biological, social, economic, and even political aspects. Insufficient intake of essential elements may result in the occurrence of hidden hunger and metabolic disorders. Some regions of the world are characterized by a lack of certain nutrients in the environment which leads to their lack in plant and animal products. The most common problem is a deficiency of iodine and selenium. To solve this problem, the government take various measures, such as direct inclusion of necessary additives in food products, as well as the modernization of technological process of crop and livestock production. In Chapter 12, Dr. Elena Chaunina and Prof. Inna Korsheva (both from Omsk State Agrarian University, Russia) analyze the provision of the population in various countries and regions with limiting nutrients. Their study specifically aims at exploring the issues of production and trade in fortified (modified) food products that can directly fill in the lack of essential elements in particular territories.

Another aspect of trade effects on the quality of nutrition is dietary diversity. In Chapter 13, Dr. Anna Veber and Dr. Zhanbota Esmurzaeva (both from Omsk State Agrarian University, Russia), Prof. Svetlana Leonova from Bashkir State Agrarian University (Russia), Prof. Elena Meleshkina from All-Russian Research Institute of Grain and Grain Products (Russia), and Prof. Tamara Nikiforova from Orenburg State University (Russia) investigate the use of elite selection cultivars of legumes as a potential way to combat undernourishment and improve dietary diversity of the population in various countries of the world. The study includes physical and chemical properties, technological characteristics, and chemical composition of new elite selection cultivars of pea ("Pisum arvense", the harvest of 2018, Bashkir Scientific and Research Institute of Agriculture) and haricot bean (harvest of 2018, Omsk State Agrarian University). The authors demonstrate that these cultivars can be used for the enrichment of foodstuff and the creation of new functional foods since most of the samples have increased phytochemical capacity and high protein concentration.

In Chapter 14, Prof. Inna Simakova from Saratov State Vavilov Agrarian University (Russia) and Roman Perkel from Peter the Great Saint Petersburg Polytechnic University (Russia) compare the biological value and safety of hydrogenated fat containing trans-isomers of oleic acid and palm oil based fat. The authors assess the potential of replacing hydrogenated fats by palm oil in the production of special fat products. Hematological and histological studies are carried out in a form of biological experiment on animals (white rats). The study reveals the explicit negative effect of trans-isomers even with a relatively low concentration of trans-isomers in a diet. Pathological changes are not observed in animals when palm-based fat is introduced into their ration. The findings suggest that palm oil along with its fractions may be considered as an alternative to hydrogenated fats in the production of margarine, cooking, baking, and deep-frying fats. The use of palm oil in the production of special fats of increased hardness (spreads, confectionery, waffles and fillings, and chocolate coating) requires the application of modern methods for modifying triglyceride composition of fats – biocatalytic interesterification and fractionation.

Dozens of thousands of tons of fruit and vegetables are lost annually during harvesting, transportation, and storage. Meanwhile, there is a problem of insufficient consumption of fruit and vegetables in the diet of modern people which results in an increase in the occurrence of alimentary-dependent diseases.

Preface

One of the possible solutions to these two interrelated problems is the development of a technology of processing of substandard raw materials directly at the harvesting site. In Chapter 15, Prof. Inna Simakova, Dr. Victoria Strizhevskaya, Prof. Igor Vorotnikov (all three from Saratov State Vavilov Agrarian University, Russia), and Dr. Fedor Pertsevyi from Sumy National Agrarian University (Ukraine) develop the technology of dehydration of fruit and vegetables applicable in a field. The economic effect of the proposed solution is contingent on the reduction of losses at the stage of cleaning and saving water resources and saving transportation and storage costs.

Part 4: Regional Aspects of Food Security and Trade in Agricultural Products

It is generally believed that free trade plays a vital role in stabilizing food supplies and food prices since abundant foods stocks in some countries coexist with shortages in some others. Contemporary global trade system, however, is becoming increasingly distorted by unfair and inefficient policies in many countries, creating both winners and losers among not only small developing economies, but also largest producers of food and agricultural products. One of the recent examples of such distortion is US-China trade tensions and potential tariff escalations where the agricultural sector is the most vulnerable. By raising import tariffs on food and agricultural products in response to protectionist policies, the countries may face a situation of rising prices for consumers, limited market access for producers, and increasing pressures on food security. In Chapter 16, Dr. Vasilii Erokhin from Harbin Engineering University (China) develops the theme of the effects of globalized agricultural trade on food security with a critical focus on the importance of balancing trade liberalization and protectionism.

The theme of food security and self-sufficiency as a basis for national security and sovereignty is developed further in Chapter 17. In the case of Russia, Dr. Kirill Zemliak, Dr. Anna Zhebo, and Dr. Aleksey Aleshkov (all three from Khabarovsk State University of Economics and Law, Russia) reveals the role of food security in ensuring economic, social, and political security and sovereignty of a state. Special attention is paid to the state of agriculture in Russia as a source of raw materials for ensuring food security, problems of its development, and ways to solve them. In Chapter 18, another team of experts from Khabarovsk State University of Economics and Law (Russia) (Dr. Alexander Voronenko and Dr. Sergei Greizik) and Mikhail Tomilov from Economic Research Institute of the Far Eastern Branch of the Russian Academy of Science present the analysis of agricultural regulations in the frames of the World Trade Organization (WTO) along with an overview of control measures in the sphere of food security in bilateral and multilateral trade unions. Main attention is given to the associations in the Asia-Pacific Region (APR). The study is concluded with an overview of interregional cooperation between Russia and China in the sphere of agriculture and analysis of its impact on food security of the region. Recommendations for improving and establishing food security are made and future research directions are discussed.

Agricultural trade can play a significant role in helping the countries to achieve food security by increasing availability and access to food in the regions that are experiencing food insecurity, specifically, in Africa. Most of the countries in Africa have failed to achieve food self-sufficiency and rely on imports to meet the demand for food. In Chapter 19, Dr. Sinmi Abosede from Pan Atlantic University (Nigeria) discusses the role of water resources in establishing food security in Sub-Saharan Africa. Water is essential for food production and it plays an important role in helping countries achieve food security. The effect of climate change poses significant threats to agricultural productivity in Sub-Saharan Africa, where 95% of agriculture is rain-fed. Changes in weather patterns in the form of prolonged drought and severe flooding, in addition to poor water and land agricultural management practices, has resulted in significant declines in crop and pasture production in several African countries. The agricultural sector in the region faces the challenge of using the existing and scarce water resources in a more efficient way.

Chapter 20 focusses on agricultural trade and investment potential between Sub-Saharan Africa and India in order to combat food insecurity. According to Dr. Ishita Ghosh and Dr. Sukalpa Chakrabarti (both from Symbiosis International (Deemed University), India) and Dr. Ishita Ghoshal from Fergusson College (Autonomous) (India), reduction of hunger and poverty, conservation of resources, reduction in the negative impact on climate and inclusivity (gender and marginalized groups) – all of this could be augmented with collaboration through government and public-private partnerships. A sustainable approach to land policies, incentivizing international exporters, knowledge sharing in terms of technology and know-how, etc., along with a seamless interaction between financial, technical, and educational cooperation between India and Africa should help in achieving the desired synergy in sustainable agriculture that would enable the goal of food security.

Chapter 21 studies the role of agricultural production and trade integration in sustainable rural development in the specific case of Ethiopia. Ethiopia is the second-most populous country in Africa with rainfed agriculture as a backbone of its economy. Most of the population (79.3%) are rural residents. Sustainable rural development can be achieved if great attention is given to the labor-intensive sector of the country, agriculture, by improving the level of productivity through research-based information and technologies, increasing the supply of industrial and export crops, and ensuring the rehabilitation and conservation of natural resource bases with special consideration packages. The improvement in agricultural productivity alone cannot bring sustainable development unless supported by appropriate domestic and international trade. In Chapter 21, Dr. Henrietta Nagy and Ahmed Abduletif Abdulkadr (both from Szent Istvan University, Hungary), and Dr. György Iván Neszmélyi from Budapest Business School – University of Applied Sciences (Hungary) identify and examine key determinants that influence agricultural productivity to assure food security, as well as analyze domestic and foreign trade in agricultural products in Ethiopia.

In Chapter 22, Prof. Drago Cvijanović from University of Kragujevac (Republic of Serbia) and Prof. Svetlana Ignjatijević from University Business Academy in Novi Sad (Republic of Serbia) conduct an analysis of international competitiveness of niche agricultural products in case of honey production in the Republic of Serbia. Specifically, the authors determine the possibilities of honey production in the monasteries of Fruska Gora and the possibilities of increasing production and its impact on rural development. The chapter introduces the readers to the sector of honey production and sale in the monasteries of Fruska Gora, as well as to the problems related to the procurement of production material, state of marketing, and engagement of human resources. Since the monasteries represent sacred places, honey produced in such an environment can be considered a unique and special product. Prof. Cvijanović and Prof. Ignjatijević reveal the factors which have a restrictive effect on the development of the honey sector in a specified geographical area, as well as explore the significance and role of production growth in the economic development of both the sector and the area.

Chapter 23 deals with the elaboration of a conceptual framework for agricultural cooperatives in Morocco: sustainable development of rural territories. The farming cooperative associations form an effective means for the advancement of the agricultural sector, being one of the elements of agricultural policy, which play an important role in the development of agricultural production, both plant and animal, as well as in the development process in Morocco, especially for rural development, and through it, rural income of the farmers and their social statuses. In this chapter, Dr. Maria Fedorova from Omsk State

Preface

Technical University (Russia) and Prof. Ismail Taaricht from Cadi Ayyad University (Morocco) have taken the Moroccan agriculture cooperatives as a case of cooperative longevity and survival in order to observe the evolution and processes of adaptation to the distinct economic, social, and environmental demands of a broad range of member-owners. The demands of the farming community, members, and society have resulted in social and environmental factors being as much a priority as economic aspects.

Part 4 is concluded by a study of participatory poverty assessment efforts in food security and extension policy in Indonesia. Dr. Muhamad Rusliyadi from Polytechnic of Agricultural Development Yogyakarta-Magelang (Indonesia and Dr. Azaharaini Jamil from University Brunei Darussalam (Brunei Darussalam) evaluate the policies and monitor the achievement of targets and the results of the development programs. The output obtained is information that is an evaluation of how the policy was planned, initiated, and implemented. Participatory monitoring and evaluation analyze the outcome and impact of the DMP Program. PPA seeks to answer the question of whether or not the policy or program is working properly. A participatory approach may improve the outcomes in the form of a new policy model for the future. The output of the PPA process from this study is the agricultural policy formulated in terms of practical ways of approaching poverty problems from a local perspective. The success of alternative policy options applied by local government such as Physical, Human Resources, Institution development at the grassroots level should be adopted at the national level. It should represent the best example of a case of successful program implementation at the grassroots level which can then be used in formulating national policies and strategies.

This book attempts to develop conceptual approaches to improving food security in the conditions of globalized agricultural trade and emerging trade tensions between major actors, such as the USA, China, the EU, Russia, and other countries. Since studying global agricultural trade and contemporary challenges to food security has a multinational application, its outcomes will be shared with a wide international network of stakeholders, including research institutions, universities, and individual researches. The book is appropriate for government officials, policymakers, and businesses of many countries. Adaptation of research outcomes and solutions to the situation in particular countries and various collaboration formats (US-China most recent trade tension, cross-continental trade in food) will let to increase the visibility of the publication and to elaborate new practices and solutions in the sphere of developing agricultural production, increasing effectiveness of trade in agricultural commodities, and achieving food security.

Vasilii Erokhin
Center for Russian and Ukrainian Studies, School of Economics and Management, Harbin Engineering University, China

Gao Tianming
Center for Russian and Ukrainian Studies, School of Economics and Management, Harbin Engineering University, China

Acknowledgment

This book is a product of research and support of many individuals to whom the editors would like to express their sincere gratitude.

We thank all the chapter contributors representing 34 respected universities, research institutes, and organizations from 11 countries. Our acknowledgment is made to the members of the Editorial Advisory Board and international team of reviewers for sharing their precious time, professional experience, valuable recommendations, and insights. We give our special thanks to Prof. Wim Heijman from Wageningen University and Research (the Netherlands) who kindly prepared the outstanding foreword.

We sincerely thank the entire IGI Global team for assistance and support throughout the book development process.

The editors appreciate the support of the Fundamental Research Funds for the Central Universities (grant no. 3072019CFP0902, HEUCFJ170901, HEUCFP201829, HEUCFW170905, 3072019CFG0901). We wish to express our sincere thanks to Harbin Engineering University for providing us with all the necessary facilities and support.

We are truly indebted to our families and friends, who shared their continuing support and encouragement throughout the preparation of this publication.

Section 1
Trade–Related Aspects of Food Security

Chapter 1
Real Need of the World in Food:
A Future of Agricultural Production

Valentin Sapunov
Saint Petersburg State Agrarian University, Russia

ABSTRACT

Mankind has minimal areas of agricultural land that produces more food than required to feed the world's population. When allocating forces and assets within the framework of the global policy of investing in agriculture, it can be safely reduced. What is food policy in the 21st century? First of all, it is advisable to increase investments in the study of food opportunities, the development of technology for the collection and processing of aborigine animals and plants in particular territories with a further increase in investments in the methods of biological technology. It is advisable to increase the investments in industrial methods for obtaining food products from animals, plants, microorganisms, in the future – in the course of chemical industrial synthesis. Vernadsky predicted that in the future, mankind will switch to autotrophic nutrition, i.e. artificial synthesis of food from inorganic materials. Biotechnology will gradually reduce the volume of traditional agricultural production.

INTRODUCTION

What could be the prospects for obtaining human resources in the future? This problem is considered using the examples of agricultural history of the world, particularly, agricultural sectors of such big countries as Russia and China.

The three following questions are under consideration.

- During the communism era in the Soviet Union (USSR) and China, many products were chronically in short supply. Store deficiencies were strongly associated with the last years of communism. Currently, there are no collective and state farms in Russia, urban population is not engaged into agricultural production. Farming as social phenomenon is underdeveloped. Despite these, however, there are enough food products in the stores. What is the source of this abundance?

DOI: 10.4018/978-1-7998-1042-1.ch001

- At the end of the 18th century, English economist Thomas Malthus (1798) predicted a series of crises, including an agrarian crisis. Humanity is growing by geometrical progression without limit, the resources of the Earth, although abundant, are still limited. The growth of the population may cause a shortage of food products. Currently, food production on a per capita basis is bigger than ever before. In average terms, globally, there is no food deficiency. Why? What is the source of this productive force?
- Agricultural production, however, experience many challenges, including natural ones. One of the questions is why do weeds grow and spread faster than cultivated plants?

BACKGROUND

About two thousand years ago, per capita acreage of agricultural land was 120 hectares. In the modern times, there are only two hectares of land per capita. If this trend continues, in the middle of the 21st century, per capita acreage will reduce down to insignificant size. Certainly, agricultural production will never disappear entirely, but it will undergo substantial changes. Why does this happen and what predictions can be made?

There are four major ways to produce food:

- Passive production without the reproduction of the resources. It is divided into gathering, hunting, and fishing. Alternatively, the passive method is also eating the corpses of the animals killed by the predators. According to Porshnev (1963), anthropologist, this method of obtaining food prevailed among the Neanderthal people.
- Active animal breeding.
- Plant growing, that is active farming in extensive and intensive forms. Extensive farming provides for the expansion of acreage, while intensive one involves increasing yields through the use of fertilizers, breeding, and other agricultural technologies.
- Biological technology, that is, the production of food products by industrial methods at the factories and plants.

A transition to a more advanced technology is food revolution, the basis for which is the increase of world population. Until 1955, the population increased at an exponential rate (Table 1).

The increase of the population at an exponential rate required more and more food to feed the people. Under the extensive method of obtaining resources without changing the technologies, new territories were developed. Sometimes it was accompanied by the aggressive wars. The resettlement of the peoples from the North of Europe and the fall of the Roman Empire in the 5th century were associated with the need to expand the territories of extensive agricultural production for the Germanic tribes. This process was accelerated by a global cooling in the 5th century. The establishment of the Genghis Khan's empire which took place through the wars of conquest was stimulated by two circumstances. The first one is the growing population in the Golden Horde. The second one is a cold snap which reduced the efficiency of grazing. Genghis Khan's empire was not able to change food resource technology to the plant agriculture because of national mentality.

In the developed countries, there happened agricultural revolutions. The transition from animal breeding to farming allowed increasing food resources by 10-100 times without substantial increase in

Real Need of the World in Food

Table 1. Growth of the world population, people

Year	Population
1,600,000 BC	100,000
200,000 BC	1,000,000
30,000 BC	10,000,000
2,000 BC	100,000,000
1840	1,000,000,000
1900	1,700,000,000
1920	1,900,000,000
1950	2,500,000,000
1970	3,700,000,000
1990	5,300,000,000
2000	6,300,000,000
2012	7,000,000,000
2025	8,500,000,000
2050	10,000,000,000

Source: Kapitsa, 2012

the land areas under cropping. This was due to the fact that the plant mass grew 100 times faster than animal mass. The development of biotechnology will open up even greater opportunities for feeding the population of the globe. Of great importance is the fact that the rate of growth of world population is declining. Accordingly, nutritional needs will not grow. By the middle of the 21st century, the world population will reach ten billion people and will stop growing further. Accordingly, a significant increase in agricultural production is not required.

What could be the prospects for obtaining human resources in the future? Let us examine the following question. In the 1980s, the USSR State Committee on Statistics published unpleasant data for the communist regime. After many decades of violence successes of collective farm and state farm building, it turned out that almost one-third of agricultural products in the country were produced by small private households. The share of the land in private property was 0.5% of the total agricultural land in the country. This meant that labor productivity in private sector was substantially more effective compared to that in the collective-farm and public sectors. A simple question arose: why not increase the share of land allocated to private households from 0.5% to 1.5%? After Perestroika, the collective and state farms were closed, but the amount of food products in the country remained approximately on the same level. The conclusion was that all the losses that the country suffered in the bloody way of collectivization and big terror were meaningless. Soviet collective and state farms were not the enterprises for food production and supply. They were the establishments that ensured the power of the communist party in rural areas. Their loss took away almost nothing from the country in terms of providing domestic market with agricultural and food products. Household plots and gardening, some preserved collective and state farms, a few farms provided the foundation for catering. And having a base, it was already possible to import bananas, pineapples and other delicacies that did not grow in Russia (Kondratyev, Krapivin, & Savinykh, 2003). Currently, private plots in Russia obtain 3.8% of agricultural lands in the country and produce 34% of aggregated volume of food and agricultural production.

MAIN FOCUS OF THE CHAPTER

Everything Afraid of Time, Time Afraid of Pyramids

This Arabic proverb refers to the pyramids in Egypt. In fact, they, although slowly, are being destroyed. However, there is a truly constant pyramid in the Earth which is an ecological pyramid. It cannot be touched, visualized, or smelled. This is an image of pattern of biological masses of different trophy (food) levels of organisms. The pyramid is based on green plants which constitute 99% of the entire biosphere. They photosynthesize primary biological compounds. This is the first floor of the ecological pyramid. It stands on the second floor, 100 times smaller in size. These are herbivores. Further, each subsequent floor is ten times smaller than the previous one. Above the herbivores are less numerous predators. Above them are super predators, eating predators. Even higher parasites are. The man is at the top of the pyramid. One of the laws of ecology suggests that the upper floor can exist if the mass of the underlying one exceeds the next one at least ten times (Vernadsky, 2002; Sapunov, 2012). If the size of the upper floor exceeds 10% of the underlying one, local destruction begins. In other words, there should be a tenfold supply of food substrate. Being at the top of the pyramid, a man as an omnivore can eat almost any organic matter. The mass of all people living in the Earth is less than 1% of the mass of all animals. The mass of the animals is less than 1% of the mass of the plants. According to the laws of ecology, the mankind may die if its mass exceeds 10% of the one of the previous floors. This will happen if the number of people increases by three orders of magnitude, i.e. reaches seven trillion people. It is clear that this will never happen. One thing to conclude from this is that in the near future, it is not serious to talk about the lack of organic resources for the mankind. Resources are unlimited (Lomborg, 2001, 2004). Social hunger is always artificially organized as an instrument of reactionary policy.

Similar to the pyramids in Egypt, the ecological one consists of blocks or ecological niches. Each niche is occupied by a particular biological specie. No more than one and no less than one. There are no ecological niches that are unoccupied. The theory of this structure was developed by Russian scientist Vernadsky (2002). One of the important provisions of the theory suggests that the entire ecological pyramid has remained almost unchanged over the millions of years. If one block is withdrawn, it is immediately replaced by another one of the same size. Instead of the lost form, a new one appears. Instead of a destroyed ecological system, a similar one arises. But the pattern of species in the restored area may be new. These changes in the biosphere can be favorable or unfavorable for humans.

Agricultural activity is associated with the destruction of the biosphere. Burning out and uprooting the forests and plowing up the meadows damage the biosphere. For each victory over nature, it takes revenge on us, and revenge can be cruel.

Is Product Use Rational?

There is a Chinese saying "There are no inedible products, there are bad cooks". This statement seems rather categorical, but in some degree it is true. Almost all living organisms inhabiting the planet may be consumed by people as food. If not entirely, then at least some of their parts are edible. Even well-known and mastered power sources, people do not use completely. In many agricultural crops, all parts of a plant are edible, healthy, and tasty. People reject consuming many digestible products due to some prejudices or traditions. Muslims do not eat pork, while Hindus do not eat beef. Few of the Europeans will dare trying roasted dog, one of the Chinese dishes. Traditions of this kind are formed quite by chance. At the beginning of the 7th century (presumably, in 615), a Bedouin army led by Mohammed broke out a massive disease caused by pork infected with trichinosis. They were not aware of trichinae worm (Trichinella spiralis) and banned consumption of pork, especially in the then unsanitary conditions and the lack of proper control over the diet of domestic animals. Muhammad simply forbade eating pork and offered to lean on the meat of other animals, less infected with dangerous worms. When the speeches of the Prophet were canonized and published in the form of the Quran, the ban on a particular case became absolute for many Muslims. Even the fact that in the Quran the ban is not absolute, there are many comments to it, and that modern veterinary science reliably protects consumers from trichinosis, Muslims still do not consume pork. Such is the power of the traditions.

Are Weeds Available for Consumption?

Let us turn to the third question – why do weeds grow and spread better than cultivated plants? What are weeds? These are the dominant aborigine plants of original geographical area and ecological systems. Quinoa, nettle, burdock, and many others are actually edible plants. They can be consumed raw, processed into flour, or fried. The majority of biosphere products are available for consumption. Some traditions of using these plants, developed during the years of famines, are still present. In the 1970s, Russian scientist Firsov made an original experiment. During summer period, he placed chimpanzees into the wild nature of north-west Russia to the islands of Pskov region. It turned out that chimpanzees perfectly survived in the forests of northern Russia (at least, during summer period), succeeded to find edible plants and avoid poisonous ones. Chimpanzees are very close to the humans in terms of basic biological characteristics – height, weight, and nutritional needs. Where a chimpanzee can survive, a human being will survive too (Sapunov, 2012).

The Introduction of Cultivated Plants Is a Violence against Nature

According to the records of the Russian scientist Nikolay Vavilov (1987), every cultivated plant appears once in one center of origin. Then, as the civilization develops, it spreads across the globe. Potato originated and improved by breeding in Mexico. After Columbus' return to Europe, potato is cultivated there. Peter the Great tried to introduce potato in Russia. However, his attempts to force the peasants to cultivate potato instead of turnip ended in the so-called "potato riots". The situation changed only in the second half of the 19th century. The reform of 1861 reduced the country population and increased the urban population. Russia's agricultural sector required intensification. One of the promising ways was the expansion of cultivation of potatoes. From that time on, potatoes became a favorite dish in Russia and acquired the status of "second bread". So was the "potato revolution". Another Russian reformer, Nikita Khrushchev, started agricultural revolution in the XX century. He relied on the following considerations. Of all the cereals, maize has the greatest biological value. Maize also came to Europe from Mexico. In Soviet non-rational agricultural production, the average productivity of wheat was one ton per hectare (up to 3.5 tons per hectare on the best lands). Maize allowed receiving 4-5 tons per hectare at lower cost compared to wheat. In Russia, maize has traditionally been viewed not as a basic food product, but as a kind of self-indulgence, an addition to the basic diet. The Russians, unlike the Mexicans, are not accustomed to corn porridge and maize tortillas. If part of the acreage could be converted from wheat to corn, this would have a tangible effect with minimal investment. The negative results of the Khrushchev's initiative are widely known. The government system that had begun to degrade brought sensible thought to the point of absurdity. Then, all the absurdity was piled on Khrushchev, who, being a sensible man, did not recommend sowing corn outside the Arctic Circle.

What Is the Result of the Long Journeys of Plants?

To some extent, any far-away relocation of crops is a violence against nature. When nature is damaged, it responses. In the early 20th century, insects and other pests damaged 10% of the world crop. At the end of the 20th century – already 13% (Kondratyev et al., 2003; Sapunov, 2012). Thus, the whole pest control, which cost a lot of money, was lost. Nature is coming. It lives by its own laws and not by the ones the people try to impose. One of the examples is a complete failure in the fight against the Colorado beetle. There are many more defeats like that! The balance between animals, plants, herbivores, predators, and parasites is described by the rule of the ecological pyramid. Any attempt to violate this rule leads to the effective counteraction. People brought potatoes from America to Europe. At first, potatoes got accustomed to life and grew up with almost no natural enemies. However, according to the law of the pyramid, ten kilograms of animals eating this plant should be consumed per ton of mass of any plant. A free ecological niche arose, which soon was filled up with the Colorado potato beetle that had come from America. No matter what quarantine barriers people put, nature would not tolerate the emptiness. Some kind of potato eater would inevitably appear. Insects are higher organized than plants, evolve faster, better adapt to the environment. European potatoes are worse than the American ones. But the European-style Colorado potato beetle is not inferior to an American relative. The paradox is that, no matter how much crop pests destroy, it still remains much more than the world population can consume.

What Is the Fate of Super Crop?

In Russia, people usually suffered not from the lack of food products, but a poor excess to food. Even in the Soviet era, which gave birth to one of the most irrational agricultural production systems in history, there was no famine, with the exception of the especially provoked cases in 1918 and 1929.

In the 1980s, the author of this chapter worked in Stavropol region on a program to suffocate foreign weeds. The pictures observed there remained forever in memory. 70% of the farms in the region were economically inefficient. In this case, the soils of the Stavropol territory are considered among the best in the world. If it was unprofitable for them to plow and sow such high-quality land, it meant that the Soviet system of agricultural production was completely ineffective. Most of the harvest remained uncollected in the fields and collective farms' gardens. Local fruits, fragrant apples, juicy plums, etc., were more than half plowed into the ground and used as a fertilizer. At the same time, fruit trees grew on forest belts, city streets, self-sown in the fields. The notorious maize was grown on collective farm fields – and there it remained. In some cases, the crop was destroyed because there was no economic opportunity to export it.

The next stage of crop destruction is the elevators and vegetable stores. They were served by school-children, students, military soldiers, scientists, doctors, teachers, and other non-professionals. All of them were amazed at the monstrous irrationality and absurdity of organizing the storage of agricultural products. Losses accounted for at least 50%. In Soviet times, some economists suggested, at first glance, contrary to common sense, but partly fair: it was cheaper for the Soviet Union to buy agricultural products from abroad than to grow them domestically. The harvesting campaign, which in the spirit of time was called the "battle for the harvest", usually ended with abundant purchases abroad. At the same time,

Real Need of the World in Food

the volume of purchases was less than what was left in the fields in the country, crumbled on the roads, and driven away in the warehouses and vegetable stores. But in the end, there was enough food, and no one died of hunger.

International experience shows that, for example, in the USA, agricultural sector employs 1-2% of the population. Due to intensive farming, more labor is not necessary. Becoming a farmer in the USA is as difficult and prestigious as an artist. For the few farmers, the government paid extra and continues to pay so that they plow less and sow. The market is no longer able to digest existing stocks of food. The aggravation of overproduction can lead to serious social problems. Soviet ways to eliminate the surplus contradict the American mentality. Opportunities to sell the surplus to Russia, where the surplus of its own production barbarically destroyed, also provided not infinite. Of course, the US government was always ready to support the Soviet way of production that was economical and beneficial to the USA. Among the high-ranking Soviet officials, there were also quite a few who were happy about the additional opportunity to break free from the iron curtain, reveling in their significance, to go to the USA to fulfill the state mission on the purchase of grain. The general conclusion from the above is as follows. In the countries with different social regimes, for many years, food products have been produced abundantly.

Danish scientist Bjorn Lomborg (2001) proved that the mankind produced more food per capita than ever before. This is also evidenced by the Food and Agricultural Organization of the United Nations (FAO) and other agricultural organizations. The forecasts of Malthus (1798), as well as the certain "limits to growth" defined in 1972 by the Club of Rome (Meadows, Meadows, Randers, & Behrens, 1972), turned out to be untenable. This is despite the fact that the per capita acreage of cultivated land is continuously decreasing (Kovach & McGuire, 2003).

Those who visited the countries of Western Europe, especially resort areas, were struck by the number of catering establishments. In modern Russia, with its ruined agriculture and industry sold by the fishing fleet, no one dies of hunger (Kokin & Kokin, 2008; Frumin, 2006).

Where Does Hunger Come from?

When the prophet Moses led the people from Egypt, he did not immediately hurry to the Promised Land. For 40 years, they walked in the desert, although the walk from Memphis to Jerusalem takes only one week. Some Jews with a characteristic wit explained the irrational route by the fact that Moses was paid per diem. Of course, this is a joke. In the course of not entirely understandable wanderings, a good half of the people died of hunger. The only food source was "manna from heaven" (seeds of native plants propagated by the wind), and its distribution was easy to control. Having strengthened his power through the distribution system, Moses led the people to really fertile places. In modern Israel, as in most countries of the world, store shelves are full of products made in local kibbutz and farms.

In Russia, traditionally, there was no famine. After the revolution in 1917, the situation changed. With the help of hunger, the Bolsheviks strengthened their illegitimate power. At first, in 1918, food detachments cordoned off Petrograd and barred entry to peasants carrying food. As a result, the population of the city decreased from 2.4 million to 700 thousand. In order not to die of hunger, people were ready to flee anywhere, especially since there was no famine in the rest of the country – for its organization on the scale of such a state there would not have been enough food detachments. The main goal was achieved – the Bolshevik power was established in Petrograd. In the period of the New Economic Policy (suggested by the government of Trotsky and Lenin), in all parts of the country, it was possible to eat to heaps. Then, the year of 1929 was named "the year of great reforms". The war was declared with the aim of destroying the recalcitrant class of the peasantry and turning it into an amorphous mass of collective farmers. Hunger went to work again. Demonstrators under the pretext of fighting the "kulaks" seized all food from the peasants. In Ukraine, in the Central chernozem soil region, a massive famine began, which claimed the lives of nine million people. At the same time, a huge amount of food was exported abroad. The goal was achieved – the peasants were driven under the threat of starvation to the collective farms, where the economy was obviously ineffective.

After the war, the Bolsheviks easily gave cottages and household plots which created the basis for feeding the entire country. The great Austrian biologist Konrad Lorenz served on the territory of Poland in a German hospital as a doctor. Soviet troops captured him among many Germans and sent him to one of the concentration camps in Russia. As the future Nobel laureate wrote in his memoirs, the prisoners of war were placed behind barbed wire and were not fed at all for a long time. Mass mortality began. Lorenz strained his biological knowledge. Determined which of the insects living in the camp, spiders and worms that was edible, and thus saved him and some fellow suffered from death. The scientist was so strong on a diet of cockroaches and spiders, that he had the strength to perfectly learn the Russian language. A man armed with modern scientific knowledge will always find food, no less effective than a monkey in the forest does.

The blockade of Leningrad claimed the lives of at least one million people who died of starvation. In fact, the products in the warehouses were. Due to the irresponsibility and criminal negligence of the authorities, they were all lost during the raids of German Nazi army. A massive mortality of civil population took place. But some survived. So, a group of entomologists organized a farm for breeding larvae of flies in the apartment. These larvae were high in calories and even useful. Experts saved themselves and their relatives from starvation.

Hunger, though difficult, is possible even in the countries where three or four harvests are harvested per year. Relatively recent examples are the massive death of people from hunger in Ethiopia and Somalia. Only the intervention of international organizations helped to stop this genocide (Danilov-Danilyan & Losev, 2000).

How Much Does Nature Produce?

Agricultural production is unprofitable. For the production of one kilocalorie of energy contained in agricultural products, it is necessary to spend ten kilocalories (Odum, 1986). In Table 2, there are the average global data on crop yields.

It is a question of full productivity, including those parts of a body of a plant which traditionally are not eaten, but, in principle, are edible. The yield of natural biocenoses is bigger (Table 3).

Table 2. Annual average productivity of various crops, tons/hectare

Crops	Productivity
Sugar cane	17.25
Sugar beet	7.65
Figure corn	4.97
Wheat	3.44

Source: Danilov-Danilyan and Losev (2000); European Geosciences Union (n.d.)

Table 3. Annual average productivity of meadows and forests, tons/hectare

Crops	Productivity
Meadows of the average band	225
Taiga woods	400
Deciduous forests	500
Tropical forests	1,700

Source: Danilov-Danilyan and Losev (2000); European Geosciences Union (n.d.)

The total mass of products consumed by one person per year is 600 kg. With proper use of plant and animal mass, one hectare of the middle band can feed 400 people. If used correctly, one hectare of meadows in the middle lane can feed 400 people. Forests of the middle zone can feed 750 people a year from one hectare – with emphasis on vegetarian food. With an emphasis on animal food, one hectare can feed eight people per year, as the mass of animals is less than the mass of plants by approximately 100 times. Since a human being is omnivorous, the number of people who can be provided by food per hectare is from 100 to 200. One square kilometer of forest can feed at least 10,000 people. 15,000 kilometers with full use of all resources can provide food for entire population of Russia. Young forest has the biggest productivity. Cuttings in some cases may be useful, because after them there is a lot of actively growing, accumulating organic matter and releasing oxygen of young growth.

The largest increase in biomass is in the tropical zone, where it is 1,700 tons per hectare per year. When using the edible parts of animals and plants inhabiting the Brazilian tropical forest, it is possible to feed about 300 billion people, the largest possible population of the Earth ever. This is only one of the tropical massifs. It should be added that 25% of the world biomass is produced in the ocean. Seafood is used in a minimum volume, measured in fractions of a percent of the possible. Few countries of the world operate large-scale fishing fleet. Only a few hundred species of fish and marine invertebrates are caught. Hundreds of thousands of other species of aquatic plants and animals are not used. The conclusion is this. Even without artificial reproduction of fodder animals and plants, only due to the gathering put on an industrial basis, it is possible to feed any grown population of the Earth without causing significant damage to the natural environment.

SOLUTIONS AND RECOMMENDATIONS

World population increases, but the rate of growth decelerates. The maximum number is expected to reach ten billion by the middle of 21st century and stay at this level in the course of time. Resources of the Earth appeared to be sufficient to feed so many people without extensive development of agricultural production. The predictable progress deals with new technologies of intensive agricultural production, new studies, and use of natural products and biotechnologies. The ideas of environmental and agricultural crisis are weak and not proved in a scientific way. In physical terms, the deficiency of food is not a global risk. The resources-related issues may be decided based on the principle of fundamental ecology suggested by Vernadsky (2002), Vavilov (1987), Odum (1986), and other scientists.

FUTURE RESEARCH DIRECTIONS

Despite of significant progress of biological science, understanding of biological diversity is not completed. Only a little part of biological species is well studied from the point of view of ecology, genetics, and biochemistry. Future progress of biology may get new significant resources for the mankind. This study will be accompanied by the progress in biological technology and biological industry. In 20th and 21st centuries, many political, economic, and environmental problems emerged. They may be resolved in the framework of the rational world policy based on the records of natural sciences.

CONCLUSION

Currently, with minimal areas of agricultural land, people produce more food than necessary to feed themselves. The volume of production may be reduced without detriment to food security of the globe by due allocation of production forces and assets within the framework of investment policy in agricultural sector. In the XXI century, food security policy should be based on the following three pillars.

First, it is advisable to increase investments in the investigation of food production opportunities, development of technology for the collection and processing of aborigine animals and plants in each region with further increase in investments in biotechnologies.

Second, it is advisable to increase investments in industrial methods for obtaining food products from animals, plants, microorganisms, in the future – in the course of chemical industrial synthesis. Russian scientist Vernadsky (2002) predicted that in the future, people would switch to autotrophic nutrition, i.e. artificial synthesis of food from inorganic materials. Biotechnology would gradually reduce the volume of traditional agricultural production.

Third, a promising direction is the development of fish fleet and infrastructure engaged in obtaining seafood. At the same time, the majority of inhabitants of the oceans and inland waters can become objects of capture and gathering accompanying with traditional fish and aquatic invertebrates.

In general, food security policy will face the three major challenges: the source of abundance of food in global scale, the source of this productive force of the planet, and combating the weeds that grow better than cultivated plants.

REFERENCES

Danilov-Danilyan, V., & Losev, K. (2000). *Environmental Challenge and Sustainable Development*. Moscow: Progress-Tradition.

European Geosciences Union. (n.d.). *General Assembly 2017-2018*. Retrieved from http://www.egu.eu

Frumin, G. (2006). *Geo-Ecology: Reality, Pseudoscientific Myths, Mistakes, and Delusions*. Saint Petersburg: Russian State Hydrometeorological University.

Kapitsa, S. (2012). Model of Earth Population Dynamics and Future of Humankind. *Svobodnaya Mysl*, 7-8, 141–152.

Kokin, A., & Kokin, A. (2008). *Modern Environmental Myths and Utopias*. Saint Petersburg.

Kondratyev, K., Krapivin, V., & Savinykh, V. (2003). *Development Prospects of the Civilization. Multidimensional Analysis*. Moscow: LOGOS.

Kovach, R., & McGuire, B. (2003). *Philip's Guide to Global Hazards*. London: Philip's.

Lomborg, B. (2001). *The Skeptical Environmentalist: Measuring the Real State of the World*. Cambridge: Cambridge University Press. doi:10.1017/CBO9781139626378

Lomborg, B. (2004). *Global Crises, Global Solutions*. Cambridge: Cambridge University Press. doi:10.1017/CBO9780511492624

Malthus, T. (1798). *An Essay of the Principle of Population*. London: J. Johnson.

Meadows, D. H., Meadows, D. L., Randers, J., & Behrens, W. W. (1972). *The Limits to Growth*. New York, NY: Universe Books.

Odum, J. (1986). *Ecology*. Moscow: Mir.

Porshnev, B. (1963). *Modern Status of Question or Relict Hominids*. Moscow: All-Russian Institute of Scientific and Technical Information of the Russian Academy of Science.

Sapunov, V. (2012). *Environmental Challenges of the Mankind and Their Solutions*. Saarbrucken: Palmarium Academic Publishing.

Vavilov, N. (1987). *Five Continents*. Leningrad: Nauka.

Vernadsky, V. (2002). *Biosphere and Noosphere*. Moscow: Rolf.

ADDITIONAL READING

EcoLeaks. (n.d.). *The Blog of Environmental Sceptics*. Retrieved from http://ecoleaks.info/

Gause, G. (1934). *The Struggle for Existence*. New York, NY: New York Academic Press. doi:10.5962/bhl.title.4489

Sapunov, V. (2013). Prediction of Natural Disasters Basing on Chrono-and-Information Field Characters. *Geophysical Research Abstracts, 15*, EGU2013–EGU2053.

Sapunov, V. (2016). The Increase of Risk of Extreme Situations in the Zone of the Baltic Sea and their Prediction by Traditional and Nontraditional Methods. *Proceedings of the XVII International Environmental Forum "Baltic Sea Day"*. Saint Petersburg: Environmental Union.

Sapunov, V. (2017). Phytomass of Ecosystems as the Basis for Attraction of Desirable and Undesirable Organisms. *Proceedings of the 9th International Conference of Urban Pests*. Birmingham: University of Birmingham.

KEY TERMS AND DEFINITIONS

Autotrophicy: A production of organic compounds from non-organic matter.
Biotechnology: An industrial production of organic compounds.
Ecological Niche: A place of species within the biosphere.
Ecological Pyramid: A diagram, describing the relation of biological mass of the organisms of different food demands.
Eutrophication: An increase of biological mass in water.
Global Climate: A climate in global scale.
Vernadsky Law: A sum of biological mass tends to its maximization.

Chapter 2
Current Tendencies in Agricultural Trade and the Roles of Major Actors

Bistra Vassileva
https://orcid.org/0000-0002-5976-6807
University of Economics-Varna, Bulgaria

ABSTRACT

Globalization was signified as a quintessence of the last decade of the century. Its effect on international trade is outstanding because of the interconnectedness of the international markets. The current wave of globalization is driven by major technological breakthroughs in transportation and communication. Major market actors are witnessing lower trade barriers and reduced transaction costs due to numerous technological innovations. The chapter starts with a literature review on global market tendencies and their implications on agricultural trade. The first section begins with a comparative analysis of the agricultural trade flows by globalization waves. The second section describes the major actors in agricultural trade and their global market positions. In the third section, network analysis is used to visualize the connections between the major actors. Their behavioral pattern is presented as well as the key interactions are revealed. The chapter ends with recommendations aimed at the future development of agricultural trade with a focus on global intervention policies.

INTRODUCTION

Nowadays, we are witnessing a growing integration of economic and societal trends and interrelations. An extensive range of actors, such as governments, international organizations, business, labor, and civil society are engaged in various economic and financial interactions, policy debates, and challenges provoked by numerous technological breakthroughs. Le Pere (2005, p. 36) outlines the following two controversial drifts which characterize our world. On the one hand, trade, investment, technology, cross-border production systems, and flows of information and communication bring economies and societies closer together. On the other hand, the new transnational dynamics marginalize and exclude a

DOI: 10.4018/978-1-7998-1042-1.ch002

great proportion of the world's countries and people. Are we coming closer in today's globalized world or are we going apart despite the apparent increasing connectedness? From macroeconomic perspective a continuous homogenization of policies and institutions, international standards and standardization, regulatory agreements on international trade is taking place toward the establishment of a common framework of international rules, norms, and practices. The endeavor to achieve a complete standardization of international trade policy is debatable especially in today's dynamic environment which requires agile approaches to global market governance.

World economic performance regarding trade development and general economic climate improved in 2017 compared to 2016. World GDP growth has been reported as at least 3% (United Nations [UN], 2018) and is expected to remain steady in the coming years (International Monetary Fund [IMF], 2018). The growth of world merchandise trade volume is estimated at a rate of 3.6% (European Commission, 2018). Global agricultural trade has been expanding in line with these overall trends. The top five exporters (EU, USA, Brazil, China, and Canada) and the top five importers (EU, USA, China, Japan, and Canada) recorded increasing agri-food export/import values in 2017 (4.3% and 5.3%, respectively) compared to 2016 (European Commission, 2018). Strong economic growth in emerging economies has driven the demand for agricultural products globally. In emerging economies and developing countries, changes in both income and its distribution have also led to changes in consumption patterns (Food and Agriculture Organization of the United Nations [FAO], 2018).

The chapter starts with a review on global market tendencies and their implications on agricultural trade. The first section begins with comparative analysis of the agricultural trade flows during the globalization waves. Second section describes the major actors in agricultural trade and their global market positions. In third section, a network analysis is used to visualize the connections between the major actors. Their behavioral pattern is presented as well as the key interactions are revealed. The chapter ends with conclusions and recommendations aimed at future development of agricultural trade with a focus on global intervention policies.

The chapter offers a holistic approach to assessing global market tendencies and their implications on agricultural trade. This approach combines historical trends of agricultural trade at different levels of analysis with positioning matrices and indices. Data and information which is available are analyzed cross-sectionally and comparatively for decision-making purposes. Despite the apparent variety of information sources and detailed data about the agricultural trade, it was a real challenge to find the exact data which were needed to achieve the abovementioned goal.

BACKGROUND

The world is more interconnected than ever. Disruptive technologies, rapid structural changes and economic turbulence are impacting the global economy by accelerating the rise of complexity. Complexity turns out to be a major force that business must consider when developing and executing its strategies. It affects businesses both by delivering challenges and opening new opportunities which means that complexity changes radically the way business is managed. Globalization has changed the strategic context for business and nowadays it is viewed not only as a geographical expansion, but rather as a new operating theory of the world based on connectedness among pre-existing political, social, economic, cultural, and geographic boundaries (Singer, 2006, p. 51). Both connectedness and its complexity have become a source of instability and risk, as well as a driver for accelerating the reorganization of the global

Table 1. Distinguished contributions to complexity economics

Author	Contribution	Focus / Results
Arthur (1994)	Study on increasing returns	■ Economics of high technology ■ Cognition in the economy ■ Financial markets
Nelson and Winter (1982); Hodgson (1998)	Evolutionary models	■ Changes in technology and routines ■ Framework for the analysis of changes in technology ■ Steady change
Brock and Durlauf (2001)	Study of social interaction	■ Social interactions as a determinant of behavior ■ Use of parametrizations to embody social interactions ■ Baseline model of social interaction effects ■ Empirical implications of the model
Glaeser, Sacerdote, and Scheinkman (1996)	Treatment of crime	■ Model of social interactions to explain the high cross-city variance of crime rates ■ Index of social interactions
Bowles (2006)	Treatment of institutional evolution	■ Strong reciprocity – a model of cooperation and punishment
Axtell (2001)	Study of firm size	■ Firm sizes are characterized by Zipf distribution
Follmer, Horst, and Kirman (2005)	Models of financial markets	■ Models which encompass the possibility of a crisis ■ Microeconomic foundations of instability in financial markets ■ New class of models which is proposed as alternative to the models associated with the CAPM paradigm
Gintis (2009); Bowles, Choi, and Hopfensitz (2003)	Models of co-evolution	■ Agent-based simulations of a model of a semi-structured population ■ Endogenous preferences: markets and other economic institutions
Gintis (2006)	Agent-based simulation	■ Agent-based simulations of general equilibrium and barter exchange
Tesfatsion and Judd (2006)	Comprehensive overview of computational methods in complexity economics	■ ACE (Agent-based Computational Economics) modeling tools for the study of macroeconomic systems

Source: Author's development

economic landscape. More or less, complexity has become the new norm for business, requiring a new perspective (Table 1). The essence of complexity, according to Arthur (2013), is about the formation of structures and how this formation affects the objects causing it. Over the years it has been examined in different economic and cultural contexts. Although complexity is a multidisciplinary concept derived from mathematics and physics, the extra complications arising in economics because of the problem of interacting human calculations in decision-making which add a layer of complexity that may not exist in other disciplines (Rosser, 1999, p. 171).

Network theory has various implications in the field of economics (Table 2). According to Powell (1990), network forms of organization are an alternative to markets and hierarchies. This notion has been further developed by several researchers such as Podolny (1993) and Zuckerman (1999) who proposed that networks created status and category differences in global markets.

The pattern of connections between network components is almost always crucial to the behavior of the system (Newman, 2010, p. 2). The pattern of connections in a given system can be represented as a network, the components of the system being the network vertices and the connections the edges. Networks are often classified according to the characteristics of nodes and edges. When nodes of a network represent individual people, groups, organizations, or any other kinds of social actors at any

Table 2. Summary of the definitions and characteristics of the networks

Author	Definition
Powell (1990, p. 295)	Network forms of organization – typified by reciprocal patterns of communication and exchange – represent a viable pattern of economic organization.
Oh and Monge (2016, p. 2)	By definition, a network is nothing other than a collection of points linked in pairs by lines, no matter how large or complicated it is. Networks capture only the very basic relational patterns among the individual components of a whole system, and little else.
Newman (2010, p. 1)	A network is, in its simplest form, a collection of points joined together in pairs by lines.
Burt (1995); Bourdieu (1985)	Networks create social capital for individuals.
Putnam (2000); Portes and Sensenbrenner (1993)	Networks create social capital for communities.
White, Burton, & Dow (1981); Baker (1984); Granovetter (1985)	Networks embed transactions in a social matrix, creating markets.

Source: Author's development

levels, and when the links are their social relations, the network is called a social network (Wasserman & Faust, 1994).

The New Trade Theory (NTT)[1] also deals with the international patterns of trade from the perspective of the substantial economies of scale and network effects that can occur in key industries. The growth of globalization quite often is explained through the concepts of the New Trade Theory. Before 1980, mainstream international trade theory had focused on trade in different products between different countries, and comparative advantage was held to be the main driving force behind trade (Medin, 2014). Based on these assumptions the role of the industrial policy was quite limited, at best, it was presumed that only certain actors could benefit while the society as a whole will generate net losses. Another aspect of the NTT is related to the trade patterns. According to Ghironi and Melitz (2005), formal models of international macroeconomic dynamics do not usually address or incorporate the determinants and evolution of trade patterns. They suggest that modern international macroeconomics neglects to analyze the effects of macro phenomena on its microeconomic underpinnings. Similarly, much of trade theory does not recognize the aggregate feedback effects of micro-level adjustments over time (Melitz, 2003).

An interesting approach to economic complexity and networks provide a group of scientists who developed "The Atlas" project[2]. Their theoretical and methodological framework is based on the concept of product space which is defined as a network of products and paths through which productive knowledge is more easily accumulated within a country. The diversity of knowledge across individuals and their ability to combine it through complex web of interactions is considered a critical element for the prosperity of a country (Hausmann et al., 2014, p. 15). Thus, economic complexity is explained as "a measure of how intricate this network of interactions is and hence of how much productive knowledge a society mobilizes" (Hausmann et al., 2014, p. 18). Following the logic of economic complexity, in order to operate efficiently, companies should rely on a large set of complementary systems, networks, and markets (Table 3).

Communities of agricultural products such as tropical agriculture and cotton, rice and soy tend to be low in complexity. Tropical agriculture is considered a peripheral community. Community of food processing holds an intermediate position because it is connected to many products but not very sophisticated. A snapshot of agricultural products ranking and a graph presenting the Chinese export of agricultural products by countries in 2016 are presented in Table 16, Appendix 1.

Table 3. Characteristics of agricultural product communities*, SITC4 classification

Community name	Average PCI	Number of products	World Trade	World Share, %	Top 3 Countries by Export Volume	Top 3 Countries by Number of Products
Food Processing	-0.07	26	603B	2.74	DEU, ITA, USA	SRB, ESP, BEL
Cereals & Vegetable Oils	-0.34	21	295B	1.34	USA, BRA, ARG	PRY, MDA, ARG
Meat & Eggs	0.64	23	242B	1.10	USA, BRA, DEU	FRA, BEL, POL
Tropical Agriculture	-1.95	16	190B	0.86	IDN, NLD, MYS	IDN, CIV, CRI
Misc Agriculture	-0.79	22	170B	0.78	BRA, DEU, FRA	ESP, TZA, NIC
Milk & Cheese	1.14	7	134B	0.61	DEU, FRA, NLD	NLD, BLR, LTU
Cotton, Rice, Soy & Others	-2.25	18	96B	0.44	USA, IND, THA	TZA, MOZ, GRC
Fruit	-0.58	4	45B	0.21	ESP, USA, CHL	NLD, LBN, LTU
Animal Fibers	-0.85	7	12B	0.06	AUS, CHN, ITA	URY, NZL, ZAF

Note: PCI – Product Complexity Index
* According to the Atlas methodology the nearly 800 products in the SITC4 classification were grouped into 34 communities
Source: Hausmann et al. (2014, p. 55); Atlas of Economic Complexity (n.d.)

The web of global economic connections is growing deeper, broader, and more intricate (Manyika et al., 2016, p. ii). The world economy is tied by global goods trade, financial flows, and cross-border flows of data. The network of global inflows and outflows is continuing to evolve and being part of it contributes to the economic value generated by market participants (regional blocks, countries, companies, etc.). According to Manyika et al. (2014, 2016), the biggest benefits of trade flows go to countries at the center of the global network. The MGI report in 2016 found that countries at the periphery of the network of data flows stand to gain even more than those at the center. The gravity center of the global flows network fluctuates thus affecting the agricultural trade as well. The shifting balance of power has been indicated as a transition from Globalization 2.0 (Western-dominated) to Globalization 3.0 (China-dominated) (Walker, 2007). For the first time in history, emerging economies are counterparts on more than half of global trade flows, and South-South trade is the fastest-growing type of connection (Manyika et al., 2016, p. 4).

Two ties (lower and upper) and six segments of countries could be identified, Singapore being a separate case with the highest flow intensity and MGI Index value (Figure 1). The lower tier consists of countries with the flow intensity up to 150. The following three segments based on the MGI Index value within the lower tie are determined: Segment L1 – the MGI Index value ranging from 0 to 20; Segment L2 – the MGI Index value ranging from 21 to 40; Segment L3 – the MGI Index value above 40. The upper tier consists of countries with the flow intensity ranging from 151 to 250. The following three segments based on the MGI Index value within the upper tie are determined: Segment U1 – the MGI Index value ranging from 0 to 20; Segment U2 – the MGI Index value ranging from 21 to 30; Segment U3 – the MGI Index value ranging from 31 to 40. It is evident that different countries develop and follow different strategies toward their positions on global markets.

Regions, regional blocks, and countries are connecting with the global economy in a myriad way and to varying degrees. International trade relations are affected by many and various actors on the global scene which pursue different goals using numerous means to achieve them. These actors are located at different levels of the pyramid of international trade relations (Figure 2).

Figure 1. Countries positioning by MGI Connectedness Index and flow intensity [3]
Source: Author's development based on Manyika et al. (2016, p. 12)

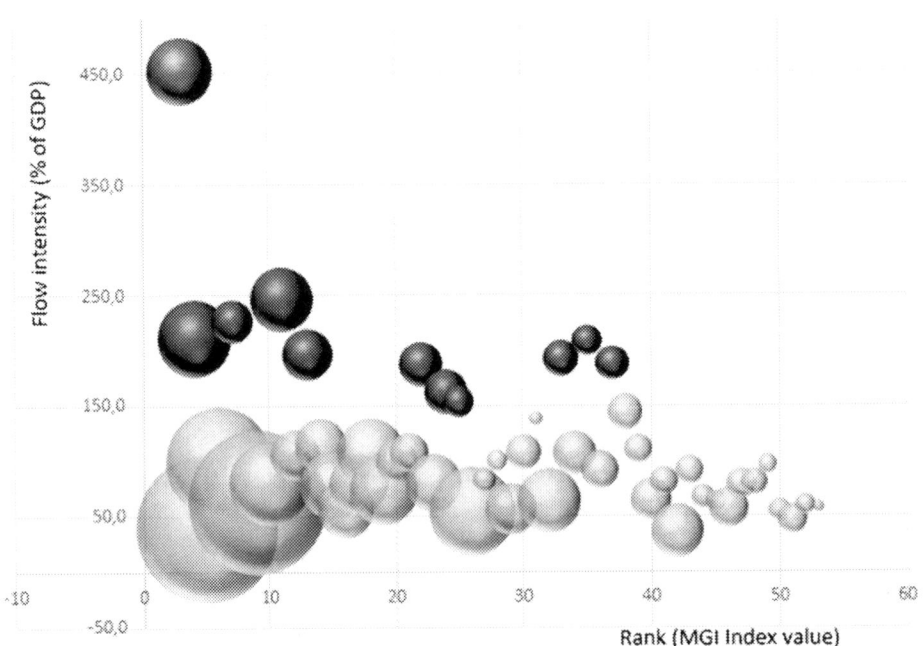

The Food and Agriculture Organization of the United Nations (FAO) plays a vital role in international agricultural trade relations. It works towards transparent and efficient global agricultural commodity markets and supports Member Nations in formulating and implementing agricultural and trade policies. Through its Strategy on Climate Change, FAO delivers transformational solutions for adaptation and mitigation in agriculture at global, national and local levels. FAO works in close interrelations with the World Trade Organization (WTO). The Organization for Economic Cooperation and Development (OECD) is another FAO's key partner. The efforts of these international organization are focused on providing support and guidance for the implementation of the following two global initiatives. The first initiative is the UN Sustainable Development Goals (SDGs) which set ambitious targets to be achieved by 2030 such as to end poverty in all its forms everywhere and to end hunger, achieve food security and improved nutrition, and promote sustainable agriculture. The second initiative is the United Nations Framework Convention on Climate Change's 2015 Paris Agreement. According to the ambitious goal set up in the agreement, 195 countries which signed it commit to take action to contain the increase in global average temperatures to well below 2°C above pre-industrial levels. Agriculture will be involved in these actions as an active part of the solution since it accounts for more than a fifth of all greenhouse gas emissions (Organization for Economic Cooperation and Development [OECD], 2017).

International, regional and national policy plays a vital role as an integrated governance process for achieving the abovementioned goals. Bazeley[4] (2005) discussed that at the macro level, there are questions about the fit of agricultural development policy to the realities of imperfect political systems that often give priority to redistribution through networks of patronage over investment in growth. At the micro-level, there are the 'how' questions of implementation in circumstances where the public sec-

Current Tendencies in Agricultural Trade and the Roles of Major Actors

Figure 2. Major actors in the hierarchy of international trade relations
Source: Author's development

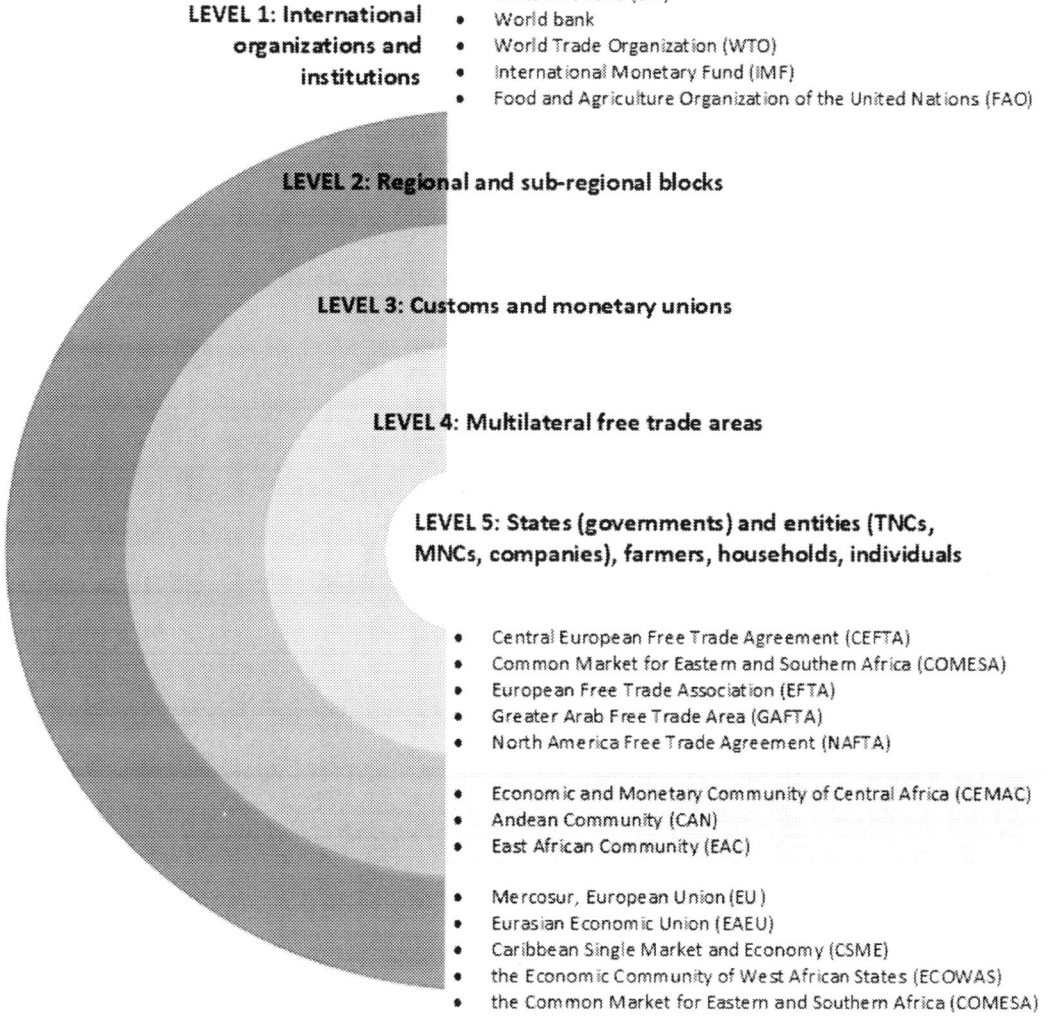

tor has low capacity. He identified three important contextual factors for analyzing agricultural policy processes as follows:

- Insufficient response in the agriculture sector (in terms of scope and scale) to technological change, institutional reforms (changing the rules of the game) and policy adjustments;
- Keeping in mind that there are two parallel processes in the agricultural sector with regards to reform: aid effectiveness and modernization of government agenda in donor countries combined with a focus on policy and institutional reform as a means to achieve change;
- Agriculture is not doing well in terms of featuring large in Poverty Reduction Strategy Papers (PRSPs)[5].

Most of the policy instruments in the field of agriculture are covered by the WTO agreements, especially the Agreement on Agriculture (AoA), which aims to limit the distortionary impact of support measures on production and trade and to establish a fair and non-discriminatory trading system that will enhance market access and improve the livelihoods of farmers around the world. Trade policies can contribute towards well-functioning international markets to which countries that experience production shortfalls due to weather shocks can resort in order to ensure food security. Global market integration can reinforce this role of trade in adaptation, as long as trade policies are combined with climate-smart domestic measures and investments (FAO, 2018). A certain balance should be kept during the implementation process of these policies in order not to distort trade. A combination of support measures (e.g. market price support and input subsidies) and investment in developing and adopting of innovative technologies in agriculture should be fostered. Adom, Djahini-Afawoubo, Mustapha, Fankem, and Rifkatu (2018, p. 301) show that climate mitigation strategies as a blend of investment in adaptation and mitigation programs in the agricultural sector and effective political institutions can boost production, especially in Africa.

MAIN FOCUS OF THE CHAPTER

Key Trends and Dynamics of Global Agricultural Trade

The food and agriculture sector is faced with a critical global challenge: to ensure access to safe, healthy, and nutritious food for a growing world population, while at the same time using natural resources more sustainably and making an effective contribution to climate change adaptation and mitigation (OECD, 2017, p. 5). According to the OECD (2015, p. 15), exports of agricultural commodities are projected to become concentrated in fewer countries, while imports become more dispersed over a large number of countries. The importance of relatively few countries in supplying global markets for some key commodities increases market risks, including those associated with natural disasters or the adoption of disruptive trade measures. Trade is expected to increase more slowly than in the previous decade. Demand growth is projected to slow considerably since the primary sources of growth in the last decade (the rising demand in China and the global biofuel sector) are no longer drivers which will support markets. Growth in food demand for virtually all agricultural commodities included in the OECD-FAO Agricultural Outlook 2017-2026 (OECD, 2017, p. 15) is anticipated to be less than in the previous decade.

Current analysis is based on a variety of secondary sources of information (Table 4).

Despite the apparent variety of information sources and detailed data about the agricultural trade, it was a real challenge to find the exact data which were needed in order to perform a cross-sectional analysis and to make comparisons by countries, products, inbound/outbound flows, etc. There was a difference in measures, the temporal dimension, the scope of gathered data provided by different sources of information. It takes a lot of time and efforts to find the information which is needed as well as it is necessary to possess certain skills to be able to extract the data from the databases because they differ by structure or from the written reports. Presented analysis makes no pretense to be comprehensive in terms of scope and details but provides an overview of the major participants on global agricultural market, key market trends, export and import flows (Figure 3).

Current Tendencies in Agricultural Trade and the Roles of Major Actors

Table 4. Secondary sources of information used for the analysis

Source of information	Type	Brief description
OECD-FAO Agricultural Outlook	Annual	It provides ten-year projections for the major agricultural commodities, as well as for biofuels and fish at national, regional and global levels.
The Agricultural Market Information System (AMIS)	Online (Agricultural Market Information System [AMIS], n.d.a)	AMIS is an inter-agency platform to enhance food market transparency and policy response for food security. It was launched in 2011 by the G20 Ministers of Agriculture.
The World Bank, DataBank, Exporter Dynamics Database	The World Bank (n.d.)	It provides interactive data cross-extraction by different indicators at Country-Product HS4-Year Level
IMF World Economic Outlook	Twice a year and online (IMF, 2019)	It presents an overview as well as more detailed analysis of the world economy; consider issues affecting industrial countries, developing countries, and economies in transition to market; and address topics of pressing current interest.
EC DG Agriculture and rural development: Monitoring EU Agri-Food Trade	European Commission (n.d.)	"Monitoring EU Agri-Food Trade" provides monthly data on EU agri-food exports and imports.
Eurostat Comext	Eurostat (n.d.a, n.d.b)	Comext is Eurostat's reference database for detailed statistics on international trade in goods. It provides access not only to both recent and historical data of the EU and its individual Member States but also to statistics of a significant number of non-EU countries.

Source: Author's development

Figure 3. The structure of Eurostat Comext database
Source: Author's development

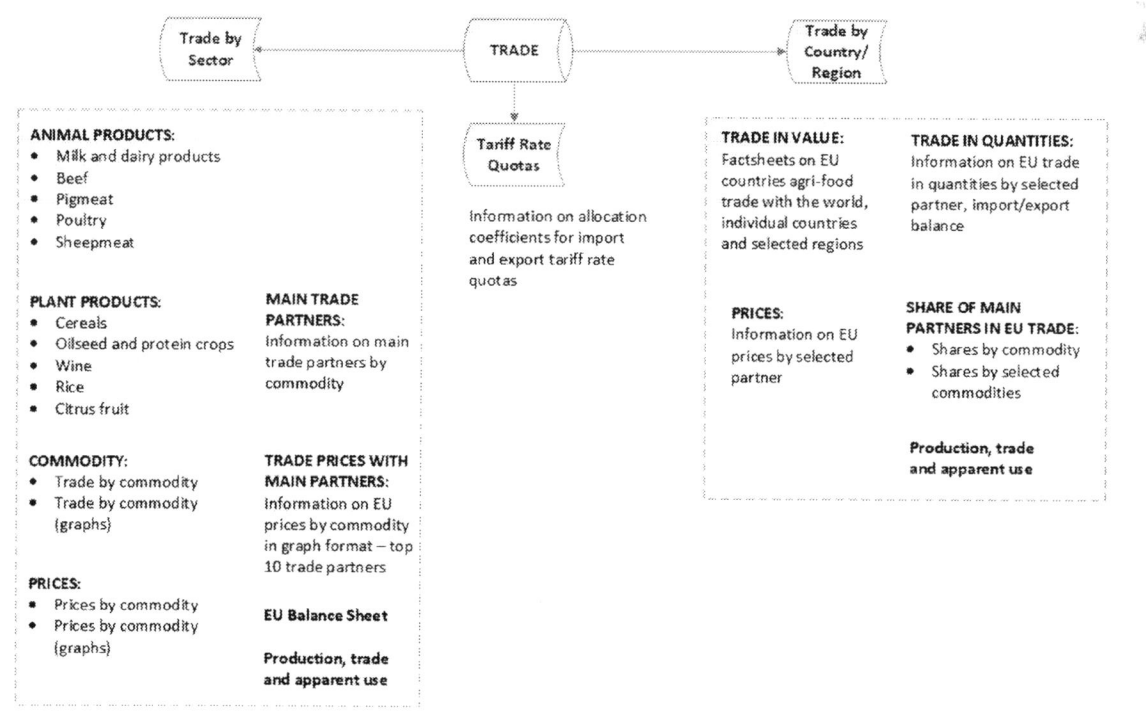

Table 5. Summary of market situation and key trends by main agricultural commodities

Commodity	Market situation	Key trends and projections
Cereals	■ Global supplies exceed overall demand ■ Lower prices compared to previous decade ■ Wheat and maize outputs increased the most	■ World trade in cereals is projected to increase 14% from the base period ■ Global cereal use is projected to grow by 13% ■ Main uncertainties will stem from the changes in demand in China and the timing of maize stocks release
Oilseeds and oilseed products	■ The main driver behind the expansion has been the growing demand for protein meals ■ Relatively tight market in view of the demand growth for vegetable oils ■ Decreased production of rapeseed in EU	■ Prices are projected to increase slightly over the next decade, except the vegetable oil prices ■ The expected expansion of soybean and palm oil production will depend on the availability of additional new land ■ Market uncertainty caused by the biofuel policies, macroeconomic environment, and weather conditions
Sugar	■ Production deficit period started in 2015 marketing year ■ Current international sugar prices are relatively high ■ Global supply shortage	■ Sugar will continue to be highly traded (33% of total production to be exported) ■ The market will continue to be influenced by production shocks, macroeconomic factors, and domestic policies ■ Exports are projected to remain concentrated (48% originating from Brazil)
Meat	■ Global meat trade recovered in 2016 ■ The imports of meat increased by China, Chile, Korea, Mexico, the EU, the Philippines, South Africa, and the UAE ■ Growth in domestic production reduced imports by the USA and Canada	■ The meat market remains relatively favorable for producers ■ Global meat production is projected to be 13% higher during the next decade ■ The traded share of meat output is expected to remain fairly constant but the share of the two largest meat exporting countries, Brazil and the United States, in global meat exports is expected to increase to around 44%
Dairy and dairy products	■ Several macroeconomic and political factors created a challenging environment for the EU dairy sector in 2015 which changed in mid-2016 ■ International dairy prices started to increase in the last half of 2016	■ There is renewed consumer enthusiasm in developed countries for butter and dairy fat over substitutes based on vegetable oil ■ More dairy products are expected to be consumed in developing countries. ■ Demand growth will support increases in dairy prices over the medium term

Source: Author's development based on OECD (2017)

Firstly, agricultural trade has increased significantly in value terms since 2000. Fast agricultural trade growth rates between 2000 and 2008 gave in to contractions during 2009-2012 and to slow-moving growth since then (FAO, 2018; OECD, 2017) (Table 5).

Conditions in agricultural markets are heavily influenced by macro-economic variables such as global GDP growth (which supports demand for agricultural commodities) and the price of crude oil (which determines the price of several inputs into agriculture) (OECD, 2017, p. 20). The macro-economic effects combined with the changes in international, regional and/or national policies (e.g. Russia's ban on imports from several countries) creates a certain instability on global agricultural market. Biofuel policies in the USA, the EU, and Indonesia, as an example, are also major sources of uncertainty because they account for a considerable share of the vegetable oil demand in these countries. The development of biofuels markets is profoundly driven by policies and crude oil prices. According to OECD (2017, p. 23), the overall growth in agricultural demand during the next decade will be mainly driven by population growth.

Secondly, the increasing importance of emerging economies has been a major development in global agricultural markets since 2000. Growing income per capita and reduced poverty advanced food consumption and imports, while increases in agricultural productivity led to growing exports. Changes in

Table 6. Import and export overview, key agricultural products in 2016-2019, million tons

	2016/2017	2017/2018	2018/2019 forecast	Change 2018/2019 over 2017/2018
Maize				
Imports (ITY)	137.80	151.83	159.51	5.06
Exports (ITY)	139.83	155.38	159.51	2.66
Wheat				
Imports (ITY)	177.53	173.44	170.50	-1.69
Exports (ITY)	176.30	176.80	170.50	-3.56
Rice				
Imports (ITY)	48.06	47.87	46.63	-2.59
Exports (ITY)	48.14	47.85	46.62	-2.56
Soybean				
Imports (ITY)	147.52	152.75	149.83	-1.91
Exports (ITY)	147.21	152.77	149.83	-1.92

Note: ITY – International Trade Year
Source: Author's development based on AMIS (n.d.b)

export patterns clearly underline the increasing importance of emerging economies in global agricultural markets (see for details next paragraph). Although traditional exporters such as the EU (Member Organization) and the USA remain at the top of the ranking in terms of the share of total export value, Brazil increased its share from 3.2% in 2000 to 5.7% in 2016. China became the fourth most important exporter, increasing its share of total export value from 3.0% in 2000 to 4.2% in 2016. Together with Brazil and China the emerging economies of India and Indonesia have increased their agricultural exports substantially. In 2016, these four countries accounted for 14.5% of global export value compared with 8.5% in 2000 (FAO, 2018). The increased participation of emerging economies in global agricultural trade reflects the speed of structural changes in these countries. That is why agricultural trade flows should be analyzed interdependently with the economic situation in emerging markets.

Thirdly, developing countries are increasingly participating in international markets. South-South agricultural trade has expanded significantly. Agricultural imports have grown faster than exports for Least Developed Countries.

The agriculture sector in Africa employs more than 60% of the total labor force and contributes about 40% to the GDP (African Development Bank, 2016). However, agricultural output in Africa has fallen from 20% of GDP in 1990 to a little below 15% in 2013.

Sub-Saharan Africa and India will account for 56% of total population growth over the next decade which determines their role to drive a large share of global demand for agricultural products. China will continue to contribute to demand for several key commodities such as meat and fish.

FAO-AMIS Database provides interactive access to data about the following agricultural products: Total cereals, Coarse grains, Maize, Wheat, Rice and Soybeans by countries (which participate in AMIS) for a period of two decades (Table 6).

Both worldwide export and import of cereals had a peak during the period from 2014-2015 to 2016-2017 which is now slowing down (Figure 4).

Figure 4. Export and import of cereals in 2000-2019, million tons
Source: AMIS (n.d.b)

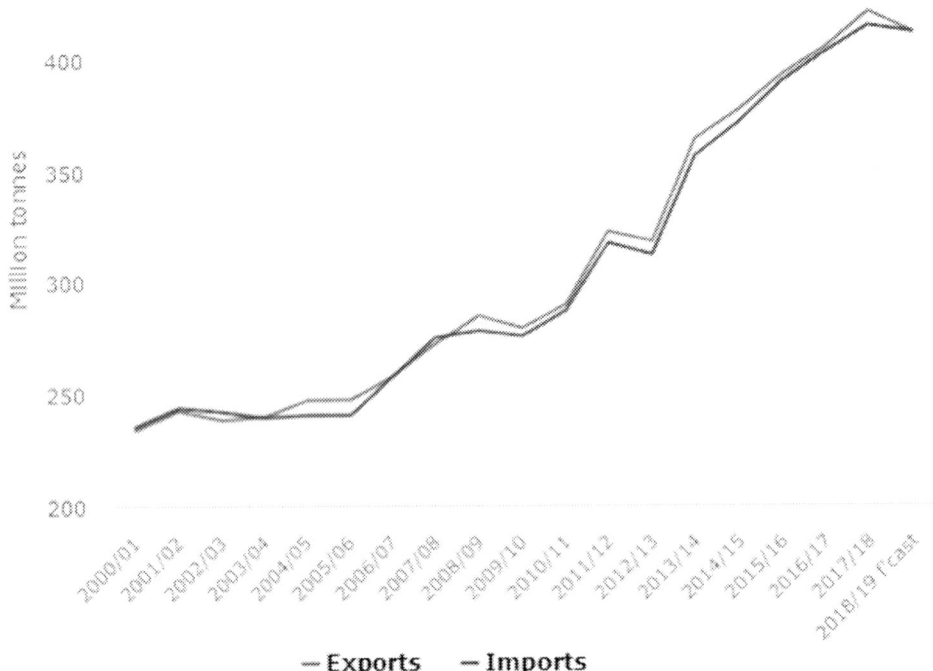

The values of total cereals exported by developed countries, BRICS, and Argentina for the period from 2001 to 2019 are presented in Table 7 and Table 8.

Global cereal production is projected to expand by 12% between the base period (2017) and 2026, mainly driven by yield growth (OECD, 2017, p. 104). Global cereal use is projected to grow by 13% to reach 2 863 Mt by 2026.

Globally, per capita food demand for cereals is anticipated to be largely flat, with growth only expected in least developed countries. Total cereals export and import values for the last two decades for the top five exporters/importers are presented in Figure 5 and Figure 6, respectively.

Figure 5. Export of cereals, top five exporters in 2001-2019, million tons
Source: AMIS (n.d.b)

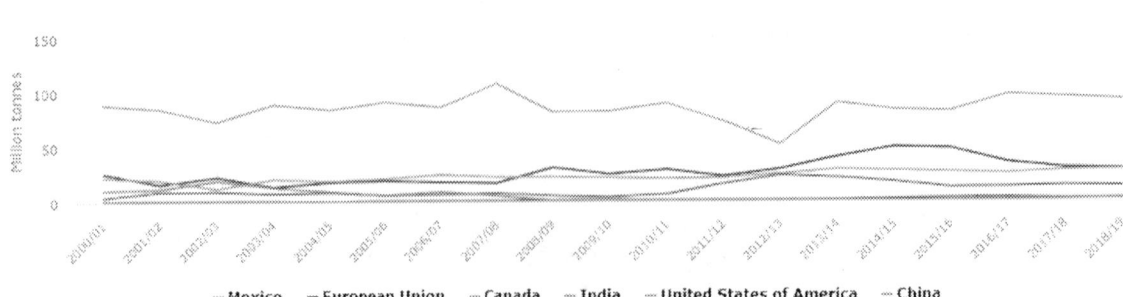

Table 7. Exports of total cereals from developed countries in 2001-2019, million tons

Years	Australia	Canada	EU	Japan	Republic of Korea	USA
2001/2002	22.20	19.15	15.91	0.60	0.41	85.64
2002/2003	15.32	10.85	22.29	0.54	0.49	73.86
2003/2004	23.04	18.88	13.31	0.57	0.21	87.91
2004/2005	22.52	17.88	17.70	0.53	0.40	84.56
2005/2006	20.77	19.59	18.16	0.50	0.17	87.31
2006/2007	14.22	23.04	16.76	0.50	0.24	87.86
2007/2008	11.35	22.44	16.28	0.52	0.07	107.78
2008/2009	18.39	21.61	30.28	0.41	0.05	79.91
2009/2010	17.97	21.16	24.71	0.43	0.06	83.26
2010/2011	23.54	21.46	28.68	0.27	0.05	91.12
2011/2012	30.72	20.98	22.39	0.48	0.05	76.05
2012/2013	27.85	23.48	28.68	0.27	0.05	53.43
2013/2014	26.23	28.07	40.48	0.27	0.05	82.02
2014/2015	23.93	29.08	48.59	0.28	0.05	83.09
2015/2016	22.18	26.34	47.74	0.24	0.05	80.27
2016/2017	32.50	24.48	35.22	0.26	0.05	100.97
2017/2018	23.71	27.75	30.41	0.26	0.11	89.76
2018/2019	16.86	30.09	28.82	0.28	0.08	95.78

Note: ITY – International Trade Year
Source: AMIS (n.d.b)

As it was mentioned above, prices on agricultural markets are subject to various effects – macro-economic forces, policies and political decisions, price levels of crude oil and other inputs to agricultural production, weather conditions, etc. During the last few years, the prices are kept well below the peaks experienced in the last decade. Average prices of cereals, meats and dairy products continued to decline, while prices of oilseeds, vegetable oils, and sugar saw a slight rebound in 2016 (OECD, 2017). For analytical and monitoring purposes several price indices were developed (Table 9).

Figure 6. Import of cereals, top five importers in 2001-2019, million tons
Source: AMIS (n.d.b)

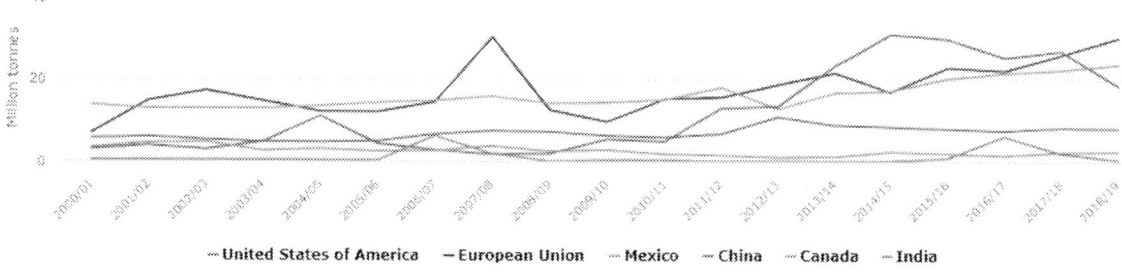

Table 8. Exports of total cereals from BRICS and Argentina in 2001-2019, million tons

Years	Argentina	Brazil	China	India	Russia	South Africa
2001/2002	21.25	5.04	9.68	9.36	7.02	1.55
2002/2003	15.77	2.45	19.19	8.15	16.24	1.20
2003/2004	17.45	7.63	15.11	8.17	6.26	1.26
2004/2005	25.65	2.98	7.29	7.99	9.16	1.23
2005/2006	19.91	2.28	8.31	5.67	12.09	2.02
2006/2007	26.17	6.16	8.46	6.90	12.29	0.68
2007/2008	26.53	12.18	5.28	6.90	13.31	0.83
2008/2009	22.58	7.83	1.30	6.01	23.58	2.63
2009/2010	20.58	8.07	1.31	5.15	21.37	1.72
2010/2011	27.45	15.37	1.19	8.01	4.50	2.64
2011/2012	34.07	11.46	0.94	16.44	27.58	2.52
2012/2013	36.56	29.01	1.04	24.90	15.45	2.40
2013/2014	17.76	24.44	0.87	21.92	25.62	2.31
2014/2015	26.81	23.23	0.57	14.81	30.79	2.13
2015/2016	31.01	37.61	0.59	11.68	34.82	0.98
2016/2017	38.35	14.01	1.43	13.70	36.04	0.87
2017/2018	42.83	32.68	2.32	13.34	52.93	2.40
2018/2019	42.77	25.66	2.51	14.12	44.50	2.11

Note: ITY – International Trade Year
Source: AMIS (n.d.b)

Table 9. Summary of the main price indices on global agricultural market

Price Index	Structure	Brief description
FAO Food Price Index	It consists of the average of five commodity group price indices weighted with the average export shares of each of the groups (for 2002-04).	It is composed of 55 commodity quotations and updated monthly.
Cereals Price Index	It is compiled using the grains and rice price indices weighted by their average trade share for 2002-04.	The grains price index consists of the IGC wheat price index and 1 maize export quotation. The rice price index is composed of the average prices of 16 rice quotations.
Oils and Fats Price Index	It consists of an average of 10 different vegetable oils weighted with average export trade shares of each oil for 2002-04.	
Sugar Price Index	It is an index form of the International Sugar Agreement prices with 2002-04 as base.	
Meat Price Index	It is computed from average prices of four types of meat, weighted by world average export trade shares for 2002-04.	Quotations include two poultry products, three bovine meat products, three pig meat products, and one ovine meat product.
Dairy Price Index	It consists of butter, skim and whole milk powder, cheese, and casein price quotations.	The average is weighted by world average export trade shares for 2002-04.

Source: Author's development based on AMIS (n.d.c)

Figure 7. IGC Grains and Oilseeds Index, April 26, 2018, to April 26, 2019
Source: AMIS (n.d.c)

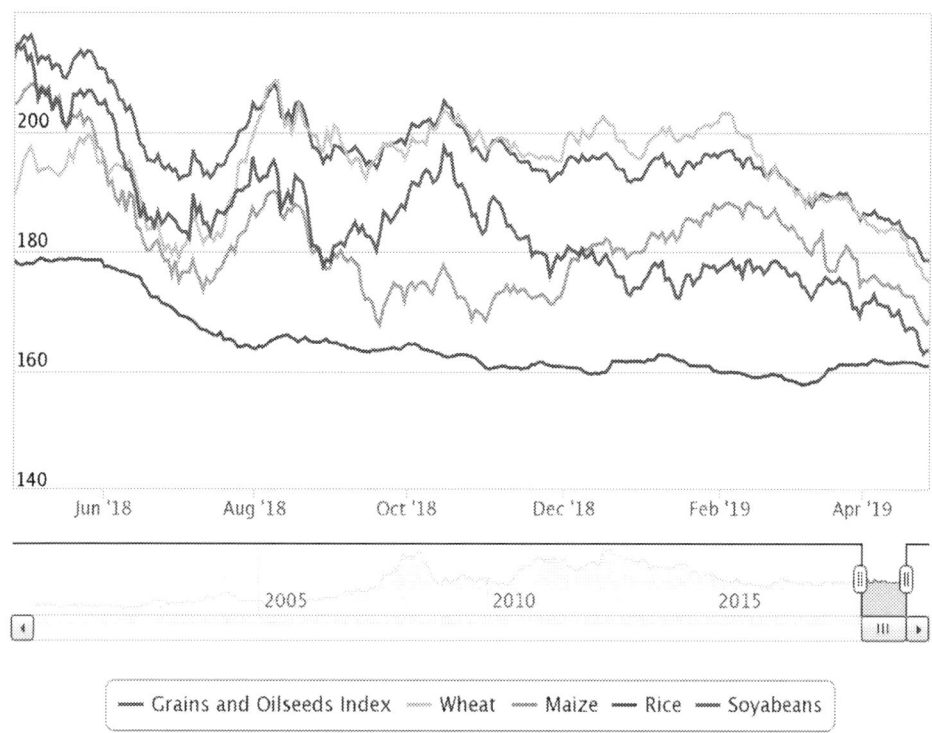

The daily IGC Grains and Oilseeds Index is comprised of the AMIS commodities plus barley, sorghum, and rapeseed/canola. With January 2000 taken as its base, component weightings are based on their five-year average share of the total trade of all commodities considered. The sub-indices for wheat, maize, rice, and soybeans are based on daily price quotations from several official and trade sources (Figure 7).

International commodity prices might indicate changes in supply and demand in major producing and consuming countries, or signal policy actions, such as a tightening of trade measures or changes in governmental purchase and stocking regimes, which have proved to be important drivers of food prices and food price volatility at global level. It should be taken into account that speculation via index funds has unjustifiably affected commodity prices (Kleinau & Lin-Hi, 2014).

The major trends of global agricultural trade presented above reveal some major problems. They can be summarized as follows. The first group of problems is policy-related. Despite the efforts of world organizations on a global scale, there is an insufficient response in terms of scope and scale to technology development, institutions and policy in the field of agriculture. The interventional policies are focused predominantly on agriculture leaving out of scope the non-agricultural determinants of agricultural productivity and growth. Policy and institutional reforms are needed in order to reconfigure, balance and sequence the efforts across sectors in order to achieve particular productivity and growth in one sector – agriculture. Additionally, financial instruments should be used in a more effective way. FDI should be seen as complementary instead of substitute efforts to domestic investment in R&D for agriculture (Adom et al., 2018). According to Cling (2014) the failure of the Doha Round of multilateral trade negotiations to reach its ambitious agenda derives from the discrepancy between the governance of world trade and the new power relationship prevailing in the world economy, with new emerging powers (China, India, etc.) rapidly increasing their share of world trade. Severe trade unbalances are caused by the multilateral trade liberalization conducted within GATT followed by WTO as well as the Marrakesh agreements. The inertia of the ongoing structural changes confirms the conclusion for expected increase of pressure for a profound reform of the governance of world trade in the next few years.

The second group of problems concerns the major actors and their relationships especially in the field of innovation. Lambrecht, Kuhne, and Gellynck (2015) show that across subsectors, different players in innovation networks play different roles. The authors observed that the majority of farmers see their relationships more as a necessity for the farm to be able to function, than as an opportunity for innovation. There is a strong preference toward symmetric relationships with similar companies on horizontal level when a collaboration for innovation is needed. A specific aspect of this group of problems is associated with the fragmentation process within the value chain. Future research on developing alternative solutions to these problems is needed, e.g. the implementation of the concept of "International Production Networks" (IPNs) in agriculture.

The third group of problems deals with the effects of the restructuring of world agricultural trade. Emerging countries are improving their position quickly in agricultural products and food industries, especially in cereals, meat, sugar, etc. for Brazil (and Argentina to a lesser extent) because they possess abundant fertile land resources (Pouch, 2012). A very limited number of developing countries is benefitting from this trend, mostly China and India. Regarding the exports of agricultural products over the last decades, developing countries' exports have fluctuated because of the fluctuations of commodity prices, but the global market shares of developing regions (Africa, South America excluding Brazil, Asia excluding China) has remained pretty stable. The share of developing countries (China and India especially) in world imports has also increased massively, while the share of industrialized countries has declined. It should be taken into account that the growth of intra-regional trade in Asia is constantly increasing that is questioning the globalization trends.

Comparative Analysis of Market Positions of the Major Agri-Food Exporters and Importers

Since 2013, the major player in the global agri-food market is EU which took over the leading position from the USA. Its share in export-import flows of agricultural products is approximately 40%. The USA is the second strongest worldwide exporter and importer with approximately 10% share (Table 10). Brazil takes the third position in the export ranking with a big gap after the USA with an export value (€72 billion) approximately half of the value of the EU export (€138 billion).

Six out of ten countries/member organizations are both top exporters and importers of agricultural products. The first six places of exporters are equally distributed between developed countries (EU28, USA, and Canada) and emerging markets (Brazil, China, and Argentina). The ratio between developed countries and emerging markets for the top six importers of agricultural products is 4:2. China and Mexico are the biggest importers among the emerging markets (Table 11). EU28, USA, China, and Canada have a strong domestic production of agricultural products but at same time are both top exporters and importers. Brazil and Argentina are primarily suppliers of agricultural products. Japan, Hong Kong SAR, Republic of Korea, and Russia are net purchasers on world agricultural markets.

Brazil has the strongest annual increase among the top five exporters (12.5% in 2017 compared to 2016), followed by the EU28 (5.34%). Chinese import shows similar growth rate of 10.75% (2017 com-

Table 10. Top ten exporters/importers of agricultural products in 2016, share of total export/import value

Exporter	Share, %	Importer	Share, %
EU	41.1	EU	39.1
USA	11.0	USA	10.1
Brazil	5.7	China	8.2
China	4.2	Japan	4.2
Canada	3.4	Canada	2.7
Argentina	2.8	Mexico	2.0
Australia	2.5	China, Hong Kong SAR	1.9
Indonesia	2.4	India	1.9
Mexico	2.3	Republic of Korea	1.9
India	2.2	Russian Federation	1.9

Source: Author's development based on FAO (2018, pp. 5-6).

Table 11. Top five importers and exporters of agricultural products in 2015-2017, annual growth rate

Exporter	2016/2015, %	2017/2016, %	Importer	2016/2015, %	2017/2016, %
EU	1.55	5.34	EU	-1.75	4.46
USA	2.34	0.76	USA	1.82	2.68
Brazil	-4.48	12.5	China	-5.10	10.75
China	4.35	2.08	Japan	-2.04	4.17
Canada	-2.5	2.56	Canada	no change	no change

Source: Author's development based on Eurostat (n.d.a)

pared to 2016). It is obvious that the export/import growth rates for emerging markets are higher than the rates for developed markets.

The business climate in developed countries is excellent because of the very low values of the business and country risk level (Table 12 and Table 13).

The values of the business and country risk level in emerging countries is reasonable or fairly high, except Argentina showing a high country risk level. Top 10 exporters and importers of agricultural products differ a lot regarding their GDP per capita, especially India and Indonesia. It was mentioned above, that India will be a major player on the agricultural market because of the size of its population and the growth rate of it. These results support the statement about the dependence of the agri-food market situation on macro-environmental factors and policy issues.

EU28 Profile as a Main Exporter/Importer on Global Agri-Food Market

Agriculture and the food-related industries and services together provide almost 44 million jobs in the EU, including regular work for 22 million people within the agricultural sector itself (European Commission, 2015). The output of the EU agricultural sector was estimated at €427 billion in 2017, which is a strong increase compared to the 2016 value of €406 billion (European Commission, 2018).

Table 12. Comparative assessment of the top ten exporters of agricultural products in 2019

Country	Population, millions of people	GDP per capita, USD	Country risk assessment	Business climate
EU	512.6			
USA	325.9	59,792	A2	A1
Brazil	207.7	9,896	B	A4
China	1,390.1	8,643	B	B
Canada	36.7	45,095	A2	A1
Argentina	44.1	14,463	C	B
Australia	24.8	55,693	A2	A1
Indonesia	262.0	3,876	A4	A4
Mexico	123.5	9,319	B	A4
India	1,316.9	1,976	B	B

Note: Country risk assessment and Business climate are measured as risk level. The legend for the risk level is as follows: A1 – Very Low; A2 – Low; A3 – Satisfactory; A4 – Reasonable; B – Fairly High; C – High; D – Very High; E – Extreme.

There are no data available for EU since it is a Member Organization.

Source: Eurostat (n.d.a)

Table 13. Comparative assessment of the top 10 importers of agricultural products in 2019

Country	Population, millions of people	GDP per capita, USD	Country risk assessment	Business climate
EU	512.6			
USA	325.9	59,792	A2	A1
China	1,390.1	8,643	B	B
Japan	126.7	38,449	A2	A1
Canada	36.7	45,095	A2	A1
Mexico	123.5	9,319	B	A4
China, Hong Kong SAR	7,4	46,080	A2	A1
India	1,316.9	1,976	B	B
Republic of Korea	51,5	29,938	A2	A1
Russian Federation	144.0	10,956	B	B

Note: Country risk assessment and Business climate are measured as risk level. The legend for the risk level is as follows: A1 – Very Low; A2 – Low; A3 – Satisfactory; A4 – Reasonable; B – Fairly High; C – High; D – Very High; E – Extreme.

There are no data available for EU since it is a Member Organization.

Source: Eurostat (n.d.a)

The value of EU agri-food exports in January 2019 increased for the 4th consecutive year to reach a level of €11.2 billion. Agri-food imports also grew to a record of €10.8 billion, leading to the monthly trade value covering €22 billion, compared to €21 billion in January 2018. Altogether, between 2002 and 2018, EU trade in agricultural products doubled, equivalent to an average annual growth of 5.0% (Figure 8).

Figure 8. Export of EU-28 to extra-EU of agri-food products, € million
Source: European Commission (2019d)

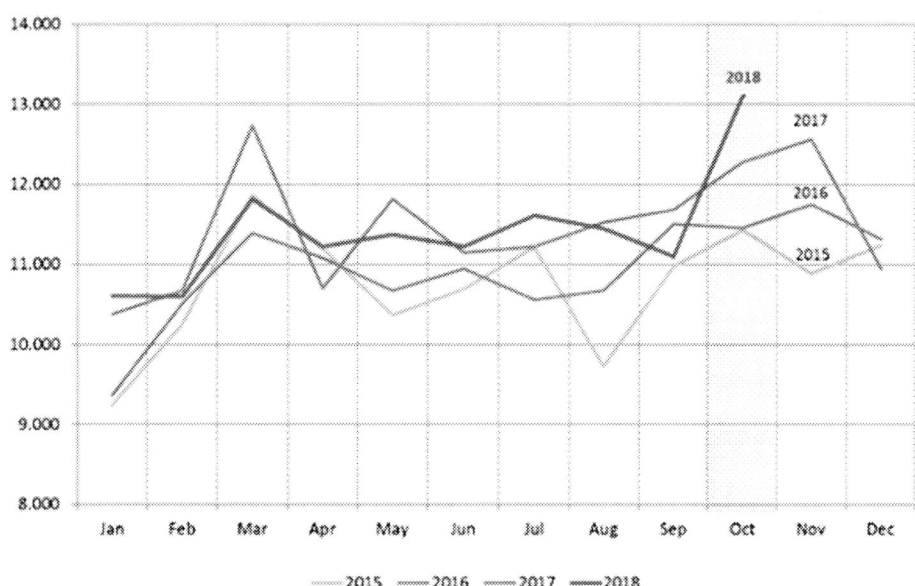

Trade surplus in January 2019 reached €0.4 billion, compared to €0.3 billion a year ago. Comparative statistics for the highest increases and decreases for EU agri-food trade on monthly basis is provided in Table 14.

The ambitious trade agenda of EU as stated by Phil Hogan, Commissioner for Agriculture and Rural Development (European Commission, n.d.), is aimed at supporting the EU farmers and food producers to make full use of the opportunities of international markets while recognizing the need to provide sufficient safeguards for more sensitive sectors. The EU trade agenda is grounded on the recent successes of EU in negotiations with Canada, Japan, and Mexico.

Figure 9. EU-28 exports, imports, and trade balance of agricultural products in 2002-2018
Source: European Commission (2019b)

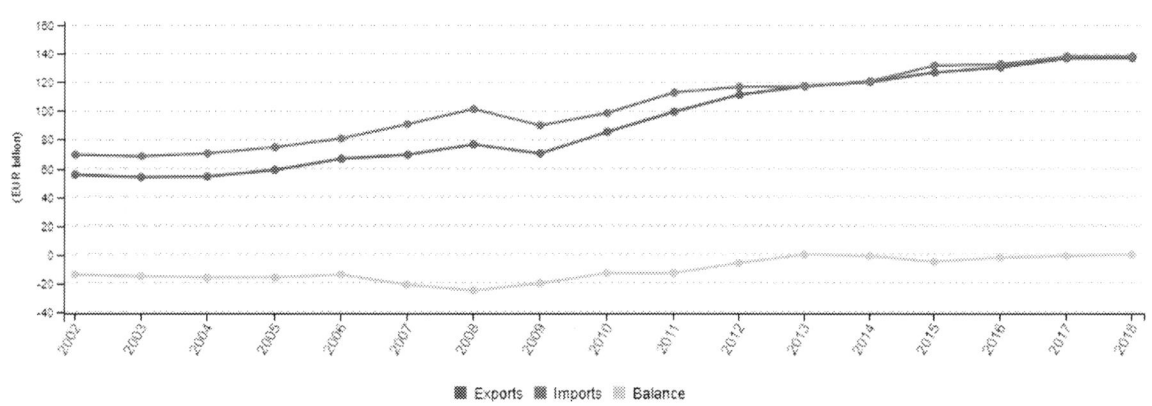

Table 14. Highest increases and decreases for EU agri-food trade in January 2019 compared to January 2018, export/import values

Exporters				Importers			
Highest	**€ million**	**Lowest**	**€ million**	**Highest**	**€ million**	**Lowest**	**€ million**
USA	+191	Hong Kong SAR, China	-44	USA	+291	Brazil	-145
China	+91	Turkey	-29	Ukraine	+246	Indonesia	-97
Switzerland	+39	Angola	-26			Malaysia	-68
Philippines	+29	Saudi Arabia	-24				
Russia	+29	Jordan	-23				

Source: European Commission (2019c)

According to the Eurostat data, exports and imports of agricultural products between the EU and non-member countries in 2018 accounted for 7.0% of total EU international trade. This is 0.4% less than in 2017. Data on trade in agricultural products is central for two important EU policies: Common Agricultural Policy (CAP) and the common trade policy, which manages trade relations with non-EU countries.

The value of trade (imports plus exports) of agricultural goods between the EU-28 and the rest of the world was €275 billion in 2018. It is almost evenly divided between exports at €137 billion and imports at €138 billion which accounts for only a small trade deficit. Between 2002 and 2018, trade measured in value more than doubled, equivalent to an average annual growth of 5.0%, with exports (5.8%) growing faster than imports (4.3%) (Figure 9).

The volume of total agricultural trade shows that in 2018 the EU-28 imported 151 million tons of agricultural products, while it exported only 99 million tons. Between 2002 and 2018, the total trade volume had an average annual growth rate of 2.1%. Here too, exports (3.1%) grew faster than imports (1.5%). The average annual increase in prices for exports (2.6%) was lower than for imports (2.8%) (Figure 10).

The USA was the main recipient of EU-28 exports of agricultural goods, with 16% of the total (Table 15). It was followed by China (8%), Switzerland (6%), Japan and Russia (both 5%), and Norway (4%). Brazil and the USA (both 9%) were the main origins of agricultural imports. They were followed by Norway and China (both 5%), Argentina, and Ukraine (both 4%). China, the USA, and Norway appear as one of the top six partners for both exports and imports. For the USA, this is also the case for each of the three product groups discussed below.

SOLUTIONS AND RECOMMENDATIONS

The growth in agriculture trade is projected to slow to about half the previous decade's growth rate (OECD, 2017). Food imports are becoming increasingly important for food security, particularly in Sub-Saharan Africa, North Africa, and the Middle East. Climate change will have an increasingly adverse impact on many regions of the world, especially many countries in Africa, Asia, and Latin America which already suffer from poverty, food insecurity and various forms of malnutrition (FAO, 2018). Agriculture in these regions will be negatively affected.

Figure 10. EU-28 exports and imports of agricultural products by product category in 2018
Source: Eurostat (n.d.a)

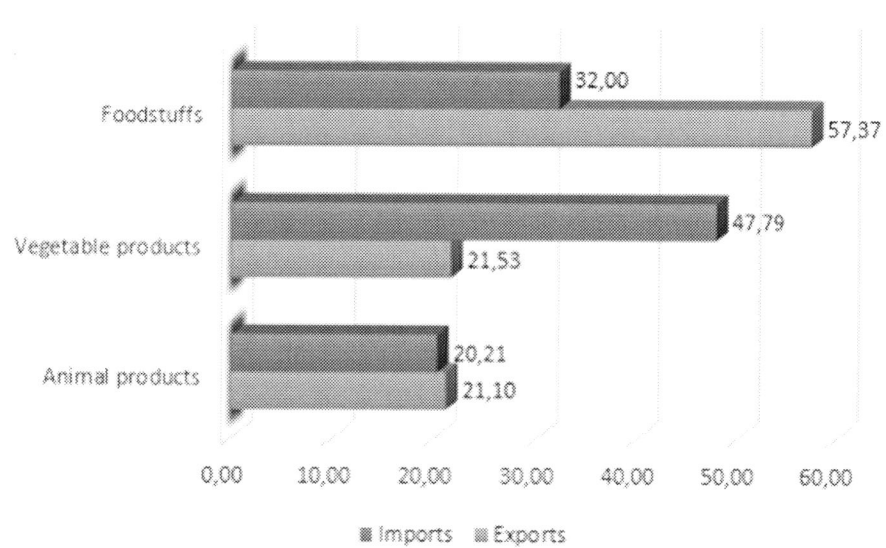

Table 15. EU-28 exports and imports of agricultural products by main partner in 2018

Countries	Total agricultural production		Animal production		Vegetable production		Foodstuffs	
	export	import	export	import	export	import	export	import
USA	16	9	10	4	12	11	20	9
China	8	5	14	9	0	4	7	5
Switzerland	6	0	5	0	9	0	6	6
Japan	5	0	8	0	0	0	4	0
Russia	5	0	0	0	6	0	6	0
Norway	4	5	0	24	6	0	4	0
Republic of Korea	0	0	4	0	0	0	0	0
Hong Kong SAR, China	0	0	4	0	0	0	0	0
Saudi Arabia	0	0	0	0	5	0	0	0
Algeria	0	0	0	0	5	0	0	0
Brazil	0	9	0	0	0	9	0	12
Argentina	0	4	0	5	0	0	0	7
Ukraine	0	4	0	0	0	7	0	0
New Zealand	0	0	0	5	0	0	0	0
Iceland	0	0	0	4	0	0	0	0
Indonesia	0	0	0	0	0	4	0	0
Turkey	0	0	0	0	0	4	0	0
Cote d'Ivoire	0	0	0	0	0	0	0	7
Other	57	64	55	49	57	62	54	55

Source: Eurostat (n.d.a)

Many initiatives have been already launched and many things have been already done. In order to accomplish the ambitious goals toward sustainability, poverty reduction, food security, etc. long-term planned collaborative and continuous efforts from all actors on different hierarchical levels are required. These efforts and activities should be effectively governed and orchestrated by the international bodies and institutions in close interrelationship with national policy-makers. In short, a feasible policy and a different paradigm is needed on how to reform agriculture in order to have a real impact on growth and people's wellbeing in a long terms and in a sustainable way.

Generally, agricultural trade has proven to be more resilient to macroeconomic fluctuations than trade in other goods but due to climate change, it is expected that the global picture of agriculture in the future will be quite different. Climate change can widen the economic gap between developed and developing countries which means that the countries will face a disproportionate risk levels.

The policy-makers should do their best to improve the sensibility of agricultural policy and agricultural sector as a whole to technological change (scope and scale) and institutional reforms (changing the rules of the game). Agricultural policy should be adjusted to the different needs of the various market participants – regions, countries, farmers, etc.

FUTURE RESEARCH DIRECTIONS

The main challenge during the writing process was to find the proper framework to integrate the right data processed in the right way to analyze them and to draw conclusions leading to useful recommendations. It is almost impossible to go through the available data and policy documentation. Effective actions require effective data-driven decision-making process. There are so many indicators, measures, metrics, interactive database platforms which sometimes perplex the analysts. A common framework of indicators aimed at policy-makers is needed to support them how to plan and how to implement structural adjustments in the agricultural sector, how to improve regional agricultural innovation systems, how to react pro-actively and timely on macro-economic and market changes, how to plan the agricultural production mix in mid and long term. Future research could investigate more deeply the potential for regional trade integration, which is reinforced by international production networks. Regional trade agreements might be an increasing alternative to multilateral trade agreements.

CONCLUSION

Global production has continued to increase to meet demand and trade has expanded significantly, with its composition and pattern following changes in demand and the emergence of new agricultural exporters and importers. The increased importance of emerging economies such as Brazil, China, India, Indonesia, and Russia has been a major development in world agricultural markets. Changes in trade patterns also include increased trade between developing countries. (FAO, 2018). Understanding the dynamics and trends that drive changes in the pattern and composition of agricultural trade is key for analysis of the effects of climate change in world agricultural markets and the linkages between trade and food security.

Despite the notion that agricultural trade has proven to be more resilient to macroeconomic fluctuations than trade in other goods, for some countries, additional policy reforms and investments could significantly adjust future production. Given relatively high protection in the farm sector, agricultural trade growth could be boosted by further market liberalization.

Agricultural sector faces many challenges and difficulties, such as populations growth, urbanization increases, and incomes grow, which guarantee that it will be under mounting pressure to meet the demand for safe and nutritious food.

REFERENCES

Adom, P. K., Djahini-Afawoubo, D. M., Mustapha, S. A., Fankem, S. G., & Rifkatu, N. (2018). Does FDI Moderate the Role of Public R&D in Accelerating Agricultural Production in Africa? *African Journal of Economic and Management Studies*, *9*(3), 290–304. doi:10.1108/AJEMS-07-2017-0153

African Development Bank. (2015). *African Development Report 2015. Growth, Poverty and Inequality Nexus: Overcoming Barriers to Sustainable Development*. Abidjan: African Development Bank Group.

Agricultural Market Information System. (n.d.a). *About AMIS*. Retrieved from http://www.amis-outlook.org/amis-about/en/

Agricultural Market Information System. (n.d.b). *Market Database*. Retrieved from https://app.amis-outlook.org/#/market-database/supply-and-demand-overview

Agricultural Market Information System. (n.d.c). *Prices and Price Volatility*. Retrieved from http://www.amis-outlook.org/indicators/prices/en/

Aharoni, Y., & Nachum, L. (Eds.). (2000). *Globalization of Services: Some Implications for Theory and Practice*. London: Routledge. doi:10.4324/9780203465363

Arthur, W. B. (1994). *Increasing Returns and Path Dependence in the Economy*. Ann Arbor, MI: University of Michigan Press. doi:10.3998/mpub.10029

Arthur, W. B. (2013). *Complexity Economics: A Different Framework for Economic Thought*. Santa Fe, NM: Santa Fe Institute.

Atlas of Economic Complexity. (n.d.). *Global Rankings and Projections*. Retrieved from http://atlas.cid.harvard.edu/rankings/product/

Axtell, R. L. (2001). Zipf Distribution of U.S. Firm Sizes. *Science*, *5536*(293), 1818–1820. doi:10.1126cience.1062081 PMID:11546870

Baker, W. E. (1984). The Social Structure of a National Securities Market. *American Journal of Sociology*, *89*(4), 775–811. doi:10.1086/227944

Bazeley, P. (2005). *Politics, Policies and Agriculture: The Art of the Possible in Agricultural Development*. ODI. Retrieved from https://www.odi.org/events/2390-politics-policies-and-agriculture-art-possible-agricultural-development

Bourdieu, P. (1985). The Social Space and the Genesis of Groups. *Theory and Society*, *14*(6), 723–744. doi:10.1007/BF00174048

Bowles, S. (2006). *Microeconomics: Behavior, Institutions, and Evolution*. Princeton, NJ: Princeton University Press.

Bowles, S., Choi, J. K., & Hopfensitz, A. (2003). The Co-Evolution of Individual Behaviors and Social Institutions. *Journal of Theoretical Biology, 223*(2), 135–147. doi:10.1016/S0022-5193(03)00060-2 PMID:12814597

Brock, W., & Durlauf, S. (2001). Interactions-Based Models. In J. Heckman & E. Leamer (Eds.), *Handbook of Econometrics* (pp. 3297–3380). Amsterdam: Elsevier. doi:10.1016/S1573-4412(01)05007-3

Burt, R. S. (1995). *Structural Holes*. Cambridge, MA: Harvard University Press.

Cling, J.-P. (2014). The Future of Global Trade and the WTO. *Foresight, 16*(2), 109–125. doi:10.1108/FS-06-2012-0044

Ehnts, D., & Trautwein, H.-M. (2012). From New Trade Theory to New Economic Geography: A Space Odyssey. *Oeconomia, 2*(2-1), 35–66. doi:10.4000/oeconomia.1616

European Commission. (2015). *Food and Framing – Focus on Jobs and Growth*. Retrieved from https://ec.europa.eu/agriculture/sites/agriculture/files/events/2015/outlook-conference/brochure-jobs-growth_en.pdf

European Commission. (2018). *Agri-Food Trade in 2017: Another Record Year for EU Agri-Food Trade*. Retrieved from https://ec.europa.eu/info/sites/info/files/food-farming-fisheries/news/documents/agricultural-trade-report_map2018-1_en.pdf

European Commission. (2019a). *Agri-Food Trade Statistical Factsheet*. Retrieved from https://ec.europa.eu/agriculture/sites/agriculture/files/trade-analysis/statistics/outside-eu/regions/agrifood-extra-eu-28_en.pdf

European Commission. (2019b). *Extra-EU Trade in Agricultural Goods*. Retrieved from https://ec.europa.eu/eurostat/statistics-explained/index.php/Extra-EU_trade_in_agricultural_goods

European Commission. (2019c). *Monitoring EU Agri-Food Trade: Development until January 2019*. Retrieved from https://ec.europa.eu/info/sites/info/files/food-farming-fisheries/trade/documents/monitoring-agri-food-trade_jan2019_en.pdf

European Commission. (2019d). *Record-Breaking Export Performance for EU Agri-Food Products*. Retrieved from https://ec.europa.eu/info/news/record-breaking-export-performance-eu-agri-food-products-2019-jan-10_en

European Commission. (2019e). *The Common Agricultural Policy at a Glance*. Retrieved from https://ec.europa.eu/info/food-farming-fisheries/key-policies/common-agricultural-policy/cap-glance_en

European Commission. (n.d.). *Agriculture and Rural Development*. Retrieved from https://ec.europa.eu/agriculture/trade-analysis/monitoring-agri-food-trade_en

Eurostat. (n.d.a). *Comext*. Retrieved from https://ec.europa.eu/eurostat/web/international-trade-in-goods/data/focus-on-comext

Eurostat. (n.d.b). *International Trade in Goods – Overview*. Retrieved from https://ec.europa.eu/eurostat/web/international-trade-in-goods

Follmer, H., Horst, U., & Kirman, A. (2005). Equilibria in Financial Markets with Heterogeneous Agents: A Probabilistic Perspective. *Journal of Mathematical Economics, 41*(1-2), 123–155. doi:10.1016/j.jmateco.2004.08.001

Food and Agriculture Organization of the United Nations. (2018). *The State of Agricultural Commodity Markets. Agricultural Trade, Climate Change and Food Security*. Rome: Food and Agriculture Organization of the United Nations.

Ghironi, F., & Melitz, M. (2005). International Trade and Macroeconomic Dynamics with Heterogeneous Firms. *The Quarterly Journal of Economics, 120*(3), 865–915.

Gintis, H. (2006). *The Economy as a Complex Adaptive System*. Retrieved from https://www.semanticscholar.org/paper/The-Economy-as-a-Complex-Adaptive-System-Gintis/dd497297c0745c744f07abb9da632439fe748bb5

Gintis, H. (2009). *Game Theory Evolving*. Princeton, NJ: Princeton University Press. doi:10.2307/j.ctvcm4gjh

Glaeser, E. L., Sacerdote, B., & Scheinkman, J. (1996). Crime and Social Interactions. *The Quarterly Journal of Economics, 111*(2), 507–548. doi:10.2307/2946686

Granovetter, M. (1985). Economic Action and Social Structure: The Problem of Embeddedness. *American Journal of Sociology, 91*(3), 481–510. doi:10.1086/228311

Hausmann, R., Hidalgo, C., Bustos, S., Coscia, M., Simoes, A., & Yildirim, M. (2014). *The Atlas of Economic Complexity: Mapping Paths to Prosperity*. Cambridge, MA: MIT Press. doi:10.7551/mitpress/9647.001.0001

Hodgson, G. M. (1998). The Approach of Institutional Economics. *Journal of Economic Literature, 36*(1), 166–192.

International Monetary Fund. (2018). *World Economic Outlook Update*. Washington, DC: International Monetary Fund.

International Monetary Fund. (2019). *World Economic Outlook (April 2019)*. Washington, DC: International Monetary Fund.

Kleinau, C., & Lin-Hi, N. (2014). Does Agricultural Commodity Speculation Contribute to Sustainable Development? *Corporate Governance, 14*(5), 685–698. doi:10.1108/CG-07-2014-0083

Lambrecht, E., Kuhne, B., & Gellynck, X. (2015). Asymmetric Relationships in Networked Agricultural Innovation Processes. *British Food Journal, 117*(7), 1810–1825. doi:10.1108/BFJ-05-2014-0183

Le Pere, G. (2005). Emerging Markets – Emerging Powers: Changing Parameters for Global Economic Governance. *Internationale Politik und Gesellschaft, 2*, 36–51.

Manyika, J., Bughin, J., Lund, S., Nottebohm, O., Poulter, D., Jauch, S., & Ramaswamy, S. (2014). *Global Flows in a Digital Age: How Trade, Finance, People, and Data Connect the World Economy*. New York, NY: McKinsey Global Institute.

Manyika, J., Lund, S., Bughin, J., Woetzel, J., Stamenov, K., & Dhingra, D. (2016). *Digital Globalization: The New Era of Global Flows*. New York, NY: McKinsey Global Institute.

Medin, H. (2014). *New Trade Theory: Implications for Industrial Policy*. Oslo: Norwegian Institute of International Affairs.

Melitz, M. J. (2003). The Impact of Trade on Intra-Industry Reallocations and Aggregate Industry Productivity. *Econometrica, 71*(6), 1695–1725. doi:10.1111/1468-0262.00467

Nelson, R., & Winter, S. (1982). *An Evolutionary Theory of Economic Change*. Cambridge, MA: Harvard University Press.

Newman, M. (2010). *Networks: An Introduction*. Oxford: Oxford University Press. doi:10.1093/acprof:oso/9780199206650.001.0001

Oh, P., & Monge, P. (2016). *Network Theory and Models*. Retrieved from https://onlinelibrary.wiley.com/doi/full/10.1002/9781118766804.wbiect246

Organization for Economic Cooperation and Development. (2015). *OECD-FAO Agricultural Outlook 2015*. Paris: OECD Publishing.

Organization for Economic Cooperation and Development. (2017). *OECD-FAO Agricultural Outlook 2017-2026*. Paris: OECD Publishing.

Pettinger, T. (2017). *New Trade Theory*. Retrieved from https://www.economicshelp.org/blog/6957/trade/new-trade-theory/

Podolny, J. M. (1993). A Status-Based Model of Market Competition. *American Journal of Sociology, 98*(4), 829–872. doi:10.1086/230091

Portes, A., & Sensenbrenner, J. (1993). Embeddedness and Immigration: Notes on the Social Determinants of Economic Action. *American Journal of Sociology, 98*(6), 1320–1350. doi:10.1086/230191

Pouch, T. (2012). *Crisis, Market Instability and Agricultural Policy: Analysis and Prospective Elements*. Retrieved from http://www.augurproject.eu/IMG/pdf/Pouch_Traduc_rf1_WP6-2.pdf

Powell, W. (1990). Neither Market nor Hierarchy: Network Forms of Organization. *Research in Organizational Behavior, 12*, 295–336.

Putnam, R. (2000). *Bowling Alone – The Collapse and Revival of American Community*. New York, NY: Simon & Schuster.

Rosser, J. B. Jr. (1999). On the Complexities of Complex Economic Dynamics. *The Journal of Economic Perspectives, 13*(4), 169–192. doi:10.1257/jep.13.4.169

Singer, J. (2006). Framing Brand Management for Marketing Ecosystems. *The Journal of Business Strategy, 27*(5), 50–57. doi:10.1108/02756660610692716

Study.com. (2018). *New Trade Theory (NTT): Definition and Analysis*. Retrieved from https://study.com/academy/lesson/new-trade-theory-ntt-definition-analysis.html

Tesfatsion, L., & Judd, K. L. (2006). *Handbook of Computational Economics II: Agent-Based Computational Economics*. London: North-Holland.

The World Bank. (n.d.). *DataBank*. Retrieved from https://databank.worldbank.org/source/exporter-dynamics-database-%E2%80%93-indicators-at-country%5Eproduct-hs4%5Eyear-level/Type/TABLE/preview/on

United Nations. (2018). *World Economic Situation and Prospects*. Washington, DC: United Nations.

United Nations. (n.d.). *Agreement on Agriculture: Recognition of Interests in Negotiations (Article 20 & the Preamble)*. Retrieved from https://www.un.org/ldcportal/agreement-on-agriculture-recognition-of-interests-in-negotiations-article-20-the-preamble/

Walker, M. (2007). Globalisation 3.0. Prospects for a New World Order. *The Wilson Quarterly*, 16–24.

Wasserman, S., & Faust, K. (1994). *Social Network Analysis: Methods and Applications*. Cambridge: Cambridge University Press. doi:10.1017/CBO9780511815478

White, D., Burton, M., & Dow, M. (1981). Sexual Division of Labor in African Agriculture: A Network Autocorrelation Analysis. *American Anthropologist*, *83*(4), 824–849. doi:10.1525/aa.1981.83.4.02a00040

World Trade Organization. (n.d.). *Agreement on Agriculture*. Retrieved from https://www.wto.org/english/docs_e/legal_e/14-ag_01_e.htm

Zuckerman, E. W. (1999). The Categorical Imperative: Securities Analysis and the Legitimacy Discount. *American Journal of Sociology*, *104*(5), 1398–1438. doi:10.1086/210178

ADDITIONAL READING

Agricultural Market Information System. (n.d.). *Interactive World Map*. Retrieved from http://statistics.amis-outlook.org/data/index.html

Antras, P. (2003). *Firms, Contracts, and Trade Structure*. Cambridge, MA: Massachusetts Institute of Technology. doi:10.3386/w9740

European Commission. (2019). *EU Agri-Food Trade Break Record for Start of 2019*. Retrieved from https://ec.europa.eu/info/news/eu-agri-food-trade-break-record-start-2019-2019-apr-11_en

European Commission. (n.d.). *Agricultural Trade Statistics*. Retrieved from https://ec.europa.eu/agriculture/statistics/trade_en

European Commission. (n.d.). *International Trade in Goods*. Retrieved from https://ec.europa.eu/eurostat/web/international-trade-in-goods/data/database

Kearney, A. T. (n.d.). *Global Business Policy Council*. Retrieved from https://www.atkearney.com/gbpc

Leromain, E. (2017). *Essays in International Trade: International Fragmentation of Production and Trade Costs*. Paris: Paris School of Economics.

Melitz, M. (2000). *International Trade and Industry Productivity Dynamics with Heterogeneous Producers*. Ann Arbor, MI: University of Michigan.

ENDNOTES

1. New Trade Theory (NTT) is an economic theory that was developed in the 1970s as a way to predict international trade patterns. NTT came about to help us understand why countries are trade partners when they are trading similar goods and services. Paul Krugman was a leading academic in developing the New Trade Theory. He was awarded a Nobel Prize (2008) in Economics for his contributions in modelling these ideas.
2. The research on which The Atlas of Economic Complexity is based started around 2006 with the idea of the product space (Hausmann et al., 2014). The process of self-discovery is a point of view of economic development of countries as a process of discovering which products a country can master. The first edition of The Atlas was released at the Center for International Development at Harvard University Global Empowerment Meeting, on October 27, 2011. The new edition (2013) has sharpened the theory and empirical evidence of how knowhow affects income and growth and how knowhow itself grows over time.
3. The database is provided in Appendix 2. Flows value represents total goods, services, and financial inflows and outflows. Flow intensity represents the total value of goods, services, and financial flows as a share of the country's GDP.
4. Peter Bazeley is a founder and Senior Partner of the IDL group until its acquisition by AHG in 2005. He is now a freelance consultant and farmer with an extensive experience on agriculture sector reform in Africa and Asia.
5. Poverty Reduction Strategy Papers (PRSPs) are documents required by the International Monetary Fund (IMF) and World Bank before a country can be considered for debt relief within the Heavily Indebted Poor Countries (HIPC) initiative.

APPENDIX 1

Table 16. Snapshot of agricultural product rankings

Rank	HS4 Code	Product	PCI Value
973	0701	Potatoes, fresh or chilled	-0.871
1134	0702	Tomatoes, fresh or chilled	-1.420
1186	0703	Onions, shallots, garlic, leeks, and other alliaceous vegetables, fresh or chilled	-1.760
1108	0704	Cabbages, cauliflower, kohlrabi, kale and similar edible brassicas, fresh or chilled	-1.330
896	0705	Lettuce (Lactuca sativa) and chicory (Cichorium spp.), fresh or chilled	-0.641
915	0706	Carrots, turnips, salad beets (salad beetroot), salsify, celeriac, radishes and similar edible roots, fresh or chilled	-0.694
979	0707	Cucumbers, including gherkins, fresh or chilled	-0.904
1194	0708	Leguminous vegetables, shelled or unshelled, fresh or chilled	-1.860

Source: Atlas of Economic Complexity (n.d.)

APPENDIX 2

Table 17. Ranking of countries based on global flows intensity and value, 2016

Rank	Country	Score	Flow intensity % of GDP	Flow value $ billion
1	Singapore	64.2	452	1,392
2	The Netherlands	54.3	211	1,834
3	USA	52.7	39	6,832
4	Germany	51.9	99	3,798
5	Ireland	45.9	227	559
6	United Kingdom	40.8	79	2,336
7	China	34.2	63	6,480
8	France	30.1	80	2,262
9	Belgium	28.0	246	1,313
10	Saudi Arabia	22.6	106	790
11	United Arab Emirates	22.2	196	789
12	Switzerland	18.0	115	848
13	Canada	17.3	79	1,403
14	Russian Federation	16.1	57	1,059
15	Spain	14.4	79	1,105
16	Korea	14.0	107	1,510
17	Italy	13.4	74	1,587

continued on following page

Table 17. Continued

Rank	Country	Score	Flow intensity % of GDP	Flow value $ billion
18	Sweden	13.0	100	572
19	Austria	11.7	108	470
20	Malaysia	11.6	187	610
21	Mexico	10.7	80	1,022
22	Thailand	10.7	162	605
23	Kuwait	10.6	153	306
24	Japan	10.5	54	2,498
25	Kazakhstan	10.0	83	176
26	Ukraine	9.8	101	133
27	Australia	9.7	57	825
28	Denmark	8.9	108	369
29	Jordan	8.8	138	50
30	India	8.5	64	1,316
32	Czech Republic	7.5	193	397
34	Poland	7.0	107	585
35	Hungary	6.8	209	287
36	Norway	6.0	92	458
37	Vietnam	5.7	188	350
39	Finland	5.5	144	390
40	Portugal	5.5	111	255
41	Turkey	5.1	65	521
43	Israel	4.9	82	248
44	Brazil	4.5	37	869
45	Chile	4.1	92	239
47	Greece	4.1	67	160
48	New Zealand	3.9	63	130
51	Indonesia	3.4	57	504
53	South Africa	3.3	79	277
54	Philippines	3.2	81	230
64	Morocco	2.6	97	104
73	Egypt	2.2	55	158
83	Nigeria	1.9	47	268
86	Peru	1.8	60	122
118	Kenya	1.3	58	35

Source: Author's development based on Manyika et al. (2016)

Chapter 3
Food Security–Related Issues and Solutions

Olga Pasko
Tomsk Polytechnic University, Russia

Natalia Staurskaya
Omsk State Technical University, Russia

ABSTRACT

The food problem has been and has remained relevant throughout the history of mankind. At the end of 20th and the beginning of the 21st century, in the lives of many nations and countries, there have been significant changes. Health status and level of education of the population, such as, for example, food security, is the priority in many countries since, in the absence of sufficient food reserves, there is an economic and political dependence of some countries on others. Having not yet received the required amount of food, the world is faced with the problem of ensuring security in its quality. Anthropogenic pollution of the environment complicates the problem with the quality of food and the exception of harmful chemicals in food. There is a problem of using environmentally friendly agrotechnical means, ensuring the production of high yields of environmentally safe products with a desirable reduction in their cost, and shortening the time required for their production.

INTRODUCTION

The most important task of agricultural production is to increase the yield and quality of agricultural crops, as well as accelerate the ripening of food products. The use of innovative technologies in agricultural production, in particular, pre-sowing stimulation of seeds (Hozayn & Qados, 2010; Shabin, Tyshkevich, & Ershova, 2017) allows to get a crop in less time with less effort and lower costs.

The first scientific results on the stimulation of crop yields by physical factors were obtained in 1746. Dr. Mimbre from Edinburgh discovered that the treatment of myrtle plants with an electrostatic field enhanced their growth and flowering. In 1748, French abbot Jean Nole established the acceleration of seed germination after treatment with an electric field. In 1885, Finn Lemstrem described the stimula-

DOI: 10.4018/978-1-7998-1042-1.ch003

tion of growth of potatoes, carrots, and celery by 40% in eight weeks. Strawberries ripened twice as fast. Raspberry harvest doubled, that of carrots increased by 25%. Control plants of cabbage, turnip, and flax grew better than the treated ones (Briggs, 1926; Ross, 1844; Nelson, 2007).

In 1918-1921, about 500 British farmers in the acreage of about 2,000 acres studied the method of treating cotton seeds with a solution of fertilizer treated with an electric field (Nelson, 2007). It turned out that at the end of the growing season there were two or three times more boxes on the test plants than on the control plants. Positive results were obtained for sugar beets, tomatoes, and corn.

The experiments have been continuing successfully. There has been accumulated a significant scientific material on the use of physical and chemical factors to increase the yield of cultivated plants. It has been established that plant growth stimulators pose a complex effect on physiological and biochemical processes, accelerate the development of plants, and enable more rational use of agricultural lands and equipment (Wierzbowska, Cwalina-Ambroziak, Glosek, & Sienkiewicz, 2015). Their positive effect on yields and quality of agricultural products has been demonstrated in case of phytopathological state of crops in different soil and climate conditions (Kuzminykh & Pashkova, 2016; Petrichenko & Loginov, 2010).

In some of the cases, the use of stimulants allows reducing the dose of applied fertilizers and pesticides which has a positive effect on quality of agricultural and food products, as well as reduces production costs (Pigorev, Zasorina, Rodionov, & Katunin, 2011; Ponomareva & Zaharova, 2015; Demchenko, Shevchuk, & Yuzvenko, 2016). In the developed countries, application of stimulants allows increasing the productivity of particular crops by 20-30% (Danilov, 2017). Specifically, in vegetable and fruit production, as well as in ornamental horticulture, their use has become a mandatory agrotechnical technique which is employed in 50-80% of agricultural enterprises all over the world (Malevannaya, 2001; Ambroszczyk, Jedrszczyk, & Nowicka-Polec, 2016).

The use of plant growth stimulants is focused on solving a specific problem of obtaining a given volume and quality of agricultural products (Chekurov, Sergeeva, & Zhalieva, 2003). Special attention is paid to the use of environmentally friendly and non-toxic methods and substances (Colla, Rouphael, Canaguier, Svecova, & Cardarelli, 2014; Paradikovic, Vinkovic, Vinkovic Vrcek, & Tkalec, 2013; Ovcharenko, 2001; Tiwari & Dhuria, 2018). Internationally, growth and development of plants of natural and synthetic origin are used (Alexandrova, Shramko, & Knyazeva, 2010; Kocira, Kornas, & Kocira, 2013).

Natural stimulants (gibberellins, auxins, ethylene, kinins, etc.) consistently participate in the metabolism at all stages of plant life and affect budding, flowering, and fruiting. The activity of synthetic growth stimulants is similar to natural growth and is carried out by regulating the general hormonal status of plants (Roumeliotis et al., 2012). Stimulants increase plant resistance to adverse environmental conditions (drought, freezing temperature drops, etc.), damage by pests, and morbidity (Wierzbowska et al., 2015). In a number of cases, their use allow reducing the amount of applied fertilizers, herbicides, and pesticides (Kozlobaev, 2016).

However, along with a number of positive aspects, the application of plant growth stimulants has particular negative effects. These substances can accumulate in products and cause allergic reactions among employees who work directly with them, as well as among consumers. Most of the stimulants are rather expensive and difficult to prepare (Gordeeva, Shoeva, Yudina, Kukoeva, & Khlestkina, 2016). The above flaws necessitate the search for new effective, affordable, cheap, and environmentally friendly solutions, one of which is an activated water. The purpose of this study is to analyze the effectiveness of application of activated water in agriculture for the purposes of increasing the volume of crop production and solving food insecurity problem.

BACKGROUND

Fluid activation is carried out in various ways, including by passing a direct electric current (Pasko, 2010), exposure to a constant or variable electric field (Vakhidov et al., 1999), ionizing radiation, mechanical vibrations, shaking or mixing the liquid, passing it through a porous filter, bubbling with gases, heating (Hoboetc, n.d.), magnetization (Kney & Parsons, 2006), exposure to phonon oscillations of the lattice of crystalline solids, and others (Federal Service for Intellectual Property, 2009).

The term "activated substances" appeared in the scientific and technical literature after the discovery of differences in the reactivity of substances before and after any impact on them by physical factors. Activated steel has been called a substance, the stock of internal potential energy of which (as a result of any external effects) temporarily does not correspond to the thermodynamically equilibrium temperature and pressure values (Bakhir, Liakumovich, & Kirpichnikov, 1983; Pasko, Semenov, Smirnov, & Smirnov, 2007; Bakhir, Prilutsky, & Shomovskaya, 2010).

During the activation of water, a change in its physicochemical parameters, primarily, the redox potential, occurs (Figure 1).

Fluids treated with physical factors have been used in almost all areas of human activity. In addition to crop production, electrochemically activated water is successfully used in various areas (Golohvast, Ryzhakov, Chajka, & Gulikov, 2011; Shramko, Alexandrova, & Knyazeva, 2011; Kraft, 2008; Huang, Hung, Hsu, Huang, & Hwang, 2008; Cai, 2005; Devilliers & Mahe, 2007; Al-Haq, Seo, Oshita, & Kawagoe, 2002), specifically:

- agriculture to reduce the toxicity of herbicides, pesticides, and fungicides for animals and humans;
- pest control, including aphids, wilt, nematode, and whitefly;
- preparation of nutrient solutions for hydroponic cultivation of plants;
- accelerate wound healing in animals and combat hoof diseases;
- disinfecting premises containing livestock and poultry (without removing animals);

Figure 1. The change in the redox potential of tap water during the processing of physical factors (A) and after their action ends (B)
Source: Pasko (2010)

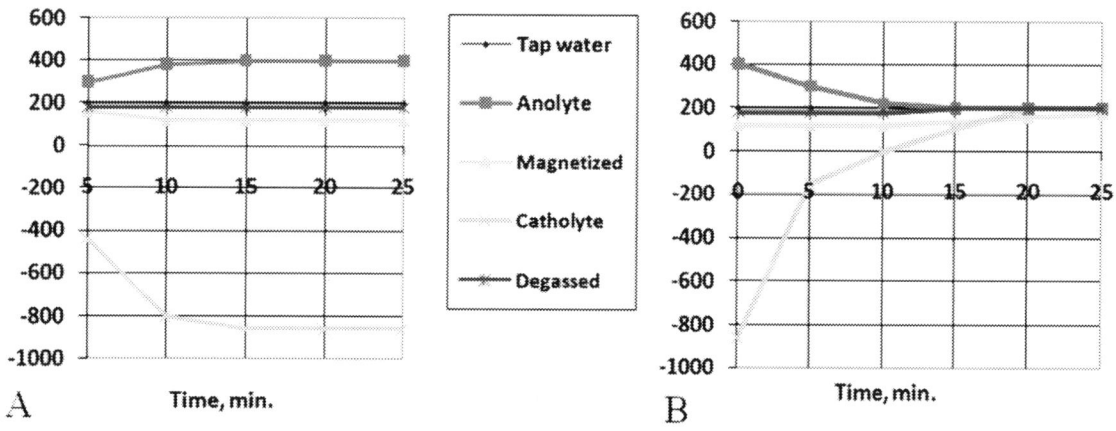

Figure 2. Overhead magnetron
Source: Medprom.ru (n.d.)

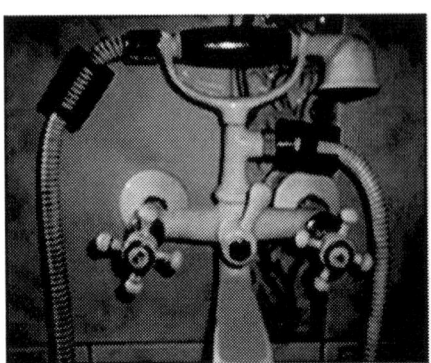  

- disinfection of drinking and recycled water, disinfection of wastewaters of agricultural complexes with simultaneous neutralization of their corrosive properties;
- disinfection of incubator and commodity eggs, incubators (anti-salmonellosis);
- soil deoxidation;
- conservation of green fodder with preservation of nutritional properties, increasing the number of fodder units, increasing the content of lactic acid, carotene, while significantly reducing the content of butyric acid in them and reducing the cost of preservatives;
- long-term storage of flowers, vegetables, fruits, and berries at a higher storage temperature.

According to the treatment method, there are the modifications of activated water (Hozayn & Qados, 2010).

Magnetic Water

Water is magnetized by affecting a stream of water moving at a speed of 1-5 m/s with a transverse magnetic field of at least 300 Oe (Klassen, 1982). At the same time, Lorentz's forces, "pulling" opposite, act on positive and negative particles. charges, the structure of water and the state of impurities in it change, it's degassing occurs. Currently, a large number of devices have been developed (Figure 2) for water magnetization (Ke La Xin, 1982; Xie, 1983).

Water Degassing

Degassed water is often used for pre-sowing seed soaking (Zelepukhin & Zelepukhin, 1973), processing cuttings before planting and watering plants, and watering birds and animals (Zelepukhin & Zelepukhin, 1987). The industrial introduction of degassed water has required the development of industrial degassers (Hoboetc, n.d.; Matusevich, Kolbatsky, Zelepukhin, Ostryakov, & Zelepukhin, 1983). Thermally activated water is usually obtained by heating to 99°C and subsequent rapid cooling in hermetic conditions (Figure 3).

Figure 3. Production of thermally degassed water under laboratory and working conditions
Source: Zelepukhin and Zelepukhin (1987)

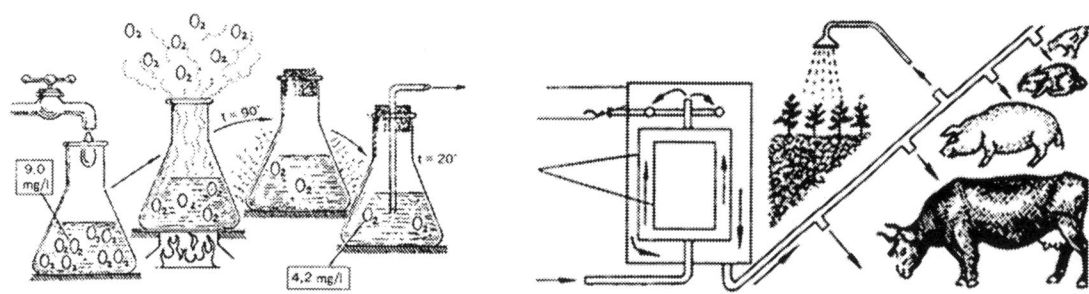

Electrochemically Activated Water

According to Gitelman (1994), the developed techniques of electroactivated aqueous solutions for the prevention and treatment of the most common human diseases are based on the two modifications of electroactivated aqueous solutions – catholyte and anolyte (Pasko & Gomboev, 2011). Catholyte has an immunostimulating and detoxifying effect and accelerates tissue regeneration. It is effective in chronic diseases, accompanied by a weakening of the immune reactivity of a body, nonhealing wounds, and ulcers. Anolyte is a strong antiseptic (Park, Hung, & Brackett, 2002; Fenner, Burge, Kayser, & Wittenbrink, 2006). It is effective in disinfecting water (including cholera vibrioes), various objects and instruments, and infections of the gastrointestinal tract (salmonellosis, dysentery), tonsillitis, and chronic tonsillitis. Anolyte is applied in surgery for the treatment of purulent wounds, abscesses, and phlegmon. Pasko (2000) successfully introduction the method of seed treatment with electrochemically activated water in the greenhouses of Russia, Kazakhstan, Kyrgyzstan, and Ukraine. Shramko, Alexandrova, and Knyazeva (2011) demonstrated high efficiency of presowing treatment of seeds of grain crops in Kuban region, Russia.

Electrolyzers: Classification and Technical Specifications

Electrolyzers are devices for unipolar electrochemical treatment of aqueous solutions of electrolytes and other liquids in order to change their acid-base and redox properties and transfer liquids to a thermodynamically no equilibrium state (Means & McMahon, 1978; Yamaguti, Ukon, Misawa, & Arisaka, 1994; Sadler & Cossich, 1997; Pasko, Semenov, & Dirin, 2000).

Different variants are known (Boltrik, 1999; Pasko, Semenov, Smirnov, & Smirnov, 2009; Yamaguti, Misawa, & Asanuma, 1999), which are conventionally divided into three types: static, immersion, and flow. In most cases, household electrolyzers are static, solutions in them are treated in individual portions, filling the cathode and anode chambers. Industrial electrolysis cells differ in significant volumes, environmentally friendly production, simplicity and durability of device operation (Figure 4).

Electrochemical household electrolysis cells, models 3002, 3003, 3006, and 3009, developed in Belarus (comply with the technical conditions of TURB 490C85159.001-2001) are currently manufactured in Russia. Manufacturer – scientific and production unitary enterprise "Akvapribor" (Belarus) (Figure 5). Water is poured into the vessel and a solution of sodium chloride in a ceramic cup.

Figure 4. Industrial electrolyzer
Source: Megakhim (n.d.)

Figure 5. Structure of an activator
Note: 1 – rectangular dielectric vessel; 2 – dielectric cover; 3 – rectangular glass made of porous ceramics; 4 – cathode made of two steel plates (food grade stainless steel); 5 – anode made of two metal plates coated with electrochemically resistant metal; 6 – power supply unit containing a step-down transformer, rectifier, and an overload protection circuit on the primary AC network; 7 – DC cable; 8 – AC cord; 9 – fork; 10 – catholyte; 11 – anolyte.
Source: School of Active Longevity (n.d.)

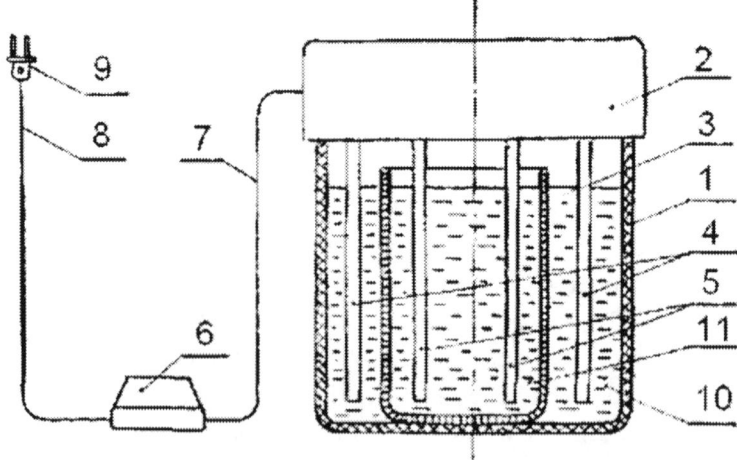

Food Security-Related Issues and Solutions

The analysis shows that the use of electrolyzers meets such tasks as resource conservation, environmental safety, ease of use, and high efficiency.

Other Species of Biologically Active Water

In the 1960s, Rodimov (1961) and Rodimov, Marshunina, Yafarov, Sadovnikova, and Labina (1975) studied biological activity of meltwater. It was suggested that its stimulating ability was determined either by a change in the isotopic composition (Mukhachev, 1975; Lobyshev & Kirkina, 2012) or structural changes associated with an increase in ice-like components (Gusev & Shpagina, 1981). Later, when more effective modifications of biologically active water were discovered, interest in research on meltwater decreased. Water stimulated by X-rays, gamma rays, short-wave radiation in the optical range and beams of rapidly charged particles can also be used as a stimulant, and since the products of radiation-chemical reactions resulting from radiolysis also exhibit biological activity.

MAIN FOCUS OF THE CHAPTER

Production Tests of Activated Water on Vegetables and Potatoes

Effect of Presowing Seed Stimulation Treatment on Tomato Yields

The authors conducted a test of pre-sowing stimulation of tomato seeds for growth, development, and yield of plants. Water treated with electricity was used as a stimulator. The area of the plot was 8 m^2. The number of plants in the plot was 20 pieces. The experiment was conducted four times. Productivity was studied by separately collecting and weighing fruits from each plot. The quality of the yields was analyzed in the agrochemical laboratory. During the elimination of the experiment, morphological parameters of plants were measured.

It was established that from the first days of development, stimulated plants differed in larger sizes – the height of seedlings, area, and the number of leaves (Table 1). Probably, these differences would be preserved and would have intensified at a later date, however, the phytotechnics adopted in greenhouses led to their leveling.

Table 1. Impact of pre-sowing treatment of tomato seeds (Verlioka variety) with electrically and chemically activated water upon the plant habit

Option	X1	X2	X3	X4	X5	X6	X7	X8
Control	60.4 ± 5.9	7.7 ± 1.0	60.5 ± 5.9	312.7 ± 10.0	76.9 ± 7.7	13.5 ± 1.5	299.0 ± 22.7	12.7 ± 1.3
Electrically and chemically activated water	68.9 ± 8.6	8.6 ± 0.8	69.0 ± 3.0	443.1 ± 8.2	78.2 ± 1.3	15.4 ± 1.3	308.8 ± 23.4	11.7 ± 1.2

Note: X1 – height of a plant (mm); X2 – width (mm); X3 – seed lobes' length (mm); X4 – seed lobes' area (mm^2) as plantlets; X5 – height of a plant (cm); X6 – number of leaves (items) as the blooming begins; X7 – height of a plant (cm); X8 – number of leaves (items) as the experiment was liquidated.

Source: Authors' development

Table 2. Impact of pre-sowing treatment with of tomato seeds (Verioka variety) with electrically and chemically activated water upon the plants' productivity

Index	X9	X10	X11	X12	X13	X14	X15
Control	12.7 ± 1.3	9.5 ± 2.0	8.4 ± 1.8	6.0 ± 2.0	1.1 ± 0.1	1.6 ± 0.1	23.4 ± 0.4
Electrically and chemically activated water	11.7 ± 1.2	11.4 ± 2.0	5.7 ± 2.0	12.8 ± 3.4	1.5 ± 0.3	4.0 ± 0.6	31.3 ± 0.5

Note: X9 – number of clusters, items; X10 – number of lowers, items; X11 – number of seed buds, items; X12 – number of fruits, items; X13 – mass of a plant, kg; X14 – mass of the fruits per plant, kg; X15 – mass of a fruit, kg

Source: Authors' development

The increase in the average mass of the fetus also indicates a more intensively passing physiological processes in stimulated plants. Note that with sudden (emergency) temperature changes in greenhouses, they showed greater stability.

Biochemical analysis revealed an improvement in the quality of the fruit under the influence of stimulation. The yield of dry matter increased by 9-13%, sugar content – by 5-9%, vitamin C – by 8-9%. The content of nitrates decreased by 7-14% relative to the control. In addition, in the plants treated with activated water, the number of non-standard fruits significantly decreased due to a weaker infection by late blight.

A significant increase in the number of brushes, flowers, ovaries, and fruits (Table 2) was found.

An average strength of the positive relationship was found between the number of vegetative organs and the mass of fruits. Steady correlations were found between the characters: number of fruits – average mass of fruits – mass of fruits from one plant; vegetative mass of a plant – mass of fruits from a

Figure 6. Verioka variety's productivity at the farm, kg/m²
Source: Authors' development

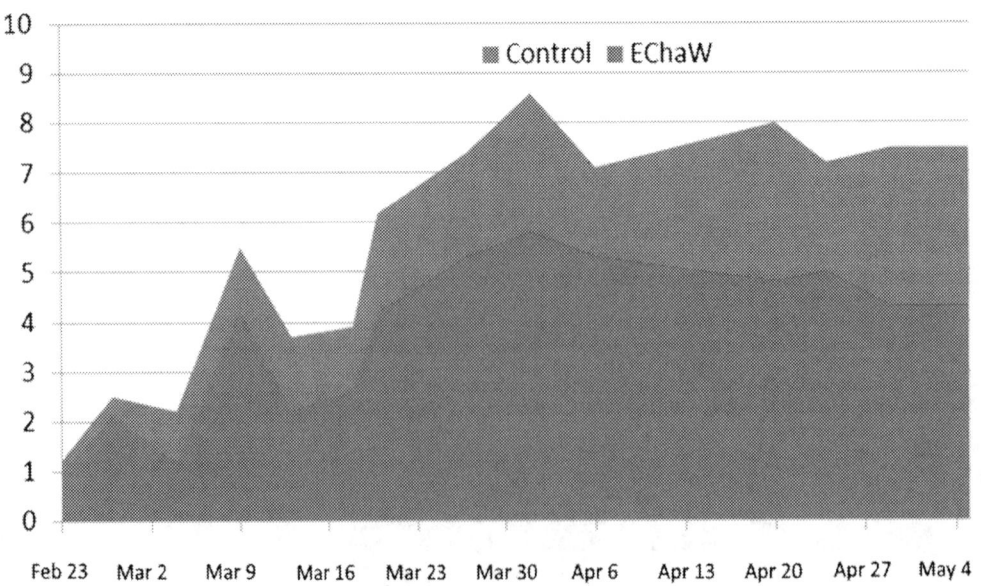

Figure 7. The structure of the crop of tomato plants (Revermun variety) in the elimination of experience and the appearance of plants in the greenhouse
Source: Authors' development

plant – number of flowers – power of initial growth (morphometric indicators at the initial stages of development). General stimulation of development processes had a positive effect on the formation of generative organs, and, consequently, on yield.

The yield of the stimulated plants during all gatherings was significantly higher than the controls (Figure 6). Consequently, pre-sowing treatment of tomato seeds did not disrupt the natural course of organ-forming processes but accelerated and intensified their passage.

The share of red fruits in the elimination of experience was 1% on control plants, 51% on stimulated plants (with a general increase in the number of fruits on a plant) (Figure 7).

In addition, it was found that pre-sowing treatment significantly influenced such indicators as the size of seedlings in the early stages of development (growth force), as well as the number of brushes in a fruit-bearing plant (Figure 8).

The estimated share of the effect of stimulation was 24-47% in terms of growth strength, and 25% in the number of brushes.

Regression analysis allowed to establish a high sensitivity of tomato plants to the action of stimulation. For all the regression equations below, the adequacy test using analysis of variance showed their adequacy at the achieved level of significance not higher than 0.02.

$X8 = 2.97 + 0.42 * X6; R^2 = 0.619; p = 0.008$
$X10 = 0.64 + 0.5 * X15; R^2 = 0.47; p = 0.05$
$X12 = 0.25 + 0.44 * X14; R^2 = 0.835; p = 0.02$
$X14 = 0.05 - 0.15 * X9 = 0.04 + 0.23 * X12; R^2 = 0.910; p = 0.05$

The obtained data allowed us to proceed to large-scale experiments. Similar processing of seeds within the greenhouse complex (the area of the protected ground is 3.3 hectares). Revermun and Tortilla varieties were used.

Starting from the very first phases, the stimulated plants differed more rapidly. They formed more powerful, leafy shoots, on which a greater number of generative organs developed. In the structure of the harvest, there was a noticeable shift towards brown fruits (Table 3). Product quality met the appropriate requirements.

Figure 8. The proportion of the effect of presowing stimulation of tomato seeds of Verlioka variety with activated water on the number of generative organs at the end of the experiment
Note: X9 – number of clusters, items; X10 – number of flowers, items; X11 – number of seed buds, items; X12 – number of fruits, items.
Source: Authors' development

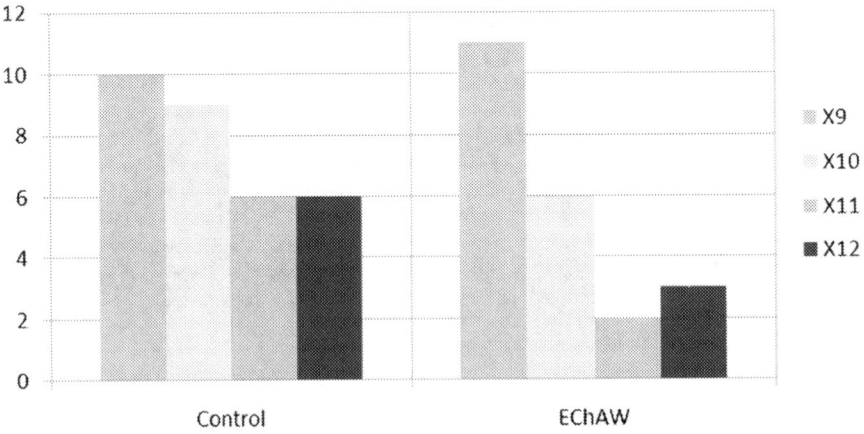

Table 3. The effect of presowing treatment of seeds of Tortilla variety on the yield of tomato plants in a large-scale experiment at Tomich state farm

Option	Yield per month, kg				
	August	September	October	total	%
Control	453	1,110	350	1,913	100
Electrically and chemically activated water	1,013	1,332	405	2,750	144

Source: Authors' development

The Effect of Presowing Stimulation of Cucumber Seeds on the Yields

The authors studied the effectiveness of pre-sowing stimulation of cucumber seeds of the hybrid Estafeta with various modifications of activated water. Experiments have shown a steady effect of increasing yields, practically in each of the 46-50 collections.

Phenological observations revealed an earlier (by 2-4 days) occurrence of flowering and fruiting phenophases in experimental plants. The yield of cucumber for three months in all experimental variants significantly exceeded the control one (Table 4).

The growth dynamics of the yield of cucumber plants is presented in Figure 9.

The maximum effect was achieved in versions with degassed water and electrochemically activated water, the minimum – with magnetized water. In some embodiments, an improvement in the quality of the fruits was noted: an increase in the content of dry matter, vitamin C, a decrease in the concentration of nitrate-nitrogen (Table 5).

Thus, from the studied modifications of activated water, degassed water and electrochemically activated water have the best stimulating properties for cucumber plants, while magnetized water is the worst.

Table 4. The effect of presowing treatment of cucumber seeds on plant yield, kg/m²

Mode of activated water	February	March	April	May-June	Total yield	Yield, %
Tap water (control)	0.22	3.56	5.03	6.23	15.0	100
Magnetized water	0.25	3.96	4.59	7.14	15.9	106
Degassed water	0.48*	4.86*	5.98*	7.63*	18.9	126
Electrically and chemically activated water	0.53*	5.37*	6.13*	7.97*	20.0*	133

Note: * the differences are reliable at a significance level of 5%
Source: Authors' development

Table 5. The effect of presowing treatment of cucumber seeds on the biochemical composition of fruits

Activated water's mode	Dry matter, %	Nitrate nitrogen, mg/kg	Ascorbic acid, mg/100	Mass of a fruit, gr
Tap water (control)	2.46	259	11.24	221.1
Magnetized water	2.91	254	12.37*	232.1*
Degassed water	3.40*	226*	12.25*	235.7*
Electrically and chemically activated water	3.80*	171*	12.65*	237.4*

Note: * the differences are reliable at a significance level of 5%
Source: Authors' development

Figure 9. The course of fruiting cucumber (Estafeta hybrid) in greenhouse conditions
Source: Authors' development

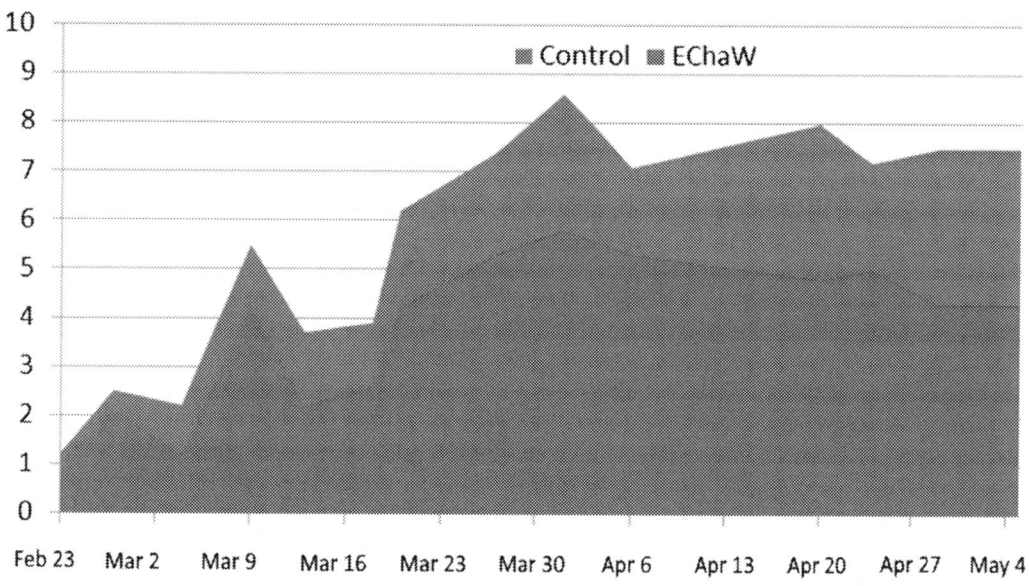

Table 6. Effect of treatment of tubers with growth and development stimulants on plant yields

Option	Tubers' mass					Yield increase, %	
	big		small		total		
	kg/m²	%	kg/m²	%	kg/m²	kg/m²	%
Control (tap water)	2.6 ± 0.3	78	0.74 ± 0.1	22	3.34	–	–
Electrically and chemically activated water	4.22 ± 0.3	92*	0.38 ± 0.1	8*	4.70	1.09	22*

Note: The differences are significant at a significance level of 5%
Source: Authors' development

The authors also conducted an additional study of the effect of irrigation of cucumber seedlings on the frost resistance of plants. Every day, for three weeks, the plants were watered with tap water (control) and electrochemically activated water. Then the plants were exposed to low positive temperatures for a day and transferred to favorable conditions. After two weeks, their viability was visually determined. The number of intact or slightly damaged plants was 36% in the control, in the variant with electrochemically activated water – 57%. Therefore, watering plants with activated water increased their resistance to low positive temperatures.

The Effect of Electrically and Chemically Activated Water on Potato Yields

For three years, the authors have been conducted an experiment to study the effect of the processing of planting potato tubers of the Ideal variety on the harvest (Figure 10). The maximum increase in total yield was obtained using electrochemically activated water (18%). At the same time, the share of large tubers increased (9%) (Table 6).

Figure 10. Field experiment with potatoes
Source: Authors' development

Food Security-Related Issues and Solutions

Table 7. The effect of the treatment of tubers and spraying of vegetative plants on their yield

Option	Tubers' yields					Increase in comparison to control level	
	large		small		total		
	kg/m^2	%	kg/m^2	%	kg/m^2	kg/m^2	%
Control	2.77 ± 0.21	77	0.84 ± 0.1	23	3.61	–	–
Sodium humate	3.18 ± 0.30	87*	0.48 ± 0.1	13*	3.66	0.05	1
Sodium humate + Electrically and chemically activated water	3.46 ± 0.27	90*	0.40 ± 0.1	10*	3.86	0.25	7*
Electrically and chemically activated water	2.94 ± 0.28	76	0.92 ± 0.2	24	3.86	0.25	7*

Note: The differences are significant at a significance level of 5%
Source: Authors' development

Additional treatment of vegetative plants with growth stimulants had a positive effect on yield (Table 7). The maximum increase in yield was observed when using sodium humate and electrochemically activated water (23%).

SOLUTIONS AND RECOMMENDATIONS

The approaches to establishing food security vary from one country to another, but they are nevertheless common in such issues as predicting and preventing food safety risks; ensurance of sustainable development of agricultural production; achievement of physical and economic accessibility of ecologically safe food for the population in accordance with rational norms of food consumption. Leading countries of the world have ensured food security through the effective government support and regulation of the entire complex of emerging problems. The market of environmentally friendly food is developing. Food security of a country may be enhanced by the improvement of the regulatory framework of agricultural production, taking into account the share of environmentally friendly food; ensuring effective monitoring of production; development of state information resources in the field of natural resource management; the use of affordable, effective stimulants for the growth and development of plants that do not leave harmful or hazardous substances in vegetable products and/or potatoes.

FUTURE RESEARCH DIRECTIONS

When planning further research, it is meant to conduct a more profound analysis on the market of biologically safe food products to range it according to its "purity" and efficiency. It is planned to make a development forecast, the species and varietal composition of the tested plants should be expanded, the plant resistance to diseases and pests should be tested in control and under the influence of presowing treatment of seeds. It seems promising to plan experiments to study the effect of activated water on the quality of crops (roots, fruits, etc.). It is possible to increase the effect when receiving electrically and chemically activated water in an electrolyzer with different chemical composition of the electrodes (with the inclusion of micro- and macro elements).

CONCLUSION

In this chapter, the authors discussed the effective stimulants for the growth and development of plants that (1) do not leave harmful or hazardous substances in plants, (2) allow increasing productivity of major crops and in such a way – the volume of production of agricultural and food products, and (3) provide environmental-friendly and healthy solution to food insecurity problem worldwide. The example used in the study is electrochemically activated water as a stimulator of plants growth, the use of which leads to a significant increase in crop yields, improving their quality, and speeding up the time to start collecting products. The experiments conducted by the authors demonstrated that the yield of tomato plants increased by 2.5 times, yield of dry matter – by 9-13%, sugar content – by 5-9%, and vitamin C – by 8-9%. The nitrate content reduced by 7-14% relative to the control. The steady effect of the growth of the yield of cucumber was established practically at 33%. An improvement in the quality of the fruits was noted due to an increase in the content of dry matter, vitamin C and a decrease in the concentration of nitrate nitrogen. Electrically and chemically activated water stimulated an increase in potato yield by 18% with a simultaneous increase in the proportion of large tubers (9%).

ACKNOWLEDGMENT

The research is carried out at Tomsk Polytechnic University within the framework of Tomsk Polytechnic University Competitiveness Enhancement Program.

REFERENCES

Al-Haq, M. I., Seo, Y., Oshita, S., & Kawagoe, Y. (2002). Disinfection Effects of Electrolyzed Oxidizing Water on Suppressing Fruit Rot of Pear Caused by *Botryosphaeria Berengeriana*. *Food Research International*, *35*(7), 657–664. doi:10.1016/S0963-9969(01)00169-7

Alexandrova, E., Shramko, G., & Knyazeva, T. (2010). New Composition of Mineral Fertilizers for Foliar Feeding of Winter Wheat. *Proceedings of Kuban State Agrarian University*, *22*(1), 71–74.

Ambroszczyk, A. M., Jedrszczyk, E., & Nowicka-Polec, A. (2016). The Influence of Nano-Gro® Stimulator on Growth, Yield and Quality of Tomato Fruit (*Lycopersicon Esculentum Mill.*) in Plastic Tunnel Cultivation. *Acta Horticulturae*, (1123): 185–192. doi:10.17660/ActaHortic.2016.1123.26

Bakhir, V., Liakumovich, A., & Kirpichnikov, P. (1983). The Physical Nature of the Phenomena of Activation Substances. *News of the Academy of Sciences of Uzbekistan Soviet Socialist Republic*, *1*, 60–64.

Bakhir, V., Prilutsky, V., & Shomovskaya, N. (2010). Electrochemically Activated Aqueous Media-Anolyte and Catholyte as a Means of Suppressing Infectious Processes. *Medical Alphabet*, *13*(3), 40–42.

Boltrik, O. (1999). *Parameters and Modes of Operation of the Electric Activator for Presowing Treatment of Seeds of Grain Crops*. Zernograd: Black Sea State Agroengineering Academy.

Briggs, L. (1926). *USDA Departmental Bulletin #1379*. Washington, DC: United States Department of Agriculture.

Cai, Z. (2005). *Characterisation of Electrochemically Activated Solutions for Use in Environmental Remediation*. Bristol: University of the West of England.

Chekurov, V., Sergeeva, S., & Zhalieva, L. (2003). New Growth Regulators. *Plant Protection Quarterly*, *9*, 20–22.

Colla, G., Rouphael, Y., Canaguier, R., Svecova, E., & Cardarelli, M. (2014). Biostimulant Action of a Plant-Derived Protein Hydrolysate Produced through Enzymatic Hydrolysis. *Frontiers in Plant Science*, *5*, 448. doi:10.3389/fpls.2014.00448 PMID:25250039

Danilov, A. (2017). Influence of Growth Promoters on Yield and Quality of Grain Crops Products. *Vestnik of Mari State University Agricultural Economics*, *9*(3), 28–32.

Demchenko, O., Shevchuk, V., & Yuzvenko, L. (2016). Investigation of the Resistance of Different Varieties of Buckwheat to Infectious Diseases after the Presowing Treatment of Seeds and Vegetating Plants with Biological Preparations. *Agrobiology*, *1*, 57–66.

Devilliers, D., & Mahe, E. (2007). Modified Titanium Electrodes. In M. Nunez (Ed.), *New Trends in Electrochemistry Research* (pp. 1–60). Hauppauge, NY: Nova Science Publishers.

Federal Service for Intellectual Property. (2009). *Patent #2349071. Russian Federation*. Retrieved from http://www.freepatent.ru/images/patents/116/2349071/patent-2349071.pdf

Fenner, D. C., Burge, B., Kayser, H. P., & Wittenbrink, M. M. (2006). The Anti-Microbial Activity of Electrolysed Oxidizing Water against Microorganisms Relevant in Veterinary Medicine. *Journal of Veterinary Medicine*, *53*(3), 133–137. doi:10.1111/j.1439-0450.2006.00921.x PMID:16629725

Gitelman, C. (1994). *Methodical Recommendations of Institutions of Uzbekistan, the Russian Federation, and Ukraine on the Use of Electroactivated Aqueous Solutions for the Prevention and Treatment of the Most Common Human Diseases*. Moscow: Espero.

Golohvast, K., Ryzhakov, D., Chajka, V., & Gulikov, A. (2011). Application Potential of Solution Electrochemical Activation. *Water: Chemistry and Ecology*, *2*, 23–30.

Gordeeva, E., Shoeva, O., Yudina, R., Kukoeva, T., & Khlestkina, E. (2016). Effect of Seed Pre-Sowing Gamma-Irradiation Treatment in Bread Wheat Lines Differing by Anthocyanin Pigmentation. *Cereal Research Communications*, *46*(1), 41–53. doi:10.1556/0806.45.2017.059

Gusev, H., & Shpagina, O. (1981). *On the Importance of Studies of the State of Water. Issues of Water Exchange and Water Status in Plants*. Moscow: USSR Academy of Sciences.

Hoboetc. (n.d.). *Degassing Is… How Degassing of Water Is Done. Methods of Degassing*. Retrieved from https://hoboetc.com/biznes/1955-degazaciya-eto-kak-provoditsya-degazaciya-vody-sposoby-degazacii.html

Hozayn, M., & Qados, A. M. S. A. (2010). Irrigation with Magnetized Water Enhances Growth, Chemical Constituent and Yield of Chickpea (*Cicer Arietinum L.*). *Agriculture and Biology Journal of North America*, *1*(4), 671–676.

Huang, Y.-R., Hung, Y.-C., Hsu, S.-Y., Huang, Y.-W., & Hwang, D.-F. (2008). Application of Electrolyzed Water in the Food Industry. *Food Control, 19*(4), 329–345. doi:10.1016/j.foodcont.2007.08.012

Ke La Xin, B. N. (1982). *Magnetization of Water*. Beijing: Measurement Press.

Klassen, V. (1982). *Magnetization of Water Systems*. Moscow: Nauka.

Kney, A. D., & Parsons, S. A. (2006). A Spectrophotometer-Based Study of Magnetic Water Treatment: Assessment of Ionic vs. Surface Mechanisms. *Water Research, 40*(3), 517–524. doi:10.1016/j.watres.2005.11.019 PMID:16386285

Kocira, A., Kornas, N., & Kocira, S. (2013). Effect Assessment of Kelpak SL on the Bean Yield (*Phaseolus Vulgaris L.*). *Journal of Central European Agriculture, 14*(2), 545–554. doi:10.5513/JCEA01/14.2.1234

Kozlobaev, A. (2016). *The Effectiveness of Growth Promoters and Micronutrients in Buckwheat*. Voronezh: Voronezh State University.

Kraft, A. (2008). Electrochemical Water Disinfection: A Short Review. *Platinum Metals Review, 52*(3), 177–185. doi:10.1595/147106708X329273

Kuzminykh, A., & Pashkova, G. (2016). Grain Yield and Quality of Winter Rye Depending on the Use of Growth Stimulants. *Vestnik of Mari State University. Chapter. Agricultural Economics, 5*(1), 26–29.

Lobyshev, V., & Kirkina, A. (2012). *Influence of Isotopic Content Variation of Water on Its Biological Activity*. Retrieved from www.biophys.ru/archive/congress2012/proc-p21-d.pdf

Malevannaya, N. (2001). Plant Growth Regulators in Agricultural Production. *Fertility, 1*, 29.

Matusevich, Y., Kolbatsky, P., Zelepukhin, I., Ostryakov, I., & Zelepukhin, V. (1983). *Copyright Certificate of the USSR #1001965 "Thermal Degasser"*. Retrieved from http://patents.su/3-1001965-termicheskijj-degazator.html

Means, E., & McMahon, R. (1978). *Patent USA 4,073,712 Electrostatic Water Treatment*. Retrieved from http://patents.com/us-4073712.html

Medprom.ru. (n.d.). *Overhead Magnetron*. Retrieved from http://medprom.ru/medprom/1394898?TESTROBOT=YES

Megakhim. (n.d.). *Electrolyzer*. Retrieved from http://www.megahim.ru/equipment/electrolysis

Mukhachev, V. (1975). *Living Water*. Moscow: Nauka.

Nelson, R. A. (2007). *Electro-Culture (The Electrical Tickle)*. Retrieved from https://pdfs.semanticscholar.org/f1a3/7c7a653e7cd6c546205989650c5c22b3864a.pdf?_ga=2.126908106.613024882.1549709931-55171684.1549709931

Ovcharenko, M. (2001). Humates Are the Activators of Productivity of Agricultural Crops. *Agrochemical Bulletin, 2*, 13–14.

Paradikovic, N., Vinkovic, T., Vinkovic Vrcek, I., & Tkalec, M. (2013). Natural Biostimulants Reduce the Incidence of BER in Sweet Yellow Pepper Plants (*Capsicum Annum L.*). *Agricultural and Food Science, 22*(2), 307–317. doi:10.23986/afsci.7354

Park, H., Hung, Y.-C., & Brackett, R. E. (2002). Antimicrobial Effect of Electrolyzed Water for Inactivating *Campylobacter Jejuni* during Poultry Washing. *International Journal of Food Microbiology*, 72(1-2), 77–83. doi:10.1016/S0168-1605(01)00622-5 PMID:11843416

Pasko, O. (2000). *Activated Water and Its Use in Agriculture*. Tomsk: Tomsk State University.

Pasko, O. (2010). Property Changes of Piped Water Treated with Different Methods. *Water: Chemistry and Ecology*, 7, 40–45.

Pasko, O., & Gomboev, D. (2011). *Activated Water and Possibilities of Its Application in Plant Growing and Animal Husbandry*. Tomsk: Tomsk Polytechnic University.

Pasko, O., Semenov, A., & Dirin, V. (2000). *Patent #2147446 Water Activation Device*. Russian Federation. Retrieved from http://www.freepatent.ru/patents/2144506

Pasko, O., Semenov, A., Smirnov, G., & Smirnov, D. (2007). *Activated Fluids, Electromagnetic Fluids, and Flicker Noise. Their Application in Medicine and Agriculture*. Tomsk: Tomsk State University of Operation Systems and Radioelectronics.

Pasko, O., Semenov, A., Smirnov, G., & Smirnov, D. (2009). *Patent #2350568 Non-Replaceable Electrolyzer*. Russian Federation. Retrieved from http://www.freepatent.ru/patents/2350568

Petrichenko, V., & Loginov, S. (2010). Influence of Plant Growth Regulators and Microelements on Productivity of Sunflower and Oil Content of Seeds. *Agrarian Russia*, 4, 24–26.

Pigorev, I., Zasorina, E., Rodionov, K., & Katunin, K. (2011). Use of Growth Regulators in the Agricultural Complex in Potato Cultivation in the Central Black Earth Region. *Agricultural Science*, 2, 15–18.

Ponomareva, Y., & Zaharova, O. (2015). The Effect of Mineral Fertilizers and Growth Regulator on Yield and Quality of Malt Barley When Drought. *Bulletin of Kostychev Ryazan State Agrotechnological University*, 27(3), 36–42.

Rodimov, B. (1961). Snow Water-Stimulator of Growth and Productivity of Animals and Plants. *Agriculture of Siberia and the Far East*, 7, 66–69.

Rodimov, B., Marshunina, A., Yafarov, I., Sadovnikova, V., & Labina, I. (1975). Biological Role of Heavy Water in Living Organisms. In *Questions of Radiobiology and Hematology* (pp. 118-126). Publishing House of Tomsk University.

Ross, U. (1844). The Patent Commissioner of the United States Report. *Journal Scientific American*, 27, 370.

Roumeliotis, E., Kloosterman, B., Oortwijn, M., Kohlen, W., Bouwmeester, H., Visser, R., & Bachem, C. (2012). The Effects of Auxin and Strigolactones on Tuber Initiation and Stolon Architecture in Potato. *Journal of Experimental Botany*, 63(12), 4539–4547. doi:10.1093/jxb/ers132 PMID:22689826

Sadler, P., & Cossich, J. (1997). *Patent UK 1487052 "Electrolytic Treatment of Drinking Water"*. Retrieved from https://patents.google.com/patent/US5807473

School of Active Longevity. (n.d.). *Activator*. Retrieved from http://www.mzk.ru/

Shabin, S., Tyshkevich, E., & Ershova, T. (2017). Influence on the Yield of Spring Wheat of Presowing Treatment of Seeds by Ozone Air Agent. *Modern Science-Intensive Technologies, 49*(1), 130–136.

Shramko, G., Alexandrova, E., & Knyazeva, T. (2011). Improving the Technology of Foliar Feeding of Winter Wheat Using Electrochemically Activated Water. *Proceedings of Kuban State Agrarian University, 33*(6), 69–72.

Tiwari, S., & Dhuria, S. (2018). Effect of Pre-Sowing Treatment on Seed Germination and Seedlings Growth Characteristics of *Albizia Procera*. *Asian Journal of Research in Agriculture and Forestry, 2*(1), 1–6. doi:10.9734/AJRAF/2018/42370

Vakhidov, V., Mamadzhanov, U., Kasymov, A., Bakhir, V., Alekhin, S., & Iskhakova, H. … Goncharov, P. (1999). *Patent #1121906 The Method of Obtaining a Liquid with Biologically Active Properties*. Russian Federation. Retrieved from http://www.bakhir.ru/inventions/03/

Wierzbowska, J., Cwalina-Ambroziak, B., Glosek, M., & Sienkiewicz, S. (2015). Effect of Biostimulators on Yield and Selected Chemical Properties of Potato Tubers. *Journal of Elementology, 20*(3), 757–768.

Xie, W. (1983). *Magnetized Water and Its Application*. Beijing: Science Press.

Yamaguti, S., Misawa, S., & Asanuma, G. (1999). *Patent US 5051161 Apparatus Producing Continuously Electrolyzed Water*. Russian Federation. Retrieved from https://patents.google.com/patent/US5051161A/en

Yamaguti, S., Ukon, M., Misawa, S., & Arisaka, M. (1994). *Patent #0612694A1 Method and Device for Producing Electrolytic Water*. Russian Federation. Retrieved from https://patents.google.com/patent/EP0612694A1/da

Zelepukhin, V., & Zelepukhin, I. (1973). Stimulation of Physiological Processes with Biologically Active Water. *Works of Works of Kazakh Agricultural Institute, 18*(14), 143–148.

Zelepukhin, V., & Zelepukhin, I. (1987). *The Key to Living Water*. Alma-Ata: Kainar.

ADDITIONAL READING

Abadias, M., Usall, J., Oliveira, M., Alegre, I., & Vinas, I. (2008). Efficacy of Neutral Electrolyzed Water (NEW) for Reducing Microbial Contamination on Minimally-Processed Vegetables. *International Journal of Food Microbiology, 133*(1-2), 151–158. doi:10.1016/j.ijfoodmicro.2007.12.008 PMID:18237810

Anderson, K., & Strutt, A. (2014). Food Security Policy Options for China: Lessons from Other Countries. *Food Policy, 49*, 50–58. doi:10.1016/j.foodpol.2014.06.008

Brooks, J. (2014). Policy Coherence and Food Security: The Effects of OECD Countries' Agricultural Policies. *Food Policy, 44*, 88–94. doi:10.1016/j.foodpol.2013.10.006

Chushkina, E., Semenenko, S., Lytov, M., & Chushkin, A. (2015). Mechanism of Biological Activity and Application Experience of Electrochemically Activated Water in Agriculture. *Scientific Journal of the Russian Research Institute of Reclamation, 20*(4), 170–185.

Commission of the European Communities. (2000). *White Paper on Food Safety*. Brussels: Commission of the European Communities.

Del Giudice, E., & Preparata, G. (1995). Coherent Dynamics in Water as a Possible Explanation of Biological Membranes Formation. *Journal of Biological Physics, 20*(1-4), 105–116. doi:10.1007/BF00700426

Dunning, B. (2005). Atoms – Giants of the Atomic World. *Science Spectra, 3*, 34–38.

Eisenberg, D., & Kauzmann, W. (1969). *The Structure and Properties of Water*. Oxford: Clarendon Press.

Federal Customs Service of the Russian Federation. (2019). *Open Data*. Retrieved from http://www.customs.ru/opendata/

Federal Service for Surveillance on Consumer Rights Protection and Human Wellbeing of the Russian Federation. (2019). *About Rospotrebnadzor*. Retrieved from https://www.rospotrebnadzor.ru/en/

Frohlich, H., & Kremer, F. (1983). *Coherent Excitations in Biological Systems*. Cham: Springer. doi:10.1007/978-3-642-69186-7

Gadhok, I. (2016). *How Does Agricultural Trade Impact Food Security?* Retrieved from http://www.fao.org/3/a-i5738e.pdf

Gil, M., Gomez-Lopez, V. M., Hung, Y.-C., & Allende, A. (2015). Potential of Electrolyzed Water as an Alternative Disinfectant Agent in the Fresh-Cut Industry. *Food and Bioprocess Technology, 8*(6), 1336–1348. doi:10.100711947-014-1444-1

Gomez-Lopez, V. M., Ragaert, P., Ryckeboer, J., Jeyachchandran, V., Debevere, J., & Devlieghere, F. (2008). Reduction of Microbial Load and Sensory Evaluation of Minimally Processed Vegetables Treated with Chlorine Dioxide and Electrolysed Water. *Italian Journal of Food Science, 20*, 321–331.

Goncharuk, V., Lapshin, V., Burdeynaya, T., Pleteneva, T., Chernopyatko, A., Atamanenko, I., ... Syroeshkin, A. (2011). Physico-Chemical Properties and Biological Activity of the Light Water. *Chemistry and Technology of Water, 33*(1), 15–25.

Government of the Russian Federation. (1993). *The Constitution of the Russian Federation*. Retrieved from http://www.constitution.ru/en/10003000-01.htm

Information and Analytical Center. (2015). *Food Security of Russia*. Retrieved from https://inance.ru/2015/11/prodbez/

Kormishkina, L., Semenova, N., & Kormilkin, E. (2017). The Solution to the Problem of Food Security and Agricultural Development in the XXI Century in European. *Agricultural Science Euro-North-East, 56*(1), 74–78.

Larionov, V. (2015). Food Security in Russia. *Food Policy and Security, 2*(1), 47–58.

Pasko, O. (2012). Presowing Treatment of Seeds with Electrochemically Activated Water. *Agricultural Science, 7*, 24–27.

President of the Russian Federation. (2009). *Decree #537 from May 12, 2009, "On the Strategy of National Security of the Russian Federation until 2020"*. Retrieved from https://rg.ru/2009/05/19/strategia-dok.html

Rodimov, B., Marshuina, A., & Ofjafarov, I. (1965). Effect of Snow Water on the Living Organisms. Agricultural Production of Siberia and the Far East. Omsk, 4, 56-57.

Shapkina, L. (2013). Food Security in the National Security System of Russia. *Bulletin of the University*, 6, 190–198.

Shcherbak, I. (2014). Role of Global Food Security in the Common Agrarian Policy of the European Union. *Bulletin of MGIMO University*, 35(2), 130–138.

Sofy, M. R., Sharaf, A. E. M., & Fouda, H. M. (2016). Stimulatory Effect of Hormones, Vitamin C on Growth, Yield and Some Metabolic Activities of Chenopodium Quinoa Plants in Egypt. *Journal of Plant Biochemistry & Physiology*, 4(1), 161.

Taranova, I. (2013). Problems of Food Safety of Russia and Its Regions in Accession to WTO. *Economics and Entrepreneurship*, 37(8), 31–36.

Yakunina, M., Fedorova, O., & Shchepakin, K. (2013). WTO and Food Security of Russia. *Proceedings of Tula State University. Economic and Legal Sciences, 2-1*, 29-41.

KEY TERMS AND DEFINITIONS

Ecological Safety: An acceptable level of negative impact of natural and anthropogenic factors.

Food Products: The products in natural or processed form, food products, food products, food water, alcoholic beverages (including beer), soft drinks, chewing gum, as well as food raw materials

Food Safety: A state of food products completely eliminating the negative impact on human health when it is consumed.

Food Security: A state of the economy and the agro-industrial complex, which, while preserving and improving the habitat, regardless of external and internal conditions, allows the country's population to smoothly obtain environmentally friendly and healthy food at affordable prices, in amounts not lower than scientifically based standards.

National Security: An ability of a nation to meet the needs necessary for its self-preservation, self-reproduction, and self-improvement with minimal risk of damage to the basic values of its current state.

Pesticides: The chemical agents used to control pests and plant diseases, as well as with various parasites, weeds, pests of grain and grain products, wood, cotton, wool, skin products, ectoparasites of domestic animals, as well as carriers of dangerous human diseases and animals.

Productivity Stimulant: The compounds causing in very low concentrations stimulation or suppression of the growth and morphogenesis of plants.

Chapter 4
World Market of Organic Food Products

Inga Ryumkina
Irkutsk State Agrarian University, Russia

Sergey Ryumkin
Novosibirsk State Agrarian University, Russia

ABSTRACT

In the past two decades, the role of international relations in various spheres has increased significantly. The world market for agricultural products is not an exception. Agricultural production is influenced by many factors, including climate, development strategies, and financing of agricultural research centers, among others. The factor of organic production should form both domestic and global markets of agricultural products and food since the health of people and the environment depends on the quality of food products. Therefore, the agrarian policy should primarily focus on the development of markets of organic food. In this chapter, the authors attempt to identify major actors in the world market of organic food products.

INTRODUCTION

The aim of this study is an analysis of current state of world food market, including healthy food and organic products. The tasks of the study:

- Review the existing literature in the sphere of organic agriculture
- Clarify the terminology and offer authors' definitions
- Assess current state of agricultural production in Russia
- Make an overview of global agricultural market
- Study the factors of development of global and domestic markets of organic products
- Offer recommendations and solutions on the topic

DOI: 10.4018/978-1-7998-1042-1.ch004

Methodology of research: review of academic literature, collection and processing of data, analysis of information, and comparison and evaluation of data.

BACKGROUND

The concept of organic agriculture existed before the invention of synthetic agrochemicals. It began taking shape in the beginning of the 1900s. The concept of organic agriculture was first introduced by Paull (2014). The era of organic farming in Europe and America started in the 1940s as a response to the dependence of agricultural production on application of insecticides and synthetic fertilizers. During the 20th century, new farming methods based on agrochemical preparations have been actively used and led to the increase in yields. The other side of this development is soil erosion, soil contamination with heavy metals, and salinization of water objects.

Albert Howard, a British botanist, is considered as one of the founders of organic agriculture. His Agricultural Testament published in 1943 had a great impact on many scientists and farmers. Howard (n.d.) described the negative impact of chemical fertilizers on health of animals and plants, proposed a system of soil fertilizers based on the use of compost from plant residues and manure, and explained the essence of organic agriculture as a maintenance of fertility of soil which is the first condition of any permanent system of agriculture. In 1939, Balfour (n.d.), influenced by the works of Howard, conducted the world's first scientific experiment on agricultural land in the UK to compare conventional and organic agriculture. After four years, her book "The Living Soil" was published, became widespread, and led to the establishment of Soil Association, one of the most famous organizations in organic agriculture today. An important contribution to the development of organic agriculture was made by Steiner who wrote the first comprehensive work on organic agriculture and advocated the development of biodynamic agriculture, a type of organic agriculture that included all the principles and standards of organic agriculture, but also affected the cosmic rhythms and spiritual aspects (Paull, 2011). The term "organic agriculture" was popularized by Rodale. In 1942, he founded Organic Farming and Gardening journal. In 1950, he also founded Prevention journal which set out the philosophy of organic agriculture (Kurochkin & Smolnyakova, 2012).

In Japan, organic agriculture began to develop about 100 years ago. An important contribution to its development was made by the Japanese philosopher Mokichi Okada. He paid special attention to the so-called "natural farming", the principles of which were largely consistent with modern organic agriculture (Howling Pixel, n.d.). Another founder of organic agriculture is Masanobu Fukuoka, a Japanese farmer. Fukuoka practiced new methods of farming which he called "non-arable, without fertilizers, without weeding, without pesticides, the method of doing nothing in subsistence agriculture". He developed the principles of natural farming which involved minimizing human intervention in farming (One-Straw Revolution. n.d.).

Until the early 1990s, almost all environmental enterprises in the EU were merged into unions which played the main role in controlling standards and providing guarantees to consumers. In 1972, there was established the International Federation of Organic Agriculture Movements (IFOAM) with a purpose to disseminate information and introduce organic agriculture in all countries of the world. In the 1990s, green movements and green philosophy acquired a global scale, environmental protection, and care for the health of their citizens have become priority areas of state policy in many countries. Currently, IFOAM unites over 750 unions and organizations from 116 countries (Sherbakova (Ponomareva), 2017).

In Russia, organic farming has deep roots in agricultural science and practice. In the XVIII century, Russian scientist Bolotov developed the principles of agricultural production. In the 1930s, Williams proposed a grass system of agriculture which was largely consistent with the principles of organic agriculture (Kovalenko, Polushkina, & Yakimova, 2017). However, the policy of intensification of agriculture which has been carried out since the beginning of the 1960s has led to significant displacement of the views of these scientists on agricultural production of Russia. In Russia, organic agriculture appeared in the 1990s when farmers began collecting and exporting mushrooms, berries, and nuts from Siberia to Western Europe. All products were certified by such European organizations as Demeter and IMO. In 1989, in Russia, there was launched the Alternative Agriculture program. That year is considered as the beginning of ecological agriculture in Russia. Due to the unavailability of the market, the program ended in failure, but in two years it brought international certification to a number of farmers.

Russia is at the beginning of the establishment of the market of environmentally friendly, safe products. The mechanism of quality control of organic products is very important. The quality control system with the help of independent certification systems which in turn is controlled by the state has proved itself all over the world (Mironenko, 2016).

In recent years, with the introduction of sanctions and counter-sanctions, the attitude towards agro-industrial production in Russia has changed. Russia's President Putin who declared that it was necessary to set a task at the national level and supply domestic market with domestically-produced food by 2050. Russia is able to become the world's largest supplier of healthy and ecologically clean food (Interfax, 2015).

The state of organic agriculture in Russia today resembles the position of organic matter in Denmark twenty years ago. Russia has the largest land resources, but organic agriculture is poorly developed. In Denmark, the share of eco-products is 8% of the total market. It is necessary to pay special attention to soil, crop rotation, and organisms living in the soil because they are the most important animals for farmers (Guseva, 2015).

Among the countries of the world, organic food market is most developed in the USA. Organic production in the USA is regulated by the Organic Food Production Act, adopted in 1990. Roos et al. (2018) describe the algorithm of interaction between farmers and the state in the manufacture of fertilizers for the development of agriculture and animal feed production. In Europe, the principles of modern organic agriculture are implemented in a number of areas, including production of organic livestock products, profitability of farming, and low use of pesticides. However, in organic farming, productivity is lower compared to that in conventional agriculture (Toleubaeva & Patlasov, 2018).

Jespersen et al. (2017) presents the basic principles of organic agriculture and analyses positive and negative contribution of organic farming to the creation of public goods. The impact of organic farming on nature and biodiversity, human and animal health and well-being is largely positive compared to conventional agriculture. However, the impacts on the environment, energy, and climate are mixed. The analysis revealed the need for further regulation and revision of organic principles and requirements (Toleubaeva & Patlasov, 2018).

Currently, the experts are interested in increasing yields in organic agriculture in order to provide livestock with more environmentally friendly feed for the growing population and reduce the negative impact on the unit of production. Strategies to increase the yield of organic products are not associated with negative side effects. Pest management through the management of ecosystem services. Some scholars believe that in addition to optimal crop rotation, biological plant protection products need to be allowed new sources of nutrients for plants, including increased nutrient recycling, and in some cases, limited use of mineral nitrogen fertilizers from renewable sources. Particularly, the foundations of organic farming and research on food markets have been considered by Howard (n.d.), Dabbert, Haring, and Zanoli (2004), Yussefi and Willer (2003), Meredith, Lampkin, and Schmid (2018), Kharitonov (2013), and Kovalenko et al. (2017).

Currently, food market of Russia is studied from all sides. In view of transformation of global economy, its digitalization, various political situations, economic crises, environmental degradation, and population growth, so to speak should not be, because this market is constantly in a dynamic state and therefore the problem of agricultural development in general and in particular, production of organic products in Russia is not fully studied. There are various reasons for this, such as weak state support for agribusiness sectors, meager funding, lack of qualified personnel in the sphere of organic farming, severe agro-climatic conditions and zones of risky agriculture, high percentage of imported decoctions, especially with Eco or Organic brands. This is only a small list of factors hindering the development of food market in Russia.

The concept of the study is to analyze the causes of hindering the development of organic agriculture. The study is aimed at sustainable development of rural areas by increasing the level of its components: ecology (organic agriculture), society (social infrastructure of rural areas), and economy (economic development of rural areas). The algorithm for the development of organic agriculture should provide for comprehensive solutions to the problems, including state support, investment, innovative technologies, and adapted and harmonized regulations for domestic producers and consumers.

MAIN FOCUS OF THE CHAPTER

World Food Market: Essence and Concepts

In the era of globalization, transformation, and digital economy, food is not only an indicator of the quality of life but also an integral part of food security of the country. Human needs for food are growing every year since the world's population has been growing steadily upward. By 2050, there will be about 10 billion people (United Nations, 2017). Accordingly, world food market should and must respond to the challenges of time, both with changes in related industries and with scientific achievements and technologies in agriculture.

The market is the point on a graph or a map (market square, central market) where the interests of consumers and producers converge (Zinchuk, 2006). The world food market is the same system of interrelations of all economic entities like any other system with its own structure and elements. The world food market is an open position of economic entities (countries participating in the world market of agricultural raw materials and food) for mutual cooperation and on voluntary principles for the transaction (buying and selling), the purpose of which is to benefit from these transactions.

The market, in general, in particular supply and demand, is a dynamic system, and is sensitive to various kinds of factors. Food market is no exception. At first glance, there is a feeling that food is essential good which means that the demand for food is inelastic. Food can also be graduated both for luxury goods and delicacies and for essential goods. On the other hand, similar food products may be considered as luxury goods and delicacies in one country, common products in another, and exotic products in other. Respectively, the costs and prices of such products vary from one country to another.

Therefore, food market is very dependent on price and non-price factors. In domestic market, there are many factors hindering the functioning of the market.

Food security is the optimal level of security and unlimited access to safe, functional foods, regardless of climatic conditions, territorial affiliation, and direct correlation according to consumption standards and diets for maintaining a normal and healthy human life (Ryumkin & Malykhina, 2018). There is not only safe food but also functionally-oriented, healthy food. The assortment offered by producers can be confused since the significance of a product lies in its purpose. For example, some products are focused on the mass purchasing power, others – on exoticism and rarity of the product, others – on environmental friendliness of products. Accordingly, a consumer needs obtain reliable information about the products with which he can identify the usefulness and significance of purchase for himself and his family. Therefore, the authors offer the classification of foodstuffs. Food products are grouped into the following classes depending on their origin and purpose:

- Plant products of high energy value (vegetables, fruits, and vegetable oil).
- Animal products of high biological value.
- Products of the auxiliary group which are not intended to be used separately (food additives, salt, and spices).
- Mixed products of high nutritional value (powdered or creamy bases and concentrates).

Food products are also differentiated according to their readiness for consumption:

- **Gastronomic:** Ready-to-eat products (sausages, cheeses, canned foods, and drinks);
- **Groceries:** Products that need to be prepared (raw vegetables and meat) (Potrebitely, n.d.).

Proper nutrition or healthy diet is a balanced diet of safe, functionally-directed, high-quality products that allow the human body to grow, develop, and stay healthy. As for healthy food products, they are classified into the following groups:

- Common products (milk, cheese, and bread);
- Healthy products, i.e. functionally-directed actions under requests of a target audience (protein bars for athletes, diabetic products, etc.).

Healthy nutrition market is divided into three major segments:

- **Organic Products:** Food Produced with an application of natural fertilizers in ecologically clean areas which does not contain artificial or unhealthy ingredients (vegetables, fruits, meat and fish, dairy products, cereals, juices, and other products)
- **Functional Foods:** Food products enriched with vitamins and useful additives which allow improving human health (dairy products, bakery products, drinks, and other food products)
- **Dietary and Diabetic Foods**: Foods designed to keep to a specific diet for medical and individual reasons (carbohydrate-free foods, low-sugar foods, low-fat foods, gluten-free foods, and other categories) (Ptukha, n.d.)

This type of products is designed for the segment of middle and high-class consumers. The introduction of new production technologies, development of modernized "smart" packaging, use of natural ingredients (environmentally friendly concentrates, natural bases, and bioproducts) increase the cost of healthy foodstuffs twofold and more. In the market of healthy food, domestic analogs of foreign products have been becoming more frequent. Suggestions of domestic producers of healthy food require extensive information support for a more trusting attitude of consumers to their products, and as a result, the expectation of increased demand for them. However, in Russia, imported products still dominate in the market (Danone, Mars, Nestle, and others). In fact, they are also pioneers in presenting new products on the Russian market of healthy foodstuffs, for example, Actimel and Activia by Danone. Trends and novelties on the Russian market come from abroad with a delay of one-three years. The reason for this is the reluctance of foreign producers to compete in the Russian market. As soon as the demand for the new product decreases, application of cutting-edge technologies allows presenting new products on the Russian product market after a certain time. Current challenge is to ensure a healthy lifestyle and natural products. This trend is called Healthonism (health and hedonism) – this is the tendency associated with the desire to combine the consumption of food, which brings pleasure, with health benefits (for example, alcohol with antioxidants, low sugar food, healthy fast food, and others) (Laboratory of Trends, 2018).

Food Market in Russia

Studying food market as final product for a consumer, it is necessary to first consider the current state and direction of development of agriculture in Russia. Agricultural production is both the final product (canned foods, beverages, meat products, yogurt, and vegetables) and intermediate and raw products (grain, flour, meat, milk, and egg).

In Russia, total volume of agricultural output amounted to RUB 5.12 trillion in 2018 and RUB 5.7 trillion in 2017 which indicates a decrease by 10.1%. In 2017, crop production in farms of all categories amounted to RUB 2.569 trillion, a decrease by 17.4% in annual terms. Livestock production declined by 3.8% down to RUB 2.551 trillion (Figure 1).

In comparable prices, crop production in 2018 decreased by 2.4% compared to 2017, and livestock production increased by 1.3%. In 2017, Russia received a record grain yield of 135.5 million tons. In 2018, gross grain harvest was 112.9 million tons which is by 17% lower than in 2017, but still by 14% higher compared to the average level in the past decade. In 2019, agricultural output is expected to grow by 1.0-1.5% (Agricultural Bulletin, 2019). There is an increase in the volume of production in 2014-2018 with the exception of 2018, where the volume of agricultural products decreased due to a number of factors, such as the rising cost of energy resources and weak financial state support.

As for agricultural imports, over the past five years, there has been a decrease in volume due to import substitution policies and imposition of sanctions on imported goods, including some agricultural products (Figure 2).

The schedule is differentiated by the sections of the Commodity Nomenclature of Foreign Economic Activity (CN FEA) which includes:

- Section I – live animals, products of animal origin;
- Section II – products of plant origin;
- Section III – fats and oils of animal or vegetable origin and their cleavage products; prepared edible fats; waxes of animal or vegetable origin;

Figure 1. Volume of crop and livestock production in 2014-2018
Source: Federal Service for State Statistics of the Russian Federation (2019)

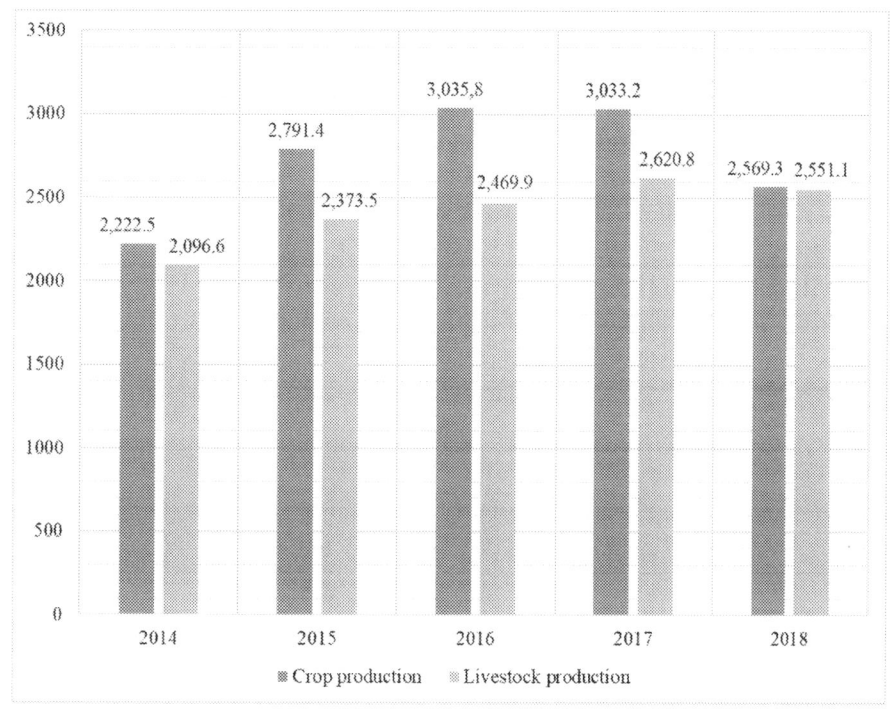

Figure 2. Imports of agricultural raw materials and food in 2014-2018, $ million
Source: AB-Center (n.d.)

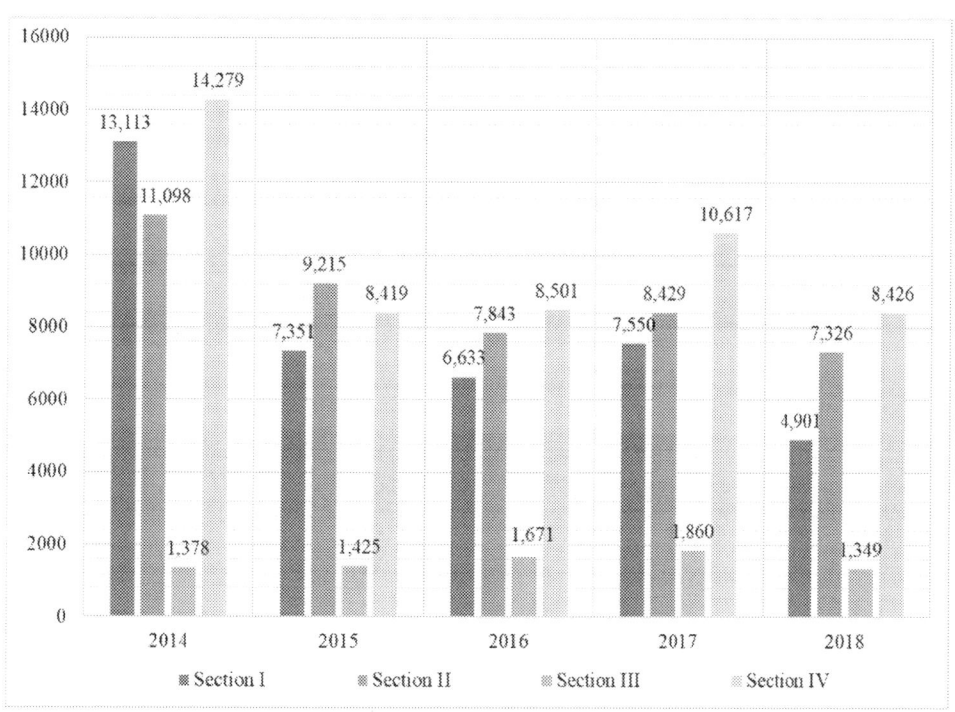

- Section IV – plant products.

There has been a general decline in the volume of imports by sections. Only in 2017, there was an increase due to the restoration of a number of supplies in sections II and IV. Further, in 2018, there is a decrease in imports due to a more trusting attitude of consumers to domestic products (AB-Center, n.d.).

In terms of the structure of per capita consumption of staple foods, the first place is taken by eggs and egg products – 279 pieces per year (Figure 3). On the second place, there is milk and dairy products which consumption has increased during the past five years. This is due to the current trend in the life of the population for healthy life and healthy diet. Milk is not only a source of calcium and other necessary micro-and macronutrients for humans. It also has an adsorbing effect, as well as beneficial properties for digestive system. Therefore, milk and, especially, dairy products (yogurts, dairy snacks, curds, and cheeses) have such a demand in the food market among the population. Bread occupies third position. Since bread and bakery products are sources of carbohydrates and energy for human vitality, demand for these products will continue increasing in the future.

The review of the current state of agriculture in Russia and demand for staple food is followed by the analysis of supply of these foods and the country's self-sufficiency in basic foodstuffs (Table 1).

For all positions, import has a positive balance. For example, import of fruit and berries prevail over export by 30 times, which is slightly above the self-sufficiency ratio (32.7%) for this position. In other words, Russia is not able to ensure self-sufficiency in fruit and berries due to climate factor. The highest self-sufficiency ratio is for potato (87%), but imports still prevail over exports.

Figure 3. Structure of per capita consumption of staple food in Russia in 2017, kilograms
Source: Federal Service for State Statistics of the Russian Federation (2019)

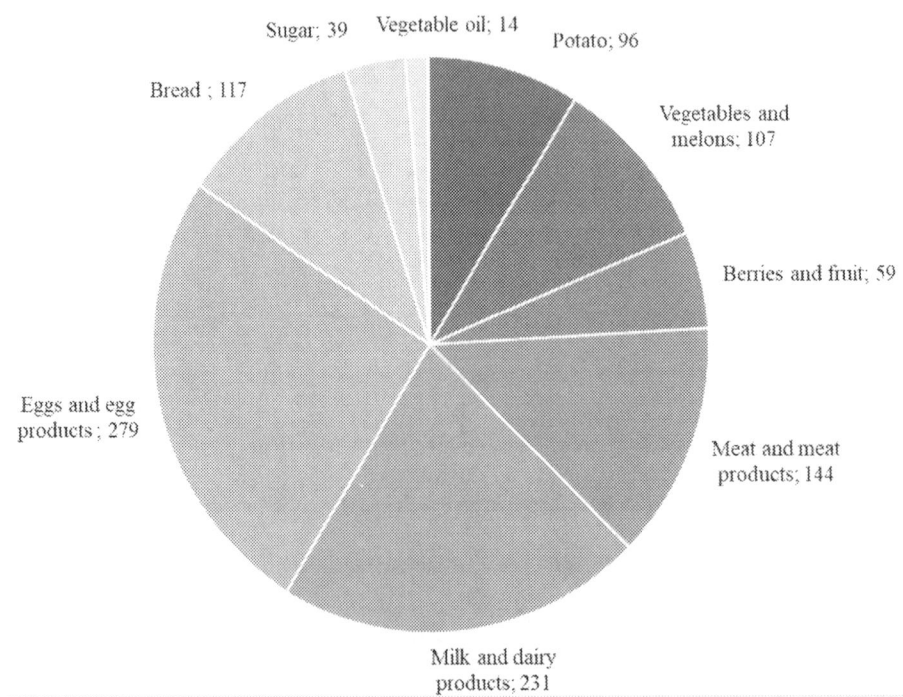

World Market of Organic Food Products

Table 1. Resources and use of main crop products in Russia in 2017, thousand tons

Parameter	Potato	Vegetables, melons	Fruits and berries
Resources:			
Stocks at the beginning of the year	18,099.9	7,129.3	1,971.9
Production	21,768.7	15,430.4	3,212.2
Import	1,499.6	2,669.9	6,677.0
Total resources	41,368.2	25,229.6	11,861.1
Usage:			
Industrial consumption	9,206.2	1,698.0	1,119.8
Losses	1,688.8	513.4	97.7
Export	246.2	360.7	210.9
Personal consumption	14,134.4	15,759.8	8,599.2
Stocks at the end of the year	16,092.6	6,897.7	1,833.5
Self-sufficiency ratio, %	87	85.86	32.7
Export / Import (+/-)	Import (+1,253.4)	Import (+2,309.2)	Import (+6,466.1)

Source: Federal Service for State Statistics of the Russian Federation (2019)

Table 2. Resources and use of main livestock products in Russia in 2017, thousand tons

Parameter	Meat and meat products	Milk and dairy products	Eggs and egg products (million pieces)
Resources:			
Stocks at the beginning of the year	804.3	1,746.0	1,316.9
Production	10,323.1	30,164.1	44,789.9
Import	1,103.1	7,129.3	1,225.7
Total resources	12,230.5	39,039.4	47,332.5
Usage:			
Production consumption	32.7	2,881.6	4,278.1
Losses	15.6	29.4	110.1
Export	307.4	607.6	720.3
Personal consumption	11,012.8	33,881.9	40,938.3
Stocks at the end of the year	862.0	1,638.9	1,285.7
Self-sufficiency ratio, %	93.3	82.0	98.8
Export / Import (+/-)	Import (+795.7)	Import (+6,521.7)	Import (+505.4)

Source: Federal Service for State Statistics of the Russian Federation (2019)

For livestock products, self-sufficiency is significantly higher compared to crop industry, particularly, 98.8% for eggs and egg products (Table 2). Imports prevail over exports. For meat and meat products and eggs and egg products, imports make up an insignificant part, but for milk and milk products, this ratio exceeds eleven times. This is explained by the large volume of imports of cheese, cottage cheese, cream, condensed milk, and powdered milk (Kozhevnikova, n.d.).

The balances of resources and used products are equal. For all positions and to varying degrees, the predominance of imports over exports is registered. For all positions, the ratio of security does not exceed 100%, which indicates food insecurity problem and the barriers to the development of agriculture.

In 2016, the market of healthy food products in Russia amounted to RUB 874.095 million. The market consists of five segments: "Best for You (BFY)", "Fortified / Functional (F/F)", "Free from (FF)", "Naturally Healthy (NH)", and "Organic". The largest market segment is NH. In 2016, its volume amounted to RUB 469.644 million. The highest growth rate was observed in F/F segment (15.1%). Over the past few years, retail has been remaining the most important sales channel for healthy food products in Russia. In 2016, 97.7% of healthy food was sold through retail channel while only 0.1% has been distributed through the specialty stores channel. Retail in Russia is represented by modern and traditional retail channels. Traditional retail is a network of hypermarkets and specialty stores (21% of healthy food sales in 2016), while modern retail is social networks and online stores (76.7% of sales). The largest market share belongs to Wimm-Bill-Dann Produkty Pitania (6.9%). Other big producers of healthy food products are Wrigley and Danone Russia Group (4.6% and 4.5% of the market in 2016, respectively). The most popular brands of healthy food are Orbit (Mars Inc.), Prostokvashino (Danone Group), and Domik v Derevne (PepsiCo Inc.) (3.8%, 2.6%, and 2.5% in 2016, respectively). By 2021, the market volume is expected to grow up to RUB 912.477 million. In 2017-2021, the annual growth rate of healthy food market is projected to be 0.9% (Discovery Research Group, 2019).

World Food Market: Analysis and Statistics

In 2018, exports of agricultural products from Russia amounted to $26 billion. The harvest obtained in 2018 allowed Russia becoming one of the leading food exporters with 39 million tons of export potential for grain crops and 34 million tons for wheat. In 2017-2018, total export of grain amounted to 52.422 million tons, including 40.449 million tons of wheat (Laboratory of Trends, 2018).

In terms of export, the largest segment in the global food market is processed food products and ready-to-use raw materials for further processing (Figure 4). This segment includes:

- Flour, starch, malt, gluten, and other products of deep processing of grain
- Finished products from meat, fish and aquatic invertebrates, canned food, and sausages
- Baking and pasta
- Canned and frozen fruit and vegetables
- Other products, including instant tea and coffee, ice cream, and sauces
- Production of eggs and eggs in the shell
- Honey

In the segment of processed food products, almost equal shares are occupied by pastries and pasta ($63 billion) and other finished products ($63 billion), while the share of vegetable and fruit products is lower ($56 billion). Products of grain processing, including flour and starch, occupy a relatively small share in the segment: below 7% or $18.5 billion. World trade in eggs and honey is very small compared to the total market volume: $5.7 billion and $2.3 billion, respectively (Mironenko, 2016).

In 2018, Deloitte conducted a study of global food market and revealed the following results:

Figure 4. Export of processed food products in 2018
Source: Mironenko (2016)

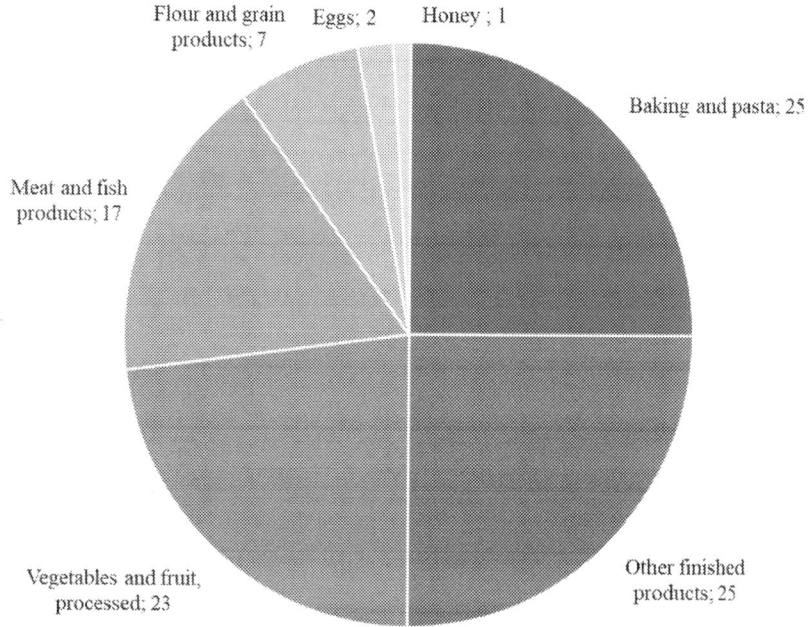

- Top five countries and regions of agricultural production: Asia (except China and India), China, Brazil, India, and the USA.
- Top five countries and regions of food consumption: China, Asia (except China and India), the USA, the EU-28, and India.
- Forecast of consumption of key agricultural products in the world (2019-2020): corn, wheat, sunflower oil, soybeans, and dairy products (Potrebitely, n.d.).

The main suppliers of high value-added products are industrialized countries. In each of the key areas of the segment (baking and pasta, fruit and vegetable processed products, other finished products), total share of the main producing countries (the USA, Germany, the Netherlands, Italy, and China) is 35-40% (Figure 5).

Meat is the second largest segment in the world market. The leading meat exporters are the USA, Brazil, and Australia (taken together, they occupy one-third of the world market). The main importers are Japan, China, and the USA. At the same time, the USA is a net exporter of meat, the volume of foreign supplies of which exceeds the level of imports by $5 billion. Another major market segment for agricultural raw materials and food is fruit and nuts. The leading suppliers in this segment are the USA (13% of the world market), Spain (9%), and Chile (5%). Key consumers are the USA (14%), Germany (9%), and the UK (5%). The USA is a net importer of fruit and nuts.

The fourth-largest segment is grain. The leading suppliers of grain to the world market are the USA (19% of the market), France (8%), and Canada (7%). The main importers are China (9%), Japan (5%), and Egypt (4%).

Figure 5. Share of leading producers of high value-added products in total world trade in 2017 Source: Mironenko (2016)

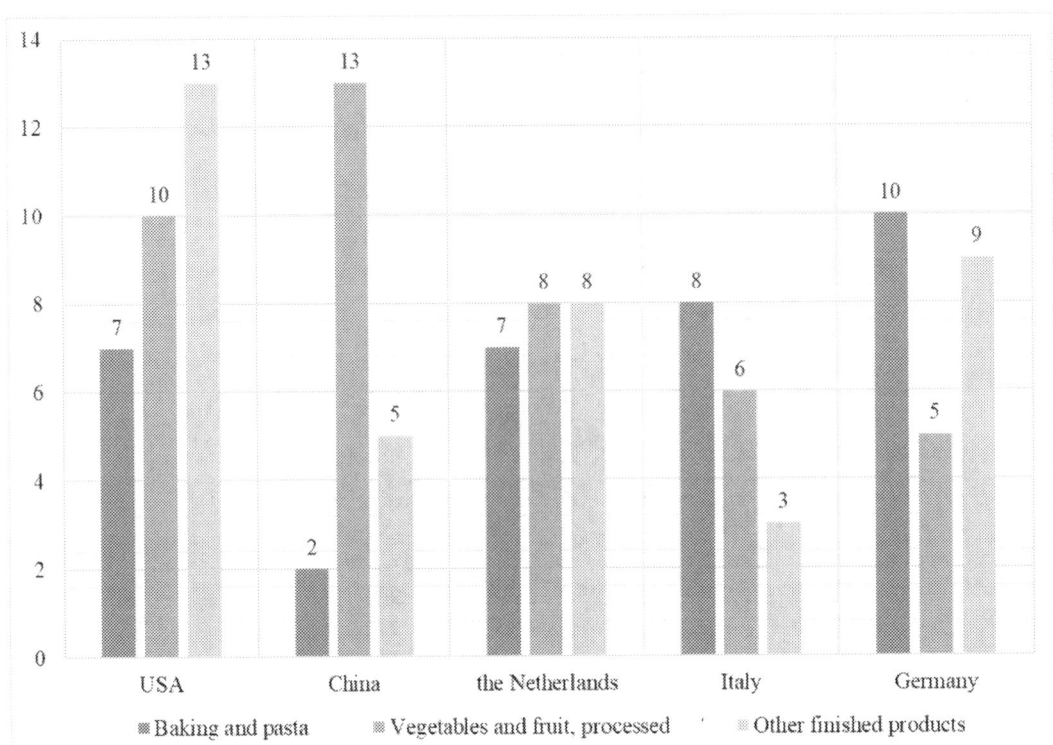

In general, the largest food exporters in the world are the USA, Brazil (tea, coffee, spices, sugar, and sugar products), Indonesia (vegetable oils), Germany (dairy products), China (fish, vegetables, and honey). The leading buyers are the USA, China, Germany, Japan, and the UK (International Independent Institute of Agrarian Policy, 2016).

For leading agricultural powers, there are several models of agro-export development:

- Agro-export as a leading export industry (20-40% of total exports – Brazil, Indonesia).
- Developed economies (8-12% of total exports – USA, Canada).
- Countries with a large population – India and China – do not have opportunities for faster growth of exports of agricultural products.
- Russia ensured a breakthrough growth in the agro-industrial sector of 10-12%, typical for the USA and Canada.

The global food and beverage market is controlled by a total of ten multinational companies. They form food basket and determine working conditions of the majority of the world's population. Nestle, PepsiCo, Coca-Cola, Unilever, Danone, General Mills, Kellogg's, Mars, Associated British Foods, and Mondelez – each company has thousands of employees, and revenues of this top ten companies make billions of US dollars annually. However, many of these companies have been associated with scandals in the field of human rights and environmental pollution.

Today, the main features of the global food market are the leading role of value-added food products, main suppliers of which are the USA and the EU. China and India are large producers and consumers at the same time due to the world's biggest population. These markets are the largest in the world in terms of sales (Pravda.ru, 2017).

World Organic Market

In the conditions of transformation of national economies, joining international unions and integration, development of smart technologies – industry 4.0, digital economy, main socio-economic indicators of the quality of human life – health, namely its nutrition, safe and functionally-directed action, the environment with the level of environmental risk for him, and the safety of life in all its interpretations are becoming increasingly important (Ryumkin, Ryumkina, Nakonechnaya, & Essaulenko, 2018).

Social well-being of a person directly depends on physiological state, which, in turn, has a direct correlation with food and food. You are what you eat makes a huge sense, that is, high-quality food has a direct impact on growth, development, and regeneration of the cells in body and brain. Accordingly, consumption patterns depend on the formation, organization, and development of organic food market. This market becomes not only a global trend, a profitable business, but also a significant social condition for health, development, and reproduction of the country's population (Ryumkin et al., 2018).

Russia has been always considered an agrarian country with the world's largest acreage of arable and agricultural land. However, the current state of agricultural sector whole requires careful study. According to Russia's President Putin, it is necessary to set the task of the national level and provide domestic market with domestic food by 2020 (Interfax, 2015). Russia is capable of becoming the world's largest supplier of healthy, environmentally friendly, high-quality food products. The demand for such products in the global market is growing (Interfax, 2015).

Development of agriculture is a priority for many economies, and the growth rate of the organic market in the world confirms this. But there are a number of constraints that suggest that the decisions of the government do not correlate with the actions to implement the task at the national level. When considering this issue, the authors revealed three categories of organic products:

- **100% Organic Product:** It consists of organic components and is grown in compliance with biotechnologies (without the use of chemical protection agents and chemicals when treating seeds and shoots; without the use of antibiotics and banning the tethered way of animals; agricultural enterprises operate in a closed-loop – farming and animal husbandry).
- **95% Organic Product:** It includes 95% of organic raw materials during processing and produced without the application of artificial additives.
- **70% Organic Product:** Produced from organic ingredients, 30% is made up of the remaining semi-natural ingredients.

There are several synonyms for these types of products – bioproducts, eco-products, and organic products. They are grown, harvested, processed, and packaged in accordance with the standards of organic farming and production adopted in Europe. Different countries use different designations: Bio in Germany and France, Organic in the UK and the USA, Eco in the Netherlands, and Nature Product in Finland) (Biomdv, n.d.).

The term "organic agriculture" or "organic farming" is commonly understood as the methods of obtaining agricultural products in which the use of artificial (synthetic) products is minimized (Novitsky, 2017). There is a term "natural products" which means the products that consist entirely or, at least, mostly of the ingredients of natural origin, with a minimum amount of chemicals, artificial fillers, and additives. Natural products include primarily organic products or "Ecological Clean Products" (ECP) which are grown on specially cleaned soil without the use of chemicals (Biomdv, n.d.). Council of the European Union (2007) defines organic production as a holistic farm management system (organization) and food production which combines the best environmental methods, high biodiversity, preserves natural resources, applies high animal protection standards in the production process using only natural substances for consumers.

Organic products are agricultural products made from plants or animals using environmentally sound technology, nature-likeness technologies without the use of pesticides and other plant protection products, chemical fertilizers, growth promoters and animal fattening, antibiotics, hormonal and veterinary drugs, GMOs, as well as based on the principles of environmental sustainability and quality production for human life and health. Under environmental sustainability, the authors understand the state of stable self-regulation of nature and the environment, the process of waste disposal without harm to ecosystems and the absence of negative physical and chemical effects on biological objects. In other words, society, as little as possible should affect the nature, to apply a negative kind of strikes and attacks against the environment, as well as to take all possible measures to protect and defend it.

Today, global organic market amounts $90 billion ($25 billion in 2003). It is expected to grow annually by an average of 15.5% in 2018-2020, whereas in 2010-2015, it increased by 8.6%. Organic farming involves 181 countries and covers over 69.7 million hectares. In 2018, the number of involved countries

Figure 6. The top ten countries with the largest number of certified organic producers in 2018
Source: FiBL (n.d.)

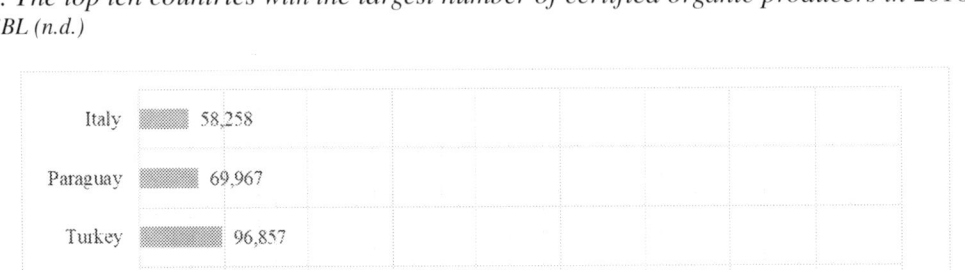

was 178, the area of farmland was 58 million hectares. In just one year, the area of land involved in organic farming increased by 20% (Ryumkin et al., 2018). The USA is a leader with a share of $40 billion, Germany ($10 billion) and France ($7.9 billion) are the second and third, respectively. China's volume is $7.6 billion. By 2024, organic food market will amount to $324 billion (Ryumkin et al., 2018). In total, there are about 3 million producers of environmentally friendly products in the world: 835,000 companies are concentrated in India, 230,410 – in Mexico, 210,352 – in Uganda. In Russia, there are only 70 certified organic producers (Figure 6).

Organic market is one of the few markets where demand exceeds supply due to the lack of land resources suitable for agricultural land and small proportion of farmers certified according to international standards. These factors not only affect demand but also constrain the formation and development of organic market. Russian organic market is at the early stage of development. The establishment and development of organic market in Russia is inadequate to its resource potential since Russia has 28 million hectares of fallow land (i.e., agricultural land that has not been used for a long time and has not been subjected to chemical treatment). Accordingly, Russia has a significant advantage over other countries. Due to this factor, it can take one of the leading places in the world organic market in combination with other important factors.

Key factors hindering the development of the domestic market are:

- High consumer value of eco-products. These products are from 50% to 400% more expensive in Russia (in the world, by 20-30%) compared to conventional food products. The cost of bioproduct is higher than the cost of a synthetic product. Not all consumers can afford buying eco-products. This sector of agriculture is directed and sells its products among the middle and upper-middle classes of the population.
- The adoption of the law on the production and turnover of organic products will make it possible to clarify the conceptual and categorical apparatus of organic agricultural products. In addition, there will be uniform norms for all producers, as well as the rules and requirements for the production of eco-products. Also, the law will protect consumers from low-quality falsified products.
- Eco-products are poorly presented in trade networks due to the low purchasing power of population.
- The risk of buying counterfeit goods and a lack of confidence in the quality of domestic products.
- Due to the small share of agricultural producers in this sector of agriculture, the issue of the logistic component has a great impact on the pricing of organic products, which in turn affects the emergence of the number of intermediaries between producer and consumer. These intermediaries sell organic products using shadow schemes (over RUB 100 billion) (Food and Agriculture Organization of the United Nations [FAO], 2018).

Today, in Russia, 385 thousand hectares of land are certified (Figure 7). The country occupies 17th place in the world by the area of certified land. Some of land plots certified for organic farming are not cultivated. In the EU, on the contrary, the area of certified land is almost equal to the number of cultivated plots (Gorbatov, 2016).

Over the decade, the area of certified agricultural land has increased by 110%. It is a good result for Russia, but there is a need to understand that to meet the demand and needs of the consumer with organic agricultural products, this is not enough. In order to reach at least the tenth place in the world, Russia needs to create a program to support and develop the production of environmentally friendly products, nature-like technologies, and smart agriculture. The number of agricultural producers should

Figure 7. Area of certified land for organic farming in Russia in 2006-2016, thousand hectares
Source: FiBL (n.d.)

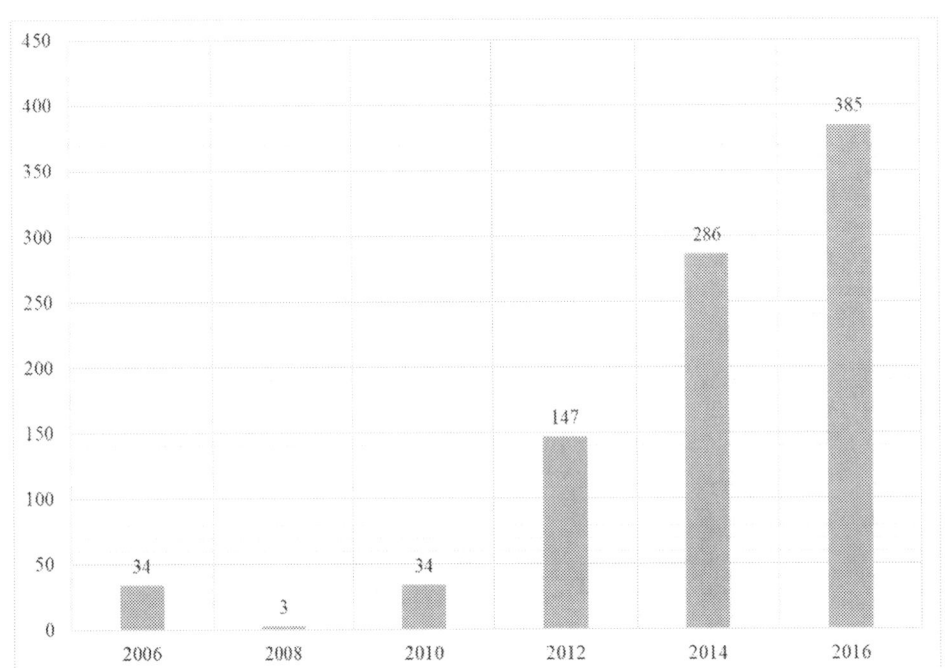

be increased at least to 2,000 units per year. Only in this case, Russia will be able to get closer to the leading countries of the world. This forecast allows judging about the existing significant gap between Russia and the rest of the world in the production of organic products (Ryumkin & Malykhina, 2017). Smart agriculture, according to the authors, is the idea of applying high technologies, using all kinds of innovative solutions and the Internet of Things in agriculture to promote and simplify the processes of management, control, regulation of production with the least losses and obtaining the maximum result, both in physical terms, in the form of safe food, and in financial indicators. And nature-likeness technologies are nature-saving and specialized technologies that allow preserving biodiversity, ecosystems, and the environment, to promote the development of biological objects, to maintain and supplement the necessary conditions for their keeping, as well as to foresee and eliminate negative impacts of various kinds. These technologies will allow optimizing production costs and reducing the level of negative impact on the environment (Ryumkin et al., 2018; Schmidt, Ryumkin, & Ryumkina, 2018).

SOLUTIONS AND RECOMMENDATIONS

Section "World Food Market: Essence and Concepts"

Promotion of new food products need simple and indisputable parameters that allow showing functionality, quality, and composition of products that do not require significant descriptions and comments. This approach is adapted to the distribution of products in the developed countries. This principle is introduced by small farms presenting their products in packages of different gradations and, accordingly, different price categories, as well as focusing consumers' attention on originality and individuality of products, such as handmade confectionery or honey with raspberry jam.

Consumers are quite picky about the characteristics of products. Therefore, to earn loyalty and conquer a consumer, a farmer needs repeating confirmation of the quality of the products, i.e. needs the creation of transparency of a brand. In other words, a producer should be opened to a buyer by:

- Masterclasses where the personnel of the farm represent the entire technological process of food production
- Exhibition stores with low prices for their products
- Fairs, where naturalness and usefulness of products are revealed

Authors' opinion is that, based on Healthonism trend, small and medium farmers should expand their product line to meet the needs of discerning and demanding consumers. For example, form sets and assorted healthy foodstuffs in combination with functional orientation for various purposes, such as weight loss, immune system maintenance, immune system restoration, rehabilitation complexes, etc.

Section "Food Market in Russia"

Depending on the target setting of agricultural producers, the factors affecting organic market have been changing. Therefore, responding to new challenges of the time, all agricultural producers regardless of the group of food products and their functional value should implement following strategies for the development of their enterprises.

- Increase of production;
- Increase of technical potential (introduction of new capacities and technologies);
- Cost reduction.

Strategy 2 requires special attention because it combines three most important technologies for the breakthrough development of agriculture. Breakthrough development is an innovative approach combined with rational management decisions. Rational management decisions are the decision-making process of correct use of factors of production (labor, land, capital, information, and entrepreneurial ability) for the production of goods and services. Innovative approach is technologies, or means of production, which allow making a big leap forward, thereby reducing the gap between the countries in the production of competitive agricultural raw materials and food. These technologies include:

- Introduction of smart farms, greenhouses, and sites for the development of crop and livestock industry, i.e. the development of digitalization in agriculture
- Development of biotechnology (biopesticides, biofertilizers, biofuels, and biologicals), i.e. development of environmental sustainability in agriculture
- Introduction of individual functional product line, i.e. development of the national idea of safe and quality food, regardless of the region, consumer segment, and price category

Since the market of healthy food is specialized, it is necessary to provide extensive information support in combination with the national priority for a healthy lifestyle.

Section "World Food Market: Analysis and Statistics"

For more dynamic growth of food exports, the attention should be paid to the following export items:

- Products with high added value
- Organic product, natural food. An important condition for entering these markets should be the Intensification of efforts by the state and business in order to harmonize standards and overcome trade barriers
- Local agricultural crops (honey, berries)

Section "World Organic Market"

International experience includes various directions in which it is possible to find the prospects for production. The main impetus of development is state support of agricultural producers. This support can take various forms:

- Adoption of the law on the production and circulation of organic products
- Federal and regional subsidies, tax benefits, and tax holidays
- Possibility of obtaining and introducing new technologies through business loans with a low-interest rate of 5-7% per annum, or through irrevocable loans
- Favorable investment (possibility to attract foreign investments) and business climate (absence of corruption)
- Protectionist measures to protect domestic producers, such as high duties on imported organic products
- An increase in the number of new qualified personnel to work with organic products by opening new training programs in sectoral universities and at the expense of the state
- Development of infrastructure and the sector as a whole

It is also necessary to take measures to eliminate the factors hindering the development of the domestic organic market:

- Reducing the cost of eco-products, through the introduction of new technologies, which will, in turn, produce products in mass mode, and thus reduce its cost
- Expanding the product range, product line for any consumer taste, as well as increasing distribution points and distribution channels for these products – online stores, specialized points of sale with qualified consultants
- Building trust in domestic producers through extensive advertising projects and PR-companies to convince consumers in the usefulness, safety, and quality of products
- Fight against corruption (legal entities and physical persons who allow falsified products to enter the market under the guise of high-quality organic products) by adopting the laws punishing this structure with high fines and criminal liability
- Elimination of intermediaries between producers and consumers of organic products, which will allow producers not to lower the price and not work at a loss to themselves, and to consumers of organic products not to buy it at an inflated price

FUTURE RESEARCH DIRECTIONS

In organic farming, there are still areas that require further study which can also affect the recommendations and conclusions. It is imperative to consider the role of international organizations in the field of food security on global and domestic levels. They play an important role in world trade. The functional characteristics of such organizations vary from the promotion of agricultural products to the market to protectionism and sanctions against a number of countries. Therefore, a study of their activities and the degree of involvement in the world trade of traditional and organic agricultural products will allow discovering the share and degree of their impact at both administrative and political level.

Some of the markets which are directly related to food trade still had been remained unexplored, specifically, market of chemical fertilizers and pesticides, market of technical means and agricultural technologies, genetic engineering products market, and logistics market and distribution channels for agricultural products. Studying these markets will allow understanding the general picture of agricultural production worldwide.

CONCLUSION

The understandings of food market vary. Basically, it is a traditional market which includes the market of agricultural raw materials and food market. Also, food market is classified into a traditional market, healthy food market, and organic market. Each of these markets has its own characteristics, target and price segment, and consumers. The factors influencing the development of the traditional food market are at the same time those which affect the other two markets, with the exception that for a healthy, functionally-directed food market, it is necessary to focus on the population observing a correct lifestyle and sport. For organic market, the main factor is land which must be certified for organic farming.

Based on the foregoing, it follows that food market in Russia does not fully meet the needs of the population for various reasons. The demand for some food products exceeds supply, and the state resolves this issue by importing food products rather than increasing the growth rate of domestic food production. Today, Russia has all the necessary prerequisites for the formation and development of the market for traditional and healthy food products with a high self-sufficiency factor, regardless of territorial affiliation, national traditions, and social-behavioral factors of consumers. Russia has the greatest export growth potential among all the major agricultural powers. The country aims to achieve $150 billion exports by 2035 (Kozhevnikova, n.d.). This is an ambitious but realistic task which will allow to fully reveal the resource potential of agriculture in Russia. This goal may be achieved through the introduction of technologies which will allow domestic producers becoming competitive in the global market.

Today, the main features of the global food market are the leading role of high value-added products (agricultural products), the main suppliers of which are the USA and the EU. The resource potential of Russia and production of organic products can become the main direction for entrepreneurs. Since it is quite difficult to compete with large agricultural holdings, production of high-quality products enables small and medium-sized businesses to be in demand in the market.

Organic market is one that combines all three vectors of sustainable rural development – economic, social, and environmental focus. In other words, it can be said that eco-market combines benefits and dividends from the economy, society, and ecology:

- Economy is represented by an increase in the volume of organic production, and as a result, employment of rural population, increase in income, import substitution, and development of exports
- Social part is represented by an increase in the level of social infrastructure in rural areas, decrease in the outflow of citizens from rural areas, and increase in the quality of life of the population
- Ecology is represented by the absence of erosion, improvement of soil water, and environment, development of ecological equipment for farms and greenhouses in favor of the protection and conservation of nature

Thus, the development of organic market will optimize the level of food, economic, social, and environmental security worldwide.

REFERENCES

AB-Center. (n.d.). *Expert-Analytical Center of Agribusiness*. Retrieved from https://ab-centre.ru/

Agricultural Bulletin. (2019). *Agricultural Production in Russia Decreased by 0.6% in 2018*. Retrieved from https://agrovesti.net/news/indst/proizvodstvo-selkhozproduktsii-v-rf-v-2018-g-snizilos-na-0-6.html

Balfour, E. (n.d.). *Towards a Sustainable Agriculture. The Living Soil*. Journey to Forever. Retrieved from http://www.journeytoforever.org/farm_library/balfour_sustag.html

Biomdv. (n.d.). *What Is Organic Food?* Retrieved from https://biomdv.ru/page/chto-takoe-organicheskie-produkty-pitaniya

Council of the European Union. (2007). *Council Regulation #834/2007 of June 28, 2007 On Organic Production and Labelling of Organic Products and Repealing Regulation (EEC) #2092/91*. Retrieved from https://eur-lex.europa.eu/legal-content/EN/TXT/?uri=celex%3A32007R0834

Dabbert, S., Haring, A. M., & Zanoli, R. (2004). *Organic Farming: Policies and Prospects*. London: Zed Books.

Discovery Research Group. (2019). *Analysis of Healthy Food Market in Russia*. Moscow: Discovery Research Group.

Federal Service for State Statistics of the Russian Federation. (2019). *Statistics*. Retrieved from http://www.gks.ru/

FiBL. (n.d.). *Research Institute of Organic Agriculture*. Retrieved from https://www.fibl.org/en/homepage.html

Food and Agriculture Organization of the United Nations. (2018). *The State of Agricultural Commodity Markets: Agricultural Trade, Climate Change and Food Security*. Rome: Food and Agriculture Organization of the United Nations.

Gorbatov, A. (2016). The Development of the Market for Organic Products in Russia. *Fundamental Research, 11*(1), 154–158.

Guseva, N. (2015). *Interview with Helena Bollesen*. Retrieved from https://lookbio.ru/obtshestvo/bio-portret/v-organike-rossiya-sejchas-kak-daniya-dvadcat-let-nazad-intervyu-s-xelenoj-bollesen/

Howard, A. (n.d.). *An Agricultural Testament*. Retrieved from http://www.journeytoforever.org/farm_library/howardAT/AT1.html

Howling Pixel. (n.d.). *Mokichi Okada*. Retrieved from https://howlingpixel.com/i-en/Mokichi_Okada

Interfax. (2015). *Putin Called Russia the World's Largest Potential Supplier of Ecologically Clean Products*. Retrieved from https://www.interfax.ru/business/482981

International Independent Institute of Agrarian Policy. (2016). *Structure of the World Food Market*. Retrieved from http://xn--80aplem.xn--p1ai/analytics/Struktura-mirovogo-prodovolstvennogo-rynka/

Jespersen, L. M., Baggesen, D. L., Fog, E., Halsnaes, K., Hermansen, J. E., Andreasen, L., ... Halberg, N. (2017). Contribution of Organic Farming to Public Goods in Denmark. *Organic Agriculture*, 7(3), 243–266. doi:10.100713165-017-0193-7

Kharitonov, S. (2013). *Organization and Economic Aspects of Development of Organic Agriculture in Russia*. Moscow: Lomonosov Moscow State University.

Kovalenko, E., Polushkina, T., & Yakimova, O. (2017). State Regulations for the Development of Organic Culture by Adapting European Practices to the Russian Living Style. Academy of Strategic Management Journal, 16(2).

Kozhevnikova, I. (n.d.). *Overview of the Russian Dairy Market*. Food Market. Retrieved from http://www.foodmarket.spb.ru/current.php?article=2279

Kurochkin, S., & Smolnyakova, V. (2012). Organic Farming. *Gardener's Bulletin*, 1, 46–49.

Laboratory of Trends. (2018). *Study of the Russian Market of Healthy Food – 2018*. Retrieved from https://t-laboratory.ru/2018/05/01/issledovanie-rossijskogo-rynka-produktov-zdorovogo-pitanija-2018/

Meredith, S., Lampkin, N., & Schmid, O. (Eds.). (2018). *Organic Action Plans: Development, Implementation and Evaluation*. Brussels: IFOAM EU.

Mironenko, O. (2016). Organic Market of Russia. Results of 2016. *Prospects*. Retrieved from http://rosorganic.ru/files/statia%20org%20rinok%20rossii.pdf

Novitsky, I. (2017). *Organic Agriculture: Profitability and Basic Principles*. Retrieved from https://xn--80ajgpcpbhkds4a4g.xn--p1ai/articles/organicheskoe-selskoe-hozyajstvo-rentabelnost-i-osnovnye-printsipy/

One-Straw Revolution. (n.d.). *About "The One-Straw Revolution."* Retrieved from https://onestrawrevolution.net/

Paull, J. (2011). Attending the First Organic Agriculture Course: Rudolf Steiner's Agriculture Course at Koberwitz, 1924. *European Journal of Soil Science*, 21(1), 64–70.

Paull, J. (2014). Lord Northbourne, the Man Who Invented Organic Farming, a Biography. *Journal of Organic Systems*, 9(1), 31–53.

Potrebitely. (n.d.). *What Are Food Products and What Applies to Them*. Retrieved from https://potrebitely.com/tovary/prodovolstvennye-tovary-eto.html

Pravda.ru. (2017). *Top 10 Companies That Control the World's Food and Drinks*. Retrieved from https://www.pravda.ru/news/economics/1329909-monopoly/

Ptukha, A. (n.d.). *Overview of the Russian Market of Healthy Food*. Step by step. Retrieved from http://www.step-by-step.ru/articles/our/RFDM%204%202017%20kartofel.pdf

Roos, E., Mie, A., Wivstad, M., Salomon, E., Johansson, B., Gunnarsson, S., ... Watson, C. A. (2018). Risks and Opportunities of Increasing Yields in Organic Farming. A Review. *Agronomy for Sustainable Development, 38*(2), 14. doi:10.100713593-018-0489-3

Ryumkin, S., & Malykhina, I. (2017). On the Issue of "Smart" Agriculture: State, Problems and Prospects of Development. *Proceedings of the XX International Conference Agrarian Science to Agricultural Production of Siberia, Mongolia, Kazakhstan, Belarus and Bulgaria*. Novosibirsk: Novosibirsk State Agrarian University.

Ryumkin, S., & Malykhina, I. (2018). Development of Agriculture as a Solution to the Food Problem of Subarctic North of Russia. *Proceedings of the Conference Arctic 2018: International Collaboration, Environment and Security, Innovation Technologies and Logistics, Legislation, History and Modern State*. Krasnoyarsk: Krasnoyarsk State Agrarian University.

Ryumkin, S., Ryumkina, I., Nakonechnaya, O., & Essaulenko, D. (2018). Prospects for the Introduction and Use of Nature-Like Technologies in Agriculture. *Proceedings of the Conference Green Economy in Agriculture: Challenges and Perspectives*. Krasnodar: Kuban State Agrarian University.

Schmidt, L., Ryumkin, S., & Ryumkina, I. (2018). NBICS-Evolution: Factors of Production in Digital Economy for Sustainable Agriculture and Rural Development. *Proceedings of the Conference Role of Agrarian Science in Sustainable Development of Rural Territories*. Novosibirsk: Novosibirsk State Agrarian University.

Shcherbakova (Ponomareva), A. (2017). Organic Agriculture in Russia. *Siberian Journal of Life Sciences and Agriculture, 9*(4), 151–173.

Toleubaeva, D., & Patlasov, O. (2018). State Regulation of the Market of Organic Products in Russia. *Human Science: Humanitarian Studies, 31*(1), 182–191.

United Nations. (2017). *Forecast of the World's Population by 2050*. Retrieved from https://www.mir-prognozov.ru/prognosis/society/oon-prognoz-naseleniya-zemli-k-2050-godu/

Yussefi, M., & Willer, H. (2003). *The World of Organic Agriculture: Statistics and Future Prospects*. Bonn: International Federation of Organic Agriculture Movements.

Zinchuk, G. (2006). Types of Food Market and Development Problems in Russia. *Russian Economic Internet Journal, 4*, 1–15.

ADDITIONAL READING

Chait, J. (2019). *What Is the Definition of an Agricultural Product?* The Balance SMB. Retrieved from https://www.thebalancesmb.com/what-is-an-agricultural-product-2538211

Egorov, A. (2014). *The Formation and Development of Organic Food Market (Case of Central Federal District)*. Moscow: All-Russian Institute of Agrarian Studies and Informatics.

Food and Agriculture Organization of the United Nations. (n.d.). *Agriculture and Food Security*. Retrieved from http://www.fao.org/3/x0262e/x0262e05.htm

Khasanova, S., & Grashorn, M. (2015). Modern Trends in Development of Organic Agricultural Production. German Experience. *Scientific Journal of Kuban State Agrarian University, 106*(2), 2–17.

Kolegov, M., & Ivanov, V. (2004). *Factors and Conditions of Production of Organic Agricultural Products in the Komi Republic (Crop Production Case)*. Syktyvkar: Ural Branch of the Russian Academy of Sciences.

Koteev, S., Yurkenayte, N., & Egorov, A. (2015). Establishment of Institutional and Legal Basis of Organic Food Market as a Key Factor of Its Development. *Agro-Food Policy in Russia, 38*(2), 35–38.

Lockeretz, W. (2007). *Organic Farming: An International History*. London: CABI. doi:10.1079/9780851998336.0000

Maretskaya, V., Omelay, A., & Topoleva, N. (2013). To the Issue of Organic Farming. *The North and the Market: Forming the Economic Order, 34*(3), 37–41.

Nakoryakova, L., Kasatkin, D., & Afanasieva, Y. (2018). *Agricultural Market Overview*. Moscow: Deloitte.

Nikitina, Z. (2008). Transition of Agricultural Enterprises Toward Ecological Production. *Agricultural Economics in Russia, 9*, 85–91.

Thwink. (n.d.). *Environmental Sustainability*. Retrieved from http://www.thwink.org/sustain/glossary/EnvironmentalSustainability.htm

KEY TERMS AND DEFINITIONS

Breakthrough Development: An innovative approach combined with rational management decisions.

Environmental Sustainability: A state of stable self-regulation of nature and the environment, the process of waste disposal without harm to ecosystems and the absence of negative physical and chemical effects on biological objects. In other words, society, as little as possible should affect the nature, to apply a negative kind of strikes and attacks against the environment, as well as to take all possible measures to protect and defend it.

Innovative Approach: The technologies, or means of production, which will allow to make a big leap forward, thereby reducing the gap between Russia and the leading countries in the production of competitive agricultural raw materials and food.

Nature-Likeness Technologies: Nature-saving and specialized technologies that allow to preserve biodiversity, ecosystems and the environment, to promote the development of biological objects, to maintain and supplement the necessary conditions for their keeping, as well as to foresee and eliminate negative impacts of various kinds.

Organic Products: Agricultural products made from plants or animals using environmentally sound technology, nature-likeness technologies without the use of pesticides and other plant protection products, chemical fertilizers, growth promoters and animal fattening, antibiotics, hormonal and veterinary drugs, GMOs, as well as based on the principles of environmental sustainability and quality production for human life and health.

Proper Nutrition: A balanced diet of safe, functionally-directed, high-quality products that allow the human body to grow, develop and stay healthy.

Rational Management Decisions: The decision-making process of correct use of factors of production (labor, land, capital, information and entrepreneurial ability) for the production of goods and services.

Smart Agriculture: An idea of applying high technologies, using all kinds of innovative solutions and the Internet of things in the agro-industrial complex to promote and simplify the processes of management, control, regulation of production with the least losses and obtaining the maximum result, both in physical terms, in the form of safe food, and in financial indicators.

World Food Market: An open position of economic entities (countries participating in the world market of agricultural raw materials and food) for mutual cooperation and on voluntary principles for the transaction (buying and selling) the purpose of which is to benefit from these transactions.

Section 2
Establishing Food Security: Policies, Practices, and Innovations

Chapter 5
Innovative Approaches to Regulation of Agricultural Production and Trade:
Protectionism-Related Effects on Food Security

Anna Ivolga
Stavropol State Agrarian University, Russia

Alexander Trukhachev
Ministry of Tourism and Recreation of Stavropol Region, Russia

ABSTRACT

The chapter studies contemporary innovative approaches to and practices of state support of agricultural production and trade in food and agricultural products. The authors attempt to discover how protectionist policies in the sphere of production and trade affect the level of food security in the conditions of expanding globalization. The chapter focuses on the investigation of advanced innovative practices of state support in the case of selected OECD countries. The authors reveal that the introduction of innovations into the system of state regulation is one of the key determinants of achieving food security in the conditions of the volatile market. Both the volume and priority directions of innovations in agricultural protectionism policies are discussed and evaluated.

INTRODUCTION

Agricultural production and trade in food and agricultural products are the two key elements in establishing food security. In most of the countries of the world, support of domestic agricultural production is considered as one of the major elements of national security (Erokhin, 2017a). Substantial resources are allocated to the support of agricultural producers, food market regulation, rural social programs, and

DOI: 10.4018/978-1-7998-1042-1.ch005

Innovative Approaches to Regulation of Agricultural Production and Trade

environment protection (Gao, Ivolga, & Erokhin, 2018). Such policies are considered as protectionist measures which directly affect food markets, agricultural production, and food security (Erokhin, 2015b).

In the context of globalizing trade in agricultural products, sustainability of domestic food market requires the introduction of innovative approaches to food security policy. While international trade in recent decades has been gradually liberalized, trade in agricultural products and food has remained among the ones most influenced by international regulations and national policies (Bozic, Bogdanov, & Sevarlic, 2011). Agricultural protectionism is focused on the selection of measures of foreign trade and economic policies to achieve the particular degree of protection of agricultural sector and domestic food market from foreign competition. In a broader dimension, protectionist policies in the form of state support of agricultural production are aimed at the development of sustainable food production and establishment of food security of a country.

A set of protectionist policies is diverse. There is a wide range of tools that affect the competitiveness of domestic farmers and food security in various ways (Erokhin, Ivolga, & Heijman, 2014). Such policies support the effective elimination of price disparity and growth of farmers' incomes. The offloading of agricultural surpluses of developed countries on the world market brings down prices and creates disincentives for local producers in many developing countries where agriculture is the main source of livelihood for a major part of the population. In the early 1990s, due to the difficult economic situation, many developing countries decreased the level of state support of agriculture and undertaken reductions in the protection of their domestic food markets, at considerable pain and effort, largely with a view to enhancing the supply of food products for their populations. Trade liberalization was successfully implemented in the countries where that process was sustained for a long period. In the short run, trade liberalization as a way to increase food security is often painful and may be even damaging for developing economies.

Liberalization froze state support and left emerging economies with very few policy instruments to protect themselves from food imports and to subsidize their agriculture (Erokhin, 2017b). As many of developing countries do not have sufficient financial capacity to support and protect domestic agricultural complex on the level comparable to that in developed economies, they have to introduce innovative practices of agricultural production and trade (Erokhin, 2017c). Despite the certain progress in economic growth during the 2000-2010s, most of the developing economies still fail to support domestic farmers on a level comparable with the developed states. In many cases, volumes of domestic support gained by farmers in emerging countries are tenfold lower than those in the developed states (Erokhin et al., 2014). Domestic production of basic agricultural products and food in many countries, including such big agricultural producers as Russia, Brazil, India, and Argentina fails to meet demand (Erokhin & Gao, 2018). Providing the population with food in sufficient quantity and variety is a challenge, which includes a range of issues of food production, import dependence and export orientation of the food market, solvency and dietary patterns of the population. Many developing economies have to rely on agricultural imports, leaving them vulnerable to global price fluctuations and affecting their export revenues, which tremendously threat food security of those nations. Innovative approach seems to be one of the solutions to food insecurity problem, and this chapter discusses the approaches which may be used in developing countries to support their agricultural production and protect domestic markets by using various trade-related tools (Erokhin, 2015a).

BACKGROUND

Protectionism in agricultural production and trade has been in the focus of many studies for rather long period of time. In relation to the state support policies implemented in the developed countries, Josling, Anderson, Schmitz, and Tangerman (2010) focused on linkages between state support, international trade in agricultural products, and food security. Schmitz, Moss, Schmitz, Furtan, and Schmitz (2010) investigated current agricultural policies in the USA and other developed countries and elaborated prognosis of agricultural policies for the next decades. Issues of state support of agriculture in the light of liberalization of international trade in food were studied by Devereux (1999), Boehringer and Rutherford (1999), and Estevadeordal, Freund, and Ornelas (2008).

Experiences of emerging countries in the sphere of state support of agriculture have been studied by Petrikov (2012, 2016), Filippov (2014), Malozemov (2014), and Visser, Mamonova, Spoor, and Nikulin (2015), among others. Anderson, Jha, and Nelgen (2013) impacted into research of political issues of agricultural protectionism and disarrays on international food market in relation to developing countries and emerging economies of Asia. Wittman, Desmarais, and Wiebe (2010) studied the specifics of food sovereignty in emerging countries in the conditions of growing degree of trade liberalization. As regards to the influences of trade liberalization on food security in developing countries, Olson (2003) investigated the alternatives to the global trade order and claimed that an expansion of the Agreement on Agriculture of the World Trade Organization (WTO) would make all countries' food security increasingly uncertain and dependent on volatile international market prices and far-flung distribution chain. Liefert and Swinnen (2002) investigated changes in agricultural markets in transition economies, and later Liefert (2004) studied food security issues in Russia in the conditions of economic growth during the early 2000s.

Despite the fact that many studies have investigated the effects of protectionism on agricultural production and trade in agricultural products, few of them have ever addressed innovations through the lens of improving food security. According to Rajalahti, Janssen, and Pehu (2008), an innovation system can be defined as a network of organizations, enterprises, and individuals that focuses on bringing new products, new processes, and new forms of organization into economic use, together with the institutions and policies that affect their behavior and performance. The innovation systems concept extends beyond the creation of knowledge to encompass the factors affecting demand for and use of knowledge in novel and useful ways. Innovation systems not only help to create knowledge; they provide access to knowledge, share knowledge, and foster learning. Berthet, Hickey, and Klerkx (2018) identified an urgent need to renew traditional approaches to regulation of agricultural production and foster more open, decentralized, contextualized, and participatory approaches to design and innovation in agricultural sector. Spielman and Birner (2008) explored the application of the innovation systems framework to the design and construction of national agricultural innovation indicators. Devaux, Torero, Donovan, and Horton (2018) studied the opportunities emerging from new and expanding markets for agricultural products and challenges to smallholder participation in those markets and identified key attributes of successful value-chain interventions. Rose and Chilvers (2018) argued that ideas of responsible innovation should be further developed in order to make them relevant and robust for emergent agri-tech, and that frameworks should be tested in practice to see if they can actively shape innovation trajectories.

The specific of the emerging markets is that they are very turbulent. There is still no clear understanding of how state support influences food security on the emerging markets in terms of four pillars of food security, i.e. availability, access, stability, and utilization. How limited tools of state support and growing openness to import of food in the conditions of trade liberalization influences domestic agricultural production, farmers and rural dwellers in the emerging countries? In all variety of conducted research, what is needed now is a rethink of international experience. The goal of the given research is an aggregation of international practices and investigation of opportunities for development of state regulation of agriculture in emerging countries with the aim to ensure food security in the conditions of import substitution and an increase of domestic agricultural production.

In this chapter, content and mechanisms of state regulation of agriculture are investigated based on the data and evaluation parameters obtained from the Organization for Economic Cooperation and Development (OECD). According to the OECD's approach, state support of agriculture includes the following elements:

- Producer Support Estimate (PSE) – annual monetary value of gross transfers from consumers and taxpayers to support agricultural producers, arising from policy measures, regardless of their nature, objectives or impacts on farm production or income.
- General Services Support Estimate (GSSE) – annual monetary value of gross transfers to services provided collectively to agriculture and arising from policy measures which support agriculture.
- Consumer Support Estimate (CSE) includes explicit and implicit transfers associated with compensation to consumers of high market prices for agricultural commodities.
- Aggregated support of agriculture includes the overall amount of transfers received by agricultural producers at the expense of taxpayers and consumers (support of market price); gross transfers from taxpayers to consumers of agricultural products, and transfers at the expense of taxpayers to services provided collectively to agriculture.

OECD unites 34 countries which aggregated share in the global GDP is about 80%. Institutional and structural shifts in the evolution of basic approaches and content of state support of agriculture in those countries reflect specific features of state regulation of agriculture in the modern era of globalization in the conditions of aggravation of global challenges. They have to be considered when correcting the measures of state support of agriculture in emerging countries in order to ensure food security. In the OECD countries, almost one-third of farmers' income on average is not actually earned in agricultural markets, but rather comes from a range of government subsidies and other support measures that restrict agricultural trade and distort markets (Secretariat of the Convention on Biological Diversity, 2005). Effects of such distortions, both positive and negative ones, have to be carefully assessed in order to balance state support, protection, and liberalization policies. The research objective is to find out the key regularities of state support of agriculture in emerging countries and to discover the opportunities of using of accumulated experience for the development of a national mechanism of state regulation of agriculture in the conditions of import substitution policy and an increase of domestic agricultural production.

For the purposes of the present research, the authors used the approach of the United Nations Department of Economic and Social Affairs (DESA) and the United Nations Conference on Trade and Development (UNCTAD), i.e. considered 17 economies (Albania, Armenia, Azerbaijan, Belarus, Bosnia and Herzegovina, Georgia, Kazakhstan, Kyrgyzstan, Macedonia, Moldova, Montenegro, Russia, Serbia, Tajikistan, Turkmenistan, Ukraine, and Uzbekistan) as emerging ones. Additionally, the authors considered seven countries (China, Brazil, India, Vietnam, Laos, Cambodia, and South Africa) recognized as emerging ones by the World Bank and the International Monetary Fund.

MAIN FOCUS OF THE CHAPTER

Global Trade in Agricultural Products and Food

Despite the resource-oriented model of export, most of the developing economies are engaged in international trade in other commodities, apart from oil, including food and agricultural products. According to Liefert and Swinnen (2002), during the 1970-1980s, most of the agricultural exports went to the Soviet Union (exports of meat by Hungary, Romania, Ukraine and Kazakhstan; grain by Hungary, Ukraine, and Kazakhstan; sugar by Ukraine; and cotton by Uzbekistan and Turkmenistan). During 1995-2015, foreign trade turnover of major food exporting countries continued to grow steadily. Specifically, in OECD countries, export volume increased from $323.7 million in 1995 to $895.8 million in 2017 (Figure 1).

During the previous decades, OECD countries have remained net importers of food and agricultural products. Import volume increased from $340.9 million in 1995 to $904.4 million in 2017 (Figure 2).

As of Liefert and Swinnen (2002), the pre-reform trade in agriculture for the emerging economies was not market driven but rather was an integral part of countries' economic planning, i.e. resulted from

Figure 1. Export of food and agricultural products in 1995-2017, $ billion
Source: Author's development based on United Nations Conference on Trade and Development [UNCTAD] (2019)

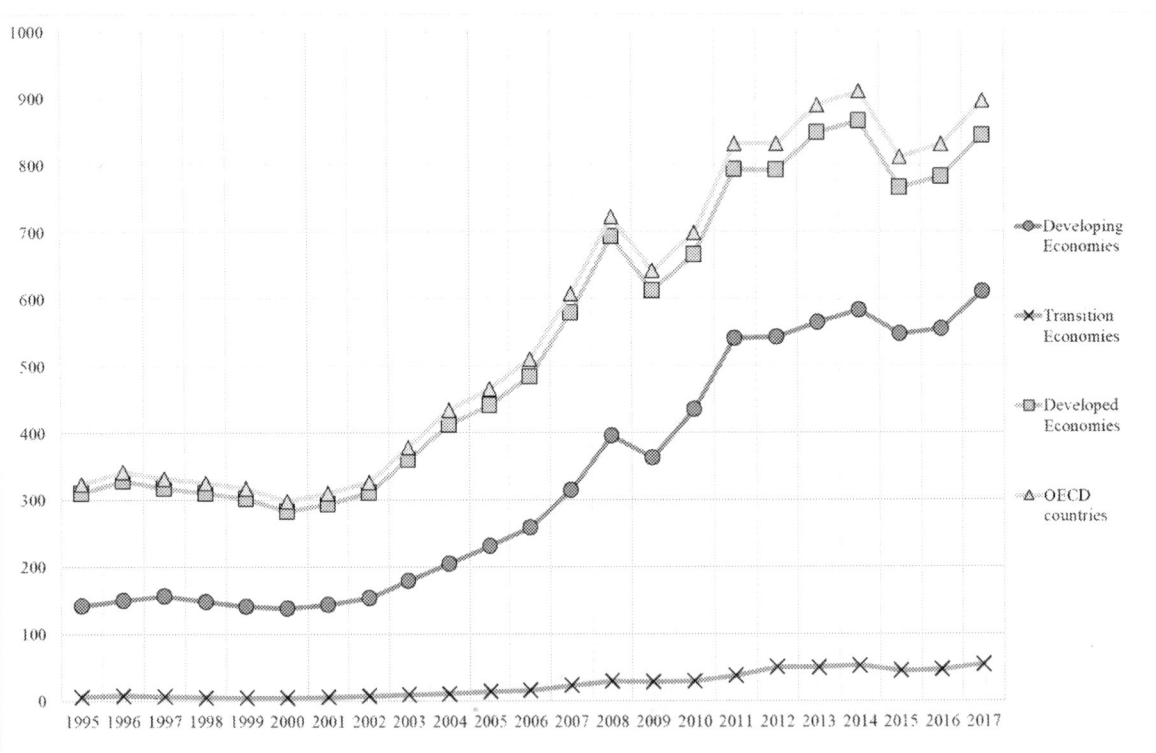

Figure 2. Import of food and agricultural products in 1995-2017, $ billion
Source: Author's development based on UNCTAD (2019)

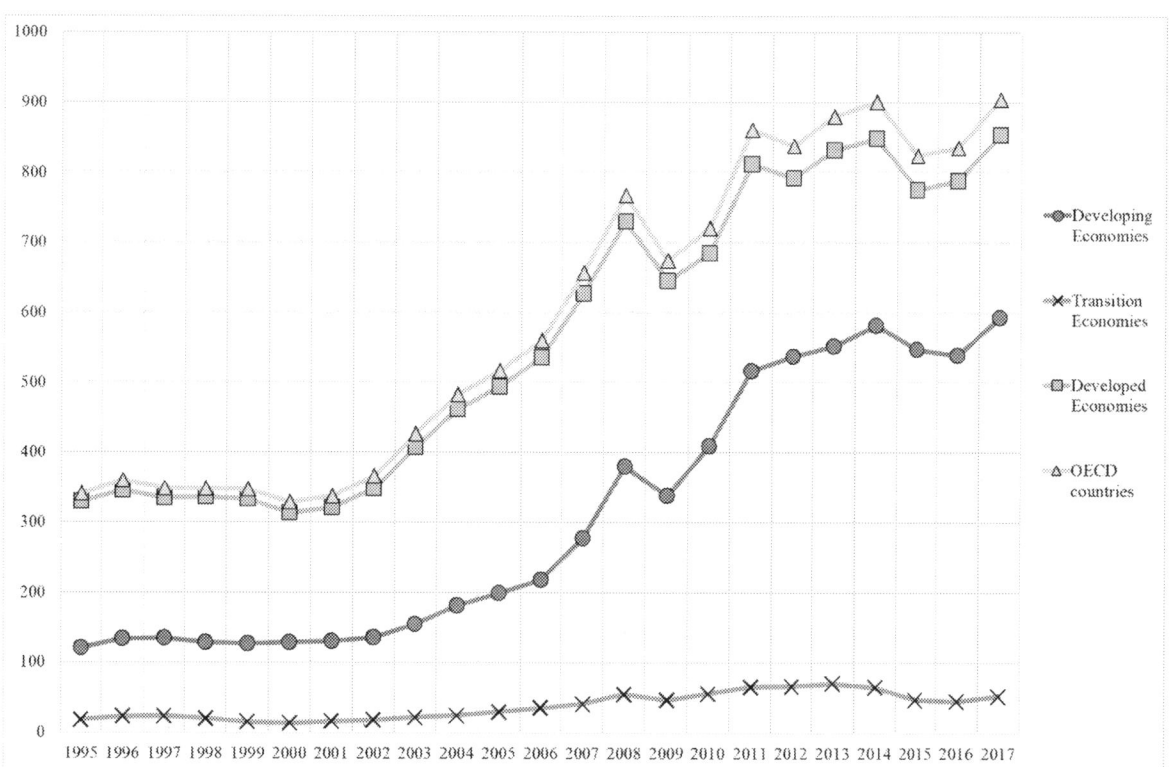

politics, not economics. When most of the emerging economies liberalized trade in the early 1990s, they experienced a collapse in the export of their agricultural products to the world and dramatic rise of import flows to their domestic food markets. Export/import ratio in foreign trade turnover (FTT) in food and agricultural products in 1995 was 25/75, while developed countries had equal proportions of export and import flows in their foreign trade. During two decades, emerging economies succeeded in balancing their foreign trade turnover to 44/56 in 2017, but import still prevails.

The share of emerging economies in world trade in food and agricultural products was 3.9% in 2017. It increased by 1.3 percentage points during two decades of transition. Share in world import was higher than the one in world export (3.5% and 4.4%, consequently). Emerging economies are oriented on the world market rather than on the internal markets. In 2017, over 68% of the total trade of the emerging economies allotted to trade with developed and developing countries, and only 31.7% – to intra-group trade. Import of food and agricultural products is even more unbalanced – over 79% of deliveries come from the developed and developing economies, not from the emerging markets.

Such a high dependency on import deliveries from developed countries along with the drop in domestic food production during the 1990-2010s have raised concerns about food security and food sovereignty in certain emerging economies in transition, especially Russia. The level of food security and changes during transition have differed importantly between the countries. However, domestic food markets in most of the emerging economies are to a greater extent influenced by internal factors (low competitiveness of food producers, their financial instability, outdated facilities, infrastructure, lowering effective demand, etc.), than external ones. Despite certain progress of reforms, volumes of GDP and foreign trade of the emerging economies are well below than those of the developed countries, while the share of foreign trade turnover in food and agricultural products in GDP is the highest among the groups of countries under consideration (4.5% for emerging economies in comparison with 4.0% for developing and 3.7% for developed countries).

Consequently, emerging economies have to concentrate on solving internal problems and increasing efficiency of domestic agricultural production. The goal is to create a profitable, market-driven agricultural economy with productivity levels and supporting infrastructure that allows it to compete effectively on the world market. In order to increase their competitiveness on the global market most of the emerging economies implement various tools of state support of agriculture.

Agricultural Protectionism

Conducted analysis reveals that a system of state regulation is a key determinant of the development of agriculture. The strategic importance of support is proved by a complete synchronization of its amount with dynamics of gross agricultural production (GAP) and aggregated support of agriculture (ASA). The increase of state support provides outrunning growth of gross agricultural production. When state support decreases and the index is below 1, gross agricultural production index changes correspondingly. It can be observed on the examples of Japan and Switzerland, where a decrease in state support ($I = 0.7$) was accompanied by a decrease in gross production ($I = 0.8$).

Analysis of state support in relation to the GDP allowed the authors to discover the tendency of its contraction in the majority of the countries over the previous 20 years. Thus, in the EU, share of state support of agriculture in the GDP decreased from 1.5% down to 0.8%; in the USA, from 0.6% down to 0.5%; in Canada, from 0.8% down to 0.4%; in Russia, from 2.6% down to 0.7%. Exceptions to this tendency are China and Indonesia, where the relative level of support increased from 1.4% up to 3.2% and from 0.8% up to 3.6%, respectively. In Brazil, in the mid-1990s, part of incomes of agricultural producers was withdrawn in the favor of the state, but currently, agriculture benefits from the support amounted to 0.4% of the GDP. In general, among OECD countries, state support of agriculture decreased twofold (from 1.5% of the GDP in 1995 down to less than 0.8% in 2017). The most critical decrease happened in those countries, where the level of support was higher, i.e. Turkey, Republic of Korea, Iceland, and Switzerland. However, it remains high, above 1% of the GDP.

The decrease of the relative level of support does not mean the agriculture now receives less. Support in absolute terms is growing. Thus, during 1995-2017, aggregated amount of state support of agriculture among the OECD countries increased by $23 billion (7.2%), including in the USA, by 195%; in Canada and Turkey, by 170%; in Norway, by 140%. Even more significant growth of the absolute amount of support occurred in the emerging economies, which are not the OECD members. In Brazil, withdrawal of incomes of agricultural producers in the amount of $4.2 billion in 1995 changed to state support in the amount of $9.4 billion in 2017. In China, the amount of support during the same period increased from only $13 billion up to $287 billion.

The variance of the importance of state support for farmers can be estimated by means of matching its volume with the gross product in agriculture. The analysis reveals that percentage ratio between the volume of state support and the gross product in agriculture in the OECD countries decreases, but fluctuates over a wide range, from negative values in Ukraine to almost 100% in Switzerland and Norway.

Table 1. Structure of aggregate state support of agriculture in the selected countries in 1995 and 2017, %

Countries	Support of producers		Support of services		Support of consumers	
	1995	2017	1995	2017	1995	2017
Australia	77	50	23	50	0	0
Brazil	176	83	-75	17	0	0
Canada	75	72	25	28	0	0
Chile	83	52	17	48	0	0
China	47	89	50	11	3	0
Columbia	91	85	9	15	0	0
EU	87	85	9	14	4	1
Iceland	87	94	9	5	4	1
Indonesia	73	89	26	6	1	5
Israel	87	80	13	20	0	0
Japan	75	85	25	15	0	0
Kazakhstan	95	76	5	23	0	0
Mexico	64	82	14	11	22	7
New Zealand	31	27	69	73	0	0
Norway	92	93	5	5	3	2
Russia	79	80	21	16	0	4
South Africa	65	57	35	43	0	0
Republic of Korea	86	87	13	13	1	0
Switzerland	82	89	6	11	12	0
Turkey	72	85	28	15	0	0
Ukraine	142	157	-42	-57	0	0
USA	53	38	9	9	38	53

Source: Author's development based on UNCTAD (2019)

Revealed differences between the countries are determined by the availability of natural resources, the role of agricultural sector in the national economy, the level of development, and goals of agrarian policy.

State regulation of agriculture is a complicated mechanism, which includes the tools of pressure on the structure of agricultural production, agricultural market, social sphere of rural territories, inter-industry, and inter-farm relations. In relation to this, the structure of state support and discovery of trends are the subjects of much attention. The structure of aggregate support of agriculture in the selected OECD countries and emerging economies in accordance with the methodical approach described above is presented in Table 1.

In 1995, support of producers dominated in the aggregated support of agriculture. Among the OECD countries, its relative share was 78%, including in the EU – 87%, in Canada and Japan – 75%, in the USA – over 50%. As for the emerging economies, this type of support also played a critical role. Over the past 20 years, the importance of the support of producers essentially decreased. Being the members of the WTO, OECD countries are obliged to limit support of producers, which distorts market patterns. That is why they have to redirect budget transfers to support of services or support of consumers. Such a tactic gives them an opportunity to save or even increase the level of support without breaking the WTO rules. Some countries almost redirected state support on the development of infrastructure, research, and development, education in the sphere of agriculture, marketing, promotion of products, and other services. Thus, in 2015, in Australia the share of such expenses in the aggregated support of agriculture is over 50%, in Chile – 48%, in New Zealand – 73%, in Canada – 28%, in Kazakhstan – 23%, and in Israel – 20%.

It is evident that usage of such capacities of support of services not restricted by the WTO is relevant for emerging countries since many of them lose a substantial portion of actual agricultural production because of poor road networks, lack of storage and other infrastructure facilities, which is resulted in the cost of production.

As for the support of consumers, this tool is implemented massively in the USA only, where the agrifood policy is focused on the expansion of domestic and external markets of agricultural raw materials and food. Over a half of the agrarian budget of the USA is spent on food aid to low-income sections of the population in purpose to promote domestic demand. The aid is implemented in the framework of the programs of distribution of food coupons (Supplemental Nutrition Assistance Program – SNAP), free lunches in schools (The National School Lunch Program – NSLP), food for women, infants, and children (The Special Supplemental Nutrition Program for Women, Infants, and Children – WIC). The state budget also finances public awareness campaign on nutrition and research in the sphere of healthy nutrition. Each dollar of food aid generates two dollars in economic activity. Each billion of dollars in the program allows creating or supporting 18,000 of workplaces, including 3,000 workplaces in agriculture.

This experience is worthwhile implementing for the development of organizational and economic mechanisms of state regulation of agriculture in emerging countries, since the tools currently implemented in the majority of them are reciprocal and directly or indirectly focused on the increase of output, i.e. on supply, without any regulations of demand. However, as the authors referred to Liefert (2004) in the beginning of the chapter, the main food security problem for emerging countries is not the low volume of output but inadequate access to food by certain socioeconomic groups. The result is that the increase of output in agriculture is not fulfilled even at the extensive support of agriculture. One of the reasons is sensibility to domestic demand. Low effective demand does not drive production increase.

Identifying the overall tendency of decrease of support in relation to the gross agricultural production (in average, in the countries under consideration it decreased from 21% in 1995 down to 17% in 2017), it is necessary to bear in mind that such average indicators submerge differences existing between the OECD countries and emerging economies. While in the former ones the level of protectionism decreased gradually, the latter ones moved up from taxation of agriculture (it still exists in Ukraine) to ensurance of an essential level of support, which just started to catch up with the OECD countries in 2014-2015.

Change in the overall level of support of producers is followed by a change of priorities in the implementation of direct and indirect subsidies. In the late 1990s, support for domestic prices was the major measure of indirect support of agricultural producers in the OECD countries. Its share was over 45% of the overall support in the EU, the USA, Canada, Israel, Turkey, and Australia, 35% in Russia, and over 90% in Japan, Republic of Korea, Kazakhstan, and South Africa.

Support of market prices ensures higher incomes for farmers because of keeping high prices for food and making consumers support farmers by paying these high prices. In the case of agricultural surpluses because of high prices, it is possible to promote export using export subsidies, which are paid from the budget, not at the consumers' expense.

The degree of support of farmers by means of such tool is characterized by the Producer Nominal Protection Coefficient (NPC), calculated as a ratio between the average price received by producers on the domestic market and the average world market price. It shows that domestic prices for agricultural products in the USA, Canada, Mexico, and Israel are close to international ones, and did not change essentially during 1995-2017.

The coefficient decreased substantially in the developed countries, where it was traditionally high. Nevertheless, domestic prices in the EU, Norway, Japan, Turkey, and the Republic of Korea surpassed international price in a range from 20% to 200% in 2017. In the emerging economies, on the contrary, the coefficient rises, which is reflected in surpassing of domestic prices over the international price in a range from 2% to 20%

Those emerging countries with financial capacities replace indirect measures of support with direct ones, which do not affect consumer prices. They are implemented by the means of direct state equalization fees; compensation of disaster damages; compensation of redeployment damage (compensation of land diversion, forced cattle slaughtering, etc.); subsidies per unit area or livestock population; payments in the form of financing of procurements (for example, subsidies for purchasing of fertilizers, agricultural chemicals, and fodder); financing of target programs. Across the OECD, direct payments to farmers increased from 19% up to 37% during 1995-2017. Payments per hectare of crop area and livestock unit reached almost a half of state support of agriculture in the USA and 2/3 of the respected payments in the EU. They are increasingly used in order to stimulate implementation of certain technologies of production, including the ones with environmental restrictions. The traditional measure of support is compensation of initial costs. Its relative share in the structure of support of producers across the OECD increased from 9.5% up to 12.0%, including in the EU from 7.0% up to 14.0%.

Agricultural insurance is also one of the major tools of support of farmers' income. International experience shows that subsidizing of insurance premium and reinsurance of risks are the common and effective ways of state support of economic stability in agriculture.

In general, the share of measures, which distort trade patterns the most, decreased by 20 percentage points in the OECD countries during 1995-2017, while in the USA and the EU – more than twofold.

Along with the support of operating activities of farmers, more and more resources of agrarian budgets of the OECD countries are spent on the realization of long-term goals, such as environmental sustainability, development of infrastructure, and implementation of innovations (Table 2). Expenses on infrastructure are essentially financed from the state budgets in Turkey, Japan, Brazil, Chile, and the Republic of Korea. Kazakhstan and Canada spend the biggest amount of support on inspection and control services. Improvement of agricultural education, development and implementation into the agricultural production of innovations are priority directions for Australia, Norway, Israel, Switzerland, the EU, and the USA. In those countries, expenses for research and education comprise from 30% up to 64% of total resources spent on support services.

Facilitating organization, financing and implementation of research and development for agriculture at the expense of state budgets, governments of most of the countries encourage innovation activities in the private sector using direct and indirect financial incentives, providing information on the results of state-funded research, developing state-private partnerships, and organizing work on information and extension support of farmers.

Over the recent years, development of information and extension services is facilitated in a way of decentralization of state services and the introduction of private entities and intermediaries in this field. Innovation brokers appeared in the EU and other countries. They formulate needs of farmers in the

Table 2. Structure of state support of services in agriculture in selected countries in 1995 and 2017, %

Countries	Research and education		Inspection and control services		Infrastructure		Other	
	1995	2017	1995	2017	1995	2017	1995	2017
Australia	75	64	5	9	14	25	5	1
Brazil	23	15	4	7	58	51	15	27
Canada	41	39	22	40	12	10	24	11
Chile	28	24	1	20	64	51	7	5
China	8	29	5	6	23	37	64	28
Columbia	25	23	3	7	71	69	0	1
EU	31	37	3	6	24	30	43	27
Iceland	35	12	10	41	20	2	35	45
Indonesia	22	11	5	3	73	72	0	5
Israel	40	44	14	16	3	30	43	10
Japan	5	14	0	1	91	82	4	3
Kazakhstan	0	12	89	62	11	22	0	4
Republic of Korea	11	25	2	6	74	49	12	20
Mexico	54	48	6	11	17	38	24	3
New Zealand	65	49	24	32	11	20	14	0
Norway	56	54	18	24	11	16	14	5
Russia	12	42	8	26	15	13	64	18
South Africa	85	41	7	16	8	36	0	7
Switzerland	28	47	3	16	14	36	56	7
Turkey	1	1	3	2	75	82	21	15
Ukraine	25	45	8	30	63	11	4	14
USA	35	29	13	16	1	26	51	30

Source: Author's development based on UNCTAD (2019)

sphere of research, help them to get an access to technologies, or develop links in value chains. Some of them are financed from the state funds and govered by a state, through regional organizations. There are pure private systems (for example, in the Netherlands and New Zealand), where farmers pay for the services and select service providers on a commercial basis. There are mixed systems, where services provided by the state entities and private consultants are paid by the farmers either in part or in whole. Finally, there are systems managed by the farmers' organizations (for example, in France and Finland) and financed by the government, farmers' organizations, and private farmers.

SOLUTIONS AND RECOMMENDATIONS

There are different perspectives, rationalities, and affordances involved in open design and innovation processes, but that these are not sufficiently yet taken into account in agricultural design processes and support tools. Transitions towards sustainable agricultural futures will require systemic approaches to design, where local solutions are also capable of contributing to larger-scale solutions - requiring both an intimate knowledge of the local context, needs and culture while also involving a range of actors and local user communities. Differences in structural content of state support of agriculture are developed under the influence of historical, political, natural, and economic conditions of various countries. They are determined by their financial capabilities and current tasks of agrarian policies, some of which are: fulfillment of economic potential of agriculture as a sector, which ensures economic growth, and employment of population; increase of environmental sustainability of production in the conditions of natural scarcity; ensurance of food security of a country; expansion of export opportunities. Nevertheless, the analysis let the authors revealing some common tendencies in a change of role and directions of state support of agriculture in the emerging countries in the light of ensurance of food security, in particular:

- While the share of support of agriculture in the GDP decreases, its absolute value increases.
- Growth rates of state support of agriculture are synchronized with the growth rates of the gross product in agriculture. During critical periods of market slump increase of direct support of agriculture helps to stabilize production and domestic food market in order to ensure food security.
- Under an influence of globalization, the structural content of state support changes: tied price support is replaced by the untied support of producers' income.
- Importance of support of services grows, primarily of innovation activities and infrastructure.

Developing countries have limited capabilities to provide the level support of their domestic farmers which would ensure achievement of target parameters of food security. Involvement into the international trade forces emerging countries opening their domestic food markets for imports. Effective protection of domestic farmers in emerging countries is impeded by the low import tariffs, which facilitate an easier market access for foreign agricultural products and food, lead to a reduction of domestic production, and in such a way threaten food security. Acting in the framework of the common directions of development of state support of agriculture, emerging countries should provide constant diversified aid to the direct producers of agricultural products. Measures of such support are increasingly frequently emerging into the tools of "precise pointing". They are aimed at smoothing negative consequences of unfavorable market environment because of objective and subjective circumstances. Account of those tendencies in the development of tools for state support in the emerging countries will let make the support more focused and effective and will help to adopt the measures of regulation of agriculture to the goals of ensurance of food security.

FUTURE RESEARCH DIRECTIONS

The vital issue for developing countries in terms of achieving food security is how to secure the sustainable development of national agriculture and agribusiness in the conditions of a growing market openness and liberalization of international trade in food, taking into consideration the incomparably lower financial capabilities. Future research in this field has to be focused on exploring the ways in which emerging countries would be able to ensure the sustainable development of agricultural production and trade. The effects of the following measures on food security have to be investigated:

state support of import substitution agricultural production; provision of environmental safety of domestic food and agricultural commodities; agricultural and food export increase once the domestic is saturated; logistical costs saving; optimization of all factors that affect competitiveness of domestic agricultural and food commodities in compliance with rational geographic distribution and specialization of agricultural production.

CONCLUSION

Many economies considered as emerging ones originated from the common socialist-type model of the economy (post-Soviet countries, Eastern European states, some of the Asian countries), where the dominant state and collective ownership of the means of production minimized the role of natural competitive factors (Kolodko, 1998). Foreign trade was controlled by the state, which resulted in various distortions, low level of diversification, specialization of export destinations and import channels due to political preferences. During early stages of transition, most of the economies of that type experienced a decline in output and export, significant increase in inflation, and decrease in living standards of the domestic population. Economic and trade liberalization rapidly embedded emerging economies into the international market, however, being considerably non-competitive, they had to specialize on low value-added exports and high value-added technological imports. For at least two decades after the beginning of market reforms in the 1980-1990s, most of the economies currently recognized as emerging ones have not been able to break the existing trade patterns.

Things are changing. Unimaginable growth of China, certain economic successes of Russia and Kazakhstan, the rapid increase in agricultural production and exports in Brazil and Argentina, and improvements in the region of the Eastern Europe demonstrated the potentials of the emerging economies on the global market. By the 2010s, particular countries and their strengthening alliances had been rather effective in increasing their shares in the global agricultural output and trade in agricultural products and food. New economic powers have arisen: China, Russia, Brazil, and other Eastern European and South-East Asian nations. The countries accelerated integration processes between themselves in order to ensure their food security, aggregate existing competitive advantages, and mitigate market risks. However, despite the evident achievements, there are still old problems and new challenges in terms of achieving food security. Unlike the past two-three decades of transition, coming years will manifest new factors: increasing competition between emerging countries and their alliances for better conditions of production and distribution of agricultural products and food; intensifying role of trade regulations (unification in the WTO framework and protectionism of particular domestic markets) and state support in achieving certain levels of food security; growing tensions between major global powers as to shape a new multi-polar world order.

The long-term outlook for agricultural development, trade in food and agricultural products, and food security in the emerging economies will depend on effective structural reforms, diversification and technical modernization of domestic production and exports, utilization of existing comparative advantages, and benefiting from effective state support of agriculture. As Von Braun, Serova, Seeth, and Melyukhina (1996) stated in relation to Russia, much would depend on how well incentives to increase agricultural production provided by a state are transmitted to the agriculture and food-processing sector and on the opening up of international trade opportunities.

REFERENCES

Anderson, K., Jha, S., Nelgen, S., & Strutt, A. (2013). Re-Examining Policies for Food Security in Asia. *Food Security*, *5*(2), 195–215. doi:10.100712571-012-0237-5

Berthet, E. T., Hickey, G. M., & Klerkx, L. (2018). Opening Design and Innovation Processes in Agriculture: Insights from Design and Management Sciences and Future Directions. *Agricultural Systems*, *165*, 111–115. doi:10.1016/j.agsy.2018.06.004

Boehringer, C., & Rutherford, T. (1999). *Decomposing General Equilibrium Effects of Policy Intervention in Multi-Regional Trade Models: Method and Sample Application*. Mannheim: Center for European Economic Research.

Bozic, D., Bogdanov, N., & Sevarlic, M. (2011). *Agricultural Economics*. Belgrade: University of Belgrade.

Devaux, A., Torero, M., Donovan, J., & Horton, D. (2018). Agricultural Innovation and Inclusive Value-Chain Development: A Review. *Journal of Agribusiness in Developing and Emerging Economies*, *8*(1), 99–123. doi:10.1108/JADEE-06-2017-0065

Devereux, M. B. (1999). Growth and the Dynamics of Trade Liberalization. *Journal of Economic Dynamics & Control*, *23*(5-6), 773–795. doi:10.1016/S0165-1889(98)00043-8

Erokhin, V. (2015a). Russia's Foreign Trade in Agricultural Commodities in Its Transition to Liberalization: A Path to Go Green. In A. Jean-Vasile, I. Andreea, & T. Adrian (Eds.), *Green Economic Structures in Modern Business and Society* (pp. 253–273). Hershey, PA: IGI Global. doi:10.4018/978-1-4666-8219-1.ch014

Erokhin, V. (2015b). Structural Changes in International Trade in Food: Competitive Growth Models for Economies in Transition. *Proceedings of the 3rd International Conference "Economic Scientific Research – Theoretical, Empirical and Practical Approaches"*. Bucharest: Academia Romana.

Erokhin, V. (Ed.). (2017a). *Establishing Food Security and Alternatives to International Trade in Emerging Economies*. Hershey, PA: IGI Global.

Erokhin, V. (2017b). Factors Influencing Food Markets in Developing Countries: An Approach to Assess Sustainability of the Food Supply in Russia. *Sustainability*, *9*(8), 1313. doi:10.3390u9081313

Erokhin, V. (2017c). Self-Sufficiency versus Security: How Trade Protectionism Challenges the Sustainability of the Food Supply in Russia. *Sustainability*, *9*(11), 1939. doi:10.3390u9111939

Erokhin, V., & Gao, T. (2018). Competitive Advantages of China's Agricultural Exports in the Outward-Looking Belt and Road Initiative. In W. Zhang, I. Alon, & C. Lattemann (Eds.), *China's Belt and Road Initiative: Changing the Rules of Globalization* (pp. 265–285). London: Palgrave Macmillan. doi:10.1007/978-3-319-75435-2_14

Erokhin, V., Ivolga, A., & Heijman, W. (2014). Trade Liberalization and State Support of Agriculture: Effects for Developing Countries. *Agricultural Economics*, *60*(11), 524–537.

Estevadeordal, A., Freund, C., & Ornelas, E. (2008). Does Regionalism Affect Trade Liberalization Toward Nonmembers? *The Quarterly Journal of Economics, 123*(4), 1531–1575. doi:10.1162/qjec.2008.123.4.1531

Filippov, R. (2014). International Experience of Subsidizing of Agriculture as Basis of Food Security of Russia. *Naukovedenie, 20*(1).

Gao, T., Ivolga, A., & Erokhin, V. (2018). Sustainable Rural Development in Northern China: Caught in a Vice between Poverty, Urban Attractions, and Migration. *Sustainability, 10*(5), 1467. doi:10.3390u10051467

Josling, T., Anderson, K., Schmitz, A., & Tangerman, S. (2010). Understanding International Trade in Agricultural Products: One Hundred Years of Contributions by Agricultural Economists. *American Journal of Agricultural Economics, 92*(2), 424–446. doi:10.1093/ajae/aaq011

Kolodko, G. (1998). *Equity Issues in Policymaking in Transition Economies*. Washington, DC: International Monetary Fund.

Liefert, W. (2004). *Food Security in Russia: Economic Growth and Rising Incomes are Reducing Insecurity*. Washington, DC: Economic Research Service, United States Department of Agriculture.

Liefert, W., & Swinnen, J. (2002). *Changes in Agricultural Markets in Transition Economies*. Washington, DC: Economic Research Service, United States Department of Agriculture.

Malozemov, S. (2014). Experience of State Support of Agriculture in Foreign Countries. *Science and Modernity, 32*(1), 136–141.

Olson, D. R. (2003). *Towards Food Sovereignty: Constructing an Alternative to the World Trade Organization's Agreement on Agriculture*. Minneapolis, MN: Institute for Agriculture and Trade Policy.

Petrikov, A. (2012). It Is Necessary to Increase Adaptation of Russian Agrarian Sector to WTO Conditions. *Economy of Agricultural and Processing Enterprises, 6*, 6–8.

Petrikov, A. (2016). Major Directions of Implementation of Modern Agrifood and Rural Policy. *International Agricultural Journal, 1*, 3–9.

Rajalahti, R., Janssen, W., & Pehu, E. (2008). *Agricultural Innovation Systems: From Diagnostics Toward Operational Practices*. Washington, DC: The World Bank.

Rose, D. C., & Chilvers, J. (2018). Agriculture 4.0: Broadening Responsible Innovation in an Era of Smart Farming. *Frontiers in Sustainable Food Systems, 2*, 87. doi:10.3389/fsufs.2018.00087

Schmitz, A., Moss, C., Schmitz, T., Furtan, W., & Schmitz, H. (2010). *Agricultural Policy, Agribusiness, and Rent-Seeking Behaviour*. Toronto: University of Toronto Press.

Secretariat of the Convention on Biological Diversity. (2005). *The Impact of Trade Liberalization on Agricultural Biological Diversity, Domestic Support Measures and their Effects on Agricultural Biological Diversity*. Montreal: Secretariat of the Convention on Biological Diversity.

Spielman, D. J., & Birner, R. (2008). *How Innovative Is Your Agriculture? Using Innovation Indicators and Benchmarks to Strengthen National Agricultural Innovation Systems*. Washington, DC: The World Bank.

United Nations Conference on Trade and Development. (2019). *Statistics Database* [Data file]. Retrieved from http://unctad.org/en/Pages/Statistics.aspx

Visser, O., Mamonova, N., Spoor, M., & Nikulin, A. (2015). "Quiet Food Sovereignty" as Food Sovereignty without a Movement? Insights from Post-socialist Russia. *Globalizations, 12*(4), 1–16. doi:10.1 080/14747731.2015.1005968

Von Braun, J., Serova, E., Seeth, H., & Melyukhina, O. (1996). *Russia's Food Economy in Transition: What Do Reforms Mean for the Long-term Outlook?* Washington, DC: International Food Policy Research Institute.

Wittman, H., Desmarais, A., & Wiebe, N. (2010). *Food Sovereignty: Reconnecting Food, Nature and Community*. Halifax, Oakland: Fernwood Publishing and Food First Books.

ADDITIONAL READING

Anderson, K., Dimaran, B., Francois, J., Hertel, T., Hoekman, B., & Will, M. (2001). The Cost of Rich (and Poor) Country Protection to Developing Countries. *Journal of African Economies, 10*(3), 227–257. doi:10.1093/jae/10.3.227

Ayres, J. M., & Bosia, M. (2011). Beyond Global Summitry: Food Sovereignty as Localized Resistance to Globalization. *Globalizations, 8*(1), 47–63. doi:10.1080/14747731.2011.544203

Boyer, J. (2010). Food Security, Food Sovereignty, and Local Challenges for Transnational Agrarian Movements: The Honduras Case. *The Journal of Peasant Studies, 37*(2), 319–351. doi:10.1080/03066151003594997

Claeys, P. (2012). The Creation of New Rights by the Food Sovereignty Movement: The Challenge of Institutionalizing Subversion. *Sociology, 46*(5), 844–860. doi:10.1177/0038038512451534

Dollar, D. (2001). *Globalization, Inequality, and Poverty since 1980*. Washington, DC: Development Research Group, The World Bank.

Dornbusch, R. (1992). The Case for Trade Liberalization in Developing Countries. *The Journal of Economic Perspectives, 6*(1), 69–85. doi:10.1257/jep.6.1.69

Edelman, M. (2014). Food Sovereignty: Forgotten Genealogies and Future Regulatory Challenges. *The Journal of Peasant Studies, 41*(6), 959–978. doi:10.1080/03066150.2013.876998

Hospes, O. (2009). Food Sovereignty: The Debate, the Deadlock, and a Suggested Detour. *Agriculture and Human Values, 31*(1), 119–130. doi:10.100710460-013-9449-3

Khanna, T., & Palepu, K. G. (2010). *Winning in Emerging Markets: A Road Map for Strategy and Execution*. Boston, MA: Harvard Business Press.

Mamonova, N. (2017). *Rethinking Rural Politics in Postsocialist Settings. Rural Communities, Land Grabbing and Agrarian Change in Russia and Ukraine*. Enschede: Ipskamp Drukkers.

Markovic, I., & Markovic, M. (2014). Agricultural Protectionism of the European Union in the Conditions of International Trade Liberalization. *Economics of Agriculture, 61*(2), 423–440.

Patel, R. (2009). Food Sovereignty. *The Journal of Peasant Studies, 36*(3), 663–706. doi:10.1080/03066150903143079

Spanu, V. (2003). *Liberalization of the International Trade and Economic Growth: Implications for both Developed and Developing Countries.* Cambridge, MA: Harvard University.

Ushachev, I. (2012). Measures to Secure Competitiveness of Russia's Agricultural Production in the Conditions of its Accession to WTO. *Economics of Agricultural and Processing Enterprises, 6,* 1–5.

Wehrheim, P., & Wobst, P. (2005). The Economic Role of Russia's Subsistence Agriculture in the Transition Process. *Agricultural Economics, 33*(1), 91–105. doi:10.1111/j.1574-0862.2005.00136.x

KEY TERMS AND DEFINITIONS

Agriculture: A pool of establishments engaged in growing crops, raising animals, and harvesting fish and other animals.

Developing Country: A country which has a less developed economy in terms of smaller gross domestic product, gross national product, and per capita income relative to other countries.

Food Market: A medium that allows buyers and sellers of agricultural raw materials, agricultural products, and food to interact in order to facilitate an exchange.

Food Security: An availability and adequate access at all times to sufficient, safe, nutritious food to maintain a healthy and active life.

Innovation System: A network of organizations, enterprises, and individuals that focuses on bringing new products, new processes, and new forms of organization into economic use, together with the institutions and policies that affect their behavior and performance.

State Support: A set of protective measures provided by a state to a certain industry or sector.

Trade Liberalization: A removal or reduction of restrictions or barriers on the free exchange of goods between nations.

Chapter 6
Technical Equipment of Agricultural Production:
The Effects for Food Security

Mikail Khudzhatov
https://orcid.org/0000-0001-6683-3206
Peoples' Friendship University of Russia (RUDN University), Russia

Alexander Arskiy
Russian Academy of Personnel Support for the Agroindustrial Complex, Russia

ABSTRACT

The guarantee of a sufficient food supply is one of the challenges in both international and national economic security. Development of world agriculture is impossible without using advanced technologies. Their application in agriculture depends on the security of agricultural producers with highly effective agricultural machinery for which agricultural engineering serves. Agricultural engineering is an important element of the agro-industrial complex of any state providing it with necessary machinery and equipment. One of the important directions of the establishment of food security is the development of agricultural engineering. In this regard, the chapter analyses the current state of the world market of agricultural machinery; develops the methodology of assessment of the competitiveness of agricultural machinery in the domestic market; and elaborates the definition of effective methods of management of logistics costs at the operation of agricultural machinery.

INTRODUCTION

Ensuring a sufficient food supply is one of the most critical challenges in establishing both international and national economic security. In this context, it is important to make a differentiation between food security and food self-sufficiency.

Food self-sufficiency is defined as the ability of a state to meet domestic food needs. The level of food self-sufficiency, calculated as the ratio of its national production to domestic consumption, var-

DOI: 10.4018/978-1-7998-1042-1.ch006

ies for different countries. It is determined by the effective public demand for food, development of the agro-industrial complex, size of its commodity resources, degree of profitability, and reliability of international food relations (Ibragimov & Dokholyan, 2010).

The concept of food security is not limited to direct provision of food to the population, although this task is its ultimate goal. The food security system also includes the establishment of strategic food stocks; formation of the optimal ratio of food for the country by means of domestic production and import; development of food base of agriculture and the network of enterprises for the processing of raw materials, as well as trade in these raw materials and foodstuffs; expansion of transport networks for the supply of raw materials to food industry and food to consumers (Shapkina, 2012).

The term "food security" is widely used in the literature. In a general sense, it includes various aspects of activities related to the development of agricultural and agro-industrial production, food supply, and provision of targeted social food aid to population. Narrowly defined, food security of a country is the level of dependence on imports of basic types of food. Anisimov, Gapov, Rodionova, and Saurenko (2019) generalized an approach to understanding food security as a state of the global economy which provided physical access to food and economic opportunity to purchase it in the required quantity for all social groups.

Food security is provided by a set of economic and social conditions associated with the development of food production and general state of the national and global economy. Food security level is ensured both by domestic food products and the availability of financial resources to import the required volume of food. Every government in the pursuance of national security interests strives minimizing the degree of potential vulnerability of food security parameters (Khudzhatov, 2018a) and stabilizing food supply amid any external fluctuations (inflation, currency deficit, violations of food imports, embargo on supplies, etc.).

Establishment of food security includes the stability of both internal (preferably) and external food sources, as well as the availability of reserve funds.

The rational level of food security involves optimal use of the agricultural potential of a country for the needs of the domestic market and intensification of foreign economic activity in the terms of import of food and raw materials (in the volume required to close food supply gaps taking into account international division of labor and situation on the global market).

Figure 1. Structure of the agro-industrial complex
Source: Authors' development based on Khudzhatov (2018b)

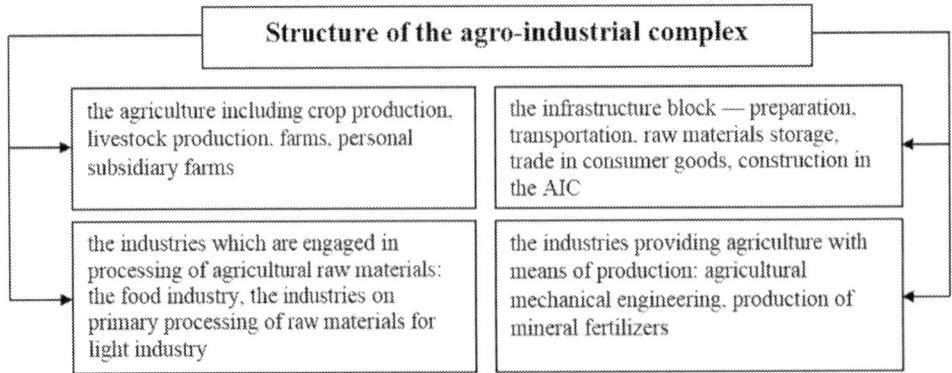

Technical Equipment of Agricultural Production

One of the key challenges in establishing food security is the development of domestic agro-industrial complex, which includes production of food and agricultural products and agricultural engineering (Figure 1).

Globally, development of agricultural production is hardly possible without using advanced technologies. Their application depends on the provision of agricultural producers with high-efficient agricultural machinery which is a major goal of agricultural mechanical engineering. It is also important to note that the global trade in agricultural machinery allows ensuring food security in the countries where domestic agricultural mechanical engineering is not capable to saturate domestic market with agricultural machinery. The above provisions determined the relevance of this chapter in the context of ensuring food security at a national level.

BACKGROUND

Agricultural mechanical engineering is an important element of agro-industrial complex of any state, providing it with necessary machinery and equipment. High mechanization of work is the key for growth of volumes and qualities of the produced agricultural products and, as a result, it has a positive impact on food security of a country.

According to food security programs of many countries, major food products have to be produced domestically. Particularly, Russia sets the following security thresholds: grain – 95%; sugar – 80%; vegetable oil – 80%; meat and meat products – 85%; milk and dairy products – 90%; potatoes – 95% (President of the Russian Federation, 2010). To be able to achieve the established levels of domestic agricultural production and in such a way ensure food security, agricultural sector has to implement advanced agricultural mechanical engineering.

Economic mechanisms, preconditions, and methodological aspects of business development in the agricultural machinery market have been studied internationally by many scholars (Bortolini, Cascini, Gamberi, Mora, & Regattieri, 2014; Huang, Yun, You, & Wu, 2011; Liu, Hu, Jette-Nantel, & Tian, 2014; Staus & Becker, 2012; Yakymenko, 2013; Yu, Leng, & Zhang, 2012; Zhovnovach, 2014).

Muzlera (2014) defined agricultural machinery market as an exchange of machinery and agricultural equipment. Among the major segments of agricultural machinery market are agricultural goods, combine harvesters, cultivators, walking tractors, plows, drill machines, and mowing machines (Bortolini, Mora, Cascini, & Gamberi, 2014). Agricultural machinery market is central for establishing national food security as it creates infrastructure required for the development of agricultural production (Morozova, Litvinova, Rodina, & Prosvirkin, 2015).

Peculiarities of business activities under the conditions of international trade integration and global competition for the enterprises operating in various countries have been examined by Antonakakis and Tondl (2014), Ganushchak-Yefimenko (2013a, 2013b), and Gong and Kim (2013). In such conditions, entrepreneurs may benefit from entering new markets of foreign countries (Bull, 2014). At the same time, globalization brings international competition to domestic markets (Wirtz, Tuzovic, & Ehret, 2015). Increasing competition is an inseparable element of integration processes in the global economy (Goncalves & Madi, 2013).

Among other scholars, Clapp and Helleiner (2012), Khafizova, Galimardanova, and Salmina (2014), Mulatu and Wossink (2014), Vosta (2014), and Yushkevych (2013) investigated major challenges and problems related to the functioning of agricultural machinery market in the conditions of international trade integration and elaborated state regulations needed to support domestic producers. In the conditions of international trade integration, national markets of agricultural machinery face intensifying competition (Solovchuk, 2015). Often, foreign competitors win the struggle and in such a way threaten national food security (Gnedenko & Kazmin, 2015). To support and protect national producers, governments implement various regulations in the sphere of agricultural machinery markets by providing tax subsidies and more favorable conditions to domestic actors (Morozova & Litvinova, 2014).

Studying and assessing competitiveness on the agricultural machinery market in relation to its influence on food security is of particular relevance today. Economic literature provides a variety of approaches to understanding competitiveness. In the most general sense, it is defined as an ability to be ahead of the others using the advantages in the achievement of the goals (Chaynikova, 2007). According to Mazilkina and Panichkina (2009), the variety of approaches to defining competitiveness may be explained by the following issues:

- Features of problem definition and research objective that require an author to focus on particular aspect of competitiveness
- Characteristics of an object under study (item, service), subject of competition (enterprise, industries, regions, national economy, state), subject to competition (demand, market, production factors), scale of activity (commodity, branch, regional, interregional, world markets)

When competitiveness is considered as a property of a product or service, its quantitative characteristic may be reflected by various parameters, quality being the most widespread one.

Similar approaches to the determination of the level of competitiveness have been elaborated by Tikhonov (1985), Fatkhutdinov (2005), Okrenilov (1998), and Rybakov (1995), who all have measured the level of competitiveness by a ratio of integrated indicators of competitiveness of the products consumed by a standard customer during a standard lifetime. There are alternative approaches to the assessment of competitiveness of technical products. According to Ferapontov (1994), competitiveness may be measured as a relation of complex indicator of quality to the actual price of its implementation.

The relationships between the impact of the production and trade of agricultural equipment on food security were studied by Gebbers and Adamchuk (2010), Auat Cheein and Carelli (2013), and Ncube, French, and Mupangwa (2018). However, in general, in the academic literature, trade in agricultural machinery remains one of the understudied aspects of food security.

Thus, at the national level, food security should be considered as a component of the aggregated concept of national economic security. One of the important directions of ensuring food security is the creation and development of agricultural engineering. In this regard, this chapter attempts analyzing current state of the global market of agricultural machinery, developing of a methodology of assessment of competitiveness of agricultural machinery in domestic market; defining effective methods of management of logistic costs associated with the operation of agricultural machinery. These research directions are one of the most relevant in terms of ensuring food security of a country by means of advanced agricultural engineering.

Technical Equipment of Agricultural Production

MAIN FOCUS OF THE CHAPTER

International Trade in Agricultural Machinery

At present stage, main development characteristics of the global economy are transnationalization and globalization which together cause the emergence of interdependence of national economies. Economic relations between the countries are beyond bilateral, and the center of gravity is displaced towards multilateral economic cooperation. In such patterns, agricultural machinery industry is being increasingly integrated to the global market and is characterized by high extent of globalization. Mechanization influences on the quantitative and qualitative growth of agricultural production. Currently, most of large producers of agricultural machinery are multinational corporations which have manufacturing and assembly enterprises around the world (Mechanical Engineering Portal, n.d.).

Agricultural machinery market is segmented based on the type of a product, function, and regions (Figure 2).

Types include tractors, harvesters, cultivation and soil separation equipment (Research and Markets, 2018). The segment of agricultural tractors is the biggest one. The share of tractors in total sales of agricultural machinery increased from 30% in 2011 up to 44% in 2018. The growth due to the increase in demand for food products, shortage of labor, and expanded government subsidies towards mechanization. The world market of tractors declined from 1.8 million machines in 2014 down to 1.68 million machines in 2015, but it is expected to grow up to 2.15 million machines in 2020 (Agriculture Equipment Market, 2019). Tractors are widely used in various spheres of agricultural production for soil cultivation, plowing, and landing.

In previous years, sales volume of equipment for plowing and cultivation has been increasing by 9%. This sector is now the fastest growing one among other types of agricultural machinery sectors. The volume has already reached $10 billion as farmers in developing countries purchase larger and more difficult equipment for soil cultivation in their attempts to increase efficiency of agricultural production.

Figure 2. Segments of the global market of agricultural machinery
Source: Authors' development based on Grand View Research (2018)

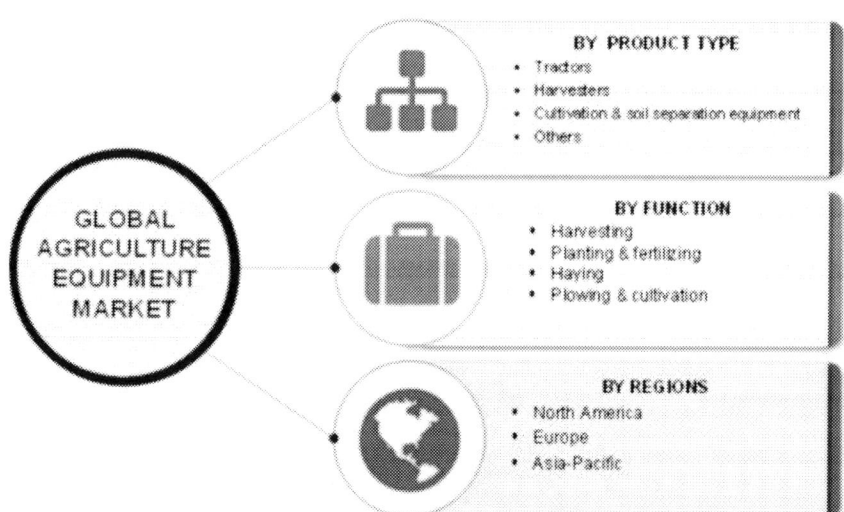

Demand for spare parts and attachable equipment grows by 5.5%. In this segment, the sales volume was $27.8 billion in 2018 (Research and Markets, 2018).

In relation to the functions of agricultural machinery, the market is segmented across harvesting, planting, fertilizing, haying, plowing, and cultivation. Harvesting machinery reduces dependence on labor, helps meeting growing demands of urban dwellers, and help breaking up soil efficiently. These factors accelerate the need for agricultural equipment for harvesting by farmers.

By regions, global agricultural machinery market spans across North America, Europe, Asia Pacific, and other regions. Global market of agricultural machinery is high fragmented due many producers and other actors, most of which are technologically advanced transnational companies which widely apply innovations to develop their production and distribution chains. As a result, smaller local suppliers experience difficulties when competing with larger actors, related particularly to quality, technology, and price. Global market of agricultural machinery is dominated by four largest companies which aggregated share of the market is over 40%. The biggest actor John Deere (USA) –18% of the global market of agricultural machinery. The other three are Case New-Holland (Italy) – 11%, AGCO (USA) – 7%, and Claas (Germany) – 4%. At the same time, there are many niche companies which focus on particular narrow segments of the market. German producers Fella, Krone, and Welger specialize in production of fodder harvesting machines. Kverneland (Norway), Kuhn (France), and Pottinger (Austria) – on soil-cultivating and fodder harvesting equipment. The USA, Germany, France, and Italy are major producers of agricultural machinery in the world (Medvedeva, 2018).

According to VDMA Agricultural Machinery Producers, the leading association of agricultural machinery producers which includes over 160 companies in Germany and worldwide, in 2108, global aggregated output of agricultural machinery reached $175 billion (Gotz, 2017). During previous eight years, the output has been increasing rapidly. Compared to 2010, in 2018, the output increased by 60% (average annual growth by 8%) (Figure 3).

The output has been driven by population growth, urbanization, and higher productivity demand amid the decrease in the acreage of agricultural land. Taken together, those factors have led to the growth in demand for agricultural machinery. Technological advancement for developing more efficient products, while keeping in mind the country-specific requirements, will provide opportunities for future growth of the sector. The key factors influencing the sale of agricultural equipment are the level of net farm income and, to a lesser extent, interest rates, and general economic conditions, availability of financing and related subsidy programs, farmland prices, and farm debt levels. Net farm income is primarily impacted by the volume of acreage planted, commodity and livestock prices, stock levels, the impacts of ethanol demand, farm operating expenses (which includes fertilizer and fuel costs), crop yields, fluctuations in currency exchange rates, tax incentives, and government subsidies. Farmers tend to postpone the purchase of equipment when the farm economy declines and increase their purchases when economic conditions improve.

Growth happened due to the increase in sales in emerging economies, primarily, China and India. By 2022, the Asia Pacific region will lead the agriculture machinery market by reaching nearly $74 billion due to growing demand for food. This may skyrocket a demand for agricultural machinery. There will be an inevitable need for agricultural equipment in the region due to these factors. Continuous adoption of technologically advanced devices due to supportive initiatives by the governments coupled with farmers' awareness programs can drive agricultural equipment market in Asia Pacific. In India, there has been noted an emergence of startups in the sphere of agricultural engineering. For instance, Agribolo located in Rajasthan is a farming services platform which provides such services as information

Figure 3. Global production of agricultural machinery, $ billion
Source: Authors' development based on Gotz (2017)

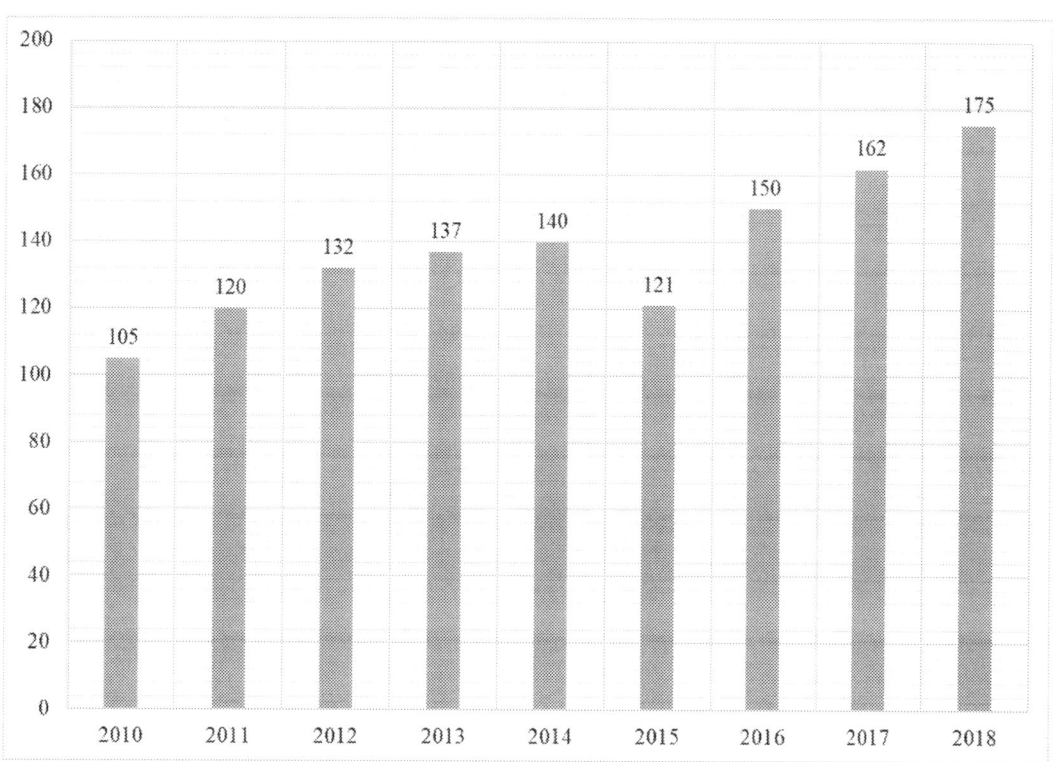

dissemination, quality input procurement, market linkages, irrigation facilities, and farming equipment (Mordor Intelligence, 2017).

Apart from China and India, the agricultural machinery markets in Thailand and Indonesia are expanding too. In the region of Central and Latin America, the sales increased in 2011-2012 due to the economic boom in Brazil and development of mechanization in agricultural production in Argentina. However, in 2014-2018, the economic decline in both Brazil and Argentina caused the decrease in the sales of agricultural machinery in the region. North America and Western Europe recorded growth below average till 2018. In both regions, demand is conditioned by technical achievements, such as increase in efficiency provided by new advanced equipment which allows farmers replacing outdated machinery in an economically effective manner. However, in the developed countries, many farmers postponed replacement of outdated equipment during economic crisis of 2008-2010, avoiding large purchases of new equipment because of uncertain economic situation. After the crisis, in 2011, the demand for new agricultural machinery emerged. Since a replacement cycle usually takes eight-nine years, the spike demand in 2011 resulted in a fact that many farmers did not seek replacement of agricultural machinery until 2018 and demand for agricultural machinery thus decreased.

With rapid technological development coupled with growing population, agricultural equipment market in North America is set to grow steadily. Moreover, there has been increase in the use of advanced agricultural equipment for more effective, reliable and time-saving production in the USA and Canada. Rapid introduction of advanced agricultural equipment in Europe can bolster the market growth in this region. Introduction of advanced equipment, such as multi-purpose tractors, will fuel the demand for agricultural equipment in Europe, thus positively impacting the growth of the market. For instance, in June 2018, SIWI Agriculture Ltd. Europe developed an automated hitch system to boost agricultural productivity and promote safety. This system enables a tractor driver to stay in a cabin instead of manually coupling and uncoupling between the implement, carriage, and the tractor.

Growth of the world population and increase in income per capita in emerging economies have resulted in the increase of demand for food products. By 2050, world population is forecasted to reach 9.5 billion people (7.4 billion in 2016). Farmers will need to increase the level of mechanization to be able to meet the growing demand. Global agricultural machinery market will grow amid the shortage of agricultural work because of mechanization and concentration of government support programs on promotion of agricultural machinery. In most of the countries worldwide, the governments will subsidize agricultural machinery production and agricultural engineering to increase production of food. The loans have been already intensively granted by banks and non-bank finance corporations to farmers to buy cars on tick. For example, in 2015, India allocated $127.9 billion for crediting of farmers, and $3.76 billion for the development of infrastructure of rural development (Ereport, n.d.).

It is supposed that the global market of agricultural machinery will experience a moderate growth by 7.3% (average) in 2018-2022. By 2022, the size of the market will reach $200 billion. Considering the growing investment opportunities in the agricultural sector, it is expected that the market of agricultural machinery will have a positive trend in 2020.

Among the agricultural machinery markets of Brazil, the USA, France, Germany, and Russia, Russian market is the one most dependent on imports. When Russia accessed the World Trade Organization (WTO), import duties on agricultural machinery were reduced by 5-10% which resulted in the increase of the openness of domestic market to foreign producers (Alekseyeva, 2018). The markets of Brazil and the USA, on the contrary, are those the most protected. Both countries are self-sufficient in the production of tractors and other agricultural machinery (Government of the Russian Federation, 2017).

It is important to note that in terms of food security it is necessary to have a methodology for assessing the competitiveness of agricultural machinery in the domestic market.

Measuring Competitiveness of Agricultural Machinery in the Domestic Market

An analysis of the existing approaches to the assessment of competitiveness of agricultural machinery allows allocating two methodologies. The first one reflects subjective preferences of consumers, while the second one reveals a position of agricultural machinery producers.

First approach is to calculate the integrated measure of competitiveness of agricultural machinery J_K:

$$JK = \frac{An}{Ac}\gamma1 + \frac{Bn}{Bc}\gamma2 + \frac{Cn}{Cc}\gamma3 + \frac{Dn}{Dc}\gamma4 + \frac{En}{Ec}\gamma5 \qquad (1)$$

A_n, A_c – cost of performance of work of this type (direct costs of funds for operation of the equipment) on new equipment and the one by a competitor, respectively;

B_n, B_c – labor productivity (activities carried out by an operator by means of a technical unit) on new equipment and the one by a competitor, respectively;

C_n, C_c – payback period of full costs of acquisition and operation on new equipment and the one by a competitor, respectively;

D_n, D_c – decrease in a loss of agricultural products on new equipment and the one by a competitor, respectively;

E_n, E_c – fuel costs related to the use of new equipment and the one by a competitor, respectively;

$\gamma_1, \gamma_2, \gamma_3, \gamma_4, \gamma_5$ – specific weight (importance) of the parameters.

According to this approach, a client chooses one or several indicators, a specific weight of the chosen indicators is accepted to a unit, and the specific weight of other indicators – to zero. If money is a major factor, then the cost index of works or a payback period are the most significant. When a cost of fuel is high and technical parameters are relatively equal to each other, a client may select the machinery with the highest efficiency in terms of fuel consumption. The main shortcoming of this approach is its subjective character as it considers only preferences of certain consumers and therefore does not provide an adequate assessment of competitiveness of agricultural machinery.

Second approach assumes an establishment of parameters which are the subjects of comparison and assessment with their quantitative expression and establishment of "ponderability" for characteristic of degree of satisfaction of a client's specific need. The specified list includes three groups of parameters: standard and legal, technical, and economic.

Standard and legal parameters establish compliance of a product to the legal standards and norms regulating the actual level and required thresholds of these parameters. Accounting of standard parameters can be provided by introduction of special indicator with two values of 1 or 0. If a machinery meets legal standards, the indicator is 1, otherwise, it is 0.

General standard and legal parameter J_{LP} represents the work of individual parameters:

$$JLP = \prod_{i=1}^{n} q_i \qquad (2)$$

q_i – individual parameter
n – number of standard parameters in the sample

If at least one of individual parameters is equal to 0 (technical tool on any indicator does not meet legal standard), then general indicator is equal to 0 too. That means a machine is not competitive in the market.

Technical parameters reflect the relation of physical volume of the products to the expenses of resources in natural measuring instruments. To find the general technical parameter J_{TP}, the importance (specific weight) of each individual parameter included the general set should be considered:

$$JTP = \sum_{i=1}^{n} \frac{bi>}{bi:} ai \qquad (3)$$

b_{io} – individual technical parameter of a machine (domestic production)
b_{ik} – individual technical parameter of a machine (foreign competitor)
a_i – specific weight of individual technical parameter
n – number of technical parameters in the sample

Economic parameters are based on actual costs and do not depend on expert assessment. The price of agricultural machinery P_C is calculated as follows:

$$PC = P_M + Po \qquad (4)$$

P_M – price of a machinery
P_o – operation price of a machinery
P_M price is determined as follows:

$$P_M = \frac{P}{n} \qquad (5)$$

P – selling price of a machinery
n – depreciation term
P_o operation price is calculated as follows:

$$Po = Et + Es + Ef + Er + Em \qquad (6)$$

E_t – transportation costs
E_s – annual charges on compensation of service staff
E_f – annual charges on fuels and lubricants
E_r – annual charges on repair
E_m – annual charges on maintenance

The aggregated economic measure E_{EP} is calculated as follows:

$$EEP = \frac{Pcd}{Pcc} \qquad (7)$$

P_{cd} – price of consumption of domestic equipment
P_{cc} – price of consumption of foreign equipment

The lower the consumption price, the higher the competitiveness of a machinery, as it means that a client has an opportunity to get a unit of quality cheaper compared to a competitor.

On the basis of the aggregated measures of standard, technical, and economic parameters, the integrated measure of competitiveness *(IC)* is determined:

$$IC = PC \times \frac{JTP}{EEP} \qquad (8)$$

Consequently, agricultural machinery of domestic production is more competitive compared to the imports if $IC \geq 1$. This approach, however, has several shortcomings:

- Aggregated measure of standard and legal parameters amounts to either 0 or 1. It has significant effect on the integrated measure of competitiveness, however, it does not make much sense measuring the competitiveness of agricultural machinery scored 0 as it does not meet required standards and legal regulations.
- Aggregated measure of technical parameters provides a weight basis for individual indicators. Measuring aggregated parameter on a basis of similar technical indicator may result in different weights for various objects.
- Specified approach ignores investment parameters as it does not consider the interests of investors, creditors, and other actors in agricultural machinery market.

Consequently, the existing approaches to the measurement of competitiveness of agricultural machinery require improvement. The specified shortcomings cause a need in the more substantiated choice by a client of particular type of equipment. The authors offer a combination of technical, economic, and investment parameters of the compared agricultural machinery options as a basis for new approach. The key technical parameters are capacity, engine capacity, reliability (time between failures), and productivity. Economic parameters include price of agricultural machinery and operation price. Among investment parameters of competitiveness, the most important ones are fuel consumption at maximum power, cost of power unit of an engine, cost of unit of mass (Zhudro, 2009).

The specified characteristics can be estimated on the basis of micro indexes of technical, economic, and investment parameters. MI_T micro index of technical parameters is determined as follows:

$$MIt = \sum_{i=1}^{n} S_i P_i \qquad (9)$$

$S_i = b_{io}/b_{iк}$, $i = 1, 2, ..., n$;

b_{io}, $b_{iк}$ – individual technical indicator of agricultural machinery (domestic vs foreign, respectively);
P_i – weight of i in relation to total n.

When calculating the general indicator for technical parameters, there is a problem with the determination of a weight base for individual indicators and discovering general parameter on its basis. To solve this problem, the authors carry out a comparison of technical parameters of agricultural machinery on the basis of convolutions of quantitative indices (Anisimov, 2009).

Formally, a convolution of technical characteristics of the two items is possible in a form of the following task. There are J items of agricultural machinery under comparison. Each of J items is characterized by a set of technical indicators. Generally, I types of indicators are established. For each parameter, a value is set: x_{ij}, $i = 1, 2, ..., I$, $j = 1, 2, ..., J$ where i – index of an indicator type, j – index of agricultural machinery.

At the first stage, it is necessary to define the coefficients B_j, $j = 1, 2, ..., J$ commensurabilities of the compared items of machinery. To find the coefficients, it is necessary to choose the reference agricultural machinery. As a standard, it is accepted a conditional machinery which has the maximum values of technical characteristics out of the machinery units under study, that is:

$$x_{i;m} = \max x_{ij}, i = 1, 2, \ldots, I, j = 1, 2, \ldots, J \tag{10}$$

At the second stage, each of the agricultural machinery units taken as a standard one is characterized by the sizes:

$$S_{ij} = x_{ij}/x_{i;m}, i = 1, 2, \ldots, I, j = 1, 2, \ldots, J \tag{11}$$

Enter parameters d_{ij} such, that:

$$d_{ij} = \begin{cases} 1, & \text{if increase } x_{ij} \text{ leads to an increase inefficiency of the machines} \\ 0, & \text{otherwise} \end{cases}$$

Then, the efficiency of each of the agricultural machinery units can be characterized by the sizes:

$$S'_{ij} = d_{ij}S_{ij}, i = 1, 2, \ldots, I, j = 1, 2, \ldots, J \tag{12}$$

Taking into account the values of the coefficients $B_j, j = 1, 2, \ldots, J$ commensurabilities of the agricultural machinery units under study are defined by a ratio:

$$Bj = \sum_{i=1}^{I} S'ij \times Pi \tag{13}$$

P_i – weight of i indicator in relation to total I.

In a basis of definition of P_i, $i = 1, 2, \ldots, I$ can be put the principle of a maximum of entropy. The essence of the principle is that at the measurement of competitiveness of agricultural machinery, the available information does not allow determining the efficiency of a machinery precisely by technical indicators and distribution of probabilities. Therefore, it is necessary to choose the steadiest one from all possible distributions. It is the distribution providing maximum of uncertainty (entropy), and, therefore, a minimum of conjectures in the developing information situation.

Practical opportunities according to the analysis of possible values of the considered weight coefficients of P_i, $i = 1, 2, \ldots, I$ are limited to pair comparison and establishment of some linear relations of an order on a set:

$$P = \{P_i\}, i = 1, 2, \ldots, I \tag{14}$$

In this regard, the problem of determination of sizes P, $i = 1, 2, \ldots, I$ comes down to the choice of a method of transformation of the preferences set in the form of the system of the relations of an order in dot estimates.

Convenient method of such transformation for the benefit of a problem of determination of weight coefficients of technical characteristics of the alternative agricultural machinery is use of the models offered by Fishbern for a priori receiving the linear restrictions of dot estimates of probabilities of events which do not contradict a system.

Information and theoretical justification of objectivity of Fishbern's estimates relies on the principle of maximum uncertainty:

$$E(p) = \prod_{i=1}^{I} p_i^{I-k+1} \tag{15}$$

p_i – probability of approach of i of an event from full group of I events reaches a maximum on Fishbern's estimates.

Taking into account that various technical characteristics of agricultural machinery units may be equal, the steadiest one (the possessing maximum of entropy of E(p)) is the uniform distribution. Then $Pi . i = 1, 2, ..., I$ are determined by a formula:

$$Pi = \frac{\sum_{j=1}^{J} S'_{ij}}{\sum_{i=1}^{I} \sum_{j=1}^{J} S'_{ij}} \tag{16}$$

Thus, in such technique of weight of P_i, $i = 1, 2, ..., I$ technical indicators and coefficients of B_j, $j = 1, 2, ..., J$ commensurabilities of the agricultural machinery are appointed not randomly, and are at a minimum of conjectures, on the basis of real characteristics of agricultural machinery units under consideration. It provides a certain objectivity to the results of comparison of agricultural machinery on their technical parameters received at its application (Khudzhatov, 2017).

MI_E micro index of economic parameters is defined as follows:

$$MIe = \frac{Pcd}{Pcc} \tag{17}$$

P_{cd} – price of domestic equipment
P_{cc} – price of competing foreign equipment
MI_I micro index of investment parameters is defined as follows:

$$MIi = \sum_{i=1}^{n} \frac{Ai>}{Ai:} ai \tag{18}$$

c_{io} – individual investment parameter of a machine (domestic production)
c_{jk} – individual investment parameter of a machine (foreign competitor)
a_i – weight of i-individual investment parameter

On the basis of micro indexes of technical, economic, and investment parameters, integrated index of competitiveness (*IIC*) of agricultural machinery is defined:

$$IIC = \frac{MIt}{MIi \times MIe} \tag{19}$$

Table 1. Key technical parameters of combine harvesters

Combine harvester	Speed of unloading, l/sec	Engine capacity, hp	Capacity of the fuel tank, l	Productivity, ha/h
Acros 585	90	300	540	4.1
John Deere W650	88	324	800	6.9

Source: Authors' development

Table 2. Weight indicators of technical parameters

Speed of unloading, P_1	Engine capacity, P_2	Capacity of the fuel tank, P_3	Productivity, P_4
0.29	0.28	0.17	0.26

Source: Authors' development

Table 3. Price of combines per year

Combine harvester	Average market price in domestic market, $	Depreciation term, years	Price per year, $
Acros 585	138,462	10	13,846
John Deere W650	200,000	15	13,338

Source: Authors' development

If $IIC \geq 1$, then technical disadvantages of domestic machinery are compensated by lower costs of acquisition and operation and more favorable conditions for investors, or the high price of consumption of this equipment and big investment expenses are compensated by its technical advantages. In both cases, domestic equipment is more competitive compared to the foreign one.

The application of the methodology is demonstrated on the cases of combine harvesters of Acros 585 (Rostselmash, Russia) and W650 (John Deere, USA). At the first stage, micro index of technical parameters is defined (Table 1).

Having taken W650 John Deere combine as a standard, the authors then define weight indicators of technical parameters (Table 2).

Further, micro index of technical parameters is calculated: $MI_T = 0.83$.

At the next stage, micro index of economic parameters is defined. It is necessary to define a price of each machinery item under study. This price consists of the price of a machine and operation price (Table 3, Table 4).

At the next stage, the price is calculated for each combine: Acros 585 – $18,723, John Deere W650 – $17,026. Micro index of economic parameters: $MI_E = 1.1$. Finally, micro index of investment parameters is defined. For this purpose, individual investment parameters of the combines are determined (Table 5).

Micro index of investment parameters in the condition of equality of specific weights of individual parameters: $MI_I = 0.84$.

Based on the calculation above, integrated index of competitiveness is determined for a domestic agricultural machinery unit in comparison with a foreign analog in domestic market:

Table 4. Operation price of combines per year, $

Combine harvester	Transportation costs	Annual charges on compensation of personnel	Annual charges on fuel and lubricants	Annual charges on repair	Annual charges on servicing	Operation price per year
Acros 585	569.5	453.1	890.2	1,304.6	1,659.5	4,876.9
John Deere W650	668.9	509.4	1,146.3	624.6	738.5	3,687.7

Source: Authors' development

Table 5. Individual investment parameters of combine harvesters

Combine harvester	Specific fuel consumption at the maximum power, kg/t	Specific cost of engine power unit, $/hp	Specific cost of mass unit, $/kg
Acros 585	4.2	461.5	8.8
John Deere W650	3.9	617.3	12.8

Source: Authors' development

$$IIC = \frac{0,83}{1,1 \times 0,84} = 0,9$$

As *IIC* < 1, Russian Acros 585 is considered as not competitive in comparison with W650 John Deere. At the same time, higher technological level of W650 John Deere provides lower price of consumption in comparison with Acros 585 and also compensates higher investment expenses. Therefore, at the expense of more considerable investment investments, John Deere is more competitive. Thus, the developed methodology for measuring competitiveness of agricultural machinery in domestic market allows considering a wide range of parameters and indicators and also revealing key problems when using agricultural machinery of different producers. At the same time, this methodology is simple in practical application as it does not require complex calculations.

Management of Logistics Costs in the Operation of Agricultural Machinery

Logistics costs is one of the main factors which affects the economic efficiency of agricultural enterprise. A relevant task is to measure the impact of the dynamics of logistics costs on the cost of agricultural products. One of the objectives of this chapter is to assess the impact of the dynamics of logistics costs on the cost of agricultural products, namely the study of the variable costs of diesel fuel during transportation of agricultural products.

The logistics costs associated with transportation of agricultural products are divided into fixed and variable ones. Variable costs include those resulting from the material costs of the motor fuel consumed by a vehicle, as well as ensuring the operation of the auxiliary units (refrigerator compartment compressor and cargo compartment climate control).

In order to measure the impact of the dynamics of logistics costs on the cost of agricultural products, the authors considered a specific type of a vehicle – grain carrier on the basis of KAMAZ 65115

Table 6. Diesel fuel costs in the selected EAEU countries in January-May 2018

Country	Diesel fuel costs in January 2018, $/l	Diesel fuel costs in May 2018, $/l	Change
Russia	0.63	0.65	+3.2%
Kazakhstan	0.48	0.51	+5.9%
Belarus	0.65	0.68	+4.5%

Source: Authors' development on Eurasian Economic Commission (2019)

(carrying capacity – 14 tons, cargo hold – 26 m^3). Consumption of motor fuel per 100 km is 27.4 liters (empty) and 32.0 liters (loaded) at air temperature above +10°C. This type of grain carrier is selected due to its average performance parameters among the trucks used in cargo transportation in agriculture. Serving the agro-industrial complex, therefore, its performance parameters provide sufficient accuracy of modeling. The average distance of transportation from agricultural enterprise to grain storage is 50 km (Arskiy, 2018a).

There is a need to assess the impact of the dynamics of the cost of motor fuel on the value of logistics costs due to their stable positive dynamics in recent years. The rise in the cost of motor fuel has a certain effect on the increase in logistics costs. A study is carried out to measure the influence of logistics costs on the cost of agricultural production and economic performance of an agricultural enterprise (Arskiy, 2018b). The authors study the cost of motor fuel in the countries of the Eurasian Economic Union (EAEU) actively engaged in farming: Russia, Kazakhstan, and Belarus. KAMAZ 65115 is the most commonly used agricultural cargo vehicle across all three countries (Table 6).

Based on the data of Table 6, the following findings are revealed:

- The dynamics of diesel fuel costs in Russia in the first half of 2018 is positive and amounts to +3.2%. The growth is due to the decline in world oil prices, which conditioned lower attractiveness of domestic market for diesel fuel producers.
- The dynamics of diesel fuel costs in Kazakhstan in the first half of 2018 is positive and amounts to +5.9%. The growth is due to the lack of sufficient refining capacity (oil refineries). Kazakhstan currently imports fuel from Russia, but there are plans to develop domestic refinery industry (there are three refining factories in the country, the fourth one will be launched in the near future).
- The dynamics of diesel fuel costs in Belarus in the first half of 2018 is positive and amount at +4.5%. Belarus imports fuel from Russia, prices in the two markets correlate closely.

There is a steady positive dynamics of changes in retail consumer prices for diesel fuel used by agricultural enterprises in their logistics. The on-going increase of logistics costs requires the assessment of their impact on the economic performance of agricultural enterprises. In this study, the authors calculate motor fuel costs associated with the transportation of grain from agricultural enterprise (field, storage) to grain elevator. The algorithm of the model (V) is as follows (Arskiy, 2016):

$$V = d_1 \times \frac{n_1}{100} + d_2 \times \frac{n_2}{100} \qquad (20)$$

Technical Equipment of Agricultural Production

d_1 – distance from agricultural enterprise (warehouse, field) to the distribution point (elevator)
n_1 – liters of motor fuel per 100 km (loaded)
d_2 – distance from agricultural enterprise (warehouse, field) to the elevator
n_2 – liters of motor fuel per 100 km (empty)

Using the parameters of KAMAZ 65115, the calculations are made based on the average cost of the volume of motor fuel spent in January 2018 ($18.76) and May 2018 ($19.38). The difference between May 2018 and January 2018 is $0.62. Based on the data presented, it is possible to determine the increase in transportation costs per one ton of grain at the rate of fourteen tons of grain transported at a time: 0.62 / 14 = $0.044. Having taken $138.5 for the cost of one ton of wheat (class 3, ex-warehouse delivery terms), the authors calculate the increase in logistics costs of motor fuel in comparison with the relative period of 2017: 0.044 / 138.5 = 0.03%. The algorithm of the model for measuring the impact of the dynamics of motor fuel costs on logistics costs (E) is as follows:

$$E = \frac{\left(\left(d_1 \times \frac{n_1}{100} + d_2 \times \frac{n_2}{100}\right) \times C_1\right) - \left(\left(d_1 \times \frac{n_1}{100} + d_2 \times \frac{n_2}{100}\right) \times C_2\right)}{\frac{V}{P}} \times 100 \qquad (21)$$

C_1 – motor fuel costs in the period under study
C_2 – motor fuel costs in the previous period
V – volume of transported cargo (tons)
P – contract value per ton of transported cargo

When $E < 1$, the influence of the actual positive dynamics of logistics costs on production costs of agricultural enterprise is insignificant.

In case of the selected EAEU countries, it is found that the dynamics of diesel fuel costs is positive. This factor determines a need to take into account the increasing costs in the logistics process of agricultural enterprise. In general, it can be concluded that there is no actual influence of motor fuel costs on the structure and volume of the corresponding logistic costs due to the meager amount of the increase in logistic costs in terms of value relative to the cost of the transported cargo. This study is based on the simplest methods of calculation when dynamics has a direct impact on the level of logistics costs. At the same time, it is necessary to practice a rational approach in forming assessments of the influence of these factors on the economic efficiency of agricultural enterprise. These estimates should also be formed taking into account government subsidies and subsidies enjoyed by some market actors while affecting the competitive market environment.

SOLUTIONS AND RECOMMENDATIONS

In general, the study of current challenges of agro-industrial complex is essential in terms of ensuring food security. Each country has to develop domestic system of agricultural engineering to obtain high performance of agricultural industry. Global market of agricultural machinery tends to grow at the expense of significant contribution of developing states which realized an importance of agricultural engineering for ensuring food security. Agricultural machinery produced in a country has to be competitive in domestic market in terms of not only technical characteristics but also economic and investment parameters. Otherwise, domestic producers may lose competition to international agro-engineering companies. Besides, measuring logistic costs associated with the operation of agricultural machinery is important. In this context, a country has to react to the world prices for oil adequately. The implementation of some of the findings of this study may allow increasing food security of a country in terms of improving technical capacity of agro-industrial sector.

FUTURE RESEARCH DIRECTIONS

Practical recommendations elaborated in this study contain some provisions which have universal character and can be implemented across various industries. The findings obtained and discussed in this chapter can become a basis for further investigations in the sphere of development of agro-industrial complex. The task for further research is the elaboration of scientific and methodical device for providing an integration of state mechanisms of stimulation of investments into agricultural sector.

CONCLUSION

The study of food security issues is essential for any country. Food security is a part of national security of a country providing its population with food products. Complex food security is impossible without development of various elements of agro-industrial complex, agricultural engineering being one of them. The scientific and methodical device of assessment of competitiveness of agricultural machinery in domestic market and calculation of logistic costs at its functioning allows defining shortcomings of agricultural engineering and ways of their correction. Implementation of the findings provides a solution of a practical problem of ensuring food security of a country by means of improvement of technical capacity of agro-industrial complex.

Thus, the study conducted in the chapter makes it possible to distinguish the link between trade in agricultural machinery and food security. Developed agricultural machinery allows ensuring food security of a country, while global production of agricultural machinery allows ensuring food security in a global scale.

REFERENCES

Agriculture Equipment Market. (2019). *Report*. Retrieved from https://www.farmmachinerysales.com.au/

Alekseyeva, Y. (2018). *Russia – Agricultural Equipment*. Retrieved from https://www.export.gov/article?id=Russia-Agricultural-Equipment

Anisimov, E., Gapov, M., Rodionova, E., & Saurenko, T. (2019). The Model for Determining Rational Inventory in Occasional Demand Supply Chains. *International Journal of Supply Chain Management*, *8*(1), 86–89.

Anisimov, V. (2009). *Optimization: An Adaptive Approach to Investment Management under Uncertainty*. Moscow: Publishing House of the Russian Customs Academy.

Antonakakis, N., & Tondl, G. (2014). Does Integration and Economic Policy Coordination Promote Business Cycle Synchronization in the EU? *Empirica, 41*(3), 541–575. doi:10.100710663-014-9254-2

Arskiy, A. (2016). Peculiarities of Calculation of Logistics Costs. Motor Fuel. *World of Modern Science, 35*(1), 34–37.

Arskiy, A. (2018a). Management of Logistics Costs of Enterprises of Agro-Industrial Complex. *Bulletin of the Moscow University of Finance and Law, 1*, 98–102.

Arskiy, A. (2018b). Assessment of Efficiency of Management Decisions in Crisis Management of Agricultural Enterprise. *Marketing and Logistics, 16*(2), 6–11.

Auat Cheein, F. A., & Carelli, R. (2013). Agricultural Robotics: Unmanned Robotic Service Units in Agricultural Tasks. *IEEE Industrial Electronics Magazine, 7*(3), 48–58. doi:10.1109/MIE.2013.2252957

Bortolini, M., Cascini, A., Gamberi, M., Mora, C., & Regattieri, A. (2014). Sustainable Design and Life Cycle Assessment of an Innovative Multi-Functional Haymaking Agricultural Machinery. *Journal of Cleaner Production, 82*, 23–36. doi:10.1016/j.jclepro.2014.06.054

Bortolini, M., Mora, C., Cascini, A., & Gamberi, M. (2014). Environmental Assessment of an Innovative Agricultural Machinery. *International Journal of Operations and Quantitative Management, 20*(3), 243–258.

Bull, B. (2014). The Development of Business Associations in Central America: The Role of International Actors and Economic Integration. *Journal of Public Affairs, 14*(3-4), 331–345. doi:10.1002/pa.1420

Chaynikova, L. (2007). *Competitiveness of an Enterprise: Scientific and Practical Study*. Tambov: Tambov State Technical University.

Clapp, J., & Helleiner, E. (2012). Troubled Futures? The Global Food Crisis and the Politics of Agricultural Derivatives Regulation. *Review of International Political Economy, 19*(2), 181–207. doi:10.1080/09692290.2010.514528

Ereport.ru. (n.d.). *World Market of Agricultural Machinery and Equipment*. Retrieved from http://www.ereport.ru/articles/commod/mirovoj-rynok-selskohozjajstvennoj-tehniki.htm

Eurasian Economic Commission. (2019). *Statistics of the EAEU*. Retrieved from http://www.eurasiancommission.org/ru/act/integr_i_makroec/dep_stat/union_stat/Pages/default.aspx

Fatkhutdinov, R. (2005). *Strategic Competitiveness*. Moscow: Economy.

Ferapontov, A. (1994). One of Variants of Mathematical Model of Indicators of Competitiveness of Technical Production. *Standards and Quality, 4*, 44–45.

Ganushchak-Yefimenko, L. (2013a). Economic Integration as a Basis for Small and Medium Enterprises Business. *Actual Problems of Economics, 141*(3), 70–77.

Ganushchak-Yefimenko, L. (2013b). Management of Innovation Potential Development of Small and Medium Business Based on Economic Integration. *Actual Problems of Economics*, *144*(6), 72–79.

Gebbers, R., & Adamchuk, V. (2010). Precision Agriculture and Food Security. *Science*, *327*(5967), 828–831. doi:10.1126cience.1183899 PMID:20150492

Gnedenko, E., & Kazmin, M. (2015). Agricultural Land and Regulation in the Transition Economy of Russia. *International Advances in Economic Research*, *21*(3), 347–348. doi:10.100711294-015-9535-y

Goncalves, J. R. B., & Madi, M. A. C. (2013). Global Economic Integration, Business Expansion and Consumer Credit in Brazil, 1994-2010. *International Journal of Green Economics*, *7*(3), 213–225. doi:10.1504/IJGE.2013.058164

Gong, C., & Kim, S. (2013). Economic Integration and Business Cycle Synchronization in Asia. *Asian Economic Papers*, *12*(1), 76–99. doi:10.1162/ASEP_a_00188

Gotz, C. (2017). *Upswing in Agricultural Machinery Industry*. Retrieved from https://lt.vdma.org/en/viewer/-/v2article/render/19870108

Government of the Russian Federation. (2017). *Order #1455-r from July 7, 2017, "Strategy of Development of Agricultural Engineering in Russia"*. Retrieved from http://government.ru/docs/28393/

Grand View Research. (2018). *Agricultural Machinery Market Analysis, Market Size, Application Analysis, Regional Outlook, Competitive Strategies, and Segment Forecasts, 2016 to 2024*. Retrieved from https://www.grandviewresearch.com/industry-analysis/agricultural-machinery-market/methodology

Huang, H., Yun, Z., You, L., & Wu, J. (2011). Forecast of Subsidy for Purchasing Agricultural Machinery Based on Life Cycle Theory in China. *Paper presented at the International Conference on Management and Service Science*, Wuhan. Academic Press. 10.1109/ICMSS.2011.5998516

Ibragimov, M.-T., & Dokholyan, S. (2010). Methodological Approaches to the Assessment of Food Security of the Region. *Regional Problems of Economic Transformation*, *26*(4), 172–193.

Khafizova, A., Galimardanova, Y., & Salmina, S. (2014). Tax Regulation of Activity of Agricultural Commodity Producers. *Mediterranean Journal of Social Sciences*, *24*(5), 421–425.

Khudzhatov, M. (2017). The Study of Differentiation of Foreign Trade Prices by Using of Dispersion Analysis. *RUDN Journal of Economics*, *25*(1), 91–101. doi:10.22363/2313-2329-2017-25-1-91-101

Khudzhatov, M. (2018a). *Enhancement of Customs Instruments of Promotion of Foreign Investments in Agricultural Mechanical Engineering in Russia*. Moscow: DPK Press.

Khudzhatov, M. (2018b). The Use of Customs Instruments for Stimulation of Foreign Investment in Agricultural Machinery in Russia. *Marketing and Logistics*, *15*(1), 58–70.

Liu, Y., Hu, W., Jette-Nantel, S., & Tian, Z. (2014). The Influence of Labor Price Change on Agricultural Machinery Usage in Chinese Agriculture. *Canadian Journal of Agricultural Economics*, *62*(2), 219–243. doi:10.1111/cjag.12024

Mazilkina, E., & Panichkina, G. (2009). *Competitiveness Management*. Moscow: Omega-L.

Mechanical Engineering Portal. (n.d.). *Analytics*. Retrieved from http://www.mashportal.ru/analytics.aspx

Medvedeva, A. (2018). *World Market of Agricultural Machinery – Stability and Need for Innovations*. Retrieved from https://www.agroxxi.ru/selhoztehnika/novosti/mirovoi-rynok-selskohozjaistvennoi-tehniki-stabilnost-i-potrebnost-v-innovacijah.html

Mordor Intelligence. (2017). *Agricultural Machinery Market – Segmented by Type (Tractors, Plowing and Cultivating Machinery, Planting Machinery, Harvesting Machinery, Haying and Forage Machinery, Irrigation Machinery), by Geography – Analysis of Growth, Trends and Progress (2019-2024)*. Retrieved from https://www.mordorintelligence.com/industry-reports/agricultural-machinery-market?gclid=CjwKCAiAiJPkBRAuEiwAEDXZZW41MxYmd5jvxtxaw-6prmgjEvU3F_UY1hMAd9lfiZtwtE8rLKLc-MxoCqwwQAvD_BwE

Morozova, I., & Litvinova, T. (2014). Russian Market of Agricultural Equipment: Challenges and Opportunities. *Asian Social Science*, *23*(10), 68–77.

Morozova, I., Litvinova, T., Rodina, E., & Prosvirkin, N. (2015). Marketing Mix in the Market of Agricultural Machinery: Problems and Prospects. *Mediterranean Journal of Social Sciences*, *36*(6), 19–26.

Mulatu, A., & Wossink, A. (2014). Environmental Regulation and Location of Industrialized Agricultural Production in Europe. *Land Economics*, *90*(3), 509–537. doi:10.3368/le.90.3.509

Muzlera, J. (2014). Capitalization Strategies and Labor in Agricultural Machinery Contractors in Argentina. *Research in Rural Sociology and Development*, *20*, 57–74. doi:10.1108/S1057-192220140000020002

Ncube, B., French, A., & Mupangwa, W. (2018). Precision Agriculture and Food Security in Africa. In P. Mensah, D. Katerere, S. Hachigonta, & A. Roodt (Eds.), *Systems Analysis Approach for Complex Global Challenges* (pp. 159–178). Heidelberg: Springer. doi:10.1007/978-3-319-71486-8_9

Okrenilov, V. (1998). *Quality Management: Scientific and Practical Study*. Moscow: Economics.

President of the Russian Federation. (2010). *Decree #120 from January 30, 2010, "Food Security Doctrine of the Russian Federation"*. Retrieved from http://www.garant.ru/hotlaw/federal/228793/

Research and Markets. (2018). *Global Agricultural Machinery Market – Industry Trends, Opportunities and Forecasts to 2023*. Retrieved from https://www.researchandmarkets.com/research/b6lzkj/global?w=5

Rybakov, I. (1995). Quality and Competitiveness in Market Relations. *Standards and Quality*, *12*, 43–47.

Shapkina, L. (2012). Regional Aspects of Management of Food Security. *Terra Economicus*, *1-2*, 128–131.

Solovchuk, K. (2015). Regulation and Support for Innovations in the Agricultural Sector of the European Union. *Actual Problems of Economics*, *165*(3), 62–68.

Staus, A., & Becker, T. (2012). Attributes of Overall Satisfaction of Agricultural Machinery Dealers Using a Three-Factor Model. *Journal of Business and Industrial Marketing*, *27*(8), 635–643. doi:10.1108/08858621211273583

Tikhonov, P. (1985). *Competitiveness of Industrial Products*. Moscow: Publishing House of Standards.

Vosta, M. (2014). The Foodstuffs Market in the CR and Its Regulation within the Framework of the EU Agricultural Policy. *Agricultural Economics – Czech, 60*, 279-286.

Wirtz, J., Tuzovic, S., & Ehret, M. (2015). Global Business Services: Increasing Specialization and Integration of the World Economy as Drivers of Economic Growth. *Journal of Service Management, 26*(4), 565–587. doi:10.1108/JOSM-01-2015-0024

Yakymenko, O. (2013). Peculiarities in Strategic Management of Enterprise Development in Agricultural Machinery Sector. *Actual Problems of Economics, 147*(9), 138–144.

Yu, X., Leng, Z., & Zhang, H. (2012). Optimal Models for Impact of Agricultural Machinery System on Agricultural Production in Heilongjiang Agricultural Reclamation Area. *Paper presented at the 24th Chinese Control and Decision Conference*, Taiyuan. Academic Press. 10.1109/CCDC.2012.6244128

Yushkevych, O. (2013). Regulation Mechanisms in the Development of Agricultural Enterprises. *Actual Problems of Economics, 147*(9), 132–137.

Zhovnovach, R. (2014). Satisfaction of Consumers' Demand as the Basis for Planning Competitiveness of Agricultural Machinery Enterprises. *Actual Problems of Economics, 155*(5), 171–180.

Zhudro, M. (2009). *Development of Economic Instruments of Increase of Competitiveness of the Use of Agricultural Technology*. Gorki: Belsha.

ADDITIONAL READING

Arskiy, A. (2017). Factor of the Economic Potential of the Customs Territory in the Anti-Crisis Management of a Trucking Enterprise. *Marketing and Logistics, 13*(5), 6–12.

Bruck, T., & d'Errico, M. (2019). Food Security and Violent Conflict: Introduction to the Special Issue. *World Development, 117*, 167–171. doi:10.1016/j.worlddev.2019.01.007

Choi, T., Chiu, C., & Chan, H. (2016). Risk Management of Logistics Systems. *Transportation Research Part E, Logistics and Transportation Review, 90*, 1–6. doi:10.1016/j.tre.2016.03.007

Colin, E. C. (2009). Mathematical Programming Accelerates Implementation of Agro-Industrial Sugarcane Complex. *European Journal of Operational Research, 199*(1), 232–235. doi:10.1016/j.ejor.2008.11.016

Erokhin, V. (2016). Development of Rural Territories in the Russian Far East and in Heilongjiang Province of China. *Agricultural Bulletin of Stavropol region, 23*(3), 256-260.

Hossain, M., Mullally, C., & Niaz Asadullah, M. (2019). Alternatives to Calorie-Based Indicators of Food Security: An Application of Machine Learning Methods. *Food Policy, 84*, 77–91. doi:10.1016/j.foodpol.2019.03.001

Kain, R., & Verma, A. (2018). Logistics Management in Supply Chain – An Overview. *Materials Today: Proceedings, 5*(2), 3811–3816.

Kou, Z., & Wu, C. (2018). Smartphone Based Operating Behavior Modelling of Agricultural Machinery. *IFAC-PapersOnLine*, *17*(51), 521–525. doi:10.1016/j.ifacol.2018.08.156

Kwesi-Buor, J., Menachof, D., & Talas, R. (2019). Scenario Analysis and Disaster Preparedness for Port and Maritime Logistics Risk Management. *Accident; Analysis and Prevention*, *123*, 433–447. doi:10.1016/j.aap.2016.07.013 PMID:27491716

Peng, W., & Berry, E. (2019). The Concept of Food Security. In Encyclopedia of Food Security and Sustainability (Vol. 2, pp. 1-7). Elsevier.

Perez-Moreno, S., Rodriguez, B., & Luque, M. (2016). Assessing Global Competitiveness under Multi-Criteria Perspective. *Economic Modelling*, *53*, 398–408. doi:10.1016/j.econmod.2015.10.030

Prosekov, A., & Ivanova, S. (2018). Food Security: The Challenge of the Present. *Geoforum*, *91*, 73–77. doi:10.1016/j.geoforum.2018.02.030

Sahnoun, H., Serbaji, M., Karray, B., & Medhioub, K. (2012). GIS and Multi-Criteria Analysis to Select Potential Sites of Agro-Industrial Complex. *Environmental Earth Sciences*, *66*(8), 2477–2489. doi:10.100712665-011-1471-4

Soltan, M., Elsamadony, M., & Tawfik, A. (2017). Biological Hydrogen Promotion via Integrated Fermentation of Complex Agro-Industrial Wastes. *Applied Energy*, *185*, 929–938. doi:10.1016/j.apenergy.2016.10.002

KEY TERMS AND DEFINITIONS

Agricultural Engineering: An area of industry engaged in the production and maintenance of equipment designed to work in agriculture.

Agro-Industrial Complex: A combination of several sectors of the economy aimed at the production and processing of agricultural raw materials and obtaining products from it, brought to the end consumer.

Competitiveness: An ability of a particular object or entity to surpass competitors in a given environment.

Entropy: A measure of the chaotic, disordered nature of a system.

Eurasian Economic Union: The union of Armenia, Belarus, Kazakhstan, Kyrgyzstan, and Russia established in 2014 for the effective promotion of free movement of goods, services, capital, and labor between participating countries.

Fishburne's Rule: An optimal distribution of the weights of the indicators from the point of view of informational entropy.

Food Security: A situation in which all people have physical and economic access to sufficient, quantitatively safe food to lead an active and healthy life at all times.

Logistics Costs: Relate to the charges for various transportation methods, including train travel, trucks, air travel, and ocean transport. Additional logistics costs include fuel, warehousing space, packaging, security, materials handling, tariffs, and duties.

Principle of Maximum Entropy: A probability distribution which best represents the current state of knowledge is the one with largest entropy, in the context of precisely stated prior data (such as a proposition that expresses testable information).

World Trade Organization: International organization established on 1 January 1995 to liberalize international trade and regulate the trade and political relations of member.

Chapter 7
Food Security Through Rational Land Management:
Innovative Aspects and Practices

Yulia Elfimova
Stavropol State Agrarian University, Russia

Anna Ivolga
Stavropol State Agrarian University, Russia

Ivan Ryazantsev
Stavropol State Agrarian University, Russia

ABSTRACT

Food security is challenged by the growth of the world population, intensification of agricultural production, depletion of scarce agricultural and environmental resources, and the consequent introduction of innovations to farming processes, one of which is land management. In this chapter, the authors discuss the role of innovative aspects and practices of land management in the establishment of food security and the provision of rational, sound, and effective administration of scarce land resources. In the case of Russia and other countries, the authors justify the necessity of rational land management focused on the innovative way of development of agricultural production.

INTRODUCTION

At the current stage of development of agricultural sector, food security is challenged by the intensification of agricultural production (Erokhin, 2017a) and introduction of innovations to farming processes (Lambin & Meyfroidt, 2011; Gomez & Ricketts, 2013), one of which is land management (Rogatnev, 2001). Recovery of agricultural sector is impossible without transition to innovative development which is the main factor of increase of effectiveness in the market economy environment (Smith, 2013; Smith

DOI: 10.4018/978-1-7998-1042-1.ch007

et al., 2010). In any country, the main issues of policy are related to the improvement of management system of social and economic development (Tilman, Balzer, Hill, & Befort, 2011).

According to Van der Molen (2017), growth of agricultural production and productivity are the measures to provide food security for the population of the world in the next decades. Land management and land administration should contribute to efficient land use and security of tenure. However, much agricultural land is not well managed and unrecorded, obstructing realizing the potential for growth (McConnell & Vina, 2018). About 38% of the Earth's land area is being used in agricultural production, with about 31% of the remaining land being under forest cover and the other half being less suitable for agricultural production due to edaphic, topographic and/or climatic factors (Keenan et al., 2015).

Land-use change is arguably the most significant driver of environmental change as it leads to many of the main areas of concern: loss of biodiversity, greenhouse gas emissions, soil degradation, and alteration of hydrological cycles. Land-use change is occurring worldwide due to human development dynamics (McConnell & Vina, 2018). It ranges from whole-scale changes in land cover to changes in the intensity of cropping on a given site, as well as changes in the type of cropping, or from crop production to conservation (Silva et al., 2017). The nature of changes in the farming technologies and practices employed can differ substantially in their effects on carbon storage, biodiversity, and hydrology (Martinelli, Batistella, Silva, & Moran, 2017). Recognizing that the issues of food (in)security are of local relevance, driven by both local, regional and global forces, that changes in land use are local in character but some of the driving forces are regional or global in nature, that food systems are influenced by land use types and changes thereof and that some actions taken to ensure/improve food security influence land use and changes thereof (Molotoks, Kuhnert, Dawson, & Smith, 2017; Chan et al., 2017; Nolasco, Soler, Freitas, Lahsen, & Ometto, 2017; Yawson et al., 2017).

Both state and society pose systematic, sensible, and purposeful influence on land relations. Land management includes the entire majority of innovative relations, i.e. social, economic, legal, and environmental, among others (Creamer et al., 2010). In order to provide sound and efficient usage of scarce land resources to improve food security in various parts of the world, this influence should be based on the cognition of objective laws (Rounsevell et al., 2010). Rational use of agricultural land is the most efficient way to increase the volume of agricultural output, intensify production with minimum harmful effects to the environment, and in such a way increase food security of population and sustainability of agricultural development (Erokhin, 2017b, 2017c; Gao, Ivolga, & Erokhin, 2018). There should be taken into account certain natural, economic, social, and political conditions in accordance with objective patterns of society and nature interaction (World Bank, 2008).

Proper and sound land management includes the following aspects:

- natural aspects that relate to the study of land operation as a component of ecosystem and environment for plants and living organisms
- social and economic aspects that reflect the effect of social processes, state policy, and social relations on land use. They all create economic aspect of using land as a resource
- technological aspects which are connected with studying technical influence on land, technology of land use, relation of sustainable land management, and technological advances
- legal aspects which are related to the study of the value of government legal affairs in organization and implementation of rational land use and protection

In regard of the above, land management is described as a system with a complex structure and is studied by various sciences of innovative technologies including the following aspects:

- political aspects that guarantee implementation of social, political, economic, and environmental tasks of a country in the spheres of food security and sustainable land use
- administrative and managerial aspects which are related to the development of a system of state and municipal bodies that manage land resources, determination of the scope of their responsibilities, and organization and coordination of their work
- legal aspects that provide rational land use and protection of agricultural land based on legal regulations stipulated in legislative acts
- scientific aspects which are related to the development of science-based guidelines in the spheres of food security and land administration taking into account technological advances
- economic aspects which determine conditions of efficient land use
- implementation aspects related to the development and implementation of economic, social, and other incentives and activities on food security, sustainable land use, and protection of agricultural land

Balanced combination of objective and subjective factors is a result of sustainability in land management. The integrated approach to land management applying innovations deals with interests of many parties of land relations. This requires an application of the systematic approach to land use and protection management and consider possible environmental and economic consequences and innovations by the control bodies (Green, Cornell, Scharlemann, & Balmford, 2005).

Thus, land management is based on general economic, ecological, and social laws intrinsic to any socio-economic formation as well as on laws of nature taking into account modern innovative technologies. Implementation of social, economic, and ecological aspects is the inner side of the content of land management. The outer side of the content of land management is a combination of economic, technical, legal, and innovative activities of the government in the spheres of establishing food security and improving land management practices (Agarkov, 2001).

Land management is implemented by legislative and executive bodies, which regulate land relations and determine the entire strategy of land use, norm setting, and law enforcement. Content of work of executive and administrative bodies in the spheres of food security and land management includes forecasting and planning of land use, setting norms and rules of land use, land distribution and redistribution, operative-managing, and control and oversight activities in land use and protection.

Land management is a complex of responsibilities of the management system that is aimed at rational use of land resources. System of land management has been changed throughout the history according to the economic framework, goals, and criteria of the social order. The system of land management should promote the interests of the country and the society, interests of the region, individual interests of the members of society, and implementation of innovation projects. In Russia, establishment of a system of land management is the main goal of land reform that determines the ways of further economic and political development of the country. In Russia, the economic conditions created innovative organization system that is characterized by abrupt transition from planned to market economy, differentiation of functions and entities of state and private regulation, development of democratization of social relations, integration into world economic, market, and information process, and integration of policy of public, social, and economic processes (All-Russian Institute of Agrarian Problems and Informatics named after A. Nikonov, 2003).

Based on the peculiarities of land use, the general system of land management for the purpose of improving of food security should include the following subsystems: legal, administrative-and-managerial, economic, social, and environmental. The backbones of the innovative land management system are the object, subject, purpose, objectives, and functions. Both the object and the subject have been created for years and are closely connected to each other. For example, management of land covered by high-rise blocks is markedly different from the management of land with low-rise buildings.

BACKGROUND

In Russia, the development of the aims, objectives, and functions of land management for the purpose of increasing agricultural output and establishing food security has been initiated rather recently (Elfimova, 2009). The object of management is agricultural land on various levels of property, including federal, regional, and municipal (administrative districts, cities, and other municipal entities), as well as land parcels of certain parties of land relations that differ by the type of use and legal status. The subject of land management is control over the use of agricultural land that provides for the variety of needs of the residents (Khlystun, 2005). Variety of needs leads to the variety of ways of land which include:

- Territorial organization of use of agricultural land, individual plots, or territories (land management, planning, zoning, etc.)
- Engineering support of land use (utilities system)
- Establishment of legal status of agricultural land (ownership, usage, tenancy, limitations, encumbrances)
- Determination of ways and types of land use (permitted use)
- Implementation of efficient environmental and economic technologies of land use
- Analysis of natural and economic condition of agricultural land
- Organization of other activities that have an impact on status and state of agricultural land

There is also a purpose and objectives that influence the development and performance of management system besides an object and a subject. An object, a subject, a purpose, and objectives of management are closely related (Elfimova, 2006a). Thus, the purpose and objectives are defined according to the conditions of the object and subject. In the process of implementation, the defined purpose and objectives create the object and the subject of management.

In the prospect of food security, the main purpose of land management is meeting the society's needs in the sufficient amount of food and agricultural products through the rational use of land resources (Erokhin, Ivolga, & Heijman, 2014; Lambin, 2012). The purpose reflects future state of land resources and process of their use. Self-management of land happens at direct influence of the society and certain individuals on land. Each of this influence has a definite purpose related to consumption of specific characteristics of land. The society is not able to control the aims of all parties of land relations because of their mass nature and variability, therefore, the management of these processes means setting general rules and limits in land use. Such limits are placed according to the existing legal rules of use regulating land relations and land-use systems. These statements are common to any form and regime land use and are extended to the entire system of land use in a country.

The purpose can have some more pronounced emphasis during a specific period: social, environmental, or various combinations of the two. In Russia, during the Soviet period, there was the proclamation of meeting social needs (Ayatskov, 2002). In the market economy, there is an emphasis on economic aspect, i.e. achievement of maximum economic benefits that usually means maximizing the income to the budget and cost reduction neglecting social and environmental components. Consequently, the main purpose of innovative land management is the creation and performance of land relations and land-use system that makes possible maximizing fiscal revenues to the federal, regional, and local budgets and providing high level of economic and environmental conditions for living, development of entrepreneurial, social, and other activities, creation of the conditions for preservation and restoration of agricultural land, and, ultimately, establishment of stable food security of a nation.

MAIN FOCUS OF THE CHAPTER

Food Security through Land Management Methods

The subjects are divided into the ones that operate at the governmental, local, and domestic levels ranging from the state as a subject of land relations to certain legal entities or citizens (Elfimova, 2006b). The main features, advantages, and disadvantages of land management modeling with the implementation of innovations are presented in Table 1.

Land management is supported by the development of the regulations, organizational structures, and procedures that enable to detect, collect, and update information on land plots. Practical solution of this issue implies the creation of a system of organization of subsystems of recognition and registration of the rights that create procedural and institutional space, where landowners and users and current and potential subjects of rights are engaged. Government authorities regulate land use relations by means of these systems, gain the required information on trends in behavior of land user, development of real estate

Table 1. Characteristics of methods applied in land management modeling

Method	Advantage	Disadvantage
Analytical method	No expenses for analysis of object's conditions and development of management model	Discrepancy of the objective and condition of the object of management.
Delphi method	Minimal expenses for analysis of object's conditions and development of management model	Inequality of the objects compared; impossibility of emphasizing some indicator from the whole scope.
Correlation method. Mathematical Modelling in Economics	Accuracy	Only comparable dimension of indicators. Complexity of the calculation, analysis of a large number of factors. Accurate relation to the time of calculations
Cluster analysis	Use of both dimensional and dimensionless number in the same model	One-time separation of a cluster – introduction of additional indicator or change in its value
Neural networks		Long observation period (big number of training pairs)

Source: Authors' development

market. This allows them to take and adjust decisions on organizational matters. Among the systems that support land management is also a subsystem of state cadastral valuation of land parcels and other real estate objects as taking organizational decisions on land use requires calculation of its economic potential. Another significant structural arrangement within the system of land resources management is a subsystem that enables massive regulation of the relationships in real estate, i.e. influencing certain classes of real estate property or land, professional groups working on the real estate market. The main element of this subsystem is taxation of real estate (including land) that implements two main functions: revenue-generation and regulation of business and social activity. Thus, several subsystems can be distinguished within the system of land management in terms of their significance in the process of making and implementing the decisions in establishing food security by means of innovative practices in land management.

Land Management Subsystems

Subsystems of legal and economic support (land registry, registration of rights to land plots, and evaluation of land plots and areas) create the necessary methodological background and infrastructure for taking and implementing organizational decisions in the sphere of land management and food security:

- Subsystem of massive regulation (land taxation, zoning, and regulation of professional participants of land management) that influences the management bodies on a large number of subjects of rights to land and land users, as well as on the operation and development of land use practices in particular territory
- Subsystem of individual regulation designed for the influence on certain land plots or certain subjects of social and economic activity

Managerial decisions within land management, especially in direct management, are implemented with a use of directive, legal, and economic methods (Figure 1).

The major methods are:

- Administrative (directive) methods which relate to the development and implementation of organizational decisions, i.e. directives. This method is based on the implementation of managerial functions by the government that are laid down in the legislation. These are directly applicable acts: removal, allocation, zoning, activities on land study and their implementation
- Legal method is employed in the case of indirect land use when the legislation and regulations in the sphere of land use force the parties of land relations to make decisions beneficial for the state
- Economic method implies the creation of economic incentives and indicators that provide implementation of state policy in the spheres of food security and land use

When managing agricultural land for the purpose of improving food security, it is necessary to apply all methods of decision-making at each level. Development of purposes and criteria, methods of implementation of each function for the creation of economically efficient system of land management and food security requires high accuracy.

Figure 1. Main methods of implementation of managerial decisions in land management
Source: Authors' development

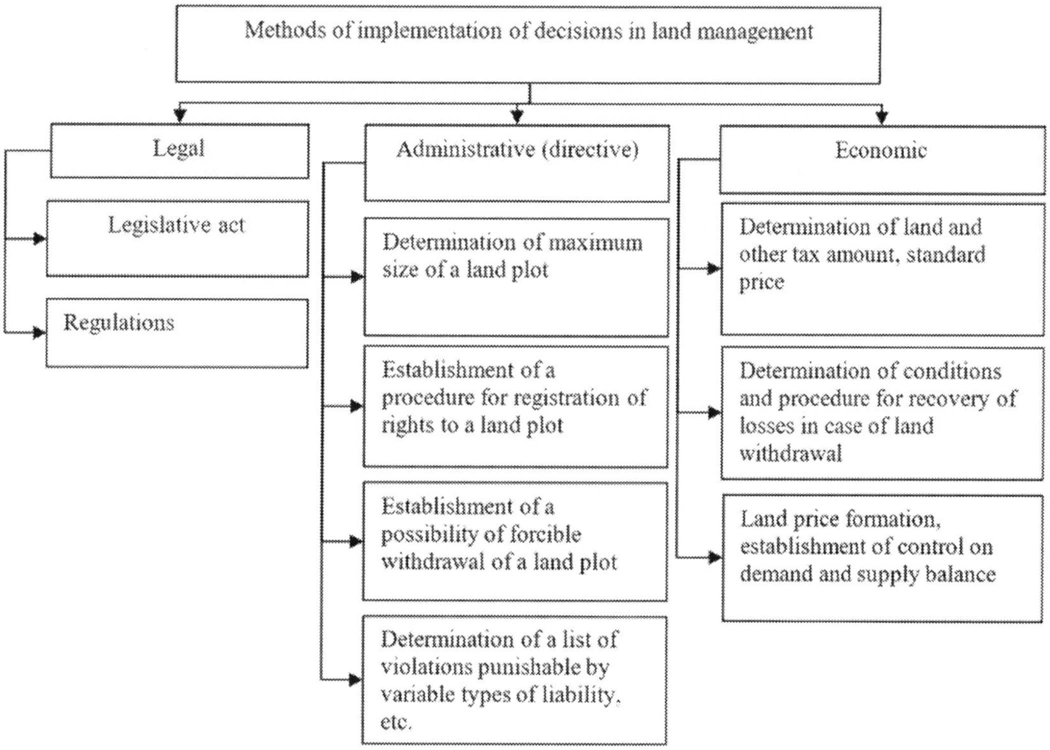

```
                    ┌─────────────────────────────────────────────────────┐
                    │  Methods of implementation of decisions in land     │
                    │  management                                         │
                    └─────────────────────────────────────────────────────┘
                           │                    │                    │
                           ▼                    ▼                    ▼
                    ┌────────────┐    ┌──────────────────────┐  ┌────────────┐
                    │   Legal    │    │ Administrative       │  │  Economic  │
                    │            │    │ (directive)          │  │            │
                    └────────────┘    └──────────────────────┘  └────────────┘
```

- Legal
  - Legislative act
  - Regulations
- Administrative (directive)
  - Determination of maximum size of a land plot
  - Establishment of a procedure for registration of rights to a land plot
  - Establishment of a possibility of forcible withdrawal of a land plot
  - Determination of a list of violations punishable by variable types of liability, etc.
- Economic
  - Determination of land and other tax amount, standard price
  - Determination of conditions and procedure for recovery of losses in case of land withdrawal
  - Land price formation, establishment of control on demand and supply balance

Principles of Land Management

The above-mentioned requirements and functions of land management provide the formation of principles that describe certain objective connection within the system. Taking into account these patterns allows not only assessing the system of management, but also foreseeing the ways for its improvement. These patterns can be divided into the formative and determinative ones.

The general (determinative) principles include the priority of state regulation of land management. The system of land management operates at three interdependent levels: federal, regional, and municipal (local) that implement similar land policy regulations and practices in an integrated and coordinated manner. Such approach provides the protection of property and tenure rights, land protection, efficient development of infrastructure, and safe operation of tax system, as well as creates the procedure of sale or leasing state and municipal agricultural land. According to Schulte et al. (2014), the relations on forming land plots are among the main troubles in terms of implementation of land legislation. At the same time, financial and time costs remain important in the implementation of land relations. This, in turn, impedes the acquisition and implementation of legitimate rights of owners, tenants, and users of land parcels, as well as creates considerable difficulties in protection of rights on agricultural land. Galinovskaya (2009) supposes that there has been an ongoing debate on the approaches to the regulation of land relations, ways of land legislation development, and its relationship to related areas of legislation since land plots have been put into civil circulation. The status of land as universal labor factor, means of production, and other social activity justifies the reasonability of this principle. State land management covers both unused lands and land parcels in use regardless the form of land rights. This follows from the territorial supremacy of state (an integral element of national sovereignty). In terms of economy, land being a source of satisfying wide range of human needs as an object of legal regulation is an object of economic activity and material basis of any production process (Brussaard, de Ruiter, & Brown, 2007; Haggblade, Hazell, & Reardon, 2010).

In Russia, the Concept of Sustainable Development of Rural Territories (Government of the Russian Federation, 2010) contains the definition of land as a territory of rural, urban, and municipal entities. Such territories have huge natural, demographic, economic, historical, and cultural potential that can provide sustainable diversified development, full employment, high level and quality of life when used fully, rationally and effectively.

According to Volkov (2001), land is a limited part of the Earth's surface with its natural and anthropogenic features and resources, characterized by the area, length, and location that is an object of some activity of research. The main elements of state land management are:

- Regional land management, development, and maintenance of state land registry, registration of rights to land, and organization of land turnover, cadastral appraisal of land, information support and personnel training, and state control.
- Differentiated approach to the organization of land of different categories and regions. In this case, legal approach to land management should be developed taking into account its economic, natural, and social features.
- Principle of sound land use. State land policy should be aimed at the development of a set of factors that provide improvement of efficiency of agricultural production and achieving certain level of food security. Thus, there could not be sustainable agricultural land use without significant capital investments and activities on land reclamation, improvement of farming practices, and plant protection. Being a holistic environmental and economic term, sound land use brings together the achievement of desirable effect of economic use of land at sufficient cost and land conservation and improvement during use. That is why the organization of land protection, when land is regarded as the most important natural resource, creation of legal, economic, organizational, technological, and other conditions for soil fertility, recovery, and improvement based on land management, is the main goal of food security policy and sustainable land use.
- Organizational coherence of land use and management. This principle means that managerial system should provide improvement of a system of objects and subjects of land relations, fulfillment of rights and obligations on use of land resources, organization of sustainable land management for smooth economic performance. In addition, the structure of land management and the number of administrative staff at all levels should be determined taking into account the quantity of work and content of managerial decisions in land management.
- Systematic improvement of functions and methods of land management. It is supported by the implementation of scientific-technical advances within the system of land management. For instance, land cadaster was developed from simple describing of church land to state land registry that includes registration of land tenure, quantitative and qualitative land inventory, soil evaluation, and cadastral valuation.

- Economic efficient combination of land management on state, regional, and municipal level. It is ensured by the vertical and horizontal separation of powers between the federal authorities, administrative regions, and municipal entities.
- Separation of powers in land management between the executive and representative bodies of the same administrative and territorial level. Legislative base should determine the responsibilities of executive and legislative authorities because the lack of distribution of functions leads to the deterioration of management and adversely affects the efficiency of the entire system of land management.
- Distribution of functions between various authorities at state, regional, and municipal levels. Implementation of this principle requires distribution of functions between various bodies involved in land management laid down in the legislation.
- Legal support of land management. Government can control land management under the conditions of building market economy only if there is a reliable legal framework. In addition, it is necessary to ensure compliance of laws and acts adopted by administrative regions and local authorities with federal legislation as well as compliance of land legislation at the federal level and the level of administrative regions with legislative framework of other branches related to land management and food security.
- Principle of sound organizational and economic balance between centralization and decentralization. This principle promotes the separation of strategic and tactical goals and current tasks in management.

The most frequent principles of land management are:

- The principle of financial and staff support of land management that implies the availability of special bodies for land management with optimal number of staff and funding, especially at municipal level. These bodies are supposed to perform most of the work on land management. This will ensure proper performance of certain types of activities on specific areas that are of particular interest to the government.
- The principle of controllability implies sound balance of managing and managed entities, their number of staff, rational level of load of the entities involved into land management, and ensuring implementation of the decisions related to land management.
- The principle of conformity of subject and object means that management structure should be created based on the specificities of land resources (of a country, region, city, etc.) as an object of management.
- The principle of variability means the ability of elements of the system (bodies involved into land management) to accommodate changes of external and internal economic, social, legal, and other conditions based on the requirements of flexibility and adaptability.
- The principle of subsidiarity implies the necessity of creation of a vertical structure of land management taking into account the compulsoriness of realization of the decisions and prescriptions of the superior authorities by the lower entities.
- The principle of economic efficiency implies that the planned result should be achieved at the lowest possible costs on land management, efficient use of labor, physical, and financial resources.

Types and content of functions (methods) of land management are determined by the legislation according to social, economic, and scientific requirements imposed on organization of sustainable land use and protection. Currently, territorial land use management is the main type of land management that provides creation of spatial conditions for efficient activity in all business spheres, registration of the rights of individuals and entities to land, accuracy and incontrovertibility of location of the boundaries of land parcels of municipal, regional, and federal administrative and territorial units, special land funds, and areas special legal regime for land use. (Schulte et al., 2014). Among the main methods of state land management are land surveying, state land cadaster, land monitoring, and state control of land use and protection.

Land Management Directions

Land management is a system of activities aimed at the regulation of land relations, organization of land use, land resources accounting and appraisal, and development of internal economic plans (Krassov, 2012). The main courses of land management include:

- development of federal and regional programs of land resources use and protection, land use schemes taking into account economic, town-planning, environmental, and other characteristics of a particular territory
- land-surveying with demarcation (restoration) of boundaries of administrative and territorial units, boundaries of land parcels belonging to the subjects of land relations according to a unified system and their technical acceptance
- development of land management projects and arrangement of existing landholdings removing inconveniences in land parcels location
- establishment and delimitation of land plots boundaries, drafting boundary plans, and preparation of documents to certificate rights to land plots
- drafting various projects including those on recovery of deteriorated land, soil protection against wind and water erosion, mudslides, landslides, under flooding, waterlogging, drying out, soil compaction, soil salinization, contamination, on improvement of agricultural land, land reclamation, conservation, and increase of soil fertility
- Justification of establishment and demarcation of the boundaries of protected areas
- Determination and marking on city and village limits
- Topographic-and geodetic surveying, soil, agrochemical, geobotanical, and historical and cultural analysis and studies
- Drafting documentation for land management that relates to the resources valuation, land use, and protection, preservation and development of traditional types of activities of small-numbered peoples and ethnic communities
- Preparation of maps and atlases on condition and use of land resources
- Land inventory and identification of unused or unsustainably used land

Land management is implemented by state design organizations, other entities and individuals, who obtain license to carry out these activities from the funds of the federal budget, regional budgets, local budgets and customers (Levushkina & Elfimova, 2012). The main courses of action are:

Food Security Through Rational Land Management

- Scientific support and forecasting of land reforms and redistribution
- Justification and implementation of unified state policy in planning, redistribution, regional use, and protection of all types of land despite the forms of use and property
- Provision of targeted use and conservation of valuable land in agricultural production
- Development and placement of environmentally and economically justified, compact landholdings of an optimal size
- Creation of territorial conditions for sound performance of agricultural and non-agricultural production
- Development of a set of measures aimed at the improvement of agricultural land, increase of soil fertility, support of sustainable landscapes, and land protection
- Land demarcation with establishment (re-establishment) and marking of administrative and territorial on the ground drafting all documentation related to land by the unified state system

Land management is implemented in all territories despite their designation purpose and form of property using scientifically grounded, broadly discussed and approved in the established procedure land surveying documentation, forecasts, programs, schemes, projects, survey data. Regime of use and designation purpose, organization of territory, package of measures aimed at support of sustainable landscape, and land protection are obligatory for land users and owners as well as for government authorities. Introduction of amendments to land surveying documentation is possible only with the authorization of the authority that approved it. Land management is implemented by the decision-making bodies of the authorities and administration, local government agencies, at the initiative of the bodies included in the State Committee on Land Resources of the Russian Federation upon a petition of landowners and users as well as the persons and corporate entities making a claim for receiving a land share. Land administration is regulated by the Constitution of the Russian Federation, legislative and other legal acts of administrative regions of Russia as well as the regulations that determine content, procedure for the elaboration, approval, and implementation of land surveying documentation. Interests of all parties of land management are considered.

State land monitoring is a system of observations of the state of land resources. It is a part of environmental monitoring and serves as a basic connection between all types of monitoring of natural resources (Rogatnev, 2003). Among the aims of land monitoring is the assessment of the state of land resources for early detection of changes, assessment of such changes, areas and development of recommendations on prevention and handling of consequences of adverse processes and information support of state land use and management, and control over land use and protection. Land monitoring is financed from the federal budget, the budgets of administrative regions, and special-purpose funds. Among the objects of land monitoring are all lands in Russia regardless the form of property, designation purpose, and pattern of land use.

The structure of monitoring activity includes seven subsystems, each of which corresponds with certain category of land and is divided into federal and regional monitoring depending on territorial coverage (Table 2).

Land monitoring is implemented by the authorities of the State Committee on Land Resources and the Ministry of Natural Resources of the Russian Federation through the allocations that come to the local budgets as land tax and other payments for land. Land monitoring includes monitoring state of land plots, fields, processes related to the changes in soil fertility (wind and water erosion, loss of humus

Table 2. Levels of settlement land monitoring and observable areas

Type of a city	Level of monitoring		
	regional	local	local detailed
Extremely large-scale regions	Area within a city boundary	Area within administrative and territorial units	Area within the boundaries of landholdings
Largest, large, and big (with administrative-territorial division)		Area within a city boundary; Area within administrative and territorial units	Area within the boundaries of landholdings
Medium, small, and urban-type settlements (without administrative-territorial division)		Area within a city boundary	Area within the boundaries of landholdings

Source: Authors' development

layer, degradation of structure), land deterioration with pesticides, toxic matters, state of riverbanks, seashores, reservoir shorelines, settlement lands, disposal sites, and parking spaces.

SOLUTIONS AND RECOMMENDATIONS

The improvement of land management in the purpose of establishment of food security should include the implementation of innovations and technologies, such as land cadaster. All observations of state of land within land monitoring by means of surveying from space crafts, high-altitude aircraft, small aircraft and ground surveys are divided by the terms and frequency into basic (at the start of monitoring), periodic (in a year and more), and operational (recording current changes). Land cadaster operates as a formal unified system of registration and drafting documentation of the rights to land parcels and connected real estate objects, and accounting and evaluation of land parcels (Elfimova & Miroshnichenko, 2011).

Cadastral and land system should include the following elements: registration of landholdings, land tenure and ownership of land plots, quantitative and qualitative recording of land resources, and soil evaluation and economic assessment of land. The potential improvements of land cadaster in view of establishing food security should include:

- Introduction of cost estimates for local budget generation
- Accomplishment of measures of economic incentives for enterprises, where environmental social and cultural activities have priority for a district or a city
- Financial incentives of reclamation and rehabilitation of deteriorated agricultural land
- Addressing environmental issues by adopting differentiated scale according to a degree of harmfulness of an enterprise, size of sanitary protection zone, costs of required activities aimed at organization of cleaner production
- Accommodation of various facilities according to optimal land-use planning
- Recovery of additional costs to landowners and land users related to less advantageous location of their land

State land cadaster should be aimed at proclaiming rights of landowners and users to land, state management of land resources, efficiency and sustainability of land rotation, and settlement of land disputes.

FUTURE RESEARCH DIRECTIONS

In the sphere of introduction of innovation practices to land management with an aim to improve food security, future research should be directed at finding the optimal ways of resolving disputes and finding the most effective ways to support rational and sustainable land management. Land disputes are settled by the competent government authorities within state land management. They are one of the ways of protection of the rights of landowners, land users, and tenants (Aleksandrov, 2001). Settlement of land disputes is the realization of legal entitlements of the state in the forms of coercive measures against a defendant into the recognition of claimant's rights to a land parcel. Legal support of land resources management should be studied in regards of the implementation of the objectives and functions which require a thorough revision of legislative framework of land management. Improvement of the system of economic relations requires improvement of the system of legal support of land resources management. The result should include the development of a new legal framework.

CONCLUSION

In addition to a possible increase in agricultural output and improvement of food security, the efficiency of land management is related to social production, therefore, land management should be a public issue and a crucial tool of any agrarian reform. Thus, the following statements can be concluded.

First, land management is the main tool of landholding (land tenure) organization in any type of business. That is, no production can be started without establishment of land management, drafting a project, its discussion, negotiation, approval, allocation of land, issuance of a document confirming rights to land ownership (land tenure). Consequently, land management is the main condition of starting any enterprise.

Second, land management leads to mutual adjustment of production and land or, in other words, territorial organization of production occurs, when specialization of a business is justified taking into account soil fertility and land location and the territory is developed. Therefore, efficiency of land management relates to the production (economy) of agricultural enterprises. This statement is supported by the fact that issues of protection and improvement of soil fertility and rational spatial planning are solved within land management. This is the main interest of state in view of establishing food security.

Third, land management creates optimal (for the given level of productive forces and production relations development) organizational and site conditions of land ownership and land tenure that is especially important at the stage of new land property formation as many joint-stock companies, associations, cooperatives, and peasant farms are continually starting and conducting their operations.

Fourth, government authorities delegate their responsibilities within land management to commercial entities. This can lead to loss of state control over implementation of policy within land use and protection. Therefore, in order to ensure state regulation of land management, improvement of quality and efficiency of land management all main activities related to protection and sound use of land resources, creation of a base for state land cadaster should be implemented by the specialized state design agencies. For other land management activities, customers may invite other legal entities and individuals (private surveyors) licensed for land survey works on a competitive basis.

ACKNOWLEDGMENT

The authors thank Prof. Vladimir Trukhachev, Academician of the Russian Academy of Sciences, Honored Worker of Science of the Russian Federation, Rector of Stavropol State Agrarian University for the continuous assistance and support provided in scientific studies and professional development.

REFERENCES

Agarkov, N. (2001). To Increase Efficiency of Production. *Economics of Agriculture in Russia, 3*, 9–10.

Aleksandrov, A. (2001). Management of Agricultural Production Requires Improvement of Legislation. *Economist, 3*, 83–89.

All-Russian Institute of Agrarian Problems and Informatics named after A. Nikonov. (2003). *Agricultural Policy and Russia's Accession to the WTO*. Moscow: All-Russian Institute of Agrarian Problems and Informatics.

Ayatskov, D. (2002). *Land Ownership in Russia: History and Present Time*. Moscow: Russian Politic Encyclopedia.

Brussaard, L., de Ruiter, P. C., & Brown, G. G. (2007). Soil Biodiversity for Agricultural Sustainability. *Agriculture, Ecosystems & Environment, 121*(3), 233–244. doi:10.1016/j.agee.2006.12.013

Chan, C., Sipes, B., Ayman, A., Zhang, X., LaPorte, P., Fernandes, F., ... Roul, P. (2017). Efficiency of Conservation Agriculture Production Systems for Smallholders in Rain-Fed Uplands of India: A Transformative Approach to Food Security. *Land (Basel), 6*(3), 58. doi:10.3390/land6030058

Creamer, R., Brennan, F. P., Fenton, O., Healy, M. G., Lalor, S., Lanigan, G. J., ... Griffiths, B. S. (2010). Implications of the Proposed Soil Framework Directive on Agricultural Systems in Atlantic Europe – A Review. *Soil Use and Management, 26*(3), 198–211. doi:10.1111/j.1475-2743.2010.00288.x

Elfimova, Y. (2006a). Land and Will. Establishment and Development of Farming in Russia: Issues of Land Use. *Russian Entrepreneurship, 9*, 172–174.

Elfimova, Y. (2006b). Methodical Fundamentals of Economic Valuation of Agricultural Land. *Achievements of Modern Life Sciences, 10*, 90–92.

Elfimova, Y. (2009). Organization of Economic Assessment of Agricultural Land. *Vestnik Universiteta, 28*, 174–175.

Elfimova, Y., & Miroshnichenko, R. (2011). Innovative Aspect in Land Management. *Social Policies and Sociology, 70*(4), 247–264.

Erokhin, V. (Ed.). (2017a). *Establishing Food Security and Alternatives to International Trade in Emerging Economies*. Hershey, PA: IGI Global.

Erokhin, V. (2017b). Factors Influencing Food Markets in Developing Countries: An Approach to Assess Sustainability of the Food Supply in Russia. *Sustainability, 9*(8), 1313. doi:10.3390u9081313

Erokhin, V. (2017c). Self-Sufficiency versus Security: How Trade Protectionism Challenges the Sustainability of the Food Supply in Russia. *Sustainability, 9*(11), 1939. doi:10.3390u9111939

Erokhin, V., Ivolga, A., & Heijman, W. (2014). Trade Liberalization and State Support of Agriculture: Effects for Developing Countries. *Agricultural Economics – Czech, 60*(11), 524-537.

Galinovskaya, E. (2009). Land Legislation: Features of Formation and Development. *Journal of Russian Law, 155*(11), 14–25.

Gao, T., Ivolga, A., & Erokhin, V. (2018). Sustainable Rural Development in Northern China: Caught in a Vice between Poverty, Urban Attractions, and Migration. *Sustainability, 10*(5), 1467. doi:10.3390u10051467

Gomez, M. I., & Ricketts, K. D. (2013). Food Value Chain Transformations in Developing Countries: Selected Hypotheses on Nutritional Implications. *Food Policy, 42*, 139–150. doi:10.1016/j.foodpol.2013.06.010

Government of the Russian Federation. (2010). *Decree #2136-r from November 30, 2010, On Approval of the Concept of Sustainable Development of Rural Areas in the Russian Federation till 2020*. Retrieved from https://rg.ru/2010/12/14/sx-territorii-site-dok.html

Green, R. E., Cornell, S. J., Scharlemann, J. P. W., & Balmford, A. (2005). Farming and the Fate of Wild Nature. *Science, 307*(5709), 550–555. doi:10.1126cience.1106049 PMID:15618485

Haggblade, S., Hazell, P., & Reardon, T. (2010). The Rural Non-Farm Economy: Prospects for Growth and Poverty Reduction. *World Development, 38*(10), 1429–1441. doi:10.1016/j.worlddev.2009.06.008

Keenan, R. J., Reams, G. A., Achard, F., de Freitas, J. V., Grainger, A., & Lindquist, E. (2015). Dynamics of Global Forest Area: Results from the FAO Global Forest Resources Assessment. *Forest Ecology and Management, 352*, 9–20. doi:10.1016/j.foreco.2015.06.014

Khlystun, V. (2005). Controversial Issues of Land Relation Development in Russia. In S. Volkov & A. Varlamov (Eds.), *Science and Education of Land Management in Russia at the Beginning of the Third Millennium* (pp. 82–91). Moscow: State University of Land Use Planning.

Krassov, O. (2012). Permitted Use and the Specific Purpose of the Land. *Environmental Law (Northwestern School of Law), 2*, 16–20.

Lambin, E. F. (2012). Global Land Availability: Malthus versus Ricardo. *Global Food Security, 1*(2), 83–87. doi:10.1016/j.gfs.2012.11.002

Lambin, E. F., & Meyfroidt, P. (2011). Global Land Use Change, Economic Globalization, and the Looming Land Scarcity. *Proceedings of the National Academy of Sciences of the United States of America, 108*(9), 3465–3472. doi:10.1073/pnas.1100480108 PMID:21321211

Levushkina, S., & Elfimova, Y. (2012). Land Resources as One of the Main Factors of a Rural Entrepreneurship. *Polythematic Online Scientific Journal of Kuban State Agrarian University, 83*(9), 606–616.

Martinelli, L. A., Batistella, M., Silva, R. F. B. D., & Moran, E. (2017). Soy Expansion and Socioeconomic Development in Municipalities of Brazil. *Land (Basel), 6*(3), 62. doi:10.3390/land6030062

McConnell, W. J., & Vina, A. (2018). Interactions between Food Security and Land Use in the Context of Global Change. *Land (Basel)*, *7*(2), 53. doi:10.3390/land7020053

Molotoks, A., Kuhnert, M., Dawson, T. P., & Smith, P. (2017). Global Hotspots of Conflict Risk between Food Security and Biodiversity Conservation. *Land (Basel)*, *6*(4), 67. doi:10.3390/land6040067

Nolasco, C. L., Soler, L. S., Freitas, M. W., Lahsen, M., & Ometto, J. P. (2017). Scenarios of Vegetable Demand vs. Production in Brazil: The Links between Nutritional Security and Small Farming. *Land (Basel)*, *6*(3), 49. doi:10.3390/land6030049

Rogatnev, Y. (2001). *Fundamentals of Planning in Land Management*. Omsk: Omsk State Agrarian University.

Rogatnev, Y. (2003). *Theoretical and Methodological Fundamentals of Land Management under Conditions of Forming of Market Relations in Western Siberia*. Omsk: Omsk State Agrarian University.

Rounsevell, M., Pedroli, B., Erb, K.-H., Gramberger, M., Busck, A. G., Haberl, H., ... Wolfslehner, B. (2012). Challenges for Land System Science. *Land Use Policy*, *29*(4), 899–910. doi:10.1016/j.landusepol.2012.01.007

Schulte, R. P. O., Creamer, R. E., Donnellan, T., Farrelly, N., Fealy, R., O'Donoghue, C., & O'hUallachain, D. (2014). Functional Land Management: A Framework for Managing Soil-Based Ecosystem Services for the Sustainable Intensification of Agriculture. *Environmental Science & Policy*, *38*, 45–58. doi:10.1016/j.envsci.2013.10.002

Silva, R. F. B. D., Batistella, M., Dou, Y., Moran, E., Torres, S. M., & Liu, J. (2017). The Sino-Brazilian Telecoupled Soybean System and Cascading Effects for the Exporting Country. *Land (Basel)*, *6*(5), 53. doi:10.3390/land6030053

Smith, P. (2013). Delivering Food Security without Increasing Pressure on Land. *Global Food Security*, *2*(1), 18–23. doi:10.1016/j.gfs.2012.11.008

Smith, P., Gregory, P. J., van Vuuren, D., Obersteiner, M., Havlik, P., Rounsevell, M., ... Bellarby, J. (2010). Competition for Land. *Philosophical Transactions of the Royal Society of London. Series B, Biological Sciences*, *365*(1554), 2941–2957. doi:10.1098/rstb.2010.0127 PMID:20713395

Tilman, D., Balzer, C., Hill, J., & Befort, B. L. (2011). Global Food Demand and the Sustainable Intensification of Agriculture. *Proceedings of the National Academy of Sciences of the United States of America*, *108*(50), 20260–20264. doi:10.1073/pnas.1116437108 PMID:22106295

Van der Molen, P. (2017). Food Security, Land Use and Land Surveyors. *Survey Review*, *353*(49), 147–152.

Volkov, S. (2001). *Land Management. Land Use Planning. Intra-Farm Land Tenure*. Moscow: Kolos.

World Bank. (2008). *Sustainable Land Management Sourcebook*. Washington, DC: World Bank.

Yawson, D. O., Mulholland, B. J., Ball, T., Adu, M. O., Mohan, S., & White, P. J. (2017). Effect of Climate and Agricultural Land Use Changes on UK Feed Barley Production and Food Security to the 2050s. *Land (Basel)*, *6*(4), 74. doi:10.3390/land6040074

ADDITIONAL READING

Anseeuw, W., Wily, L. A., Cotula, L., & Taylor, M. (2012). *Land Rights and the Rush for Land, Findings of the Global Commercial Pressures on Land Research Project*. Rome: International Land Coalition.

Baldos, U., & Hertel, T. W. (2014). Global Food Security in 2050: The Role of Agricultural Productivity and Climate Change. *The Australian Journal of Agricultural and Resource Economics*, 58(4), 1–18. doi:10.1111/1467-8489.12048

Baltissen, G., & Betsema, G. (2016). *Linking Land Governance and Food Security in Africa. Outcomes from Uganda, Ghana, and Ethiopia*. Utrecht: Utrecht University.

Bindraban, P. S., Van der Velde, M., Ye, L., Van den Berg, M., Materechera, S., Kiba, D. I., ... Van Lynden, G. (2012). Assessing the Impact of Soil Degradation on Food Production. *Current Opinion in Environmental Sustainability*, 4(5), 478–488. doi:10.1016/j.cosust.2012.09.015

Deininger, K., Byerlee, D., Lindsay, J., Norton, A., Selod, H., & Stickler, M. (2011). *Rising Global Interest in Farmland: Can It Yield Sustainable and Equitable Benefits? Agriculture and Rural Development*. Washington, DC: World Bank. doi:10.1596/978-0-8213-8591-3

Devereux, S. (2007). The Impact of Droughts and Floods on Food Security and Policy Options to Alleviate Negative Effects. *Agricultural Economics*, 37, 47–58. doi:10.1111/j.1574-0862.2007.00234.x

Erokhin, V. (2019). Trade in Agricultural Products and Food Security Concerns on Emerging Markets: How to Balance Protection and Liberalization. In I. Management Association (Ed.), Urban Agriculture and Food Systems: Breakthroughs in Research and Practice (pp. 1-27). Hershey, PA: IGI Global.

Erokhin, V., Heijman, W., & Ivolga, A. (2014). Sustainable Rural Development in Russia through Diversification: The Case of the Stavropol Region. *Visegrad Journal on Bioeconomy and Sustainable Development*, 3(1), 20–25. doi:10.2478/vjbsd-2014-0004

Erokhin, V., Ivolga, A., & Lisova, O. (2016). Challenges to Sustainable Rural Development in Russia: Social Issues and Regional Divergences. *Applied Studies in Agribusiness and Commerce – APSTRACT*, 10(1), 45-52.

Fouilleux, E., Bricas, N., & Alpha, A. (2017). Feeding 9 Billion People: Global Food Security Debates and the Productionist Trap. *Journal of European Public Policy*, 24(11), 1658–1677. doi:10.1080/13501763.2017.1334084

Kaag, M., & Zoomers, A. (Eds.). (2014). *The Global Land Grab. Beyond the Hype*. London; New York, NY: ZED Books.

Kepe, T., & Tessaro, D. (2014). Trading-off: Rural Food Security and Land Rights in South Africa. *Land Use Policy*, 36, 267–274. doi:10.1016/j.landusepol.2013.08.013

Oyekale, A. S. (2016). Assessment of Sustainable Land Management and Food Security among Climatic Shocks' Exposed to African Farmers. *Journal of Developing Areas*, 50(1), 319–332. doi:10.1353/jda.2016.0017

Verburg, P. H., Mertz, O., Erb, K.-H., Haberl, H., & Wu, W. (2013). Land System Change and Food Security: Towards Multi-Scale Land System Solutions. *Current Opinion in Environmental Sustainability*, 5(5), 494–502. doi:10.1016/j.cosust.2013.07.003 PMID:24143158

KEY TERMS AND DEFINITIONS

Agriculture: A pool of establishments engaged in growing crops, raising animals, and harvesting fish and other animals.

Agricultural Land: A land devoted to agricultural production which includes arable land, permanent cropland, and permanent pastures.

Food Security: An availability and adequate access at all times to sufficient, safe, nutritious food to maintain a healthy and active life.

Land Cadaster: A comprehensive recording of agricultural land of a country which includes details of the ownership, tenure, precise location, dimensions, and the value of individual parcels of land.

Land Management: The use of land resources, including soils, water, animals and plants, for the production of goods to meet changing human needs, while simultaneously ensuring the long-term productive potential of these resources and the maintenance of their environmental functions.

Land Tenure: A relationship, whether legally or customarily defined, among people, as individuals or groups, with respect to land.

Land Use: An activity which involves the management and modification of natural environment into semi-natural habitats such as arable fields, pastures, and managed woods.

State Support: A set of protective measures provided by a state to a certain industry or sector.

Chapter 8
Food Waste Reduction Towards Food Sector Sustainability

Giovanni Lagioia
University of Bari Aldo Moro, Italy

Vera Amicarelli
University of Bari Aldo Moro, Italy

Teodoro Gallucci
University of Bari Aldo Moro, Italy

Christian Bux
University of Bari Aldo Moro, Italy

ABSTRACT

FAO estimates on average more than 1.3 billion tons of food loss and waste (FLW) along the whole food supply chain (equivalent to one-third of total food production) of which more than 670 million tons in developed countries and approximately 630 million tons in developing ones, showing wide differences between countries. In particular, EU data estimates an amount of more than 85 million tons of FLW, equal to approximately 20% of total food production. This research presents two main goals. First, to review the magnitude of FLW at a global and European level and its environmental, social and economic implications. Second, use Material Flow Analysis (MFA) to support and improve FLW management and its application in an Italian potato industry case study. According to the case study presented, MFA has demonstrated the advantages of tracking input and output to prevent FLW and how they provide economic, social, and environmental opportunities.

DOI: 10.4018/978-1-7998-1042-1.ch008

INTRODUCTION

Since each human being needs energy and chemical products to maintain his vital signs and produce cells, skin, and bones, food plays a fundamental role in human life becoming an essential commodity for human feeding (Nebbia, 1995). However, since a huge percentage of food is wasted daily, a critical question comes to mind: is food available for all people on Earth?

According to Food and Agriculture Organization of the United Nations [FAO] (2018), nowadays the world population amounts to approximately over 7.5 billion people, divided between rural (45%) and urban (55%) population. Moreover, a huge percentage is employed in agriculture, even with a sharp decrease between 1995 (more than 40%) and 2016 (under 27%).

Worldwide, the food production value exceeds $2.3 trillion with no homogeneous distribution (Table 1). More than 10% of the global population is undernourished because of severe food insecurity. Moreover, more than 22% of children under five suffer from stunting and more than 7% from wasting. Lastly, safely managed drinking water is used by approximately 70% of population. Thus, more than 30% people cannot access drinking water (FAO, 2018; FAO, International Fund for Agricultural Development [IFAD], United Nations International Children's Emergency Fund [UNICEF], World Food Programme [WFP], & World Health Organization [WHO], 2018).

Food insecurity presents different insecurity degrees. According to its meaning, the first indicator is uncertainty about obtaining food, followed by food quality and quantity reduction and meal skipping. In this phase of moderate food insecurity, people cannot afford a healthy diet because of insufficient money or resources. However, severe food insecurity means no food for one day or more (FAO, 2018).

These indicators focus the attention on food security and require several national and international policies to be adopted. According to the 1996 World Food Summit, food security represents a situation that exists when all people, at all times, have physical, social and economic access to sufficient, safe and nutritious food that meets their dietary needs and food preferences for an active and healthy life. In order to better understand this problem, some definitions should be taken into account. Hunger represents a physical discomfort caused by luck of food and can be measured at the individual level. Underweight means individual anthropometric variables and regards two standard deviations below the global reference values. Undernutrition concerns insufficient caloric intake according to international standards. Malnutrition is related to undernutrition, obesity, and micronutrient deficiencies (Barrett, 2010).

Table 1. Worldwide food insecurity overview

Phenomena	Percentage	Million people
Undernourished people	10	750
Food insecure people	10	750
Obese people	13	975
Children under five years affected by wasting	7	50
Children under five years stunted	22	150
Children overweight under five years	5	40

Source: Authors' development based on FAO (2018)

Trying to answer the abovementioned question (Is food available to all people on Earth?), some key factors should be considered. As stated by FAO, IFAD, UNICEF, WFP, & WHO (2018), food security is a multi-layer concept based on four key pillars of equal importance, strictly linked and affected by different variables: food availability, food access, food utilization, and food stability.

The first pillar regards food availability, considering both locally produced and imported food and existing food stock. However, food availability itself cannot ensure food security. For this reason, it is necessary to take into account food access regarding both physical and economic access to food (e.g. countries purchasing power, level of income, local infrastructures or financial means) and other practical circumstances. Food utilization, the third pillar of food security, concerns the way food is handled from hygiene perspectives along the whole supply chain. Last but not least, food stability as a macroeconomic indicator for food security considers prices, political stability, local economy but also weather issues or natural catastrophes (McCarthy et al., 2018).

Related to the factors affecting food security, they can be grouped and summarized on social, economic, and environmental grounds.

In terms of social reasons, culture affects food security. First, it regards food production, processing, and storage techniques and determines food eating models. Cultures worldwide shape food quantity and quality according to social food prescriptions, historical events or political context (Alonso, Cockx, & Swinnen, 2017).

In terms of economic issues, population growth and urbanization, it is estimated that global population growth will follow an annual rate increase of approximately 1% till 2030 and a yearly increase of more than 0.5% till 2050. It means that the population will increase from 7.5 billion people to over 9 billion people in 2050 of whom more than 65% will occupy urban areas with exaggerated urbanization trends and land losses (McCarthy et al., 2018).

Lastly, food security poses environmental issues, in particular in terms of limited natural resources and negative externalities, such as greenhouse gas (GHG) emission and water pollution. Moreover, the amount of food produced each year in each country contributes to soil exploitation and agricultural production decrease. Approximately 20 million hectare of land (for instance, equal to more than 65% of Italian total area) each year are lost through soil erosion and irrigation issues. Climate shifts have catastrophic effects on economic prosperity and agricultural production entailing population displacement and resource depletion (McCarthy et al., 2018). Based on these general considerations, food insecurity seems to be mainly due to loss and distribution problems.

This research presents two main goals. First, to review the magnitude food loss and waste (FLW) at a global and European level and its environmental, social and economic implications. Second, use Material Flow Analysis (MFA) to support and improve food loss and waste management and its application in an Italian potato industry case study. According to the case study presented, MFA has demonstrated the advantages of tracking input and output to prevent FLW and how they provide economic, social, and environmental opportunities.

The chapter is organized in two different sections. In the first one, the authors provide an overview of the main differences between food loss (FL) and food waste (FW), especially in quantity and quality at different stages of the Food Supply Chain (FSC). In the second one, the main results of MFA of Italian potato industry are presented, showing the importance of FLW adding to efficiency and sustainability.

After discussing the results, the final section concludes with some indications and recommendations to pursue food sector sustainability.

BACKGROUND

Food Supply Chain and Food Loss and Waste

As stated by FAO (2018), the food available for human consumption is measured by a dietary energy supply which is on average over 2,900 kcal/day per capita. However, there is a wide difference between the countries all around the world (Table 2).

In descending order, Europe's average dietary energy supply is the highest (approximately 3,400 kcal/day per capita) followed by the Americas (more than 3,200 kcal/day), Oceania (about 3,000 kcal/day), Asia (more than 2,800 kcal/day), and Africa (less than 2,600 kcal/day). The highest percentage of food produced and consumed worldwide regards crops with relevance to harvest areas, yields, and quantities produced. The worldwide production shows commodities such as sugar cane (more than 1.8 billion tons in 2016), maize (more than 1.0 billion tons), wheat (approximately 750 million tons), rice (more than 740 million tons), and potatoes (approximately 370 million tons) as the top five items produced. However, cereals represent the biggest part of crop production (FAO, 2018).

Livestock covers a huge percentage of global food production. The highest amount of bred animals is represented by chickens (more than 22.7 billion heads in 2016), followed by cattle (approximately 1.5 million), ducks (more than 1.2 million), sheep (approximately 1.1 million) and goats (around 1.0 million). Moreover, fish provides over 20% of global average intake of animal protein.

With regard to the environment, on average agriculture exploits more than one-third of the total land area with its highest percentage in Asia (approximately 50% of total land in Asia) and lowest one in Europe (less than 25%). Furthermore, water represents a fundamental input for food production and its global demand (agriculture, municipalities, and industries) has increased over the past decade reaching a volume of over than 3,900 km^3 per year, of which approximately 70% (more than 2,700 km^3) in agriculture. Moreover, this sector contributes to GHG emissions with about 5 billion metric tons of CO_2 eq. per year and represents over one-fourth of global emissions (FAO, 2018).

These numbers provide an idea of environmental costs and economic and social implications associated with FSC and agricultural development. Thus, FLW entails not only social issues but also economic and environmental consequences.

Table 2. Worldwide dietary energy supply, crops production, and harvested area

Region	Dietary energy supply, kcal/day/capita	Production, million tons					Harvested area, million ha
		Sugar cane	Maize	Wheat	Rice	Potatoes	
Europe	3,365	5,918	118.0	252.4	4.2	117.6	186.7
America	3,265	1,032.3	548.8	121.2	36.4	44.1	298.1
Oceania	3,002	36.2	0.6	22.7	0.3	1.6	24.7
Asia	2,824	701.3	359.6	329.4	677.3	187.4	619.5
Africa	2,593	91.4	73.5	23.3	38.0	23.5	255.8
World		1,861.2	1,100.2	749.0	756.2	374.3	1384.8

Source: Authors' development based on FAO (2018)

Figure 1. FLW scheme
Source: Authors' development based in Pellegrini, Sillani, Gregori, and Spada (2019)

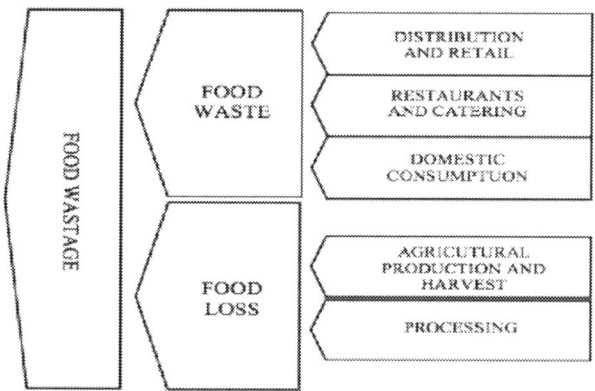

Before attempting to give a definition of FLW, it is important to remark what food is. According to Food Loss and Waste Accounting and Reporting Standard, food is any substance, whether processed, semi-processed or raw, that is intended for human consumption including drinks and any other substances used for manufacture, preparation or food treatment. Subsequently, FLW can be defined as a decrease in the quantity and quality of edible food intended for human consumption, with particular difference between FL and FW (Food Loss + Waste Protocol [FLW], 2016; Pinstrup-Andersen, 2009).

FL regards essentially food production and supply system malfunctioning. It occurs along the whole supply chain between producer and market, resulting in pre-harvest problems or harvesting, handling, storage, packaging or transportation issues. Moreover, it can be determined by policy or institutional frameworks according to different national and international waste definitions and is mainly caused by managerial and technical limitations, such as lack of efficient storage technologies, improper food handling practices or inefficient packaging.

FW refers to still edible food removal from FSC for different reasons. Generally, it refers to discarding or alternative uses of still safe and nutritious food for human consumption and mainly occurs at retail or consumer levels (FAO, 2013, 2015; Rezaei & Liu, 2017) (Figure 1).

In particular, FLW occurring at different FSC stages depends essentially on its definition. However, average values can be estimated according to FAO (2015, 2019), Moller et al. (2014), and Stenmarck, Jensen, Quested, and Moates (2016). It is stated that along FSC approximately 11-23% occurs at the agricultural production and harvest stage, 17-19% at processing, 8-17% at distribution (wholesale, retail, and food service), and 52-53% at the consumption stage (households). These little differences depend on FLW definition and analysis breakdown (Corrado et al., 2019).

Food Loss and Waste: Quantity and Costs

A shared definition of FLW is not enough to minimize this phenomenon; it must be determined in quantity and quality so as to better manage it.

In the last decades, FLW has increased on a rate of more than 50%. FAO estimates on average more than 1.3 billion tons of FWL along the whole FSC (equivalent to one-third of total food production) of which more than 670 million tons in developed countries and approximately 630 million tons in developing ones, showing wide differences between countries. In particular, EU data estimates an amount of more than 85 million tons of FLW, equal to approximately 20% of total food production (more than 170 kg of FLW out of 860 kg of food produced per capita). In descending order, domestic consumption accounts for more than 45 million tons of FLW, followed by processing (approximately 17 million tons), food service, such as restaurants and catering (more than 10 million tons), agricultural production and harvesting (more than 9 million tons), and distribution and retail (about 4 million tons). Thus, EU domestic consumption and food service account for more than 60% of FLW, while agricultural production and harvest for less than 10%. (Stenmarck et al., 2016; McCarthy et al., 2018; Boiteau, 2016).

On a global level, the average FLW associated cost is equal to approximately 1.65 Euro/kg, accounting for more than $700 billion in developed countries and less than $300 billion in developing ones. Therefore, the USA wastes more than $210 billion per year, equaling 1.3% of the U.S. GDP. In the EU, this cost goes over €140 billion, of which approximately €100 billion are generated at the household level (Stenmarck et al., 2016; Boiteau, 2016; McCarthy et al., 2018; FAO, 2019).

Figure 2 shows FLW quantity and costs divided per regions and occurring between production and retail and at the consumer stage each year.

FLW differences around the world depend on the technical, economic and social development of each country. On average, more than 160 kg/year per capita is the estimated amount of FLW occurring from the production to retailing stages, ranging from the highest value in Latin America (approximately 200 kg/year per capita) to the lowest value in South and Southeast Asia (more than 100 kg/year per capita). However, although developed countries show the highest percentage at the retail level, while developing ones at the harvest and post-harvest stage, both show on average similar FLW values at the pre-consumption stages (FAO, 2019; Pellegrini et al., 2019; Philippidis, Sartori, Ferrari, & M'Barek, 2019).

Wide difference in FLW is registered at consumer level with an average value of approximately 50 kg/year per capita. According to global data, the highest value is recorded in North America and Oceania

Figure 2. FLW quantity (kg per capita) and costs (€ per capita) per regions at different FSC stages
Source: Authors' development based in Pellegrini et al. (2019)

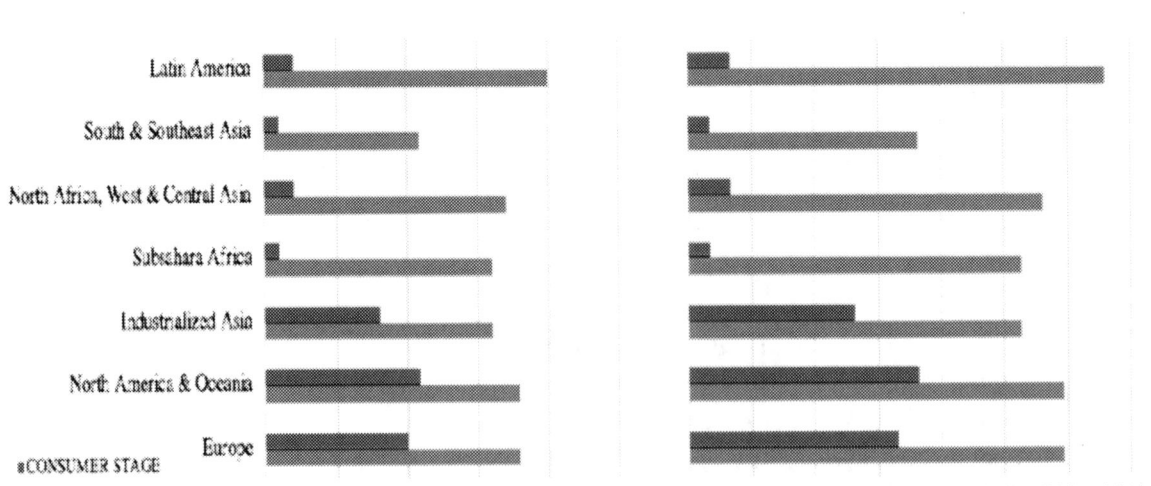

(approximately 120 kg/year per capita) and the lowest one in Sub-Saharan Africa (less than 10 kg/year per capita) (Boiteau, 2016).

FLW generally causes change worldwide, with big differences between developed and developing economies. In developed ones, the main causes for FLW relate to over-production, high consumer appearance and quality standards, higher costs for recovery and re-using than for discarding and sell-by date reached in general organized distribution. An Italian multiple-choice answer survey shows several reasons for household FLW. The highest percentage of them depends on out of date food (approximately 55% of given answers) followed by moldy food (less than 40%), badly smelling or tasting food (more than 30%), wrong planning of meals (more than 25%), too much served food (approximately 20%), unintelligible data labeling (approximately 5%), incorrect storage (less than 5%) or insufficient cooking skills (less than 5%) (Jorissen, Priefer, & Brautigam, 2015).

In developing countries, the main causes for FLW concern production phases, such as agricultural ones (too early harvesting of crops), poor storage, processing, and market facilities, and lack of infrastructure (Boiteau, 2016).

Food Loss and Waste: Quality and Composition

FLW composition differs around the world, recording deep differences between regions and FLW groups. On average and based on the total production of a single food group, roots and tubers show the highest percentage (52%) and dairy products the lowest one (15%). For various reasons and in different FSG stages, more than half of the total roots and tubers and less than 50% of fruit and vegetables produced lose their value as a human diet component.

Based on 2018 production data, Figure 3 shows the worldwide FLW composition and percentage of lost or wasted food at different FSC stages. FLW is calculated as total production of a single food group (Figure 3).

Figure 3. Worldwide FLW composition at different FSC stage
Source: Authors' development based on FAO (2019)

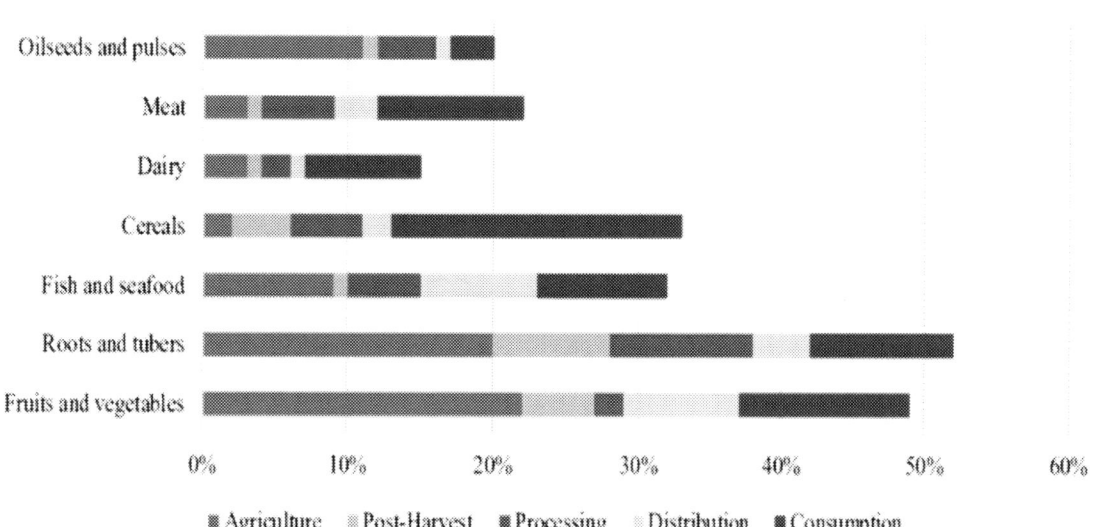

Food Loss and Waste: Accounting and Reporting

FLW minimization aims at food security, economic benefits, and environmental gains. However, its management faces several challenges according to complex FLW measurement along the FSC.

The inverted pyramid for solid waste FW management based on prevention, re-use, recycling, recovery, and disposal can be applied also to FLW (Figure 4). Generally, FLW can be used for and/or transformed in animal feed, bio-based materials, bioenergy recovery, composting or landfilling.

The first solution requires surplus food generation avoidance and prevention along the whole FSC, while re-use presupposes redeployment for human consumption through food redistribution and food banks. Recycling requires FLW usage as animal feed or composting, recovery means energy, solid or liquid material production (e.g. conversion into useable heat, electricity, biofuel, or fertilizers) and the last and residual option means that it has to be landfilled (Papargyropoulou, Lozano, Steinberger, Wright, & Bin Ujang, 2014; FLW, 2016a, 2016b).

However, FLW hierarchy implementation requires accurate accounting and reporting tools along the whole FSC.

At the production stage, farming and husbandry should be considered. Moreover, processing and manufacturing on the primary (e.g. drying, sieving, and milling) and secondary stages (e.g. mixing, cooking, and molding) as well as on the retail stage need particular attention. And finally, consumption distinguishing between household use, catering (e.g. restaurants), and institutions such as educational and medical treatment ones should be taken into account (Papargyropoulou et al., 2014). Table 3 details FLW causes along the FSC.

As stated by previous paragraphs, FLW at consumption stage deserves special mention accounting highest percentage along FSC equal to 52-53% at global level. Households FLW generation can be considered as results of food purchased, people cooking and shopping behavior and general lifestyle.

For instance, EU surveys show average households FLW accounting for more than €450 per family in 2010, however that is less if compared to 2008 (more than €550/family). The highest percentage of FLW concerns fresh products such as eggs, meat, cheese, and milk with over 35% of total purchase, fol-

Figure 4. FLW management hierarchy
Source: Authors' development based on FLW (2016a, 2016b)

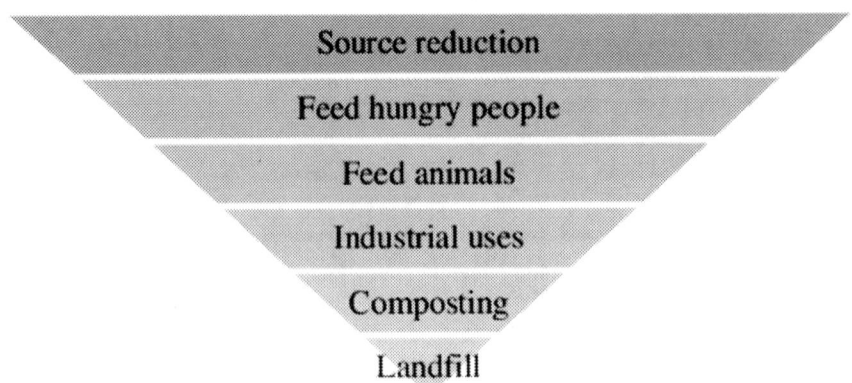

Table 3. Causes for FLW along the FSC from production to distribution

FSC stage	FLW causes
Agricultural stage (harvesting, threshing, breeding)	• Crops left in field (e.g. esthetical reasons, out of retail standard products, handling costs higher than market prices) • Crops damaged during harvesting due to poor harvesting and threshing techniques
Transport Distribution	• Poor transport infrastructure • Transport accidents and damages • Out of date products due to poor or insufficient supply programming
Primary processing Secondary processing	• Contamination and losses occurring during cleaning, classification, packaging, mixing, cooking, frying at industrial stage
Quality control	• Discharging due to out of retail standards products (e.g. esthetical or weight standards)
Storage	• Storage issues such as pests, disease, contamination due to poor storage infrastructure (e.g. lack of cooling/cold storage)
Packaging	• Packaging damages • Contaminations • Weighing, labeling, and sealing mistakes
Marketing and retail	• Publicity, selling, and distribution issues

Source: Authors' development based on Papargyropoulou et al. (2014) and Segrè and Falasconi (2011)

Table 4. External, FSC, and individual variables influencing FLW at consumption stage

Influences source	FWL causes
External variables	• Food-safety campaigns • Household make-up • Income levels • Customs and traditions • Food fashion
FSC variables	• Products availability • Packaging • Storage guidance • Date labeling • Promotions
Individual variables	• Attitude and values • Motivation • Habits • Knowledge and skills • Facilities and resources

Source: Authors' development based on Moller et al. (2014)

lowed by bread (less than 20%), fruit and vegetables (approximately 15%), cold cuts and salads (about 10%), pasta and quick-frozen (less than 5%) (Segrè & Falasconi, 2011).

In terms of FLW causes occurring at the consumption stage, it is possible to distinguish between external variables related to social and cultural issues, FSC variables regarding problems and malfunctioning along the supply chain and individual variables related to personal habits, attitudes, and values (Table 4).

Several impacts are associated with FLW from economic to environmental and social ones (Figure 5).

Figure 5. FLW economic, social, and environmental impacts
Source: Authors' development based on Barilla Center for Food and Nutrition [BCFN] (2012)

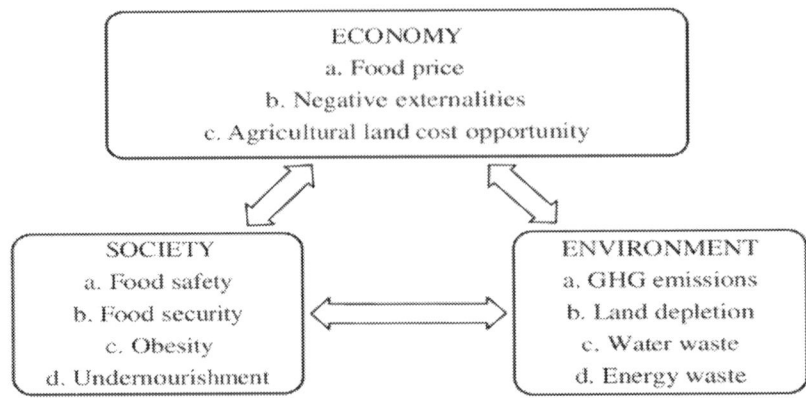

MAIN FOCUS OF THE CHAPTER

Potato FSC Case Study in Italy

As mentioned above, FLW data available are not homogenous and present several differences. Based on Italian FLW equaling on average to 150 kg/per capita per year (approximately 5% of Italian waste generation), in Italy approximately more than 9 Mt of food are wasted yearly corresponding to more than €14.5 billion loss (€1.65/kg). However, references record lower values. For instance, Segrè and Azzurro (2016), Segrè and Falasconi (2011), Segrè, Falasconi, and Politano (2016), BCFN (2012), and Pellegrini et al. (2019) estimate that Italy wastes less than 3% (approximately 2 Mt) of its national food production each year, losing approximately €3.5 billion and human feed for approximately three-quarters of its population. In particular, the highest FLW costs are associated with meat wastes for over €440 million (Bräutigam, Jörissen, & Priefer, 2014; Priefer, Jörissen, & Bräutigam, 2016).

Figure 6. Italian FLW composition
Source: Authors' development based on Segrè and Falasconi (2011)

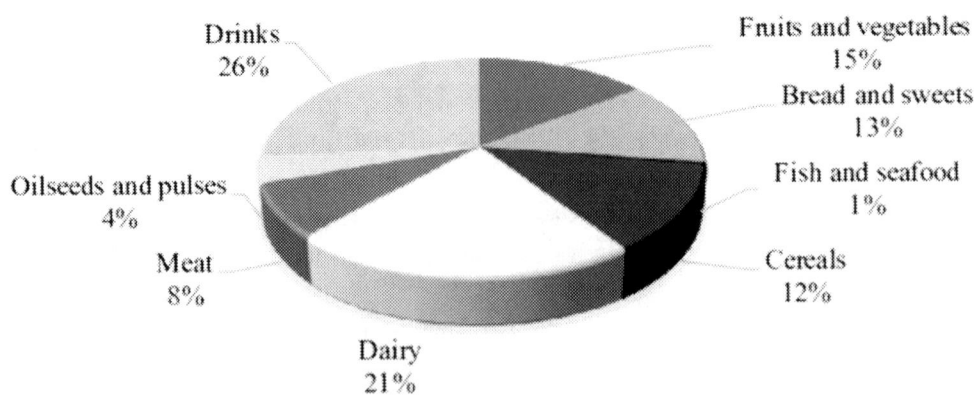

Figure 7. Italian FLW composition at different FSC stages
Source: Authors' development based on FAO (2019) and Segrè and Falasconi (2011)

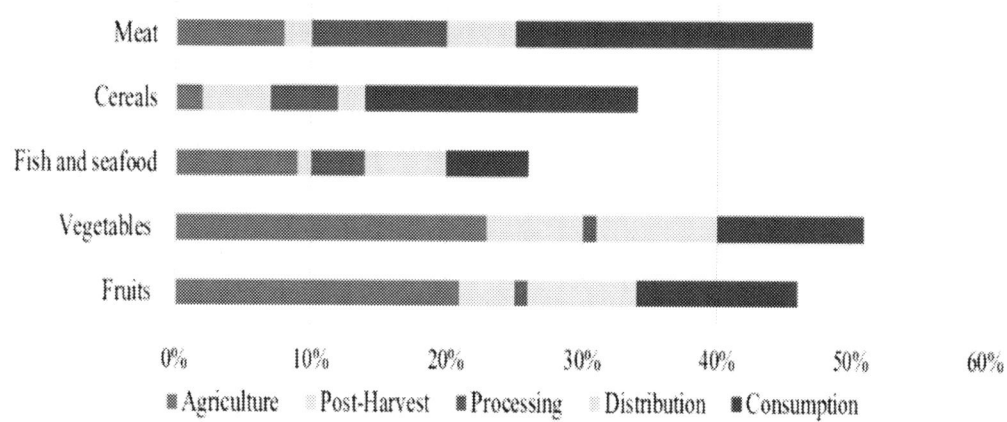

With regard to the Italian FLW composition, drinks and dairy together account for 47%, followed by fruits and vegetables (15%), bread and sweets (13%), cereals (12%), meat (8%), oils and vegetable fats (4%), and fish (1%) (Figure 6).

To detail: with regard to the hypothesis in which the Italian FLW follows the same trends of the global FLW (Figure 3), it is possible to calculate the Italian FLW percentages along the FSC (Figure 7).

In particular, between 2% and 12% of the FLW occurs in the agricultural stage as harvest land loss (e.g. less than 2% of citrus fruits, 5% of potatoes and legumes, and more than 12% for tomatoes) (BCFN, 2012).

The case study presented regards potato industry. The choice is based on the following considerations: potato is the fourth most important staple food worldwide after maize, wheat and rice mainly due to its starch content which is the primary energy source in the human diet. It is cultivated in about 130 countries, mostly in developing ones even if, in the last years, its production has recorded a continuously growing also in developing countries. It has become a crucial food both for its direct and indirect use in human diet in the European countries (Eriksson, Carlson-Nilsson, Ortíz, & Andreasson, 2016; Eurostat, 2019, FAO, n.d.). Moreover, vegetables are one of the most wasted food categories worldwide, recording high percentage of FLW along FSC. Moreover, potatoes as raw materials can be transformed in several final products, giving the chance to compare costs and wastes along different FSCs.

Worldwide, potato production (PP) is estimated to be approximately less than 390,000 kt with several changes in global markets and trades. Until the 1990s, the highest percentage was consumed in Europe, North America, and USSR countries. However, recent statistics show a sharp increase of PP in Asia, Africa, and Latin America from 30 million tons in the 1960s to more than 165 million tons in 2007. Moreover, China shows the biggest PP and approximately one-third of all potatoes is harvested between China and India (PotatoPro.com, n.d.). China accounts for about 100,000 kt, followed by India (approximately 50,000 kt) and Russia (approximately 30,000 kt). Table 5 shows PP, harvested area and yield, while Figure 8 shows PP worldwide.

Fresh and processed potatoes global consumption is on average approximately less than 35 kg/capita each year at consumption stage. European PP accounts for more than 50,000 kt in 2018, less with more

Figure 8. Potato production worldwide, tons
Source: Authors' development based on FAO (2018)

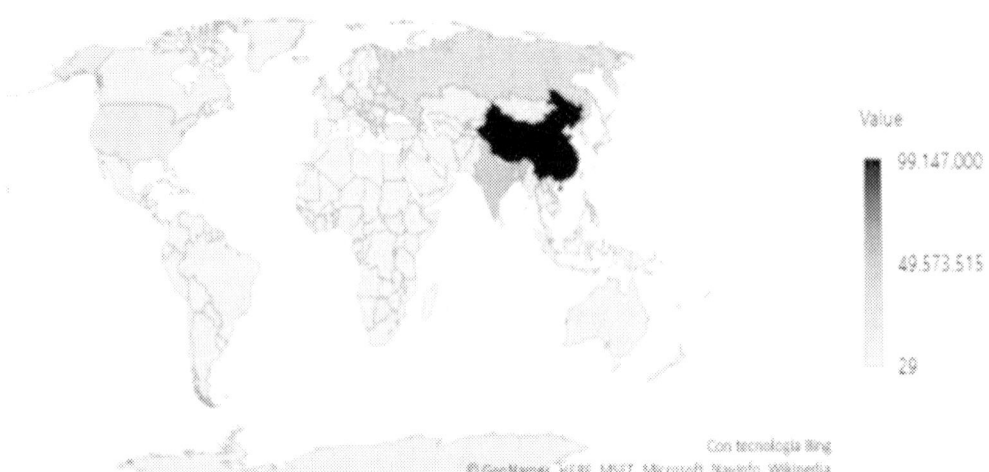

Table 5. Potato production, harvested area, and yield

Phenomena	Quantity
World Potato Production (2017)	374,252,073 tons
World Potato Harvested Area (2017)	19,302,600 ha
World Potato Yield (2017)	201,108 hg/ha

Source: Authors' development based on PotatoPro.com (n.d.)

than 60,000 kt if compared to the 2017 PP. Italian PP registers more than 1,300 kt in 2018 (approximately 2.5% of European PP) (Eurostat, 2019).

In particular, the potato FSC registers a sharp increase requiring quality standards such as dimension, shape, peel characteristics, sugar in raw content, and taste. However, the one can distinguish between potatoes for industrial use and potatoes for consumption.

Available data state that the Italian potato FSC depends approximately on 85% imported raw material, followed by 10% national produce and less than 5% is selected for planting. Moreover, Italian harvested land is approximately over 60,000 ha and PP accounts for more than 1,300 kt.

Potato consumption (PC) shows some important trends. According to an Italian consumer survey, more than 55% of consumers habitually consume them and approximately 25% is very interested in. This means that less than 2% of the population does not consume potatoes, thus becoming one of the most preferred products by consumers. Furthermore, several varieties of potatoes can be distinguished. The highest percentage of PC is registered by yellow flesh potatoes (approximately 50%), followed by new potatoes, white flesh potatoes and lastly frozen potatoes. However, their use changes according to consumer preferences. More than 60% of consumers eat baked potatoes, approximately 55% – French fries, about 50% boiled ones, and less than 30% for other preparation, showing baked and French potatoes as the preferred ones in a double-choice answer survey (Coltura & Cultura, n.d.; ISMEA, 2014; Tirelli, n.d.).

With regard to the market, Table 6 shows important details according to consumers purchase preferences.

Table 6. Consumer purchase preferences

Potato-based final product	%	kt
In jute bags	33	429
Packaged potatoes (with quality label)	25	325
Loose potatoes	24	312
Frozen potatoes	15	195
Ready for consumption (peeled, cooked)	3	39
Total	100	1,300

Source: Authors' development based on Tirelli (n.d.)

Considering that Italian PP is approximately more than 1,300 kt in 2018, it is possible to estimate and share its national transformation according to consumer purchase preferences. Applying these percentages to Italian PP, it is possible to calculate how PP is divided between industrial and consumer end-use.

After having stated PP and PC quantity and quality and remembering potatoes importance in daily diet, in consumer purchase preference and in FLW phenomena, it is possible to create a Material Flow Analysis (MFA) along the whole FSC to measure FLW. In particular, three types of industrial processing can be taken into account: frozen pre-fried potatoes, chips and dehydrated potatoes.

SOLUTIONS AND RECOMMENDATIONS

The authors have chosen to use the MFA for analyzing the potato industry. MFA is a systematic assessment of the state and change of materials flow and stocks defined in space and time. It connects sources, pathways, and intermediate and final sinks of materials. Since MFA is related to the matter of conservation law, MFA results can be controlled by mass balance comparing all inputs, stocks, and outputs (Brunner & Rechberger, 2017; Lagioia & Camaggio, 2002; Zaghdaoui, Jaegler, Gondran, & Montoya-Torres, 2017; Kytzia, Faist, & Baccini, 2004).

MFA is the main method applied in the following case study and takes into account three specific typologies of processed potatoes: pre-fried, chips, and dehydrated potatoes, which represent the highest quota of processed potatoes. According to reference literature, the main type of processed potatoes are pre-fried ones (e.g. linear sticks, zig-zag sticks, sliced potatoes), potatoes for direct consumption (e.g. chips), and dehydrated ones.

The functional unit of MFA is one ton of final product and boundaries are from the agricultural to the processing stage, without considering distribution/retail and consumption ones. However, according to different technical routes, MFA tries to summarize common sections for both pre-fried potatoes and chips and one common supply chain for dehydrated ones. MFA first deals with the agricultural stage, then the processing technologies and particular issues along the supply chain, focusing its attention to FLW along the FSC.

According to Wang et al. (2016), it is possible to calculate resources (seeds, fertilizers, and energy, without considering land and water) and waste produced along whole FSC (boundaries from agricultural stage to storage, without considering retail and consumption stages). Figure 9 illustrates in details the whole MFA from the agricultural to storage stage for pre-fried potatoes (in circles) and chips (in rectangles), showing global input-output. The agricultural stage, reception, preparation for cutting, cutting, washing, drying, burning, pre-frying and frying are common for both industrial processes. After frying, separate processes are represented (cooling-down and freezing for pre-fried potatoes and salting for chips) (Figure 9).

Starting from the agricultural stage, seeds, fertilizers, and energy are required as input. However, this stage produces land loss (5% of tubers required). To obtain a functional unit, an amount of 1.7-2.0 tons of tubers are needed in the pre-fried potatoes process, while it is 3.5-4.0 tons in the chips process. During reception and preparation for the cutting phase, approximately 1 m³ of water are required in both processes. Moreover, about 50 kg of uncalibrated tubers are wasted while more than 0.2-0.4 tons of skins and scraps are produced. Later, during the cutting, washing, drying, and burning phases, additional water is required (5 m³ for pre-fried potatoes and more than 15-20 m³ for chips) with little waste.

Figure 9. MFA for pre-fried potatoes and chips (1 ton)
Source: Authors' development

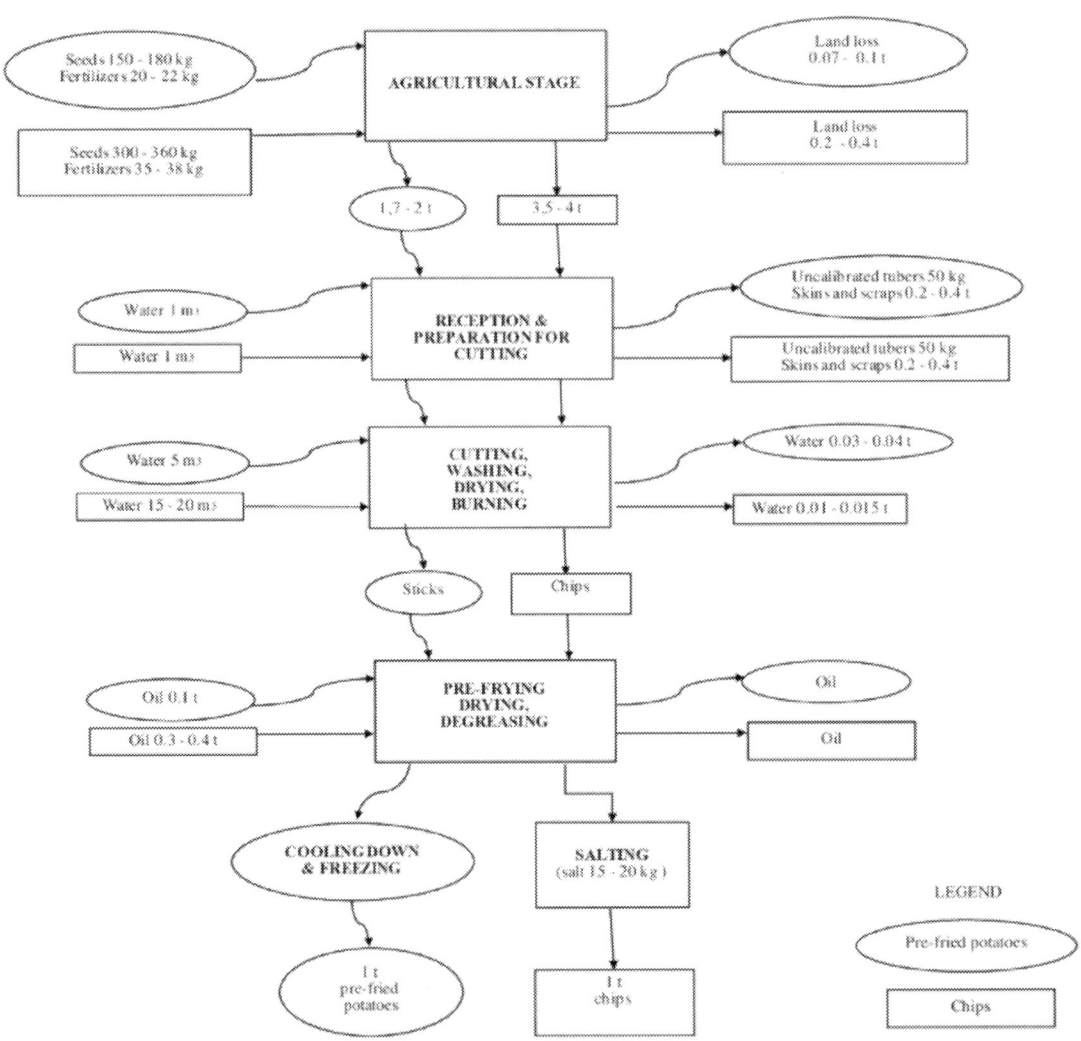

Food Waste Reduction Towards Food Sector Sustainability

Subsequently, sticks (1.4-1.5 tons) and chips are obtained. Sticks should be about 10 cm in length and 6-12 mm in depth while chips should be 1.0-1.7 in depth.

During the pre-frying, drying, and degreasing phase, oil is required in a range of 0.1-0.4 tons. After that, the process differs between pre-fried potatoes and chips. In the first case, a cooling-down and freezing phase is required, while in the second one salting (15-20 kg of salt) is needed.

Moreover, it is possible to calculate resources (seeds, fertilizers, and energy, without considering land and water) and waste produced along the whole FSC (boundaries from agricultural stage to storage, without considering retail and consumption stages) for dehydrated potatoes (Figure 10).

As stated in Figure 10, to obtain 1 ton of dehydrated potatoes (functional unit), 55-60 kg of potato seeds, 70-80 kg of fertilizers, tubers in a range of 6.3-7.3 tons, 30-35 tons of water and 40 GJ of energy, of which 3 GJ for electricity production and 37 GJ for the production of 15-20 tons of steam are required.

Figure 10. MFA for dehydrated potatoes (1 ton)
Source: Authors' development

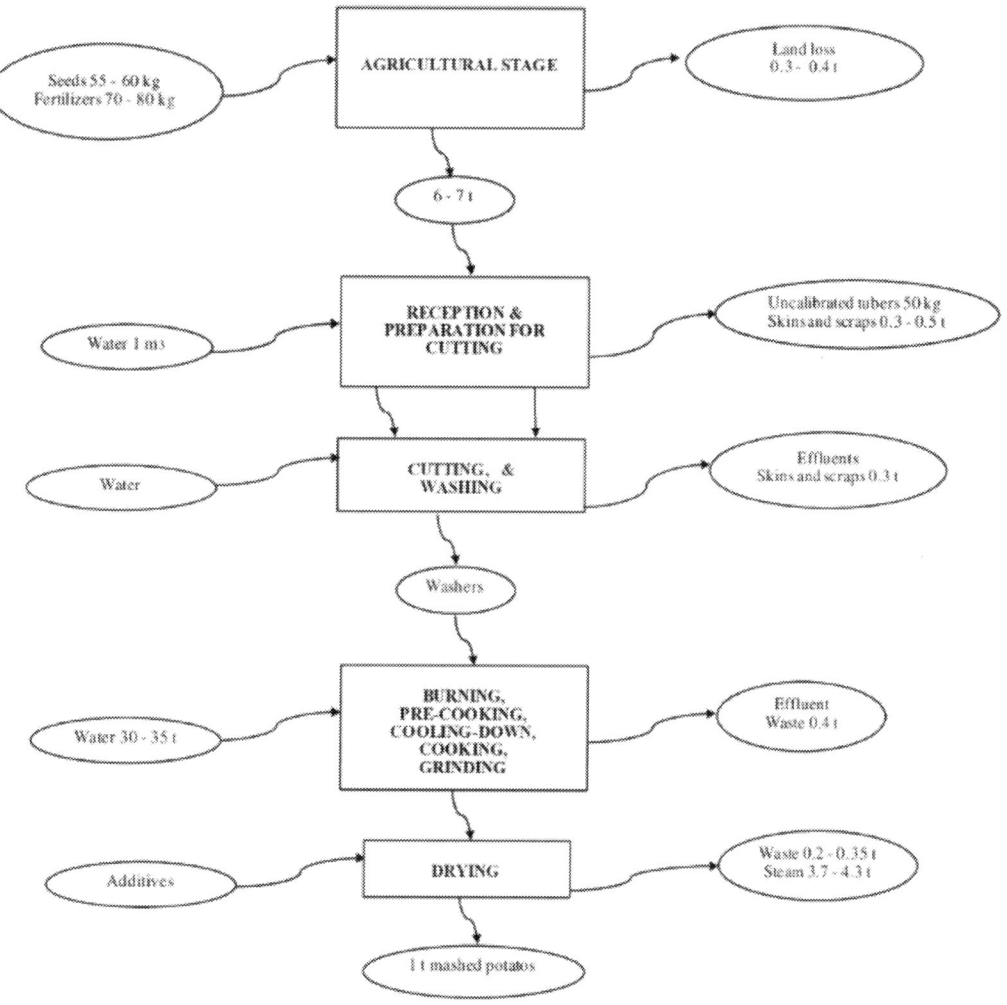

As output, approximately 0.3-0.4 tons of land loss, 50 kg of uncalibrated tubers, 1.20-1.55 tons of skins, scraps, and waste and water, oil and steam are produced as waste and FLW.

FUTURE RESEARCH DIRECTIONS

As mentioned in the previous paragraphs, the authors have chosen MFA to analyze the Italian potato industry. In particular, this tool is applied to describe and evaluate material and substance balances in a well-defined system and can be useful to define a systematic assessment of the state and change of material flows and stocks in space and time connecting sources, pathways and final sinks of materials. Moreover, since MFA is related to the matter of conservation law, its results can be controlled by mass balance comparing all inputs, stocks and outputs (Brunner & Rechberger, 2017; Zaghdaoui et al., 2017; Kytzia et al., 2004; Wang et al., 2016).

In particular, the present research uses MFA to support and improve FLW management and its application in an Italian agri-food sector towards its sustainability and main results demonstrated MFA as an important tool able to quantify and qualify natural resources utilization and economic costs associated to 1 t of final product. Moreover, MFA results are useful in FLW context to account energy and material lost or waster along FSC when food stops to be "feed" and begins to be "waste". Being MFA comparable, this tool provides useful information in order to understand if FLW associated to potato-based final products change according to new or old technologies, or to compare FLW associated to different vegetables or food FSCs.

In such context, being availability of data and FLW measurement tools European priority, authors have proposed MFA as one of the possible options for its accounting and reporting and as a basis for the construction of others environmental indicators such as carbon and water footprint or for the application of Life Cycle Assessment (LCA) methodology. Thus, its application could be improved within future studies and researches and proposed to further policy implementations, giving the chance to minimize FLW towards food safety and food security better management (Brunner & Rechberger, 2017; Caldeira, Corrado, & Sala, 2017; Zaghdaoui et al., 2017).

CONCLUSION

Table 7 shows FLW along the FSC according to the potatoes transformation quota in Table 6. FLW (kt) displays total quantity along the FSC, in a range of 585 kt and 845 kt (45-65% of global production).

According to MFA methodology and Tables 6-7, potatoes FLW costs have been calculated.

With regard to pre-fried, chips, and dehydrated products, FLW occurs in a range of 585-845 kt. In terms of monetary costs and based on Istituto di Servizi per il Mercato Agricolo Alimentare (2019), FLW occurs in a range of €111.2-160.2 million (for €0.19/kg) and €274.9-€397.2 million (for €0.47/kg). Overall, FLW in PP accounts in a range of 3-12% of the Italian FLW or 0.7-2.7% considering €14.5 billion as Italian FLW associated cost.

The contribution of the MFA is a very effective tool to track input and output and in balanced mass and energy flows. As cited above, wasting food has a negative impact on many levels: ethical and social level (throwing away the food that other people could eat), economic level (people have paid for this food), environmental impacts that are underestimated (for example, CO_2 emissions derived from the improper

Table 7. FLW (kt) along FSC for different potato-based final products

Potato-based final product	Total (kt)
In jute bags	193-279
Packaged potatoes (with quality label)	146-211
Loose potatoes	140-203
Frozen potatoes	88-127
Ready for consumption (peeled, cooked)	18-25
Total	585-845

Source: Authors' development

management of organic waste). So, analyzing the factors influencing the wastage can help for better management of the whole supply chain. MFA is possible to include into the wider concept of sustainable development whose application can improve FSC management and drive the food sector towards sustainability goals. Furthermore, MFA can be integrated in the new paradigm of Circular Economy (CE). The CE concept has been introduced by Ellen MacArthur Foundation. It is an economic system where products and services are traded in closed loops or cycles. A circular economy is characterized as an economy which is regenerative by design, with the aim to retain as much value as possible of products, parts, and materials. This means that the aim should be to create a system that allows for the long life, optimal reuse, refurbishment, remanufacturing and recycling of products and materials (Ellen MacArthur Foundation, 2019).

It means that the CE paradigm can lead FLW to economic opportunities such as reuse, recycling, energy efficiency, and nutrient recovery against landfill and disposal (Garcia-Garcia, Stone, & Rahimifard, 2019; Lemaire & Limbourg, 2019; Principato, Ruini, Guidi, & Secondi, 2019; Segneanu et al., 2018). According to FLW hierarchy (Figure 4), FLW can be first reused for human consumption or secondly destined to animal feed and, if not possible, it can follow several alternative pathways such as:

- Recycling for industrial use;
- Anaerobic digestion;
- Composting;
- Combustion for energy recovery.

Several policies have adopted the CE paradigms. In particular, in Europe, European Commission (2015) proposed a transition to a more circular economy where products, materials, and resources value is maintained in the economy for as long as possible and waste generation is minimized.

This document dedicates one paragraph to FLW as an EU increasing concern. In detail, it states that FLW produces environmental impacts and financial losses for both consumers and companies and considers social implications such as limited food donation of still edible food. Moreover, it considers FLW as part of 2030 Sustainable Development Goals (SDGs), especially in terms of "Zero Hunger". Moreover, it states that FLW is still hard to quantify since there is no harmonized and reliable method to measure food waste in the EU giving public authorities the chance to assess its scale, origins, and trends. However, FLW minimization requires actions by EU Member States, regions, cities, and businesses along the whole FSC. For this reason, the EU Commission has tried to address FLW policies such as the elaboration of a common EU methodology to its measurement, the creation of an EU platform dedicated to FLW and the application of actions regarding as instance "best before" marking. The case study analyzed in Italy shows the importance of FLW accounting towards efficiency and sustainability. This is because it has become possible to develop action plans, guidelines, recommendations to address policymakers, restaurants, hotels, producers, and anyone interested in helping to reduce food waste to new managerial policy.

EU interventions against FLW can be summarized with particular attention to:

- Development of EU common methodology to measure FLW and define relevant indicators;
- Improvement of data marking use by FSC actors and its understanding by consumers (e.g. "best before" label);
- Platform involving Member States and Stakeholders to support SDGs achievement and share best practices;
- Clarification of EU legislation relating waste, food, and feed;
- Food waste facilitation and former food-stuff and by-product use.

According to EU policies and other studies and researches, and since the available amount of data on FLW is low in quality and quantity, data collection should become policy priority. Table 8 registers FLW legislative frameworks opportunities and limits according to European Parliament (2008) and Technopolis Group (2016).

When considering FLW prevention policies, some possible options can be taken into account, such as reduction in portion sizes, tailored hygiene rules, best before dates for products and better estimations of the number of day-to-day customers. However, several lacks can be registered: lack of rules on portion sizes, hygiene, best before dates and lack of accurate FLW definition. Moreover, in terms of FLW recovery and re-use and related separate collection, the lack of an efficient collection infrastructure determines an expansive collection process. One more option is FLW reuse through donation. However, it is limited by VAT and liability rules with negative economic incentive on food donation and competitive implications to anaerobic digestion, incineration, composting and animal feed. In terms of FLW recycling

Table 8. FLW legislative framework opportunities and limits

FLW hierarchy stage	Opportunities	Limits
FLW prevention	Reduction in portion sizes Tailored hygiene rules Best before practices Estimation of day-to-day customers	No legislation on portion sizes Strict hygiene rules Problems in estimation
FLW collection		Lack of collection infrastructure Expansive collection
FLW reuse	Food donations Food banks	VAT Limited liability rules
FLW recycling	Animal feed Composting	Limits in catering separation as only biotic materials can be processed Cannibalism and safeguard of animal health
FLW incineration	Biofuels generation	CO_2 emissions

Source: Authors' development based on Technopolis Group (2016)

which regards food conversion into animal feed and composting, the limits are represented by lack of efficient conversion plants. Last option regards incineration, being worldwide used as a biofuel generation source. Its limit is represented by environmental impacts such as GHG and related CO2 emissions.

In conclusion, according to typologies and the FSC stage, different FLW quality can be reported with different FLW destinations, such as natural fertilizer (left in field), animal feeding, energy recovery, food banks and composting.

REFERENCES

Alonso, E. B., Cockx, L., & Swinnen, J. (2017). *Culture and Food Security*. Leuven: Centre for Institutions and Economic Performance.

Barilla Center for Food and Nutrition. (2012). *Lo sprecoalimentare: cause, impatti e proposte*. Parma: Barilla Center for Food and Nutrition.

Barrett, C. B. (2010). Measuring Food Insecurity. *Science*, *327*(5967), 825–828. doi:10.1126cience.1182768 PMID:20150491

Boiteau, J. M. (2016). *Food Loss and Waste in the United States and Worldwide*. Retrieved August 20, 2019, from https://www.worldhunger.org/food-loss-and-waste-in-the-united-states-and-worldwide/

Bräutigam, K. R., Jörissen, J., & Priefer, C. (2014). The Extent of Food Waste Generation across EU-27: Different Calculation Methods and the Reliability of their Results. *Waste Management & Research*, *32*(8), 683–694. doi:10.1177/0734242X14545374 PMID:25161274

Brunner, P. H., & Rechberger, H. (2017). *Handbook of Material Flow Analysis. For Environmental, Resource and Waste Engineers*. London: CRC Press.

Caldeira, C., Corrado, S., & Sala, S. (2017). *Food Waste Accounting – Methodologies, Challenges and Opportunities*. Brussels: Publication Office of the European Union.

Coltura & Cultura. (n.d.). *Un Volume, una Coltura*. Retrieved from https://www.colturaecultura.it/download

Corrado, S., Caldeira, C., Eriksson, M., Hanssen, O. J., Hauser, H.-E., Van Holsteijn, F., ... Sala, S. (2019). Food Waste Accounting Methodologies: Challenges, Opportunities, and Further Advancements. *Global Food Security*, *20*, 93–100. doi:10.1016/j.gfs.2019.01.002 PMID:31008044

Ellen MacArthur Foundation. (2019). *What Is the Circular Economy?* Retrieved from https://www.ellenmacarthurfoundation.org/circular-economy/what-is-the-circular-economy

Eriksson, D., Carlson-Nilsson, U., Ortíz, R., & Andreasson, E. (2016). Overview and Breeding Strategies of Table Potato Production in Sweden and the Fennoscandian Region. *Potato Research*, *59*(3), 279–294. doi:10.100711540-016-9328-6

European Commission. (2015). *Communication from the Commission to the European Parliament, the Council, the European Economic and Social Committee and the Committee of the Regions. Closing the Loop – An EU Action Plan for the Circular Economy*. Retrieved from https://eur-lex.europa.eu/legal-content/EN/TXT/?uri=CELEX:52015DC0614

European Parliament. (2008). *Directive 2008/98/EC of the European Parliament and of the Council of 19 November 2008 on Waste and Repealing Certain Directives*. Retrieved from https://eur-lex.europa.eu/legal-content/EN/TXT/?uri=celex%3A32008L0098

Eurostat. (2019). *Crop Production in EU Standard Humidity*. Retrieved from https://data.europa.eu/euodp/en/data/dataset/u33K8Gi1MFYGN7HyHUNhg

Food and Agriculture Organization of the United Nations. (2013). *Food Wastage Footprint. Impacts on Natural Resources*. Rome: Food and Agriculture Organization of the United Nations.

Food and Agriculture Organization of the United Nations. (2015). *Global Initiative on Food Loss and Waste Reduction*. Rome: Food and Agriculture Organization of the United Nations.

Food and Agriculture Organization of the United Nations. (2018). *World Food and Agriculture. Statistical Pocketbook 2018*. Rome: Food and Agriculture Organization of the United Nations.

Food and Agriculture Organization of the United Nations. International Fund for Agricultural Development, United Nations International Children's Emergency Fund, World Food Programme, & World Health Organization. (2018). The State of Food Security and Nutrition in the World 2018. Building Climate Resilience for Food Security and Nutrition. Rome: Food and Agriculture Organization of the United Nations.

Food and Agriculture Organization of the United Nations. (2019). *Save Food: Global Initiative on Food Loss and Waste Reduction*. Rome: Food and Agriculture Organization of the United Nations.

Food and Agriculture Organization of the United Nations. (n.d.). *Crops*. Retrieved from http://www.fao.org/faostat/en/#data/QC

Food Loss + Waste Protocol. (2016a). *Food Loss and Waste Accounting and Reporting Standard*. Retrieved from https://flwprotocol.org/

Food Loss + Waste Protocol. (2016b). *Guidance on FLW Quantification Methods*. Retrieved from https://flwprotocol.org/wp-content/uploads/2016/05/FLW_Protocol_Guidance_on_FLW_Quantification_Methods.pdf

Garcia-Garcia, G., Stone, J., & Rahimifard, S. (2019). Opportunities for Waste Valorisation in the Food Industry – A Case Study with Four UK Food Manufacturers. *Journal of Cleaner Production, 211*, 1339–1356. doi:10.1016/j.jclepro.2018.11.269

Istituto di Servizi per il Mercato Agricolo Alimentare. (2014). *Patate: nel 2014 la produzione italiana cresce del 20%*.

Istituto di Servizi per il Mercato Agricolo Alimentare. (2019). *Osservatorio patate. Prezzi all'origine. Trend annui*. Retrieved from http://www.ismeamercati.it/flex/cm/pages/ServeBLOB.php/L/IT/IDPagina/4845#MenuV

Jorissen, J., Priefer, C., & Brautigam, K.-R. (2015). Food Waste Generation at Household Level: Results of a Survey among Employees of Two European Research Centers in Italy and Germany. *Sustainability, 7*(3), 2695–2715. doi:10.3390u7032695

Kytzia, S., Faist, M., & Baccini, P. (2004). Economically Extended-MFA: A Material Flow Approach for a Better Understanding of Food Production Chain. *Journal of Cleaner Production, 12*(8-10), 877–889. doi:10.1016/j.jclepro.2004.02.004

Lagioia, G., & Camaggio, G. (2002). *La trasformazione industrial della patata. Dal tubero al fast food*. Bari: Progedit.

McCarthy, U., Uysal, I., Badia-Melis, R., Mercier, S., Donnell, C. O., & Ktenioudaki, A. (2018). Global Food Security – Issues, Challenges and Technological Solutions. *Trends in Food Science & Technology, 77*, 11–20. doi:10.1016/j.tifs.2018.05.002

Moller, H., Hanssen, J., Gustavsson, J., Ostergren, K., Stenmarck, A., & Dekhtyar, P. (2014). *Report on Review of (Food) Waste Reporting Methodology and Practice*. Krakeroy: Ostfold Research.

Nebbia, G. (1995). *Lezioni di Merceologia*. Rome: Laterza & Figli Spa.

Papargyropoulou, E., Lozano, R., Steinberger, J. K., Wright, N., & Bin Ujang, Z. (2014). The Food Waste Hierarchy as a Framework for the Management of Food Surplus and Food Waste. *Journal of Cleaner Production, 76*, 106–115. doi:10.1016/j.jclepro.2014.04.020

Pellegrini, G., Sillani, S., Gregori, M., & Spada, A. (2019). Household Food Waste Reduction: Italian Consumers' Analysis for Improving Food Management. *British Food Journal, 121*(6), 1382–1397. doi:10.1108/BFJ-07-2018-0425

Philippidis, G., Sartori, M., Ferrari, E., & M'Barek, R. (2019). Waste not, Want not: A Bio-Economic Impact Assessment of Household Food Waste Reductions in the EU. *Resources, Conservation and Recycling, 146*, 514–522. doi:10.1016/j.resconrec.2019.04.016 PMID:31274960

Pinstrup-Andersen, P. (2009). Food Security: Definition and Measurement. *Food Security, 1*(1), 5–7. doi:10.100712571-008-0002-y

PotatoPro.com. (n.d.). *The Potato Sector*. Retrieved from https://www.potatopro.com/world/potato-statistics

Priefer, C., Jörissen, J., & Bräutigam, K.-R. (2016). Food Waste Prevention in Europe – A Cause-Driven Approach to Identify the Most Relevant Leverage Points for Action. *Resources, Conservation and Recycling, 109*, 155–165. doi:10.1016/j.resconrec.2016.03.004

Principato, L., Ruini, L., Guidi, M., & Secondi, L. (2019). Adopting the Circular Economy Approach on Food Loss and Waste: The Case of Italian Pasta Production. *Resources, Conservation and Recycling, 144*, 82–89. doi:10.1016/j.resconrec.2019.01.025

Rezaei, M., & Liu, B. (2017). Food Loss and Waste in the Food Supply Chain. *Nutfruit*, 26-27.

Segneanu, A.E., Grozescu, I., Cepan, C., Cziple, F., Lazar, V., & Velciov, S. (2018). Food Security into a Circular Economy. *HSOA Journal of Food Science and Nutrition, 4*, 38.

Segrè, A., & Azzurro, P. (2016). *Spreco alimentare: dal recupero alla prevenzione. Indirizzi applicativi della legge per la limitazione degli sprechi*. Milan: Fondazione Giangiacomo Feltrinelli.

Segrè, A., & Falasconi, L. (2011). *Il libro nero dello spreco in Italia: il cibo*. Milan: Edizioni Ambiente.

Segrè, A., Falasconi, L., & Politano, A. (2016). *Crisi dei prezzi agricoli, sostenibilità e sprechi alimentari*. Retrieved from https://www.researchgate.net/publication/228837174_Crisi_dei_prezzi_agricoli_sostenibilita_e_sprechi_alimentari

Stenmarck, A., Jensen, C., Quested, T., & Moates, G. (2016). *Estimates of European Food Waste Levels*. Stockholm: IVL Swedish Environmental Research Institute.

Technopolis Group. (2016). *Regulatory Barriers for the Circular Economy. Lessons from Ten Case Studies*. Amsterdam: Technopolis Group.

Tirelli, D. (n.d.). *Richieste del consumatore*. Retrieved from https://www.colturaecultura.it/capitolo/richieste-del-consumatore

Wang, W., Jiang, D., Chen, D., Chen, Z., Zhou, W., & Zhu, B. (2016). A Material Flow Analysis (MFA)-Based Potential Analysis of Eco-Efficiency Indicators of China's Cement and Cement-Based Materials Industry. *Journal of Cleaner Production, 112*(1), 787–796. doi:10.1016/j.jclepro.2015.06.103

Zaghdaoui, H., Jaegler, A., Gondran, N., & Montoya-Torres, J. (2017). Material Flow Analysis to Evaluate Sustainability in Supply Chains. *Proceedings of the 20th IFAC World Congress*. Toulouse: The International Federation of Automatic Control.

ADDITIONAL READING

Aramyan, L., Valeeva, N., Vittuari, M., Gaiani, S., Politano, A., & Gheoldus, M. ... Hanssen, O.J. (2016). Marked-Based Instruments and Other Socio-Economic Incentives Enchancing Food Waste Prevention and Reduction. Wageningen: Wageningen UR.

Aschemann-Witzel, J., & Peschel, A. O. (2019). How Circular Will You Eat? The Sustainability Challenge in Food and Consumer Reaction to either Waste-to-Value or yet Underused Novel Ingredients in Food. *Food Quality and Preference, 77*, 15–20. doi:10.1016/j.foodqual.2019.04.012

Boulding, K. (1996). The Economics of the Coming Spaceship Earth. In H. Jarrett (Ed.), *Environmental Quality in a Growing Economy* (pp. 3–14). Baltimore: Johns Hopkins University Press.

De Marco, O., Lagioia, G., Amicarelli, V., & Sgaramella, A. (2009). Constructing Physical Input-Output Tables with Material Flow Analysis (MFA) Data: Bottom-Up Case Studies. In S. Sangwon (Ed.), *Handbook on Input-Output Economics in Industrial Ecology* (pp. 161–187). Amsterdam: Springer Netherlands. doi:10.1007/978-1-4020-5737-3_9

European Food Banks Federation. (2016). *Circular Economy in Favour of the Most Deprived. Preventing Food Waste through Food Redistribution*. Brussels: European Food Banks Federation.

Fresco, L. O. (2009). Challenges for Food System Adaptation Today and Tomorrow. *Environmental Science & Policy, 12*(4), 378–385. doi:10.1016/j.envsci.2008.11.001

Gustavsson, J., Cederberg, C., Sonesson, U., Van Otterdijk, R., & Meybeck, A. (2011). *Global Food Losses and Food Waste. Extent, Causes and Prevention*. Rome: Food and Agriculture Organization of the United Nations.

International Food Policy Research Institute. (2018). *2018 Global Food Policy Report*. Washington, DC: International Food Policy Research Institute.

Jeffries, N. (2018). *A Circular Economy for Food: 5 Case Studies*. Retrieved from https://medium.com/circulatenews/a-circular-economy-for-food-5-case-studies-5722728c9f1e

Kaza, S., Yao, L. C., Bhada-Tata, P., & Van Woerden, F. (2018). *What a Waste 2.0: A Global Snapshot of Solid Waste Management to 2050*. Washington, DC: World Bank. doi:10.1596/978-1-4648-1329-0

Lagioia, G., Calabrò, G., & Amicarelli, V. (2012). Empirical Study of the Environmental Management of Italy's Drinking Water Supply. *Resources, Conservation and Recycling*, *60*, 119–130. doi:10.1016/j.resconrec.2011.12.001

Renner, G. T. (1947). Geography of Industrial Localization. *Economic Geography*, *23*(3), 167–189. doi:10.2307/141510

Rood, T., Muilwijk, H., & Westhoek, H. (2017). *Food for the Circular Economy*. The Hague: PBL Netherlands Environmental Assessment Agency.

Saint Ville, A., Po, J. Y., Sen, A., & Quiñonez, H. M. (2019). Food Security and the Food Insecurity Experience Scale (FIES): Ensuring Progress by 2030. *Food Security*, *11*(3), 483–491. doi:10.100712571-019-00936-9

Tseng, M.-L., Chiu, A. S. F., Chien, C.-F., & Tan, R. R. (2019). Pathways and Barriers to Circularity in Food Systems. *Resources, Conservation and Recycling*, *143*(1), 236–237. doi:10.1016/j.resconrec.2019.01.015

KEY TERMS AND DEFINITIONS

Circular Economy: A concept that entails gradually decoupling economic activity from the consumption of finite resources, and designing waste out of the system.

Food Loss: Food that gets spilled, spoilt or otherwise lost, or incurs reduction of quality and value during its process in the food supply chain before it reaches its final product stage.

Food Sector: A collection of all activities that facilitate the consumption and supply of food products and services across the world.

Food Supply Chain: The processes that describe how food from a farm ends up on a table of a consumer, including the processes of production, processing, distribution, consumption, and disposal.

Food Waste: Food that completes the food supply chain up to a final product, of good quality and fit for consumption, but still does not get consumed because it is discarded, whether or not after it is left to spoil or expire.

Material Flow Analysis: A systematic assessment of the flows and stocks of materials within a system defined in space and time.

Sustainability: A concept focuses on meeting the needs of the present without compromising the ability of future generations to meet their needs.

Chapter 9
Comparative Analysis of Food Security and Tort Liability Doctrines in Russia and China:
Harm Caused by Poor-Quality Agricultural Products

Dmitry Bogdanov
 https://orcid.org/0000-0002-9740-9923
Kutafin Moscow State Law University, Russia

Svetlana Bogdanova
 https://orcid.org/0000-0001-7159-5310
Bauman Moscow State Technical University, Russia

ABSTRACT

Provision of the population with environmentally friendly and safe agricultural products is an important challenge in the developed states. This chapter analyzes the issues of food safety and quality. The indemnification caused by low-quality products stimulates producers to ensure the quality and safety of food resources. The institute of indemnification caused by low-quality agricultural products is analyzed in the chapter. Special attention is paid to the issues of consumer protection in the legislation of Russia and China.

INTRODUCTION

Providing the population with environmentally friendly and safe agricultural products is an important problem worldwide. Solution of global food insecurity problems require coordinated actions of the states based on uniform legal approaches. This study focuses on the overview and comparison of approaches to establishing food security in the countries which are considered as large producers of agricultural products and consumers of food, specifically, Russia and China.

In Russian, National Security Strategy was accepted in 2009 (President of the Russian Federation, 2009). On the basis of the Strategy, there was approved Food Security Doctrine (President of the Russian Federation, 2010), which declared provision of the population with safe agricultural products as a strategic objective of food security. Proceeding from the Doctrine, the key components of food security are sufficient volume of food products and safety of agricultural and food resources. Such understanding of the term "food security" is coordinated with the international terminology. In this context, many studies have been focusing on the investigation of food safety concept, analysis of criteria of safety, study of a relationship between the concepts of safety and quality of products, and their standard fixing on both national and international levels.

The indemnification mechanism which is caused by low-quality products stimulates producers to ensure quality and safety of food resources. The issue of indemnification caused by low quality of agricultural products is studied in this chapter. Special attention is paid to the protection of consumers' rights in Russia and China. There should be resolved a problem of mass infliction of harm by the uniform (replaced) products offered by various producers (mass products torts). Within this delict, there is a problem of interaction of a collective victim and a collective (uncertain) offender. New models of responsibility, for example, responsibility according to a share in the market (market-share liability), model of alternative responsibility (alternative liability), are applied. In this chapter, the prospects of application of these models of responsibility in Russia and China is considered. There are made suggestions for improvement of the legislation and practice of its application.

BACKGROUND

Food Security Legislation

Globally, every year, over 420,000 people die and almost 600 million people become sick by reason of consuming contaminated food. For food to be edible, it must be safe (Food and Agriculture Organization of the United Nations [FAO], 2019). Currently, supplying the population with ecologically clean and safe agricultural products is an important task in most of the countries.

In 2015, the Strategy of National Security of Russia was approved (President of the Russian Federation, 2015), which included some of the parameters of Food Security Doctrine (President of the Russian Federation, 2010). The strategic goal of food security is to supply the population of the country with safe agricultural products. In paragraph 54 of the Doctrine (President of the Russian Federation, 2010), it is stated that food safety is ensured, among other things, by improving the system of technical regulation, sanitary and phytosanitary supervision, and control in the sphere of food safety for human health.

Russia's legislation in the sphere of food safety includes, but is not limited to, the following major regulatory acts:

- Constitution of the Russian Federation (Government of the Russian Federation, 1993a);
- Federal Law "On Protection of Consumers' Rights" (Government of the Russian Federation, 1992);

- Federal Law "On Quality and Safety of Food Products" (Government of the Russian Federation, 2000b);
- Federal Law "On Sanitary and Epidemiological Wellbeing of Population" (Government of the Russian Federation, 1999);
- Federal Law "On Technical Regulation" (Government of the Russian Federation, 2002);
- Federal Law "On Plant Quarantine" (Government of the Russian Federation, 2014);
- Federal Law "On Veterinary Medicine" (Government of the Russian Federation, 1993b);
- Decree of the President of the Russian Federation (2010) "On the Approval of Food Security Doctrine of the Russian Federation";
- Decree of the President of the Russian Federation (2015) "On National Security Strategy of the Russian Federation"
- Decree of the Government of the Russian Federation (2011) "On the Approval of the Rules for the Implementation of State Veterinary Supervision at Border Crossing Points of the Russian Federation";
- Decree of the Government of the Russian Federation (2005) "On the Regulations on the Use of Means and Methods of Control in the Implementation of the Passage of Persons, Vehicles, Cargoes, Goods, and Animals Across the State Border of the Russian Federation";
- Decree of the Government of the Russian Federation (1997) "On the Regulations on the Examination of Low-Quality and Dangerous Food Raw Materials and Food Products, their Use or Destruction";
- Decree of the Government of the Russian Federation (2000a) "On the Organization and Conduct of Quality, Food Safety and Public Health Monitoring";
- Order of the Federal Agency on Technical Regulating and Metrology (2011) "On the Introduction of Interstate Standard."

National legislation in the sphere of food and nutrition security can no longer solve the arising problems to the necessary and sufficient extent on a unilateral basis. Different legislative approaches may create obstacles to trade in food. The need to tackle global food insecurity problems requires concerted actions of many states based on common legal approaches.

International Acts

Consideration of provisions of international acts and standards in the development of national legislation is required. International acts in this sphere include, but are not limited to:

- Convention on Biological Diversity (United Nations, 1992)
- International Plant Protection Convention (FAO, 1951)
- International Standards for Phytosanitary Measures (FAO, 2006)

Russia's Involvement in International Cooperation

Food Security Doctrine of the Russian Federation (President of the Russian Federation, 2010, paragraph 12) emphasizes a need to harmonize food safety metrics with international requirements. As a part of a set of measures to implement the provisions of the Doctrine, a comprehensive program has been developed to ensure Russia's involvement in international cooperation in the field of agriculture, fisheries, and food security (Government of the Russian Federation, 2010). The program determines key directions and forms of international cooperation. Among the key tasks of the program are the establishment of food safety and quality of food and agricultural products. The Program also points to the need for a full-fledged representation of Russia in the Food and Agriculture Organization of the United Nations (FAO), as well as the establishment of agricultural attaché positions in Ukraine, Belarus, Kazakhstan, Germany, the Netherlands, France, Italy, Israel, China, Turkey, India, Saudi Arabia, Iran, Australia, Brazil, Argentina, Ecuador, the USA, Canada, Egypt, and Morocco. The program calls for the intensification of international cooperation of Russia on a multilateral basis as part of specialized international organizations and institutions, including organizations and agencies of the United Nations system (UN World Food Program) and the FAO, as well as in the framework of international conventions, agreements, memoranda, and other treaties, international conferences, forums, and multilateral events, such as G8, G20, the L'Aquila Food Security Initiative, and the Global Partnership for Agriculture and Food Security.

In particular, cooperation between the FAO and Russia covers a wide range of initiatives. The FAO collaborates with Russia in the areas of food security; agriculture, including knowledge sharing and technical assistance in areas such as land management, veterinary medicine, plant protection, and agricultural engineering; forestry and fisheries; food safety; and participation in the work of the Codex Alimentarius Commission. In 2015, the FAO launched its Communications Offices in Moscow. FAO's structure includes Agriculture and Consumer Protection Department (AG), which, among other things, aims to ensure safety and quality of food (FAO, n.d.). The Codex Alimentarius Commission was established by the FAO and the World Health Organization (WHO) in the 1960s to protect health of consumers and promote fair trade in food products. The Commission plays a central role in the implementation of the Joint FAO/WHO Food Standards Program. The Codex Alimentarius or Food Code is a collection of standards, guidelines, norms, and rules approved by the Codex Alimentarius Commission (CAC) (FAO & World Health Organization [WHO], 2019). Since the establishment of the World Trade Organization (WTO), the standards developed by the Commission have played a special role in the WTO law because they are a benchmark for international food standards. Despite their advisory nature, they acquire a special status under the Sanitary and Phytosanitary Measures (SPS) Agreement and the Agreement on Technical Barriers to Trade (TBT).

Cooperation in the Framework of the Eurasian Economic Union (EAEU)

The Eurasian Economic Union (EAEU) was established with the aim of comprehensive modernization, cooperation, enhancing the competitiveness of national economies, and creating conditions for sustainable development in order to raise the living standards in the member states. The member states of the EAEU are Armenia, Belarus, Kazakhstan, Kyrgyzstan, and Russia.

According to Article 13 of the "Agreement on Common Principles and Rules of Technical Regulation in the Republic of Belarus, the Republic of Kazakhstan and the Russian Federation" (Eurasian Economic Commission [EEC], 2010), the Commission of the Customs Union decided to adopt the Technical Regulations of the Customs Union "On Food Safety" (EEC, 2011). According to Article 3 of the Technical Regulations (EEC, 2011), the objects of technical regulation are food products and processes of their production (manufacturing), storage, carriage (transportation), sale, and disposal associated with food product requirements.

The Treaty on the Eurasian Economic Union (EEC, 2014, paragraph 1, Article 53) confirmed that products put into circulation within the EAEU must be safe. The rules and procedures for food safety assurance and food products handling are determined by the technical regulations of the EAEU. Technical regulations have direct force in the territory of the EAEU. Article 25 "Principles of the Customs Union Functioning" stipulates that within the Customs Union, there is a single mode of trade in goods with third parties.

Supreme Eurasian Economic Council (2018b) formulated the main directions of international activities of the EAEU in 2019. To wit, it is planned to develop and deepen interactions between the EAEU and the EU at the expert level on the issues of legislative regulation, standardization, technical regulation, and consumers' rights protection. Decision of the Supreme Eurasian Economic Council (2018a) emphasized the need for the EAEU to interact with China as a key partner.

Food safety issues in China are governed primarily by the Law "On Food Safety" (Government of the People's Republic of China, 2009) and the Law "On Quality and Safety of Agricultural Products" (Government of the People's Republic of China, 2006). However, despite the fact that from November 1, 2006, there is a specialized regulatory act on agricultural products in effect in China, some issues are regulated by the Law "On Food Safety."

MAIN FOCUS OF THE CHAPTER

Food Quality and Food Safety

Federal Law "On Quality and Safety of Food Products" (Government of the Russian Federation, 2000b) has clearly enshrined the definitions of food quality and food safety. Food quality is a set of characteristics of food products capable to meet human needs for food under normal conditions of their use. Food safety is a state of reasonable confidence that food products are not harmful under normal conditions of their use and do not pose risks to the health of present and future generations.

According to Article 469 of the Civil Code of the Russian Federation (Government of the Russian Federation, 1996) and Article 4 of the Law "On Protection of Consumers' Rights" (Government of the Russian Federation, 1992), quality is determined by contract conditions and mandatory regulatory requirements. Terms related to quality are also defined in GOST ISO 9000-2011 (Federal Agency on Technical Regulating and Metrology, 2011) which is identical to ISO 9000:2005 "Quality Management Systems – Fundamentals and Vocabulary" (International Organization for Standardization, 2005). In Section 3.1 of the Standard, along with the term "quality", other related terms are listed, namely: "requirement", "grade", "customer satisfaction", "capability", and "competence." Food safety is defined as a state of reasonable confidence which is based on the producer's information that a product is not harmful and hazardous under normal conditions of use. According to the Ministry of Health and Social Development of the Russian Federation (2008), harm to health is a violation of the anatomical integrity or physiological function of human organs and tissues as a result of physical, chemical, biological exposures, disease, disability, or death.

Federal Law "On Quality and Safety of Food Products" (Government of the Russian Federation, 2000b) does not contain a criterion for food safety but indicates what food products, materials, and goods are considered to be poor-quality and hazardous. The following food products, materials, and goods shall be excluded from circulation: incompliant with the requirements of regulatory documents; having apparent signs of poor quality; those which do not correspond to the information provided and in respect of which there are reasonable suspicions about their falsification; those without specified shelf life (for food products, materials, and goods with a mandatory shelf life requirement) or those with expired date; those which do not have a label containing information required by law or regulatory documents, or those for which there is no such information.

According to Food Security Doctrine (President of the Russian Federation, 2010), to ensure food safety, it is necessary to monitor a compliance with the requirements of Russia's legislation in the spheres of agriculture, fishery products, and food, including imported ones, at all stages of their production, storage, transportation, processing, and sale.

EEC (2011, Article 4) determines food safety is a state of food products with no evidence of unacceptable risk associated with harmful effects on humans and future generations. EEC (2011, Chapter 2) also establishes safety requirements to food products, namely: requirements for ensuring the safety of food products during their production (manufacturing); requirements for water supply of food production (manufacturing) processes; safety requirements for edible raw materials used in food production; requirements for the use of technological equipment and utensils in the process of production (manufacture) of food; requirements for storage and disposal of food production (manufacturing) waste; requirements for the processes of storage, carriage (transportation) and sale of food products; requirements for food recycling processes; requirements for the processes of obtaining non-processed food products of animal origin.

European Parliament (2002) enshrines the fundamental norms on the issues of one of the independent branches of the European Community legislative regulation – "food legislation", also referred to as (in English sources) "food law." Based on European Parliament (2002), high level of protection of human health and consumer interests in relation to food products can be guaranteed by food safety assurance. A product is understood as any processed, partially processed, or non-processed food item. In this context, European Parliament (2002) uses the definition of food safety, not food quality. To ensure food safety, all aspects of a continuous food production chain shall be controlled from primary production and feed production to the sale or supply of food products to a customer. Safe product is a product that does not contain hazards to human life and health, i.e. a product not expected to produce adverse effects on health, the risk of which must be avoided (prevented). Hazard is understood as biological, chemical, or physical processes present in food products or feed that can have detrimental health effects, or as a condition of these food products or feed that can have similar impact. For these purposes, European Parliament (2002) calls for risk analysis which includes three interrelated components: risk assessment, risk management, and risk communication. Furthermore, European Parliament (2002) establishes the presumption of unsafe food. According to European Parliament (2002, Section 4), food shall be deemed to be unsafe if it is injurious to health of a consumer and subsequent generations and has probable cumulative toxic effects (European Parliament, 2002).

In China, the legislation distinguishes between the concepts of quality and safety of food products. Only the definition of safe food products is directly enshrined. Government of the People's Republic of China (2009, Chapter 10) states that safe food products are non-toxic, harmless foods that meet the established requirements for nutritional values, which cannot lead to acute, subacute, and chronic diseases. Food safety is assured through the monitoring and risk assessment, development of standards, prescription of control measures covering various stages of food production, storage, transportation, sales, as well as inspection of raw materials, semi-finished products, and finished goods. It is interesting to note that the above-mentioned regulatory acts contain constituent elements of tortious conduct in the sphere of food security and enshrine the relevant penalties in the form of both administrative fines and compensation for harm caused to consumers.

Thus, China's legislation on food safety includes special provisions for civil torts aimed at protecting the life and health of final consumers of food. Given that China's tort liability law is new and largely reflects advanced scientific experience, it seems appropriate to conduct a comparative analysis of Russian and Chinese legislation for its adequacy to new challenges in the sphere of food security, one of which is the rapid development of 3D-printing technology. This technology is a new challenge in legal support of food security. Currently, 3D is one of the fastest-growing technologies is the technology in printing. This technology is an example of developing additive manufacturing technology concept that differs from traditional subtractive manufacturing. The additive technology is based on a special bonding technique, when a 3D printer through sequential bonding of ingredients (powders, metal, polymers, etc.), performs layer-by-layer printing of a new 3D object. Such printing requires appropriate software and preliminary creation of a digital 3D model of a future object (Computer-aided design files – CAD-files) which can be obtained, for example, by 3D scanning.

3D-printing technology is already beginning to have a serious impact on the development of society. Goldman Sachs (2013) listed 3D-printing technology among the eight technologies that radically transform the existence of mankind and change the vision of human capabilities and limits. 3D-printing technology poses serious challenges not only in the economic sphere but also to the legal system, which lags behind the rapid scientific and technological progress in its evolution. Lawmakers and law enforcers will soon have to answer the questions put by the new industrial revolution. For example, one of the fundamental challenges to the legal system is the problem of 3D-printing of food.

According to Tran (2016), 3D printer will soon become a common kitchen device representing the development of a "smart kitchen" concept (Lupton & Turner, 2018). Active deployment of 3D-printing technology in food production will have serious socio-economic consequences due to the phenomenon of "consumer co-creation", that is, consumer involvement in the process of producing various welfare items, including foods. Market segment associated with consumer-centric mass innovation is emerging (Li et al., 2014). Owing to 3D printing, people get the possibility to create freely whatever they wish, this opens the door to the next wave of "home innovation" (Desai & Magliocca, 2014; Dubuisson, 2014).

However, decentralization of production associated with the new technology deployment carries the risk of losing control over food quality and safety. The technology of 3D printing of food products creates two main challenges that require an adequate legal solution. First, the issue of ensuring proper quality and safety of printed foods. Second, their proper labeling, i.e. informing consumers that a particular food product is not organic but synthetic and was produced by 3D printing (Tran, 2016a). From the long-term risk assessment standpoint, there is no any empirical data on the possible effects, including impact on health of future generations, of the replacement of natural food with 3D-printed food. Besides, food can be printed from either organic or synthetic ingredients. Tran (2016b) and Endres (2000) suggest that the problems of safety and labeling of food products manufactured using 3D-printing technology should be addressed in a way similar to genetically modified foods.

The lack of empirical data on the negative impact of consumption of GMO-containing food persuaded the EU countries to base their legislation on the precautionary principle. The principle allowing for taking preventive measures, "precautions" in the absence of precise scientific data was first formulated in the Rio Declaration on Environment and Development (United Nations Conference on Environment and Development, 1992). The precautionary approach was further developed in the Cartagena Protocol on Biosafety to the Convention on Biological Diversity (Secretariat of the Convention on Biological Diversity, 2000).

Russia is not a party to the Cartagena Protocol, but the President of the Russian Federation (2013) indicates the need to improve the regulation of transboundary movement of genetically modified organisms and accede to the Cartagena Protocol. Russian legal system has been improving continually to prepare an effective response to the global challenge associated with the production and circulation of GMO-containing food products. Currently, such products are subject to appropriate state registration (Government of the Russian Federation, 2013). In China, genetically modified foods are also subject to appropriate labeling (Government of the People's Republic of China, 2009, Article 69).

Modern science is not fully confident in the safety of food produced by 3D-printing technology, especially that of products printed from synthetic ingredients. Therefore, it seems reasonable to extend the precautionary approach to foods produced by 3D printing. Consumers are entitled to know that they purchase goods manufactured by 3D-printing technology, that is, exercise their right to be informed about the product (Government of the Russian Federation, 1992, Article 10; Government of the People's Republic of China, 2013, Article 8) and the right to safety of the goods (Government of the Russian Federation, 1992, Article 7; Government of the People's Republic of China, 2013, Article 7). This implies the need to establish mandatory requirements for equipment and ingredients used in production of food by 3D printing. Such food products shall be the subjects to mandatory labeling.

3D Printing of Food Products

Food must be safe for both present and future generations. In this regard, there is a need to develop an effective model of civil liability for the harm caused by food produced using 3D-printing technology, especially, when the harm is inflicted to the public at large, and the tortfeasor identity is unknown, i.e. in a "mass products tort" situation. The tortfeasor may be unidentifiable due to the fact that the harm to health is caused by a separate component, ingredient that is brought on the market by dozens, hundreds of manufacturers, and then massively used by people in 3D printing of food.

Tran (2016a) conceded that strict (no-fault) joint and several liability could be established for all persons who participated in production and sale of hazardous foods or their components (product liability). A well-known case of Summers v. Tice is casually mentioned in that context (Tran, 2016a). In this regard, it is necessary to comparatively analyze the development trends of product liability doctrine in order to develop an effective liability model that can be used in assigning the obligation to compensate harm caused by food products produced by 3D printing, including their components.

The active use of 3D-printing technology in everyday life, its application for production of goods and food products can generate a wave of lawsuits related to the situation of massive infliction of harm to health of an indefinite number of people ("public at large"). Pelley (2018) and Joob and Wiwanitkit (2017) have already reported higher risk of respiratory, cardiovascular, and oncological diseases due to expanded use of 3D-printing technology since the operation of 3D printers is accompanied by the emission of harmful substances and gases that get into the respiratory tract and blood. In Russia, this tort is covered by the Government of the Russian Federation (1996, articles 1095-1098) which imposes no-fault liability for the harm resulting from the product design flaw, formulation flaw, or other defects in the goods both on a producer and a seller of such goods regardless of any contractual relations with a victim (Government of the Russian Federation, 1996, Article 1095).

The establishment of liability for harm caused by the defects in goods (product liability) regardless of the tortfeasor's fault is a characteristic of many legal systems, for example, the ones in the USA (Beck & Jacobson, 2017) and China (Li & Jin, 2014). The laws governing product liability in China are, in general, similar to those applied in Russia. China's approach is more in line with the European experience of legislative regulation in this sphere than that of the USA. Unlike Russia, in China, one should differentiate between no-fault liability standard for producers and fault liability standard for sellers which is established by the Government of the People's Republic of China (2010, articles 41 and 42). The Russian approach establishing strict no-fault liability for both a producer and a seller appears to be more successful and fair for a victim. However, in China, a seller may also be a subject to strict (no-fault) liability if he fails to point to a particular producer or supplier of the goods. In accordance with the Government of the People's Republic of China (1993, Article 41), a producer is exempted from liability if he can prove the presence of any of the following circumstances: the product was not put into circulation; the defect that caused the damage occurred after the product was put into circulation; the level of science and technology at the time of putting the product into circulation did not allow to identify the defect.

In the Russian legal system, the attitude towards the goods manufacturer is even more stringent, since the Government of the Russian Federation (1992, paragraph 4, Article 14) reads that a producer (performer) is responsible for the harm caused to life, health or property of a consumer through the use of materials, equipment, tools, and other means necessary for the production of goods (performance of works, provision of services), regardless of whether the level of scientific and technical knowledge allowed them to reveal their specific properties or not. The Russian approach in this matter seems to be more equitable since if a businessman (manufacturer) benefits from putting into circulation goods with dangerous properties, he cannot be exempt from liability solely on the grounds that the level of scientific and technical knowledge did not allow to identify the dangerous properties of the goods. If a businessman puts into circulation a dangerous product to earn profit, then the risk of harm to the injured should be placed on the businessman. Otherwise, the law will become antisocial.

3D Food Printing Technology

A victim in both Russia and China must prove the fact of harm and justify the amount of harm and the presence of causal relationship between the tortfeasor's behavior and the occurrence of harm to the victim. Thus, regardless of the nature of tort liability (fault or regardless of fault), a victim must reasonably point out the particular tortfeasor who caused the particular harm. A victim must prove the presence of a physical connection between the tortfeasor's behavior and the occurrence of harm to a victim (Jones, 2007). This is the basic postulate of individualistic understanding of tort liability, individualized causation.

Rapid development of technologies and consequent evolution of social and economic sphere of society give rise to a crisis in the individualized causation theory of civil law. The individualized causal relationship is replaced by the concept of socialized causation. A serious challenge to the principle of individualized causation was posed by the development of liability for harm caused by defects in goods, works or services (product liability) in situations of harm to public at large from similar goods produced by various manufacturers (mass products torts). According to Gifford (2005), in their mutual interaction, a collective victim and a collective or indefinite tortfeasor created a fundamental challenge to the traditional requirement of individual causation in tort law.

Socialization of tort liability is manifested in the strengthening of its distributional orientation. For example, according to the Government of the People's Republic of China (2010, Article 1), one of its goals is to ensure social harmony and stability. The function of tort law is to distribute losses and balance social interests through the redistribution of social benefits. This function of tort law is implemented, inter alia, through the establishment of joint and several liability, mixed liability, and shared liability (Li & Jin, 2014).

The concept of socialized causation is aimed at ensuring equitable distribution of harm (damages) by shifting the burden of proof from the victim to the alleged tortfeasor who has put the product with hazardous properties into circulation. One of the first attempts to de-individualize causal relationship in situations involving the infliction of harm by product deficiencies was the concept of market-share liability. The leading precedent of this concept is judgment in the case of Sindell v. Albott Laboratories. Essentially, liability, in this case, was imposed not for the fact of causing harm to a specific victim, but for the fact of putting hazardous goods in the circulation. For example, according to Rosenberg (1984), market-share liability is a form of proportional liability associated with creating the risk of harm to the victim. A similar position was expressed by Robinson (1982) and Wright (1985).

Tran (2016b) analyzed the problem of compensation for harm caused by 3D-printed goods and their components and conceded that strict (no-fault) joint and several liability can be established for all persons who participated in production and sale of hazardous foods or their components (product liability), mentioning the well-known precedent in the case of Summers v. Tice. Snider (2019) believes that liability of manufacturers of 3D printed goods and their components shall be constructed using market-share liability model. In the case of Summers v. Tice, the California Supreme Court formulated the concept of alternative liability or alternative causation. The Court, later constructing the market-share liability model, considered it as based on an extended interpretation of alternative liability concept. Geistfeld (2006) also expressed an opinion about the doctrinal unity of market-share liability and alternative liability for a more successful use of the first in judicial practice. "Ratio decidenti" of this judgement was included in The American Law Institute (1965, Part 433) and later in HeinOnline (n.d., Part 28), according to which, if the plaintiff makes a claim to several persons and can prove that each of them by his behavior put a victim before the risk of causing harm, and such behavior did cause harm, but a victim, reasonably and expectedly, was not able to prove who exactly caused the harm, then the burden of proving the absence of causal relationship is placed on the defendants.

The concept of alternative causation is also known to the European doctrine. Thus, according to Item 1 of Art. 3:103 (Alternative causes) of the Principles of European Tort Law (Busnelli et al., 2005), in case of multiple actions, when each of them separately could be a sufficient cause of harm, but it remains unclear which of them actually caused it, each of these actions is considered as a cause to that extent which corresponds to the probability of causing harm to the victim by such action.

Following the recent reform, Chinese tort liability legislation also explicitly provides for an alternative liability model. For example, according to the Government of the People's Republic of China (2010, Article 10), if two or more persons are involved in behavior that endangers personal safety or property security of another person, and such behavior subsequently caused harm to the victim, and a specific tortfeasor is unidentifiable, then all persons who created such endangerment are jointly and severally liable. This legal phenomenon is characterized as a manifestation of the tendency to socialize tort liability.

Apparently, alternative liability (causation) model will be actively used in the future for cases of causing harm by goods manufactured using the technology of three-dimensional printing. In this matter, the Chinese legislation reflects the advanced world experience and is more effective and equitable in comparison with the Russian legislation. The tortfeasor may also be unidentifiable due to the fact that harm to health of victims will be caused by a separate component, ingredient used along with others in production of goods by three-dimensional printing. In this regard, of interest are the rules of the French Civil Code (FCC) (Government of France, 2016). In FCC Article 1245-5, a producer is understood not only as a producer of the final goods (product) but also as producer of raw materials or components (ingredients) of a product. According to FCC Article 1245-7, in case of damage caused by a defect in goods (product) incorporated into another product, a producer of a separate component and a person who performed such incorporation shall be jointly and severally liable for the harm caused. Thus, a liability in solidum is directly established both for the OEM manufacturer and the person who manufactured its separate parts (components). Apparently, to ensure equitable liability for harm, it is necessary to establish the same liability in solidum both in relation to persons who have manufactured the hazardous final product using 3D-printing technology and those who have manufactured the pertinent printing components which predetermined the final product properties harmful to consumers' health.

Punitive Damages as a Tool to Ensure Food Security

In Russia and China, tort liability is not a mono but a multifunctional legal phenomenon. For the Russian and Chinese legal systems, the key value lies in compensatory-restorative function of tort liability which is expressed in the full compensation conception of damages (harm). However, considering liability from the standpoint of a monofunctional phenomenon primitives the doctrine. Tort liability cannot be placed on the Procrustean bed of corrective justice and exclusively compensatory orientation since it is a multifunctional phenomenon. In civil law, liability solves the tasks of not only compensation but also a fair distribution of the burden of adverse consequences (distributive justice), as well as retribution for the wrongs committed (retributive justice). Thus, the entire triad of private justice in the sphere of liability is implemented: corrective, distributive, and retributive (Bogdanov, 2016).

The theoretical postulate that civil liability solves the tasks of not only compensation but also a fair retribution for the wrongs committed, as well as prevention and control of antisocial behavior is typical of North American civilians. This is due to the fact that the doctrine of punitive damages is actively used in their law enforcement practice (Markel, 2009).

A decade ago, Koziol (2008) analyzed the prospects for punitive damages in Europe and revealed that the concept of punitive damages did not correspond to the European legal tradition. However, he drew attention to the fact that the new edition of "The Rome II" (European Parliament, 2007) had softened the approach to overcompensation damages. Finally, Koziol (2008) concluded that in the sphere of private law, it was possible to use the construction of punitive damages, for example, in cases of non-pecuniary damage or in respect of difficult-to-prove economic losses.

A decade after the release of Koziol's publication, there happened a revolutionary departure from the classical understanding of the civil liability functional orientation in Italy, since the Italian Court of Cassation in the case of AXO Sport, SpA v. NOSA Inc (July 5, 2017) had formulated a legal position that the task of civil liability is not only to provide compensation to the victim but also to prevent unlawful conduct and to impose punishment upon the tortfeasor. The Court of Cassation indicated that punitive damages are not ontologically incompatible with the Italian legal system. The law enforcement practice of other European countries, for example, France and Spain, is also developing in this direction (Venchiarutti, 2018).

Government of the Russian Federation (1996) is actually familiar with the phenomenon of punitive (non-compensatory) damages established, for example, in part 2, clause 2, article 15. If a person who infringed the right has received any income as a result of infringement, a person whose right was infringed is entitled to demand, along with other lost profit damages, compensation in an amount not less than the income received. Thus, the law established the possibility of seizing unduly received benefits of the tortfeasor regardless of the losses that were incurred by the victim.

Zhang (2011) points out that China's legal system also explores the category of punitive damages which combine the functions of compensation, punishment, and prevention. The recovery of punitive damages in China is possible only in cases expressly provided by law. For example, Government of the People's Republic of China (2010, Article 47) establishes the possibility to recover punitive damages in favor of the victim if the manufacturer or seller deliberately brought into circulation (produced or sold) defective goods. Recovery of punitive damages is possible if the defect of the goods caused the death or harm to health of the victim.

Punitive damages are also provided for in the legislation aimed at ensuring food products safety, in particular, Government of the People's Republic of China (2009, Article 148) sets forth that in the case of production or sale of food products not conforming to food safety standards, in addition to compensation for the inflicted harm, a consumer is entitled to demand that a producer or a seller pays a penalty equal to tenfold cost of the goods or threefold amount of harm.

SOLUTIONS AND RECOMMENDATIONS

For the Russian legal system, China's experiences of introducing a model of socialized (alternative) tort liability, as well as experience in local enshrinement of punitive damages doctrine, are of great interest. Apparently, for China, it would be useful to establish a strict (no-fault) liability standard for food sellers and to exclude instruction to take into account the level of science and technology at the time of putting the product into circulation from tort liability exemption factors. The development of 3D food printing technology predetermines the need for both legal systems to take into account the legal experience of France regarding the joint and several liability of both a producer of the final product (food) and a producers of its individual component.

FUTURE RESEARCH DIRECTIONS

Food quality and safety shall be considered from both short-term and long-term perspective. Thus, according to the Government of the Russian Federation (2000b, Article 1), food safety is a state of reasonable confidence that food products under normal conditions of their use are not harmful and do not pose hazard for the health of present and future generations. The new revision of China's law on food safety states that it is adopted in order to ensure food safety and protect health and life of the population.

Therefore, the issue of ensuring the safety of food products manufactured using 3D-printing technology shall be considered as one of the strategic goals in the development of the legal system.

As a rule, the legal system lags behind technological development. Rapid development of technology, first of all, entails transformations in socio-economic relations, to which the legal system has to respond in the "catch-up" development model. Tran (2016a) points out that the legal system lags in regulating relations connected with the use of 3D-printing technology since special legal provisions are not yet available.

Further research is needed to improve the legislation in the field of quality and safety of 3D-printed food products, as well as compensation for damage caused by such products.

CONCLUSION

This comparative legal study suggests that the convergence of Russia-China legal experience in the sphere of consumer protection and food safety can increase the effectiveness of legislation in both Russia and China. It will allow the legal systems of Russia and China to respond effectively to new challenges associated, inter alia, with the development of 3D food printing technology.

It can be asserted that, to secure consumer interests in the sphere of food security, China's legislation has developed effective remedies that are not only aimed at imposing punishment for already committed torts, but are also able to prevent new torts and control unlawful encroachments in food security, since the punitive damages doctrine is an effective economic tool for the general prevention of torts, creating economic incentives for businessmen to ensure compliance with safety standards in production and sale of foods.

In the Russian legal system, such special provisions are not yet available, so it seems appropriate to take into account China's experience in order to improve Russia's legislation in the sphere of consumer protection and safety assurance in production and circulation of food.

REFERENCES

Beck, J. M., & Jacobson, M. D. (2017). 3D Printing: What Could Happen to Product Liability When Users (and Everyone Else in Between) Become Manufactures. *Minnesota Journal of Law, Science & Technology, 18*, 143–150.

Bogdanov, D. (2016). *Evolution of Civil Liability from a Position of Justice: Comparative Legal Aspect.* Moscow: Prospect.

Busnelli, F. D., Comande, G., Cousy, H., Dobbs, D. B., Dufwa, B. W., & Faure, M. G. ... Widmer, P. (2005). Principles of European Tort Law. Vienna: Springer.

Desai, D. R., & Magliocca, G. N. (2014). Patents Meet Napster: 3D Printing and the Digitization of Things. *The Georgetown Law Journal, 2*, 1691–1715.

Dubuisson, T. (2014). *3D Printing and the Future of Complex Legal Challenges: The Next Great Disruptive Technology Opportunity or Threat?* Retrieved from https://papers.ssrn.com/sol3/papers.cfm?abstract_id=2718113

Endres, B. A. (2000). "GMO": Genetically Modified Organism or Gigantic Monetary Obligation? The Liability Scheme for GMO Damage in the United States and the European Union. *Loyola of Los Angeles International and Comparative Law Review, 22*, 453–462.

Eurasian Economic Commission. (2010). *Agreement on Common Principles and Rules of Technical Regulation in the Republic of Belarus, the Republic of Kazakhstan and the Russian Federation.* Retrieved from http://www.eurasiancommission.org/en/act/texnreg/Pages/acts.aspx

Eurasian Economic Commission. (2011). *Technical Regulations of the Customs Union TR CU 021/2011 "On Food Safety."* Retrieved from http://www.eurexcert.com/TRCUpdf/TRCU-0021-On-food-safety.pdf

Eurasian Economic Commission. (2014). *Treaty of the Eurasian Economic Union.* Retrieved August 14, 2019, from https://docs.eaeunion.org/en-us

European Parliament. (2002). *Regulation (EC) #178/2002.* Retrieved from https://eur-lex.europa.eu/legal-content/EN/ALL/?uri=celex%3A32002R0178

European Parliament. (2007). *Regulation (EC) #864/2007.* Retrieved from https://eur-lex.europa.eu/legal-content/en/ALL/?uri=CELEX%3A32007R0864

Federal Agency on Technical Regulating and Metrology. (2011). *Order #1575 from December 22, 2011, "On the Introduction of Interstate Standard."* Retrieved from http://docs.cntd.ru/document/902387886

Food and Agriculture Organization of the United Nations. (1951). *International Plant Protection Convention.* Retrieved from https://www.ippc.int/en/publications/1997-international-plant-protection-convention-new-revised-text/

Food and Agriculture Organization of the United Nations. (2006). *International Standards for Phytosanitary Measures.* Retrieved from http://www.fao.org/3/a0450e/a0450e.pdf

Food and Agriculture Organization of the United Nations. (2019). *FAO's Work on Food Safety.* Retrieved from http://www.fao.org/publications/highlights-detail/en/c/1180272/

Food and Agriculture Organization of the United Nations, & World Health Organization. (2019). *CODEX. Protecting Health, Facilitating Trade.* Rome: Food and Agriculture Organization of the United Nations.

Food and Agriculture Organization of the United Nations. (n.d.). *Russian Federation.* Retrieved from http://www.fao.org/countryprofiles/index/en/?iso3=RUS

Geistfeld, M. A. (2006). The Doctrinal Unity of Alternative Liability and Market-Share Liability. *University of Pennsylvania Law Review, 155*(2), 447–501. doi:10.2307/40041311

Gifford, D. G. (2005). The Challenge to the Individual Causation Requirement in Mass Products Torts. *Washington and Lee Law Review, 62*(3), 873–935.

Goldman Sachs. (2013). *The Search for Creative Destruction.* New York, NY: Goldman Sachs.

Government of France. (2016). *French Civil Code 2016.* Retrieved from https://www.trans-lex.org/601101/_/french-civil-code-2016/

Government of the People's Republic of China. (1993). *Product Quality Law of the People's Republic of China.* Retrieved from http://english.mofcom.gov.cn/article/policyrelease/Businessregulations/201303/20130300046024.shtml

Government of the People's Republic of China. (2006). *Law "On Quality and Safety of Agricultural Products."* Retrieved from https://www.fsvps.ru/fsvps-docs/ru/importExport/china/files/china_law_quality.pdf

Government of the People's Republic of China. (2009). *Law "On Food Safety."* Retrieved from https://www.fsvps.ru/fsvps-docs/ru/importExport/china/files/zakon1.pdf

Government of the People's Republic of China. (2010). *Tort Law of the People's Republic of China.* Retrieved from https://www.wipo.int/edocs/lexdocs/laws/en/cn/cn136en.pdf

Government of the People's Republic of China. (2013). *Law "On Protection of Consumers' Rights."* Retrieved from https://chinalaw.center/civil_law/china_consumer_rights_protection_law_revised_2013_russian/

Government of the Russian Federation. (1992). *Law #2300-I from February 7, 1992, "On Protection of Consumers' Rights."* Retrieved from http://base.garant.ru/10106035/

Government of the Russian Federation. (1993a). *Constitution of the Russian Federation.* Retrieved from http://www.constitution.ru/en/10003000-01.htm

Government of the Russian Federation. (1993b). *Law #4979 from May 14, 1993, "On Veterinary Medicine."* Retrieved from http://base.garant.ru/10108225/

Government of the Russian Federation. (1996). *Law #14 from January 26, 1996, "Civil Code of the Russian Federation."* Retrieved from https://legalacts.ru/kodeks/GK-RF-chast-2/

Government of the Russian Federation. (1997). *Decree #1263 from September 29, 1997, "On the Regulations on the Examination of Low-Quality and Dangerous Food Raw Materials and Food Products, Their Use or Destruction."* Retrieved from https://legalacts.ru/doc/postanovlenie-pravitelstva-rf-ot-29091997-n-1263/

Government of the Russian Federation. (1999). *Law #52 from March 30, 1999, "On Sanitary and Epidemiological Wellbeing of Population."* Retrieved from http://base.garant.ru/12115118/

Government of the Russian Federation. (2000a). *Decree #883 from November 22, 2000, "On the Organization and Conduct of Quality, Food Safety and Public Health Monitoring."* Retrieved from https://legalacts.ru/doc/postanovlenie-pravitelstva-rf-ot-22112000-n-883/

Government of the Russian Federation. (2000b). *Law #29 from January 2, 2000, "On Quality and Safety of Food Products."* Retrieved from http://base.garant.ru/12117866/

Government of the Russian Federation. (2002). *Law #184 from December 27, 2002, "On Technical Regulation."* Retrieved from http://base.garant.ru/5139626/

Government of the Russian Federation. (2005). *Decree #50 from February 2, 2005, "On the Regulations on the Use of Means and Methods of Control in the Implementation of the Passage of Persons, Vehicles, Cargoes, Goods, and Animals Across the State Border of the Russian Federation."* Retrieved from http://pravo.gov.ru/proxy/ips/?docbody=&nd=102090891

Government of the Russian Federation. (2010). *Decree #1806 from October 18, 2010, "On the Approval of the Comprehensive Program for the Participation of the Russian Federation in International Cooperation in the Field of Agriculture, Fisheries and Food Security."* Retrieved from https://rulaws.ru/goverment/Rasporyazhenie-Pravitelstva-RF-ot-18.10.2010-N-1806-r/

Government of the Russian Federation. (2011). *Decree #501 from June 29, 2011, "On the Approval of the Rules for the Implementation of State Veterinary Supervision at Border Crossing Points of the Russian Federation."* Retrieved from http://pravo.gov.ru/proxy/ips/?docbody=&nd=102148718

Government of the Russian Federation. (2013). *Decree #839 from September 23, 2013, "On State Registration of Genetically Modified Organisms Intended for Release into the Environment, as well as Products Obtained Using Such Organisms or Containing Such Organisms, Including Specified Products Imported into the Territory of the Russian Federation."* Retrieved August 14, 2019, from http://www.consultant.ru/document/cons_doc_LAW_152217/

Government of the Russian Federation. (2014). *Law #206 from July 21, 2014, "On Plant Quarantine."* Retrieved from http://base.garant.ru/70699630/

HeinOnline. (n.d.). *Restatement, Third, Torts: Liability for Physical and Emotional Harm.* Retrieved from https://home.heinonline.org/titles/American-Law-Institute-Library/Restatement-Third-Torts-Liability-for-Physical-and-Emotional-Harm/?letter=R

International Organization for Standardization. (2005). *ISO 9000:2005 "Quality Management Systems – Fundamentals and Vocabulary."* Retrieved from https://www.iso.org/standard/42180.html

Jones, M. A. (2007). *Textbook on Torts.* Oxford: Oxford University Press.

Joob, B., & Wiwanitkit, V. (2017). Estimation of Cancer Risk Due to Exposure to Airborne Particle Emissions of a Commercial Three-Dimensional Printer. *Indian Journal of Medical and Paediatric Oncology: Official Journal of Indian Society of Medical & Paediatric Oncology, 38*(3), 409. doi:10.4103/ijmpo.ijmpo_118_17

Koziol, H. (2008). Punitive Damages – A European Perspective. *Louisiana Law Review, 68*(3), 741–764.

Li, P., Mellor, S., Griffin, J., Waelde, C., Hao, L., & Everson, R. (2014). Intellectual Property and 3D Printing: A Case Study on 3D Chocolate Printing. *Journal of Intellectual Property Law and Practice, 9*(4), 322–332. doi:10.1093/jiplp/jpt217

Li, X., & Jin, J. (2014). *Concise Chinese Torts Laws.* Heidelberg: Springer-Verlag. doi:10.1007/978-3-642-41024-6

Lupton, D., & Turner, B. (2018). Both Fascinating and Disturbing': Consumer Responses to 3D Food Printing and Implications for Food Activism. In T. Schneider, K. Eli, C. Dolan, & S. Ulijaszek (Eds.), *Digital Food Activism* (pp. 151–167). London: Routledge.

Markel, D. (2009). Retributive Damages: A Theory of Punitive Damages as Intermediate Sanction. *Cornell Law Review, 94*, 239–340.

Ministry of Health and Social Development of the Russian Federation. (2008). *Order #194 from April 24, 2008, "On the Approval of Medical Criteria for Determining the Severity of Harm Caused to Human Health."* Retrieved from http://base.garant.ru/12162210/

Pelley, J. (2018). Safety Standards Aim to Rein in 3-D Printer Emissions. *ACS Central Science, 4*(2), 134–136. doi:10.1021/acscentsci.8b00090 PMID:29532010

President of the Russian Federation. (2009). *Decree #537 from May 12, 2009, "On the National Security Strategy of the Russian Federation till 2020."* Retrieved from https://www.garant.ru/products/ipo/prime/doc/95521/

President of the Russian Federation. (2010). *Decree #120 from January 30, 2010, "On the Approval of Food Security Doctrine of the Russian Federation."* Retrieved from http://base.garant.ru/12172719/

President of the Russian Federation. (2013). *Decree #2573 from November 1, 2013, "Fundamentals of the State Policy in the Field of Ensuring Chemical and Biological Security of the Russian Federation for the Period up to 2025 and Beyond."* Retrieved from https://www.garant.ru/products/ipo/prime/doc/70423098/

President of the Russian Federation. (2015). *Decree #683 from December 31, 2015, "On the National Security Strategy of the Russian Federation."* Retrieved from http://base.garant.ru/71296054/

Robinson, G. O. (1982). Multiple Causation in Tort Law: Reflections on the Des Cases. *Virginia Law Review, 68*(4), 713–769. doi:10.2307/1072725

Rosenberg, D. (1984). The Causal Connection in Mass Exposure Cases: A "Public Law" Vision of the Tort System. *Harvard Law Review, 97*(4), 849–929. doi:10.2307/1341021

Secretariat of the Convention on Biological Diversity. (2000). *Cartagena Protocol on Biosafety to the Convention on Biological Diversity*. Montreal: Secretariat of the Convention on Biological Diversity.

Snider, M. (2019). *Asbestos and Additive Manufacturing: Addressing Early Concerns Surrounding Manufacturing 3D-Printing Technology Using Asbestos Litigation as a Model.* Retrieved from https://papers.ssrn.com/sol3/papers.cfm?abstract_id=3343881

Supreme Eurasian Economic Council. (2018a). *Decision #3 from May 8, 2018, "On the Approval of the Agreement on Trade and Economic Cooperation between the Eurasian Economic Union and Its Member States, from One Side, and the People's Republic of China, from Another Side."* Retrieved from https://docs.eaeunion.org/docs/ru-ru/01418737/scd_10052018_3

Supreme Eurasian Economic Council. (2018b). *Decision #19 from December 6, 2018, "On Major Directions of International Activity of the Eurasian Economic Union in 2019."* Retrieved from https://docs.eaeunion.org/docs/ru-ru/01420197/scd_07122018_19

The American Law Institute. (1965). *Restatement of the Law*. Washington, DC: American Law Institute Publishers.

Tran, J. (2016a). 3D-Printed Food. *Minnesota Journal of Law, Science & Technology, 17*, 855–880.

Tran, J. (2016b). Press Clause and 3D Printing. *Northwestern Journal of Technology and Intellectual Property, 14*(1), 75–80.

United Nations. (1992). *Convention on Biological Diversity*. Retrieved from https://www.cbd.int/convention/text/

United Nations Conference on Environment and Development. (1992). *Rio Declaration on Environment and Development*. Retrieved from http://www.unesco.org/education/pdf/RIO_E.PDF

Venchiarutti, A. (2018). The Recognition of Punitive Damages in Italy: A Commentary on Cass Sez Un 5 July 2017, 16601, AXO Sport, SpA v NOSA Inc. *Journal of European Tort Law, 9*(1), 104–122. doi:10.1515/jetl-2018-0105

Wright, R. W. (1985). Causation in Tort Law. *California Law Review*, *73*(6), 1735–1828. doi:10.2307/3480373

Zhang, M. (2011). Tort Liabilities and Torts Law: The New Frontier of Chinese Legal Horizon. *Richmond Journal of Global Law and Business*, *10*(4), 415–495.

ADDITIONAL READING

Bogdanov, E. (2008). The Influence of the Scientific Views of Leon Duguit on the Development of Civil Law. *Journal of Russian Law*, *6*, 32–38.

Chuyko, N. (2016). Application of Precautionary Principle when Settling a Dispute on Trade of Genetically Modified Products within the WTO. *International Economic Law*, *1*, 47–56.

Colby, T. B. (2003). Beyond the Multiple Punishment Problem: Punitive Damages as Punishment for Individual, Private Wrongs. *Minnesota Law Review*, *87*, 583–678.

Ellis, D.D., Jr. (1982). Fairness and Efficiency in the Law of Punitive Damages. *South California Law Review*, *56*.

Liu, C. (2018). Socialized Liability in Chinese Tort Law. *Harvard International Law Journal*, *59*, 16–44.

Oberdiek, J. (2008). Philosophical Issues in Tort Law. *Philosophy Compass*, *3/4*(4), 734–748. doi:10.1111/j.1747-9991.2008.00156.x

Osborn, L. S. (2014). Regulating Three-Dimensional Printing: The Converging Worlds of Bits and Atoms. *The San Diego Law Review*, *51*, 553–621.

Thomas, K. (2014). The Product Liability System in China: Recent Changes and Prospects. *The International and Comparative Law Quarterly*, *63*(3), 755–775. doi:10.1017/S0020589314000219

Zweigert, K., & Koetz, H. (1998). *An Introduction to Comparative Law*. Oxford: Oxford University Press.

KEY TERMS AND DEFINITIONS

3D Printing: A production of bulk products by applying the material using the print head, nozzle, or other printer components.

Codex Alimentarius (Food Code): A collection of standards, guidelines, norms, and rules approved by the Codex Alimentarius Commission.

Ensurance of Food Safety: A monitoring of compliance with the requirements of Russia's legislation in the spheres of agriculture, fishery products, and food, including imported ones, at all stages of their production, storage, transportation, processing, and sale.

Food Quality: A set of characteristics of food products capable to meet human needs for food under normal conditions of their use.

Food Safety: A state of reasonable confidence that food products are not harmful under normal conditions of their use and do not pose risks to the health of present and future generations.

Harm to Health: A violation of the anatomical integrity or physiological function of human organs and tissues as a result of physical, chemical, biological exposures, disease, disability, or death.

Safe Product: A product that does not contain hazards to human life and health, i.e. a product not expected to produce adverse effects on health, the risk of which must be avoided (prevented).

Tort: An act of injury or damage to a person or property that is covered by a law, so that a person can start a court action.

Chapter 10
Agrarian Reforms of the 20th Century in Russia:
Impacts on Agricultural Sector and Food Security

Olga Pasko
Tomsk Polytechnic University, Russia

Natalia Staurskaya
Omsk State Technical University, Russia

Alexey Gorodilov
Tomsk Polytechnic University, Russia

Alexander Zakharchenko
https://orcid.org/0000-0002-1559-7925
Institute of the Problems of Northern Development, Russian Academy of Sciences, Russia

ABSTRACT

Current political and economic reforms, as well as the development of market relations and private property rights, need a retrospect to the experience of the past. An ambitious reform implemented by Russian public entities in the early 20th century was a result of a compromise between the government, society, and individuals. The goals of the reforms offered by Pyotr Stolypin were similar to those of the contemporary ones. Stolypin's reforms aimed at the substitution of group type of land use by public property. The reforms were not evolutional but were motivated by the explosive political and social-economic situation. Another agrarian reform took place in the early 1990s in the Soviet bloc, including the USSR. It aimed at state land property and a centrally planned agrarian economy, the domination of big manufacturers like collective and communal farms, and state pricing control. Despite similar basic principles, the states chose different strategies for the implementation of agrarian reforms.

DOI: 10.4018/978-1-7998-1042-1.ch010

INTRODUCTION

In modern conditions of political and economic transformations and development of market relations and property rights, studying previous experience acquires particular importance. Reforms are important and momentous events in the history of any country as they stipulate national development strategies for many years. From a scientific point of view, reform is a reorganization of any aspect of social life without destroying the foundations of its social structure (History of Russia, n.d.). In agriculture, it is meant to be a complex of measures undertaken by a state in order to redistribute land ownership and increase production efficiency.

In the history of many countries, reforms were carried out for the benefits of upper-class' interests without consideration of common people. Such reforms gave rise to protracted conflicts, transformed the structure and efficiency of national economies, dramatically changed the ways of life of many people.

In this regard, it is important to study both achievements and failures of reforms not only from the point of view of the public good but also the interests of individuals. The history of the Soviet Union (USSR) and modern Russia provides extensive material for such a study from the reforms of Petr Stolypin to current agrarian transformations. In the majority of previous studies, the analysis has been carried out from a unified position, critical discussion is lacking. Most of the publications cover only particular aspects of agrarian reforms, such as law, economics, land management, agricultural production, and social sphere. Statistic data allowing to evaluate the progress and effectiveness of agrarian reforms is heterogeneous and difficult to compare. There are complications in the form of well-defined ideological approaches regarding strategies and means of conducting reforms in the early and mid-XX century. The examples are the theses about the need for an exclusively revolutionary way of conducting reforms in Russia or about the advantages of small farms over large ones.

Development of agricultural sector in the countries of the former USSR represents the unique historical experience. It started from similar positions but used different strategies and after 25 years came to different results. A comparison of the experience of 15 countries makes it possible to identify both general patterns and national (regional) features. Such reviews are now missing in the literature.

The purpose of this study is to analyze the strategies, course, and effectiveness of major agrarian reforms in Russia in XX century on the basis of common approaches. To achieve this goal, the authors attempt describing the reasons of agrarian reforms; revealing similarities and differences in carrying out reforms in different time periods and different territories of the USSR; analyzing the effects and major consequences of reforms.

BACKGROUND

A large-scale agrarian reform started in Russia in the beginning of XX century based on a compromise between the government and public society. Its goal was to increase manufacturing of high-quality products and minimize living and materialized labor, as well as reduce damage to the environment. To achieve the goal, the following tasks were supposed to be accomplished:

- Radical change in land relations
- Improvement of the organization of production and ensuring stable agricultural production
- Development of optimal conditions for life and work of rural population

Figure 1. Stolypin during inspection of the farm gardens in Moscow suburbs in 1910 and during the reception of a report in Pristanny village
Source: Volgareva and Pasko (2014)

- Equal distribution of human resources across the territory of Russia, including those people who were meant to protect eastern parts of the country from China and other countries of Asia
- Rational use of natural and production resources, particularly, land.

The reform started in 47 territories of the European part Russia. It dealt with political, economic, social, technical, technological, scientific, educational, and other aspects. The reform aimed at dynamic development of financial sector, infrastructure (road and housing construction), education (new universities, including engineering and pedagogical ones), medicine (universities, hospitals, and pharmacies), among other sectors. Transformation of agricultural and social institutions was supported by the Government. Government expenditures on land management grew exponentially from 2.3 million rubles in 1906 up to 14.1 million rubles in 1914.

The reform was inspired and implemented by Petr Stolypin, Russian Prime Minister at that time (Figure 1). Leading international experts contributed to the development and implementation of Stolypin's reform. Stolypin as its ideologist took full responsibility for its implementation. Stolypin declared that all the concerns of the Government were directed towards the implementation of progressive reforms. The relentless development of cities was supposed to go hand in hand with the economic advancement of rural life. The government promoted the penetration into the consciousness of the broad masses of the great truth that the people alone in work could find salvation.

The essence of both Stolypin's and modern agrarian reforms was transformation and abolishment of collective forms of land use (communities, collective farms) through the transition to an individual form of ownership (private owner, farmer) by radical liberal or administrative-command methods. In the center of the Government's concerns was the success of the institution of small-scale land ownership. The real progress of farming can be accomplished only in case of private land ownership, which develops in the owner a consciousness of both rights and obligations (Chramkov, 1994). According to Stolypin, it is necessary to give freedom, but at the same time, it is necessary to create citizenship and make the people worthy of their freedom (Bestuzheva-Lada, 2011). Both reforms were caused by not the consistent evolutionary progress in the development of economic relations, but tense political and socio-

Figure 2. Development of new territories: (a) explorers before going to work; (b) dam under construction; (c) wooden well; (d) farmyard of the immigrant family.
Source: Volgareva and Pasko (2014)

economic situation in Russia. In the course of the reform, land was transferred to private ownership. In reality, private and individual property was created at the expense of collective property. This allowed for a rapid growth in agricultural production. The reform started in 1906 and ended in 1917 when it was annulled by the Land Management Commissions (Volgareva & Pasko, 2014).

The historical need for large-scale agrarian reform was caused by many reasons, including low tax-capacity of agricultural producers and relative overpopulation of the European part of Russia (23 million people by 1900). Between 1907 and 1912, the amount of cash loans and non-repayable benefits amounted to 21.2 million rubles (Kofod, 1914). Almost 250 thousand households received aid. Loans were issued for such activities as drainage; irrigation; strengthening the banks of rivers, ravines, and loose sand; afforestation; cultivation of orchards, vineyards, and hop fields; clearing land; arrangement of farms in possessory estates and manor settlements in the resettlement of peasants within the allotment of their lands; purchase of livestock (Figure 2)

The majority of the peasants who arrived in Tomsk region in 1906-1914 were poor people and had no own funds (Shilovsky, 2003). This caused the need for their crediting from the state. For three years, the number of loans increased dramatically (Table 1).

Agrarian Reforms of the 20th Century in Russia

Table 1. Dynamics of state loans for immigrants in 1906-1909 in Tomsk region

Year	Number of Loans	Total Amount, rub
1906	2,754	106,857
1907	9,819	417,333
1909	47,725	2,291,704

Source: Volgareva and Pasko (2014)

During the years of the reform, the emergence and strengthening of credit cooperatives contributed to the growth in the number of credit unions. In 1911-1913, the number of credit unions in Tomsk region increased from 97 up to 297, their capital – from 835 thousand rubles up to 2,423 thousand rubles. The money provided was not enough for a complete device, but they helped the new settlers, who added the money brought to reach out to their first harvest (Tyukavkin, 1986; Rogachevskaya, n.d.). Large sums were allocated by the state for the organization of transportation, maintenance of agricultural machinery warehouses and equipment. In 1905-1911, total expenditures increased fivefold (Stolypin & Krivoshein, 1911). Construction of Trans-Siberian Railway contributed to the development of the region (Kulomzin, 1903) (Figure 3)

State Archive of Tomsk region stores many cases containing design estimates for the construction of roads, bridges, dams, as well as their repair and maintenance. They detail the name of the work and labor, the rates for building materials and labor, as well as their cost (Stolypin & Krivoshein, 1911).

Figure 3. Construction of Trans-Siberian Railway
Source: Volgareva and Pasko (2014)

Figure 4. Population growth in Tomsk region
Source: Directorate (1914)

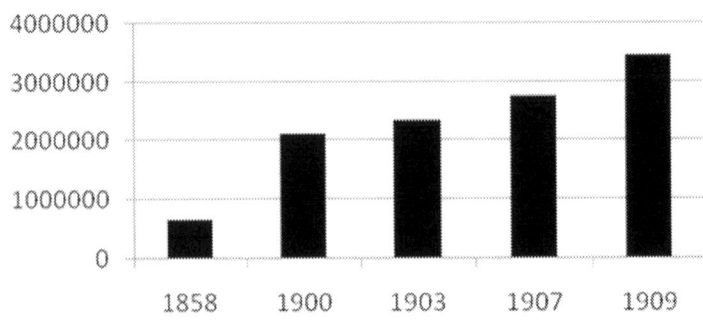

At the beginning of XX century, Tomsk region was one of the largest in the empire (today, its territory of Novosibirsk, Tomsk, Kemerovo, and Altai regions and the Republic of Altai) (General Directorate for Land Management and Agriculture [Directorate], 1914).

In August-September 1910, Stolypin and Krivoshein visited six districts of four provinces and regions of Western Siberia. In the report, they noted that annually Siberia provided hundreds of thousands of new plowmen with resettlement, and the acreage increased with amazing speed. In 1909, it amounted to 6 million tithes or about 0.6 tithes per capita in rural areas. In the European part of Russia, the ratio was 0.7 (Shilovsky, 2012).

In 1913, the area of the province was 78.9 million acres, the population – 3.57 million people (Directorate, 1914). The was one of the centers of the migration movement. Between 1890 and World War I, its population almost doubled. During 1893-1906, the resettlement process in Tomsk region was spontaneous and rather sluggish. In 1907-1908, it grew rapidly, but in 1909 declined (Figure 4).

Growing shortage of land in Europe is considered among the reasons for a radical change of residence in the beginning of XX century. Russia's Minister of Agriculture (1894-1905) Yermolov argued that there was no absolute shortage of land, but lack of land for preserving traditional forms of extensive farming (Belyanin, 2014). The government managed to achieve a skillful combination of interests of individual farmers and the state (Resettlement Directorate, 1911): "The government does not invite anyone to be resettled. The Government's help is provided only to those who, before moving with their family to new lands, will search for land for resettlement, examine it and, by order, enlist a free section of it themselves or through a trusted walker. The land for relocation is harvested annually much less than that required for the device of all those who wish to relocate. Therefore, anyone who has decided to relocate should first carefully consider this matter. Everybody has a lot of ways to improve his household at home. Only those who cannot help themselves in such ways, and who are firmly resolved by hard work and patience to overcome all the hardships of the first time in a new place, can resettlement be beneficial".

MAIN FOCUS OF THE CHAPTER

Land Resources and Land Management

Families with young workers preferred to move to Tomsk region. The resettlement department built in advance wells, ferries, and dirt roads. The work required a great deal of effort, but it progressed rather successfully. In 1914, the region experienced lack of fertile lands, which under the pressure of the annual migration camp was quickly spent in the order of the land allotment always haste for accommodating the immigrants (Dorofeev, 2014).

In the Description of Tomsk resettlement area (reference book for immigrants), published in 1911, it was noted that in Altai and in the area of the railway, all sections were dismantled, and free plots could be found in the north of Tomsk and Kainin counties only.

Amid the outbreak of the First World War, exhaustion of convenient fertile lands slowed down the pace of reform. The relative shortage of land resources and the lack of specialists led to the fact that the quality of the land was estimated not by peasants, but by workmen (Dorofeev, 2007), who were poorly versed in business. The work was done poorly with many shortcomings.

Land relations in Tomsk region differed in many ways from those established in the European part of Russia (Directorate, 1914).

- There was no landlord tenure.
- "Office" land tenure was distributed (Figure 5).
- Siberian old-timers often encountered communal land tenure.
- There were still areas for potential relocation, despite the fact that thanks to Stolypin's agrarian reform, hundreds of thousands of people moved to the region.

Figure 5. Land plots in Tomsk region in 1912
Source: Directorate (1914)

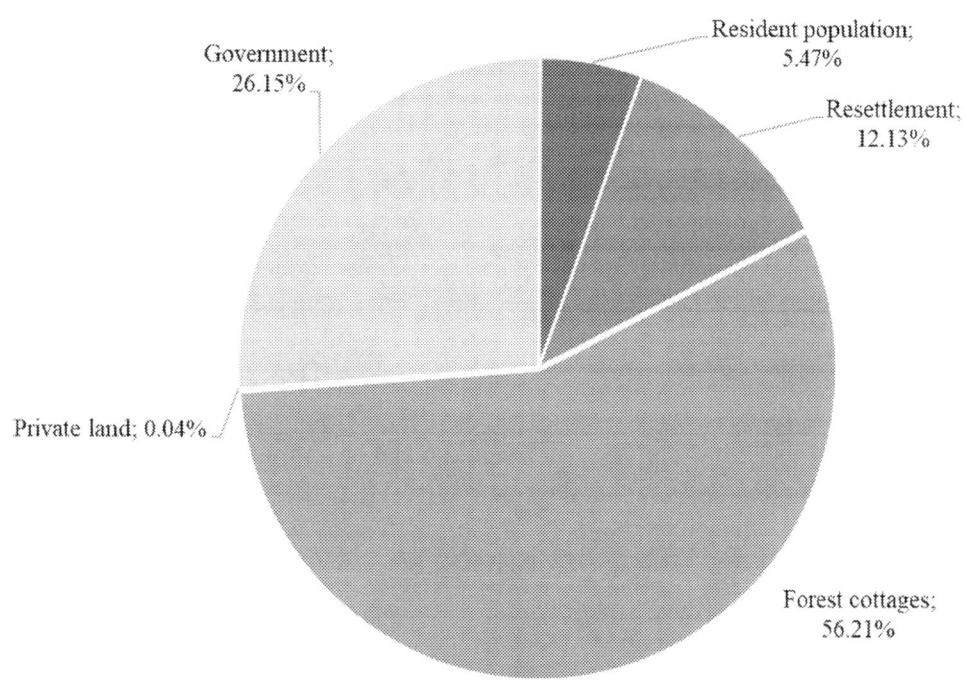

Figure 6. (a) migrants in the area of Chulym; (b) Zaimka on the bank of the river Kuendat, Tomsk region
Source: Volgareva and Pasko (2014)

a   b

According to the Atlas of Asian Russia (Directorate, 1914), over 3.3 million tithes were assigned to the land plots of local people, 7.4 million tithes under resettlement plots, and 34.4 million tithes as forest dachas of a single possession of the treasury.

During Stolypin's reform, there was a massive settlement of the northern areas of Tomsk region, particularly, along Chulym River (Figure 6).

Each farmer had a five-wall house, a bath heated by wood nine months a year; farm buildings made out of local tree species. Each owner kept cattle (over 15-20 heads of cattle, 40-60 heads of sheep and goats, and horses) and poultry. Major crops were rye, wheat, oats, and peas (Museum of Local History in Chulym, n.d.).

Hospitals and canteens were established in the resettlement centers (Figure 7).

Agricultural Production

Tomsk region was called the breadbasket of Siberia. Crop production and livestock breeding contributed the most to the gross regional product. The province was the leader in Russia in the number of domestic animals (2.5 million heads of cattle, 7% of all horses in Russia). The region held the first place in haymaking – over 300 million tons annually. In 1912, there were 2,609 butter-making factories. Agricultural products were distributed across Russia and exported to Europe (Goryushkin, Bocharova, & Nozdrin, 1993).

Production of wheat expanded rapidly. The growth of agricultural land under wheat markedly outpaced the growth of settlement, which indicated a greater activity of migrants compared with local dwellers. Each year, resettlement gave Siberia hundreds of thousands of new plowmen, and the acreage grew. In 1909, it amounted to 6 million tithes (Stolypin & Krivoshein, 1911). The economic development of the territories led to an increase in yields (Figure 8). The immigrants brought to Siberia many new varieties of cultivated plants, especially fruit, vegetable, and flower crops, which markedly enriched the possibilities of crop production.

As a result of an increased production, trade emerged with the help of entrepreneurs in rural areas. One of the immigrants, Mr. Novikov, got successfully engaged in arable farming, cattle breeding, and

Figure 7. Typical design of the hospital building at medical centers of Tomsk resettlement area: (a) 1913; (b) new temple in Siva; (c) rural family
Source: Volgareva and Pasko (2014)

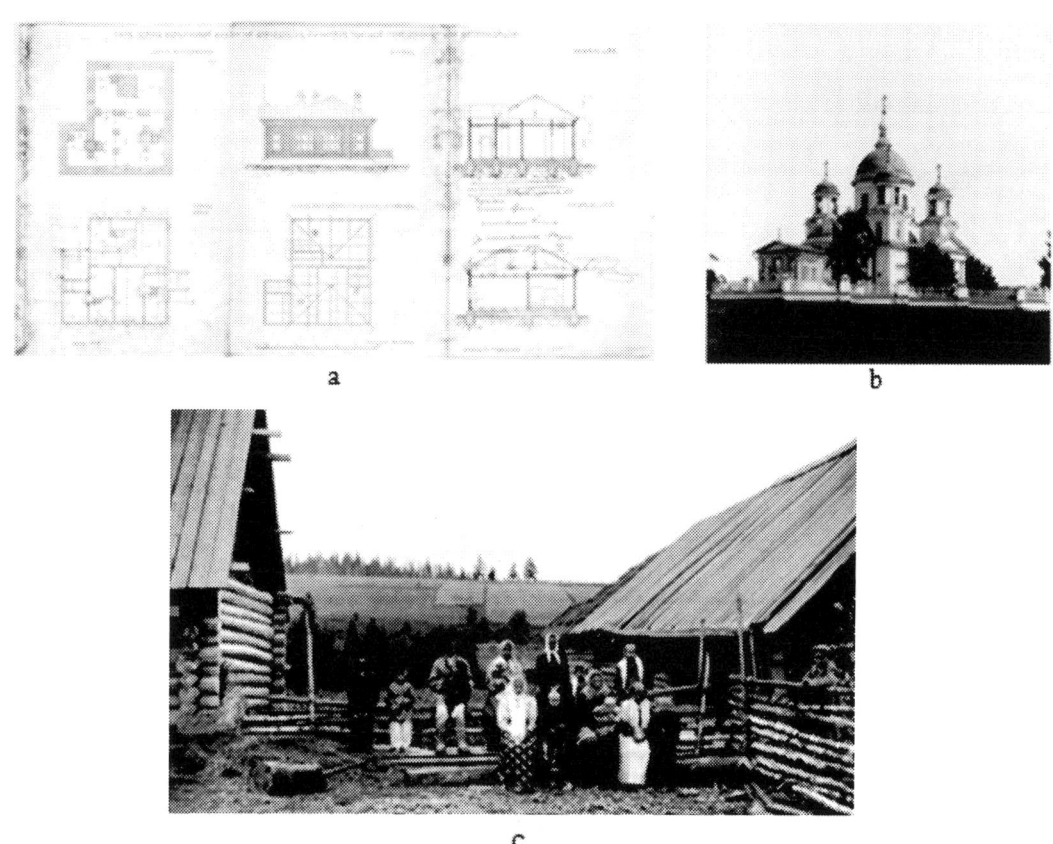

Figure 8. Growth of acreage and yield in the Tomsk province: (a)- area under crops, million acres; (b) – ehe harvest of grain, million pounds.
Source: Atlas of Asian Russia (1914)

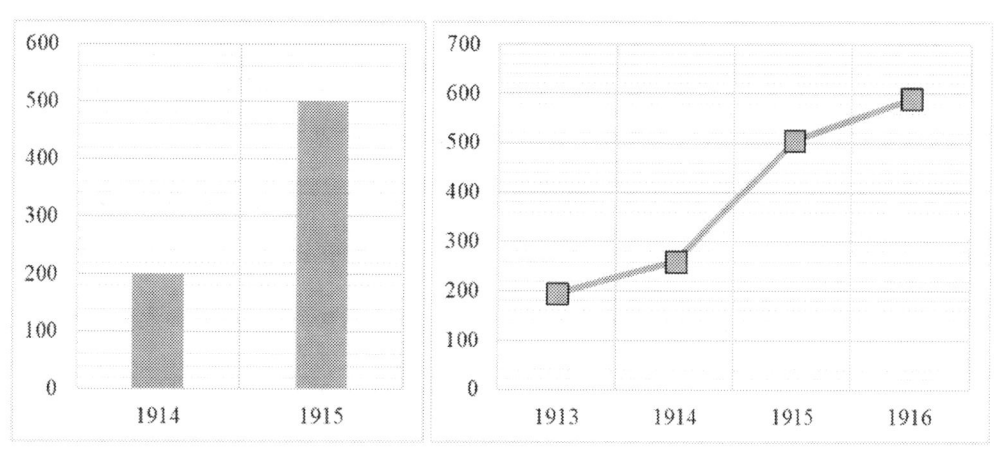

trade. He managed 80 workers, 40 clerks, and hundreds of local laborers in the areas of manufacturing, rural trade, buying up agricultural goods and furs. Mr. Novikov was the owner of numerous inns, as well as vehicles – horses, riverboats, etc

Development of agricultural production resulted in the full supply of bread to Siberia and the possibility of its supply to the European part of Russia. Grain surplus exceeded 5.6 thousand tons of grain. Its appearance in the new market plunged producers of Kuban and Central Black Earth grain, suppliers of cheap grain from the USA, and the introduction of "internal customs". As a result of a sharp increase in the tariff for the transit of Siberian bread, its price also increased. There was a conflict between powerful supporters (merchants Stakheyev and Putilov) and opponents (Senator Khvostov) of Siberian agriculture. According to Mr. Krivoshein, Chief Governor of Land Management and Agriculture at that time, Siberian bread did not compete with Russian landowners, and Siberian trade could not be delayed even if it threatened competition since Siberia was an integral part of Russia. In 1913, unfair tariffs were abolished, and Siberia received an opportunity to increase production of grain.

In twenty years, population of the region increased by 2.6 times. Tomsk became one of the richest and beautiful cities in Siberia (Figure 9).

Figure 9. (a) Millionnaya Street; (b) market in Tomsk
Source: Directorate (1914)

a

b

Figure 10. Professors of Tomsk Institute of Technology
Source: Directorate (1914)

Complexity, flexibility, and a vision of the future were the major factors of the reform. Two examples are medicine and education. The development of production and construction required qualified specialists. Establishment of Tomsk Institute of Technology (Figure 10) and Imperial Siberian University provided an opportunity to obtain higher education for people living in Siberia. The industrialists of Siberia and the Urals made an invaluable contribution to the development of young universities. In the future, they hoped for a decent return on investment as a result of the scientific exploration of natural resources.

Due to the enormous, well-thought-out systematic work on the resettlement of the peasants, their competent motivation, and the rendering of various assistance to them, the growth of agricultural and industrial production in the region and in the country accelerated significantly (Vinogradov, 1914). This allowed Russia producing locomotives, military and commercial ships, agricultural equipment, as well as increase exports. According to Teri (1914), the increase in agricultural output was achieved without the aid of expensive foreign labor, as it was the case in Argentina, Brazil, the USA, and Canada. Agricultural production not only satisfied the increasing domestic demand of growing population but also allowed Russia to significantly expand exports. Within a decade, expenditures on education increased twofold (compared to the expenditures on national defense – growth by 68.2%). According to Rybas and Tarakanova (1991), if the trend continued, by the middle of XX century Russia would dominate in Europe in economic and financial areas.

Agrarian Reform of 1965

In the first half of XX century, Russia participated in two world wars, experienced revolution, and civil war. Military and social turbulences postponed agrarian reforms for several decades. Only after the restoration of the national economy and the end of the World War II, the agenda focused on the provision of population with grain and other food products. Modernization of agricultural production and industry emerged as the key factors of new agrarian policy. Material incentives were widely used: approval of purchase prices for ten years and their increase for state and collective farms; surcharge to the purchase price of 50% for over-fulfillment of the plan; payment of guaranteed salaries to collective farmers (instead of working days); removal of restrictions on subsidiary farming; powerful financial investments in the material and technical equipment of collective and state farms (History of Russia, n.d.). At the same time, fixed wages for collective farmers remained. On the one hand, it provided collective farmers with social guarantees and a sense of security, on the other, it resulted in the leveling and de-motivation.

Agrarian reform of 1965 contributed to the growth of profitability of agri-industrial sector (state farms increased their output by 22%, collective farms – by 34%) (Markov, 1995). However, there was a reduction in the area of arable land; increase in product losses (20-40%), and aggravation of environmental problems. This led to a subsequent drop in economic indicators (Figure 11) (Skiba, 1988).

Among the causes of economic decline are an attempt to solve major problems by superficial changes; refusal to change the fundamentals of the country's economic basis (Belousov, 1972); contradictions in the Communist Party about the scenario of the reform; priority of the ideology over the economy; restriction of the independence of collective and state farms; full directive management by the Ministry of Agriculture; lack of necessary funds due to the launch of a number of long-term programs (economic development of Siberia and the Far East, modernization of the army, provision of loans to other countries) (Belousov, 1972). Due to lack of funds, implementation of the agrarian reform was halted.

Figure 11. The impact of economic reform on the development of the USSR
Source: History of Russia (n.d.)

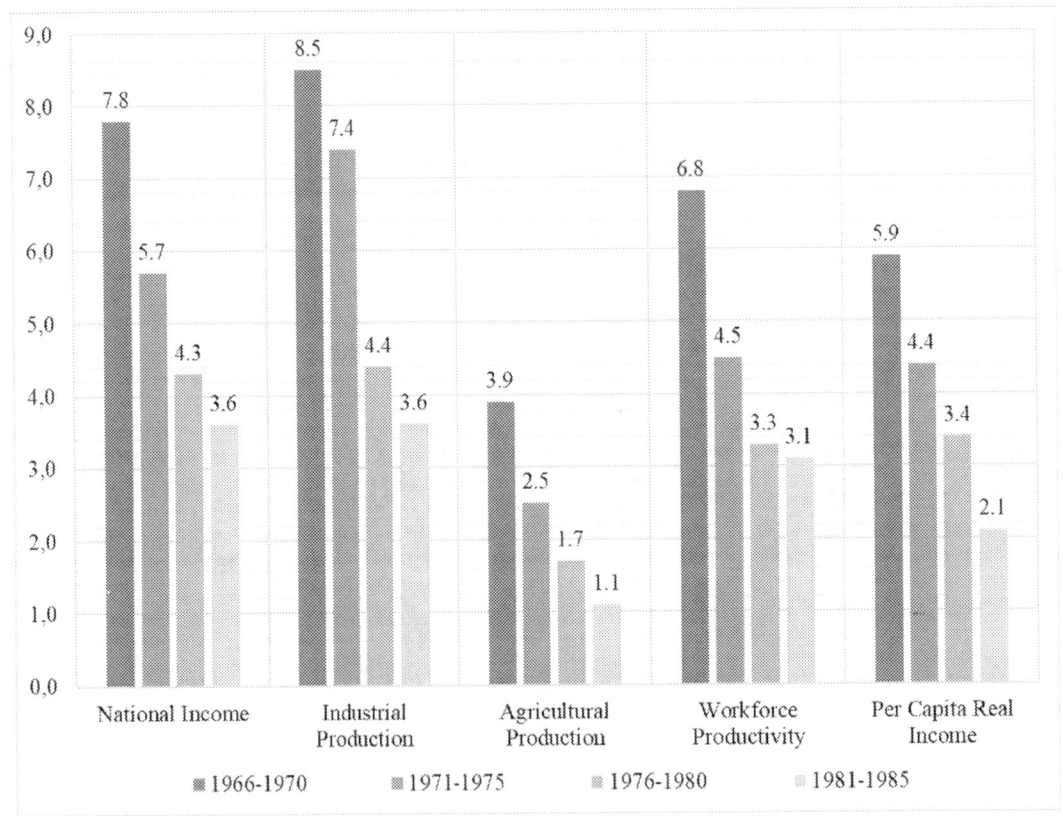

Agrarian Reform of the 1990s

The problems not solved by the reform of 1965 were actualized in the early 1990s. Radical changes in the field of agrarian relations took place in 1990 (Organisation for Economic Cooperation and Development, 1997). They were associated with the adoption of the laws that secured the right of land ownership to citizens and their associations. The second stage of land reform (1991-1993) marked the beginning of mass privatization of agricultural land, the reorganization of collective and state farms based on the Decree of the President of Russian Federation (1993) "On the Regulation of Land Relations and Development of Agrarian Reform in Russia". Despite a set of new laws, public concern about land privatization and farm restructuring, the implementation of agricultural sector reforms in the former socialist countries of Europe was much slower than planned (The World Bank, 1992; The International Bank for Reconstruction and Development, 1992). The reasons were political and legal uncertainty (Lerman, Csaki, & Moroz, 1998), the lack of favorable conditions, high risks and the lack of management tools (Csaki, 2000), as well as inadequate mechanisms for restructuring farms.

The third stage of the reform took place in 2001-2002. Agrarian reform in Russia allowed the development of peasant (farmer) farms. A significant number of personal subsidiary farms were created, which did not require registering. The products produced and sold by them were not subjects to taxation.

In 2015, the Federal Law of the Russian Federation "On the Amendments to the Federal Law on the Development of Agriculture" (Government of the Russian Federation, 2015) was adopted. The law guaranteed state support to both large agricultural producers and individual entrepreneurs in rural areas. Support emerged as the major direction in the development of agricultural sector in Russia.

In the course of modern agrarian reform, the foundations were laid for the market mechanism of economic management and the system of a mixed economy in the agro-industrial complex. Legal foundations have been created for solving economic, social, and environmental problems of agri-industrial sector. Development of agribusiness proceeded amid the emergence of various forms of entrepreneurial activity. The role of local authorities has increased.

At the same time, there were negative aspects of the reform, particularly, insufficient financing of economic entities of all forms of ownership (Pasko, 2013a); low payback of investments in production and social spheres; wear and tear of technical equipment (Pomelov, Pasko, & Baranova, 2015); degradation of land resources (Bogdanov, Posternak, Pasko, & Kovyazin, 2016); aggravation of environmental and social problems; rising unemployment in rural areas and outflow of rural population to the cities. The growth of labor productivity was accompanied by a reduction in labor force and an increase in unemployment. Increase in livestock productivity was associated with the reduction in the number of herds; an increase in general agricultural support – with decrease in its efficiency (Uzun & Lerman, 2017). The agricultural market is dominated by vertically integrated structures with many branches, which have no analogs in developed countries (Uzun, 2012). After the accession to the World Trade Organization (WTO), Russia harmonized its support policy with international regulations. Among the problems, there were the fall in consumer demand with the loss of income and the liberalization of foreign trade (Nikolsky, 2016). Import of meat increased from 290 thousand tons in 1992 to 968 thousand tons in 1997, poultry meat – from 46 thousand tons to 1,149 thousand tons, animal oils – from 25 thousand tons to 190 thousand tons, sunflower oil – from 174 thousand tons to 326 thousand tons, sugar – from 2,137 thousand tons to 3,709 thousand tons. Dumping import pricing ruined the level of wholesale and retail prices on the domestic market and led to the bankruptcy of many agricultural producers. The reform resulted in the substantial decrease of food self-sufficiency of Russia (Adukov, n.d.). To maintain the standard of living of the population, some of the products are imported. In the past five years, there has been a clear decrease in its volume due to the policy of import substitution (Federal Customs Service of the Russian Federation, 2018). In 2015, the volume of imported products was estimated at $26.58 billion, two times higher compared to 2013. The volume of domestic agricultural output amounted to 5,037 billion rubles (Expert and Analytical Center for Agribusiness, n.d.).

One of the main grain crops imported by Russia is rice (30.0% of consumption). Grapes (45.3%), as well as pears and apples (29.4%), make up a large share in crop imports. Among animal products, cheese, butter, and cottage cheese have the highest share, while the share of poultry and pork meat is decreasing (National Rating Agency, 2016). Major suppliers of agricultural products to Russia are Australia, the USA, the Netherlands, Germany, Belarus, Poland, Brazil, and Argentina.

Since 2014, in retaliation to economic sanctions, Russia banned import of some agricultural products from a number of countries. Exports to the EU decreased as Russian producers failed meeting high quality and safety requirements set by European regulations (Koptseva & Kirko, 2017). Instead, import substitution program and development of agricultural production allowed increasing food exports to the countries of Asia and Africa. Russia supplies grain, food industry waste and ready-made animal feed (about 30% of the aggregated supply).

Currently, Russia is the world's third largest exporter of grain (31 million tons), primarily, to Turkey, Egypt, Saudi Arabia, and some countries of North Africa. Russia plans to enter the markets of Southeast Asia and Latin America. Apart from grain, Russia exports fats and oils of plant and animal origin (1.5 million tons, about a quarter of the aggregated national output of oil). Fats and oils are exported to Algeria, Turkey, Egypt, China, Belarus, Kazakhstan, Azerbaijan, and Uzbekistan. Among exclusive Russian export brands are honey and caviar (Sugar.ru, 2016). Russia is one of the leading suppliers of fish, live animals, meat and poultry meat, dairy products, vegetables, and fruits. The government aims to get 10-15% of the global market of these products by 2020 (Agronews, 2013) provided that the following problems related to foreign trade in agricultural products are resolved: insufficient state support, economic sanctions, and restricted regulations of the WTO (Mitin & Rozhdestvensky, n.d.; Frolova & Boiko, 2015; Uzun & Lerman, 2017). Fundamental challenges are ensuring food security of the country, social orientation of the economy, attraction of investment to increase productive capacity of agri-industrial sector.

Thus, historical experience indicates that the success of the agrarian reform is due to its complexity and comprehensiveness, financial support, use of innovative technologies and tools, common political, economic, and social goals of all social groups.

The Course of Agrarian Reform in the Post-Soviet Area

Agrarian reforms were implemented in all countries of the former USSR and characterized by similar economic conditions (Alam, et al., 2005; Csaki & Jambor, 2009; Lerman & Sedik, 2008). Among the similarities are state ownership of land and centralized planning of agricultural production, predominance of large producers in the face of collective and state farms, state control of pricing and prices. All former Soviet countries transitioned from socialist model to market economy. Despite very similar starting positions, the countries have varied in the principles and strategies of agrarian reforms. They have been carried out in the four formats, including "shock therapy" (Abazova & Kulova, 2004).

Format 1

Countries – Latvia, Lithuania, and Estonia. The reform aimed at the combination of restitution and privatization of land by rural residents. The process went particularly rapidly in Latvia. The reforms were supported by investments from Finland, Sweden, the USA and Germany to financial sector, trade, agricultural production, and industry. Baltic countries have made significant progress in the transition period from a planned distribution to open and liberal market economy, as well as in conducting agrarian reforms (Baumane & Pasko, 2014).

Format 2

Countries – Armenia, Moldova, Russia, and Ukraine. The reform aimed at the creation of conditions for equal development of various forms of management; development of a mixed economy based on diversity and equality of land ownership; introduction of private property on land; and development of agricultural land market.

In Russia, private property and land transactions are recognized by law. In reality, the allocation of land plots in the form of land shares is organized within agricultural organizations. "Farmer will feed Russia" slogan has not justified itself. Farmers fail competing with large agricultural enterprises, they have to sell their products at low prices. Many citizens registered as farmers and thus obtained land plots and agricultural machinery which were later sold out. As a result, the acreage of agricultural land reduced (Pasko, 2013b, 2014). To a considerable extent, the reform was carried out in a formal way, in the absence of effective market mechanisms, and with minor regulations from the government.

In 1992, Russia recorded the highest per capita volume of grain and grain products, including output, exports, and imports (Figure 12).

Figure 12. Per capita volume of grain and grain products, including output, exports, and imports, tons
Source: Russian Century (2011); Pikabu (n.d.)

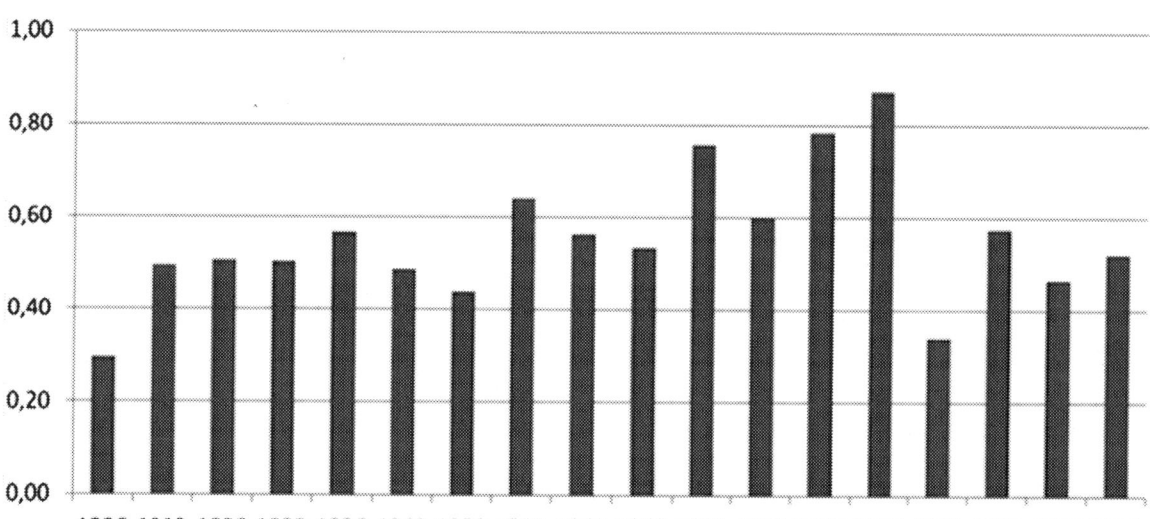

Figure 13. (a) grain output, million tons; (b) meat output in slaughter weight, million tons, in Russia in 1990-2016
Source: Russian Century (2011); Pikabu (n.d.)

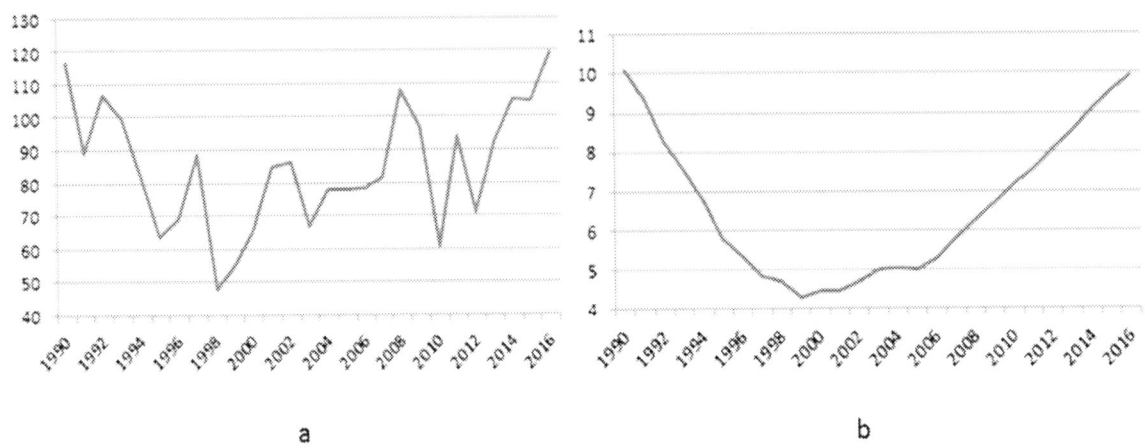

a   b

The growth of volume was mainly due to the imports since the output of grain reduced from 118 million tons to 112 million tons. Seven years after the start of the reform, grain output reached its minimum of 48 million tons. Similarly, by 2000, production of meat had decreased sharply. Only in 2016, the output of both grain and meat returned to the pre-reform level (Figure 13).

Ukraine also legally recognized private property on land, but imposed a moratorium on sale and purchase operations. Moldova introduced private property on land, but have not adopted the legislation to regulate land market.

Format 3

Countries – Uzbekistan and Tajikistan. Development of the land market was based on the legislative consolidation of state property on land. Private property on land is not allowed (Lerman & Sedik, 2008). Land plots are granted to peasant farms in life-long inherited possession, a system of state orders for agricultural products operates. The reform aims at the optimization of work of large collective farms and establishment of the conditions for their self-financing and self-sufficiency (Lerman, 2008).

Format 4

Countries – Belarus, Georgia, Kazakhstan, Kyrgyzstan, and Turkmenistan. Among the objectives are the development of private property on land while preserving most of the agricultural land in state ownership (Bluashvili & Sukhanskaya, 2015; Akhramovich, Chubrik, & Shymanovich, 2015; Arakelyan, 2013). In Belarus and Kazakhstan, private property is allowed only on household plots and their sale. In Turkmenistan, private property on land is permitted, but without the right to sell. In Kyrgyzstan, private property on land is not allowed, but the rights to use land may be marketed.

The Results of Agrarian Reform in the Countries of the Former USSR

The results of the agrarian reform in the countries of the former USSR are heterogeneous. Since that time, four economies has experienced an increase in the volume of agricultural output (Armenia, Georgia, Kyrgyzstan, and Uzbekistan). In Moldova and Ukraine, there was a sharp drop in the level of agricultural production (Lerman & Cimpoies, 2006; Urutyan, Yeritsyan, & Mnatsakanyan, 2015).

Economic problems were common to all former Soviet republics (Syzdykov, Aitmambet & Dautov, 2015). In 1992, GDP in Estonia decreased by 25.5%, in Latvia – by 34.0%, in Lithuania – by 39.3%, in Russia – by 13.8%, in Belarus – by 9.6%, in Turkmenistan – by 5.3%. Crisis in agricultural sector was caused by unresolved socio-economic problems, absence of a full-fledged agricultural policy, and underdeveloped legal framework and market infrastructure. Until 1996, the decline continued an all post-Soviet countries except the Baltic states and Armenia. In 2001, the level of agricultural production in the countries of the Commonwealth of Independent States (CIS) amounted to an average of 68% of the 1991 level, including the excess in Armenia (12%) and Kyrgyzstan (2%) and the backlog in Moldova (56% of the pre-reform level), in Russia (65%), in Ukraine (61%) (Lerman, Sedik, Pugachov, & Goncharuk, 2007), in Azerbaijan (64%) (Food and Agriculture Organization of the United Nations [FAO], 2012).

In the 1990s, the fall in GDP caused impoverishment of the population. In 2000, 41.2% of the population was considered poor in Russia, 44.4% in Belarus (Buchenrieder, Hanf, & Pieniadz, 2009), and 36.4% in Kazakhstan (Dudwick, Fock, & Sedik, 2007). Since 2000, there has been a rise in the economy in all countries of the former USSR (Arakelyan, 2013; Erjavec et al., 2017; Khalilov, Shalbuzov, & Huseyn, 2015).

It should be emphasized that contradictory processes are taking place in the agricultural sector in the post-Soviet space, including concentration of production in central and southern regions, reduction of acreage, and depopulation of rural territories. Since the beginning of the reform, Russia has lost 40 million hectares of arable land (one-third of the total area), mainly in the non-chernozem zone and in the Urals and Siberia (Rylko, Khotko, Abuzarova, Yunosheva, & Glazunova, 2015). State support is mainly allocated to large agricultural holdings. There are processes that are actually opposite to the successful Stolypin's reform.

Was there an alternative? Some experts noted that Soviet Union had to adopt market economy anyway. It is believed that for decades, Soviet system has been living beyond its means. High level of social welfare and standard of living in rural areas were maintained by budgetary transfers that masked a gross lack of profitability and inefficiency in agricultural sector dominated by collective farms.

The countries of Eastern Europe experienced similar processes in the 1990s. A transition from socialist planning to market economy has led to fundamental changes in political and social institutions and economic conditions (Csaki, 2000; Lerman, Csaki, & Feder, 2004; Liefert & Liefert, 2012; Moroz, Stratan, Ignat, & Lucasenco, 2015; Nivievskyi, Stepaniuk, Movchan, Ryzhenkov, & Ogarenko, 2015). Similarly, as a result of land reform, there were massive changes in land ownership along with fragmentation of agricultural fields (Csaki & Jambor, 2009; Lerman, 2009). The areas of abandoned lands are growing at a high rate, they are overgrown, cultivated and planted again. Large areas turn into meadows and forests (Baumane & Pasko, 2014; Pasko, 2013a).

Despite common history, common starting point, and common aspirations, the transition of countries to a more efficient economic system took place in various ways. The implementation strategies for agrarian reform and the possibilities for their implementation differed significantly. As a result, there was a significant discrepancy between the leaders and the laggards which may increase in the future.

Development Prospects

Further development of agrarian reforms in the post-Soviet space depends on a combination of external and internal factors. The factors that impede the implementation of the reform include low productivity and yield of the cult, old equipment and machinery; outdated or missing irrigation system; insufficient development of the transport system and logistics; insurance and credit systems; bureaucracy, excessive government regulation, and corruption; lack of investors; need to develop rural infrastructure; insufficient scientific support and a shortage of skilled labor (Bruinsma, 2009).

All post-Soviet countries except Belarus, Moldova, and Ukraine, have more or less significant potential for increasing land area for agricultural purposes. Belarus, Moldova, and Ukraine after the Chernobyl accident in 1986 excluded some areas from agricultural use and are faced with the problem of preserving productive potential of soil (Drozdz & Jurkenaite, 2017). The advantage of Russia and Belarus is a developed industry for the production of fertilizers and agricultural machinery.

These shortcomings can be overcome by pursuing a sustainable policy for the development of the agro-industrial sector with appropriate financing. This will allow realizing the unused potential of the domestic market, increasing exports, improving investment environment, increasing yields and product quality. Among the risks of the development of the agri-food sector should be noted the dependence on imported products in the domestic market; urbanization and emigration of youth, aging of rural population; soil erosion and imbalance of nutrients.

SOLUTIONS AND RECOMMENDATIONS

Further implementation of the agrarian reform in the post-Soviet space should be based on solving the problems related to the withdrawal of the industry from the economic crisis. It is necessary to take into account rich successful experience of Stolypin's reform, start with in-depth scientific analysis and generalization of international experience, involvement of leading specialists in the planning and carrying out of events and fully provided with financial resources. There are required an inventory of agricultural land with the definition of the characteristics of particular land areas and development of optimal cultivation technologies; introduction of innovative means and technologies in the production and processing of agricultural products (remote sensing, precision farming, agro-landscape farming, etc.).

Similarly, the analysis of the financial and economic state of each economy should be carried out and adequate measures (external management, division, consolidation, etc.) should be taken. The government should ensure sufficient funding along with the development of effective economic instruments for improving economic efficiency of the entire agro-industrial complex. The complexity of the reform requires the creation of clear legal and organizational conditions, ensuring reliable social protection of rural residents, and the establishment of an effective market infrastructure.

In particular, at the state level it is necessary to control the combination of all forms of management with the right of choice by producers; to form procurement prices taking into account real costs; to protect the domestic market of agricultural products from imports through customs duties and quotas; to carry out targeted investment in manufacturing and processing industries. There is an acute task of creating a system for product sales, marketing and management, and a wholesale food market. Thus, only an integrated approach to reforming a complex and diverse system of the agro-industrial complex can bring positive changes.

The experience of Russian agrarian reforms shows the damage that the national economy brings to the realization of erroneous principles, such as self-organization and self-regulation of agricultural production and the market when private land ownership is introduced; unregulated "one-time" opening of borders for the import of foreign products; rapid replacement of public production with private production; reassessment of farm opportunities; the need for fragmentation of agricultural land.

The experience of Russia and other countries of the former USSR proves the need for an evolutionary method of conducting agrarian reforms, ensuring food security of the state in the conditions of international competition, a thorough and comprehensive analysis of the agrarian sector and all productive forces, developing optimal scenarios for its consistent development and ensuring financial support.

FUTURE RESEARCH DIRECTIONS

It seems promising to continue monitoring the development of the agro-industrial complex of the post-Soviet countries. The unique historical situation, in which fifteen states started practically from similar positions and significantly differed in the mechanism and effectiveness of the reform, will allow discovering the most effective mechanisms and adapting them for specific economic conditions of different countries.

CONCLUSION

An analysis of the reforms of XX century in Russia and the countries of the former USSR demonstrates that they have been carried out successfully in case of comprehensive analysis of the current situation, providing the necessary funding, organizational, personnel, technical, technological conditions, monitoring and correction of the current situation. The economic efficiency of agrarian reforms is associated with a whole range of factors, and not only with land ownership.

Regardless of particular regulations imposed by the government, the influence of traditions on land relations is strong. Historical experience coupled with political will, consistency, competence, realism, sufficient funding, and competent operational management are key success factors of any land reform.

At present, the agricultural sector is emerging from a protracted crisis in Russia. One of the prerequisites for this was the assignment of agriculture to priority sectors. The process is contradictory. It combines opposing trends – reducing labor and rising unemployment; increasing livestock productivity and reducing the number of herds; increasing overall support for agriculture and reducing its effectiveness. There are also problems that have not been solved in the course of previous reforms, such as a strong depreciation of fixed assets, shortage of qualified personnel, and underdeveloped infrastructure. Increasing the efficiency of agricultural production requires constant monitoring and ensuring sufficient funding. Establishing of food security along with the development of food imports are promising development strategies for any country.

ACKNOWLEDGMENT

This study is carried out at Tomsk Polytechnic University within the framework of Tomsk Polytechnic University Competitiveness Enhancement Program.

REFERENCES

Abazova, F., & Kulova, A. (2004). Major Opportunities to Overcome Crisis in Agriculture in the Conditions of Market Transformation. *Advances in Current Natural Sciences, 8,* 127–128.

Adukov, R. (n.d.). *Agrarian Reforms and Rural Economy Development in Russia.* Retrieved from https://www.adukov.ru/articles/agrarnye_reformy/

Agronews. (2013). *How Russia Loses the Market of Organic Products.* Retrieved from http://agronews.ua/node/35624

Akhramovich, V., Chubrik, A., & Shymanovich, G. (2015). *AGRICISTRADE Country Report: Belarus.* Retrieved from http://www.agricistrade.eu/document-library

Alam, A., Murthi, M., Yemtsov, R., Murrugarra, E., Dudwick, N., Hamilton, E., & Tiongson, E. (2005). *Growth, Poverty, and Inequality: Eastern Europe and the Former Soviet Union.* Washington, DC: The World Bank. doi:10.1596/978-0-8213-6193-1

Arakelyan, M. (2013). *CIS Frontier Countries: Economic and Political Prospects.* Frankfurt am Main: Deutsche Bank.

Baumane, V., & Pasko, O. (2014). Comparison of Land Reform of Latvia and Russia in Conditions of Transition Period. *International Scientific Journal, 1,* 40–44.

Belousov, R. (1972). *History of National-State Construction in the USSR.* Moscow: Politizdat.

Belyanin, D. (2014). Factors of Peasant Migration to Siberia in the Second Half of XIX – Beginning of XX Century. *Bulletin of Tomsk State University, 378,* 100–108.

Bestuzheva-Lada, S. (2011). The Reformer of All Russia. *Smena, 6,* 16–27.

Bluashvili, A., & Sukhanskaya, N. (2015). *AGRICISTRADE Country Report: Georgia.* Retrieved from http://www.agricistrade.eu/document-library

Bogdanov, V., Posternak, T., Pasko, O., & Kovyazin, V. (2016). The Issues of Weed Infestation with Environmentally Hazardous Plants and Methods of Their Control. *IOP Conference Series: Earth and Environmental Science, 43.* 10.1088/1755-1315/43/1/012036

Bruinsma, J. (2009). *The Resource Outlook to 2050: By How Much Do Land, Water, and Crop Yields Need to Increase by 2050?* Retrieved from http://www.fao.org/fileadmin/templates/esa/Global_persepctives/Presentations/Bruinsma_pres.pdf

Buchenrieder, G., Hanf, J. H., & Pieniadz, A. (2009). 20 Years of Transition in the Agri-Food Sector. *German Journal of Agricultural Economics, 58*(7), 285–293.

Chramkov, A. (1994). *Land Reform in Siberia (1896-1916) and Its Impact on the Situation of the Peasants. Allowance.* Barnaul: Altai State University.

Csaki, C. (2000). Agricultural Reforms in Central and Eastern Europe and the Former Soviet Union: Status and Perspectives. *Agricultural Economics, 22*(1), 37–54. doi:10.1111/j.1574-0862.2000.tb00004.x

Csaki, C., & Jambor, A. (2009). *The Diversity of Effects of EU Membership on Agriculture in New Member States*. Budapest: FAO Regional Office for Europe and Central Asia.

Dorofeev, M. (2007). Land Reform in Siberia 1896-1916. Aspects of the Relations between the Authorities and the Peasantry. *Tomsk State University Journal, 3*, 69–73.

Dorofeev, M. (2008). Reverse Migration from Siberia (End of XIX – Beginning of XX Century): The Climatic Aspect of the Problem. *Tomsk State University Journal, 309*, 75–79.

Drozdz, J., & Jurkenaite, N. (2017). Agri-Food Sector Potential in the Chosen CIS Countries. *Zagadnienia Ekonomiki Rolnej, 352*(3), 103–115. doi:10.30858/zer/83035

Dudwick, N., Fock, K. M., & Sedik, D. (2007). *Land Reform and Farm Restructuring in Transition Countries. The Experience of Bulgaria, Republic of Moldova, Azerbaijan, and Kazakhstan*. Washington, DC: The World Bank. doi:10.1596/978-0-8213-7088-9

Erjavec, E., Volk, T., Rac, I., Kozar, M., Pintar, M., & Rednak, M. (2017). Agricultural Support in Selected Eastern European and Eurasian Countries. *Post-Communist Economies, 29*(2), 216–231. doi: 10.1080/14631377.2016.1267968

Expert and Analytical Center for Agribusiness. (n.d.). *Agriculture of Russia*. Retrieved from https://ab-centre.ru/page/selskoe-hozyaystvo-rossii

Federal Customs Service of the Russian Federation. (2018). *Foreign Trade Statistics of the Russian Federation*. Retrieved from http://www.customs.ru/index.php?%20option=com_content&view=article&id=26274:20

Food and Agriculture Organization of the United Nations. (2012). *The State of Food and Agriculture: Investing in Agriculture for a Better Future*. Rome: FAO.

Frolova, N., & Boiko, T. (2015). Agriculture in the Context of Russia's Membership in WTO: Present-Day State, Problems, Perspectives. *Problems of Modern Economics, 56*(4), 312–315.

General Directorate for Land Management and Agriculture. (1914). *Atlas of Asian Russia*. Saint Petersburg: A.F. Marks.

Goryushkin, L., Bocharova, G., & Nozdrin, G. (1993). *Experience of Traditional Agriculture in Siberia (Second Half of XIX – Beginning of XX Century)*. Novosibirsk: Novosibirsk State University.

Government of the Russian Federation. (2015). *Federal Law #11-FZ from February 12, 2015, "On the Amendments to Article 14 of the Federal Law "On Development of Agriculture.""* Retrieved from http://base.garant.ru/70866588/

History of Russia. (n.d.). *Economic Reform of 1965*. Retrieved from https://istoriarusi.ru/cccp/ekonomicheskie-reformi-1965-goda.html

Khalilov, H., Shalbuzov, N., & Huseyn, R. (2015). *AGRICISTRADE Country Report: Azerbaijan*. Retrieved from http://www.agricistrade.eu/document-library

Kofod, A. (1914). *Land Management in Russia*. Saint Petersburg: Selsky Vestnik.

Koptseva, N., & Kirko, V. (2017). Development of the Russian Economy's Agricultural Sector under the Conditions of Food Sanctions (2015-2016). *Journal of Environmental Management and Tourism*, *8*(1), 123–131.

Kulomzin, A. (1903). *Siberian Railway: Past and Present. Historical Essay*. Saint Petersburg: State Printing House.

Lerman, Z. (2008). Agricultural Development in Central Asia: A Survey of Uzbekistan, 2007-2008. *Eurasian Geography and Economics*, *49*(4), 481–505. doi:10.2747/1539-7216.49.4.481

Lerman, Z. (2009). Land Reform, Farm Structure, and Agricultural Performance in CIS Countries. *China Economic Review*, *20*(2), 316–326. doi:10.1016/j.chieco.2008.10.007

Lerman, Z., & Cimpoies, D. (2006). Land Consolidation as a Factor for Rural Development in the Republic of Moldova. *Europe-Asia Studies*, *58*(3), 439–455. doi:10.1080/09668130600601933

Lerman, Z., Csaki, C., & Feder, G. (2004). *Agriculture in Transition: Land Policies and Evolving Farm Structures in Post-Soviet Countries*. Lanham, MD: Lexington Books.

Lerman, Z., Csaki, C., & Moroz, V. (1998). *Land Reform and Farm Restructuring in Moldova – Progress, and Prospects*. Washington, DC: The World Bank. doi:10.1596/0-8213-4317-3

Lerman, Z., & Sedik, D. (2008). *The Economic Effects of Land Reform in Tajikistan*. Budapest: FAO Regional Office for Europe and Central Asia.

Lerman, Z., Sedik, D., Pugachov, N., & Goncharuk, A. (2007). *Rethinking Agricultural Reform in Ukraine*. Halle: Leibniz Institute of Agricultural Development in Transition Economies.

Liefert, W. M., & Liefert, O. (2012). Russian Agriculture during Transition: Performance, Global Impact, and Outlook. *Applied Economic Perspectives and Policy*, *34*(1), 37–75. doi:10.1093/aepp/ppr046

Markov, A. (Ed.). (1995). *History of the National Economy*. Moscow: Law and Right, Unity.

Mitin, A., & Rozhdestvensky, V. (n.d.). *The Analysis of State of Domestic Agriculture in Conditions of Fierce and International Competition*. Retrieved from http://bmpravo.ru/show_stat.php?stat=323

Moroz, V., Stratan, A., Ignat, A., & Lucasenco, E. (2015). *AGRICISTRADE Country Report: Moldova*. Retrieved from http://www.agricistrade.eu/document-library

Museum of Local History in Chulym. (n.d.). *The History of Settlement of Chulym Land*. Retrieved from https://vmuseum.ucoz.ru/index/istorija_zaselenija_zemli_chulymskoj/0-9

National Rating Agency. (2016). *Agriculture of Russia in 2015 and First Half of 2016*. Retrieved from http://www.ra-national.ru/sites/default/files/analitic_article/Сельское%20хозяйство%20и%20семена%202016%202_0.pdf.

Nikolsky, S. (2016). *Agrarian Reform of the 1990s: Regional Models and Ideological Foundations*. Retrieved April 27, 2019, from http://lawinrussia.ru/content/agrarnaya-reforma-90-h-godov-regionalnye-modeli-i-ih-ideologicheskie-osnovaniya

Nivievskyi, O., Stepaniuk, O., Movchan, V., Ryzhenkov, M., & Ogarenko, Y. (2015). *AGRICISTRADE Country Report: Ukraine*. Retrieved from http://www.agricistrade.eu/document-library

Organisation for Economic Cooperation and Development. (1997). *Agricultural Outlook*. Paris: OECD.

Pasko, O. (2013a). Economic Development and Perspectives of Cooperation Between the USA, Europe, Russia, and CIS States. *Cibunet Publishing*, *1*, 45–60.

Pasko, O. (2013b). The Use of Agricultural Land in Tomsk Region. *Agrarian Science*, *6*, 9–10.

Pasko, O. (2014). Dynamics of Changes in Agricultural Land in Tomsk Region. *Lucrari Stintifice*, *33*, 28–33.

Pikabu. (n.d.). *How the Standard of Living Has Changed in Russia over the Past 130 Years*. Retrieved from https://pikabu.ru/story/kak_menyalsya_uroven_zhizni_v_rossii_za_poslednie_130_let_5695251

Pomelov, A., Pasko, O., & Baranova, A. (2015). Comparative Analysis of Land Management in the World. *IOP Conference Series: Earth and Environmental Science*, *27*. 10.1088/1755-1315/27/1/012040

President of the Russian Federation. (1993). *Decree of the President of the Russian Federation #1767 from October 27, 1993, "On the Regulation of Land Relations and Development of Agrarian Reform in Russia*. Moscow: Kremlin.

Resettlement Directorate. (1911). Resettlement beyond the Urals in 1911: A Reference Book for Walkers and Settlers with a Road Map of Asian Russia. Saint Petersburg: Printing House "Village of the Messenger".

Rogachevskaya, M. (n.d.). *P.A. Stolypin: Agrarian Reform and Siberia*. Retrieved from http://econom.nsc.ru/eco/arhiv/ReadStatiy/2002_09/Rogachevska.htm

Russian Century. (2011). *Figures: Who Lost and Who Won from the Collapse of the USSR*. Retrieved from http://www.ruvek.ru/?module=articles&action=view&id=6031

Rybas, S., & Tarakanova, L. (1991). *The Reformer: The Life and Death of Peter Stolypin*. Moscow: Nedra.

Rylko, D., Khotko, D., Abuzarova, S., Yunosheva, N., & Glazunova, I. (2015). *AGRICISTRADE Country Report: Russia*. Retrieved from http://www.agricistrade.eu/document-library

Shilovsky, M. (2003). *Siberian Relocation. Documents and Materials*. Novosibirsk: Novosibirsk State University.

Shilovsky, M. (2012). Branch Records as a Source on the History of Land Management in Tomsk Province in the Beginning of XX Century. *Tomsk State University Journal. History (London)*, *18*(2), 74–78.

Skiba, I. (1988). *Agro-Industrial Complex: Economic Reform and Democratization: On the Experience and Problems of Economic Reform and Strengthening of Democratization in Agri-Industrial Complex of a Country*. Moscow: Politizdat.

Stolypin, P., & Krivoshein, A. (1911). *Trip to Siberia and Volga Region*. Retrieved from https://xn--90anbaj9ad0j.xn--80asehdb/documents/poezdka-v-sibir-i-povolzhe-zapiska-pa-stolypina-i-av-krivosheina-spb-1911/

Sugar.ru. (2016). *Ten Biggest Agricultural Producers in the World*. Retrieved from http://sugar.ru/node/14299

Syzdykov, R., Aitmambet, K., & Dautov, A. (2015). *AGRICISTRADE Country Report: Kazakhstan.* Retrieved April 27, 2019, from http://www.agricistrade.eu/document-library

Teri, E. (1914). *Russia in 2014. Economic Review.* Retrieved from http://www.mysteriouscountry.ru/wiki/index.php/%D0%AD%D0%B4%D0%BC%D0%BE%D0%BD_%D0%A2%D1%8D%D1%80%D0%B8_%D0%A0%D0%9E%D0%A1%D0%A1%D0%98%D0%AF_%D0%92_1914_%D0%B3._%D0%AD%D0%BA%D0%BE%D0%BD%D0%BE%D0%BC%D0%B8%D1%87%D0%B5%D1%81%D0%BA%D0%B8%D0%B9_%D0%BE%D0%B1%D0%B7%D0%BE%D1%80/%D0%9F%D1%80%D0%B5%D0%B4%D0%B8%D1%81%D0%BB%D0%BE%D0%B2%D0%B8%D0%B5

The International Bank for Reconstruction and Development. (1992). *Food and Agricultural Policy Reforms in the Former USSR. An Agenda for the Transition Country Department III Europe and Central Asia Region.* Washington, DC: The International Bank for Reconstruction and Development.

The World Bank. (1992). *International Agriculture and Trade Reports.* Washington, DC: The World Bank.

Tyukavkin, V. (1986). Agrarian Resettlement in Russia in the Era of Imperialism. In *Socio-Demographic Processes in Rural Areas of Russia (XVI – beginning of XX century)* (pp. 214-225). Academic Press.

Urutyan, V., Yeritsyan, A., & Mnatsakanyan, H. (2015). *AGRICISTRADE Country Report: Armenia.* Retrieved from http://www.agricistrade.eu/document-library

Uzun, V. (2012). Russian Policy of Agriculture Support and the Necessity of Its Modification after WTO Accession. *Issues of Economics, 10*, 132–149.

Uzun, V., & Lerman, Z. (2017). Outcomes of Agrarian Reform in Russia. In S. Gomez y Paloma, S. Mary, S. Langrell, & P. Ciaian (Eds.), *The Eurasian Wheat Belt and Food Security: Global and Regional Aspects* (pp. 81–101). Amsterdam: Springer. doi:10.1007/978-3-319-33239-0_6

Vinogradov, P. (1914). *The Memorial Book of Tomsk Province in 1914.* Tomsk: Printing House of Provincial Administration.

Volgareva, G., & Pasko, O. (2014). *Land Reforms at the Beginning and the End of XX Century in Russia.* Tomsk: Demos.

ADDITIONAL READING

Allina-Pisano, J. (2004). Sub Rosa Resistance and the Politics of Economic Reform: Land Distribution in Post-Soviet Ukraine. *World Politics, 56*(4), 554–581. doi:10.1353/wp.2005.0001

Belyanin, D. (2011). Resettlement of Peasants in Siberia During the Stolypin Agrarian Reform. *Russian History (Pittsburgh), 1*, 86–95.

Bloch, P. C. (2002). Kyrgyzstan: Almost Done, What Next? *Problems of Post-Communism, 49*(1), 53–62. doi:10.1080/10758216.2002.11655970

Gaidar, E. (1996). *Economic Reforms and Hierarchical Structures.* Moscow: Berator.

Gizatullin, H., & Antonov, D. (2009). The Model of Full Cost of the Enterprise-the Basis for Determining the Effective System of Taxation. *Journal of Economic Theory*, *4*, 16.

Gumenyuk, A. (2009). Man and Reform in the USSR, 1953-1985 Years. *Bulletin of Saratov University. Series. History and International Relations*, *9*(2), 92–102.

Kabytov, P. (2009). Stolypin Program of Reformation of Russia. *Bulletin of Samara State University*, *73*, 87–92.

Katzenellenbogen, A. (2014). The Conflict of Change in the Soviet Economy and the Post-Stalin Era. *Review of the Russian Economy*, *35*(4), 373–399. doi:10.2307/128437

Kosygin, A. (1965). *On the Improvement of Industry Management, Improvement of Planning and Strengthening of Economic Stimulation of Industrial Production*. Moscow: Politizdat.

Leonard, C. S. (2000). Rational Resistance to Land Privatization: The Response of Rural Producers to Agrarian Reforms in Pre- and Post-Soviet Russia. *Post-Soviet Geography and Economics*, *41*(8), 605–620. doi:10.1080/10889388.2000.10641160

Mogilevsky, K., & Soloviev, K. (2009). Stolypin's Project of Transformation of Russia. *Polity*, *52*(1), 151–166.

Patsiorkovsky, V., O'Brien, D., & Wegren, S. (2005). Land Reform and Land Relations in Rural Russia. *Eastern European Countryside*, *11*, 5–17.

Plotnikov, V. (2010). Agrarian Reform in Russia: Stolypin's Experience and Modern Problems of Farming. *Agribusiness: Economics and Management*, *12*, 3–10.

Sitnin, V. (2007). The Main Ideas of the Economic Reform of 1965 and the Reasons for Its Failures. *Economic Policy*, *2*, 80–96.

Vedeneev, Y. (1994). *Institutional Reform of the State Administration of Industry*. Moscow: Politizdat.

KEY TERMS AND DEFINITIONS

Agrarian Reform: A set of measures taken by the state to redistribute land ownership in favor of direct producers, increase their interest in the results of labor, increase in production.

Agro-Industrial Complex: A set of industries (enterprises) whose activities are directly or indirectly aimed at the production of food or other products produced from agricultural raw materials.

Economic Crisis: A sharp deterioration in the economic state of the country, manifested in a significant decline in production; violation of existing production relations; bankruptcy of enterprises; and rising unemployment. The result of the economic crisis is a decline in the living standards of the population and a decrease in the real gross national product.

Economic Reforms: The changes in the business system, economic management, ways and methods of economic policy implementation. Economic reforms are carried out under conditions when low efficiency of the economic system is revealed, economic crises occur, market does not sufficiently satisfy the needs of people, country lags behind in its development from other countries.

Efficiency of Agricultural Production: An effectiveness of the financial and economic activities of an economic entity in agriculture, ability to ensure the achievement of high rates of productivity, efficiency, profitability, and product quality. Criterion is the maximum production of agricultural products at the lowest cost of living and materialized labor. It is measured using a system of indicators, including labor productivity, capital productivity, cost, profitability, crop yields, and animal productivity.

Post-Soviet Space: Also known as the republics of the former USSR, the CIS countries and the Baltic states, are independent states that left the Soviet Union during its collapse in 1991. The post-Soviet states are the subject of various studies in the field of geography, history, politics, economics, and culture.

Province: In Russia, an administrative division, part of the empire, entrusted to the governor.

State Policy: A set of the basic principles, norms, and activities for the implementation of state power.

Stolypin's Agrarian Reform: The reform of peasant allotment land ownership in Russia. It is named after its ideologist Petr Stolypin. Permitting the exit from the peasant community to the farms and cuts, strengthening the Peasant Bank, compulsory land management and strengthening the resettlement policy (relocating the rural population of the central regions of Russia for permanent residence to sparsely populated districts – Siberia, the Far East) were aimed at eliminating peasant land shortage, intensifying economic activities peasantry on the basis of private ownership of land, and increasing the marketability of the peasant economy.

Chapter 11
Economic Aspects of Agricultural Public Policy as a Key Factor of Establishing Food Security:
Retrospectives of Post-Soviet Russia

Stanislav Lipski
https://orcid.org/0000-0003-1283-3723
State University of Land Use Planning, Russia

Olga Storozhenko
Bauman Moscow State Technical University, Russia

ABSTRACT

In Russia, food security is ensured by sustainable development of domestic agriculture and related industries. Arable lands, the key agricultural resource in Russia, account for about 9% of the world's total. This study investigates changes in public policy related to agricultural lands in post-Soviet period, namely, arguments for land redistribution; privatization that covered over 60% of agricultural lands and resulted in appearance of land shares owned by about 12 million rural citizens barely understanding what to do with their land shares; post-privatization issues and problems concerned with the involvement of agricultural and other lands in economic activity; implementation of public economic policy measures aimed to resolve the above-mentioned issues (transfer of unclaimed land shares to municipalities); current transformation of ownership structure of agricultural lands; specifics of demarcation of un-privatized lands between federal, regional, and local authorities.

DOI: 10.4018/978-1-7998-1042-1.ch011

INTRODUCTION

Food security is ensured by sustainable development of domestic agriculture and related industries. In Russia, food security is defined as a share of domestically-produced food in total volume of domestic product turnover (95% for grain and potato, 90% for milk, 85% for meat, and 80% for sugar). The main factors for the implementation of such thresholds are resource potential and institutional environment. Meanwhile, key agricultural resource is agricultural land. The total world's area of agricultural land is about five billion hectares including 1.4 billion hectares of arable lands (Loyko, 2009). In addition, there are about 0.5 billion hectares of lands that are appropriate for agricultural purposes, but their development is limited by severe climate conditions, geographic terrain, and lack of water supply. Meanwhile, rapid growth of population makes food supply a very challenging task. According to Pitersky (1999), available land resources are enough to provide food for about twelve billion people. As a result, the availability of land is a key factor of food security and competitiveness of domestic farmers in the global market. Russia accounts for about 9% of world's arable land (Figure 1). The economic policy of the country is aimed at the ensurance of food security by the improvement of soil fertility and increase of arable land.

In Russia, allocation of agricultural land is not homogeneous. Most of the land is concentrated in the south-western part of the country (Figure 2), therefore, further analysis is aimed to study the results of state economic policy in agriculture implemented at federal and regional levels.

During the Soviet regime of state land property, all lands were withdrawn from property ownership relations. As a result, there were the references to the definition of real estate in neither economic theory nor legislation. Along with economic and social transformations in post-Soviet Russia, land property rights have been transformed. Today, the monopoly of state land property has been abolished while private and municipal ownership has emerged. Moreover, after a complex process of privatization, land

Figure 1. Top countries in arable land acreage, billion hectares
Source: Loyko (2009); Federal Service for State Registration, Cadastre and Cartography [Rosreestr] (n.d.)

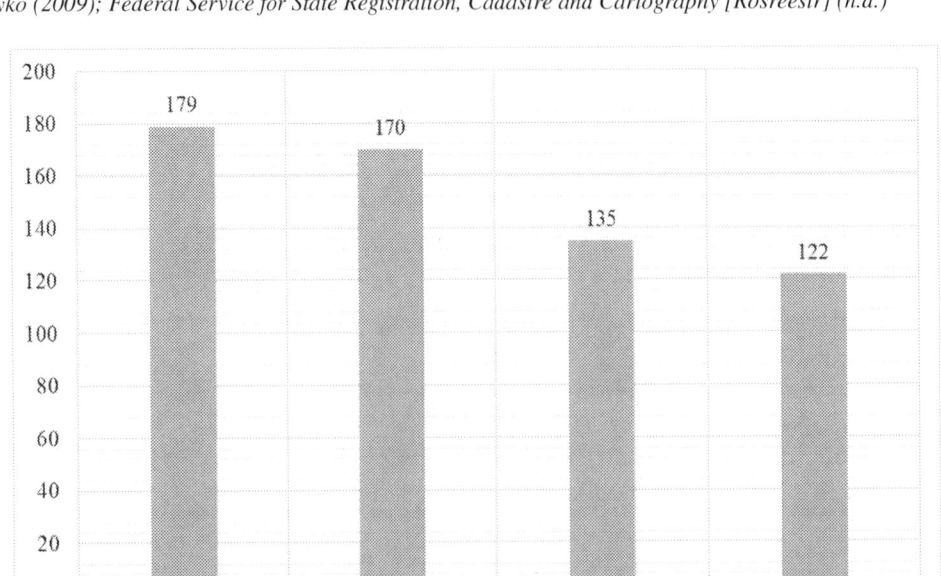

Figure 2. Allocation of agricultural land in Russia
Source: Authors' development based on Rosreestr (n.d.)

has been acknowledged as real estate and thus engaged in commercial turnover. Non-privatized lands are still under the process of allocation between federal, regional, municipal, and state forms of property (Figure 3).

Agricultural land has been affected by privatization the most (about 96.5% of all privatized land is that of agricultural purpose). Privatization was implemented in three ways: transfer of land ownership rights to the individuals using the related lands under other conditions (privatization based on the fact of use); transfer of ownership for new land plots to individuals and legal entities; transfer of agricultural lands to collective ownership of agricultural farms and individuals engaged in agriculture ("mass" privatization). The major part of lands was transferred to private ownership during the 1990s based on the third way of privatization. In 2001, individuals and legal entities owned 129 billion hectares (7.6% of national land fund). In 2017, the share of private property increased up to 133 million hectares (7.8%).

Agricultural policy has reached its goals in the fulfillment of privatization and reorganization of agricultural entities. In general, however, it was controversial and inconsistent. Privatization of agricultural land has resulted in numerous institutional issues that impacted food security of the country as well as foreign agricultural trade. Nowadays, Russian agriculture is targeted at the increase of agricultural exports which is achievable by the exploitation of available land resources. At the same time, withdrawal of abandoned lands from private owners is still a challenging task.

BACKGROUND

This study focuses on the features of modern agricultural policy of Russia in the light of ensuring food security. The issues of land management have been widely addressed by many scholars, land surveyors, and experts in the field of agricultural and land law.

Figure 3. Transformation of land ownership in post-Soviet Russia
Source: Authors' development

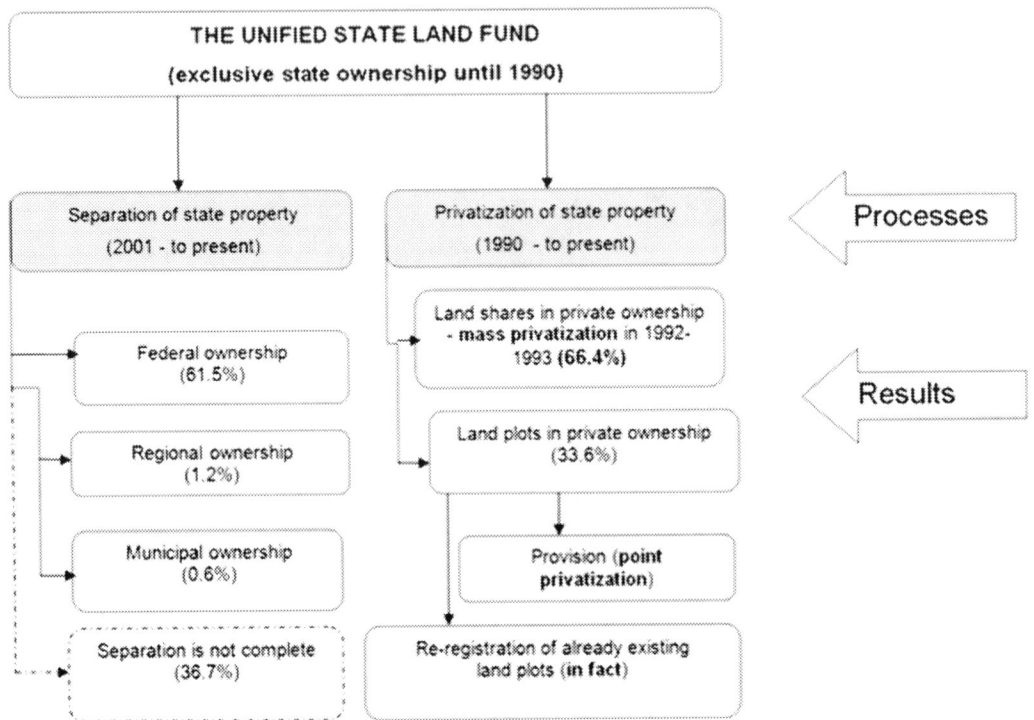

Positive influence of organizational changes in agriculture and land privatization on the performance of agricultural producers were investigated by Krylatykh (1997, 1998) and Rabinovich, Fedoseev, and Ignatiev (1995). One of the first comparisons of land privatization in Russia as well as others agricultural resources with experience of other countries were fulfilled by Serova (2000) who positively evaluated the effect of reforms and pointed out problematic issues concerned with the use of lands transferred to private owners and investments in agricultural industry. Petrikov (2000, 2016) and Ushachev (2003) claimed that privatization resulted in the concentration of lands in big agricultural enterprises and insufficient state regulation of agricultural industry. Statistical analysis of agricultural sector and land market were investigated by Shagaida and Alakoz (2017), Shagaida and Fomin (2017), Kresnikova (2008), and Uzun and Shagaida (2015). The issues of state policy in the sphere of agricultural land management were investigated by Ogarkov (2000) and Pankova (2000) who generally supported the relevance of privatization but questioned its methods. Land privatization was also investigated by Wegren (2012) who indicated the specific features of agricultural reforms in Russia in the 1990s.

The issues of rational use of privatized agricultural land were studied by Alakoz, Kiselev, and Shmelev (1999), Alakoz, Vasiliev, Kiselev, and Pulin, (2001), Leppke (1998, 2000), Loyko (2001), Komov (1995), Komov and Aratskiy (2000), and Khlystun (2005, 2011, 2012a, 2012b, 2015). In the status of government officials, the above-mentioned scholars were engaged in the establishment of land privatization tools. As a result, the results of their studies were used during the reforms and thus opposed radical ways of privatization. However, recently, many experts criticize the results of privatization and current land relations. Vershinin and Petrov (2015), Volkov and Khlystun (2014), and Volkov and Lipski (2017) support the negative influence of lands privatization on rational use and protection of land and investigate the weak points of land management (mosaic location of land areas, irrational demarcation of land plots, and intersection of land plots) as a consequence of economically and technologically unreasonable allocation of land plots during the reform. Legal issues of agricultural land ownership, land shares status, and limitation of private landowners' rights were investigated by Bogolubov (1998) and Ustyukova (2007).

The above-mentioned studies have explored the related issues separately in a view of economic aspects in agriculture, legal aspects of land and private law related to the privatization and further land use, and land management improvements or provision of food security. Additionally, current wide-scale changes concerned with agricultural lands ownership rights (withdrawal of the lands privatized in the 1990s from private owners to municipal authorities and distribution of non-privatized lands between different public entities) have not been studied due to their novelty. Thus, the studies of the issues concerned with economic policy in agricultural field that ensures national food security should consider that some aspects still need diversified investigations. Therefore, this study explores the following issues:

- Arguments for privatization of agricultural land
- Procedure of privatization which covered over 60% of agricultural land, its objectives and results
- Engagement of agricultural lands in economic activities
- Implementation of state policy aimed at the solution of the above-mentioned problems, the results of recommended solutions, and analysis of current situation with unclaimed land shares
- Legal status of lands not involved in privatization and current issues of land ownership
- Consistency of economic policy related to agricultural land in post-Soviet period

Some of the results explored in this study have been published at specialized articles as well as presented by the authors at various conferences (Lipski, 2002, 2005, 2014, 2015), however, a complex and systematized form of the study including the analysis of agricultural lands use is presented for the first time.

MAIN FOCUS OF THE CHAPTER

Privatization of Agricultural Land

In Soviet period, all lands in Russia were owned by the state, withdrawn from turnover, and not considered as real estate. Land reform of the 1990s was aimed to change that situation. Initially, land reform mostly targeted agricultural land, its main purpose was to solve food insecurity problem (Bogolubov, 1998). Among other goals, there were:

- Encourage of economic interest in rational use of agricultural land and motivation of landowners to get rid of unused lands (motivation tools included land taxes and permission to sell off unused privatized lands)
- Facilitate the development of new economic relations by the establishment and differentiation of land property rights aimed at increase of economic performance
- Minimize land degradation and improve rational and ecological use of lands based on the solicitous approach of private owners to their property

- Meet the demand of urban population for land plots intended to be used as gardens and kitchen-gardens (Alakoz et al., 1999; Krylatykh, 1998)

Traditionally, the transfer of state-owned lands to private individuals has been widely used as a part of public policy. Specifically, the monarchs passed the lands to feudal lords in exchange for military and other duties. In Russia, historically, accession of new territories, political organization, and nationalization of lands often took place concurrently with privatization. In the times of Ivan the Terrible, lands of disgraced landlords were confiscated and granted to guardsmen. Catherine the Great granted the conquered lands to her favorites and loyal noblemen. In the beginning of the 1990s, land privatization had totally different ground. The purpose was to ensure fundamental changes in agriculture and irreversibility of those changes at state level. Specifically, privatization aimed to provide the following:

- Transformation of land ownership rights and fast allocation of lands to private entities and individuals that meant to drive reforms in agriculture
- Accelerated reorganization of collective and state farms (administration of the farms assumed to become dependent on agricultural workers – new owners of agricultural lands previously belonged to farms; it is important to note that only agricultural lands were the subject of privatization while other collective and state farms lands such as forests and ravines were excluded from privatization)
- Improvement of rational use of agricultural land (land shares in less effective farms meant to be acquired by more effective agricultural entities since it would be more profitable for the former to sell shares off than to continue to using them)

There were limited number of options to carry out privatization with the above-mentioned objectives:

- **Restitution:** Return of collective farms' lands back to the individuals (or their successors) whom these lands belonged to before the nationalization during the Revolution in 1917. That option was rejected, the former owners and their successors were assigned to get land ownership rights only on a general basis (Government of the Russian Federation, 1990). The nationalization happed more than 70 years ago, most of the lands owners already passed away. The return of the lands to the successors of second and third generation was considered very complicated and unfair. Besides, restitution could have negative effect on societies in the republics of Russia, where the entire nations suffered from nationalization. Nevertheless, restitution was successfully implemented in Bulgaria, Czechoslovakia, East Germany, Slovenia, and Estonia.
- **Land Sharing:** Transfer of ownership rights on equal basis. Land sharing can be carried out by sharing the lands between all urban and rural citizens similar to privatization of non-agricultural factories by the issue of privatization shares (vouchers) or by distribution of lands between citizens engaged in agriculture, including retired people. In Russia, land sharing involved all rural citizens. In a similar way, privatization was carried out in Albania, Hungary, Belarus, Kazakhstan, Latvia, Lithuania, and Ukraine (Serova, 2000).
- **Spot Privatization:** Cost-reimbursable transfer of lands to particular individuals or farms (similar to ongoing privatization of property). The option does not allow mass privatization and results in the growth of corruption.

- **Privatization "Upon The Use":** Transfer of land ownership rights to individuals granted the lands upon the other rights (in the Soviet Union, it was allowed to have lands at own use excluding the rights to sell, inherit, etc.). The way was used mostly for social purpose regarding small land plots used by private individuals.

The majority of private-owned lands were allocated during 1992-1993 (President of the Russian Federation, 1991; Government of the Russian Federation, 1991, 1992). In the 1990s, privatization was characterized by the following features:

- Foundation of collective-shared form of land ownership for the most of the country's agricultural lands except the lands excluded from privatization or the lands that are subject of privatization under special terms. The above should result in transfer of land ownership rights from state farms to the workers in the form of land shares. The size of land share depends on the total area of agricultural lands of a particular state farm and number of people pretending to get shares (current and retired workers);
- Intention to combine land privatization of state farms (state agricultural enterprises) and collective farms with prescriptive reorganization of the above-mentioned farms. It was supposed that agricultural workers with acquisition of ownership right for agricultural lands would participate in the reorganization of the farm structure and legal form while new executives of reorganized farms would contribute to efficient problem solving of privatization process and legalization of land use agreements between farms and land shares owners. In the reality, none of the above has been fulfilled;
- Large scale of privatization: about 12 million people became the owners of 115 million hectares of agricultural lands for 1-2 years that amounted to 61.8% of total area of agricultural lands belonging to collective and state farms at the beginning of 1992;
- Free of charge transfer of lands from state to private ownership;
- Combination of land privatization with privatization of other property of agricultural enterprises (individuals together with land shares were granted the property shares exchangeable for land shares);
- Lack of demarcation borders for privatized lands: privatization covered agricultural lands only (arable lands, gardens, and grasslands) that were not delimited from other lands belonging to agricultural enterprise (forests, ravines, and swamps). The purpose of the above was to privatize lands applicable for agricultural production but it resulted in the difficulty to determine the specified object for privatization;
- Significant share of retired workers among private owners. As a result, in ten years after privatization, 60-70% of land share owners were out of agricultural production and unable to exploit lands independently;
- Controversial interpretation of institutional nature of land shares that generated doubts either that tool should be considered as short-term instrument ensuring transformation of management in agricultural industry or as long-term perspective development of agricultural industry economy. The legal acts regulating land shares allocation did not presume temporarily nature of land shares aimed at the reorganization of agricultural enterprises. On the contrary, the attitude of legislators and management was built up upon the idea of land shares consolidation in hands of effective

owners by means of purchase or rent and further allocation of land shares into land plots, but that had not happened;
- Insufficiency of legal regulation: procedure of land shares allocation had been changed several times during privatization period that influenced the size of shares. After two months of privatization, the local administration was empowered to differentiate the average size of granted lands stated by President of the Russian Federation (1992) depending on the population density in the region. In addition, the expansion of the official list of individual's categories entitled to get free land shares according to Government of the Russian Federation (1992) resulted in decrease of land share size as the total area of lands covered by privatization had not been changed. Most of legal acts which administrated privatization and use of land shares were acknowledged inconsistent and contradictory. Unfortunately, the gaps of federal land legislation had barely been improved amid the changes in government approach to privatized lands and their owners (Lipski, 2014).

The opportunities of land shares owners were broadened by the changes in agricultural policy (President of the Russian Federation, 1993). During the first two years of privatization, the owners were limited to use land shares for organization of agricultural farm, acquisition of part of restructured farm, and direct sale of shares to the related agricultural farm or its workers. After two years, they were allowed to sell, rent, mortgage, inherit, and exchange land shares for property shares without quitting up the agricultural farm (President of the Russian Federation, 1993). After three years, the owners were granted the right to transfer land shares upon rental and long-life rental agreements. Additionally, they could manage their shares without consent of other shareholders (President of the Russian Federation, 1996).

The government administration approach was aimed to ease operational procedures with land shares deals and enable consolidation of land shares in hands of reorganized and newly-established agricultural enterprises or individuals intended to set up an agricultural farm or lease land plots. The above objective had not been reached as 46% of land shares in the 1990s were rented by agricultural farms, or former owners of those lands (Lipski, 2005). The main reason for that was the doubt in the legitimacy of land shares deals that was not reinforced by legal norms as required by Government of the Russian Federation (1993). According to the Constitution of Russian Federation, procedures related to land use are the subject to federal regulation. As a result, adoption of legislation without federal law made regulation of the related issues unlawful (Government of the Russian Federation, 1993).

Transfer of land shares was completed in middle of the 1990s (Table 1).

Russian regions are ranged based on the share of agricultural land transferred to private owners in the total amount of agricultural land in the region. The highest parameters are demonstrated by agricultural regions in the European part of Russia and south of Siberia. In the regions where agricultural land is scarce (Moscow Oblast, Leningrad Oblast, Republic of Chuvashia), the ratio is about 50%. The lowest share of privatized lands is indicated in the North and Far East of Russia. The size of land share allocated to private owners was dependent upon the availability of agricultural lands in the region as well as number of potential shareholders. Nationally, land share average is about 9.7 hectares and it varies from 1.5 hectares to 34.3 hectares from one region to another. In 38 regions out of 74, the land share size is between 5-10 hectares.

Prior the adoption of the law On the Turnover of Agricultural Land (Government of the Russian Federation, 2002), privatized lands had controversial status due to the fact that their transfer to private owners was not ensured by the law, but prescribed by the Constitution. Thus, land privatization aimed

Table 1. Results of agricultural land privatization in the mid-1990s

Region	Shares, thousand	Total area of land shares, thousand hectares	Average land share, hectares	Share of privatized lands in total acreage, %
Volgograd Oblast	308.9	6,471.4	20.9	73.9
Belgorod Oblast	297.1	1,582.2	5.3	73.8
Kursk Oblast	295.4	1,801.1	6.1	73.7
Agin-Buryat Autonomous District	24.3	724.4	29.8	73.3
Omsk Oblast	316.3	4,831.6	15.3	71.9
Lipetsk Oblast	202.2	1,396.9	6.9	71.3
Yaroslavl Oblast	108.0	790.0	7.3	69.3
Tambov Oblast	257.6	1,899.7	7.4	69.2
Ryazan Oblast	208.2	1,749.3	8.4	69.0
Bryansk Oblast	243.3	1,302.1	5.4	68.6
Ulyanovsk Oblast	166.8	1,517.1	9.1	68.5
Tver Oblast	192.9	1,652.0	8.6	67.7
Kurgan Oblast	207.0	3,015.7	14.6	67.5
Orenburg Oblast	353.0	7,249.7	20.5	66.9
Rostov Oblast	476.0	5,676.4	11.9	66.4
Tula Oblast	163.0	1,314.8	8.1	66.4
Nizhny Novgorod Oblast	329.5	2,077.0	6.3	66.2
Orel Oblast	180.8	1,372.5	7.6	66.1
Stavropol Krai	363.8	3,821.8	10.5	66.0
Saratov Oblast	322.1	5,513.4	17.1	64.3
Voronezh Oblast	431.4	2,605.8	6.0	63.9
Republic of Mordovia	172.7	1,060.9	6.1	63.8
Kirov Oblast	211.2	2,120.9	10.0	63.7
Samara Oblast	228.8	2,553.3	11.2	63.5
Smolensk Oblast	170.2	1,329.8	7.8	63.1
Krasnodar Krai	658.8	2,980.4	4.5	63.1
Pskov Oblast	136.9	966.4	7.1	62.9
Novosibirsk Oblast	261.6	5,235.5	20.0	62.3
Ust-Ordyn Buryat Autonomous District	51.5	532.4	10.3	61.3
Republic of Udmurtia	160.6	1,160.5	7.2	61.3
Kaluga Oblast	102.3	843.0	8.2	60.9
Vologda Oblast	155.5	872.7	5.6	60.0
Penza Oblast	200.5	1,829.4	9.1	60.0
Ivanovo Oblast	63.7	495.2	7.8	59.5
Kostroma Oblast	74.1	580.2	7.8	56.6
Republic of Mary-El	100.5	444.2	4.4	56.5
Republic of Tatarstan	464.3	2,557.7	5.5	56.2

continued on following page

Table 1. Continued

Region	Shares, thousand	Total area of land shares, thousand hectares	Average land share, hectares	Share of privatized lands in total acreage, %
Altay Krai	362.3	6,176.0	17.0	56.0
Arkhangelsk Oblast	75.4	409.9	5.4	55.8
Kaliningrad Oblast	58.3	452.9	7.8	55.7
Chelyabinsk Oblast	235.8	2,850.5	12.1	55.6
Krasnoyarsk Krai	191.1	3,009.9	15.8	55.1
Komi-Perm Autonomous District	34.0	189.7	5.6	54.5
Novgorod Oblast	70.7	448.1	6.3	53.9
Vladimir Oblast	102.7	544.1	5.3	53.1
Republic of Chuvashia	257.2	525.6	2.0	50.6
Moscow Oblast	243.1	901.4	3.7	50.2
Republic of Khakassia	41.5	951.2	22.9	49.5
Perm Oblast	152.7	1,254.9	8.2	49.2
Irkutsk Oblast	93.9	949.8	10.1	49.2
Sverdlovsk Oblast	182.5	1,286.8	7.1	48.9
Tomsk Oblast	53.9	669.9	12.4	48.8
Chita Oblast	94.0	3,227.4	34.3	48.7
Tyumen Oblast	156.1	1,623.7	10.4	47.8
Primoskiy Krai	82.6	770.5	9.3	47.2
Leningrad Oblast	110.1	376.2	3.4	47.0
Kemerovo Oblast	123.8	1,252.3	10.1	47.0
Republic of Adygea	53.4	169.6	3.2	46.7
Amur Oblast	85.9	1,211.2	14.1	44.3
Republic of Buryatia	70.7	1,138.8	16.1	36.1
Republic of Karachay-Cherkessia	34.4	203.2	5.9	30.3
Jewish Autonomous Oblast	16.0	153.1	9.6	28.5
Astrakhan Oblast	87.7	823.4	9.4	26.2
Sakhalin Oblast	10.9	46.2	4.2	25.2
Republic of Altay	22.1	348.8	15.8	19.6
Murmansk Oblast	3.3	5.1	1.5	18.9
Koryak Autonomous District	2.9	8.3	2.9	18.5
Republic of Karelia	12.7	37.4	2.9	17.3
Khabarovsk Krai	12.4	91.2	7.4	13.3
Khanty-Mansi Autonomous District	3.3	80.5	24.4	12.6
Kamchatka Oblast	5.8	39.4	6.8	9.1
Komi Republic	3.8	23.4	6.2	5.6
Magadan Oblast	0.1	1.6	16.0	1.2

Source: Rosreestr (n.d.)

to fasten redistribution of lands among land users became the obstacle for that redistribution once it was easier to get lands for the establishment or expansion of agricultural farm from state or municipality.

There were several ways to strengthen cooperation between owners of land shares and agricultural enterprises. Some of regions followed their own way of privatization, for example, Nizhny Novgorod and five other regions. "Nizhny Novgorod model" proposed the reorganization of agricultural enterprises that had to be established by the owners of land and property shares and distribution of lands and property of agricultural enterprises by means of closed internal auctions with further public announcement of the distribution results. The lands and property could be acquired only for the establishment or expansion of agricultural enterprises or entities engaged in agricultural activities (Government of the Russian Federation, 1994). The majority of the regions, however, followed federal instructions and sub-laws that had insufficient legal power and were grounded on the recommendations instead of regulating legislation (Government of the Russian Federation, 1995). None of the privatization initial targets was completely reached due to the above-mentioned issues and insufficiency of legislation ground.

Inconsistency of legal regulation related with agricultural and land reforms had caused continuous disputes about ownership rights status of land shares. Due to the fact that there was no clear legal definition of land share, there were several points of view defining land share as a part of collective farm land's fund; a right of individual to allocate land plot; a share of property collectively owned by farmworkers. Krasnov (1993), Syroedov (1997), Kozir (1998), and Krassov (2000) specified land share as an obligatory right of individuals to allocate land shares into their own land plots and stated that ownership rights would arise after allocation of land plot.

The variety of practical interpretations leaded to the case that ownership rights for a particular land plot belonged simultaneously to individuals (land shares owners with related certificates) and farms which are legal entities owned the lands based on the certificate for perpetual use from Soviet period or agreements with land shares owners confirming contribution of that land shares to the capital equity of reorganized farms. In the 1990s, contribution of land shares to capital equity was allowed by direct transfer of land share as well as the right to exploit it. In the former case, land share owner loses ownership rights in exchange for agricultural farm stocks. In the latter case, land share owner kept ownership rights and granted the right to exploit the lands for three years with possible further prolongation upon party's agreement. It was not always clear from the registration documents what is contributed: land share or a right to exploit it. Juridical practice added uncertainty to this issue by opposite court's decisions in different regions. Thus, Moscow arbitrage court accepted the presence of registration company documents confirming contribution of land shares to capital equity as sufficient proof of ownership rights transfer from individuals to farms even in the cases of violations of contribution procedure. At the same time, the court of North-West administrative region requested the farms to present clear and detailed evidence of legality of shares contribution to capital equity otherwise the ownership rights were acknowledged to belong to individuals. The legal inconsistency issues of agricultural and land reforms were widely studied by Ustyukova (2007).

After fifteen years of legal regulation improvements, there are still points of view stating that land share is a right to request allocation of land plot while the ownership rights for lands belong to legal entities of reorganized agricultural farms (Evsegneev, 2013); land shares as collectively owned property were merged by legal entities of the farms after aggregation of land shares (Sheynin, 2011); there was legally structured shared property relationship between land shares owners and farms (Bystrov, 2000). The suggestions to return land shares to the state (Savchenko, 2001), to limit ownership rights to one year (Stroev & Volkov, 2001), to cancel land shares (Tarasov & Volodin, 1998) had contributed to the idea that land shares owners could not been fully considered as private lands owners.

Meanwhile, the reorganization of collective and agricultural farms that was initiated by the privatization barely brought new economic and legal sense to the modernization of agricultural entities.

One of the expected outcomes of privatization is redistribution of land shares among land share owners and concentration of land shares in the hands of the most effective and productive farms or individuals. In reality, by the end of the 1990s, 5.5 billion of land shares owners leased their land shares to the same agricultural farms they got lands shares from. Additionally, nearly two billion people contributed their land shares or the right to use them to capital equity, about 300 thousand people allocated land shares into land plot for individual or commercial use (Lipski, 2002). Approximately one-third of all privatized lands had been exploited and are still used by the farms without registration of ownership rights.

The most disputable issue is the way and terms of privatization that provoked deep conflict between key stakeholders: individuals, new owners of lands that could not organize proper, effective, and profitable use of lands, from one side, and farms that used those lands for years before and have to continue to exploit them for agricultural purposes, from another side (Lipski, 2015).

There are two alternatives for future development of land shares issues:

- Social scenario means that land shares are to be used as long-term market tool in agriculture. At the same time, regulation of land shares has to correspond to legal regulation of shared property. Such scenario is in favor of individual owners because they could get economic effect from the lease of lands. The lease rate was dependent on financial and economic results of the farms and was exchanged for grain and services, but it was feasible return from land (Stroev, 2001);
- Economic (production) scenario that implies liquidation of land shares as a temporarily tool used for the reorganization of farms (liquidation can be accomplished step by step limiting opportunities of land shares owners or by on-time withdrawal of all lands shares). This scenario was mostly in support of agricultural entities.

In the 1990s, the state program was based on social scenario. At the same time, there was no federal law ensuring economic policy in agriculture. Formally, workers of collective and agricultural farms (including retired workers) benefitted from privatization of agricultural lands as they got ownership for most part of lands exploited by the farms. In reality, however, the benefit was questionable and, in some way, negative for the workers as the farms faced some difficulties after establishment of land shares. As a result, in late 1990s, scholars started to explore problematic issues of land shares.

Problematic Issues of Land Shares and Possible Solutions

In 2002, legal status of land share was defined as a share in collective property right for lands plots related to agricultural lands (Government of the Russian Federation, 2002). The initial version of the document had changed nothing in the above-mentioned conflict (land shares owners and farms). In addition, the law kept land shareowners' rights related to land shares use unchanged: they were able to inherit, sell, grant, exchange, use as mortgage, contribute to capital equity of legal entity or use them in other way taking into consideration that priority right to buy shares belonged to co-owners. There was only one restriction exposed by the law related to new lease agreements as up to 46% of all land shares were under lease.

The law forbidden to lease land shares but did not put any restrictions upon the lease of allocated land plots based on one or several lands shares. Adoption of the Law "On Agricultural Lands Use" clarified legislation issues related to the privatized agricultural land. Slight restrictions of land shares owners' rights did not influence its general ideology of social scenario of land shares development.

Main issues of shared property status influencing agricultural production had not been changed at all but had been legally disputed:

- Continuing conflict of interest between individuals and farms;
- Decision-making related to the use of shared land plot (for example, lease of land plot to agricultural entity and terms of the lease) should be managed upon mutual consent of co-owners but considering that up to 2018, only 20% quorum from total amount of owners was enough for decision-making, the accepted decisions were the subject of continuous reconsiderations;
- Disputes about land shares ownership rights as a result of land shares commercial circulation and allocation into land plots by individuals decided to quit shared property ownership;
- No penalties for non-used part of privatized (split into shares) lands similar to land plot. There are tools ensuring ceasing of ownership rights for the land plots granted or acquired as land plots in case if they are not used (Khlystun & Alakoz, 2016; Volkov & Lipski, 2017). In case of collectively shared property, the above tools cannot be used due to several reasons, first, the object of privatization is not specified as privatization was concerned with agricultural lands while other lands of former state and collective farms continue to be state-owned, secondly, it is impossible to identify and personify for shared property lands whose lands shares related to unused lands;
- **Unclaimed Lands Shares:** Owners of about two billion of shares are not known. Huge number of unclaimed lands is the result of agrarian policy. During the initial twelve years after the start of privatization, state administration approach was aimed to make any deal with land shares as easier as possible so to reach the goal of land shares concentration in hands of effective owners (individuals or reorganized farms). In 2005, new version of federal law limited the owners' rights in the sphere of land shares use (Government of the Russian Federation, 2002) to the following options: inherit, refusal and transfer to another co-owner or farm exploiting the lands. Another option to use land shares was allocation land shares into lands plots, but that option was hardly implemented.

Unclaimed Land Shares

Restrictions of lands shares owners' rights imposed in 2005 resulted in the fact that over 20% of land shares remained unclaimed. It means a share has an owner as well as an owner has the documents confirming the ownership, but the owner failed to realize in some way the property right and did not know where the share was located. Kovalev (2009), Lukianchikova (2016), and Rumyantsev (2012) reported that most of lands referred to shares were used by farms without any registration. The total amount of unclaimed lands was quite stable during the 1990s, but starting from 2012, shares have been acknowledged as municipal property. In the early 1990s, the area of unclaimed land shares was estimated at 26.5 billion hectares (Garankin & Komov, 2005), while in the early 2000s, it was 22-25 billion hectares (Khlystun, 2012b). The number of land shares owners declined by 30% from 12.0 billion of people to 8.5 billion of people.

Figure 4. Change in the scope of land rights
Source: Authors' development based on Lipski (2014)

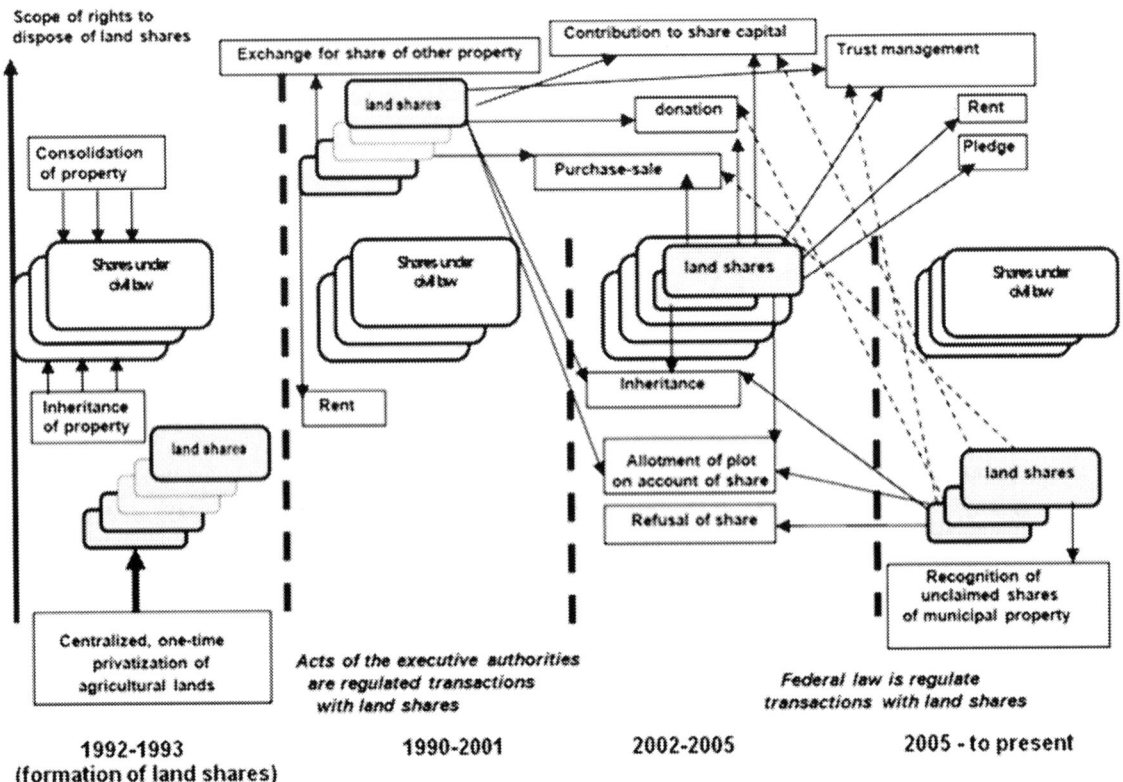

Amendments added to the law "On Agricultural Lands Use" in 2010 were aimed to solve the problem of unclaimed land shares by empowering municipal administration with a right to identify unclaimed shares, list them, and recognize them as municipal property after certain procedures, including court process. The law specified that lands were referred as unclaimed if their owners died and nobody claimed for succession; indisposed for over three years; were not mentioned in municipal administration documents as an object of privatization.

The issue of unclaimed lands has been actual for more than twenty years. Such a period is sufficient for land shares owners to decide how to realize their rights, but it is important to note the shift from social to economic approach in federal legislation policy adopted in 2005 and 2010. As first step, the legislation limited land shares owners in disposing land shares by leaving them only one option to transfer them to farms in any possible form, and then after five years, the legislation approved withdrawal of unused land shares.

Rights of land shares owners correlate with the rights of co-shared property owners regulated by civil law (Figure 4). Fragmentary lines related to landowners' rights indicate significant limitation of the rights. Starting from 2005, the owners of land shares could dispose them only to farms or individuals exploiting the lands related to the shares. The restriction forced the owners to dispose their land shares

Table 2. Total area of land shares owned by individuals in 2012-2016, billion hectares

	2012	2013	2014	2015	2016
Total owned by individuals	97.6	94.9	92.3	89.3	88.4
Change in % to the previous year		97.2	97.3	96.7	99.0
i.e.:					
unclaimed		15.3	16.6	18.1	18.5
Change in % to the previous year			108.5	109.0	102.2
Recognized as municipal property			3.7	4.4	5.5
Change in % to the previous year				118.9	125.0

Source: Authors' development based on Russian Research Institute of Information and Technical and Economic Studies on Engineering and Technical Support of Agriculture [Rosinformagrotekh] (2018)

to farms using that lands was the first step to abolish land shares and further, it was supported by the withdrawal of unclaimed lands to municipal property.

As a result, in 2013, about 15.3 billion hectares of land shares (16.1% of total land shares owned by individuals) were recognized as municipal property by court procedures. From that year, the area of lands acknowledged as unclaimed increased up to 18.5 billion hectares that resulted in the increased area of lands withdrawn to municipal property upon court decision.

The highest rate of settled lawsuits concerned with the recognition of unclaimed land shares as municipal property is indicated at Voronezh, Belgorod, Amur, and Novgorod oblasts (Table 3). To some extent, this can be explained by the fact that the governors of those territories are particularly experienced in the establishment of economic policy in agriculture.

Notwithstanding the adopted legislative measures, less than 30% of unclaimed land shares are acknowledged as municipal property. There are several reasons for that:

- Organizational and technical flaws of local administration including violation of the established procedures;
- Lack of mutual decision of co-owners concerned with demarcation of land plots formed on account of unclaimed land shares;
- Court doubts in legitimacy of unclaimed land shares recognition: there are some deputes related to three years of non-use of land shares (in case of inheritance of land shares, the period should be counted from the date of succession); additionally, an owner which share is recognized as unclaimed due to fact of non-use has the right to argue the decision and follow the procedure to exclude the share from the list of unclaimed ones;
- Possible conflict of interests between municipal administration and farms exploiting the related lands that can face extra expenses (lands taxes or release of some land plots in favor of other individuals) due to the recognition of unclaimed land shares as municipal property.

The transfer of ownership rights for unclaimed land to municipal administration was aimed to distribute lands to effective farmers in a form of auctions.

Table 3. Unclaimed land shares

	Total area, thousand hectares		Share of property recognized at court, %
	Recognized by municipal administration as unclaimed	Recognized at court as municipal property	
Voronezh Oblast	270.42	204.67	75.7
Belgorod Oblast	40.83	21.88	53.6
Amur Oblast	385.76	220.41	57.1
Novgorod Oblast	230.14	129.57	56.3
Lipetsk Oblast	86.87	39.22	45.1
Penza Oblast	244.49	103.03	42.1
Tambov Oblast	127.56	52.79	41.4
Yaroslavl Oblast	58.19	24.05	41.3
Orel Oblast	240.77	98.06	40.7
Kaluga Oblast	183.30	71.80	39.2
Tula Oblast	85.20	28.20	33.1
Jewish Autonomous Oblast	42.90	14.20	33.1
Republic of Mordovia	293.66	91.96	31.3
Republic of Chuvashia	231.48	71.87	31.0
Kostroma Oblast	203.30	58.94	29.0
Republic of Buryatia	656.76	189.60	28.9
Vologda Oblast	523.05	149.23	28.5
Perm Krai	683.20	190.10	27.8
Tver Oblast	77.78	21.57	27.7
Krasnodar Krai	46.62	12.58	27.0
Kursk Oblast	228.18	61.33	26.9
Kirov Oblast	327.84	70.00	21.4
Saratov Oblast	663.00	134.50	20.3
Ulyanovsk Oblast	290.80	58.30	20.0
Pskov Oblast	298.04	53.51	18.0
Zabaykalskiy Krai	1,012.02	165.59	16.4
Smolensk Oblast	480.76	75.86	15.8
Samara Oblast	276.35	42.06	15.2
Republic of Kalmykia	40.98	5.90	14.4
Vladimir Oblast	117.69	13.43	11.4
Khabarovsk Krai	30.01	3.30	11.0
Republic of Bashkortostan	588.94	63.57	10.8
Kemerovo Oblast	401.26	42.58	10.6
Arkhangelsk Oblast	110.62	11.58	10.5
Primorsky Krai	244.65	25.74	10.5
Ivanovo Oblast	229.16	22.39	9.8

continued on following page

Table 3. Continued

	Total area, thousand hectares		Share of property recognized at court, %
	Recognized by municipal administration as unclaimed	Recognized at court as municipal property	
Kurgan Oblast	901.50	84.90	9.4
Orenburg Oblast	1,294.44	114.91	8.9
Republic of Mary-El	105.18	7.76	7.4
Republic of Adygea	5.76	0.36	6.2
Republic of Udmurtia	210.12	12.84	6.1
Bryansk Oblast	306.00	17.90	5.8
Kaliningrad Oblast	14.72	0.86	5.8
Novosibirsk Oblast	1,572.91	80.99	5.1
Chelyabinsk Oblast	136.52	5.75	4.2
Astrakhan Oblast	279.60	8.19	2.9
Leningrad Oblast	47.20	1.34	2.8
Komi Republic	24.50	0.50	2.0
Tomsk Oblast	35.65	0.62	1.8
Volgograd Oblast	330.70	5.20	1.6
Omsk Oblast	444.76	5.04	1.1
Sakhalin Oblast	18.08	0.12	0.6
Irkutsk Oblast	463.13	2.00	0.4

Source: Authors' development based on Ministry of Agriculture of the Russian Federation (n.d.)

Non-Privatized Lands (Demarcation of State-Owned Lands)

Demarcation of state-owned lands between federal, regional, and municipal authorities is another tool aimed to change the structure of land ownership. The active phase of that process started in the 2000s, not the 1990s, due to the insufficiency of federal legislation. About a half of the entire territory of the country was demarcated before 2013. Since that year, the area of demarcated lands has increased (55% in 2014, 61% in 2016, and 63% in 2017). The analysis of the land fund indicates that the major part of demarcated lands (970.8 billion hectares) has been transferred to federal ownership that accounts for 61.5% of land fund, while municipal authorities acquired about 9.9 billion hectares (0.5% of land fund). The reason for that was demarcation of forest land fund that constitutes two-thirds of total land fund. These lands can be owned only by federal authorities. The area of lands that was a subject to demarcation decreased by 238.4 billion hectares in 2012-2016 (Table 4). Federal lands dominated despite the increase of demarcated lands owned by regional and municipal authorities.

Regional ownership constitutes the major part of demarcated agricultural lands, while the percentage of lands in federal property decreased (Table 5). The share of demarcated lands in municipal property has increased due to the withdrawal of unclaimed land shares. Agricultural lands have been demarcated to a lower extent in 2012-2016 and indicated decrease in the percentage of non-demarcated lands.

Table 4. Demarcated lands in state property in Russia, billion hectares

	2012	2013	2014	2015	2016	2016 to 2012, %
State and municipal property, including	1,576.7	1,576.8	1,576.9	1,577.3	1,579.1	100.1
Federal	745.4	770.4	862.1	945.1	970.8	130.2
Regional	8.8	9.3	12.6	12.9	18.2	206.8
Municipal	3.9	5.1	6.5	8.0	9.9	253.8
Non-demarcated state property	818.6	792.0	695.7	611.3	580.2	70.9
% of non-demarcated lands	51.9	50.2	44.1	38.8	36.7	70.7

Source: Authors' development based on Rosreestr (n.d.)

Table 5. Demarcated agricultural lands in state property in Russia, billion hectares

	2012	2013	2014	2015	2016	2016 to 2012, %
State and municipal property, including	260.4	258.0	258.3	258.0	255.4	98.1
Federal	8.3	8.5	8.5	8.7	5.8	69.9
Regional	7.7	7.9	11.3	11.4	11.2	145.5
Municipal	3.4	4.6	5.8	7.3	9.1	267.6
Non-demarcated state property	241.0	237.0	232.7	230.6	229.3	95.1
% of non-demarcated lands	92.5	91.8	90.1	89.4	89.8	97.1

Source: Authors' development based on Rosreestr (n.d.)

It is expected that demarcation of lands of non-agricultural purpose will be completed in three-five years while the demarcation of agricultural non-privatized lands will go on for uncertain period.

Transformation of Land Ownership in Post-Soviet Russia

In Russia, 92.2% of lands are in state or municipal property while the share of private-owned lands is about 7.8% (133.1 billion hectares) (Table 6).

The highest area of privatized lands (over 50%) is registered in the regions where agriculture is well developed (Figure 5).

It is necessary to note that the ratio of private and public owned lands has not changed recently. The slight decrease of private-owned lands was registered in 2014 after the accession of Crimea. At the same time, the distribution of private-owned lands between individuals and legal entities for the specified period has shifted in the favor of the latter ones. The acreage of lands owned by individuals decreased by 11.3 billion hectares, while the acreage of lands owned by legal entities increased by 15 billion hectares. Thus, it is possible to conclude that privatization of state and municipal lands has stopped and turned to the redistribution of lands between individuals and legal entities. Despite the fact that about 1.5 billion hectares of lands is redistributed from individuals to legal entities annually, the area of lands owned by individuals exceeds the area of lands owned by enterprise fivefold.

Table 6. Land fund of Russia based on ownership rights

Period	Ownership					
	billion hectares			percentage		
	State and municipal	Individuals	Legal entities	State and municipal	Individuals	Legal entities
1990	1,709.8	0.0	0.0	100.0	0.0	0.0
...						
2005	1,580.4	124.2	5.2	92.4	7.3	0.3
2006	1,580.0	123.8	6.0	92.4	7.2	0.4
2007	1,577.6	125.1	7.1	92.3	7.3	0.4
2008	1,576.9	124.3	8.6	92.2	7.3	0.5
2009	1,576.3	123.2	10.3	92.2	7.2	0.6
2010	1,576.4	121.4	12.0	92.2	7.1	0.7
2011	1,576.7	119.6	13.5	92.2	7.0	0.8
2012	1,576.8	118.3	14.7	92.2	6.9	0.9
2013	1,576.9	117.0	15.9	92.2	6.8	0.9
2014	1,577.3	115.4	17.2	92.2	6.7	1.0
2015	1,579.1	115.3	18.1	92.2	6.7	1.1
2016	1,579.4	114.1	19.1	92.2	6.7	1.1
2017	1,579.4	112.9	20.2	92.2	6.6	1.2

Source: Rosreestr (n.d.)

Figure 5. Share of privatized lands in Russian regions
Source: Authors' development based on Rosreestr (n.d.)

SOLUTIONS AND RECOMMENDATIONS

Economic aspects of agricultural policy ensuring food security are specified as follows:

- About 88.3 billion hectares of agricultural lands (45.2%) are in shared ownership of 8.5 billion of individuals that became the owners of land shares during privatization in the early 1990s
- Existence of the conflict between individuals (landowners) who are not able to exploit them and farms that seek to exploit the lands
- Special procedure related to the exploitation of agricultural lands (a decision is taken upon mutual consent of co-owner)
- Economic inefficiency of land shares owners' stimulation to allocate land shares into land plots
- Disputes concerning land shares status of owners which allocated land shares into lands plots or decided to refuse from shared ownership
- The issue of unclaimed land shares (about two billion of shares) that has remained unsolved even after the adoption of withdrawal procedure to municipal property in 2012
- Necessity to adopt the measures ensuring the right of municipal authorities to buy land shares from individuals that enable to solve the issue of unclaimed land shares

FUTURE RESEARCH DIRECTIONS

This study analyses economic issues of agricultural sector assuming that privatization of agricultural land took place, mainly completed, and accepted by the majority of individuals and enterprises involved in agriculture. The study allows identifying positive and negative effects of privatization and determining the measures aimed to solve food insecurity problem. The study, however, has not covered the following issues:

- Analysis of performance of agricultural production when exploiting own (private) lands and rented lands, as well as rational use of lands by private owner and renter
- Lands that are not used in agriculture – the total area is over 28 billion hectares in 2015 (15% of agricultural lands). Those lands were acquired as capital investment for the construction or further resale in case of land price growth and are subject to identification, legal and fair withdrawal, and further transfer to new owners or renters who are expected to ensure their effective utilization. Additionally, taxation measures aimed to force landowners to exploit their lands in agricultural production or concede to others need diversified investigations
- Ineffectiveness of court decisions related to unclaimed lands issues

The above-mentioned issues related to economic aspects of public policy in agriculture should be studied as key factors affecting food security of a country.

CONCLUSION

In general, privatization of agricultural lands in Russia had positive effects. Harmonization of legal status of lands with economic relations in agriculture has been settled. Privatization has ensured the engagement of lands to economic activities which helped to meet demand for land plots that exceeded the supply at the beginning of land reform and adjust the fair market price for land.

Meanwhile, there were some negative issues concerned with land shares:

- Unclaimed land shares issue; complicated procedure related to the exploitation of privatized agricultural lands (upon mutual consent of co-owners); huge number of owners who do not have any idea what to do with acquired ownership; contradiction between individuals (new owners of lands) having no experience how to manage the lands or facing difficulties related with lands management and agricultural farms that have to exploit the related lands (used in the soviet time by collective or state farms) to meet agricultural industry needs
- Land shares tool was aimed to enable consolidation of agricultural lands in the hands of effective farmers that ensure their rational use, but in practice, it became a barrier for private agriculture development and rational use of agricultural lands due to the unwillingness of private owners to cultivate and develop the lands as well as invest in improvement of land fertility
- Corruption in agriculture related to illegal procedures of land allocation.

The demarcation of state-owned lands between federal, regional, and municipal authorities highly likely will be completed during next three-five years for non-agricultural lands, while the settlement period for agricultural non-privatized lands is uncertain

REFERENCES

Alakoz, V., Kiselev, V., & Shmelev, G. (1999). *Reasons for Land Reform in Russia*. Moscow: Interdesign.

Alakoz, V., Vasiliev, I., Kiselev, V., & Pulin, A. (2001). *Joint Shared Land Property: Theory, Data, Practice*. Moscow: Construction Formula.

Bogolubov, S. (1998). *Ecological Law: Manual*. Moscow: Norma-Infra.

Bystrov, G. (2000). Land Reform in Russia: Legal Doctrine and Practice. *State and Law, 4*, 49–50.

Evsegneev, V. (2013). Unresolved Issues of Privatization of State-Owned Land in Russia. *Agrarian and Land Law, 103*(7), 100–108.

Federal Service for State Registration, Cadastre and Cartography. (n.d.). *State (National) Reports on Land Use in the Russian Federation in 1992-2015*. Moscow: Federal Service for State Registration, Cadastre and Cartography.

Garankin, N., & Komov, N. (2005). Rent as a Key Issue of Civil Turnover of Land. *Land Management. Land Monitoring and Cadaster, 12*, 54–56.

Government of the Russian Federation. (1990). *Law #374-1 from November 23, 1990, "On Land Reform."* Retrieved from https://base.garant.ru/10107009/

Government of the Russian Federation. (1991). *Decree #86 from December 29, 1991, "On the Reorganization of Collective and State Farms."* Retrieved from http://base.garant.ru/10104699/

Government of the Russian Federation. (1992). *Decree #708 from September 4, 1992, "On the Privatization and Reorganization of Agro-Industrial Enterprises and Organizations."* Retrieved from http://pravo.gov.ru/proxy/ips/?docbody=&nd=102018291&rdk=&backlink=1

Government of the Russian Federation. (1993). *The Constitution of the Russian Federation.* Retrieved from http://www.constitution.ru/en/10003000-01.htm

Government of the Russian Federation. (1994). *Decree # 874 from July 27, 1994, "On the Reforming of Agricultural Enterprises with Account of the Experience of Nizhny Novgorod Oblast."* Retrieved from http://www.consultant.ru/document/cons_doc_LAW_4234/

Government of the Russian Federation. (1995). *Decree #96 from February 1, 1995, "On the Procedure of Execution of Rights of the Owners of Land Shares and Property Shares."* Retrieved from http://www.consultant.ru/document/cons_doc_LAW_5724/

Government of the Russian Federation. (2002). *Federal Law of the Russian Federation #101 from July 24, 2002, "On the Turnover of Agricultural Land."* Retrieved from http://base.garant.ru/12127542/

Khlystun, V. (2005). Structural Transformation and Development of Land Relations. *Farming, 3,* 20–21.

Khlystun, V. (2011). On State Land Policy. *Economy of Agricultural and Processing Enterprises, 10,* 1–4.

Khlystun, V. (2012a). Agrarian Transformations in Post-Soviet Russia (20th Anniversary of Agricultural Reform). *Economy of Agricultural and Processing Enterprises, 6,* 17–21.

Khlystun, V. (2012b). Land Relations in Agricultural Industry of Russia. *Domestic Notes, 6,* 78–84.

Khlystun, V. (2015). Quarter of a Century of Land Transformations: Intentions and Results. *Economy of Agricultural and Processing Enterprises, 10,* 13–17.

Khlystun, V., & Alakoz, V. (2016). Tools of Circulation of Unused Agricultural Land. *Economy of Agricultural and Processing Enterprises, 11,* 38–42.

Komov, N. (1995). *Management of Land Resources in Russia: Russian Model of Land Management and Land Tenure.* Moscow: RUSSLIT.

Komov, N., & Aratskiy, D. (2000). *Methodology of Land Management at Regional Level.* Nizhny Novgorod: Nizhny Novgorod Branch of the Russian Presidential Academy of National Economy and Public Administration.

Kovalev, A. (2009). Theoretical and Practical Issues of Economic Turnover of Unclaimed Land Shares. *Bulletin of Notarial Practice, 5,* 44–47.

Kozir, M. (1998). On Conceptual Development of Land Reform in Russia at Modern Stage. *Economy of Agricultural and Processing Enterprises, 6,* 14.

Krasnov, N. (1993). Land Reform and Land Law in Modern Russia. *State and Law, 12,* 3–13.

Krassov, O. (2000). *Land Law: Manual.* Moscow: Jurist.

Kresnikova, N. (2008). Tools of Agricultural Land Turnover. *Land Management. Land Monitoring and Cadaster, 4,* 52–54.

Krylatykh, E. (1997). Formation and Development of Economic Regulation of Land Relations. *Forecasting Issues*, *1*, 31–39.

Krylatykh, E. (1998). Development of Land Relations in Agricultural Sector and Rural Areas. *Scientific and Technological Development in Agricultural Industry*, *4*, 30–35.

Leppke, O. (1998). Economic and Legal Aspects of State Policy Related to Rational Use and Preservation of Land Resources (Analysis and Solutions). *International Agricultural Journal*, *3*, 3–6.

Leppke, O. (2000). *Scientific and Organizational-Economic Regulation of Land Relations in Agro-Industrial Industry in a Transition Economy (Methodology and Regional Level)*. Moscow: Agripress.

Lipski, S. (2002). Specifics of Land Reform at Modern Stage. *Economist*, *10*, 77–87.

Lipski, S. (2005). Land Shares: Actual Issues and Perspectives. *Real Estate and Investments. Legal Regulation*, *4*, 86–90.

Lipski, S. (2014). *Land Relations and State Land Policy in Modern Russia (Theory, Methodology, Practice)*. Moscow: State University of Land Use Planning.

Lipski, S. (2015). Private Ownership for Agricultural Lands: Advantages and Disadvantages (Experience of Two Decades). *Studies on Russian Economic Development*, *26*(1), 63–66. doi:10.1134/S1075700715010074

Loyko, P. (2001). *Agricultural Reform: Issues and Practice*. Moscow: AS Plus.

Loyko, P. (2009). *Land Management: Russia and the World (Future Outlook)*. Moscow: State University of Land Use Planning.

Lukianchikova, S. (2016). «Exceptional Privatization»: How to Avoid Legal Violations upon Acquisition of Unclaimed Land Shares on a Preferential Basis. *Legal Issues of Real Estate*, *2*, 37–40.

Ministry of Agriculture of the Russian Federation. (n.d.). *Reports on the Conditions and Use of Agricultural Lands in 2010-2016*. Moscow: Ministry of Agriculture of the Russian Federation.

Ogarkov, A. (2000). Agriculture and Its Resource and Production Potential. *Economy of Agricultural and Processing Enterprises*, *5*, 7–9.

Pankova, K. (2000). Land Shares and Collective Land Use. *Agrarian Science*, *1*, 3–5.

Petrikov, A. (2000). Agricultural Reform in Russia and Agricultural Policy Issues. *Economy of Agricultural and Processing Enterprises*, *12*, 7–8.

Petrikov, A. (2016). The Main Directions of the Modern Agri-Food and Rural Policy. *International Agricultural Journal*, *1*, 3–9.

Pitersky, V. (1999). *Strategical Potential of Russia. Natural Resources*. Moscow: Geoinformmark.

President of the Russian Federation. (1991). *Decree #323 from December 27, 1991 "On Urgent Measures for the Implementation of Land Reform in Russian Socialist Federative Soviet Republic."* Retrieved from http://www.consultant.ru/document/cons_doc_LAW_206/

President of the Russian Federation. (1992). *Decree #213 from March 2, 1992, "On Establishment of the Norm of Free Land Allotment to the Individuals."* Retrieved from http://www.consultant.ru/document/cons_doc_LAW_365/

President of the Russian Federation. (1993). *Decree #1767 from October 27, 1993, "On the Regulation of Land Relations and Agrarian Reform in Russia."* Retrieved from http://www.consultant.ru/document/cons_doc_LAW_2601/

President of the Russian Federation. (1996). *Decree #337 from March 7, 1996, "On the Implementation of Citizens' Constitutional Rights for Land."* Retrieved from https://base.garant.ru/10105753/

Rabinovich, B., Fedoseev, I., & Ignatiev, A. (1995). *Land Reform in Russia: Principles and Implementation*. Moscow: Stars and Co.

Rumyantsev, F. (2012). On Unclaimed Land Shares. *Economy and Law, 5*, 111–115.

Russian Research Institute of Information and Technical and Economic Studies on Engineering and Technical Support of Agriculture. (2018). *Report on the Conditions of Rural Areas in the Russian Federation in 2016*. Moscow: Russian Research Institute of Information and Technical and Economic Studies on Engineering and Technical Support of Agriculture.

Savchenko, E. (2001). Contemporary Issues of Land Relations Regulation. *International Agricultural Journal, 1*, 3–6.

Serova, E. (2000). Agricultural Reforms in Transition Economies: Mutual Objectives and Diverse Tools. Moscow: Encyclopedia of Russian Villages.

Shagaida, N., & Alakoz, V. (2017). *Land for People*. Moscow: Center for Strategic Research. doi:10.2139srn.3090400

Shagaida, N., & Fomin, A. (2017). Improvement of Land Policy in the Russian Federation. *Moscow Economic Journal, 3*, 1–26.

Sheynin, L. (2011). *Real Estate: Legislative Gaps*. Moscow: Delovoy Dvor.

Stroev, E. (Ed.). (2001). *Mixed Agrarian Economy and Russian Village (Middle of the 1980-1990s)*. Moscow: Kolos.

Stroev, E., & Volkov, S. (2001). *Land Issues in Russia in the Beginning of the XXI Century (Challenges and Solutions)*. Moscow: State University of Land Use Planning.

Syroedov, N. (1997). Notarization of Land Shares Transactions. *Bulletin of Legal Information. Land and Law, 3*, 6.

Tarasov, N., & Volodin, V. (1998). Land Relation Issues in Agricultural Cooperatives – Collective Farms. In *Proceedings of the Conference "Land Relations in Agro-Industrial Industry of Russia."* Uglich.

Ushachev, I. (2003). To Improve State Agricultural Policy. *Agro-Industrial Complex: Economy. Management, 11*, 3–9.

Ustyukova, V. (2007). Citizens Collective Ownership Rights for Land Plots of Agricultural Purpose: Myth or Reality? *Ecological Law, 2,* 19–25.

Uzun, V., & Shagaida, N. (2015). *Agrarian Reform in Post-Soviet Russia: Tools and Results.* Moscow: Delo.

Vershinin, V., & Petrov, V. (2015). Development of Tools Ensuring Agricultural Circulation of Unused Agricultural Land. *International Agricultural Journal, 5,* 9–11.

Volkov, S., & Khlystun, V. (2014). Land Policy: How to Improve Effectiveness? *International Agricultural Journal, 1-2,* 3–6.

Volkov, S., & Lipski, S. (2017). Legal and Land Use Planning Measures for Involvement of Unused Agricultural Land into Economic Circulation and for Ensuring Their Effective Use. *Land Management. Land Monitoring and Cadaster, 145*(2), 5–10.

Wegren, S. (2012). Institutional Impact and Agricultural Change in Russia. *Journal of Eurasian Studies, 3*(2), 193–202. doi:10.1016/j.euras.2012.03.010

ADDITIONAL READING

Khlystun, V., Volkov, S., & Komov, N. (2014). Land Resources Management in Russia. *Economy of Agricultural and Processing Enterprises, 2,* 41–43.

Kulik, G. (2013). Preservation and Rational Use of Lands and Natural Resources Ensure Food Security of Russia. *Agricultural and Food Policy of Russia, 3,* 17–24.

Lerman, Z., & Shagaida, N. (2007). Land Policies and Agricultural Land Markets in Russia. *Land Use Policy, 24*(1), 14–23. doi:10.1016/j.landusepol.2006.02.001

Petrikov, A. (2016). Food Security and Import Substitution as Key Factor of Russian National Security: Development, Priorities and Perspectives. *Scientific Publications of Free Economic Society of Russia, 199,* 437–444.

KEY TERMS AND DEFINITIONS

Agricultural Lands: The lands used for agricultural activities: arable lands, hayfields, pastures, deposits, gardens, and fruit gardens.

Food Security: A share of domestically-produced food in total volume of domestic product turnover.

Lands: The lands engaged or applicable for economic activities characterized by natural and historical features.

Land Plot: An individual and demarcated part of lands.

Land Share: A share in shared property rights for land plots of agricultural purpose originated from the process of land privatization in the 1990s.

Land Sharing: A transfer of ownership rights on equal basis.

Privatization: A transformation of ownership ensuring transfer of state and municipal property to individuals.

Spot Privatization: Cost-reimbursable transfer of lands to particular individuals or farms.

Section 3
Agricultural Trade and Quality of Nutrition

Chapter 12
Agricultural Trade and Quality of Nutrition:
Impacts on Undernourishment and Dietary Diversity

Elena Chaunina
Omsk State Agrarian University, Russia

Inna Korsheva
Omsk State Agrarian University, Russia

ABSTRACT

Proper nutrition is not only a biological but also a social, economic, and political issue. Insufficient intake of essential elements may result in the occurrence of hidden hunger and metabolic disorders. Some regions of the world are characterized by a lack of certain nutrients in the environment which leads to their lack in plant and animal products. The most common problem is a deficiency of iodine and selenium. To solve this problem, the government takes various measures, such as direct inclusion of necessary additives in food products, as well as the modernization of technological process of crop and livestock production. In this chapter, the authors analyze the provision of the population in various countries and regions with limiting nutrients. The study specifically aims at exploring the issues of production and trade in fortified (modified) food products that can directly fill in the lack of essential elements in particular territories.

INTRODUCTION

One of the important challenges faced by the world community today is the question of providing a sufficient amount of affordable, high-quality, and safe food to the growing population of the planet which is expected to reach ten billion by 2050. This must be done against the background of the growing shortage of fresh water and land suitable for agricultural use, soil degradation, and the reduction of biological diversity. The ongoing climate change aggravates food insecurity problem. Achieving food

DOI: 10.4018/978-1-7998-1042-1.ch012

security requires an integrated approach that takes into account all existing forms of undernourishment of the population, as well as the stability and efficiency of production of agricultural goods and food products and international trade in food.

BACKGROUND

In 2018, about 113 million people in 53 countries around the world experienced severe food shortages. Despite the fact that the number of people confronted with food crises has slightly decreased, the total number of people who do not receive adequate nutrition is still above 100 million. In addition, over the past three years, the scope of countries and continents that have been engulfed by the food crisis has increased. Nearly two-thirds of the total number of people experiencing acute hunger inhabit eight countries: Afghanistan, the Democratic Republic of Congo, Ethiopia, Nigeria, Sudan, South Sudan, Syria, and Yemen. In 17 countries, number of people experiencing acute hunger either remained the same or increased (Food and Agriculture Organization of the United Nations [FAO], International Fund for Agricultural Development [IFAD], United Nations International Children's Emergency Fund [UNICEF], World Food Programme [WFP], & World Health Organization [WHO], 2018).

Proper nutrition is not only a biological but also a socio-economic, as well as political issue. A balanced diet is an essential condition for public health. To maintain the normal flow of energy and plastic and catalytic processes in the body, a certain amount of nutrients is required. Some of these substances can be synthesized in the body, but a large proportion of them are obtained from the outside with food and thus is indispensable. A balanced diet provides the optimal ratio for the human body in the daily diet of proteins, fats, carbohydrates, and biologically active substances.

The recommended ratio of carbohydrates, proteins, and fats is 4:1:1, respectively. Proteins are of paramount importance for vital activity. Without them, life, growth, and development of a body are impossible. It is a plastic material for the formation of cells and intercellular substance. Proteins are constituents in hormones, immune bodies, and enzymes. They are involved in metabolism of vitamins and minerals and delivery of oxygen, lipids, carbohydrates, vitamins, hormones, and drugs by blood. The earliest manifestation of protein deficiency is the reduction of protective properties of an organism in relation to the action of adverse environmental factors. With a lack of proteins, the processes of digestion, blood formation, and activity of the endocrine glands and nervous system are disturbed, the growth and development of an organism are inhibited, the mass of muscles and liver is reduced, and trophic disorders of the skin appear.

Fats supply a body with energy, polyunsaturated fatty acids, phospholipids, and sterols. Insufficient intake of these substances may result in the impairment of the function of central nervous system, skin, kidney, and organs of vision, as well as to a decrease in the body resistance.

Carbohydrates are the main source of energy. They are necessary for ensuring metabolism. Carbohydrates stimulate absorption of proteins, contribute to normal activity of liver, muscles, nervous system, heart, and other organs.

Functioning as biological regulators of metabolism and constituents of enzymes, vitamins provide normal course of biochemical and physiological processes in a body.

Minerals are vital to human body and, along with other food components, are an essential part of the diet.

Major sources of nutrients for humans are the products of animal and vegetable origin which are conventionally divided into several main groups:

- Dairy products
- meat
- Fish, eggs, and products made from them
- Bread, pasta and bakery products, cereals, and sugar
- Oils and fats
- Herbs, vegetables, berries, and fruit
- Coffee, tea, cocoa, and spices

These product groups are distinguished by their composition and predominant nutrients.

The qualitative composition of the products is of great importance. Insufficient intake of certain indispensable elements can lead to "hidden hunger" and disorder of metabolic processes. Monotonous eating can also lead to the phenomenon of "hidden hunger." Food security implies that the population has access to a sufficient amount of safe and high-quality food, satisfying its nutritional needs to ensure an active and healthy life, and also takes into account food preferences of people. However, in low- and middle-income countries, about 13% of the population is undernourished. About two billion people worldwide have a shortage of nutrients. A lack of zinc, vitamin A, and iron leads to growth retardation, anemia, immunity, and cognitive functions impairment. Products of animal origin are rich in such substances as vitamin B12, riboflavin, calcium, iron, zinc, and essential fatty acids which are difficult to obtain in sufficient quantities by consuming mainly vegetable food. This is especially important for children.

From the products of animal origin, people consume 39% of protein and 18% of calories of their daily consumption, but these are average data for different countries. Poor people in low- and middle-income countries often do not consume sufficient amounts of products of animal origin, while others – especially in high-income countries, and increasingly – in middle-income countries – consume them in excessively.

Over the past three decades, consumption of meat, milk, and eggs in low- and middle-income countries has tripled. Population growth, urbanization, increased incomes, and globalization stimulate agricultural development. According to FAO (2019), demand for meat in these countries will increase by another 80% by 2030 and more than double by 2050. Agricultural producers increase their capacity to meet the growing demand for food, as well as update technologies due to the changing food preferences of an increasingly affluent and urbanized population in a globalized economy.

Some regions are characterized by a lack of certain nutrients in the environment, which leads to a decrease in their content in plants, as well as in products of animal origin produced in the area. The population of these regions has a number of specific symptoms that reduce the quality of life and lead to various diseases. The most frequent deficiency of such microelements as iodine and selenium is a global problem. To solve this problem, a number of measures are taken at national level, such as the direct inclusion of necessary additives in food products, as well as the modernization of the technological process of crop and livestock production.

The simultaneous presence of multiple forms of malnutrition is frequent. This is typical for poor people in middle- and low-income countries. Obesity in high-income countries is also concentrated among the poor. The coexistence of various forms of malnutrition can occur not only at national level, but also in households, and can also occur in the same person at different periods of their life.

Agricultural Trade and Quality of Nutrition

This may seem paradoxical, but food insecurity can contribute to overweight and obesity. Fresh and high-quality products are often expensive. Thus, people choose less expensive products that usually contain large amounts of calories and few nutrients. This is ubiquitous but is especially characteristic of urban dwellers in upper-middle and high-income countries.

According to FAO (2019), in 2019, food imports in the world will decrease by 2.5%. Lower costs on food purchase are expected mainly in developed countries. In low-income countries, whose currency weakens against the US dollar, the major currency in international trade, the volume of imported food may also decrease. About half of the expected reduction is accounted for by coffee, tea, cocoa, and spices, while the costs on sugar and cereals, despite the decline in international prices for the latter, will generally remain unchanged. A positive phenomenon is the expected decline in prices for vegetable oils, which, as a rule, occupy a large share of imported goods.

The new food forecast presents the first FAO's supply and demand forecasts for 2019/2020 (FAO, 2019). It offers updated information on new developments in the map of world food production and trade. It is noteworthy that India and Russia consolidate their recent ascent to the top status as the world's largest sugar producers and wheat exporters, overtaking Brazil and the USA, respectively. While sugar consumption per capita is growing, in developed countries, it is stabilized due to the fact that it is a subject to increased attention from the regulators and a change in consumer preferences. This leads to the fact that international prices are set below production costs. In this regard, Brazil is expected to use almost two-thirds of its sugar cane crop for ethanol production, compared with 53% in 2018 (FAO, 2019).

The emergence and rapid spread of African swine fever in China, where half of the world population of pigs is located, will have a significant impact on world meat markets. According to forecasts, pork imports will grow by 26%. Growth in imports of other meats, including cattle and poultry, is also expected.

MAIN FOCUS OF THE CHAPTER

In modern conditions, the structure and patterns of nutrition the population has changed in many ways. Various pathological conditions associated with a deficit of various macro- and micronutrients in the diet have become widespread. The nutritional disorders of the population are largely due to the crisis in the production and processing of food raw materials and food products, the deterioration of the economic opportunities of the majority of the population and low purchasing power. There is an acute problem of food quality, as well as a low level of food culture of the population.

Meat products are always poor in micronutrients, which has been especially aggravated in recent years. Enrichment with vitamins, microelements, phytocomplexes, and other biologically active substances significantly increases their biological value (Bou, Guardiola, Barroeta, & Codony, 2005). Therefore, the consumption of fortified products is affordable for the population and a safe way to prevent the shortage of the essential substances with which these products are fortified. Many countries have succeeded in correcting nutrition and improving public health with fortified foods. Food fortification should not be a separate independent procedure, but a part of national (regional) programs related to nutrition and health of certain groups of citizens or population. In many developed countries, food fortification is regulated at the state level.

In Russia, the elimination of micronutrient deficiencies through food fortification is provided for by the Foundations of State Policy of the Russian Federation in the Sphere of Healthy Nutrition of the Population till 2020 (Government of the Russian Federation, 2010) and a number of national programs, including "Overcoming Iron Deficiency", "Overcoming Iodine Deficiency", "Overcoming Selenium Deficiency", "Food Vitaminization", "Diabetes", and others.

Food fortification is a serious interference with the traditionally established structure of human nutrition, so it can only be done according to scientifically based and proven principles, including the following:

- Micronutrients, the shortage of which really exists, is quite widespread and dangerous to health, should be used for food fortification. The most common deficiencies in the world are a lack of vitamin A, iodine, and iron. In Russia, these are B and C group vitamins, folic acid, carotene, iodine, iron, zinc, and calcium.
- First of all, it is necessary to fortify mass-consumption products that are available for all groups of children and adults and are regularly used in everyday food (flour and bakery products, milk and dairy products, salt, sugar, beverages, and baby foods).
- Food fortification should not change the organoleptic characteristics of products and reduce their shelf life.
- When fortifying food products, it is necessary to take into account the possibility of chemical interaction of the fortifiers between themselves and with the components of the product to be fortified. It is necessary to choose such combinations, forms, their safety in the course of production and storage.
- Regulated (guaranteed by a producer) content of vitamins and minerals in food product fortified with them should provide 30-50% of the average daily requirement at the usual level of consumption of the product.
- The amount of micronutrients additionally introduced into products must be calculated taking into account their possible natural content in the original product or raw materials used for its manufacture, as well as losses during the production and storage process in order to ensure their content at a level not lower than the regulated level during the whole shelf life of the fortified product.
- The amount of the fortifier must be at a level that will not be exceeded by adding small amounts of this fortifier to other sources.
- The additional cost of the fortified product must be acceptable to the consumer.
- Introduced substances must be biologically available in the product.
- The regulated content of ingredients in the products fortified with them should be indicated on the individual packaging of this product and be strictly controlled.
- The effectiveness of fortified products and their harmlessness should be convincingly confirmed by testing on representative groups of people.

The process of food fortification is rather complicated since a number of factors should be taken into account:

- Compatibility of fortifiers introduced among themselves. For example, ascorbic acid contributes to a better absorption of iron, the presence of vitamin E in the product increases the activity of vitamin A, calcium has a blocking effect on the absorption of iron. Ascorbic acid destabilizes folic acid and cyanocobalamin.
- Compatibility of fortifiers and a carrier. For example, in products containing a large amount of dietary fiber, it is impractical to introduce iron salts or other trace elements, since dietary fibers are able to bind them tightly, disrupting absorption in the gastrointestinal tract.

- Impact of processing, including heat, treatment of products on the fortification efficiency. For example, it is advisable to fortify flour and bread with B vitamins, since they relatively well tolerate the effects of high temperature in the baking process, while ascorbic acid is significantly less resistant. The inclusion of small amounts of ascorbic acid in vitamin and vitamin-mineral mixtures for the fortification of flour has purely processing goals: it accelerates aging of flour and improves its baking properties.

The category of fortified products includes:

- Specialized products for children, pregnant and lactating women, athletes, elderly people, and people of extreme professions (submariners, climbers, astronauts, etc.).
- Specialized food products are developed for healthy people who have certain physiological needs related to the functional state of the body or lifestyle.
- Specialized baby foods include products for artificial nutrition and complementary foods that are necessary to ensure the full physical and mental development of the child, especially when there is insufficient breastfeeding. Products for pregnant women, lactating women, and elderly people are designed to ensure appropriate adjustment of their physiological status.
- Specialized products are also a necessary element of nutrition for athletes, extreme activities, accompanied by high energy consumption, hypoxia, physical, and psycho-emotional stress. At the same time, there is an increased need of the body for energy, food, essential, and minor substances which are problematic to compensate with conventional traditional products.
- Therapeutic-prophylactic and prophylactic products – products for people working in hazardous industries, living in environmentally unfavorable conditions, having certain diseases or predisposed to them (diabetes, obesity, atherosclerosis, etc.).

Food products intended for therapeutic and prophylactic nutrition are dietary foods. Dietary foods can be used by healthy people for the prevention of nutrition-related diseases, etc. Functional foods are foods that contain ingredients that benefit human health by improving many physiological processes in a body. There are intended for healthy people and risk groups. Additional (functional) ingredients that give functional properties to products should be beneficial to health, safe, natural, not reduce nutritional value, and taken orally.

The volume of intake of functional ingredients must be medically agreed. Currently, in the EU and the USA, there is a provision that functional foods having the ability to improve health state do not have to meet full medical requirements. The production of livestock products of a functional purpose is of strategic importance in ensuring health of the population. Functional and special-purpose products on the USA market account for 40% of the total production, in Europe – exceed 20%.

World production of edible eggs and poultry meat provides more than 30% of the population's need for natural foods of animal origin. Complete protein, optimal fatty acid, vitamin and mineral composition of chicken eggs contribute to the constant increase of their production and consumption. In Russia, there are few functional products, for example, poultry farms have started supplying eggs fortified with iodine, selenium, carotenoids, unsaturated fatty acids, etc. With organic forms of selenium and iodine, real prerequisites have been created for solving the problem of their deficiency in the human diet through the consumption of eggs fortified with these elements. Due to less toxicity and prolonged action, organic forms are more preferable to meet the need for this microelement (Fisinin, Yegorov, Yegorova, Rozanov, & Yudin, 2011; Komarova, Ivanov, & Nozhnik, 2012). The authors' study on the fortification of edible eggs with iodine, selenium, and carotene revealed the positive effect of additives on the nutritional value of eggs.

The use of Monclavit-1 has increased the iodine content in eggs to 33.3%. Moreover, according to Korsheva (2017), chicken eggs fortified with iodine with the application of Monclavit-1 almost completely retain it during cooking and can be considered functional products that provide human body with necessary micronutrients. As a result, due to higher productivity and livestock survival, as well as the realizable value of eggs fortified with iodine, in experiment group, the profitability of production was higher by 4.2% compared to that in control group. The introduction of organic selenium (Sel-Plex) into the diet increased the content of this element in eggs by an average of 21.1% and increased the profitability of production by 2.3% (Korsheva & Trotsenko, 2015). The use of carotenoids in feeding birds also had a positive effect on the content of vitamins in eggs – the amount of carotenoids increased by 74.7%.

Thus, today, fortified mass-consumption products are means capable of countering the development of a deficiency of essential substances (vitamins, BAS, microelements) in human body. Requirements for fortified products are well developed and approved. Therefore, the consumption of fortified products is affordable for the population and a safe way to prevent the shortage of the essential substances with which these products are fortified. Functional products are fortified products, the effect of which on human health has been proven in scientific research. The established requirements for these products continue to be improved since currently they have already been developed and approved by experts.

SOLUTIONS AND RECOMMENDATIONS

Regardless of the degree to which undernourishment is expressed, its effects have a significant impact on health, well-being, and quality of life of the population. Undernourishment leads to a decrease in labor productivity and economic efficiency and to an increase in medical expenses.

In all countries of the world, one can find various forms of hunger and undernourishment. Although poverty is the main cause of undernourishment, programs to increase incomes or to provide food to the poor are not the only way to remedy the situation. The solution of the problem is possible through an integrated approach with the application of political measures, implementation of social and educational actions, improvement of trade relations, and new methods of production of agricultural products.

FUTURE RESEARCH DIRECTIONS

The study conducted by the authors in this chapter is not enough to make final conclusions. Future research should confirm the existing solutions and open up new possibilities for solving the issue of the availability of a balanced diet.

To combat undernourishment, it is necessary to direct research to expand the range of food fortifiers, including through the use of specialized feed additives, to determine the most effective economic models of their use in agriculture, and to increase the range of functional nutrition and sustainability of micro additives to cooking.

In order to increase the availability of high-quality food for the population, studies on the possibility of the development of international trade relations in terms of food products turnover on special conditions will be promising, allowing for more comfortable conditions on the market.

CONCLUSION

The shortage of food has accompanied he mankind throughout its history. Food insecurity problem is one of the most ancient global problems of mankind. There are many causes of hunger in the world and they are often interrelated. Previously, the main reason was considered to be high birth rate of the population, leading to overpopulation and reduction in planted acreage. But the results of the studies showed that the causes of hunger and malnutrition differed not only across the countries and regions but also at the level of a family. There are social differences and customs of food distribution by gender and age.

According to Sen (1998), one of the main causes of hunger in the world is not the absence of natural prerequisites for food production and high population density, but poverty. Therefore, starving regions of the world coincide geographically with the regions of the spread of poverty. Hunger and poverty prevail mainly in African countries that lie south of the Sahara, in some regions of East Asia and Latin America, as well as in most countries of Southeast Asia.

For the developed countries, the phenomenon of hunger as a whole is no longer typical. But along with this, chronic deficiency of vitamins and microelements in the diet has become widespread in the world. Life in ecologically unfavorable cities and nervous overloads require an increased consumption of vitamins from a modern person. In addition, the adopted food technologies do not contribute to the preservation of the most valuable nutrients in products. The severity of the situation becomes even more tangible in a cold climate, economic poverty, lack of certain elements in the soil and water. Pregnant women and children are particularly affected by the deficits.

Developed countries are taking emergency measures to combat hidden hunger. Mandatory fortification of basic foods is already embodied in various kinds of national programs fixed by laws. The positive results of such prevention are obtained.

Thus, the problem of hunger, undernourishment, and dietary diversity is relevant today for both a single state and for all of humanity as a whole. It is not only about the need to increase food production, but also to ensure access to food, eliminate the shortage of quality food supply and the development of market institutions.

REFERENCES

Bou, R., Guardiola, F., Barroeta, A. C., & Codony, R. (2005). Effect of Dietary Fat Sources and Zinc and Selenium Supplements on the Composition and Consumer Acceptability of Chicken Meat. *Poultry Science*, *84*(7), 1129–1140. doi:10.1093/ps/84.7.1129 PMID:16050130

Fisinin, V., Yegorov, I., Yegorova, T., Rozanov, B., & Yudin, S. (2011). Enrichment of Eggs with Iodine. *Poultry and Poultry Products*, *4*, 37–40.

Food and Agriculture Organization of the United Nations. (2018). The State of Food Security and Nutrition in the World 2018. Building Climate Resilience for Food Security and Nutrition. Rome: Food and Agriculture Organization of the United Nations.

Food and Agriculture Organization of the United Nations. (2019). *Food Outlook – Biannual Report on Global Food Markets*. Rome: Food and Agriculture Organization of the United Nations.

Government of the Russian Federation. (2010). *Decree #1873 from October 25, 2010, "Foundations of State Policy of the Russian Federation in the Sphere of Healthy Nutrition of the Population till 2020."* Retrieved from https://rmapo.ru/medical/58-osnovy-gosudarstvennoy-politiki-rossiyskoy-federacii-v-oblasti-zdorovogo-pitaniya-naseleniya-na-period-do-2020-goda.html

Komarova, Z., Ivanov, S., & Nozhnik, D. (2012). Production of Table Eggs with a Predominantly Functional Properties. *Scientific Journal of Kuban State Agrarian University*, *81*(7), 476–485.

Korsheva, I. (2017). Production of Iodine-Rich Edible Eggs. In *Proceedings of the Conference Biotechnology: Current State and Future Development*. Moscow: BioTech World.

Korsheva, I., & Trotsenko, I. (2015). The Effectiveness of Sel-Plex Usage for Production of Selenium Enriched Eggs. *Omsk Scientific Bulletin*, *144*(2), 199–201.

Sen, A. (1998). *The Possibility of Social Choice*. Nobel Prize Organization. Retrieved from https://www.nobelprize.org/prizes/economic-sciences/1998/sen/lecture/

ADDITIONAL READING

Ebdon, L., Pitts, L., Cornelis, R., Crews, H., Donard, O. F. X., & Quevauviller, P. (2001). *Trace Element Speciation for Environment, Food, and Health*. Cambridge: Royal Society of Chemistry.

Nys, Y., Bain, M., & Van Immerseel, F. (2011). *Improving the Safety and Quality of Eggs and Egg Products*. Oxford: Woodhead Publishing. doi:10.1533/9780857093912

Ponomarenko, Y. (2015). Chlorella Enriched in Iodine and Selenium in the Diets of Chickens-Broilers. *Proceedings of the VIII International Research and Practice Conference*. Munich: Strategic Studies Institute.

Ponomarenko, Y. (2015). Chlorella Enriched in Iodine and Selenium in the Diets of Laying Hens. *Proceedings of the IV International Scientific Conference*. Chicago: Strategic Studies Institute.

KEY TERMS AND DEFINITIONS

Balanced Diet: A physiologically complete nutrition, satisfying the needs of a body in metabolizable energy and nutrients.

Biologically Active Substances: The chemicals that in low concentrations may affect metabolic processes in a body.

Feed Ingredients: The products of plant, animal, and synthetic origin used in animal feeding.

Fortified (Functional Foods): The food products produced using a special technology and able to fill the body's needs for individual nutrients.

Hidden Hunger: An inadequate intake of certain nutrients with a sufficient level of nutrition in general.

Nutrients: The body's digestible food components – proteins, fats, carbohydrates, vitamins, and minerals.

Nutrition Structure: The ratio of individual food groups in a diet.

Specialized Fortifiers: Specialty feed additives designed to increase the content of certain nutrients in foods of animal origin.

Chapter 13
Agricultural Trade and Undernourishment, Nutrition, and Dietary Diversity:
The Use of Elite Selection Cultivars of Legumes

Anna Veber
https://orcid.org/0000-0003-0715-0426
Omsk State Agrarian University, Russia

Svetlana Leonova
Bashkir State Agrarian University, Russia

Elena Meleshkina
All-Russian Research Institute of Grain and Grain Products, Russia

Zhanbota Esmurzaeva
https://orcid.org/0000-0002-0471-1921
Omsk State Agrarian University, Russia

Tamara Nikiforova
Orenburg State University, Russia

ABSTRACT

The results described in this chapter are of the investigation based on the collaborative research of scientists from the three Russian universities (Omsk State Agrarian University, Bashkir State Agrarian University, and Orenburg State University) which started in 2014. The authors assess various indicators of food safety. The study includes physical and chemical properties, technological characteristics, and chemical composition of new elite selection cultivars of pea ("Pisum arvense", the harvest of 2018, Bashkir Scientific and Research Institute of Agriculture) and haricot bean (harvest of 2018, Omsk State Agrarian University). Most of the samples have increased phytochemical capacity and high protein concentration (21.15-22.49% in haricot bean; 19.38-23.75% in pea). The authors demonstrate that these cultivars can be used for the enrichment of foodstuff and the creation of new functional foods.

DOI: 10.4018/978-1-7998-1042-1.ch013

INTRODUCTION

In the previous decade, factual nutrition of the world population can be characterized as unsatisfactory. Nutrition is the intake of food considered in relation to the dietary needs. Proper nutrition (an adequate, well-balanced diet combined with regular physical activity) is a cornerstone of good health (Food and Agriculture Organization of the United Nations [FAO], 2013). The priority of state policy is ensurance the right of people for balanced nutrition. The disturbance of food status causes health disorders and leads to the development of foodborne diseases, such as atherosclerosis, hypertension, obesity, diabetes mellitus, osteoporosis, arthralgia, and malignant tumors, among others. The maintenance of health along with the prevention of foodborne diseases seem to be the aims of utmost importance for the governments worldwide (Tutelyan, n.d.)

According to Tutelyan (2005), globally, there is a decrease in consumption of meat, dairy, and fish products, vegetable oils, vegetables, and fruit. Meanwhile, there is an increase in consumption of fat, sugar, confectionery, bakery products, bread, and potato. Despite the abundance of feedstuffs and low level of hunger in the world, there is a steady deficiency in consumption of quality protein, essential nutrients, vitamins, and food fibers. The problem is particularly relevant in Russia, where over 80% of people struggle the deficiency of crude protein consumption. The average norm of protein consumption is 90-100 grams per day including 60-70% of animal protein. According to the standards of physiologic requirements in energy and feedstuffs for various groups of population in Russia (Government of the Russian Federation, 2008), animal proteins should provide at least 50% of a daily diet of an adult individual, while plant proteins should make the remaining 50%. For adults, daily physiologic need for proteins is 65-117 grams for men and 58-87 grams for women. Currently, the deficiency of animal proteins varies from 15% to 20% of the recommended norm. It does not meet the recommendations of the World Health Organization (WHO) and the standards of physiologic requirements in energy and feedstuffs in Russia (Government of the Russian Federation, 2008).

Globally, the general deficiency of protein is estimated at 10-25 million tons per year. About half of the global population suffers from a lack of protein. Apart from proteins, the deficiency in other nutrients (fats, carbohydrates, vitamins, and microelements) leads to weaker immunity, increase in disease vulnerability, and disturbance of physical and mental development. The need for protein is an evolutionarily developed dominant in human nutrition caused by the need of providing optimum physiologic level of irreplaceable amino acids intake. Two groups of proteins are defined in the structure of world resources of food protein, i.e. plant protein and animal protein. The sources for animal protein are meat and meat products, dairy, and fish and fish products which can be absorbed by up to 93-96% by a human organism. They include the full amount of irreplaceable amino acids sufficient for protein biosynthesis in a body. Plant proteins are derived from oil, grain and leguminous crops, root and tuber crops, nuts, vegetables, and melon cultures. In plant proteins, there is a deficiency of irreplaceable amino acids. Also, they contain inhibitors of proteinase that reduce digestion of proteins. Plant protein is absorbed by up to 62-80% by a human organism. Despite the high biological value of plant proteins, the correlation of plant and animal proteins makes 80% and 20%, respectively, in the world nutrition balance.

Beans and cereal crops grain are the traditional sources of protein ingredients. Over the times, grain crops have been particularly important in the formation of protein sources. However, their widespread application is limited by the content of gluten which may cause food allergy among certain groups of people. One of the solutions of this problem is production of soybeans and other crops, such as pea, chickpea, lupine, vetch, lentil, haricot bean, and other leguminous crops (Table 1).

Table 1. Global production of leguminous crops

Country	Volume of production, million tons	Share in the global production, %	Average annual increase during previous 50 years	Main crops
World, total	75.0	100.0	1.3	
India	17.6	24.0	1.5	Chickpea, haricot bean, Cajanus
Canada	5.4	7.0	12.0	Pea, lentil
Myanmar	5.0	7.0	7.0	Chickpea, haricot bean, Cajanus
China	4.5	6.0	0.4	Bean, haricot bean, pea
Nigeria	3.9	5.0	7.5	Vigna
Brazil	3.0	4.0	2.8	Bean, haricot bean, pea
Australia	3.0	4.0	18.5	Lupine, lentil, chickpea
Ethiopia	2.6	3.0	2.6	Bean, haricot bean, pea, chickpea
Russia	2.3	3.0	2.3	Pea, chickpea, vetch
USA	2.0	3.0	3.5	Haricot bean, pea
Tanzania	1.7	2.0	6.4	Haricot bean
Nigeria	1.6	2.0	17.6	Bean
Mexico	1.3	2.0	4.2	Haricot bean, chickpea, vetch
Turkey	1.3	2.0	2.0	Haricot bean, chickpea, lentil, vetch

Source: International Independent Institute of Agrarian Policy (2017)

India is the world's biggest producer of leguminous crops. Canada and Myanmar take the second and third places in the ranking, respectively. Russia's share in the global market of leguminous crops is only 3%. In Russia, there are three major regions for production of leguminous crops:

- Central Federal District, the leading territory in growing forage and universal leguminous crops – pea, soybean, and lupine.
- Volga Federal District, which specialized in cultivation of forage, food, and universal crops, in particular, pea, lentil, chickpea, and vetch.
- Far Eastern Federal District, the major territory in Russia for soybean farming.

In 2018, production of soybean, lentil, pea, and haricot in Russia increased compared with 2016. Particularly, in 2018, Russian farmers got the record yield of soybean since 1990. The volume of production reached 3.9 million tons (in 1990, there were only 0.7 million tons). According to Nekrasov (2019), total production of grain and leguminous crops in Russia in 2018 made 112.9 million tons. In 2019, production of grain and leguminous crops is expected to reach 118.0 million tons.

Such processes as overpopulation, limitation of land resources and acreage, and economic development require more resources. Therefore, the production of leguminous crops takes considerable place in food security worldwide. The necessity of solution of food insecurity problems along with the need to ensure economic growth and development in emerging economies (BRICS countries, in particular), increase trade cooperation in agriculture. The authors agree with Bazga (2015) who stated that food-secure countries are those which agricultural potential allow supplying agricultural raw materials and food in the volume bigger than domestic market demands.

In the previous decade, BRICS has acquired the status of the powerful economic block with over 40% of world population and over 20% of world GDP. Taken together, BRICS countries produce about one-third of global output of grain and leguminous crops. BRICS includes the world's biggest exporters of leguminous crops (Russia, India, and Brazil), as well as net importers (China and South Africa). The structure of export and import shows the potential opportunities for the development of trade within the alliance (Popova & Klimenko, 2018). Wheat, rice, corn, and barley are also among potential crops to be traded between BRICS countries.

Based on the discussion above, there are three issues which are taken into consideration in this chapter: the size of the markets, the variety of sources of plant protein, and poor development of the market. There is a lack of correct motivation of domestic demand. This demand will allow increasing the deliveries of different types of leguminous crops and products of their processing between BRICS countries. It will promote diversification of sales markets and bring mutual benefits to all member countries. Due to such cooperation, the level of self-dependence of BRICS countries on leguminous crops and food security can be increased. Environmental situation in the planet will improve, the share of plant protein in a diet will grow.

One of the paramount values of leguminous crops is the maintenance of soil fertility in its natural way by means of soil enrichment with available nitrogen forms and the activity of tuberous bacteria. It is also necessary to preserve ecosystem biodiversity and maintain underground biodiversity (Zotikov, Naumkina, & Sidorenko, 2014). High yields of leguminous crops provide extra economic abilities for food industry. As such, leguminous crops are suggested to be the ones which can create a sustainable model of organic agriculture (Shcherbakova (Ponomareva), 2017) and contribute to regional and global food security (Halimi, Barkla, Mayes, & King, 2019).

BACKGROUND

According to Karmas and Harris (1988), there are over 13,000 species of legumes in the world, but only twenty of them are consumed by people. Those crops contain a high proportion of proteins, fats, dietary fibers, carbohydrates, vitamins, and minerals (Prodanov, Sierra, & Vidal-Valverde, 2004). The nutrition content varies depending on the species, their habitats, climate, and type of soil (Bishnoi & Khetarpaul, 1993). High concentration of protein makes leguminous crops a source of a balanced diet (Muimba-Kankolongo, 2018). Soybean, chickpea, lentil, haricot bean, and other crops provide people with the largest amount of plant protein. This is testified by the chemical analysis of bean grain and its components in percent per dry matter basis (Table 2).

Among a variety of leguminous crops, soybean is the most widespread one. It is a traditional source of ingredients used in the production of high-protein food, for instance, isolates, concentrates, textured soybean products, and different types of soybean flour or semolina. Also, it is used in the production of food products for vegetarians (Skripko, Isaycheva, & Pokotilo, 2015; Migina, 2016).

One of the ways to increase the amount of high-protein products is the cultivation and research of new selection cultivars of leguminous crops. They are less widespread in food industry compared to soybean, but they have well-balanced amino acid composition and low activity of inhibitors. Kazantseva (2016) used chickpea grains of "Krasnoutsky 28" and "Privo 1" cultivars as the unconventional material source

Table 2. Chemical compound of bean grain, % per dry matter

Crop	Protein	Starch and other carbohydrates	Fiber	Fat	Leach
Pea	26.4	60.1	6.8	2.1	2.7
Chickpea	23.6	59.4	8.3	5.2	3.3
Vetchling	25.8	60.9	5.3	1.5	2.7
Lentil	28.9	60.9	4.4	1.9	3.2
Haricot bean	27.7	60.2	5.2	2.4	4.1
Bean	28.0	54.1	10.6	2.1	3.5
Mung bean	23.5	42.3	3.8	2.0	3.5
Soybean	34.9	24.6	4.3	17.3	5.0

Source: Skurikhin and Tutelyan (2002)

of plant origin. The grains were processed for healthy foodstuffs with high protein content as they had low activity level of trypsin inhibitors, increased phytochemical potential, and quality protein (Table 3).

Methionine and cysteine, valine, isoleucine, and threonine are the limiting amino acids for chickpea grains of "Privo 1" cultivar. Methionine and cysteine are the limiting amino acids for "Krasnoutsky 28" cultivar (Table 4).

Shelepina (2010) and Shelepina and Parshutina (2013) carried out complex assessment of technological, biochemical, and quality indicators of crops. Comparative study of the composition of protein of domestic pea selection ("Temp" and "Spartacus" cultivars and selection lines of "Amius-98-891" and "Amikh-99-1132") demonstrated the identity of proteins in pea grain structures (Shelepina, 2010) and the identity of the content of all irreplaceable and replaceable amino acids (Shelepina & Parshutina, 2013). Both protein full value and its utilization were testified by the content of methionine (1.06-1.10%), tryptophan (1.12-1.14%), lysine (7.06-7.36%), and isoleucine and leucine (12.86-13.46%). The results indicated insufficient balance of protein in the studied pea cultivars according to the content of irreplaceable amino acid valine. The amount of threonine, isoleucine, and leucine exceeded the quantity of similar amino acids in "FAO protein." Therefore, "Krasnoutsky 28" and "Privo 1" pea cultivars process high-quality protein. They can be used not only in forage production but in foodstuff production by processing of grain into protein foodstuffs.

In Russia, there is a sustainable increase in the volume of pea yield. Russia has emerged to a strong competitor to Canada and Australia in supply of pea crops to India. Russia's domestic pea cultivars "Yamal", "Aksay Usatyi", "Pharaoh", "Spartacus", "Taloved", "Chishminsky 95", "Chishminsky 229", "To the Memory of Hangildin" have high phytochemical potential and compete with European cultivars

Table 3. Phytochemical potential of chickpea grain

Chickpea cultivar	Weight fraction, %					
	Moisture	Protein	Fat	Leach	Starch	Fibre
Krasnoutsky 28	9.00 ± 0.50	24.00 ± 0.20	3.70 ± 0.30	3.30 ± 0.03	46.00 ± 1.30	3.60 ± 1.10
Privo 1	8.50 ± 0.50	20.07 ± 0.20	4.30 ± 0.30	3.15 ± 0.03	41.80 ± 1.30	3.90 ± 1.10

Source: Kazantseva (2016)

Table 4. Amino acid composition of chickpea grain protein, grams per 100 grams of protein

Amino acid	Krasnoutsky 28	Privo 1
Irreplaceable amino acids		
Valine	5.45	4.00
Isoleucine	4.29	3.39
Leucine	7.38	8.06
Lysine	5.59	6.16
Methionine	0.92	0.75
Threonine	3.98	3.74
Tryptophan	1.02	1.00
Phenylalanine	5.10	5.27
Replaceable amino acids		
Alanine	4.87	4.74
Arginine	8.33	7.91
Asparagine acid	11.04	10.59
Histidine	2.89	2.38
Glycine	4.29	3.87
Glutamine acid	15.40	17.69
Proline	4.88	5.45
Serotonin	4.87	4.94
Tyrosine	2.12	2.48
Cysteine	0.45	0.33

Source: Kazantseva (2016)

grown in France and Canada. They are pest-resistant and highly productive. Among the territories of Russia, the Republic of Bashkortostan is the leader in production of pea. High protein pea cultivars for foodstuff production were selected at Bashkir Scientific and Research Institute of Agriculture. These cultivars have high commercial and culinary qualities, a lack of bean flavor being one of them. Over ten pea cultivars with relevant characteristics were included in the State Register of Selection Achievements of the Russian Federation (Vakhitova, 2015; Akhmadullina, 2018).

Haricot is less popular and therefore less cultivated compared to pea, soybean, and chickpea. Among the countries of the world, Kyrgyzstan is one of the biggest producers and exporters of haricot. The country supplies about twenty cultivars of haricot. Bodoshov (2015) found that the level of protein content in haricot cultivars in Kyrgyzstan differed from 20% to 30%. There was also sufficient amount of replaceable and irreplaceable acids. The major cultivars of Kyrgyzstani selection are "Sugar", "Boxer", "Chinese Woman", "Tashkent", "Goose Feet." They all have high concentration of amino acid. In food industry, it is recommended to use "Sugar" cultivar due to its rich content of amino acid (Bodoshov, 2015). Kazydub and Marakaeva (2015) developed the adaptive haricot cultivars with high yield characteristics and better consumers' qualities.

The preservation and enhancement of natural nutrition value of initial raw materials in the production of socially important foodstuffs, for instance, bread, cereals, macaroni, confectionery, dairy drinks, etc.

is the next main method of protein deficiency correction. The process of germination is one of ways of enhancement of nutrition value of plant native raw materials. Antipova, Grebenshchikov, Mishchenko, Osipova, and Tychinin (2017), Kazymov and Prudnikova (2012), and Dotsenko, Bibik, Lyubimova, and Guzhel (2016) studied the process of germination of leguminous cultivars on the example of lentil, soybean, and mung bean grains ("Amurskaya", "Octyabrskaya-70", "Harmony", "Sonata", "Dauriya", and others). They proved the necessity of germination process as an essential stage in the production of consumers' foodstuffs with balanced combination of protein and amino acids. It was testified by the protein value assessment before and after germination. The amino acid composition of proteins in lentil, soybean, and mung bean was improved approximately to "FAO protein" model after the germination. In average, the total amount of protein boosted by 13-15%. There was an increase in components and bioavailability of vitamins and minerals in germinated grain. The decrease in an anti-alimentary factor of leguminous cultivars was indicated (Antipova et al., 2017; Kazymov & Prudnikova, 2012; Dotsenko et al., 2016). The prospects of use of germinated leguminous raw materials in production of foodstuffs for different purposes are confirmed by Relf (2015). The use of new selection cultivars in combination with nature-like technologies allows increasing protein content at the minimum material and technical expenses. Therefore, it will help to produce balanced foodstuffs of high quality and to solve a problem of protein deficiency.

In spite of the large number of studies devoted to the characteristics of leguminous raw materials and foodstuffs (Matejčeková, Liptáková, & Valík, 2017; Petruláková & Valík, 2015), most of them deal with leguminous crops as cereal materials (Ponomarev, 2011), ingredients in confectionery production (Tsareva, 2007; Antipova & Presnyakova, 2008), and water retraining and high protein additives in meat industry (Giro & Chirkova, 2007; Kolesnikova, 2006; Razvedskaya, 2001; Ryazanova, 2014). In this chapter, the authors evidenced inconsiderable number of studies of technological properties of haricot and pea cultivars in the public catering sector. Raw materials for production of drinks on leguminous basis with a universal basic composition allowing expanding the range of technologies of ready-made foodstuffs are considerably less studied. There are insufficient results related to particular pea and haricot cultivars. The ways and criteria for production of high-protein foodstuffs are not specified. Therefore, this chapter aims at studying biological and nutrition value of haricot of the selection of Omsk State Agrarian University cultivated in the territory of Omsk Oblast, Russia and pea of the selection of Bashkir Scientific and Research Institute of Agriculture cultivated in the territory of the Republic of Bashkortostan, Russia. Identifying the differences of technological properties of pea and haricot cultivars may be the key to the development of the production technology of high-protein foodstuffs.

MAIN FOCUS OF THE CHAPTER

In 2018, the following selection cultivars were included in the study: "Lukeriya", "Omichka", "Nerussa" (all three from Omsk State Agrarian University), "Chishminsky 95", "Chishminsky 229", and "To the Memory of Hangildin" (all the remaining three from Bashkir Scientific and Research Institute of Agriculture). A sample of one hundred sorted grains was used per each cultivar. They were cleaned from extraneous impurity, dirty grains, and bruises by hand.

Indicators of Technological Properties of Grain

The analysis of economically valuable properties of domestic selection shows that new cultivars are characterized by large weight per 1,000 grains: 21.1-28.7 grams (haricot) and 77.0-88.0 grams (pea). The mass of haricot and pea grains from the plant differs from 232 grams to 406 grams and from 240 grams to 277 grams, respectively.

According to sensory indicators, the studied selection cultivars of haricot and pea correspond to the requirements of the standards "Pea. Requirements for State Purchases and Deliveries" (Government of the Russian Federation, 1991) and "Food Beans. Specifications" (Government of the Russian Federation, 1976). The selection cultivars are specified in accordance with the requirements of interstate standard "Food Beans. Specifications" (Government of the Russian Federation, 1976) and considering characteristics of grains. Therefore "Omichka" cultivar belongs to "white haricot" type, "white oval" subtype. "Nerussa" cultivar is a "white haricot" type, "pearl barley" subtype. "Lukeriya" cultivar is "concolorous" type, "other concolorous" subtype. The studied cultivars are of food type, the first subtype according to the requirements of interstate standard "Pea. Requirements for State Purchases and Deliveries" (Government of the Russian Federation, 1991). Taking into account bean cotyledons, the selection cultivars relate to the first class.

Consumer Qualities of New Selection Cultivars

"Omichka" haricot cultivar, "To the Memory of Hangildin", "Chishminsky 95", and "Chishminsky 229" pea cultivars demonstrated the best consumer qualities. They were assessed according to their sensory and cooking characteristics. These indicators were developed in the Laboratory of Technological Assessment of Agricultural Crops of the All-Russian Research Institute of Plant Industry. The studied haricot and pea cultivars refer to Class I (excellent) due to their cooking test.

The sensory assessment of boiled leguminous grains show that "Nerussa" and "Omichka" haricot cultivars, "Chishminsky 95" and "Chishminsky 229" pea cultivars have better taste. It is best to use them in cooking of snacks with protein content, for instance, hummus. On the contrary, "Lukeiya" haricot cultivar and "To the Memory of Hangildin" pea cultivar are not characterized by strong leguminous taste. The use of protein and carbohydrate compositions of these crops as plant additives does not influence significantly on the taste of a ready-made foodstuff. They can be recommended for the production of beverages on grain basis of emulsion type.

"Lukeriya" grains are bigger than those of "Omichka" and "Nerussa" cultivars. They are monochromatic in black color which intensity changes during heat treatment. "Omichka" and "Nerussa" grains have the same white color before and after heat treatment. The studied pea cultivars are of yellow color of different tones (with the cotyledons translucent through grain peel). The intensity of color does not change during heat treatment.

Biological Safety and Phytochemical Potential of Haricot and Pea Grains

Haricot and pea cultivars of domestic selection show the lack of extraneous impurities and contamination by pests and mycotoxins. The content of toxic elements, radionuclides, and pesticides meets the technical regulations of the Eurasian Economic Union "On Safety of Grains" (Eurasian Economic Commission,

Table 5. Characteristics of phytochemical potential of haricot and pea cultivars samples (2018)

Leguminous crop cultivar	Content							
	Zinc, mg/kg	Calcium, %	Ferrum, mg/kg	Protein, %	Fat, %	Starch, %	Fiber, %	leach%
"Food Beans. Specifications" standard (Government of the Russian Federation, 1976), haricot cultivar	32.1	0.150	59.0	21.0	2.0	43.8	12.4	3.6
"Omichka", haricot cultivar	36.60 ± 11.40	0.03 ± 0.04	80.00 ± 23.00	21.44 ± 0.83	1.4	42.5	14.2	3.5
"Lukeriya", haricot cultivar	20.90 ± 4.44	0.03 ± 0.04	57.00 ± 19.00	19.40 ± 0.75	1.4	44.5	16.3	3.6
"Nerussa Standard", haricot cultivar	30.90 ± 9.90	0.03 ± 0.04	64.00 ± 19.00	20.87 ± 0.75	1.7	42.0	14.0	4.2
Pea polished, whole grains, Class I, "Polished Pea. Specifications" standard (Government of the Russian Federation, 1968)	31.80	0.115	68.0	20.5	2.0	44.9	11.2	2.8
"Chishminsky 95", pea cultivar	21.16 ± 4.44	0.02	0.0	23.00 ± 0.75	2.1	46.5	6.8	3.6
"Chishminsky 229", pea cultivar	20.90 ± 4.44	0.02	0.0	18.63 ± 0.75	1.6	46.0	7.0	3.6
"To the Memory of Hangildin", pea cultivar	21.00 ± 4.44	0.170	0.0	22.28 ± 0.75	1.9	44.3	6.1	3.0

Source: Skurikhin and Tutelyan (2002)

2011). According to chemical composition, new selection haricot and pea cultivars refer to the group of socially important foodstuffs which provide people with protein, main macro- and microelements, including zinc, ferrum, and calcium (Table 5)

Zinc content is 20.9-36.6 mg/kg in "Omichka" and "Lukeriya" haricot cultivars. This is higher than in "Nerussa Standard" and "Food Beans. Specifications" (Government of the Russian Federation, 1976) haricot cultivars. Zinc content in pea samples of Bashkir Scientific and Research Institute of Agriculture varies from 20.9 mg/kg to 21.16 mg/kg. The average content of ferrum microelement in haricot cultivars samples of selection of Omsk State Agrarian University varies from 57.0 mg/kg to 80.0 mg/kg depending on a cultivar. It exceeds the parameters of "Nerussa Standard" cultivar. In the studied pea cultivars, no ferrum microelement is revealed. As for protein content, it varies from 18.63% to 23.00% in the cultivars of Bashkir Scientific and Research Institute of Agriculture. The protein content of "Lukeriya" and "Omichka" haricot cultivars is 19.40-21.44%. It exceeds the standard indicator of "Nerussa" cultivar (20.87%). In addition, the studied samples of haricot and pea cultivars of domestic selection are characterized by low fat content, high ash, and food fiber contents.

Inhibitors Activity of New Haricot and Pea Cultivars

Natural biologically active anti-alimentary agents, such as phytates, lectins, condensed tannins, inhibitors of trypsin, and α-amylase decreased nutrition value of haricot and pea protein. It makes difficult to use them and stipulates the search for technological solutions of their processing.

Proteinases, the main regulators of proteins activity, are localized in grains and can perform the function of reserved proteins. There are inherent and induced inhibitors proteinases in haricot and pea grains which differ in chemical composition, level of activity, and substrate peculiarity (one-centered and two-centered). Each of these proteinases has its function. They can be free and bound with enzyme in a cell (Petibskaya, 1999). The effect of proteinases inhibitors is ambiguous. On the one hand, they are able to make stable, reversible, substrate matrixes and to lead to suppression of proteolytic enzymes activity. As a result, there is incomplete digestion of food proteins and decrease in their absorption by the body (Valuyeva & Mosolov, 2011). On the other hand, the studies carried out by Abu-Afife, Bushuyeva, and Kolomiychuk (2005) show positive influence of trypsin inhibitors on a human body in post-operative period.

There seems to be enough methods, ways and technological solutions of activity reduction of anti-alimentary agents (Valueva & Mosolov, 1995; Gorlov, Slozhenkina, Danilov, Semenova, & Miroshnik, 2018), including germination. Germination process allows reducing anti-alimentary agents and increasing phytochemical potential of grains (Shaskolskiy & Shaskolskaya, 2007). Samofalova (2010) studied germinated leguminous grains and found out the modification of leguminous chemical composition. Currently, heat treatment and germination of leguminous crops are widely used; however, reduction of inhibitors activity occurs as sporadic episodes.

Despite the revealed advantage of particular cultivars on a number of qualitative characteristics, it is necessary to consider the activity of inhibitors proteinases (which are important for consumers) to discover the best cultivars of haricot and pea for food production. This part of the chapter presents the results of the study of activity of inhibitors of proteinases within new haricot and pea cultivars before and after germination. The research allows discovering cultivars differences in the proteolytic enzymes activity, which hydrolyzes BAPNA substrate, as well as the activity of trypsin inhibitors and urease enzyme in haricot and pea cultivars. The authors employed the method of Erlanger with modifications (Solomintsev & Mogilny, 2009) to determine proteinases and hydrolyzing Nα-Benzoyl-DL-arginine 4-nitroanilide hydrochloride (BAPNA, Sigma, USA). The activity of trypsin inhibitors was identified according to the method of Gofman and Vicebly with modifications. Specification was carried out corresponding to the determination of enzymatic activity. The buffer solution contained 1 mg/ml of trypsin. When determining the activity of urease in pH units on pH-meter, the measurement pH phosphate buffer solution was employed (pH = 6.86). This indicator changes in the result of urease influence on the urea in

Table 6. Proteolytic activity of selection haricot and pea cultivars before and after germination

Leguminous crop cultivar	Proteolytic activity of samples without heat treatment, E		Proteolytic activity of samples after heat treatment, E		Proteolytic activity of self-proteinases, E	
	b.g.	a.g.	b.g.	a.g.	b.g.	a.g.
"Omichka" haricot cultivar	166.10	83.05	120.30	60.15	45.80	91.60
"Lukeriya" haricot cultivar	184.50	102.50	117.30	48.87	65.29	143.30
"Nerussa Standard" haricot cultivar	165.00	71.73	119.80	39.93	35.00	105.00
"Chishminsky 95" pea cultivar	242.20	80.73	140.30	46.67	98.48	196.96
"Chishminsky 229" pea cultivar	233.90	64.97	146.60	41.25	110.48	187.86
"To the Memory of Hangildin" pea cultivar	200.72	132.43	151.10	75.50	67.72	169.29

Note: b.g. – before germination; a.g. – after germination
Source: Authors' development

Table 7. Urease enzyme activity within haricot and pea cultivars before germination

	Cultivar					
	Haricot			Pea		
	Omichka	Lukeriya	Nerussa	Chishminsky 95	Chishminsky 229	To the Memory of Hangildin
Urease enzyme activity, pH-units	0.011	0.020	0.014	0.000	0.000	0.000

Source: Authors' development

the buffer. The germination of haricot cultivars samples was conducted in the range of automatic control of humidity from 40% to 90%, temperature of 24°C, and the period of 17-20 hours. The pea cultivars samples were germinated in the range of automatic control of humidity from 40% to 90% and the temperature of 21°C. The period of germination was 13-15 hours before seedling emergence of about 50-70 mm (Algazin, Vorobiev, Zabudsky, & Zabudskaya, 2016). The results of proteolytic activity and activity of urease enzyme testify the differences between selection haricot and pea cultivars samples (Table 6).

The finding about the activity of trypsin inhibitors due to the data of proteolytic activity of heated samples is that the higher indicator of absorbance, the lower inhibitors' activity is (Table 7). The samples of "To the Memory of Hangildin" pea cultivar and "Omichka" and "Lukeriya" haricot cultivars showed low inhibitors' activity before and after germination. The proteolytic activity within trypsin inhibitors of germinated haricot and pea grains decreased by 2.0-2.5 times and 3.0-4.0 times, respectively. In the process of germination of "Omichka", "Lukeriya", and "Nerussa" haricot cultivars grains, the proteolytic activity of self-proteinases increased by 2.00, 2.19, and 3.00 times respectively. Samples of "To the Memory of Hangildin", "Chishminsky 95", "Chishminsky 229" pea cultivars showed the proteolytic activity in the process of germination. It increased by 1.4-2.5 times compared to the rest period of cultivars grains.

The activity of urease enzyme before germination is about zero in the studied selection pea cultivars of Bashkir Scientific and Research Institute of Agriculture. The results correspond to the study of other pea cultivars made by Vozhyan, Taran, Yakobutsa, and Avadrniy (2013), whose data of urease enzyme activity allows concluding about traces quantity of this inhibitor within all pea cultivars irrespectively. The activity of urease enzyme before germination varies from 0.011 pH to 1.020 pH units in "Omichka", "Lukeriya", and "Nerussa" haricot cultivars. The study of urease activity in haricot after germination shows traces quantity of this inhibitor ferments. The methods of germination employed in the study allow increasing proteolytic activity of self-proteinases and decreasing inhibitors proteinases to 50-65%. That will have positive effect on protein absorption. Considering the results of phytochemical potential, the assessment of consumers' qualities of new selection cultivars, the activity of inhibitors, and the methods of germination, "Lukeriya" haricot cultivar and "To the Memory of Hangildin" pea cultivar were chosen for research of biological value of grains.

Biological Value of Protein in "Lukeriya" Haricot Cultivar and "To the Memory of Hangildin" Pea Cultivar

The most important factors for the choice of raw material for foodstuffs are the qualitative and quantitative compositions of protein. To assess the quality of a protein, different indicators are used including

Agricultural Trade and Undernourishment, Nutrition, and Dietary Diversity

Table 8. Amino-acid composition of the FAO protein formula

Amino acids	Content of protein, mg/g
Leucine	61.0
Isoleucine	30.0
Valine	40.0
Lysine	48.0
Threonine	25.0
Tryptophan	6.6
Methionine + cysteine	23.0
Phenylalanine + tyrosine	41.0
Histidine	16.0

Source: FAO (1991)

amino-acid score and protein digestibility-corrected amino acid score (PDCAAS). PDCAAS was recommended for protein quality assessment by FAO/WHO (Molchanova, 2013). Since 1989, it has been used especially in assessment of plant protein. According to Schaafsma (2012), the method of rectified calculation of biological value of foodstuffs and dietary intakes by protein digestibility-corrected amino acid score (PDCAAS) allows having the results close to the data of clinical researches conducting of which is sometimes irrelevant from the economic and ethical points of view. Tavano, Neves, and Da Silva (2016) studied in vitro methods to determine protein digestibility for chickpea proteins and concluded that the use of in vitro methods for calculating PDCAAS might be promising and deserved more discussions. The influence of processing on lentil protein quality was discovered by Nosworthy et al. (2018). A significant correlation was found between in vivo and in vitro methods of red and green lentils protein digestibility.

The research of amino-acid composition conducted by the authors made it possible determining that the proteins of "Lukeriya" haricot cultivar and "To the Memory of Hangildin" pea cultivar contain all irreplaceable amino acids after germination. These cultivars are characterized by high biological value (Veber, Leonova, Kazydub, Simakova, & Nadtochii, 2019).

The amino-acid score and protein digestibility-corrected amino acid score (PDCAAS) of "Lukeriya" haricot cultivar and "To the Memory of Hangildin" pea cultivar were calculated before and after germination (Molchanova & Suslyanok, 2013). For the calculation of PDCAAS, it is necessary to multiply the least datum for limiting acid by the accepted or established indicator of protein absorption (Molchanova & Suslyanok, 2013). The authors used the FAO protein formula to make a calculation of amino-acid score (Table 8).

The irreplaceable amino acid limiting for this protein was discovered along with the definition of amino-acid score. The limiting amino acid showed the minimum score (Baranovskiy, 2008). In the process of germination, there is a change in quality and quantity concentration of separate amino acids corresponding to the whole protein (Table 9). That is testified by the data of amino-acid score. The studied cultivars contain sufficient amount of replaceable and irreplaceable amino acids including deficient ones: lysine, threonine, tryptophan, threonine, and methionine + cysteine. Therefore, the protein of selection haricot and pea cultivars can be considered complete in optimum quality and amino acids correlation.

Table 9. Amino-acid score of leguminous grains before and after germination

Amino acids	Amino-acid score corresponding to the FAO protein, %			
	"To the Memory of Hangildin" grain before germination	"To the Memory of Hangildin" grain after germination	"Lukeriya" grain before germination	"Lukeriya" grain after germination
Leucine + isoleucine	77.8	80.0	86.0	102.0
Valine	77.0	96.3	82.0	98.0
Histidine	95.0	139.0	121.0	144.0
Lysine	96.4	130.0	124.0	101.0
Threonine	99.0	193.0	96.0	175.0
Tryptophan	93.0	97.0	110.0	131.0
Methionine + cysteine	60.0	75.0	84.0	105.0
Phenylalanine + tyrosine	105.0	125.0	129.0	148.0
PDCAAS	53.0	66.0	67.0	77.0

Source: Authors' development

"Lukeriya" haricot cultivar and "To the Memory of Hangildin" pea cultivar have all irreplaceable amino acids in the protein content but their number differs

The minimum amino-acid index calculated with amino-acid score method demonstrated the results of 0.75 and 0.98 after germination and 0.60 and 0.82 before germination of "To the Memory of Hangildin" haricot cultivar and "Lukeriya" pea cultivar, respectively (Table 10). PDCAAS was calculated due to the index of protein absorption by 88% for pea cultivars and 78% for haricot. So it made 66% for the pea and 67% for the haricot before germination and 53% and 64% after germination, respectively.

The foodstuffs containing high-quality proteins and having PDCAAS indexes of about 1.0 are considered to be complete in terms of providing of a dietary reference intake of proteins. The presented

Table 10. Indicators of amino-acid index according to protein absorption

Amino acids	Amino-acid index			
	"To the Memory of Hangildin" grain before germination	"To the Memory of Hangildin" grain after germination	"Lukeriya" grain before germination	"Lukeriya" grain after germination
Leucine + isoleucine	0.78	0.80	0.86	1.02
Valine	0.77	0.93	0.82	0.98
Histidine	0.95	1.39	1.21	1.44
Lysine	0.96	1.30	1.24	1.01
Threonine	0.99	1.93	0.96	1.75
Tryptophan	0.93	0.97	1.10	1.31
Methionine + cysteine	0.60	0.75	0.84	1.05
Phenylalanine + tyrosine	1.05	1.25	1.29	1.48
PDCAAS	53.00	66.00	67.00	77.00

Source: Authors' development

research data show that the use of germination allows decreasing of proteinases inhibitors and increasing nutrient value of leguminous. The data also prove the idea of selection cultivars proteins being the closest to the complete protein in composition. The results of the previous studies have become the basis of new ways of foodstuffs production, for instance, hummus and bread.

Use of Germinated Seeds of "Lukeriya" Haricot Cultivar and "To the Memory of Hangildin" Pea Cultivar for Compensation of Protein Deficiency in Food Products

The use of preliminary germinated grains of domestic selection haricot and pea cultivars is suggested to solve the problem of deficient consumption of the balanced protein. The studies, conducted by Veber et al. (2019) allowed developing technological processing ways of germinated grains of new selection haricot and pea cultivars into foodstuffs with high phytochemical potential. One of them is to determine the influence of concentration of flour from the seeds of domestic selection haricot upon rheological qualities of dough from a composite mixture (wheat and haricot) by Mixolab apparatus (protocol "Chopin+") (Veber, Maradudin, Strizhevskaya, Romanova, & Simakova, 2019). Another way is to work out a series of foodstuffs for healthy nutrition. For instance, the use of powders component from the germinated seeds of "Omichka" and "Lukeriya" haricot cultivars in the technology of a sour milk product increased the nutrition value and sensory indicators of samples compared to the quality-controlled yoghurt from cow milk (Veber, Kazydub, Leonova, & Zhiarno, 2017; Buyakova, Da Costa, Veber, De Castro, & Kazydub, 2015; Buyakova et al., 2015; Zhiarno, Zabodalova, Veber, Petushkova, & Kazydub, 2017). Standard technological processing of "Fa-sol" ice-cream included addition plant fat (coconut oil), haricot dispersion from the germinated seeds of "Omichka", coconut in coconut water, preliminary activated biomass of bifidus bacteria lyophilized according to the standard, and others. These ingredients provide experimental samples with high sensory qualities: more homogeneous structure without curds of fat and ice compared to the ice-cream on the basis of rice, soybean and coconut milk; delectable sweet and sour flavor; light taste and bland coconut aroma. Physical and chemical analysis showed the moisture content of 71.0% and protein content above 3.5%. The content of fat is not less than 12.0% in a ready-made foodstuff. The quantity of viable cells of probiotic cultures in a product makes not less than $1*10^6$ CFU/g for the end of an expiration date. This indicator testifies high pro-biotic properties of the developed product (Vorobiev, 2017).

The problem of protein deficiency can be also solved by the development of such foodstuffs as hummus and bread from the germinated seeds of leguminous. The authors worked out the technology of hummus production which includes the following stages: seeds separation, cleaning, germination, seeds grinding, heat treatment, mixture of ingredients, beating for 3-5 minutes and packing. Preliminary heated samples of haricot and pea seeds were put in a germination apparatus which allowed maintaining optimum moisture and temperature conditions for equal germination. The germination of haricot cultivars samples was conducted in the range of automatic control of moisture from 40% to 90% and the temperature of 24°C in the period of 17-20 hours. The pea cultivars samples were germinated in the range of automatic control of moisture from 40% to 90% and the temperature of 21°C. The period of germination took from 13 to 15 hours before seedling emergence of about 50-70 mm. Heat treatment was carried out at a temperature of 100°C during 40-60 min. Then the seeds were cooled down to 30-45°C and ground to homogenous substance. In addition, the compound included oil of chia seeds, garlic, paprika, and lemon juice in correlation per 1,000 kg (Table 11).

Table 11. Formula of hummus production from the germinated seeds of "Lukeriya" and "To the Memory of Hangildin"

Ingredients	Weight, kg
Germinated seeds of "Lukeriya" and "To the Memory of Hangildin"	768.3
Ground seeds of chia	45.0
Oil of chia seeds	45.0
Salt	11.5
Plant oil	95.0
Garlic	8.2
Paprika	2.0
Lemon juice	25.0

Source: Authors' development

The use of ground seeds of "Lukeriya" haricot cultivar and "To the Memory of Hangildin" pea cultivar which were germinated and heated increases biological and nutrient value of a foodstuff. The seeds contain a complete plant protein, all irreplaceable amino acids, increased amount of ferrum, iodine, calcium, zinc, and other important nutrients.

Introduction of ground seeds of chia meets daily nutrient requirements because they are a plant source of rich Omega-3 and Omega-6 fat acids, food fibers (20-40%). The seeds also contain significant amount of protein (up to 26%), vitamins and mineral matters, oxidation retarders (meletin, kaempferol, and alpha of tocopherol). Chia seeds implementation in foodstuffs has positive influence upon ill-health prevention of catarrhal and virus diseases, cancer pathologies, and atopic diseases due to their antihistamine effect. Scientists from all over the world proved chia seeds advantageous influence on humans' digestive organs and their rejuvenated impact (Figure 1).

Figure 1. Hummus from the germinated seeds of "To the Memory of Hingildin" pea and "Lukeriya" haricot cultivars
Source: Authors' development

Table 12. Sensory indicators of hummus from the germinated seeds of "To the Memory of Hingildin" pea cultivar and "Lukeriya" haricot cultivar

Indicators	Specification
Taste and aroma	Clean, typical for ingredients, without extraneous taste
Color	Creamy, with light red tone
Texture	Homogenous, sticky to the extent, spreading

Source: Authors' development

Table 13. Physical and chemical indicators of hummus from the germinated seeds of "To the Memory of Hingildin" pea cultivar and "Lukeriya" haricot cultivar

Indicators	Specification
Content of minerals (Ca), mg per 100 g	72.9
Content of minerals (Fe), mg per 100 g	8.2
Protein, not less than %	18.0
Fat, to a maximum of %	25.0
Carbohydrates, %	20.0
Fibers, %	3.1

Source: Authors' development

Sensory indicators of hummus from the germinated seeds of "To the Memory of Hingildin" pea cultivar and "Lukeriya" haricot cultivar are presented in Table 12.

Physical and chemical indicators of hummus from the germinated seeds of "To the Memory of Hingildin" pea and "Lukeriya" haricot cultivars are presented in Table 13.

The analysis of physical and chemical indicators shows that a haricot foodstuff has high biological and nutrition value. The caloric value of 100 g of the hummus from the germinated seeds of "Lukeriya" haricot and "To the Memory of Hangildin" pea cultivar makes not less than 369 kcal / 1544.9 kJ. Daily consumption of hummus would provide dietary reference intake of protein up to 34% (Veber et al., 2019). Bread is a foodstuff of the most social significance. Bread is vital for the world essential population's nutrition.

The authors suggest the following technology of making high protein bread to increase biological and nutrition value. The technology consists of two stages.

During Stage 1, the seeds of "Lukeriya" haricot and "To the Memory of Hangildin" pea cultivars are germinated according to technology presented in this chapter. Then, the seeds are cleaned by drinking-water during about 20-30 minutes, dried up to 30% of moisture, cooled to 18-20°C, ground to particles of 0.3-0.5 mm, and used for cooking of a liquid acid mixture. This mixture contains germinated seeds, wheat flour of Class I, compressed yeast, and drinking water. The mixture is aged for about 30-40 minutes at the temperature of 31-35°C for obtaining moisture at the rate of 70-75%.

Stage 2 includes kneading dough. Chia seeds oil and salt are added to the liquid acid mixture at the rate of 2.0% and 1.5% to the amount of flour in the dough, respectively. Drinking-water is then adjusted to the moisture of dough not more than 48%. Kneaded dough is fermented at the temperature of $35\pm2°C$ during 20-35 minutes. Then, the dough is separated into parts according to the weight of a loaf and

Table 14. Sensory indicators of enriched bread compared to the model

Indicators	Traditional bread	Enriched bread
Color of crumb	Light	Light yellow, equal color
Crumb characteristics	Elastic, soft	Elastic, soft
Crumb vesiculation	Developed, cells of medium size	Developed, cells of medium size
Aroma	Bread aroma	Thick bread aroma
Flavor and taste	Usual	Pleasant; typical for bread from germinated seeds; without extraneous taste; more apparent taste compared to the model; easily chewed crumb
Crust rise, mm	100.0	105.0
Crust thickness, mm	1.0	1.5
Crust specification	Dome-shaped top crust, perfect smooth surface, without bubbles, cracks, breaks	
Shape	Regular	
Crust color	Light yellow	Brown

Source: Authors' development

Table 15. Sensory indicators of enriched bread compared to the model

Indicator	Traditional bread	Enriched bread
Crumb acidity, °C	2.60	3.00
Dough acidity, °C, ≤	2.70	3.50
Weight fraction of moisture in dough, %, ≤	46.00	48.00
Crumb moisture, %	43.50	45.00
Crumb vesiculation, %	64.00	72.00
Protein, %	6.70	8.30
Fat, %	0.69	0.99
Sugar, %	1.63	2.79
Food fibers, %	2.19	2.97
Ash, %	1.28	2.80

Source: Authors' development

rested for 40-50 minutes at the temperature of 40°C and relative moisture of 80±2%. The shaped loaves are baked at the temperature of 210-230°C. The term of bread preservation is 36 hours after baking, the temperature is about 20-25°C.

Sensory indicators of bread enriched with germinated seeds of "Lukeriya" haricot and "To the Memory of Hangildin" pea cultivars, in comparison with the bread, baked according to traditional technology (a model) are presented in Table 14.

The analysis shows high sensory indicators of bread cooked according to the developed technology (with the use of germinated seeds of "Lukeriya" haricot and "To the Memory of Hangildin" pea cultivars) (Table 15).

Figure 2. Bread samples worked out according to the presented technology
Source: Authors' development

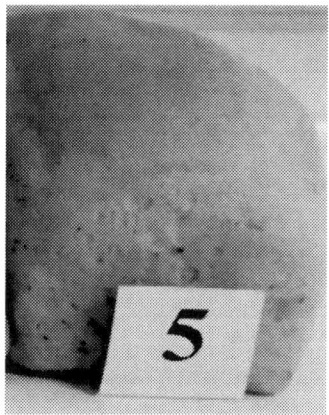
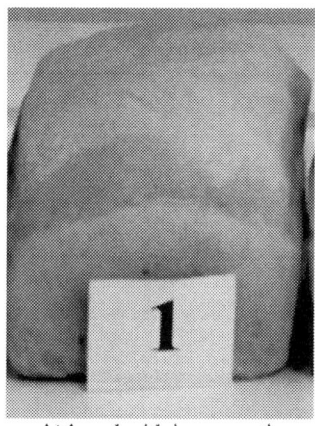
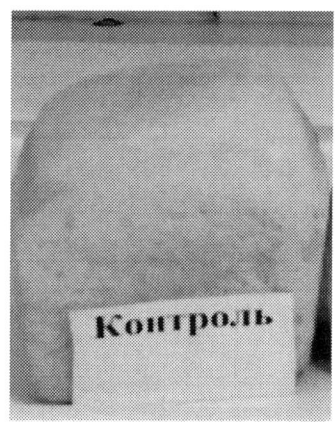

a) bread with incorporation of the germinated seeds of 'Lukeriya' haricot cultivar

b) bread with incorporation of the germinated seeds of 'To the memory of Hangildin' pea cultivar

c) model

The analysis of physical and chemical indicators allows concluding about high nutrition and biological value, increasing consumers' qualities of enriched bread. The distinctive criterion of crumb vesiculation and moisture makes it possible to expect high absorption of enriched bread by the body and its physiological value (Figure 2).

SOLUTIONS AND RECOMMENDATIONS

The studied samples of selection haricot and pea cultivars are characterized by high protein content, food fibers, ash, and starch. Haricot cultivars in the selection of Omsk State Agrarian University are better in microelements composition compared to the pea cultivars of Bashkir Scientific and Research Institute of Agriculture. However, the nutrition value of pea remains high because of zero urease activity and comparably low activity of trypsin inhibitor. The germination methods for the studied haricot and pea cultivars allow decreasing of trypsin inhibitors activity from 50% to 60% and can be used for further development of technologies of complex raw material foodstuffs production. According to the results of the studies, it is recommended to use the germinated seeds of "Lukeriya" haricot and "To the Memory of Hangildin" pea cultivars for the technology of high protein foodstuffs production (hummus and bread with high protein content). Consumers' oriented foodstuffs developed on the basis of domestic selection cultivars allow solving the problem of protein deficiency and deficiency of other important nutrients.

FUTURE RESEARCH DIRECTIONS

Further research will be oriented towards the expansion of a line of the multicomponent, balanced, specialized, and functional food products. They will contain a set of properties and structure for different population groups according to their genetic peculiarities and nature and climate zones. These products will help to increase the level of food security. Another potential direction is the study of physiologically functional ingredients from regional cultivars of leguminous crops of domestic selection. The results of the research can be used for the development of actions allowing strengthening the ties between the universities, individual researchers, and experts worldwide. The latter include food producers and sellers, the authorities in the field of food safety and security, and consumers.

CONCLUSION

The lack of gene modifications, high nutrition and biological value, low prime cost, and high level of profitability, the outputs and possible increase in added value make seeds of domestic selection pea and haricot cultivars excellent export item which has a potential for expanding a raw material resource base for production of foodstuffs with increased protein content.

REFERENCES

Abu-Afife, S., Bushuyeva, N., & Kolomiychuk, S. (2005). The Use of Trypsine Inhibitors at Sclera Reinforcement within Children and Teenagers with Progressive Myopia. *Journal of Ophthalmology, 2*, 11–14.

Akhmadullina, I. (2018). *Enhancement of the Techniques of Primary Seed Breeding and Pea Cultivation for Seed Purpose in the CIS-Urals Region of the Republic of Bashkortostan*. Kazan: Kazan State Agrarian University.

Algazin, D., Vorobiev, D., Zabudsky, A., & Zabudskaya, E. (2016). *Patent #1600896 "Device for Growing Plants."* Retrieved from https://yandex.ru/patents/doc/RU160896U1_20160410

Antipova, L., Grebenshchikov, A., Mishchenko, A., Osipova, N., & Tychinin, N. (2017). Histochemical and Physiological-Biochemical Properties of Germinated Seeds Lentils as a Source of Nutrients. *Technologies of Food and Processing Industry of AIC – Healthy Food, 4*, 69-79.

Antipova, L., & Presnyakova, O. (2008). Opportunities of Chickpea and Lupin Utilization in Condensed Milk Analogues Technology. *Storage and Processing of Farm Products, 4*, 50–52.

Baranovskiy, A. (2008). *Dietology: Guidance*. Saint Petersburg: Piter.

Bazga, B. (2015). Food Security Component of Sustainable Development – Prospects and Challenges in the Next Decade. *Procedia Economics and Finance, 32*, 1075–1082. doi:10.1016/S2212-5671(15)01570-1

Bishnoi, S., & Khetarpaul, N. (1993). Effect of Domestic Processing and Cooking Methods on In-Vitro Starch Digestibility of Different Pea Cultivars (*Pisum Sativum*). *Food Chemistry, 47*(2), 177–182. doi:10.1016/0308-8146(93)90240-G

Bodoshov, A. (2015). Amino-Acid Composition of Haricot Grains Cultivated in Kyrgyzstan. *Young Scientist, 104*(24), 94–96.

Buyakova, A., Da Costa, R. M., Veber, A., De Castro, J. C. M., & Kazydub, N. (2015). *Patent #2616864 "Process of Production of Fermented Milk Drink."* Retrieved from https://findpatent.ru/patent/261/2616864.html

Buyakova, A., Veber, A., De Castro, J. C. M., Da Costa, R. M., Kazydub, N., & Staurskaya, N. (2015). *Patent #2599569 "Functional Alimentary Product from Sprouted Kernels."* Retrieved from https://xn--90ax2c.xn--p1ai/catalog/000224_000128_0002599569_20161010_C1_RU/viewer/

Dotsenko, S., Bibik, I., Lyubimova, O., & Guzhel, Y. (2016). Kinetics of Biochemical Processes of Germination of Soybean Seeds. *The Bulletin of KrasGAU, 1*, 66–74.

Eurasian Economic Commission. (2011). *Technical Requirements of the Customs Union #015/2011 "On Safety of Grain."* Retrieved from http://docs.cntd.ru/document/902320395

Food and Agriculture Organization of the United Nations. (1991). *Protein Quality Evaluation*. Rome: Food and Agriculture Organization of the United Nations.

Food and Agriculture Organization of the United Nations. (2013). *Dietary Protein Quality Evaluation in Human Nutrition*. Rome: Food and Agriculture Organization of the United Nations.

Giro, T., & Chirkova, O. (2007). Meat Foodstuffs with Plant Ingredients for Functional Nutrition. *Meat Industry, 1*, 43–46.

Gorlov, I., Slozhenkina, M., Danilov, Y., Semenova, I., & Miroshnik, A. (2018). Use of a New Food Ingredient in the Production of Meat Products of Functional Purposes. *Bulletin of Lower Volga Agrouniversity Complex, 52*(4), 219–229.

Government of the Russian Federation. (1968). *Interstate Standard "Polished Pea. Specifications."* Retrieved from http://docs.cntd.ru/document/1200022305

Government of the Russian Federation. (1976). *Interstate Standard "Food Beans. Specifications."* Retrieved from http://docs.cntd.ru/document/1200023726

Government of the Russian Federation. (1991). *Interstate Standard "Pea. Requirements for State Purchases and Deliveries."* Retrieved from http://docs.cntd.ru/document/gost-28674-90

Government of the Russian Federation. (2008). *Norms of Physiological Requirements in Energy and Feedstuffs for Various Groups of Population in the Russian Federation*. Retrieved from http://docs.cntd.ru/document/1200076084

Halimi, R. A., Barkla, B. J., Mayes, S., & King, G. J. (2019). The Potential of the Underutilized Pulse Bambara Groundnut (*Vigna Subterranean (L.) Verdc.*) for Nutritional Food Security. *Journal of Food Composition and Analysis, 77*, 47–59. doi:10.1016/j.jfca.2018.12.008

International Independent Institute of Agrarian Policy. (2017). *Global Market of Legume Crops*. Retrieved from http://xn--80aplem.xn--p1ai/analytics/Mirovoj-rynok-bobovyh-kultur/

Karmas, E., & Harris, R. S. (1988). *Nutritional Evaluation of Food Processing*. Amsterdam: Springer Netherlands. doi:10.1007/978-94-011-7030-7

Kazantseva, I. (2016). *Scientific Foundation and Elaboration of Technological Solutions of Complex Processing of Chickpea Seeds for the Development of Healthy Food Products for People in Russia*. Saratov: Yuri Gagarin State Technical University of Saratov.

Kazydub, N., & Marakaeva, T. (2015). *Comparative Assessment of Economically Valuable Qualities of Haricot Samples (Phaseolus Vulgaris L.) and Development of New Selection Material on Their Basis in the Territory of the Southern Forest-Steppe of Western Siberia.* Omsk: Kant.

Kazymov, S., & Prudnikova, T. (2012). Germination Influence on Amino Acids Composition of Mash Beans. *News of Institutes of Higher Education. Food Technology, 5-6*, 25–26.

Kolesnikova, N. (2006). *The Development of Technology and Assessment of Consumers' Qualities of Foodstuffs on the Basis of Haricot for Pupils.* Krasnodar: Kuban State Technological University.

Matejčeková, Z., Liptáková, D., & Valík, L. (2017). Functional Probiotic Products Based on Fermented Buckwheat with Lactobacillus Rhamnosus. *Lebensmittel-Wissenschaft + Technologie, 81*, 35–41.

Migina, E. (2016). Prospective Usage of Soybean Seeds and Products of Their Processing in Development of New Feed Additives. *Young Scientist, 125*(21), 284–288.

Molchanova, E. (2013). *Physiology of Nutrition.* Saint Petersburg: Piter.

Molchanova, E., & Suslyanok, G. (2013). Quality Assessment and Food Proteins Value. *Storage and Processing of Farm Products, 1*, 16–22.

Muimba-Kankolongo, A. (2018). Leguminous Crops. In A. Muimba-Kankolongo (Ed.), *Food Crop Production by Smallholder Farmers in Southern Africa* (pp. 173–203). Amsterdam: Elsevier. doi:10.1016/B978-0-12-814383-4.00010-4

Nekrasov, R. (2019). *The Results of Plant Breeding Branch and Engineering and Technical Services in 2018, Realization of Tasks of the State Programme for 2013-2020.* Moscow: Ministry of Agriculture of the Russian Federation.

Nosworthy, M. G., Medina, G., Franczyk, A. J., Neufeld, J., Appah, P., Utioh, A., ... House, J. D. (2018). Effect of Processing on the In Vitro and In Vivo Protein Quality of Red and Green Lentils (*Lens Culinaris*). *Food Chemistry, 240*, 588–593. doi:10.1016/j.foodchem.2017.07.129 PMID:28946315

Petibskaya, V. (1999). Inhibitors of Proteolytic Enzymes. *News of Institutes of Higher Education. Food Technology, 5-6*, 6–10.

Petruláková, M., & Valík, L. (2015). Legumes as Potential Plants for Probiotic Strain *Lactobacillus Rhamnosus GG*. *Acta Universitatis Agriculturae et Silviculturae Mendelianae Brunensis, 63*(5), 1505–1511. doi:10.11118/actaun201563051505

Ponomarev, S. (2011). *The Development of Resource Saving Technology of Byproducts of Pea Processing.* Orenburg: Orenburg State University.

Popova, I., & Klimenko, E. (2018). Prospects for the BRICS Countries Cooperation in the Field of Food Security in the Grain Production Sector. *Technico-Tehnologicheskie Problemy Servisa, 43*(1), 43–48.

Prodanov, M., Sierra, I., & Vidal-Valverde, C. (2004). Influence of Soaking and Cooking on the Thiamin, Riboflavin and Niacin Contents of Legumes. *Food Chemistry, 84*(2), 271–277. doi:10.1016/S0308-8146(03)00211-5

Razvedskaya, L. (2001). *The Development of the Technology of Production of Plant Preserves with Addition of Milk and Products of Leguminous Crops Processing*. Krasnodar: Kuban State Agrarian University.

Relf, D. (2015). *Sprouting Seeds for Food*. Retrieved from http://pubs.ext.vt.edu/content/dam/pubs_ext_vt_edu/426/426-419/426-419_pdf.pdf

Ryazanova, K. (2014). Convenience Meat Product with Fillings. In *Proceedings of the Conference "Youth. Science. Future – 2014."* Magnitogorsk: Magnitogorsk State Technical University.

Samofalova, L. (2010). *The Scientific Foundation of Application of Germinated Seeds of Biobated Plants in the Production of Plant Basis and Substitutes of Dairy Products for Special Purpose*. Orel State Technical University.

Schaafsma, G. (2012). Advantages and Limitations of the Protein Digestibility-Corrected Amino Acid Score (PDCAAS) as a Method for Evaluating Protein Quality in Human Diets. *British Journal of Nutrition*, *108*(S2), 333–336. doi:10.1017/S0007114512002541 PMID:23107546

Shaskolskiy, V., & Shaskolskaya, N. (2007). Antioxidant Activity of Germinated Seeds. *Bread Products*, *8*, 58–59.

Shcherbakova (Ponomareva), A. (2017). Organic Agriculture in Russia. *Siberian Journal of Life Sciences and Agriculture*, *9*(4), 151–173.

Shelepina, N. (2010). Main Trends in Pea Grains Cultivation. Security and Quality of Goods. *Proceedings of IV International Scientific and Practical Conference*. Saratov: KUBiK.

Shelepina, N., & Parshutina, I. (2013). Comparative Characteristics of Amino-Acid Composition of Wheat Flour and Embryonic Foodstuffs from Pea. *Proceedings of the III International Conference "New in Technologies and Techniques of Food for Special Purpose with Medical and Biological Approach."* Voronezh: Voronezh State University of Engineering Technologies.

Skripko, O., Isaycheva, N., & Pokotilo, O. (2015). Preparation of Protein-Vitamin-Mineral Products with Use of Soy for a Healthy Nutrition. *Food Industries*, *5*, 34–37.

Skurikhin, I., & Tutelyan, V. (2002). *Chemical Composition of Foodstuffs in Russia: Guidance*. Moscow: DeLi Print.

Solomintsev, M., & Mogilny, M. (2009). Determination of Proteinase Inhibitors Activity in Food Products. *News of Institutes of Higher Education. Food Technology*, *1*, 13–16.

Tavano, O. L., Neves, V. A., & Da Silva, S. I. J. (2016). In Vitro Versus in Vivo Protein Digestibility Techniques for Calculating PDCAAS (Protein Digestibility-Corrected Amino Acid Score) Applied to Chickpea Fractions. *Food Research International*, *89*(1), 756–763. doi:10.1016/j.foodres.2016.10.005 PMID:28460976

Tsareva, N. (2007). *The Use of Foaming Qualities of Leguminous Crops in the Technology of Whipped Cottage Cheese Deserts. Orel*. Orel State Technical University.

Tutelyan, V. (2005). Nutrition and Health. *Food Industries*, *5*, 6–7.

Tutelyan, V. (n.d.). *Priorities of Government Policy in Healthy Nutrition in Russia on the Federal and Regional Levels.* Retrieved from http://pfcop.opitanii.ru/articles/state_feed_prioritets.shtml

Vakhitova, R. (2015). *The Formation of Pea Harvest Depending of the Elements of Cultivation Technology in the Cis-Ural region of the Republic of Bashkortostan.* Ufa: Bashkir State Agrarian University.

Valuyeva, T., & Mosolov, V. (1995). Proteins-Inhibitors of Proteolytic Enzymes of Plants. *Applied Biochemistry and Microbiology, 31*(6), 579–589.

Valuyeva, T., & Mosolov, V. (2011). Inhibitors of Proteolytic Enzymes while Plants Abiotic Stress. *Applied Biochemistry and Microbiology, 47*(5), 501–507.

Veber, A., Kazydub, N., Leonova, S., & Zhiarno, M. (2017). The Process of Biologically Active Component Derivation from Germinated Haricot Seed for Further Development. *Bread Products, 17*(6), 35–38.

Veber, A., Kazydub, N., Leonova, S., Zhiarno, M., Nadtochii, L., Vorobiev, D., & Fialkov, D. (2019). *Patent #2685911 "Process of Foodstuff from Haricot Production."* Retrieved from http://www1.fips.ru/registers-doc-view/fips_servlet?DB=RUPAT&DocNumber=2685911&TypeFile=html

Veber, A., Leonova, S., Kazydub, N., Simakova, I., & Nadtochii, L. (2019). Special Legume-Based Food as a Solution to Food and Nutrition Insecurity Problem in the Arctic. In V. Erokhin, T. Gao, & X. Zhang (Eds.), *Handbook of Research on International Collaboration, Economic Development, and Sustainability in the Arctic* (pp. 570–592). Hershey, PA: IGI Global. doi:10.4018/978-1-5225-6954-1.ch027

Veber, A., Maradudin, M., Strizhevskaya, V., Romanova, H., & Simakova, I. (2019). The Research of Functional and Technological Qualities of Composite Mixtures from Wheat and Haricot of Domestic Selection. *Food Industries, 3*, 45–49.

Vorobiev, D., Zhiarno, M., Leonova, S., Veber, A., Kazydub, N., & Zabudsky, A. … Ponomareva, E. (2017). *Patent #2676166 "Process of Ice-Cream Production."* Retrieved from https://findpatent.ru/patent/267/2676166.html

Vozhyan, V., Taran, M., Yakobutsa, M., & Avadrniy, L. (2013). The Nutrition Value of Pea, Soybean, Haricot Cultivars and Antinutritional Stuffs in Them. *Scientific and Industrial Journal. Leguminous and Cereal Foodstuffs, 5*(1), 26–29.

Zhiarno, M., Zabodalova, L., Veber, A., Petushkova, Y., & Kazydub, N. (2017). *Patent #266119 "Process of Production of Fermented Product."* Retrieved from https://findpatent.ru/patent/266/2661119.html

Zotikov, V., Naumkina, T., & Sidorenko, V. (2014). Role of Leguminous Crops in Economy of Russia. *Farming, 4*, 6–8.

ADDITIONAL READING

Evteev, S., & Perelet, R. (n.d.). *Report of the International Committee on Environment and Development (ICED).* Retrieved from http://xn--80adbkckdfac8cd1ahpld0f.xn--p1ai/files/monographs/OurCommon-Future-introduction.pdf

Fabbri, A. D. T., & Crosby, G. A. (2016). A Review of the Impact of Preparation and Cooking on the Nutritional Quality of Vegetables and Legumes. *International Journal of Gastronomy and Food Science, 3*, 2–11. doi:10.1016/j.ijgfs.2015.11.001

Gorlov, I., Nelepov, Y., Slozhenkina, M., Korovina, E., & Simon, M. (2014). The Development of New Products with Germinated Chickpea for Special Purpose. *All about Meat, 1*, 28-30.

Government of the Russian Federation. (2010a). *National Standard of the Russian Federation "Public Catering Service. Method of Sensory Evaluation of Catering Products Quality."* Retrieved from http://docs.cntd.ru/document/1200069392

Government of the Russian Federation. (2010b). *Order #1873-r from October 25, 2010, "Foundations of State Policy of the Russian Federation in the Sphere of Healthy Nutrition of Population till 2020."* Retrieved from https://www.gnicpm.ru/UserFiles/osnovi_zdor_pitania_do_2020.pdf

Government of the Russian Federation. (2012). *Order #717 from July 14, 2012, "On State Program of Development of Agriculture and Regulation of the Markets of Agricultural Products, Raw Materials, and Food."* Retrieved from http://base.garant.ru/70210644/

Government of the Russian Federation. (2015). *National Standard of the Russian Federation "Organic Foods. Terms and Definitions."* Retrieved from http://docs.cntd.ru/document/1200113488

Government of the Russian Federation. (2016). *Decree #1378-r from June 30, 2016, "Strategy of Development of Food and Food Processing Industry in the Russian Federation for a Period till 2020."* Retrieved from http://static.government.ru/media/files/65bZISIOP6bA0VSJ67GnnpKIhhoHhxgP.pdf

State Commission of the Russian Federation on Probation and Treatment of Selection Inventions. (n.d.). *Plant Cultivars Included in State Register of Selection Achievements.* Retrieved from https://reestr.gossort.com/reestr

World Health Organization. (1974). *Energy and Protein Requirements.* Geneva: World Health Organization.

World Health Organization. (2015). *Second International Conference on Nutrition. The Report of FAO-WHO Secretariat.* Rome: World Health Organization; Food and Agriculture Organization of the United Nations.

KEY TERMS AND DEFINITIONS

Energy Value: An amount of energy in kilocalories which is contained in foodstuffs. It helps to provide physiological functions of a human body.

Macronutrients: The nutrient materials (proteins, fat, and carbohydrates) which are essential for a human body and measured in grams. They provide a human body with energy, elastic, and other needs.

Micronutrients: The nutrient materials (vitamins, minerals, and microelements) which are contained in food in small amounts – milligrams and micrograms. They are not energy sources but active in food absorption, functions regulations, growth process, adaptation, and development of a body.

Nutrients: The substances that an organism must obtain from its surroundings for growth and sustenance of life. They are used as energy sources, sources or precursor substrates for construction, growth, and renovation of organs and muscles, physiologically active substances which regulate vital activity, and determine value of foodstuffs.

Nutrition Value: A measure of the quantity or availability of nutrients found in materials ingested and utilized by humans as a source of nutrition and energy.

Organic Agriculture: A production system which improves ecosystem, preserve soil fertility, protects humans' health. Considering local climates, it preserves biodiversity in the territory and does not use components which can be harmful for environment.

PDCAAS: A method of evaluating the quality of a protein based on both the amino acid requirements of humans and their ability to digest and recommended by FAO/WHO.

Proteins: High molecular and nitrogen-containing biopolymers which consist of L-amino acids. Proteins perform a vast arrays of functions: energetic, catalyzing metabolic reactions, elastic, harmonic, protection, regulatory, and transport.

Protein Quality: Defined by the presence of a full complex of irreplaceable amino acids in appropriate correlation with each other and replaceable amino acids. In the process of oxidation 1 g of protein makes 4 kcal.

Chapter 14
Imports and Use of Palm Oil as a Way to Increase Safety of Food Fats

Inna Simakova
https://orcid.org/0000-0003-0998-8396
Saratov State Vavilov Agrarian University, Russia

Roman Perkel
https://orcid.org/0000-0003-1435-1330
Peter the Great Saint Petersburg Polytechnic University, Russia

ABSTRACT

The authors compare the biological value and safety of hydrogenated fat containing trans-isomers of oleic acid and palm oil-based fat. The chapter assesses the potential of replacing hydrogenated fats by palm oil in the production of special fat products. Hematological and histological studies are carried out in a form of biological experiment on animals (white rats). The study reveals the explicit negative effect of trans-isomers even with a relatively low concentration of trans-isomers in a diet. Pathological changes are not observed in animals when palm-based fat is introduced into their ration. The findings suggest that palm oil along with its fractions may be considered as an alternative to hydrogenated fats in the production of margarine, cooking, baking, and deep-frying fats. The use of palm oil in the production of special fats of increased hardness (spreads. confectionery, waffles and fillings, and chocolate coating) requires the application of modern methods for modifying triglyceride composition of fats – biocatalytic interesterification and fractionation.

INTRODUCTION

Ensuring a sufficient amount of food fats in the diet is one of the most important factor in establishing food security. Consumption of fats affects food security status as fats are the every-day components of the diets and food products. Special fat products play an important role in the production of catering

DOI: 10.4018/978-1-7998-1042-1.ch014

products and home cooking as they provide taste, aroma, formation of the structure of a product, oxidative stability during storage, and other characteristics of culinary products. They also have a significant impact on the quality of finished products (O'Brien, 2007).

Recently, natural vegetable oil has been increasingly displacing cooking fats based on hydrogenated fats which are characterized by a high content of trans-isomers of fatty acids. Palm oil occupy a significant part of consumed vegetable oils today (AB-Center, n.d.) due to thermal stability and physicochemical properties which make it possible to consider palm oil as an alternative to partially hydrogenated fats.

Until recently, the technology of partial hydrogenation has been intensively used in many countries being the most important industrial method of stabilizing vegetable oils to oxidation and obtaining plastic food fats for margarine products on their basis. The technology of partial hydrogenation is specifically aimed at increasing the content of trans-isomers of oleic acid in fat which makes it possible to regulate the ratio of melting point and structural characteristics of resulting solid hydrogenated fats. However, in recent years, numerous studies have revealed negative effect of oleic acid trans-isomers in a body. According to the World Health Organization (WHO) (n.d.), human body should receive no more than 1% of daily norm of total energy consumption from trans-fats (about 2-3 grams of trans-fats). Based on these recommendations, the content of trans-isomers in fat products has been legally limited (no more than 2%) in the EU since 2010 and in Russia and the USA since 2018.

Therefore, improvement of the quality and safety of edible fats and regulation of the content of trans-isomers are relevant and require close attention of scientists, producers, and government agencies.

BACKGROUND

For almost 100 years and until very recently, selective hydrogenation has been the most important industrial method of stabilizing vegetable oils and obtaining plastic edible fats for margarine products on their basis. The technology of partial hydrogenation is specifically aimed at increasing the content of trans-isomers in the fat in order to provide necessary ratio between melting point and structural characteristics of the resulting solid hydrogenated fat.

The formation of trans-isomers is associated with a complex of reactions that occur during the partial hydrogenation of vegetable oils (Gassenmeier & Schieberle, 1994; Schmidt, 2000). As a result of the addition of hydrogen to double bonds in molecules of unsaturated fatty acids, the content of polyunsaturated fatty acids (PUFAs) in fat decreases while the content of oleic and stearic acids increases with simultaneous formation of trans- isomers of oleic acid. For this reason, various producers periodically brought fats containing 15% or more trans-isomers to the market (Table 1).

The content of trans-fatty acids in fast-food products reaches one-third of total fatty acids and accounts for a significant proportion of a daily diet (Table 2).

In addition, the process of formation of trans-isomers was studied in detail during deodorization of oils at the refining stage. Non-hydrogenated refined vegetable oils contain a small amount of trans-isomers (0.5-2.0%) depending on the degree of their unsaturation and the exposure conditions used for their processing (Schwarz, 2000; Schmidt, 2000).

In frying technology, the process of formation of trans-isomers has not been investigated sufficiently. Beatriz, Oliveira, and Ferreira (1994), Gamel, Kiritsakis, and Petrakis (1999), and Sebedio et al. (1996) demonstrated that when using non-hydrogenated oils that did not contain trans-isomers of oleic and linoleic acids, their concentration in the frying medium was low. Therefore, if they present in fried foods

Imports and Use of Palm Oil as a Way to Increase Safety of Food Fats

Table 1. Content of trans-isomers in various fats

Food products	Content of trans-isomers, %
Milk fat	2.3-8.6
Beef fat	2.0-6.0
Hydrogenated fats	35.0-67.0
Raw vegetable oils	<0.5
Refined vegetable oils	<1.0
Soft margarines	0.1-17.0
Cake margarine	20.0-40.0
Cooking fats	18.0-46.0
Spreads	1.5-6.0

Source: Kulakova, Viktorova, and Levachev (2008)

in significant quantities, they are not formed as a result of frying but because of actual presence in the original fats. Trans-isomers can migrate to non-hydrogenated oils from foods or semi-finished products previously fried in hydrogenated oils and regenerated in catering establishments (Romero, Cuesta, & Sanchez-Muniz, 2000).

Hutchinson (1984), Shapiro (1995), Gans and Lapane (1995), Mozaffarian, Katan, Ascherio, Stampfer, and Willett (2006) concluded that trans-isomers disrupted the work of enzymes and cell membranes and increased the level of cholesterol in blood. Their action increases the risk of cancer, cardiovascular disease, and diabetes. Although no direct relationship between the content of trans-isomers and the incidence of cardiovascular disease has been revealed, the studies in Canada demonstrated that breast milk of 25% of women contains more than 10% of trans-isomers. In the USA, this problem came to the forefront only when statistics put obesity, colon cancer, and cardiovascular diseases in first place.

It has been proved that trans-isomers of unsaturated fatty acids obtained by industrial processes differ from natural trans-isomers contained in animal fats (up to 8% in milk fat) by the position of trans-double

Table 2. Content of trans-fatty acids in some fast-food products

Food products	Content of trans-isomers of unsaturated fatty acids		
	Grams per portion	g/100 g	% of total fatty acids
French fries, crisps	4.7-6.1	4.2-5.8	28.0-36.0
Fish burger	5.6	3.4	28.0
Pizza	1.1	0.5	9.0
Popcorn	1.2	3.0	11.0
Pie	3.9	3.1	28.0
Donuts	2.7	5.7	25.0
Cakes	1.8	5.9	26.0
Muffin	1.7	2.7	16.0
Pastry	0.7	1.3	14.0

Source: Bezuglov and Konovalov (2009)

bond in fatty acid chain. Observations of their influence on human body have made it possible to conclude that there is neither lower safe and upper tolerant border nor adequate level of daily consumption. Particularly, Bezuglov and Konovalov (2009) registered similar effects of trans-isomers of natural origin and trans-isomers formed during the hydrogenation process on a body. Without characteristics of natural cis-unsaturated fatty acids, trans-fatty acids compete with normal fatty acids for enzymes of biosynthesis and metabolism of these acids. In animal experiments, feeding of trans-fatty acids led to the changes in gene expression (Bezuglov & Konovalov, 2009).

In 2009, the World Health Organization (WHO) recommended all industrial trans-fats to be removed from food products (Uauy et al., 2009). The reason was that in view of insufficient number of clinical data, normalization of natural trans-fats remained an open question. On the basis of WHO recommendations since July 19, 2010, EU countries have established strict standards for the content of trans-isomers of unsaturated fatty acids in fatty products – no more than 2%.

Rationing of the trans-isomers of oleic acid requires a fundamental revision of the technology for producing the solid component of edible fats. The issue is particularly acute for those countries where the main raw materials for margarine products are liquid vegetable oils (USA and Russia). In accordance with the Eurasian Economic Commission (2011), since January 1, 2018, in Russia, the content of trans-isomers has been strictly specified limited by 2% for all types of anhydrous and emulsified edible fats. In the USA, where the hydrogenation of soybean and cottonseed oils has been an obligatory stage in the production of margarines and special fats for about 100 years, a detailed system for obtaining fat sets of the required composition by combining five bases of different depth of hydrogenation has been developed (O'Brien, 2007). The refusal to use hydrogenated fats leads to the fact that the composition indicators, the nature of crystallization, and the curve of the solid phase of special fat products change accordingly.

At the World Conference of Fat and Oil Industry Specialists in Istanbul in 2006, the US experts reported that they were reviewing margarine product formulations in order to reduce the content of trans-isomerized acids to meet nutrition labeling guidelines established by the U.S. Food and Drug Administration (FDA) (List, 2006). As of January 1, 2006, trans fatty acid (TFA) content must be listed on nutrition labels as a separate line, but foods containing less than 0.5 grams TFA/serving (14 grams) may be declared as zero. The American Soya Association appealed to farmers to take part in a five-year research program to produce products without trans-isomers, including sowing of special varieties of genetically modified soybeans and the use of innovative methods and special emulsifiers for modifying oils and fats. Since June 16, 2015, the FDA has offered enterprises to stop using partially hydrogenated trans-isomerized fats in food for three years (National Academy of Sciences, 2005; Morris et al., 2003).

The difficulties of a complete transition to innovative methods for modifying fatty raw materials (interesterification, fractionation) explain the restrain attitude of domestic producers to the introduction of strict standards for the content of trans-isomers in margarine products. In this regard, for the production of culinary, baking, and deep-frying fats, it is appropriate to use natural vegetable oils. Thus, enterprises producing special fats and oils should find possibilities for technical modernization to stop using hydrogenation technology. For a number of years, producers in Russia have been increasing purchases of palm oil in order to reduce the consumption of hydrogenated fats used in the production of margarine, spreads, and other edible fats, as well as to reduce the content of trans-isomers and the cost of production of oil and fat products. In 2018, the volume of palm oil imports was about one million tons which had led to the reduction in the use of domestic sunflower oil and required certain measures to ensure food security of the country.

This study aims at the justification of the practicability of usage of palm oil for the complete replacement of hydrogenated fats in the formulations of special-purpose fats. The following objectives have been accomplished:

- In an animal experiment, the authors have assessed safety and biological value of palm oil in comparison to fat based on hydrogenated fats.
- There have been evaluated the prospects for the use of palm oil instead of hydrogenated fats in the production of special fat products in related food industry sectors.

MAIN FOCUS OF THE CHAPTER

Study Materials and Methods

The subject of the study included:

- Frying fats containing different levels of trans-isomers of fatty acids: palm oil fat and partially hydrogenated fat
- Statistically interesterified fats of known fatty acid and triglyceride composition used for modeling a fatty bases of special fatty products.

The triglyceride composition of statistically interesterified fat has been calculated. If two groups of fatty acids ("s" for saturated and "u" for unsaturated ones) are considered, then the concentration of the corresponding triglycerides can be expressed as follows:

$S_3 = s^3 * 10^{-4}\%$

$S_2U = 3s^2u * 10^{-4}\%$

$SU_2 = 3su^2 * 10^{-4}\%$

$U_3 = u^3 * 10^{-4}\%$

where

s, u – molar concentrations of acids "s" and "u" in a mixture of fats, %;
S_3, S_2U, SU_2, S_3 – molar concentrations of the corresponding monoacid and di-acid triglycerides in the interesterified fat, %.

The fatty acid composition of fats was determined by gas-liquid chromatography of methyl esters of fatty acids (Government of the Russian Federation, 1999a, 1999b). The content of trans-isomers of oleic acid was determined according to the Government of the Russian Federation (2003b). The melting point of fats was determined according to the Government of the Russian Federation (2003a). The solid phase content (SFC) of fat at temperatures 10-35°C was determined by NMR method using PC-20 device of Brucker (International Organization for Standardization, 2008; Government of the Russian Federation, 2008a).

The effects on animals during long-term consumption of palm oil and hydrogenated fats containing different levels of trans-isomers were studied by pathomorphological, histological, and hematological methods. Animal studies were carried out on the basis of a certified vivarium and scientific-technological center of the Veterinary Hospital and the laboratories of the Department of Morphology, Animal Pathology and Biology of Saratov State University, Russia and conducted in accordance with the "Rules for Carrying out Work on Experimental Animals" (Loskutova, 1980). All experimental studies were performed on groups of clinically healthy rats, formed according to the method of analogs: same breed, same sex, same age, and same weight. The animals were fed for 40 days. During the experiment, rats were kept in individual cells (ten animals each). Before introducing the studied fat into the diet, the animals were kept in quarantine for 21 days and transferred to an experimental diet in accordance with the experiment schedule (Loskutova, 1980).

Three groups of rats participated in the experiment: a control group of rats that received the usual full-fledged diet, which included 50 g of mixed fodder, as well as carrot and hay additives (Government of the Russian Federation, 1977). Experimental groups of rats received the usual full-fledged ration, part of which was replaced with frying fat: Group 1 – fat based on palm oil; Group 2 – fat based on hydrogenated fat (Simakova, Perkel, Lomnitsky, & Terentiev, 2015) (Table 3).

Table 3. Content of basic elements in the diet of laboratory animals

Food items	Net weight, g	Proteins, g	Fats, g	Carbohydrates, g	Energy value, kcal
Control group					
Mixture of cereals	15.0	2.1345	3.51000	3.0525	147
Bread	4.0	0.9760	0.48800	3.3320	
Oatmeal	3.0	0.7290	0.14700	0.4512	
Milk	8.0	0.2800	0.28000	1.0000	
Meat	5.0	1.2500	0.75000	0.0000	
Juicy foods	20.0	2.4100	0.50000	7.5000	
Fish oil	0.1	0.0000	0.08990	0.0000	
Beefish flour	0.7	0.0039	0.00021	0.0000	
Total		8.6800	5.76000	15.3300	
Fat consuming group					
Mixture of cereals	15.0	2.1345	3.51000	3.0525	132
Oatmeal	3.0	0.7290	0.14700	0.4512	
fat based on palm oil / hydrogenated fat	2.0	0.0000	1.95000	0.0000	
Milk	8.0	0.2800	0.28000	1.0000	
Meat	5.0	1.2500	0.75000	0.0000	
Juicy foods	10.0	1.2000	0.25000	7.5000	
Beefish flour	0.7	0.0039	0.00021	0.0000	
Total		5.5900	6.80000	12.0300	

Source: Authors' development

Table 4. Content of vitamins and minerals in the diet of laboratory animals

Vitamins, mg				Minerals, mg			
A	D	E	K	Fe	Mg	K	Ca
0.90	3.00	0.60	2.00	0.25	1.00	12.00	40.00

Source: Authors' development

Table 5. Fatty acid composition of the studied fats

Fatty acid	Length of the chain and the number of double bonds	Mass fraction of fatty acid,% of total fatty acids of frying fa	
		hydrogenated fat	palm oil fat
Lauric	12.0	-	0.2
Myristic	14.0	0.3	1.1
Palmitic	16.0	18.7	44.0
Palmitoleic	16.1	0.2	0.1
Stearic	18.0	16.5	4.5
Oleic acid	18.1	50.0	39.2
including trans- isomers	18.1	25.9	2.2
Linoleic	18.2	13.2	10.1
Linolenic	18.3	-	0.4
Arachic	20.0	-	0.4

Source: Authors' development

Vitamins and minerals were introduced into the food in the form of a premix (Table 4).

The autopsy was performed with detailed logging and photographing of the material. Patho-morphological changes were studied on the material of 30 slaughtered animals. An autopsy was performed in the first two hours after slaughter. For histological examination, pieces of liver, spleen, and aorta were taken. To fix the pathological material, a 10% solution of neutral formalin was used (Merkulov, 1969; Skopichev, 2004).

Sections were obtained on a freezing microtome of model 2515 (Reichert Wien). The histological sections were stained with Ehrlich hematoxylin and eosin followed by microscopy. The morphological structure of the organs was studied in 30 fields of view of the microscope in different histological sections. Histological examination of the manufactured preparations was carried out under different magnifications, with detailed logging and photographing of the sites studied. Microphotography of histological specimens was performed using a CANON Power Shot A460 IS camera.

Comparative Analysis of the Effect of Partially Hydrogenated and Palm Oil Fats on a Body

The studied fatty-acid composition of fats is demonstrated in Table 5.

Palm oil based fat contained about 2% of trans-isomerized fatty acids formed as a result of thermal exposure during high-temperature deodorization of raw materials. Hydrogenated fat based fat contained about 25% of trans-isomers mainly formed during hydrogenation (Table 6).

Table 6. General blood analysis

Descriptors	White rat (normogram, average)	Control	Group 1 (palm oil based fat)	Group 2 (hydrogenated fat)
Hemoglobin, gr/dm^3	150.0	145.0	120.0	135.0
Hematocrit, %	46.0	47.2	39.2	43.5
Erythrocytes, 10^{12}/dm^3	8.3	7.6	6.0	6.0
Leukocytes, 10^9/dm^3	15.5	14.6	16.7	17.3
Platelets, 10^9/dm^3	400.0	610.5	629.3	709.5
ESR, mm/hr	-	1.0	1.0	2.0
Basophils, %	0.5	0.0	2.0	2.0
Eosinophils, %	3.0	3.0	3.0	6.0
Myelocytes, %	-	0.0	0.0	0.0
Juvenile, %	-	0.0	0.0	1.0
Stab. neutrophils, %	2.5	6.0	8.0	9.0
Segmented neutrophils, %	26.5	32.0	19.0	25.0
Lymphocytes, %	65.0	56.0	64.0	52.0
Monocytes, %	3.0	3.0	4.0	5.0
Anisocytosis, %	-	10.4	12.5	14.01
Color index	0.8	1.1	1.0	1.00

Source: Authors' development

Table 7. Biochemical analysis of rat blood

Descriptors	White rat (normogram, average)	Control	Group 1 (palm oil based fat)	Group 2 (hydrogenated fat based fat)
Bilirubin total, mcM/dm^3	7.2	3.8	8.4	6.0
Cholesterol, mcM/dm^3	4.4	3.4	6.5	5.6
Sodium, mcM/dm^3	17.1	42.5	59.6	102.5
Potassium, mcM/dm^3	4.4	5.3	2.1	6.1
Alkaline phosphatase, mcM/dm^3	-	37.2	84.3	62.8

Source: Authors' development

The results of biochemical analysis of rat blood are presented in Table 7.

The results of studies presented in Table 6 and Table 7 indicate changes in blood of experimental groups of animals. The most significant changes are observed in a group of rats, in the diet of which hydrogenated fat based fat was present. The changes were indicated by a shift in the leukocyte formula, an inflammatory process, allergic reactions, changes associated with intracellular and intercellular exchange processes, and disorders of cellular immunity of the body.

To confirm the conclusions based on the results of a blood test, histological examination of the organs and tissues obtained after the autopsy of rats in each group was performed (Figure1).

Figure 1. Aortic and liver tissues
Source: Authors' development

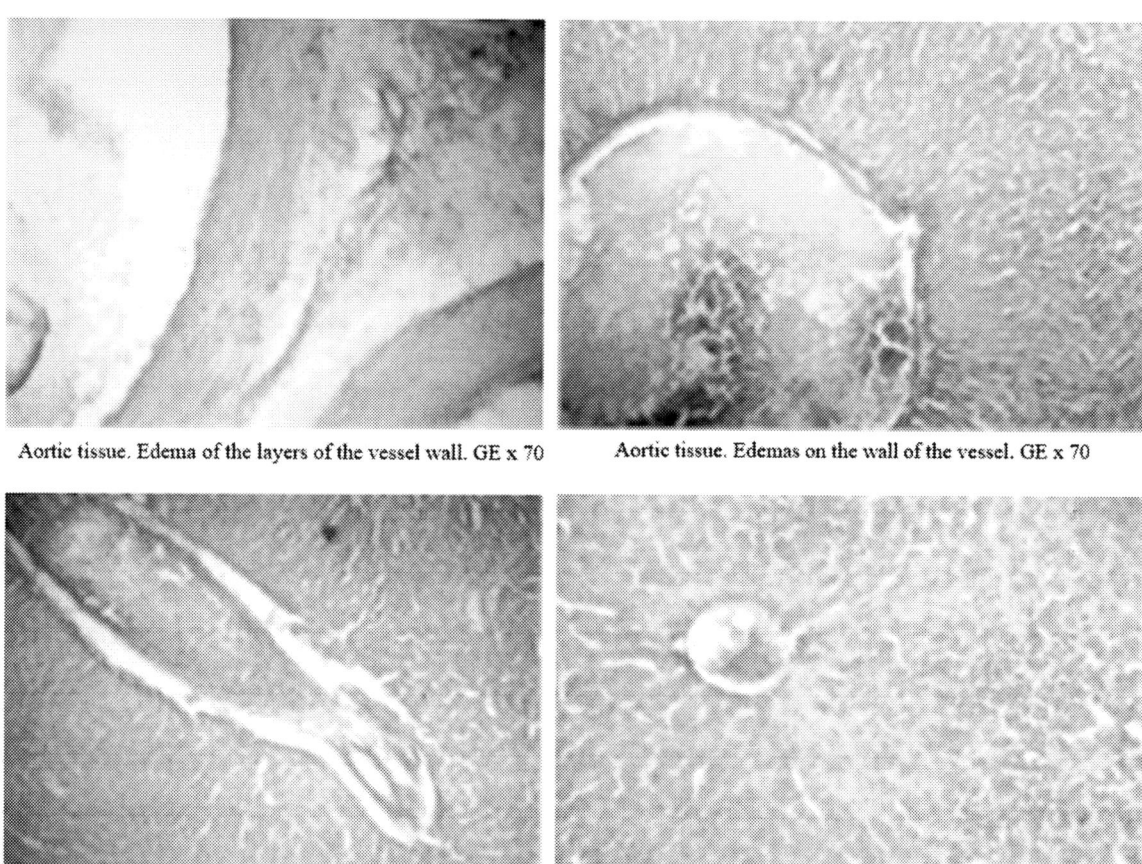

Aortic tissue. Edema of the layers of the vessel wall. GE x 70

Aortic tissue. Edemas on the wall of the vessel. GE x 70

Liver tissue. Light coloration of perivascular sites, hyperemia. GE x 100

Liver tissue. Tinctorial properties are not violated. GE x 100

The results of histological examination of internal organs of animals (white rats) after feeding them with experimental diets during four weeks are presented in Table 8.

In authors' opinion, trans-isomers should have caused disturbances in the cardiovascular system. During external examination of slaughtered rats, no apparent pathological changes are detected. When analyzing the results of histological studies, it is found that in the experimental Group 2, myocardia`s coronary arteries have uneven intimal surface with some elements of proliferation process. The walls of the vessels are slightly thickened, there is accumulation of edematous fluid around them. The nuclei of individual myocytes are round-oval, clearly contoured. The cytoplasm is fine-grained; the transverse striation of the fibers is slightly preserved. Separate fibers are fragmented. The walls of blood vessels of animals from the Control group are not thickened, they are sufficiently smooth.

In Group 2, areas of depression are identified in spleen. The walls of arterial vessels are thickened; the intima is in a condition of proliferation. Vessels are surrounded by edematous fluid. There are the sites of accumulation of erythrocytes outside the vascular bed. The intima of the animals from the Control group is smooth and even.

Table 8. The results of histological studies of internal organs of white rats

Organ	Control	Results of histological examination after feeding with experimental diets	
		palm oil fat	hydrogenated fat
Aorta	Normal	Minor proliferative processes	Pathology: granular dystrophy, proliferation, and edematous phenomena
Spleen	Normal	Normal	Pathology: enlargement of the spleen
Liver	Normal	Normal	Pathology: changes are characteristic for granular dystrophy of hepatocytes and perivascular edema

Source: Authors' development

Vacuolar dystrophy and perivascular edema are observed in brain of experimental animals of Group 2. The vessels are hyperemic; the walls of the vessels are thickened. In some cases, there is proliferation of intima. The walls of the vessels of the animals from the Control group are sufficiently smooth.

In Group 2, discomplexation of the beam structure is observed in liver; hepatocytes are enlarged in size; cytoplasm has a fine-grained appearance and contains drops of protein origin; nuclei are poorly contoured and located acentrically. Vessels are hyperemic, the walls are thickened, surrounded by edematous fluid. Animals from the Control group have normal, smooth walls of blood vessels.

In group 2, in kidneys of experimental animals, there is a collection of edematous fluid in the Shumlyansky's capsule; cytoplasm of the epithelium of the renal tubules contains inclusions of protein origin. Intima of blood vessels has some elements of proliferation. In the Control group of animals, the vessels are rounded, with an even surface of the intima.

The animals of Group 1 were fed a diet with the addition of frying fat based on palm oil with a low content of trans-isomers of oleic acid. At the same time, special attention was paid to the study of changes in the liver and gastrointestinal tract. Studies have shown no changes in liver, stomach, small intestine; minimal changes have been detected in large intestine only – an increase in the number of goblet cells along with edema.

The study concludes that there is a negative effect of trans-isomers on animals under conditions of normal provision of a body with essential linoleic acid, even with a relatively low concentration of trans-isomers in the fatty part of the diet. When these fats are used for deep-frying, as the oxidation products accumulate, negative effect of trans-isomerized fats on safety and quality indicators of products increases. The authors' findings show the unacceptability of the use of hydrogenated fats in the composition of deep-frying and cooking fats. Therefore, to achieve the required thermal stability, it is necessary to use mixtures of natural oils and fats with a reduced content of PUFAs. Palm oil based fat can be used as a limited alternative to partially hydrogenated fats as it does not have an explicit negative effect on a body.

SOLUTIONS AND RECOMMENDATIONS

The quantity of solid glycerides at any given temperature as determined by SFC is responsible for many of the product characteristics including general appearance, ease of packing, organoleptic properties (flavor release, coolness, and thickness), ease of spreading, and oil exudation. For example, in a table spread, the SFC between 4°C and 10°C determines the ease of spreading of the product from a refrigerator; a SFC of not greater than 32% at 10°C is essential for the production of a spread having a yield value of

Table 9. Melting point and solid phase content (by NMR method) in structured cooking and confectionery fats and palm oil

Indicator	Value of the indicator in various fats			
	Milk fat substitute	General-purpose cooking oil	Chocolate coating fat	Palm oil
Melting point, °C	33-36	below 36	below 36	31-38
Mass fraction of solid phase SFC, %, at temperature, °C:				
10	31-37	42-46	93-96	40-55
15	27-29	34-38	86-92	27-40
20	18-25	30-34	80-86	15-28
25	10-12	16-24	61-67	7-19
30	4-8	10-14	34-38	5-14
35	below 5	5-7	4-8	2-12

Source: Authors' development

1,000 g/cm^2 at 4°C. The ease of packing is governed by the SFC at 15°C and a high SFC is desirable for foil wrapping (Palm Oil Research Institute of Malaysia, 1993).

The melting point and solid phase content (by NMR method) in palm oil at a temperature between 10°C and 35°C are given in Table 9. Palm oil and its mixtures with liquid vegetable oils are characterized by an unfavorable melting point and hardness ratio, they are characterized by the so-called post-crystallization, that is, slow freezing in equipment for cooling, structuring, and packaging. After such a delayed crystallization, an additional amount of heat is released from the finished prepacked product, which leads to disruption of the product structure and leakage of the liquid oil phase. For this reason, the traditional technology provides the use of natural palm oil mainly in mixtures with components of increased hardness.

Typically, such mixtures used to contain partially hydrogenated fats with the desired optimal performance. The almost complete abandonment of the use of hydrogenated fats requires the development of fundamentally new technologies and formulations for a wide range of food fats in related sectors of the food industry.

Culinary and confectionery fats according to their characteristics and purpose can be divided into two groups. In deep-frying oils and cooking fats used for frying, thermal stability and, in some cases, melting points, is a determining factor. Structural characteristics of such products, for example, solid-phase curve, are of secondary importance.

In the fatty bases of margarine and special fats for baking, toppings, and chocolate coating, the determining factor is the solid fat content at temperatures of 10-35°C, on which the melting temperature and crystal structure of the finished product depend. Table 9 shows melting characteristics of the most important types of structured cooking and confectionery fats, obtained as a result of research products of advanced domestic and foreign firms.

The main component of the fatty bases of this product, providing the required structure, character of crystallization and melting curves depending on temperature, are medium melting triacylglycerols (triglycerides). This term refers to triglycerides with a melting point of 35-45°C, containing triglycerides of lauric and myristic acids (from coconut and palm kernel oils) and disaturated triglycerides of palmitic and stearic acids.

Table 10. Composition of fatty foods for healthy nutrition

Fatty acid unsaturation	Ratio of fatty acids, in % of the total caloric intake of a daily diet	Ratio of fatty acids, in % to the total fat content
Saturated	below 10	38.5-33.3
Monounsaturated	10	38.5-33.3
Polyunsaturated	6-10	23.0-33.3
Ratio of polyunsaturated acids ω6 / ω3	5-15	21.6/1.4-28.4/5.6

Source: Government of the Russian Federation (2008b)

To ensure biological usefulness of special fats, there should be ensured a compliance with the current recommendations of the Government of the Russian Federation (2008b) on the fatty acid composition of edible fats (Table 10).

In order to increase the oxidative stability of culinary and confectionery fats, it is advisable to choose a variant with a minimum content of polyunsaturated fatty acids (20-25%) and maximum content of saturated fatty acids (no more than 38%). In particular, in special fats, the content of linolenic acid (ω3) should be no more than 1%.

The need to improve the safety of culinary products for consumers requires immediate reduction in the content of trans-isomers in special fats, margarines, and spreads to meet European standards.

FUTURE RESEARCH DIRECTIONS

Consequently, it is necessary to develop new scientific principles for constructing fat sets in order to obtain special fats with given characteristics. As was mentioned above, in order to obtain fatty bases with the required characteristics, it is necessary to use mixtures of palm oil with other fats of increased hardness that do not contain trans-isomers. This problem is solved partially by using lauric oils (coconut, palm kernel). However, the widespread use of lauric oils in domestic products has several disadvantages, including a rather high cost, reduced storage stability, high waste during processing, soapy taste during hydrolysis, and formation of eutectic mixtures with palm oil (Palm Oil Research Institute of Malaysia, 1993; Herrington, 1965). In addition, it is impossible to manufacture all domestic products only from mixtures of palm and lauric oils. Therefore, the production of the entire range of margarine products without selectively hydrogenated fats and without lauric oils is a serious scientific and practical task.

The problem of excluding trans-isomers from the structure of dietary fats can be solved at the molecular level depending on the type of product. The preservation of the structural characteristics of such mixtures is achieved by replacing medium melting triglycerides containing trans-isomers of oleic acid with triglycerides of a different fatty acid composition with a similar melting point, stable fine-crystalline β´-structure and a fairly high crystallization rate. These requirements suggest using fats with a high content of disaturated glycerides of palmitic and stearic acids, which plays a decisive role ratio trisaturated glycerides to disaturated (S_3/S_2U), the ratio of symmetrical and asymmetrical disaturated triglycerides (SUS/SSU) and the ratio of palmitic and stearic acids (S/P). In particular, to create a microcrystalline β´-structure that is stable during storage, the concentration in the palmitic acid fat should be at least 11% (O'Brien, 2007).

Imports and Use of Palm Oil as a Way to Increase Safety of Food Fats

The solution of these structural problems is impossible without the use of the interesterification process of high-melting natural and fully hydrogenated vegetable oils with liquid vegetable oils. Interesterification refers to the exchange reaction of fatty acid residues between ester groups within molecules and between molecules of fat mixtures. It has been established that in the presence of special alkaline catalysts (sodium alcoholates) the exchange of fatty acid residues does not depend on the structure of the fatty acid and its position in the triacylglycerol molecules. As a result, fatty acid exchange within 30-40 min reaches an equilibrium state in which fatty acid residues are distributed in molecules in accordance with their concentration, and the content of individual triacylglycerol groups is determined by the formulas of mathematical statistics. As a result, the interesterification of mixtures of the same fatty acid composition achieves the same molecular composition of the interesterified fat corresponding to the statistical distribution of fatty acids in a triglycerides.

Statistically, interesterified fats, similar in physicochemical parameters, can be obtained from various fats and oils by selecting mixtures of the same group fatty acid composition, which is explained by the similar properties of mixed-acid triglycerides of palmitic and stearic, as well as oleic and linoleic acids.

The authors obtained systematic data characterizing the relationship of physicochemical parameters of interesterified fats with their group fatty acid and triglyceride composition. As the content of high-molecular saturated fatty acids (palmitic and, especially, stearic) increase from 25% to 45%, the melting point of interesterified fat increases from 28°C to 41°C. At the same time, mass fraction of the solid phase at 15°C (SFC_{15}) rises from 8% to 42% (Perkel, 1990).

In laboratory experiments and industrial production, the authors achieved a degree of interesterification of fatty mixtures on alkaline catalysts – sodium alcoholates – at 92-95%, which made it possible to consider the obtained interesterified fats as mixtures with a known triglyceride composition and use them to model the molecular composition of edible fats that do not contain trans-isomerized fatty acids (Perkel, 1990).

It is known that fat in the first approximation is a heterogeneous two-phase system, the liquid phase of which is formed by molten triglycerides. The solid phase of this system (SFC) is formed by triglycerides, which are insoluble or partially soluble in the liquid phase at a given temperature. At each temperature, SFC composition is different depending on melting temperature and solubility in the liquid phase of the system. For example, at 35°C, SFC contains only the highest melting trisaturated glycerides of palmitic and stearic acids (S_3). At 20°C, SFC contains, in addition to trisaturated, also asymmetric disaturated glycerides (SSU). At 15°C, the composition of the SFC also includes symmetric disaturated glycerides (SUS).

Considering the fraction of solid glycerides (SG) of fat as a solute and liquid phase of fat as a solvent, it is possible (although approximately) to consider phase equilibria in narrow temperature ranges as equilibria in a two-component system and apply to them the ratios valid for ideal solutions, in particular, the Raoult law (Bailey, 1950; Kireev, 1975). A theoretical consideration of the problems based on the application of that law showed that at each certain temperature, there should be a linear relationship between the molar fraction of solid glycerides at this temperature and the mass fraction of the solid phase in the fat. Thus, at each certain temperature, there should be a linear relationship between the molar fraction of S_t at this temperature and the mass fraction of the solid phase SFC_t in fat (Perkel, 1990).

Processing of the experimental data shows that the triglyceride composition of interesterified fat and the SFC curve obtained by the NMR method are related by approximate linear dependences of the indicated type. The parameters of the approximate linear dependences determined for interesterified fats containing mixed palmitic and stearic acid triglycerides:

Table 11. Indicators of interesterified fat containing a total of 49.2% palmitic and stearic fatty acids

Indicator	Value indicator	
	calculated data	experimental data
Melting point, °C	42.8	41.5
SFC, %, at temperatures °C:		
10	48.8	59.5
15	-	49.2
20	-	38.2
25	11.4	27.1
30	-	18.5
35	-	11.5

Source: Authors' development

$$t_{mp} = 8 \ln S_3 + 24.0°C$$

$$SFC_{35} = (S_3 - 0.5)\%$$

$$SFC_{15} = (S_3 + SSU + SUS) - (3...5)\%$$

The value 3-5% characterizes the solubility of the solid phase in the liquid phase of the fat mixture with a relatively low content of the solid phase. With an increase in the solids content and a corresponding decrease in the liquid phase, this value decreases.

Table 11 shows the experimentally determined melting point and the SFC curve of interesterified fat containing, according to the calculation, 11.9% of trisaturated and 36.9% of disaturated glycerides.

The obtained dependences are, of course, especially approximate due to the different melting points of individual TG and their unequal contribution to the general physicochemical parameters of interesterified fats. Nevertheless, they are very useful for practical purposes, as they allow to roughly calculate the technological characteristics of interesterified fat of known acid composition and, conversely, to calculate the required acid composition of the fat mixture undergoing the interesterification from the desired SFC fat curve.

For example, by interesterification of a mixture of 50-60% high-melting fraction of palm oil (palm stearin), with 50-40% sunflower oil containing 40-42% of saturated fatty acids, one can obtain interesterified fat, by its structural characteristics rather closely reproducing the characteristics of natural palm oil, and the nature of crystallization is much more favorable for special fats. At the same time, the fatty acid composition of the obtained interesterified fat better meets the requirements for optimum fatty acid composition (Table 10) recommended for healthy feeding. Using this advanced technology, the volume of imports of palm oil and palm stearin can be almost halved. Further studies have shown that these dependencies, with certain limitations, are also applicable to the calculation of mixtures containing natural and interesterified oils and fats.

Comparing the obtained dependences with the characteristic of special fats (Table 9), it can be concluded that cooking oils should contain no more than 6% trisaturated and 25-30% disaturated triglycerides. More solid fats for confectionery products should contain 50-80% disaturated glycerides of palmitic and stearic acids, in general symmetrical type.

Currently, advanced domestic enterprises adapted the process of interesterification of fats on both alkaline and biological catalysts-enzymes, which allow a controlled process of interesterification, affecting exchange of fatty acids only in the extreme (1st and 3rd) positions of triglycerides.

The cost of production of interesterified fats is about 1.5 times lower than the cost of hydrogenating vegetable oils.

Fats of the desired fatty acid and triglyceride composition can be obtained by mixing natural and interesterified fats and oils. The structural characteristics of the interesterified fats can be further improved by fractionation.

CONCLUSION

Palm oil and its fractions can be considered as a real alternative to hydrogenated fats in the production of certain types of oil and fat products – margarines, cooking, baking, and deep-frying fat. The use of palm oil for the production of special fats of increased hardness (for spreads, confectionery, waffles and fillings, and chocolate coating) requires the use of modern methods for modifying the triglyceride composition of fats – biocatalytic interesterification and fractionation. The interesterification of palm stearin mixtures with sunflower oil makes it possible to almost halve the purchases of palm fats. The use of deepest hydrogenation of liquid vegetable oils (the so-called hydrogenation to full saturation) allows us to obtain deeply hydrogenated fats that do not contain unsaturated fatty acids and their trans-isomers. Interesterification of mixtures of such saturated fats with liquid vegetable oils results in a mixture of triglycerides that are quite similar to the properties of palm oil. In the future, the development of new technologies must lead to lower purchases of palm oil and the maximum use of domestic vegetable oils.

REFERENCES

AB-Center. (n.d.). *Expert-Analytical Center of Agribusiness*. Retrieved from https://ab-centre.ru/

Bailey, A. (1950). *Melting and Solidification of Fats*. New York, NY: Interscience Publishers.

Beatriz, M., Oliveira, P., & Ferreira, M. (1994). Evolution of the Quality of the Oil and the Product in Semi-Industrial Frying. *Grasas y Aceites, 45*(3), 113–118. doi:10.3989/gya.1994.v45.i3.982

Bezuglov, V., & Konovalov, S. (Eds.). (2009). *Lipids and Cancer. Essays on Lipidology of Oncological Process*. Saint Petersburg: Prime-Evroznak.

Eurasian Economic Commission. (2011). *Technical Regulations of the Customs Union TR CU 021/2011 "On Food Safety."* Retrieved from http://www.eurexcert.com/TRCUpdf/TRCU-0021-On-food-safety.pdf

Gamel, T. H., Kiritsakis, A., & Petrakis, C. (1999). Effect of Phenolic Extracts on Trans-Fatty Acid Formation during Frying. *Grasas y Aceites, 50*(6), 421–425. doi:10.3989/gya.1999.v50.i6.689

Gans, K. M., & Lapane, K. (1995). Trans-Fatty Acid and Coronary Disease: The Debate Continues. 3. What Should We Tell Consumers? *American Journal of Public Health, 85*(3), 411–412. doi:10.2105/AJPH.85.3.411 PMID:7892934

Gassenmeier, K., & Schieberle, P. (1994). Formation of the Intense Flavor Compound Trans-4,5-Epoxy-(E)-2-Decenal in Thermally Treated Fats. *Journal of the American Oil Chemists' Society, 71*(12), 1315–1319. doi:10.1007/BF02541347

Government of the Russian Federation. (1977). *Order #755 from August 12, 1977, "On the Measures on Further Development of Organizational Forms of Work with the Use of Experimental Animals."* Retrieved from https://docplayer.ru/31723947-Ministerstvo-zdravoohraneniya-sssr-prikaz-12-avgusta-1977-g-n-755.html

Government of the Russian Federation. (1999a). *National Standard "Vegetable Oils and Animal Fats. Determination by Gas Chromatography of Constituent Contents of Methyl Esters of Total Fatty Acid Content."* Retrieved from http://docs.cntd.ru/document/gost-r-51483-99

Government of the Russian Federation. (1999b). *National Standard "Vegetable Oils and Animal Fats. Preparation of Methyl Esters of Fatty Acids."* Retrieved from http://docs.cntd.ru/document/gost-r-51486-99

Government of the Russian Federation. (2003a). *National Standard "Margarines, Cooking Fats, Fats for Confectionery, Baking and Dairy Industry. Sampling Rules and Methods of Control."* Retrieved from http://docs.cntd.ru/document/1200036186

Government of the Russian Federation. (2003b). *National Standard "Spreads and Melted Blends. General Specifications."* Retrieved from http://docs.cntd.ru/document/1200032516

Government of the Russian Federation. (2008a). *National Standard "Animal and Vegetable Fats and Oils and Their Derivatives. Determination of Solid Fat Content. Pulsed Nuclear Magnetic Resonance Method."* Retrieved from http://docs.cntd.ru/document/1200074555

Government of the Russian Federation. (2008b). *Norms of Physiological Requirements in Energy and Feedstuffs for Various Groups of Population in the Russian Federation*. Retrieved from http://docs.cntd.ru/document/1200076084

Herrington, E. (1965). *Zone Melting of Organic Compounds*. Moscow: Peace.

Hutchinson, R. (1984). *The Carp Strikes Back. Henlow Camp*. Beekay Publishers.

International Organization for Standardization. (2008). *ISO 8292-1:2008 "Animal and Vegetable Fats and Oils – Determination of Solid Fat Content by Pulsed NMR – Part 1: Direct Method."* Retrieved from https://www.iso.org/standard/41256.html

Kireev, V. (1975). *Course of Physical Chemistry*. Moscow: Chemistry.

Kulakova, S., Viktorova, E., & Levachev, M. (2008). Trans-Isomers of Fatty Acids in Food Products. *Oils and Fats, 85*(3).

List, G. (2006). Reformulation of Food Products for Trans-Fatty Acid Reduction: A U.S. Perspective. *Proceedings of World Conference and Exhibition on Oilseed and Vegetable Oil Utilization*. Istanbul: Exponet.

Loskutova, Z. (1980). *Vivarium*. Moscow: Medicine.

Merkulov, G. (1969). *Course of Pathology Techniques*. Saint Petersburg: Medicine.

Morris, M. C., Evans, D. A., Bienias, J. L., Tangney, C. C., Bennett, D. A., Aggarwal, N., ... Wilson, R. S. (2003). Dietary Fats and the Risk of Incident Alzheimer Disease. *Archives of Neurology, 60*(2), 194–200. doi:10.1001/archneur.60.2.194 PMID:12580703

Mozaffarian, D., Katan, M. B., Ascherio, A., Stampfer, M. J., & Willett, W. C. (2006). Trans-Fatty Acids and Cardiovascular Disease. *The New England Journal of Medicine, 354*(15), 1601–1613. doi:10.1056/NEJMra054035 PMID:16611951

National Academy of Sciences. (2005). *Dietary Reference Intakes for Energy, Carbohydrate, Fiber, Fat, Fatty Acids, Cholesterol, Protein, and Amino Acids*. Washington, DC: National Academies Press.

O'Brien, R. (2007). *Fats and Oils: Formulating and Processing for Applications*. Saint Petersburg: Profession.

Palm Oil Research Institute of Malaysia. (1993). *Selected Readings on Palm Oil and Its Uses. Bandar Baru Bangi*. Palm Oil Research Institute of Malaysia.

Perkel, R. (1990). *Research, Technology Development and Organization of Production of Interesterified Edible Fats*. Saint Petersburg: All-Russian Research Institute of Fats.

Romero, A., Cuesta, C., & Sanchez-Muniz, F. J. (2000). Trans-Fatty Acid Production in Deep Fat Frying of Frozen Foods with Different Oils and Frying Modalities. *Nutrition Research (New York, N.Y.), 20*(4), 599–608. doi:10.1016/S0271-5317(00)00150-0

Schmidt, S. (2000). Formation of Trans-Unsaturation during Partial Catalytic Hydrogenation. *European Journal of Lipid Science and Technology, 102*(10), 646–648. doi:10.1002/1438-9312(200010)102:10<646::AID-EJLT646>3.0.CO;2-2

Schwarz, W. (2000). Formation of Trans-Polyalkenoic Fatty Acids during Vegetable Oil Refining. *European Journal of Lipid Science and Technology, 102*(10), 648–649. doi:10.1002/1438-9312(200010)102:10<648::AID-EJLT648>3.0.CO;2-V

Sebedio, J. L., Dobarganes, M. C., Marquez-Ruiz, G., Wester, I., Christie, W. W., Dobson, F., ... Lahtinen, R. (1996). Industrial Production of Crisps and Prefried French Fries Using Sunflower Oils. *Grasas y Aceites, 47*(1-2), 5–13. doi:10.3989/gya.1996.v47.i1-2.836

Shapiro, S. (1995). Trans-Fatty Acid and Coronary Disease: The Debate Continues. 2. Confounding and Selection Bias in the Data. *American Journal of Public Health, 85*(3), 410–413. doi:10.2105/AJPH.85.3.410-a PMID:7892932

Simakova, I., Perkel, R., Lomnitsky, I., & Terentiev, A. (2015). Clinical Studies on the Safety of Frying Fats Containing Trans-Isomers of Oleic Acid. *Scientific Review (Singapore), 2*, 52–56.

Skopichev, V., Eisymont, T., Alekseev, N., Bogolyubova, I., Enukashvili, A., & Karpenko, L. (2004). *Animal Physiology and Etiology*. Moscow: Kolos.

Uauy, R., Aro, A., Clarke, R., Ghafoorunissa, R., L'Abbe, M., Mozaffarian, D., & Tavella, M. (2009). WHO Scientific Update on Trans-Fatty Acids: Summary and Conclusions. *European Journal of Clinical Nutrition, 63*(S2), S68–S75. doi:10.1038/ejcn.2009.15

World Health Organization. (n.d.). *Population Nutrient Intake Goals for Preventing Diet-Related Chronic Diseases*. Retrieved from https://www.who.int/nutrition/topics/5_population_nutrient/en/

ADDITIONAL READING

Arshad, F., & Sundram, K. (2003). Trans Fatty Acids – An Update on Its Regulatory Status. *Palm Oil Developments, 39*, 16–21.

Husum, T. L., Pedersen, L. S., Nielsen, P. M., Christensen, M. W., Kristensen, D., & Holm, H. C. (2003). Enzymatic Interesterification: Process Advantages and Product Benefits. *Palm Oil Developments, 39*, 7–10.

Minal, J. (2003). An Introduction to Random Interesterification of Palm Oil. *Palm Oil Developments, 39*, 1–6.

Minal, J. (2003). Application of Palm-Based Interesterified Fats. *Palm Oil Developments, 39*, 11–15.

Schneider, C., Boeglin, W. E., Yin, H., Ste, D. F., Hachey, D. L., Porter, N. A., & Brash, A. R. (2005). Synthesis of Dihydroperoxides of Linoleic and Linolenic Acids and Studies on Their Transformation to 4-Hydroperoxynonenal. *Lipids, 40*(11), 1155–1162. doi:10.100711745-005-1480-3 PMID:16459928

KEY TERMS AND DEFINITIONS

Biological Studies: The studies conducted on laboratory animals (most often, on white rats) to determine the effect of certain food components, toxicants, food additives, the mode and duration of feeding experimental animals with experimental diets, on a living organism.

Fatty Acid Composition: A composition of fatty acids in mass percentage or molar percentage relative to the amount of fatty acids of the test oil. Fatty acids are carboxylic acids, differ in chain length, the number of double bonds. Some of the fatty acids (only in some oils) contain additional functional groups (for example, ricinol hydroxy acid in castor oil).

Fractionation: A technological process of dividing the original fat into several parts (fractions) differing in their melting point and other characteristics.

Hydrogenated Fat: A fat obtained in technological process hydrogenation.

Hydrogenation: A technological process in which hydrogen gas is added to the double bond in the fatty acid molecule in the presence of special catalysts (most often nickel-based) and the double bond is transformed into a single one. Reducing the number of double bonds lowers the degree of the fatty acid unsaturation, increases its resistance to oxidation by atmospheric oxygen. During hydrogenation, not only double bonds can be converted into single ones, but unsaturated fatty acids in cis-configurations can be partially converted into unsaturated trans-acids.

Interesterification: A technological process in which fatty acid residues are exchanged within molecules and between molecules of triacylglycerols (triglycerides) in the presence of special catalysts. The interesterification is called biocatalytic when the enzyme serves as a catalyst for the process.

Palm Oil: A vegetable oil that is solid at room indoor temperature and is extracted from the fruits of tropical palm trees of African origin Elaeis guineensis Jacq.

Trans-Isomers of Unsaturated Fatty Acids: The features of the spatial structure of unsaturated fatty acids relative to the double bond plane. If both parts of the fatty acid (before and after the double bond) are located on the same side of the double bond plane, this is the cis-isomer; if on either side of the double bond plane, it is the trans-isomer. Trans-isomers differ from cis-isomers in their biological effects on a body.

Chapter 15
Resource–Saving Technology of Dehydration of Fruit and Vegetable Raw Materials:
Scientific Rationale and Cost Efficiency

Inna Simakova
https://orcid.org/0000-0003-0998-8396
Saratov State Vavilov Agrarian University, Russia

Victoria Strizhevskaya
Saratov State Vavilov Agrarian University, Russia

Igor Vorotnikov
Saratov State Vavilov Agrarian University, Russia

Fedor Pertsevyi
Sumy National Agrarian University, Ukraine

ABSTRACT

Thousands of tons of fruit and vegetables are lost annually during harvesting, transportation, and storage. Meanwhile, there is a problem of insufficient consumption of fruit and vegetables in the diet of modern people which results in an increase in the occurrence of alimentary-dependent diseases. One of the possible solutions to these two interrelated problems is the development of a technology of processing of substandard raw materials directly at the harvesting site. This study aims at the development of the technology of dehydration of fruit and vegetables applicable in a field. The economic effect of the proposed solution is contingent on the reduction of losses at the stage of cleaning and saving water resources and saving transportation and storage costs.

DOI: 10.4018/978-1-7998-1042-1.ch015

Resource-Saving Technology of Dehydration of Fruit and Vegetable Raw Materials

INTRODUCTION

Contemporary globalization processes emerge new challenges to food security. Concentration of people in big cities aggravates food supply problems. Since 1950, the share of urban population in the world has increased from 30% up to 54% and is expected to reach 66% by 2050. Megalopolises with population over ten million people are at particular risk in terms of stable food supply due to the complex logistics, high intensity of economic activity, cascade effects in case of failures in functioning of separate elements of infrastructure, critical dependence on food and agricultural products produced outside urban areas.

There are fears of increased food insecurity in different countries including developed ones. One of the reasons of hunger are losses of food and agricultural raw materials which is a complex problem worldwide. For example, in the EU and Russia, major factors of food losses are losses during cleaning (about 11%), consumption (10%), and processing and packaging (4%). Annual losses of vegetables and melon are about 3.5% of the total output. Particular, in 2017, Russia's domestic production vegetables, cucurbits, fruit, and berries totaled 18.6 million tons. At a stage of cleaning, 559.26 thousand tons were lost, not to mention processing and other stages. Reduction of losses is one of the factors of food security improvement and decrease in adverse environmental effects of agriculture and food processing.

Food and Agriculture Organization of the United Nations (FAO) focuses on the development of policies to reduce food losses and spoilage. The methodology elaborated by the FAO as part of a global initiative to reduce food losses and damage in food preservation is the basis of many reports developed to analyze critical points in food value chains and identify possible solutions and strategies to reduce food loss. Considerable losses of food during storage, transportation, and retail trade have different nature and different reasons in developing and developed countries. In developing world, major losses happen during harvesting, transportation, pre-processing, and storage of agricultural products and raw materials due to underdeveloped technologies, lack of availability of expensive modern equipment, as well as organizational issues. In the developed countries, significant losses (up to 30%) happen in retail trade and at end users. In a number of developed countries, governments respond to food losses problem by the implementation of large-scale program actions.

Another global challenge to food security is the problem of hidden hunger associated with the intensification of technological processes and the saturation of food market with refined food. The issue has been becoming common and increasingly relevant for both poorer and richer states. Hidden hunger causes chronic deficiency of vitamins and minerals in diet. Life in environmentally neglected urban zones requires higher consumption of vitamins. Modern technologies of food production do not promote preservation of the most valuable products. Decrease in volume of consumed food and its partial replacement by industrially developed foodstuff lead to the development of year-round deficiency of minor food components in diet. The effect has been registered in many countries worldwide. It may provoke development of a large number of alimentary and dependent metabolic disorders and diseases. The problem becomes even more adverse in the conditions of frigid climate, poverty, stress, location, and shortage of particular elements in soil and water. Hidden hunger has particular negative effect on pregnant and lactating women and children.

Indirectly, the problem of hidden hunger is influenced by the emergence of agro-holdings which strive to every intensification of agricultural production. This process leads to sharp polarization of rural people in employment opportunities and income level. It provokes structural unemployment in rural areas and deterioration in social status of rural dwellers. Rural unemployment is associated with increasing transaction expenses owing to territorial distribution of population and places of application of work; small amount, fragmentariness, and isolation of local labor markets; lateness in the expansion of consumer innovations in rural areas; and rather poor quality of social infrastructure. It leads to decrease in the level of social stability in rural areas and involves outflow of population to the cities.

Large-scale losses of food at storage, transportation, and retail trade, as well as hidden hunger problem, require the search of essentially new technological solutions of food production and processing to achieve the sustainable development and food security goals. This study aims at the development of resource-saving technology of dehydration of fruit and vegetable raw materials and economically expedient design solutions of processing of sub-standard raw materials in field conditions focused on the reduction of food losses during harvesting and storage.

BACKGROUND

In large urban zones, the issue of stable food supply can be solved by means of the development of infrastructure of agricultural production. In developing countries, such modern technologies as vertical farms and robotic greenhouse complexes are not widespread due to their high cost in the conditions of extensive development of agricultural sector and food processing industry.

Dynamic character of business models and technologies implemented in agricultural production and processing is one of the reasons of such social and economic problems as bankruptcy of agricultural enterprises, unemployment in rural areas, reduction of cropland, and degradation of rural infrastructure. Development strategies of large agricultural corporations often create conditions unfavorable for small farmers. Due to higher flexibility at the expense of application of advanced technologies (agrochemicals of new generation, genetically engineered and modified organisms, robots, among others), faster attraction, and geographical redistribution of capital, large multinational corporations are able to solve environmental, social, regulatory, and other problems in more effective manner. Therefore, in the aspiration to gain new market niches and reduce competitive pressure, they are able to operate without profit in the form of hidden dumping, carrying costs of the large-scale modernization programs for later periods, using low-interest rates, and state support whereas small and medium producers, particularly, individual farms, are forced to curtail activity without maintaining competitive pressure.

Technological gap between large agro holdings and small farms is particularly sharp because of low availability of credits to small agribusiness and high risks of investment. Other reason why smaller farms lose to holdings is a lack of advanced processing technologies. Operation of large agro holdings allow solving the problem of hunger, but not that of hidden hunger. The priority direction of development in processing of food raw materials allowing to solve hidden hunger problem is the development of new types of products. Due to their physical and chemical structure, such products are capable to fill the gap of the substances required for the maintenance of human health.

Bioflavonoids and other minor components of natural food are necessary for the increase in the efficiency of immune system of human body, decrease the impact of harmful effects of toxic components in a daily food allowance. Fruit, vegetables, and berries contain a complex of various biologically active agents, sources of vitamins C, P, and E, some vitamins of group B, V-carotene, a number of minerals, carbohydrates and phytoncides available in fresh only (Bessonov, Knyaginin, & Lipetskaya, 2017). Until recently, it has been believed that processing, including canning, allows preserving food substances. It is known that different ways of influencing a product do not allow achieving equal preservation of biologically valuable substances. For example, conservation of vegetables, fruit, and berries promotes only elimination of seasonality in their consumption and also allows supplying industrial centers and remote areas of the planet, for example, the Arctic.

The search for technological solutions of processing of fruit and vegetables without losing their nutritional value continues. There have been explored various ways of processing, including those by ionizing beams, radiation in inert gases, vacuum, low temperatures, antioxidants, conservation by currents of ultrahigh and microwave oven frequency, radiation by ultraviolet rays, and aseptic conservation. In Russia, the Ministry of Health approves various products processed by ionizing radiation. It is possible to apply ionizing radiation, short period of storage of fruit and vegetables which considerably depend on microorganisms. The extension of storage periods even for several days is important. For example, wild strawberry can be kept during 4-5 days in refrigerator and 10-12 days with the use of additional radiation. It is possible to extend storage period of irradiated sweet cherry and red tomatoes twofold. Essential lack of conservation by ionizing radiation is that during processing a product changes its chemical composition and hydrolysis processes. Enzymes are not inactivated which leads to the deterioration in taste, smell, and consistency. The amount of vitamins in comparison with initial phase decreased.

In recent years, much attention has been paid to the selection of the modes of radiation of foodstuff, not defiant changes of organoleptic properties. The most perspective way is radiation in inert gases, vacuum, at low temperatures, and with use of antioxidants. For the extension of storage period of potatoes and some of the vegetables, the allowed norms of ionizing radiation should not exceed 0.10-0.12, which allows complete suppression of germination in onions, garlic, and potatoes during storage. Such way of conservation, however, still has not been applied widely. Its impact on human health and degree of resistance of microorganisms to ionizing radiation has been studied comprehensively, the changes occurring in irradiated foodstuff has been investigated. In some countries, such processing is recognized as unsafe and its application is forbidden.

Conservation by currents of ultrahigh and ultrahigh (microwave oven) frequency is based on the creation of the movement of charged particles in a product. The temperature increases above 100°C. The product is corked in a tight container, placed in ultrahigh-frequency waves, and heated up to boiling during 30-50 seconds. Unlike thermal sterilization, microwave heating happens at the same time in all points, at one speed, warming up heat conductivity of a product does not influence. Microorganisms disappear much quicker than at thermal sterilization due to the oscillating motions of particles in cells. There happens not only allocation of heat but also polarization which influences vital signs of microorganisms. This method is widely used in fruit and vegetable processing for the sterilization of fruit, berries, and vegetable juices.

Radiation by ultraviolet rays (UVR) is a radiation by invisible part of light beams with the length of the waves of 60-400 nanometers. It affects microflora of food. The most effective action is posed by the beams with the length of the waves of 255-280 nanometers. Nucleonic acids and nucleoproteins in microbial cells adsorb UVR that leads to denaturation changes in these substances. Resistance of microorganisms to UVR varies, specifically, bacteria are more sensitive than mold. UVR is used for sterilization of a surface of meat hulks and sausages as their penetration does not exceed 0.1 mm. UVR can be used for sterilization of refrigerators and warehouses. However, this way of conservation requires implementation of tight security measures as UVR are dangerous to people, they affect eyes and skin.

The above-described ways of processing have not been widely used, therefore, the bulk of vegetables used for the production of tinned products and for the extension of seasonality of their processing are processed during summer and autumn in the form of fresh, dried, frozen, semi-finished products, natural purees and juices based on direct extraction, preserved in aseptic way, and concentrated juice and puree.

In aseptic conservation, short-term (dozens of seconds) sterilization of a product in a thin stream layer at high temperature is carried out in combination with rapid cooling, bottling in sterile conditions into a pre-sterilized container with a hermetic seal in aseptic conditions. It allows obtaining high-quality semi-finished products that practically do not differ in its consumer properties (color, taste, smell, vitamin composition, etc.) from non-sterilized one (Barkhatov, Lisitsky, & Kozachenko, 1993). The main advantage of aseptic conservation is the possibility of packaging in sterile conditions into sterile containers of any capacity and its hermetic sealing in aseptic conditions. It allows storing semi-finished product up to a year at a temperature between 0°C and 25°C without sudden fluctuations in temperature and relative humidity of 75% without loss and continue to use it for various types of canned products until new season (Barkhatov et al., 1993).

The most popular way to preserve fruits, vegetables, and berries is freezing. Today, the market of frozen vegetables and berries is one of the largest and fastest-growing segments of food market. In Russia, the share of this segment in food market is 16-17%, which is rather low compared to many developed countries. For example, in the USA, the share of frozen food market is 71%. In Russia, annual average per capita consumption of frozen vegetables is only one kg, while in the developed countries, this parameter reaches 4-6 kg. Russian market of frozen vegetables and berries has almost tripled during previous four years. In the 2000s, the growth rate of the market was about 15%, while in the past few years it has been ranging from 20-25% to 30-40% (Boltavin, 2006). The forecasts expect the continuation of dynamic growth of this market in the future (Vlahovich, 2008).

Freezing, however, is not the best way to preserve natural structure of food product. It partially breaks the intracellular structure of a product which negatively affects its nutritional value. Since there is a redistribution of moisture, tissues are injured by ice crystals. When freezing fruit, berries, and vegetables, it is almost impossible to achieve maximum reversibility of the phenomena. As water turns into ice, it leads to the compression of fibers and cells which causes additional water outflow and damages outer layers. Increased volume of central freezing layers leads to increased internal pressure in a product. Dense and inelastic outer ice layer is not able to withstand internal pressure and frozen product breaks. In addition to the loss of nutrients, freezing is an energy-intensive process, as it requires maintenance of low temperature during the entire storage cycle.

One of the most promising ways of preserving food products is dehydration which can dramatically reduce the cost of storage, transport, and losses and ensure long-term preservation of organoleptic characteristics of a product. Mass exchange and thermal processes are often accompanied by oxidation reactions, changes in structural and physical properties, formation of polymorphic forms and crystalline hydrates which leads to a partial or complete loss of nutritional value. One of the reasons is the omission of manufacturers and technologists of one of the most important properties of raw materials – thermolability, that is, the instability of essential substances under temperature changes. This leads to intracellular interactions during dehydration and loss of vitamin C, bioflavonoids, and aromatic substances. Loss of nutritional biological value may continue during storage of dehydrated product.

The study of dehydration process of freshly harvested crops should be based on the development of new ways to improve the efficiency of dehydration process using non-traditional energy sources and new types of heat generators. At any scale of drying technologies development, the implementation of a number of technical and economic parameters, such as minimum possible energy consumption, maximum homogeneity of dehydration, minimum time to reach a given humidity, and some other characteristics of dehydration, is considered fundamental (Table 1, Table 2).

Table 1. Comparative characteristics of various methods of dehydration

Type of dehydration	Principle of heat convection	Parameters of final product	Energy consumption, kW•h/kg
Infrared	Heat convection by IR rays	The product quality is as close as possible to the quality of freeze-drying. Up to 90% of the initial product properties are preserved and microbial contamination is reduced.	0.9 – 1.0
Sublimation	Removal of moisture in two stages: sublimation of ice from frozen product and heat finish drying in vacuum	The shape, color, and organoleptic properties are preserved with minimal losses of bioactive substances, the recoverability is 85-95%.	2.7 – 3.0
SHF	Heat generators are water dipoles contained in raw materials that are placed in a microwave electromagnetic field	Uniform heating, almost independent of thermal conductivity of drying material. The most promising combined drying: convection pre-drying and microwave finish drying. No specific effect of the microwave field on the product was found.	1.6 – 1.8
Convective	Heat transfer to raw material with drying agent (heated air or steam-gas mixture)	Reduced product thermal conductivity at the end of drying significantly lengthens the process, degrading the quality of the finished product. Achievement of stable quality is possible by correct cutting and blanching. Convective method produces 90% of dried products.	1.8 – 3.0

Source: Authors' development

The most effective and safest method of dehydration is IR exposure since it has an optimal dehydration time, minimum specific area of evaporating moisture, and product recoverability at the level of freeze-drying. In addition, this method is environmentally safe and allows obtaining products with a prolonged shelf life.

Chekrygina, Bukreev, and Eremin (2002) proposed a method for drying and disinfection of fruit and berries. It includes four stages: heating by low-frequency currents up to a temperature of 55-65°C until electrolytic disinfection and further drying with IR and microwave energy with a power flow density of

Table 2. Parameters of various methods of dehydration

Parameter	Methods			
	Infrared rays	Convective	Sublimation	Microwave heating
Time of dehydration, hours	4-6	8-10	10-20	up to 4
Specific area occupied by evaporated moisture m²/kg	0.04	0.07	0.26	0.18
Recoverability	85-95%	60-70%	85-95%	85-95%
Residual achieved humidity, %	3-4	8.0	3.5	2.5-4.0
Environmental safety of production	safe	safe	unsafe (freon)	unsafe (SHF)
Ability to store	more than 1 year	0.3 – 0.5 years	more than 1 year	more than 1 year

Source: Authors' development

not more than 0.2 W/cm² (stage 2), 0.3 W/cm² (stage 3), and 0.4 W/cm² (stage 4). The disadvantages of this method are long period of drying and multi-stage process.

Ivanov and Sapunov (2003) developed alternative method of drying of fruit and vegetables based on blanching of raw materials produced by microwave radiation with a capacity of 0.5-3.0 W/g and a pressure of 200-400 mm Hg at simultaneous centrifugation at a speed of 250-500 rpm. Drying is carried out by supplying microwave energy with a specific radiation power of 1.0-0.25 W/g at a pressure of 30-100 mm Hg. The disadvantages of this method are the inability to obtain dried berries of a certain shape due to significant compression of raw material during centrifugation. The essence of the method is that raw materials are sorted, washed, cut into pieces, blanched, placed on pallets, and placed in a chamber in which raw materials are heated up to 70°C in vacuum, the value of which is cyclically changed in the range of 0.00-0.04 MPa. The moisture obtained during drying is collected for its subsequent processing. The disadvantages of this method are long drying time and low-quality indicators of the final product.

Ermolaev, Fedorov, Sosnina, and Lifentseva (2015) proposed a method of vacuum drying of fruit and berries on the basis of pressure reduction in the chamber to 10-30 kPa during the first stage of dehydration. This leads to self-freezing of a product and sublimation of the resulting ice. After two hours, the chamber pressure is increased up to 3-5 kPa and infrared heating lamps are turned on maintaining a drying temperature of 70-80°C. This invention shortens vacuum drying process while maintaining high quality of dried product.

Antipov, Zhuravlev, Vinichenko, and Kazartsev (2015) offered a technological line for the production of dried berries and powder from them. Production line contains a consistently arranged truck, dump, scraper conveyor, washing machine, inspection conveyor, dryer, cooling chamber, pneumatic conveyor, cooling bin, sifter, screw conveyor, and filling machine. The line has additional installation: calibration vibrating table after dump, fumigation chamber after washing machine; shredder after dryer which is used as vacuum drying chamber. The authors argue that this invention will improve the quality of dried products, increase the shelf life of the finished product, and reduce their cost.

Analyzing the above methods, it should be noted that:

- Energy consumption of the above methods is quite high, especially when vacuum is used;
- None of the given methods can be offered for small businesses;
- The methods do not allow processing of substandard raw materials which are unstable in transportation.

Despite the variety of patents for dehydration of vegetables, fruit, and berries, most manufacturers use convective or combined heating of the working environment for dehydration. For example, ZHARKO SPb company produces drying chambers with convective and infrared heating and blowing. Among the advantages is rapid dehydration process, but it is likely to overheat a product in different layers and other defects.

One of the priorities is establishment and development of small and medium enterprises for production of dried fruit and vegetables. For low-power enterprises, the most promising method is IR dehydrogenation in modes that preserve native component by 80-90%. It allows solving food problems: concentration of minor components per kg of food substance increases, while cellulose, hemicellulose, and protopectin remain in native state thus providing a substrate for microorganisms. From a technical point of view, IR dehydrogenation allow solving the problem of high energy consumption inherent in other dehydration methods. The advantages include a small area occupied by the equipment as the heating elements are located parallel to the product under process.

MAIN FOCUS OF THE CHAPTER

In this study, the analysis of the assortment, functional ingredients, main processing methods, and ways of snacks processing recommended as healthy nutrition, different producers, and using mainly regional raw materials has been carried out (Table 3).

Natural sources of ingredients of monocomponent and multicomponent snacks demonstrating functional properties are vegetable raw materials (fruit, berries, nuts, and vegetables) which have food fibers, oligosaccharides, antioxidants, vitamins, and mineral substances. Vegetable raw materials include spinach, carrots, radish, beet, tomato, and pumpkin. Fruit snacks (chips) include oranges, apples, pears, and bananas. They are ranked as high-quality food products with high dietary and flavor properties and useful components (fructose, glucose, apple acid, cellulose, pectin, and iron). Among monocomponent and multicomponent snacks, it is necessary to mention Brainfood snacks which ingredients are different and made in the form of a mixture. Apple-banana-spinach snack is in the form of solid foam.

Fruit raw materials, vegetables, and berries come fresh and undergo traditional hydro-mechanical processing. Cutting is made in various ways, generally in the form of circles for fruit snacks and medium-size cubes for candied fruits. Berries are mainly dried. Sibirskiye Prostory company produces vegetable snacks in the form of straws. In other cases, vegetable chips are produced in the form of segments or slices. It means that such snacks find their application when cooking the first and second courses, namely, being specialized for junk foods. During mechanical action on raw materials, there is a loss of vitamin value. Therefore, it is necessary to consider that when cutting straws loss of nutrients is much higher. The loss is lower when cutting circles, segments, and drying. Therefore, snack mixture by Brainfood has positive sides as the nutritious quality of such snacks is higher compared to monocomponent snacks. It is up-to-date to produce snacks in the form of a mix as they are less commonly presented in the market.

Generally, various ways of drying, such as sublimation by Brainfoods, are applied to produce snacks. At the same time, drying is followed by the processes of warm transfer and mass transfer. Their intensity and depth have significant effect on chemical composition, structure, and physical and organoleptic properties of a product. Only Yablokoff disclosure all of its data, while other producers do not specify time and temperature of drying. The only information provided is that drying takes place at low temperature, presumably, 49°C. The type of sublimation drying of snack products has been specified by Brainfood. It means that data on production of snacks are not authentic. The existing mode of drying by Yablokoff makes it possible stating that at drying temperature above 77°C, duration of processing is reduced minimum to three hours. Other firms presumably use low temperature (minimum value of temperature of process equal to 49°C). At such temperature, standard value of mass fraction of moisture is reached in 20 hours of processing (Demidov, Voronenko, & Bazhanova, 2015; Doymaz, 2007).

Sublimation drying is a hi-tech process which allows keeping up to 95% of nutrients, vitamins, and microelements in their initial form, as well as preserving natural smell, taste, and color of a product. It is one of the most important advantages of sublimation. The way allows avoiding destruction of the structure of a product and restoring sublimated products as they have porous structure. This fact is remarkable because sublimated products are fully suitable for baby and dietary nutrition. However, energy consumption on the organization of sublimation process in vacuum exceeds the cost of thermal drying

Table 3. Comparison of the assortment, functional ingredients, and major processing methods

Product / group of products	Producer	Functional ingredients	Main processing methods	Processing regime
Oranges	Fruits, Saint Petersburg	Circle-shaped oranges	Traditional hydro mechanized processing (washing, cleaning, cutting), drying	Kind and parameters of drying are not stated
Apples		Circle-shaped apples		
Peas		Pea pieces		
Bananas		Banana in the form of oblong slices		
Chestnut –Cranberry – Pea snack	Brainfood, Moscow	Chestnut in the form of halves, pea in the form of segments, cranberry	Nuts' mechanized processing (cleaning from the nutshell) hydro mechanized processing (washing, cleaning, pea cutting), drying	Sublimation drying / Parameters of drying are not stated
Apple – Banana – Spinach snack		All components (apple – banana – spinach) in the form of homogenized mix / solid foam	Traditional hydro mechanized processing (washing, cleaning, apple cutting, crushing in the mashed mass, formation), drying	
Apple – Strawberry snack		Mix: apple in the form of oblong slices, strawberry	Traditional hydro mechanized processing (washing, cleaning, cutting for apple), drying	
Cherry snack		Cherry cut into circles	Traditional hydro mechanized processing (washing, cleaning, cutting), drying	
Apple chips	Yablokoff, Moscow	Circle-shaped apples and peas	Hydro mechanized processing (removal of a core, cutting into circles), drying	Kind of drying is not known / Drying parameters: temperature (60-80°C), drying time – 2.5-3.5 hours
Pea chips				
Carrot snack	C-Fruit Siberia, Omsk	Carrots in the form of straws	Hydro mechanized processing (washing, cleaning, cutting in the form of straw), drying	Drying under moderate temperature regime (according to producer's data), parameters are not stated
Horseradish snack		Horseradish in the form of straws		
Sugar beet snack		Beet in the form of straws		
Candied fruit "Cherry-taste beet"		Beet in the form of a cube and cherry (small cube)		
Candied cherry		Whole cherry		
Pumpkin snack	Ecofarmer, Krasnodar	Pumpkin in the form of slices	Hydro mechanized processing (washing, cleaning, trimming, extracting seeds out of pumpkin, cutting in slices), drying	Drying at low temperature, parameters are not stated
Tomato snack		Tomato in the form of segments	Hydro mechanized processing (washing, trimming, cutting in the form of segments), drying	

Source: Authors' development

Figure 1. Carbohydrate content in snacks
Source: Authors' development

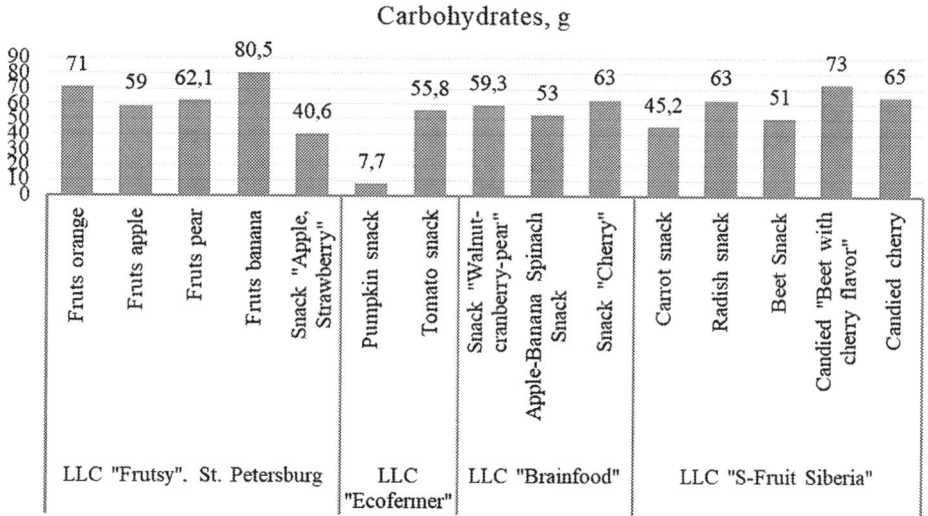

by 15-20 times. Besides, vacuum sublimation dryers are characterized by high cost, heavy operating costs, and complexity of service (Heredia, Barrera, & Andrés, 2007; Atanazevich, 2000).

In this study, the authors also conducted the analysis of nutrition value declared by producers. The analysis of data on protein content in snack products of different producers demonstrated that among vegetable snacks, more proteins are contained in tomato chips and horseradish and beet snacks. Among monocomponent, it is necessary to distinguish banana fruits from fruit snacks. Among multicomponent snacks, more proteins are contained in walnut snacks consisting cranberry and pear mixture. At the same time, it is necessary to understand that protein content is insignificant and fluctuates in the range from 1% to 9%. Content of fats in various snacks is unequal. Higher content of fats is registered in Brainfood multicomponent snack because walnut contains saturated and nonsaturated fatty acids – palmitic, linoleic, olein, linolenic (according to the producer) are contained in the mixture in the amount of 17.4%. In multicomponent snack (apple, banana, spinach) and tomato chips, fat content is below 2.4%. Other snacks except banana fruit contain minimum quantity of fat.

Most of the carbohydrates are found in banana fruits, candied beetroot with cherry flavor, and candied cherry. Comparing vegetable and fruit snacks, carbohydrates prevail in fruit snack foods due to the fact that fruit snacks contain predominantly glucose and fructose in contrast to vegetable snacks. Regarding multicomponent snacks, they are inferior to fruity in carbohydrate content since it depends on the characteristics of drying and the content of the component. Candied snacks predominate in carbohydrates due to sugars introduced into technological process during processing (Figure 1).

Brainfood multicomponent snacks are those with the highest calorie value as they contain walnuts, cranberries, pears, and bananas. Comparing the calorie value of vegetable snacks, it should be emphasized that candied beet with cherry flavor is higher-calorie product compared to Beet snack. Radish snack is one of the highest in calorie value while pumpkin snack is one of the lowest (Figure 2).

Yablokoff did not provide nutritional value of its snacks, only calorie content, but indicated that the products are obtained by infrared drying at a temperature of 60-80°C during 2.5-3.5 hours.

Figure 2. Caloric value of snack foods
Source: Authors' development

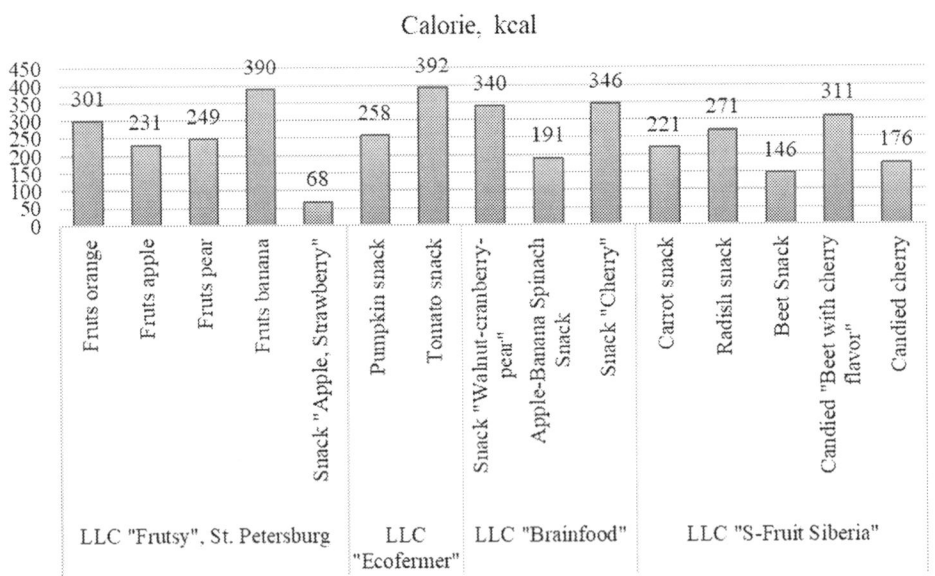

Protein content in snacks from vegetables and multicomponent snack foods is higher than that in fruit snacks. Multicomponent snacks and tomato chips have high content of fat. Fruit and vegetable snacks have high content of carbohydrates. Analysis of the calorie content demonstrated that multi-component and banana snacks are those with the highest caloric content. Other fruit snacks are low-calorie. At the same time, manufacturers probably do not take into account the digestibility of various carbohydrates. Simple sugars (glucose, fructose, and sucrose) are absorbed in the body, while complex sugars (cellulose and protopectin) are not. Therefore, there is a question if calorie calculation is correct. Some producers indicate the nutrient content values according to which moisture content in a product is 45%. It contradicts the data on dry vegetables and that is why producers misinform consumers about the true nutrient content.

Analysis of labeling data for snack products showed that product information meets the requirements for products intended for the mass segment, but not a functional product and/or a product for a healthy diet. Nutriciological analysis of the preservation of essential and minor components in industrial products offered for a healthy diet is not specified. It is not possible to draw conclusions about the replenishment of the need for necessary substances.

Such conclusions can be made on the basis of the processing modes used. However, the modes of the main process are not fully spelled out, for example, Fruits company has not provided information about drying. Sibirskiye Prostory and Ecofarmer companies also provide incomplete information indicating only that drying is conducted at low temperature. According to this information, it is impossible to understand whether all essential components are really preserved in a product.

Therefore, effective technologies are needed to ensure the safety of minor components in food products (flavonoids and their glycosides, indoles, exogenous peptides, organic acids, and phenolic compounds) with specific biological effects on various functions of individual metabolic systems and the body. One of the potential solutions to this problem is the creation of technologies to preserve natural properties of a product. Focusing on the FAO appeal and the global challenges to food security to reduce food loss and damage, the authors proposed a technological solution that allows processing fruit and vegetable raw materials at harvesting site, including substandard. The focus of the study is increase of the bioavailability of minor components in food by means of gentle processing methods, one of which is IR treatment (dehydrogenation). Not only the preservation of sensory, physical, and chemical properties is achieved but also the ability to increase the availability of substrate for intestinal microorganisms. The work was carried out at Saratov State Agrarian University and the Center for Collective Use (CCU) by scientific equipment in the field of physicochemical biology and nanobiotechnology "Symbiosis" of the Research Institute of Biochemistry and Physiology of Plants and Microorganisms of the Russian Academy of Sciences (IBFRM RAS). Fresh and dehydrated oranges were selected as the objects of the study. Dehydration was carried out by the method of resonant IR-drying at a wavelength of near and middle infrared range of 1.8-3.0 microns. The temperature was lowered from intense (67-75°C) to moderate (32-35°C). In resonant infrared drying, there was used Sator installment equipped with ceramic shell emitters. The study was conducted by comparing the data of the content of bioflavonoids and vitamin C in fresh oranges and dehydrated (immediately after dehydration and during storage for 12 months). The authors also studied a combined snack consisting of chopped tomato, onion, parsley, basil, and coriander, pressed and hydrated.

The analysis was performed by the method of reversed-phase HPLC on a DionexUltimate 3000 chromatograph (ThermoScientific, USA) using a Luna 5u C18 (2) 100A column, 5 μm 4.6 mm × 150 mm (Phenomenex, USA), serial number 125617-12. Components were identified by comparing the retention times of standard flavonoid samples (rutin as a hydrate (≥94%, Sigma-Aldrich, USA), quercetin as a dihydrate (97%, Alfa Aesar, UK), naringin (≥95%, Sigma-Aldrich, USA), apigenin (≥97%, Sigma-Aldrich, USA), and naringenin (≥95%, Sigma-Aldrich, USA).

Analysis of the composition of vitamins was performed by the method of reversed-phase HPLC on a DionexUltimate 3000 chromatograph (ThermoScientific, USA) using a Luna 5u C18 (2) 100A column, 5 μm 4.6 mm × 150 mm (Phenomenex, USA), serial number 125617-12. Analysis time – 15 minutes. The extracts were chromatographed under isocratic elution (Solvent A – methanol, qualification (Ultra) gradient HPLC grade (JTBaker, the Netherlands), solvent B – acetone nitrile qualification HPLC grade (Panreac, Spain), in the ratio 80:20. Speed 1 ml/min flow volume, sample volume 20 μl. Chromatograph control and data analysis was performed by Chromeleon version 7.1.2.1478 (ThermoScientific, Dionex, USA). The detection was performed at the following wavelengths: A, E – 265 nm.

Quantitative calculation of the content of vitamins was performed according to the ratio of the peak areas of the standard and sample.

Analysis of the composition of vitamins was performed by the method of reversed-phase HPLC on a DionexUltimate 3000 chromatograph (ThermoScientific, USA) using a Luna 5u C18 (2) 100A column, 5 μm 4.6 mm × 150 mm (Phenomenex, USA), serial number 125617-12. Analysis time – 25 minutes in a water-acetonitrile gradient.

SOLUTIONS AND RECOMMENDATIONS

A comparative analysis of the content of ascorbic acid (vitamin C) in freshly squeezed orange juice, the residue after extraction and dehydrated orange shows that this method of dehydration allows preserving vitamins (Figure 3). The presence of vitamin C 1.9269 mg per 1 g of dehydrated oranges was registered,

Figure 3. Vitamin C content, mg per g (for juice and residue after pressing mg per 10 g)
Source: Authors' development

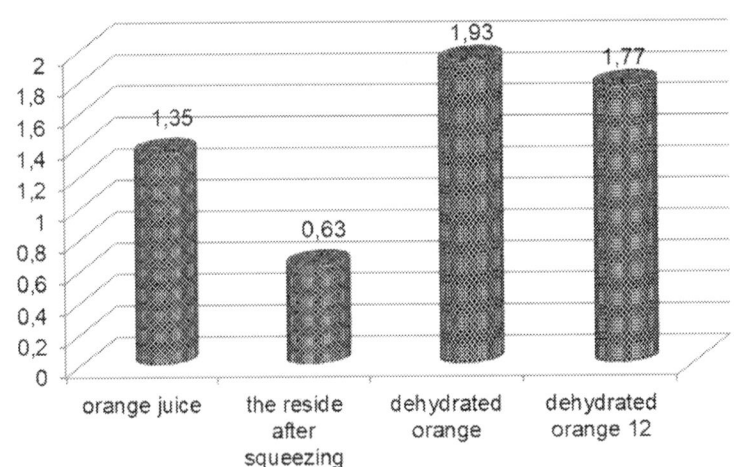

which is identical to the content of 10 g of fresh orange. The loss of vitamin C during storage of dehydrated orange for 12 months is 8%.

Analysis of chromatograms of orange juice, chips, and fresh orange residue after separation of juice shows that they all have a similar profile, but differ significantly in the content of certain components. A group of peaks corresponding to polyphenolic compounds, in particular, flavonoids, is observed in the chromatograms of all studied samples. Thus, the most representative flavonoid found in the extracts is pruning which is monoglycoside-naringenin-7-O-glucoside (Figure 4). Retention time (15.00 minutes) and UV-visible spectrum coincide with those of the component obtained as a result of partial acid hy-

Figure 4. Chromatogram of orange chips extract, integrating at a wavelength of 342 nm
Source: Authors' development

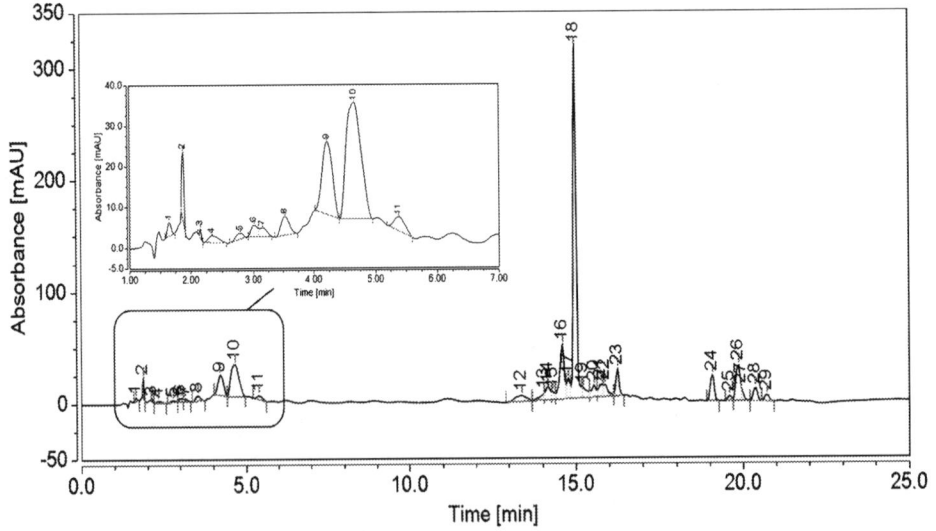

drolysis of naringin sample. The component has an almost identical absorption spectrum with prunin, however, it is characterized by a slightly shorter retention time (14.60 minutes) than the standard naringin sample (14.80 minutes). It suggests that this component is a polymer form of naringin with a low degree of polymerization (dimer, trimer, etc.).

In combined senecah confirmed the safety of chlorophyll (Figure 5). This minor component improves detoxification of the body by quickly removing waste and regulating the level of fluid. In addition, preliminary studies have shown the advantage of chlorophyll in accelerating metabolism which leads to weight loss.

Figure 5. Chromatogram of chlorophyll peak
Source: Authors' development

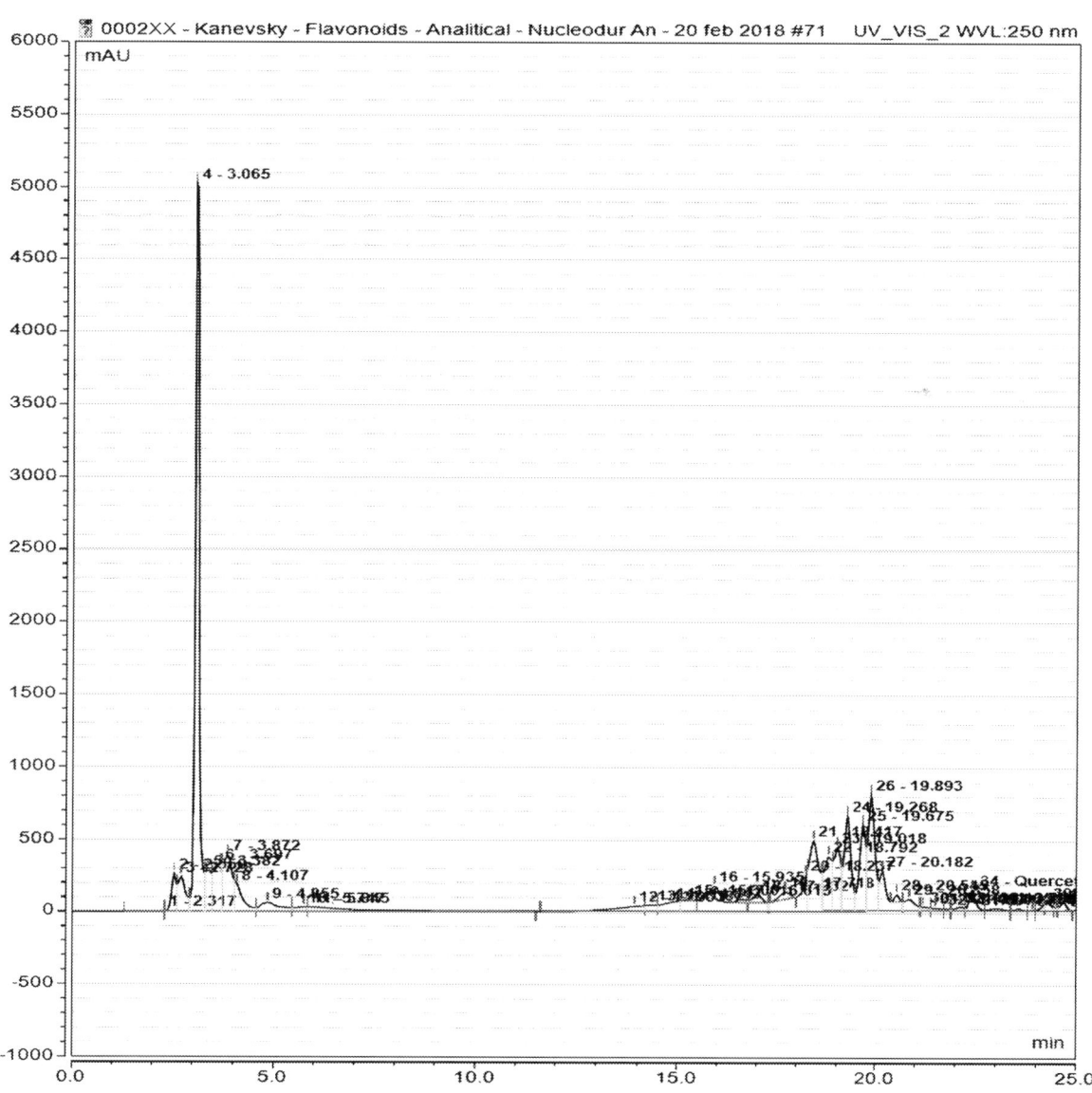

Table 4. Quercetin content in samples

Sample	Peak area, m unit*min	Quercetin content, mkg/g
Damp sample	3.132	6.10
Dry sample	27.619	44.86

Source: Authors' development

Table 5. Phenolic compounds content in samples

Sample	Total area peaks, m unit. * min	Phenolic connections content, mg/g (in terms of quercetin)
Damp sample	188.10	2.97
Dry sample	1,657.21	26.24

Source: Authors' development

In presented dry and raw (damp) samples, there is found a wide range of phenolic compounds, including flavonol quercetin. However, its content is low compared to other components. The chlorophyll peak is presumably most expressed. In the damp (raw) raw materials, extract concentration of compounds is lower (Table 4).

Most part of the extract is made by hydrophobic compounds. However, there is a hydrophilic fraction which can be represented by glycosides of flavonoids and free phenolic acids.

The quantitative analysis of content phenolic compounds in a sample (in terms of quercetin) showed the following results (Table 5).

The data on the prescription mixture consisting of tomato, onion, dill, parsley green, basil, and coriander after drying show that vitamin K remains and its content in the dried-up mixture is higher than in damp mixture on 1 g of raw materials. It is proved that the used energy effective and delicate technology of dehydration with the use of long-wave resonant IK-radiation allows keeping the vitamins and other biologically active agents which are contained in a product.

Most small farms and individual farms are not able to organize their own processing facilities. In addition, on-site collection of raw materials is always non-conforming raw materials unsuitable for the long-term storage, losing the quality of fresh raw materials during transportation. It is expensive to keep vegetable and fruit storage facilities while it is convenient to store the packaged dry product in a sealed package and does not require specially prepared premises. The idea is to create a mobile workshop, equipped with resonant IR dryers, collecting evaporated moisture in tanks and used for further washing of raw materials, leaves for the time of collection, thereby minimizing the loss of material and technical resources. A user only needs connect 380 V, water after washing can be used for watering. At the following stage of work, the calculation of the economic efficiency of the mobile workshop for drying of fruits and vegetables was carried out.

The annual output is equal to the product of the designed variable capacity by the number of shifts of the enterprise's work per year.

$$B_{year} = M \cdot K_{СТ} = 120 * 250 = 30{,}000 \text{ packs} \tag{1}$$

Annual production output will be 30 thousand packs. Annual commodity output is equal to the product of annual output at wholesale prices.

$$TP_{year} = B_{year} * Ц_{опт} \tag{2}$$

Annual commodity products when connected to a substation:
$TP_{year} = 30 * 46.87 = $ RUB 1,406.1 thousand.
Annual commodity products during the operation of the generator of current on solar oil:
$TP_{year} = 30 * 56.78 = $ RUB 1,703.4 thousand.

Cost of Establishing a Mobile Shop

Capital investments in the creation of production facilities consist of the cost of buildings' construction and equipment cost. The cost of buildings' construction of the main production purpose is calculated by multiplying their areas by the cost of building 1 m² of buildings.

Construction = 18 m² * RUB 25 thousand = RUB 450 thousand.

The cost of equipment specified in the specification is calculated at current wholesale prices for equipment.

Cost of equipment:
Drying chambers – RUB 3-900 thousand.
Vegetable cutting machine – RUB 1-100 thousand.
Other equipment (production table, table with washing tub, sealing machine, rack) – RUB 27 thousand.
Total: RUB 1,027 thousand.
K invest = 450 + 1,027 = RUB 1,477 thousand.

Depreciation (A)

Depreciation rates for equipment – 15%. Depreciation is calculated by producing the cost of equipment by the depreciation rate: 1,027.0 * 0.15 = RUB 154.05 thousand. Accrual of depreciation for the annual output: equipment depreciation (dehydrated product) – RUB 154.05 thousand.

Maintenance and Routine Repairs of Equipment (Po)

Deductions for equipment repair accept 20% of its value. The calculation is carried out similarly to the previous section. Deductions for equipment repair: equipment repair (dehydrated product) – RUB 205.4 thousand. The cost prices are calculated in accordance with costing items for the annual production volume.

Raw and Packaging Materials

Since the production of products is carried out from substandard horticultural raw materials, costs are not provided (Table 6).

Table 6. Calculation of the cost of packaging materials

Name	Quantity per shift	Price per unit, RUB	Amount, RUB	Amount per year, RUB thousand
Packaging	120	3	360	90.0

Source: Authors' development

Energy Resource (Ce)

The cost of 1 kW/h of electricity is taken as RUB 9.0 (Table 7, Table 8).

Wage

For the production, one worker per shift is required. Operation mode is single 8-hour shift, 250 shifts are assumed annually. Monthly salary of one employee is RUB 11,280. Annual salary: wages (dehydrated product) – RUB 135.36 thousand. Salary accruals (LF) (30% of salary): dehydrated product – 40.61.

Production Cost

Production cost (SPr) defined as the sum of all previous cost items. Other expenses take 5% of the amount of the previous cost items (Table 9).

Commercial expenses (Ce) take 5% of the product cost. Total cost (Tc) is defined as the sum of production costs and selling expenses.

Table 7. Electricity cost for the annual output when connected to the substation

Energy resource	Dehydrated product
Power consumption per shift, kW/h	60.0
Electricity consumption per year, thousand kW/h	15.0
Price 1 kW/h of electricity, RUB	9.0
Electricity cost, RUB thousand	135.0

Source: Authors' development

Table 8. Cost of energy for the annual production with the use of diesel fuel generator

Energy resource	Dehydrated product
Cost of diesel fuel in the shift, RUB	1,171.80
Cost of diesel fuel per year, RUB thousand	292.95

Source: Authors' development

Table 9. Total annual cost of production

Product	Operation of a substation, RUB thousand	Operation of diesel fuel generator, RUB thousand
Depreciation	154.05	154.05
Deductions for equipment repair	205.40	205.40
Cost of packaging materials	90.00	90.00
The cost of energy	135.00	292.95
Wages	135.36	135.36
Salary accruals	40.61	40.61
Other expenses	38.00	45.90
Production cost	798.40	964.30
Selling expenses	39.90	48.20
Total cost	838.30	1,012.50

Source: Authors' development

Product Pricing

The price is determined taking into account the normative profitability of products (40%) and VAT (20%). Dehydrated product when connected to the substation = 27.9 *·1.4 * 1.2 = RUB 46.87. Dehydrated product when operating on the diesel current generator = 33.8 *·1.4 * 1.2 = RUB 56.78.

Production Efficiency and Sales

The results of the calculations indicate that the effect of the production and sale of annual volume of dehydrated product when the current generator runs on diesel fuel is higher than when connected to the substation and amounted to RUB 690.9 thousand (profit from sales). The construction of a mobile plant for drying fruit and vegetables requires capital investments in the amount of RUB 1,477 thousand. The payback period of capital investments, respectively, with two options of work, is 3.3 years and 2.7 years. The profitability of capital investments is 30.8% and 37.4%, respectively. Efficiency of using fixed assets is higher at the second variant of the mobile production facility of RUB 1.15. This also applies to the indicator of efficiency of labor resources use (labor productivity) – RUB 1,703.4 thousand (Table 10).

At the same time, the parameters of economic efficiency (product profitability and sales profitability) are almost the same for both options of the mobile shop. These figures have a high enough value for the food industry, therefore, production and sales of products are effective from an economic point of view.

FUTURE RESEARCH DIRECTIONS

The proposed technology is simple and allows solving the following tasks on a global scale:

- Obtaining a universal sustainable model for processing of fruit and vegetable raw materials and large-scale reduction of losses at the stages of harvesting and storage.

Table 10. Efficiency of production and sales of products at estimated selling prices

Parameters	Operation of a substation	Operation of diesel fuel generator
Selling price of 1 product packaging, RUB	46.87	56.78
Production and sales per year, thousand packs	30.00	30.00
Revenue from sales for the year, RUB thousand	1,406.10	1,703.40
Total cost of 1 product package, RUB	27.90	33.80
Total cost of annual production, RUB thousand	838.30	1,012.50
Profit from sales, RUB thousand	567.80	690.90
Net profit, RUB thousand	454.20	552.70
Product profitability, %	67.70	68.20
Return on sales, %	40.40	40.60
Capital productivity, RUB	0.95	1.15
Capital intensity, RUB	1.05	0.87
Labor productivity, RUB thousand	1,406.10	1,703.40
Return on capital investments, %	30.80	37.40
Payback period of capital investments, years	3.30	2.70

Source: Authors' development

- Solution to the problems of logistics in promoting the product from producer to consumer, including the storage of raw materials.
- Obtaining products that can be used as an independent product, or as an integral component of food systems.
- Ensuring the safety of the beneficial properties of vegetables and fruit, solving the problem of hidden hunger.

Usage of IR-dehydrogenation allows keeping native components in vegetables and fruit at a level of 80-90% which, in turn, provides concentration of minor components per kg of food substance, allows cellulose, hemicellulose, and protopectin remaining in native state, and thus providing a substrate for microorganisms. Besides, it allows solving complex problems of insufficient consumption of fresh fruit and vegetables and preservation of nutrients during long storage period. Newly developed products are subjected to minimal temperature and time exposure, IR dehydrogenation does not cause a significant effect on the cellular wall of fruits, vegetables, and berries, and thus the technology allows obtaining products of a combined oncology protective and antioxidant action.

Although the production technology with preserved minor components is simple, it requires further studies. It consists of a standard hydro-mechanical treatment of washing, cleaning, and grinding of raw materials. Dehydration is conducted by means of IK resonant drying without compulsory convection. Resonance allows preventing the interaction of molecules of osmotically-bound and physic-mechanically bound moisture prevailing in fruit raw materials with substances in an intact cellular structure.

It is on the basis of these considerations that the most intensive influence of temperature 65-70°C happens in the first hour and allows removing the surface moisture by intensive evaporation. In this case, the main moisture remains in excess. Then the temperature decreases to 45-50°C and most of the moisture removal occurs in this range of 2-3 hours, the last stage of drying occurs at a temperature of 35-38°C, and the remainder of the osmotically retained water is removed. The final moisture of the product is 8-9%. The dehydrated product is cooled in air to 20°C and packed in a sealed package.

The given technologies are applied at different periods of harvesting and a farmer can hand over surplus and non-conforming raw materials for processing or use the opportunity to recycle self-contained for further realization.

CONCLUSION

The authors have proposed a universal technical and economic model of processing of fruit and vegetable raw materials in field conditions. It assumes creating a mobile workshop equipped with resonant IR dryers with devices for collecting evaporated moisture used for washing raw materials. This model is expected to reduce the loss of fruit and vegetables at the stages of harvesting and storage and provide the population with the products with prolonged shelf life and preserved profile of minor components to solve the problem of hidden hunger. The calculation of economic efficiency has shown that there is a potential associated with:

- the possibility of saving water resources, due to the collection of evaporated moisture 96%;
- saving energy consumption (for IR-drying is 3-4 times less energy than for other methods of drying);
- processing of fruit and vegetable raw materials on the spot before receiving the product allowing to get 30% of additional profit.

The production of food products with a prolonged implementation period from substandard and perishable raw materials allows solving the problem of logistics in promoting the product from producer to consumer, including storing raw materials, which also solves the problem of reducing losses and damage to food raw materials, and ensures sustainable development of agricultural production and improvement of food security.

REFERENCES

Antipov, S., Zhuravlev, A., Vinichenko, S., & Kazartsev, D. (2015). *Patent #2548209 "Vacuum Dryer of Continuous Operation with Ultra-High Frequency Power Supply."* Retrieved from https://findpatent.ru/patent/254/2548209.html

Atanazevich, V. (2000). *Drying Food. Reference Manual*. Moscow: DeLi.

Barkhatov, V., Lisitsky, V., & Kozachenko, Z. (1993). Aseptic Preservation of Fruit and Vegetable Puree of Semi-Finished Products from Low-Acid Raw Materials. *News of Institutes of Higher Education. Food Technology, 3-4*, 51–52.

Bessonov, V., Knyaginin, V., & Lipetskaya, M. (Eds.). (2017). *Nutritiology-2040. The Horizons of Science through the Eyes of Scientists*. Saint Petersburg: Center for Strategic Research.

Boltavin, A. (2006). Market for Frozen Convenience Foods from Vegetables. *Ice Cream and Frozen Products, 5*, 16–18.

Chekrygina, I., Bukreev, V., & Eremin, A. (2002). *Patent #2194228 "Method of Drying and Disinfection of Fruits and Berries."* Retrieved from http://www.freepatent.ru/patents/2194228

Demidov, S., Voronenko, B., & Bazhanova, I. (2015). Kinetics of Infrared Drying Shredded Carrots. *Scientific Journal NRU ITMO. Series. Processes and Food Production Equipment, 3*, 158–163.

Doymaz, I. (2007). Air-Drying Characteristics of Tomatoes. *Journal of Food Engineering, 78*(4), 1291–1297. doi:10.1016/j.jfoodeng.2005.12.047

Ermolaev, V., Fedorov, D., Sosnina, O., & Lifentseva, L. (2015). *Patent #2541395 "Method for Vacuum Drying of Fruit and Berries."* Retrieved from http://www.freepatent.ru/patents/2541395

Heredia, A., Barrera, C., & Andrés, A. (2007). Drying of Cherry Tomato by a Combination of Different Dehydration Techniques. Comparison of Kinetics and Other Related Properties. *Journal of Food Engineering, 80*(1), 111–118. doi:10.1016/j.jfoodeng.2006.04.056

Ivanov, V., & Sapunov, G. (2003). *Patent #2195824 "Method of Drying of Fruits and Vegetables."* Retrieved August 11, 2019, from http://allpatents.ru/patent/2195824.html

Vlahovich, S. (2008). Frozen Food Market Today and Forecasts of Its Development for the Future. *Ice-Cream and Frozen Products, 12*, 28–31.

ADDITIONAL READING

Golovkin, N. (1984). *Refrigeration Food Technology*. Moscow: Consumer Goods and Food Industry.

Kutsakova, V., Rogov, I., Frolov, S., & Filippov, V. (2001). *Examples and Tasks of Food Technology Refrigeration. Part 1: Theoretical Basis of Canning*. Moscow: Kolos.

Makarova, N., & Zyuzina, A. (2011). Investigation of Antioxidant Activity of Juice Production Semis with DPPH Method. *Food Processing: Techniques and Technology, 22*(3), 1–5.

Nariniyants, G., Pacyuk, L., Kostromina, N., Lukashevich, O., & Medvedeva, E. (2005). Technology of Aseptic Canning of Fruit Convenience Foods for Baby Food. *Food Industries, 3*, 20–21.

Roberfroid, M. B. (1999). Functional Foods and the Intestine: Concepts, Strategies and Examples. In L.A. Hanson & R.H. Yolken (Eds.), Probiotics, Other Nutritional Factors, and Intestinal Microflora (pp. 203-216). Lippincott-Raven, PA: Lippincott-Raven Publishers.

Sacilik, K., Keskin, R., & Elicin, A. K. (2006). Mathematical Modelling of Solar Tunnel Drying of Thin Layer Organic Tomato. *Journal of Food Engineering, 73*(3), 231–238. doi:10.1016/j.jfoodeng.2005.01.025

Zholik, G., & Kozlov, N. (2004). *Technology of Processing of Plant Raw Material: A Tutorial*. Gorki. Belarus State Agricultural Academy.

KEY WORDS AND DEFINITIONS

Bioactive Substances: The chemicals that have high physiological activity at low concentrations in relation to certain groups of living organisms or to certain groups of their cells.

Dehydration of Vegetables or Fruits: Maximum removal of moisture from the product, including osmotic moisture.

Economic Efficiency: A ratio between the obtained results of production – products and services, on the one hand, and the expenditures of labor and means of production, on the other.

Material and Technical Resources: A set of objects of labor (raw materials, materials, fuel, etc.) and tools (machines and equipment), processing objects of labor.

Minor Food Components: Natural food components of an established chemical structure (vitamin-like compounds, some minerals, indole compounds, flavonoids, isoflavones, phytosterols, etc.) present in food in milligrams and micrograms, which play an important and proven role in the adaptation reactions of the body and maintaining health, but not essential nutrients.

Mobile Workshop: A mobile plant processing plant raw materials, including substandard, working independently.

Quercetin: A natural biochemical substance of the flavonoid group, a strong antioxidant.

Section 4
Regional Aspects of Food Security and Trade in Agricultural Products

Chapter 16
Emerging Trade-Related Threats to Food Security:
Evidence From China

Vasilii Erokhin
https://orcid.org/0000-0002-3745-5469
Harbin Engineering University, China

ABSTRACT

It is generally believed that free trade plays a vital role in stabilizing food supplies and food prices since abundant foods stocks in some countries coexist with shortages in some others. Contemporary global trade system, however, is becoming increasingly distorted by unfair and inefficient policies in many countries, creating both winners and losers among not only small developing economies, but also largest producers of food and agricultural products. One of the recent examples of such distortion is US-China trade tensions and potential tariff escalations where the agricultural sector is the most vulnerable. By raising import tariffs on food and agricultural products in response to protectionist policies, the countries may face a situation of rising prices for consumers, limited market access for producers, and increasing pressures on food security. In this chapter, the author develops the theme of the effects of globalized agricultural trade on food security with a critical focus on the importance of balancing trade liberalization and protectionism.

INTRODUCTION

The majority of scholars agree that liberalized trade plays a vital role in stabilizing food supplies and food prices since abundant foods stocks in some countries coexist with shortages in some others (Erokhin, 2017, 2018). Unforeseen price surges can push millions of people in developing countries into poverty, aggravating income inequalities and threatening social cohesion. Malnutrition can result in maternal mortality and stunted growth of children, reducing their learning ability and lowering their productivity in adulthood. Government are pursuing various policy options for ensuring food security: expanding investment in agriculture; encouraging climate-friendly technology; restoring degraded farmland; improving

DOI: 10.4018/978-1-7998-1042-1.ch016

post-harvest storage and supply chains; and even indulging in the promotion of niche products. They face special challenges as their growing middle class shifts from traditional staples to more nutritious products such as meat, fish and dairy whose higher resource intensity requires expansion of domestic agricultural capacity or greater reliance on imports. But importing countries also worry about unreliability of world markets in times of need and promote self-sufficiency in food along with trade policy restrictions.

Liberalization of international trade in food and agricultural products can affect the availability of certain foods by removal of barriers to imports, but also to foreign investments in the development of domestic production (Erokhin & Ivolga, 2012; Erokhin, Ivolga, & Heijman, 2014). Thow and Hawkes (2009) found that availability of processed food has risen in developing countries after foreign direct investment by multinational food companies. Thus, changes in trade policies have facilitated the rising availability and consumption of foods. Dorosh (2004) found the positive contribution of trade liberalization to food security and stabilization of supply and food market. Protection policies, including administrative restrictions of food imports with a view to increasing food self-sufficiency of the country, on the contrary, may cause unnecessary social costs and place food self-sufficiency into conflict with the goals of food security and poverty reduction. Warr (2011) studied the effects of an import ban on rice introduced in Indonesia and concluded that it let to achieve the required level of food self-sufficiency, but at the expense of reducing food security.

Despite a generally accepted belief that free trade plays a vital role in stabilizing food supplies and food prices since abundant foods stocks in some countries coexist with shortages in some others, liberalization of international trade has become a significant source of tension in contemporary agricultural change (Lee, 2007). In the conditions of globalization, where liberalization of food trade and the reduction of administrative protection of food producers are mandated by the rules of the World Trade Organization (WTO), many countries have lost a part of their sovereignty over food policies (Lawrence & McMichael, 2012). Some of them (primarily, developing ones) have become food dependent, others managed to benefit from easier access to foreign markets and unified framework of global trade in food. In general, globalization has refocused attention from trade-based food self-sufficiency to availability-based food security. According to Ghosh and Ghoshal (2017), structural changes in global food market have put in place concerns and challenges related to food prices, stability of food supply, and trade-economic aspects of food accessibility and affordability.

This study of trade-related threats to food security is focused on the specific case of China. Special concerns of China's food security are based on the facts that rural people still earn over a half of their income from farming, share of food in total spending is high, and food price is one of the critical elements of consumer price index. The core to understand China's food security policy is controllability perceived by policymakers in three aspects. First is to avoid over-dependence on foreign trade by means of the development of domestic production and stockholding and regulations on domestic utilization of staple foods. Second is to avoid external-induced shocks to domestic food market by practicing domestic food price and border interventions and food stock measures. Third is to ensure social stability by means of preventing of combating income disparity, preventing volatility of food prices, and targeting support policies to rural and urban poor. As China's agriculture is an integral element of globalized food market, food and trade policies of China and other market actors affect food prices bringing about price volatility and food security concerns since many people still spend a substantial part of their income on food and related goods. This chapter overviews China's foreign trade in agricultural commodities and food and assesses possible effects of rising protectionism in global trade in food.

Emerging Trade-Related Threats to Food Security

BACKGROUND

Before the economic reforms of 1980s, China's foreign trade had been restricted down to exporting surplus raw materials and simple manufactured goods to cover payments for imported goods (Sun & Heshmati, 2010). Since the start of economic reforms, domestic agricultural sector has achieved remarkable growth in terms of production of major crops, vegetables, fruits, meat, and aquatic products.

In the previous decades, there has occurred the substantial increase in productivity in agricultural sector. However, taking into account scarce land resources coupled with the world's biggest population, China remains heavily dependent on agricultural imports. With the integration of China's agriculture into the global economy and accession to the WTO, agricultural production and trade are now facing competition from home and abroad (Panitchpakdi & Clifford, 2002; Cheng, 2007). The growth of import is conditioned by the restructuring of domestic demand in favor of qualitative food products of high nutrient value amid the progressive degradation of agricultural lands and shrinking possibilities for agricultural production in China. Zhou (2010) records growth of consumption of food products of higher quality, nutrient value and price in China (meat and meat products, milk and dairy products, and seafood) – to the prejudice of cheaper and fewer nutrient crops. Consequently, Chinese meat and dairy producers demand more crops as fodder for agricultural animals, which increases the load on agricultural lands and accelerates their degradation (Liu & Diamond, 2005). Zhang and Chen (2016) identify four key pressures on China's agricultural sector, namely, widening gap between demand and supply on the domestic food market, emerging threat to sustainability of China's agricultural sector due to degradation of limited arable land and heavy use of fertilizers, prevalence of cash crops production and consequent distortion of domestic agricultural market and income growth for farmers, and expansion of overseas operations of large state-owned agricultural as domestic competition intensified.

The country heavily depends on food imports being the world's top importer of soybeans, cotton, palm oil, and sugar. About 80% of consumed soybean and other agricultural products, such as milk and sugar, are imported to China. Soybean imports significantly increased from 0.3 million tons in 1995 to 95 million tons in 2017, presently accounting for two-thirds of the world's soybean market (Cui & Shoemaker, 2018). China is one the world's biggest importer of palm oil, sugar, meat, fish, fruits, and feeding stuff for animals. Supply and demand are roughly in balance for grains such as rice, corn, and wheat. The World Bank (2018) forecasts that due to positive population growth coupled with growing incomes and changing diets in China, the demand for oilseeds, meats, milk, and sugar will continue to increase. According to Nath et al. (2015), by 2030, China will need to increase agricultural lands under crops by 21% to be able to meet the growing internal demand for food. According to Gao (2017), China's demand for cereals will rise up to 700 million tons in 2020, including imports of 100 million tons. By 2022, China will double import of feed grain, beef, and pork, as well as increase purchases of soybeans by 40%. Huang, Wei, Cui, & Xie (2017) predict China's overall food self-sufficiency to reduce to 91% by 2025. Economic development along with progressing urbanization and transformation of food consumption patterns are likely to bring increased demand for major crops, including corn, rice, soybeans, and wheat (Gro Intelligence, 2018). Particularly, by the 2030s, the demand for corn is projected to rise up to 300 million tons per year (currently, it is about 250 million tons), for soybeans – up to 180-190 million tons (about 100 million, respectively).

In view of expected rise in consumption of food, Zhu (2016); Yu, Elleby, and Zobbe (2015), Yu, Feng, Hubacek, and Sun (2016), and Zhou (2010) consider high dependence of China on imports as a potential threat to food security. Zhang and Chen (2016) show that China's grain imports have continued to grow in spite of increased domestic grain production in the 2010s. In 2004-2014, grain imports nearly tripled, and imports of rice, wheat, and corn increased eightfold. Zhang (2016) points out that large-scale agricultural imports, particularly, soybeans from the USA, has almost destroyed domestic soybean production sector causing China to lose control of its supply chain to foreign producers and suppliers. In a situation of expanding trade tensions between the USA and China and, potentially, between China and other countries, such over-reliance on foreign food supply equips China's counterparts with a power to influence on China's trade policies by pressing its food market and threatening food security of its population.

MAIN FOCUS OF THE CHAPTER

Agricultural Sector

By now China has achieved self-sufficiency on major kinds of food and agricultural products on the level of about 91-97% of domestic consumption (Mahendra Dev & Zhong, 2015; Kravchenko & Sergeeva, 2014). Agricultural sector has been developing rapidly. The growth has been affected by many factors, the most important of which have been institutional innovations in rural areas, technological breakthrough, market reforms, and agricultural investments. Institutional reforms started in 1978 with the introduction of responsibility system, according to which land plots were attached to particular rural households depending on the number of workers. At the start of the reform, such solution allowed substantial increasing of labor productivity by 40-50% (Fan, 1997). In the late 1980s, introduction of land contracts and land tenure rights both encouraged agricultural investments.

Due to the scarcity of fertile land, agricultural production has been intensifying. There has been developed a system of agricultural science, technologies, and innovations. China has established the world's largest network of information consulting offices in rural areas (Huang & Yang, 2017). Government invested in the development of irrigation and flood prevention, construction of transport infrastructure in rural areas, and development of distribution infrastructure. All those efforts allowed involving small farms into network relationship with food processing enterprises, distributors, and consumers. Along with the expansion of irrigation, investments in the improvement of soil fertility allowed increasing productivity. Yields have increased steadily for all major crops, including rice, corn, wheat, and soybeans (Park, Jin, Rozelle, & Huang, 2002).

Significant portions of China's yield gains have been obtained by means of overuse of pesticides and fertilizers, which, in turn, caused problems with water and air pollution. The volume of contaminated soil in China is increasing while the environmental remediation industry is still in its infancy (Kennedy, Zhong, & Corfee-Morlot, 2016). The yield of soybean is lower than that of other major crops, about 1/3 of wheat and 1/4 of rice and corn. With China's shift to importing rather than growing soybean, a total of 50 million hectares of fertile cropland (40% of China's total arable land) freed up for growing other higher yielding crops (Cui & Shoemaker, 2018). A common way to boost domestic agricultural production is to expand farm acreage. China's total farm area has increased since the 1960s, particularly, for corn, but currently, only 12.8% of the total national land area is available for agricultural production (Chen, 2007).

Despite the overall growth, China's land and water issues remain the primary constraints to the expansion of agricultural production. Pressures from increased urbanization has prevented expansion in the arable area, and competition for land is high. China's agricultural sector is dominated by small households (Ma, 2011), which provide the bulk of domestic product in agriculture. Land area per capita is only 0.08 ha, which is far below the world average of 0.22 ha per capita (Nath et al., 2015). As of 2016, the number of households is above 200 million, and the average size of a household is 0.66 ha (264 ha in the USA and 37 ha in the EU, respectively) (Gao, Ivolga, & Erokhin, 2018).

Foreign Trade

Over the past decade, the total agricultural trade deficit has increased significantly to $40 billion. Today, China is one of the major global producers of agricultural products. Despite the world's biggest population of over 1.3 billion people, limited agricultural resources, and domination of smallholders in

Table 1. China's agricultural export in 1995-2017, $ billion

Product	1995	2000	2005	2010	2015	2016	2017
Live animals	0.50	0.38	0.33	0.45	0.60	0.65	0.56
Meat and edible meat offal	0.99	0.72	0.71	0.89	1.01	0.86	0.90
Prepared and preserved edible meat offal	0.33	0.49	1.19	1.46	1.76	1.65	1.87
Milk, cream, and milk products	0.03	0.05	0.10	0.06	0.06	0.08	0.09
Eggs and eggs' yolks	0.03	0.03	0.08	0.14	0.20	0.19	0.19
Fresh, chilled, or frozen fish	0.91	1.31	3.02	5.74	7.53	7.74	7.75
Dried, salted, or smoked fish	0.13	0.11	0.21	0.35	0.47	0.49	0.51
Crustaceans, mollusks, and aquatic invertebrates	1.02	0.84	1.12	2.70	5.32	5.48	4.47
Fish, prepared and preserved aquatic invertebrates	0.79	1.39	3.18	4.40	6.25	6.29	7.68
Rice	0.02	0.56	0.22	0.42	0.27	0.38	0.60
Maize	0.01	1.05	1.10	0.03	0.01	0.01	0.02
Cereals, excluding wheat, rice, barley, and maize	0.04	0.03	0.05	0.09	0.05	0.04	0.05
Meal and flour of wheat and flour of meslin	0.06	0.05	0.09	0.12	0.07	0.06	0.08
Cereal preparations, flour of fruits or vegetables	0.15	0.12	0.32	0.67	0.89	0.90	0.90
Vegetables	1.32	1.27	2.56	6.03	6.81	7.80	7.61
Prepared and preserved vegetables, roots, tubers	1.20	1.02	2.18	4.48	6.65	7.10	7.87
Fruits and nuts	0.44	0.35	0.91	2.41	4.88	5.21	5.06
Preserved fruits and fruit preparations	0.34	0.53	1.15	2.21	2.93	2.87	4.05
Sugar, molasses, and honey	0.29	0.18	0.28	0.73	1.01	1.14	1.16
Confectionery sugar	0.09	0.14	0.34	0.68	1.08	1.09	1.13
Coffee and coffee substitutes	0.01	0.02	0.05	0.14	0.45	0.95	0.43
Cocoa	0.03	0.02	0.06	0.10	0.11	0.09	0.06
Chocolate and food preparations with cocoa	0.01	0.01	0.05	0.11	0.33	0.34	0.32
Tea and mate	0.28	0.36	0.50	0.83	1.49	1.60	1.73
Spices	0.19	0.15	0.42	0.77	0.93	0.97	1.08
Feeding stuff for animals	0.35	0.30	0.50	1.98	2.68	2.77	2.66

Source: United Nations Conference on Trade and Development [UNCTAD], 2019

agriculture, China is largely self-sufficient in food, except few agricultural commodities. With the fast development of foreign trade, trade in agricultural products has also grown rapidly. China's major export agricultural products are vegetables (fresh, prepared, and preserved), fish (fresh, chilled, frozen, dried, etc.), fruits (fresh, preserved, and prepared), crustaceans, mollusks, and aquatic invertebrates (Erokhin & Gao, 2018) (Table 1).

Table 2. China's agricultural import in 1995-2017, $ billion

Product	1995	2000	2005	2010	2015	2016	2017
Live animals	0.04	0.05	0.11	0.27	0.55	0.39	0.36
Meat of bovine animals, fresh, chilled or frozen	0.01	0.01	0.01	0.08	2.32	2.52	3.07
Meat and edible meat offal	0.09	0.64	0.61	2.18	4.45	7.60	6.38
Milk, cream, and milk products	0.06	0.21	0.41	1.80	2.63	2.71	3.89
Butter and other fats and oils derived from milk	0.00	0.01	0.03	0.09	0.27	0.30	0.50
Cheese and curd	0.00	0.00	0.03	0.11	0.35	0.42	0.50
Fresh, chilled, or frozen fish	0.39	0.77	2.29	3.32	3.61	3.84	4.43
Dried, salted, or smoked fish	0.02	0.04	0.03	0.02	0.02	0.03	0.06
Crustaceans, mollusks, and aquatic invertebrates	0.18	0.39	0.52	0.98	2.70	3.06	1.80
Fish, prepared and preserved aquatic invertebrates	0.01	0.01	0.02	0.09	0.23	0.17	2.00
Wheat and meslin	2.03	0.15	0.76	0.31	0.89	0.80	1.03
Rice	0.43	0.11	0.20	0.25	1.47	1.59	1.83
Barley	0.24	0.31	0.43	0.54	2.86	1.14	1.82
Maize	0.82	0.00	0.00	0.37	1.11	0.64	0.60
Cereals, excluding wheat, rice, barley, and maize	0.06	0.00	0.00	0.04	3.02	1.49	1.12
Meal and flour of wheat and flour of meslin	0.01	0.01	0.01	0.01	0.02	0.01	0.05
Cereal preparations, flour of fruits or vegetables	0.02	0.02	0.06	0.22	0.86	0.98	1.03
Vegetables	0.08	0.09	0.53	1.56	2.72	1.95	2.12
Prepared and preserved vegetables, roots, tubers	0.01	0.04	0.06	0.12	0.28	0.27	0.25
Fruits and nuts	0.08	0.37	0.63	2.06	5.87	5.72	6.24
Preserved fruits and fruit preparations	0.00	0.01	0.07	0.27	0.58	0.66	0.84
Sugar, molasses, and honey	0.91	0.15	0.42	0.99	1.98	1.35	1.31
Confectionery sugar	0.03	0.03	0.03	0.06	0.20	0.20	0.21
Coffee and coffee substitutes	0.01	0.01	0.03	0.10	0.46	0.90	0.45
Cocoa	0.05	0.04	0.11	0.28	0.36	0.35	0.31
Chocolate and food preparations with cocoa	0.01	0.03	0.06	0.16	0.52	0.34	0.35
Tea and mate	0.00	0.00	0.01	0.06	0.12	0.12	0.17
Spices	0.01	0.01	0.01	0.03	0.06	0.06	0.06
Feeding stuff for animals	0.42	0.91	1.31	3.30	4.97	3.61	3.93
Margarine and shortening	0.01	0.11	0.02	0.07	0.29	0.31	0.42

Source: UNCTAD, 2019

China has succeeded to ensure self-sufficiency on major kinds of food and agricultural products. Nevertheless, the country still depends on food imports being the world's top importer of soybeans, cotton, palm oil, and sugar. About 80% of consumed soybean and other agricultural products, such as milk and sugar, are imported to China. Soybean imports significantly increased from 0.3 million tons in 1995 to 95 million tons in 2017, presently accounting for two-thirds of the world's soybean market (Cui & Shoemaker, 2018). China is one the world's biggest importer of palm oil, sugar, meat, fish, fruits, and feeding stuff for animals (Table 2).

China-USA

China's largest supplier of agricultural products is the USA which accounted for about 20% of the value of China's agricultural imports during 2015-2017. In 1995-2017, the combination of top five suppliers of agricultural products to China is rather stable. In 2017, however, Brazil has overtaken the USA to become the leading partner of China in agricultural trade (21.05% of China's total agricultural import in 2017, or $28.88 billion) (Spring, 2018; Spring & Polansek, 2018). The share of the USA in China's agricultural import declined from 21.66% in 2016 ($21.58 billion) to 19.09% ($21.66 billion) in 2017.

Trade conflicts between the USA and China have escalated recently. China has threatened to impose a 25% tariff on 128 products, including agricultural commodities and food products, in response to a U.S. proposal to impose a 25% tariff on imported products from China (Taheripour & Tyner, 2018a). In April 2018, China began imposing tariffs on U.S. goods including soybeans, dairy, and pork, and more tariffs were added in July 2018 (Filloon, 2018). Zhang (2016) finds that Chinese policymakers are convinced that over-reliance on the USA for agricultural imports could pose a threat to not only the country's food security but also national security.

The U.S. economy is external-oriented, but to a lesser degree compared to that of China. In the USA, export-GDP ratio is 11.9%, while that in China is 19.2%. Such trade position of China in relation to the USA limits the response resources in trade tensions. By raising import tariffs on food and agricultural products in response to recent U.S. protectionist policies, China may face a situation of food insecurity. According to Zhang (2016), the USA supplied 36% of total China's oilseeds imports, 42% of cereals imports, 30% of cotton imports, and 25% of meat imports. Gale, Hansen, and Jewison (2015) report that the USA is also one of China's top three destinations for agricultural exports. Bilateral agricultural trade has experienced phenomenal growth from 2001, the year of China's accession to the WTO. U.S. imports of agricultural products from China increased from $1 billion in 2001 to almost $7 billion in 2017; whereas its agricultural exports to China skyrocketed from about $2.2 billion in 2001 to over $21.6 billion in 2017.

In 2017, China was the USA's fourth-largest supplier of agricultural imports. Particularly, China is one of the major suppliers of canned fruits, farmed fish and shellfish, apple juice, garlic, mushroom, pet food, noodles, tea and spices to the USA (Table 3).

China is a major agricultural export market for the USA that has grown tremendously over the past two decades. In fact, China was the second-largest export market for the USA in 2017 behind Canada, accounting for 14% of the U.S. agricultural exports (Bakst, 2018). In terms of value, the largest China's agricultural import from the USA is soybeans totaling over $12.3 billion in 2017 (United States Department of Agriculture, 2018). The USA is also one of the major China's suppliers of meat and edible meat offal, cereals, fish, fruits and nuts, and feeding stuff for animals (Table 4).

Table 3. China's agricultural export to the USA in 1995-2017, $ billion

Product	1995	2000	2005	2010	2015	2016	2017
Live animals	0.002	0.002	0.015	0.020	0.022	0.036	0.027
Meat and edible meat offal	0.001	0.002	0.008	0.005	0.002	0.002	0.004
Prepared and preserved edible meat offal	0.000	0.003	0.027	0.024	0.014	0.015	0.007
Eggs and eggs' yolks	0.002	0.003	0.002	0.004	0.007	0.007	0.007
Fresh, chilled, or frozen fish	0.092	0.216	0.689	1.260	1.243	1.155	1.191
Dried, salted, or smoked fish	0.001	0.004	0.030	0.051	0.069	0.057	0.054
Crustaceans, mollusks, and aquatic invertebrates	0.159	0.129	0.151	0.407	0.514	0.533	0.423
Fish, prepared and preserved aquatic invertebrates	0.021	0.170	0.408	0.835	1.292	1.226	1.452
Rice	0.000	0.000	0.000	0.000	0.001	0.001	0.003
Cereal preparations, flour of fruits or vegetables	0.008	0.005	0.013	0.037	0.059	0.065	0.065
Vegetables	0.027	0.030	0.135	0.314	0.379	0.396	0.364
Prepared and preserved vegetables, roots, tubers	0.095	0.061	0.184	0.385	0.432	0.616	0.687
Fruits and nuts	0.011	0.014	0.049	0.101	0.146	0.148	0.130
Preserved fruits and fruit preparations	0.017	0.057	0.178	0.413	0.591	0.547	0.591
Fruit and vegetable juices	0.004	0.038	0.182	0.412	0.330	0.316	0.324
Sugar, molasses, and honey	0.016	0.021	0.027	0.021	0.026	0.045	0.047
Confectionery sugar	0.004	0.017	0.072	0.142	0.151	0.140	0.142
Coffee and coffee substitutes	0.000	0.000	0.002	0.023	0.025	0.026	0.030
Cocoa	0.023	0.018	0.036	0.002	0.016	0.017	0.024
Chocolate and food preparations with cocoa	0.000	0.000	0.004	0.006	0.010	0.010	0.010
Tea and mate	0.018	0.015	0.032	0.063	0.105	0.111	0.099
Spices	0.009	0.015	0.049	0.095	0.106	0.104	0.120
Feeding stuff for animals	0.006	0.016	0.047	0.427	0.439	0.421	0.434

Source: UNCTAD, 2019

Tariff escalations between such large actors as China and the USA will initiate trade diversion and change the global agricultural market substantially. Taheripour and Tyner (2018b) evaluated the global economic impacts of Chinese tariffs on their imports of U.S. soybeans, corn, wheat, sorghum, and beef. In the case of soybean, it is demonstrated that the total decline in U.S. soybean exports is not as large as the decline in Chinese imports, as exports to some other regions increase. Chinese imports from the USA will fall by 48%, but U.S. global exports will fall 24%. Exports to other countries make up about half of the loss in Chinese exports. Brazil and other exporters capture more of the Chinese market, and the USA takes some of the markets that other exporters give up (Taheripour & Tyner, 2018b). In general, the 25% Chinese tariff could reduce exports of U.S. soybeans to China by about 17 MMT in the long run. In an attempt to compensate the losses on China's market, the USA would export more to the EU and the rest of the world. Taheripour and Tyner (2018b) predict total U.S. soybean exports to the world to drop by 14 MMT and China's soybean imports from the USA to drop by 32.6 MMT. In this case, U.S.

Table 4. China's agricultural import from the USA in 1995-2017, $ billion

Product	1995	2000	2005	2010	2015	2016	2017
Live animals	0.012	0.011	0.017	0.034	0.011	0.005	0.016
Meat of bovine animals, fresh, chilled or frozen	0.001	0.003	0.000	0.000	0.000	0.000	0.025
Meat and edible meat offal	0.069	0.417	0.247	0.405	0.519	1.347	1.160
Milk, cream, and milk products	0.015	0.021	0.059	0.165	0.250	0.232	0.368
Cheese and curd	0.000	0.000	0.001	0.013	0.054	0.041	0.060
Fresh, chilled, or frozen fish	0.037	0.046	0.307	0.563	0.757	0.674	0.902
Dried, salted, or smoked fish	0.002	0.001	0.000	0.000	0.001	0.002	0.000
Crustaceans, mollusks, and aquatic invertebrates	0.030	0.033	0.035	0.157	0.325	0.351	0.122
Fish, prepared and preserved aquatic invertebrates	0.000	0.000	0.001	0.001	0.003	0.005	0.295
Wheat and meslin	0.679	0.028	0.104	0.031	0.185	0.208	0.391
Maize	0.786	0.000	0.000	0.348	0.121	0.056	0.160
Cereals, excluding wheat, rice, barley, and maize	0.000	0.000	0.000	0.000	2.472	1.262	0.959
Meal and flour of wheat and flour of meslin	0.002	0.003	0.001	0.001	0.000	0.000	0.000
Cereal preparations, flour of fruits or vegetables	0.002	0.003	0.007	0.014	0.052	0.046	0.040
Vegetables	0.004	0.018	0.024	0.047	0.052	0.049	0.048
Prepared and preserved vegetables, roots, tubers	0.002	0.032	0.030	0.069	0.150	0.149	0.125
Fruits and nuts	0.007	0.049	0.106	0.382	0.523	0.573	0.760
Preserved fruits and fruit preparations	0.001	0.002	0.019	0.090	0.094	0.117	0.162
Fruit and vegetable juices	0.001	0.003	0.004	0.018	0.015	0.020	0.019
Sugar, molasses, and honey	0.001	0.005	0.015	0.037	0.063	0.049	0.068
Confectionery sugar	0.001	0.005	0.002	0.004	0.014	0.016	0.015
Coffee and coffee substitutes	0.001	0.001	0.005	0.010	0.013	0.017	0.022
Cocoa	0.000	0.003	0.005	0.024	0.012	0.009	0.001
Chocolate and food preparations with cocoa	0.003	0.014	0.006	0.009	0.028	0.026	0.028
Tea and mate	0.000	0.000	0.001	0.002	0.001	0.001	0.004
Feeding stuff for animals	0.041	0.120	0.097	1.004	2.745	1.500	0.921
Margarine and shortening	0.000	0.003	0.002	0.001	0.015	0.017	0.016
Edible products and preparations	0.006	0.048	0.095	0.151	0.421	0.443	0.515
Oil seeds and oleaginous fruits	0.039	1.185	3.164	11.359	12.437	13.923	14.004

Source: UNCTAD, 2019

soybean exports go up by 1.9 MMT to the EU and by 10.7 MMT to the rest of the world. The proposed Chinese tariffs will not significantly affect the global markets for corn, wheat, and sorghum or cause significant changes in the total U.S. exports of these products. However, the tariffs decrease U.S. exports of corn, wheat, and sorghum to China by 42% (0.11 MMT), 82% (0.74 MMT), and 13% (0.68 MMT), respectively (Taheripour & Tyner, 2018b).

SOLUTIONS AND RECOMMENDATIONS

One of the possible ways to reduce resistance is to implement joint investment projects in agricultural sector instead of direct land leases. The most promising areas are production of soybeans, cash crops, and meat, including pork and poultry. China should develop investment collaboration with other countries in the improvement of technical crops, feed production, soil fertility, production of organic fertilizers, and environmental protection. There should be established joint agricultural parks and zones for the development of collaboration in the spheres of crop production (grain, soybeans, oilseed, vegetables, and fruits) and animal husbandry (production of meat and milk with a focus on organic farming).

Second way is a long-term oriented solution to diversify supply chains of high value-added agricultural products. Currently, China heavily depends on the USA on a number of cash crops, soybeans, and other high value-added agricultural products. Development of overseas agricultural production is an alternative to, first, reduce imports spending on value-added foods. China has been diversifying where it obtains crops that Chinese farmer are not able grow in sufficient quantities domestically. China's agricultural expansion has resulted in 6.6 million hectares of land being acquired around the world, primarily, in Southeast Asia, Africa, and Latin America. China should continue supporting its agricultural companies to grow into internationally competitive corporations that can address growing concerns about food security of the nation. At the same time, China should also continue establishing overseas agri-business and investment associations to enhance coordination and cooperation among Chinese companies to improve risk management and conflict resolution.

Collaboration should be promoted within the Belt and Road Initiative, particularly, with such countries land-abundant countries as Russia, Ukraine, Kazakhstan, Belarus, and countries of Central Asia (Mariani, 2013). China has already become the most important trade and investment partner for those countries, including Russia and the economies of Central Asia (Golam & Monowar, 2018). Some barriers, however, should be removed, including production and processing, technical barriers, and institutional constraints. Production and processing constraints include low yields resulting in low production volumes, low quality and insufficient quantity of planting materials and inputs, underdeveloped system of irrigation, logistical constraints, and outdated processing and packaging practices. Technical barriers include the lack of adequate sanitary and phytosanitary capacities, mismatch between national and China's quality standards, and lack of compliance with specific China's regulations, including labeling and packaging requirements. Finally, institutional constraints are related to weak inland transport infrastructure, lack of export infrastructure, cumbersome customs procedures, and low spending on agricultural research and extension. Enabling those countries to convert their natural advantages into competitive advantage and to become significant sources of agricultural imports for China requires complementary investments in farming, processing, logistics, trade, and transport infrastructure.

FUTURE RESEARCH DIRECTIONS

Although the current China-USA trade dispute continues to evolve, it is valuable to understand the potential negative impact on global agricultural trade and to be informed of possible consequences of tariff escalation and trade distortions on global market share reallocations. Threats of trade tensions on agricultural trade shook the agricultural production globally, that is why a deeper analysis is required in the sphere of the trade policy impact on not only agricultural trade, but also changes in agricultural production, producer prices, and welfare. It is important to determine the actual level of security of domestic food market with an account of the internationally recognized reference nutrient intakes. There should be monitoring of the level of security of certain market over the certain period of time, which allows comparing it with both the official statistics reports and levels of food self-sufficiency and assessing influences of various foreign trade policies on food security.

To expand overseas agricultural investments, China strategically places Agriculture Technology Demonstration Centers in the countries where agricultural productivity is low and has room to improve, as well as in places where China already owns land or has a high probability of acquiring it. Further investigation is required to reveal the effects of such platforms not only on local food supplies but also on deliveries of agricultural commodities and food products to China. It is necessary to identify the determinants of competitiveness for the selected agri-food value chains in countries – food producers other than the USA (Latin America, Southeast Asia, Russia) and suggest policy reforms and investments that could facilitate expansion of agricultural imports from those countries to China.

CONCLUSION

The majority of scholars agree that liberalized trade plays a vital role in stabilizing food supplies and food prices since abundant foods stocks in some countries coexist with shortages in some others. Unforeseen price surges can push millions of people in developing countries into poverty, aggravating income inequalities and threatening social cohesion. Malnutrition can result in maternal mortality and stunted growth of children, reducing their learning ability and lowering their productivity in adulthood. Government are pursuing various policy options for ensuring food security: expanding investment in agriculture; encouraging climate-friendly technology; restoring degraded farmland; improving post-harvest storage and supply chains; and even indulging in the promotion of niche products. They face special challenges as their growing middle class shifts from traditional staples to more nutritious products such as meat, fish and dairy whose higher resource intensity requires expansion of domestic agricultural capacity or greater reliance on imports. But importing countries also worry about unreliability of world markets in times of need and promote self-sufficiency in food along with trade policy restrictions.

International trade has an important role to play in increasing food self-sufficiency, which is a proportion of imported food in domestic consumption, but also in improving food security. Transforming paradigm of international economic relations causes rising influence of integration and trade liberalization on domestic food markets of developing and emerging countries. In the conditions of globalizing agricultural trade, adaptation of domestic food markets should go in the two directions: promotion of external opportunities and overcoming of internal bottlenecks. Current global trade systems are becoming increasingly distorted by unfair and inefficient policies in many countries, creating both winners and losers. One of the recent examples of such distortion is US-China trade tensions and potential tariff escalations where agricultural sector is the most vulnerable. By raising import tariffs on food and agricultural products in response to protectionist policies, China may face a situation of raising prices for consumers, limited market access for producers, and increasing pressures on food security.

Over the past few decades, China has experienced a transition from a food shortage to an achievement of security status on the provision with food staples. However, despite the remarkable achievements, China still faces both domestic problems in the sphere of sustainability of its achieved food security (instability of agricultural production growth rates, low income in agriculture, rural poverty, etc.) and new external challenges (trade policy of major food exporters, particularly, the USA, and volatility of global markets). The country depends on food imports of soybeans, meat, and many other agricultural products. The analysis of China's foreign trade in food and agricultural products and food security status of the country on major kinds of agricultural commodities demonstrates that import volumes grow as domestic demand is restructured in favor of qualitative food products of high nutrient value amid the progressive degradation of agricultural lands and shrinking possibilities for agricultural production in China. Currently, the majority of such value-added food products is imported from the USA. In the situation of trade tensions and tariff escalations between the USA and China, the agricultural sector is one of the most vulnerable. By raising import tariffs on food and agricultural products in response to the recent US protectionist policies, China may face a situation of food insecurity.

The possible solution to the improvement of food security is the intensification of domestic agricultural production and, primarily, diversification of imports. China needs to learn from its decades of experience in international agricultural cooperation and initiate the new model to keep up with global changes. It aims to further liberalize China's agricultural sector to enhance the country's food security. Based on the principle of mutual beneficial cooperation, China will give economic and technological support to develop the agricultural sector in neighboring countries. China needs to enhance connectivity with neighboring countries, establish more cross-border trade centers and free trade zones, and improve environmental conditions for cross-border investment. China should fund agricultural development in developing countries and also sign bilateral agricultural cooperation agreements.

ACKNOWLEDGMENT

This chapter is supported by the Fundamental Research Funds for the Central Universities (grant no. 3072019CFP0902, HEUCFJ170901, HEUCFP201829, HEUCFW170905, 3072019CFG0901).

REFERENCES

Bakst, D. (2018). *Agricultural Trade with China: What's at Stake for American Farmers, Ranchers, and Families.* Retrieved from https://www.heritage.org/agriculture/report/agricultural-trade-china-whats-stake-american-farmers-ranchers-and-families

Chen, J. (2007). Rapid Urbanization in China: A Real Challenge to Soil Protection and Food Security. *Catena, 69*(1), 1–15. doi:10.1016/j.catena.2006.04.019

Cheng, G. (2007). China's Agriculture within the World Trading System. In I. Sheldon (Ed.), *China's Agricultural Trade: Issues and Prospects* (pp. 81–104). Beijing: International Agricultural Trade Research Consortium.

Cui, K., & Shoemaker, S. (2018). A Look at Food Security in China. *NPJ Science of Food, 2.* doi:10.103841538-018-0012-x

Dorosh, P. A. (2004). Trade, Food Aid and Food Security: Evolving Rice and Wheat Markets. *Economic and Political Weekly, 36*(39), 4033–4042.

Erokhin, V. (2017). Factors Influencing Food Markets in Developing Countries: An Approach to Assess Sustainability of the Food Supply in Russia. *Sustainability, 9*(8), 1313. doi:10.3390u9081313

Erokhin, V. (2018). Contemporary Foreign Trade Policy of China in the Region of Central and Northeast Asia. In A. C. Ozer (Ed.), *Globalization and Trade Integration in Developing Countries* (pp. 27–54). Hershey, PA: IGI Global. doi:10.4018/978-1-5225-4032-8.ch002

Erokhin, V., & Gao, T. (2018). Competitive Advantages of China's Agricultural Exports in the Outward-Looking Belt and Road Initiative. In W. Zhang, I. Alon, & C. Lattemann (Eds.), *China's Belt and Road Initiative: Changing the Rules of Globalization* (pp. 265–285). London: Palgrave Macmillan. doi:10.1007/978-3-319-75435-2_14

Erokhin, V., & Ivolga, A. (2012). How to Ensure Sustainable Development of Agribusiness in the Conditions of Trade Integration: Russian Approach. *International Journal of Sustainable Economies Management, 2*(1), 12–23. doi:10.4018/ijsem.2012040102

Erokhin, V., Ivolga, A., & Heijman, W. (2014). Trade Liberalization and State Support of Agriculture: Effects for Developing Countries. *Agricultural Economics – Czech, 60*(11), 524-537.

Fan, S. (1997). Production and Productivity Growth in Chinese Agriculture: New Measurement and Evidence. *Food Policy, 22*(3), 213–228. doi:10.1016/S0306-9192(97)00010-9

Filloon, W. (2018). *These Are All the Foods Being Affected by Trump's Trade War.* Eater. Retrieved from https://www.eater.com/2018/7/18/17527968/food-tariffs-trump-canada-china-mexico-eu

Gale, F., Hansen, J., & Jewison, M. (2015). *China's Growing Demand for Agricultural Imports.* Washington, DC: U.S. Department of Agriculture, Economic Research Service.

Gao, T. (2017). Food Security and Rural Development on Emerging Markets of Northeast Asia: Cases of Chinese North and Russian Far East. In V. Erokhin (Ed.), *Establishing Food Security and Alternatives to International Trade in Emerging Economies* (pp. 155–176). Hershey, PA: IGI Global.

Gao, T., Ivolga, A., & Erokhin, V. (2018). Sustainable Rural Development in Northern China: Caught in a Vice between Poverty, Urban Attractions, and Migration. *Sustainability, 10*(5), 1467. doi:10.3390u10051467

Ghosh, I., & Ghoshal, I. (2017). Implications of Trade Liberalization for Food Security under the ASEAN-India Strategic Partnership: A Gravity Model Approach. In V. Erokhin (Ed.), *Establishing Food Security and Alternatives to International Trade in Emerging Economies* (pp. 98–118). Hershey, PA: IGI Global.

Golam, M., & Monowar, M. (2018). Eurasian Economic Union: Evolution, Challenges and Possible Future Directions. *Journal of Eurasian Studies, 9*(2), 163–172. doi:10.1016/j.euras.2018.05.001

Gro Intelligence. (2018). *China's Road Map to Food Security.* Retrieved from https://gro-intelligence.com/insights/chinas-roadmap-to-food-security

Huang, J., Wei, W., Cui, Q., & Xie, W. (2017). The Prospects for China's Food Security and Imports: Will China Starve the World via Imports? *Journal of Integrative Agriculture, 16*(12), 2933–2944. doi:10.1016/S2095-3119(17)61756-8

Huang, J., & Yang, G. (2017). Understanding Recent Challenges and New Food Policy in China. *Global Food Security*, *12*, 119–126. doi:10.1016/j.gfs.2016.10.002

Kennedy, C., Zhong, M., & Corfee-Morlot, J. (2016). Infrastructure for China's Ecologically Balanced Civilization. *Engineering*, *2*(4), 414–425. doi:10.1016/J.ENG.2016.04.014

Kravchenko, A., & Sergeeva, O. (2014). China Policy in the Area of Food Security: Modernization of Agriculture. *Pacific Rim: Economics, Politics. Law*, *32*(4), 57–65.

Lawrence, G., & McMichael, P. (2012). The Question of Food Security. *International Journal of Sociology of Agriculture and Food*, *2*(19), 135–142.

Lee, R. (2007). *Food Security and Food Sovereignty*. Newcastle upon Tyne: Centre for Rural Economy, University of Newcastle upon Tyne.

Liu, J., & Diamond, J. (2005). China's Environment in a Globalizing World. *Nature*, *435*(7046), 1179–1186. doi:10.1038/4351179a PMID:15988514

Ma, L. (2011). Sustainable Development of Rural Household Energy in Northern China. *Journal of Sustainable Development*, *5*(4), 115–124.

Mahendra Dev, S., & Zhong, F. (2015). Trade and Stock Management to Achieve National Food Security in India and China? *China Agricultural Economic Review*, *7*(4), 641–654. doi:10.1108/CAER-01-2015-0009

Mariani, B. (2013). *China's Role and Interests in Central Asia*. London: Saferworld.

Nath, R., Luan, Y., Yang, W., Yang, C., Chen, W., Li, Q., & Cui, X. (2015). Changes in Arable Land Demand for Food in India and China: A Potential Threat to Food Security. *Sustainability*, *7*(5), 5371–5397. doi:10.3390u7055371

Panitchpakdi, S., & Clifford, M. L. (2002). *China and the WTO: Changing China, Changing World Trade*. Singapore: J. Wiley & Sons.

Park, A., Jin, H., Rozelle, S., & Huang, J. (2002). Market Emergence and Transition: Arbitrage, Transaction Costs, and Autarky in China's Grain Markets. *American Journal of Agricultural Economics*, *84*(1), 67–82. doi:10.1111/1467-8276.00243

Spring, J. (2018). *Brazil Record Soy Exports to China Could Expand Further – Official*. Retrieved from https://www.reuters.com/article/brazil-agriculture-trade/brazil-record-soy-exports-to-china-could-expand-further-official-idUSL4N1XP5YI

Spring, J., & Polansek, T. (2018). *Trump Trade War Delivers Farm Boom in Brazil, Gloom in Iowa*. Retrieved from https://www.reuters.com/article/us-usa-trade-china-brazil-insight/trump-trade-war-delivers-farm-boom-in-brazil-gloom-in-iowa-idUSKCN1ML0E7

Sun, P., & Heshmati, A. (2010). *International Trade and Its Effects on Economic Growth in China*. Bonn: Institute for the Study of Labor.

Taheripour, F., & Tyner, W. E. (2018a). Impacts of Possible Chinese 25% Tariff on U.S. Soybeans and Other Agricultural Commodities. *Choices*, *33*(2).

Taheripour, F., & Tyner, W. E. (2018b). *Impacts of Possible Chinese Protection on US Soybeans*. West Lafayette, IN: Purdue University.

The World Bank. (2018). *Central Asia. China (and Russia) 2030 – Implications for Agriculture in Central Asia*. Washington, DC: The World Bank.

Thow, A. M., & Hawkes, C. (2009). The Implications of Trade Liberalization for Diet and Health: A Case Study from Central America. *Globalization and Health*, 28(1), 5. doi:10.1186/1744-8603-5-5 PMID:19638196

United Nations Conference on Trade and Development. (2019). *Statistics Database* [Data file]. Retrieved from http://unctad.org/en/Pages/Statistics.aspx

United States Department of Agriculture. (2018). *Top U.S. Agricultural Exports in 2017*. Retrieved from https://www.fas.usda.gov/data/top-us-agricultural-exports-2017

Warr, P. (2011). *Food Security vs. Food Self-Sufficiency: The Indonesian Case*. Canberra: The Australian National University.

Yu, W., Elleby, C., & Zobbe, H. (2015). Food Security Policies in India and China: Implications for National and Global Food Security. *Food Security*, 7(2), 405–414. doi:10.100712571-015-0432-2

Yu, Y., Feng, K., Hubacek, K., & Sun, L. (2016). Global Implications of China's Future Food Consumption. *Journal of Industrial Ecology*, 20(3), 593–602. doi:10.1111/jiec.12392

Zhang, H. (2016). Food in Sino-U.S. Relations. From Blessing to Curse? In F. Wu & H. Zhang (Eds.), *China's Global Quest for Resources. Energy, Food and Water* (pp. 100–118). London: Routledge.

Zhang, H., & Chen, G. (2016). China's Food Security Strategy Reform: An Emerging Global Agricultural Policy. In F. Wu & H. Zhang (Eds.), *China's Global Quest for Resources. Energy, Food and Water* (pp. 23–41). London: Routledge.

Zhou, Z. (2010). Achieving Food Security in China: Past Three Decades and Beyond. *China Agricultural Economic Review*, 2(3), 251–275. doi:10.1108/17561371011078417

Zhu, Y. (2016). International Trade and Food Security: Conceptual Discussion, WTO and the Case of China. *China Agricultural Economic Review*, 8(3), 399–411. doi:10.1108/CAER-09-2015-0127

ADDITIONAL READING

Afonso, O. (2001). *The Impact of International Trade on Economic Growth*. Porto: University of Porto.

Anderson, K., Jha, S., Nelgen, S., & Strutt, A. (2013). Re-examining Policies for Food Security in Asia. *Food Security*, 5(2), 195–215. doi:10.100712571-012-0237-5

Anderson, K., Martin, W., & van der Mensbrugghe, D. (2010). China, the WTO and the Doha Agenda. In D. Greenaway, C. Milner, & S. Yao (Eds.), *China and the World Economy* (pp. 1–20). London: Palgrave Macmillan. doi:10.1057/9781137059864_1

Beloglazov, G. (2007). Food Security of the People's Republic of China and its Russian Vector. *Russia and Pacific RIM, 3,* 75–83.

Cass, D. Z., Williams, B. G., & Barker, G. R. (2003). *China and the World Trading System: Entering the Millennium.* New York, NY: Cambridge University Press. doi:10.1017/CBO9780511494482

Chang, X., DeFries, R. S., Liu, L., & Davis, K. (2018). Understanding Dietary and Staple Food Transitions in China from Multiple Scales. *PLoS One, 13*(4), e0195775. doi:10.1371/journal.pone.0195775 PMID:29689066

Dohmen, H. (1976). China's Foreign Trade Policy. *Inter Economics, 11*(7), 197–201. doi:10.1007/BF02929005

Estevadeordal, A., Freund, C., & Ornelas, E. (2008). Does Regionalism Affect Trade Liberalization Toward Nonmembers? *The Quarterly Journal of Economics, 124*(4), 1531–1575. doi:10.1162/qjec.2008.123.4.1531

Frankel, J. A., & Romer, D. H. (1999). Does Trade Cause Growth? *The American Economic Review, 89*(3), 379–399. doi:10.1257/aer.89.3.379

Ghose, B. (2014). Food Security and Food Self-Sufficiency in China: From Past to 2050. *Food and Energy Security, 3*(2), 86–95. doi:10.1002/fes3.48

Hamrin, C. L., & Zhao, S. (1995). *Decision-Making in Deng's China: Perspectives from Insiders.* Armonk, NY: M.E. Sharpe.

Henneberry, S. R., & Diaz, C. C. (2015). Food Security Issues: Concepts and the Role of Emerging Markets. In A. Schmitz, P. L. Kennedy, & T. G. Schmitz (Eds.), *Food Security in an Uncertain World* (pp. 63–79). Bingley: Emerald Group Publishing Limited. doi:10.1108/S1574-871520150000015005

Huang, J., & Rozelle, S. (2006). The Emergence of Agricultural Commodity Markets in China. *China Economic Review, 17*(3), 266–280. doi:10.1016/j.chieco.2006.04.008

Kneller, R., Morgan, C. W., & Kanchanahatakij, S. (2008). Trade Liberalization and Economic Growth. *World Economy, 31*(6), 701–719. doi:10.1111/j.1467-9701.2008.01101.x

Lamaj, J. (2015). The Impact of International Trade and Competition Market on Developing Countries. *Proceedings of the International Conference Managing Intellectual Capital and Innovation for Sustainable and Inclusive Society.* Bari: Management, Knowledge and Learning; Technology, Innovation and Industrial Management.

Lardy, N. R. (2002). *Integrating China into the World Economy.* Washington, DC: Brookings Institution Press.

Lee, J. W. (1995). Capital Goods Import and Long-Run Growth. *The Developing Economies, 48*(1), 91–110. doi:10.1016/0304-3878(95)00015-1

Levchenko, A., & Zhang, J. (2016). The Evolution of Comparative Advantage: Measurement and Welfare Implications. *Journal of Monetary Economics, 78,* 96–111. doi:10.1016/j.jmoneco.2016.01.005

Liapis, P. (2011). *Changing Patterns of Trade in Processed Agricultural Products.* Paris: OECD Publishing.

Liberthal, K. (2004). *Governing China: From Revolution Through Reform*. New York, NY: W. W. Norton.

Lu, N. (1997). *The Dynamics of Foreign-Policy Decisionmaking in China*. Boulder, CO: Westview Press.

Omoju, O., & Adesanya, O. (2012). Does Trade Promote Growth in Developing Countries? Empirical Evidence from Nigeria. *International Journal of Development and Sustainability, 1*(2), 743–753.

Perez Motta, E. (2016). *Competition Policy and Trade in the Global Economy: Towards an Integrated Approach*. Geneva: International Center for Trade and Sustainable Development, World Economic Forum.

Raynolds, L., Murray, D., & Wilkinson, J. (2007). *Fair Trade: The Challenges of Transforming Globalization*. London: Routledge. doi:10.4324/9780203933534

Schmitz, A., & Meyers, W. H. (Eds.). (2015). *Transition to Agricultural Market Economies. The Future of Kazakhstan, Russia and Ukraine*. Boston, Oxfordshire: CABI. doi:10.1079/9781780645353.0000

Spoor, M., & Robbins, M. J. (Eds.). (2012). *Agriculture, Food Security, and Inclusive Growth*. The Hague: Institute of Social Sciences.

Wagner, J. (2007). Exports and Productivity: A Survey of the Evidence from Firm Level Data. *World Economy, 30*(1), 60–82. doi:10.1111/j.1467-9701.2007.00872.x

Windfuhr, M., & Jonsen, J. (2005). *Food Sovereignty: Towards Democracy in Localized Food Systems*. Heidelberg: FIAN ITDG Publishing. doi:10.3362/9781780441160

Wittman, H., Desmarais, A., & Wiebe, N. (2010). *Food Sovereignty: Reconnecting Food, Nature and Community*. Oakland, CA: Food First Books.

Wu, Y. (2006). *Economic Growth, Transition and Globalization in China*. Cheltenham: Edward Elgar Publishing Limited.

Yanikkaya, H. (2003). Trade Openness and Economic Growth: A Cross-Country Empirical Investigation. *Journal of Development Economics, 72*(1), 57–89. doi:10.1016/S0304-3878(03)00068-3

Zhou, Z., Liu, H., Cao, L., Tian, W., & Wang, J. (2014). *Food Consumption in China: The Revolution Continues*. Cheltenham: Edward Elgar Publishing. doi:10.4337/9781782549208

KEY TERMS AND DEFINITIONS

Belt and Road Initiative: A development strategy proposed by the Chinese government in 2013 and focused on the improvement of connectivity and collaboration among the countries of Eurasia through the increase of China's role in global affairs.

Food Market: The supply and demand of agricultural commodities and food products within a single country (domestic) or between countries (international).

Food Security: A condition when people have access to sufficient amounts of safe and nutritious food and are therefore consuming the food required for normal growth and development, and for an active and healthy life.

Food Self-Sufficiency: An extent to which a country can satisfy its food needs from its own domestic production.

Foreign Trade: The system of international commodity-money relations composed of foreign trade activities of all countries worldwide.

Tariff Escalation: A situation where the import duties on components or raw materials are lowest and move progressively higher on semi-finished goods upwards to the finished goods.

Trade Liberalization: A process of reduction or elimination of constraints in the sphere of international trade, including reduction or removal of customs tariffs, import quotas, abolishment of multiple exchange rates, and simplification of administrative requirements to import and export operations.

Trade Protectionism: A set of measures to limit imports or promote exports by putting up barriers to trade that countries use to limit unfair competition from foreign industries.

Trade War: A practice to counterbalance tariff rates in retaliation for the tariff rates of another country in an effort to gain trade advantages.

Chapter 17
Food Security and Self-Sufficiency as a Basis for National Security and Sovereignty:
Evidence From Russia

Kirill Zemliak
Khabarovsk State University of Economics and Law, Russia

Anna Zhebo
https://orcid.org/0000-0003-3142-2188
Khabarovsk State University of Economics and Law, Russia

Aleksey Aleshkov
https://orcid.org/0000-0003-3853-4772
Khabarovsk State University of Economics and Law, Russia

ABSTRACT

The study discusses one of the global problems of mankind—ensuring food security for the population. The historical context of the food problem, the formation of the concept of food security, the approaches of the world community and individual countries to its provision and evaluation are considered. The case of Russia reveals the role of food security in ensuring economic, social, and political security and sovereignty of a state. Special attention is paid to the state of agriculture in Russia as a source of raw materials for ensuring food security, problems of its development, and ways to solve them. The place of Russia in ensuring the food security of the world is shown.

DOI: 10.4018/978-1-7998-1042-1.ch017

INTRODUCTION

Food problem has a global relevance. It is caused by the mismatch between the growing demand of the population for food products and the capacities of agricultural production. The food problem is multi-dimensional as it combines the following aspects:

- Hunger and malnutrition
- Structure and quality of food
- Health status of population
- Food supplies
- Irregularity in the distribution of food
- Balance between food supply and population needs
- Food prices

Food problem is characterized by increasing severity at the national and global levels. As a result of the significant increase in world prices for major crops and food products, global status of food security deteriorated during the economic crisis in 2008. That fact was stated by international organizations as the onset of the world food crisis. The impact of this crisis has seriously affected low-income countries. These countries are net food importers in international trade (Mintusov, 2016).

The main causes of food shortages in the world are:

- High level of energy intensity of agro-industrial production, which was especially evident in developed countries, major food exporters
- Growing production of biofuels from oilseeds and grain
- Rapid industrial development and consumption growth in China and India
- Growth of intensity in use of natural resources, in a number of regions it reaches the lowest level
- Reduction of agricultural acreage as a result of urbanization and industrialization
- Growth of environmental pollution due to the growth of industrial waste, pesticides, and fertilizers
- Impact of global warming that caused a series of crop failures
- Growth of stocks under the influence of speculative turnover in the world market of food products, the increase in the volume of fixed-term insurance and speculative operations (Mintusov, 2016)

Hunger and malnutrition are linked primarily to local poor production conditions, war and civil strife, poverty and, therefore, to economic inaccessibility of food products. In addition, global climate change and extreme climate events are among the main causes of serious food crises (Food and Agriculture Organization of the United Nations (FAO), 2018b).

Over the past few decades, the world has made some progress in addressing food security problem. In particular, there is a steady decline in the number of undernourished people. Globally, it is up to 820.8 million people (10.9% of the world population). In Russia, it is up to 3.6 million people (2.5% of population of the country) (FAO, 2018a).

Along with the global problem of hunger, there is a problem of overweight. In 2016, the number of people with obesity in the world amounted to 672.3 million people (13.2% of the world population), in Russia – 29.3 million people (25.7% of the population of the country) (FAO, 2018b). In addition to the above, obesity is found among people in the regions where people experience malnutrition. This can be explained by the imbalance in the diets, the lack in a number of nutrients and biologically active substances. Thus, about 2 billion people in Russia experience a deficit of one or more minor-nutrient elements (FAO, 2013). In particular, 14.1% of children in Russia are deficient in vitamin A, while 11.7% of adults are deficient in zinc (FAO, 2015a; FAO, 2016). The increase in the prevalence of obesity in some regions is associated with an increase in per capita income and purchasing power for higher-calorie food products in combination with less active lifestyles of people (FAO, 2016). It is rather relevant for other regions with low income and consumption of cheap products with a high content of fat and sugar.

BACKGROUND

There are three stages in the development of the food security concept in the world (Revenko, 2015). Before the raw materials' crises of the 1970s food security was interpreted exclusively at the national level in connection with the desire of countries to independence from the external environment. After the raw materials' crises of the 1970s, the concept of "world food security" appeared. The fight against hunger and the solution of the food problem on a global scale acquired the status of a global problem of mankind. The Rome Declaration on World Food Security (FAO, 2014), adopted at the World Food Summit in 1996, defined poverty (not demographic and resource imbalances) as a threat to food security at both the national and global levels (Revenko, 2015).

The beginning of the 1970s was marked by significant fluctuations in global cereal production. As a result of the reduction of the world's grain reserves and the lack of food in the markets, food prices increased. That fact actualized the problem of poverty and hunger in the world. To solve this problem, in 1973, a specialized Committee on World Food Security was established within the framework of the Food and Agriculture Organization of the United Nations (FAO). In the 1980s, there was developed the Concept of World Food Security. In 1974, at the World Food Conference in Rome, representatives of 134 countries formulated the concept of "food security" (Ushachev, 2014). Initially, the term meant the availability at all times of adequate world food supplies sufficient to sustain steady expansion of food consumption and to offset fluctuations in production and prices (FAO, 1981).

In the 1980s, on the rise of Green Revolution, food production in the world increased. However, despite the sufficiency of total food supply and price stability, some vulnerable groups of population were deprived of physical or economic access to food (Belugin, 2017). Therefore, in 1983, the definition of food security, as proposed by the FAO, was clarified and formulated as ensuring that all people at all times had both physical and economic access to the basic food that they needed (FAO, 1981).

The Rome Declaration on World Food Security, adopted at the World Food Summit in 1996, referred to the obligation of any country to ensure the human right to access to adequate and safe food (FAO, 1996). The solution to the problem of food and water supply is declared as one of the Millennium Development Goals proclaimed in 2000 by the United Nations (UN) (United Nations, 2000). The UN World Food Council defines food security as a policy of achieving the highest level of food self-sufficiency through integrated efforts to increase the production of necessary food products, to improve the quality of food supply and consumption, to solve the problem of hunger and malnutrition.

A modern definition of food security was adopted at the World Summit on Food Security in Rome in 2009. It is a situation when all people, at all times, have physical, social and economic access to sufficient, safe and nutritious food that meets their dietary needs and food preferences for an active and healthy life (FAO, 2009; FAO, 2018a). According to this definition, the following dimensions of food security should be emphasized:

- Availability
- Access
- Utilization
- Stability (FAO, 2006)
- Food safety (Mintusov, 2016; Belugin, 2017)

The FAO assesses food security at both global and regional (national) levels through a set of indicators developed in accordance with the guidelines of the Committee on World Food Security (Belugin, 2017). Seven key indicators are used to measure the level of world food security. The most widely used is the ratio of grain reserves to consumption of grain. It is for this reason that food insecurity is often linked to the development of grain production. There are indicators that show the state of grain markets in the main exporting and importing countries. Based on this, the ratio of grain supply is estimated across the five main grain exporters (Australia, Argentina, the EU, Canada, and the USA) against the required amount, as well as the proportion of rolling stocks to total consumption. Indicators of changes in grain production in developing countries (India and China are not reported) are used to measure grain security in importing countries. Finally, average annual export prices are monitored for wheat, soy, and maize (Mintusov, 2016).

In total, there are 43 (31 core and 12 supporting) indicators used by the FAO to assess the status of food security at the national level. They are divided into four groups: availability; access; stability; use/usefulness. There are also additional indicators: GDP per capita, prevalence of anemia among pregnant women and children under five years, prevalence of iodine and vitamin A deficiency, average consumption of fat per capita, fat consumption per capita, proportion of systematically overeating people, and others (Belugin, 2017). An alternative measure is the Global Food Security Index (GFSI), a global ranking of countries in terms of food security and the effectiveness of public institutions to ensure it, compiled by The Economist Intelligence Unit (2018). It provides an assessment of food security in 113 countries around the world. For this purpose, 28 indicators are divided in three group: economic availability, provision and sufficiency, quality and safety.

MAIN FOCUS OF THE CHAPTER

International Approaches to Food Insecurity Problem

Food insecurity problem is studied in close connection with food self-sufficiency, economic and physical availability of food, and safety of food products (Mintusov, 2016). International trade plays an important role in combating food insecurity contributing to the redistribution of resources and products between producers and suppliers of agricultural products worldwide and meeting the needs of domestic markets in food. The largest exporters and importers of food products are the EU, the USA, and China. A group of fast-growing exporters include such developing countries as Brazil, India, Indonesia, and Thailand. The share of the EU in the global market of agricultural products and food is decreasing. The countries

Table 1. World's largest exporters and importers of agricultural products in 2010-2017

Region	Volume in 2017, $ billion	Share of the global agricultural market, %			
		2000	2005	2010	2017
Export					
EU (28)	647	42.0	44.4	39.4	37.4
USA	170	10.1	9.7	9.4	10.0
Brazil	88	2.8	4.1	5.1	5.1
China	79	3.0	3.4	3.8	4.6
Canada	67	6.3	4.9	3.8	3.9
Indonesia	49	1.4	1.7	2.6	2.8
Thailand	43	2.2	2.1	2.6	2.5
Australia	40	3.0	2.5	2.0	2.3
India	39	1.1	1.2	1.7	2.3
Argentina	36	2.2	2.3	2.6	2.1
Import					
EU (28)	649	42.7	45.3	40.3	36.6
China	183	3.3	5.0	7.8	10.3
USA	161	11.6	10.6	8.4	9.1
Japan	79	10.4	7.3	5.6	4.4
Canada	39	2.6	2.4	2.3	2.2
Republic of Korea	35	2.2	1.9	1.9	2.0
India	33	0.7	0.8	1.3	1.9
Russia	30	1.3	1.9	2.6	1.7
Mexico	29	1.8	1.8	1.7	1.6
Hong Kong SAR, China	29	-	-	-	-

Source: Authors' development based on World Trade Organization [WTO] (2019)

of Asia are emerging as exporters and importers of food. Food imports in Asia have been growing even in the times of crisis of world food prices and the economic downturn (Table 1).

China is the largest importer of food worldwide, its trade balance in agricultural products is negative. The role of Japan in global food imports is declining. Instead, Russia and the Republic of Korea have been becoming prominent importers of food (Table 2).

The level of food security depends on the degree and role of state regulation in agriculture. The higher the level of government regulation, the higher the degree of food self-sufficiency of a country (Revenko, 2003). According to Reinert (2015), even the most efficient national agricultural sector cannot survive without subsidies and protectionism. Even in the most developed countries, economic performance and profitability in agriculture are relatively low. However, the strategic importance of agriculture is beyond any problems caused primarily by natural factors (Lomakin, 2017).

The major trade-related approaches to the solution of food insecurity problem include state support (economic and non-economic measures), development of agricultural production, protection of domestic markets, and promotion of domestic products to the world market (Lomakin, 2017). For instance, in

Table 2. World's largest exporters and importers of food products in 2010-2017

Region	Volume in 2017, $ billion	Share of the global food market, %			
		2000	2005	2010	2017
Export					
EU (28)	560	44.1	46.2	40.5	38.3
USA	138	12.6	9.1	10.1	9.4
Brazil	78	3.0	4.5	5.4	5.3
China	69	3.2	3.6	4.0	4.7
Canada	49	4.1	3.6	3.3	3.3
Indonesia	38	1.3	1.4	2.3	2.6
Argentina	35	2.7	2.7	3.0	2.4
India	35	1.3	1.3	1.6	2.4
Australia	32	2.9	2.5	2.0	2.2
Mexico	32	1.9	1.7	1.6	2.2
Import					
EU (28)	559	43.5	46.8	41.4	37.4
USA	139	11.1	10.1	8.5	9.3
China	113	2.0	3.0	5.2	7.6
Japan	67	10.5	7.4	5.6	4.5
Canada	35	2.6	2.4	2.4	2.4
Republic of Korea	28	1.7	1.6	1.7	1.9
Hong Kong SAR, China	28	-	-	-	-
Russia	28	1.5	2.2	3.0	1.9
Mexico	25	1.8	1.9	1.7	1.7
India	25	0.5	0.7	1.1	1.7

Source: Authors' development based on WTO (2019)

Brazil, development of agricultural production and exports is due to the changes in government regulations. Three decades ago, the policies focused on the intervention procurement, control of prices, high import duties, and export restrictions. Today, it provides for more concessional lending, project financing, sectoral research and development, including export operations for the private sector (Lomakin, 2017). Another example is China, where the development of agricultural production has been promoted by the policy of credit and tax benefits, as well as the establishment of free economic zones in rural and coastal regions of the country. This allowed attracting foreign investments and advanced technologies to agricultural sector. Much attention was paid to the pricing policy during the reform period – purchase prices for agricultural goods increased by more than 50% (Mintusov, 2016).

Internationally, the approaches to the development of national and supranational markets of food and agricultural production are based on differentiated systems of subsidies, government regulations, and foreign trade restrictions. For example, in Finland, domestic food market is protected by not only import duties but also government regulations in the form of equalization and countervailing taxes. This allows the farmers selling their products at the prices higher than those in the world market. Sales taxes, compensation fees, and excise duties (up to 80% of all customs duties and import duties) are implemented but do not affect the market in a significant way. Such a system of regulation creates the conditions for the effective development of agricultural production in harsh climate. As a result, Finland has achieved high level of self-sufficiency in major food products: wheat – 177%, barley – 108%, oats – 174%, sugar – 95%, milk – 122%, and beef – 109% (Mintusov, 2016).

Another approach to combating food insecurity is rather costly means of commodity intervention and price regulation. State support for the competitiveness of the agricultural sector in the world market is implemented in the form of budget expenditures. These expenditures correspond to significant share of farm income. Such policies are implemented in the EU (Mintusov, 2016).

Agricultural policies of the developed countries, for example, Canada, are based on various benefits, grants, and subsidies. A key role belongs to tax benefits that are used to promote the development of major areas of research and development (Mintusov, 2016). In the USA, agriculture is one of the most important areas of national economy. Agricultural mechanization started in the 1930s. In the 1950s, the government limited food production and exported grain on favorable terms for buyers. By the 1960s, the USA achieved self-independence in agricultural products and food, and food exports exceeded imports (Lomakin, 2017). Currently, the promotion of the US food products to foreign markets is rather noteworthy. During 1980-2012, the USA increased its agricultural exports from $41.2 billion to $141.3 billion. The export intensity of agriculture amounted to 31.8% which was more than twice above the average for other industries (14.0%) (United States Department of Agriculture, 2010; Lomakin, 2017). In the midst of the 2008 crisis, the US government allocated $131.5 billion to support farmers (Lomakin, 2017). The country is characterized by a stable share of federal budget spending on agriculture at 4.4% with a steady increase in the absolute amount of these expenditures in the period of 1962-2015 (The White House, 2016). The amount of funds received by the U.S. Department of Agriculture (USDA) allows implementing a variety of programs to support agriculture, ensure farmers' profitability, and stabilize domestic food market. Government programs are presented as a set of measures that provide agricultural producers with protection against risks, such as loss of income, limited access to credit, and impoverishment due to natural disasters (Shields, 2015). In the USA, food security is considered as a support for the stability of domestic food sales; a form of food aid programs for the poor people; and stimulation of agricultural exports and the use of food supplies for foreign policy (United States Government, 1985).

According to the current global trend of declining tariff restrictions on imports, there are widely used various non-tariff regulatory measures, such as quantitative restrictions and special protective measures. The widespread use of non-tariff measures is partly related to the strong monopolization of sales in the global food market, where state trading firms and transnational corporations play the major role. They prefer regulating their supplies through administrative non-tariff measures that allow domination on food markets regardless of the price level and financial capabilities of competitors (Lomakin, 2017).

The review of international practices in sphere of food security allows summarizing the following key trends:

- In the developed countries, the establishment of agro-industrial sector (including its industrial basis) has been completed, thus the access to the market of ready-made food through the retail network has accelerated
- Vertically integrated chains play important role in production and distribution of agricultural products

- Concentration and specialization in agriculture takes place in parallel with the process of formation of zones for specialized industries in domestic or world markets
- Specialization in agriculture has emerged as a result of the development of infrastructure for transportation and storage of agricultural products
- Food industry is based on the application of modern technologies in the sphere of processing of agricultural raw materials, particularly, large-scale use of modern precision equipment, fine chemistry, and electronics
- There is an active transition from multi-sectoral structures to national, specialized ones; single production chains that cover all processes (from production to consumption) are being developed (Mintusov, 2016)

Food Security and Sovereignty and National Security and Independence

Meeting the food needs of the population in a particular country is directly related to the mutual conditionality of food security and national security. There are opinions that food sovereignty does not contribute to the aspect of trade relations, hinders economic development as it leads to the loss of the effects of industrial specialization, and creates barriers to the spreading of advanced technologies of agriculture and food production.

The rigid model of food security, characterized by the maximum closure of the domestic market to food imports, inevitably poses the problem of full food needs at the expense of domestic production. Depending on resource and technological self-sufficiency, each sovereign state decides the issue of openness of the domestic market in the food sector in different ways.

However, any degree of dependence of domestic consumption from foreign supplies provides the basis for the risks and threats to national security, as foreign exporters may cut off their supply at any time, thereby creating a threat to social stability and internal sustainability of the state through rising prices and reducing food supply.

Economic history experiences different examples when provisions served as an instrument of political and economic pressure: in 1933 – the UK banning on grain imports from the USSR; in 1980-1981 – banning on grain exports to the USSR introduced by the USA in connection with the entry of Soviet troops into Afghanistan; since 2014 – food embargo in response to anti-Russian sanctions after the annexation of Crimea; since 2018 – establishment of high import duties on food products in the framework of the USA-China trade war, etc.

For Russia, the role of domestic production in food supply is currently increasing, including the decline in the possibility of importing products caused by sanctions, reduction in foreign exchange earnings, weakening of the national currency and rising prices (Botkin, Sutygina, & Sutygin, 2016).

The link between food security and food sovereignty as well as national sovereignty is obvious. The level of food sovereignty is considered to be high when a complete stop of food imports will not lead to a food crisis. This is the optimum of food sovereignty, which allows receiving benefit from participation in the international division of labor, but does not create catastrophic risks to social stability and national sovereignty.

Thus, food security policy is inextricably linked to the national security and sovereignty of the country. With this in mind, a number of countries have integrated food sovereignty into their national constitutions or laws (first, in 2008, Ecuador, and then Venezuela, Mali, Bolivia, Nepal, Senegal, and Egypt) (Pena, 2008; Wittman, Desmarais, & Wiebe, 2010).

Food Security Policy of Russia

In Russia, the foundations of national food security policy are laid by the presidential decrees "On the Approval of Food Security Doctrine of the Russian Federation" (President of the Russian Federation, 2010) and "On National Security Strategy of the Russian Federation" (President of the Russian Federation, 2015), as well as the orders and resolutions of the Government of the Russian Federation.

In Russia, food security is recognized among the directions of ensuring national security and sovereignty of the country, the important component of demographic policy, the necessary condition for the improvement of the quality of life by guaranteeing high standards of life support. The main tasks of ensuring food security in Russia are:

- Detection and prevention of internal and external threats to food security, minimization of their negative consequences due to the constant readiness of the system of providing citizens with food products, the establishment of strategic food stocks
- Sustainable development of domestic production of food and agricultural raw materials sufficient to ensure food independence of the country
- Attainment and maintenance of physical and economic availability of safe food products in the volumes and assortment which correspond to the established norms of healthy nutrition necessary for active and healthy life
- Safety of food products in the domestic market

Russia's Food Security Doctrine (President of the Russian Federation, 2010) defines food security as a status of national economy which ensures food independence of Russia, guarantees physical and economic accessibility of food products that meet the requirements of Russia's legislation on technical regulation for every citizen of the country in the amounts not less than the established norms of healthy nutrition necessary for active and healthy life. In Russia and other countries of the Commonwealth of Independent States (CIS), including Azerbaijan, Belarus, Kazakhstan, Tajikistan, Turkmenistan, and Uzbekistan, food security is considered as food independence or food self-sufficiency (FAO, 2015b). The definition of food independence refers to sustainable domestic food production in volumes not less than the established thresholds of its share in the commodity resources of the domestic market of relevant products. The Concept of Improvement of Food Security of CIS Member States adopted in 2011 defines food security of CIS countries as the status of the economy, in which own annual production of vital food products is not less than 80% of the annual needs of the population in these food products in accordance with physiological standards of nutrition (CIS Council of Heads of Government, 2010).

Russia's Food Security Doctrine introduces the concept of food security into the national legislation, sets the criteria of food security as the share of domestic agricultural and fish products in the total volume of commodity resources, and stipulates the thesis on the availability of food in the domestic market (Yakutin, 2014). In economic terms, food availability is understood as the opportunity provided by the level of income of the population, the purchase of food products at prevailing prices in volumes and assortment that are not less than the established rational norms of consumption. In physical terms, food availability is the level of development of the commodity distribution infrastructure, in which all settlements of the country are provided with the possibility of purchasing food products or catering in volumes and assortment, which are not less than the established rational norms of food consumption. The Doctrine does not contain provisions on sectoral priorities and structural development of domestic

Figure 1. Share of imports in the retail trade in food products in Russia in 2005-2018, %
Source: Authors' development based on Unified Interdepartmental System of Information and Statistics (2019)

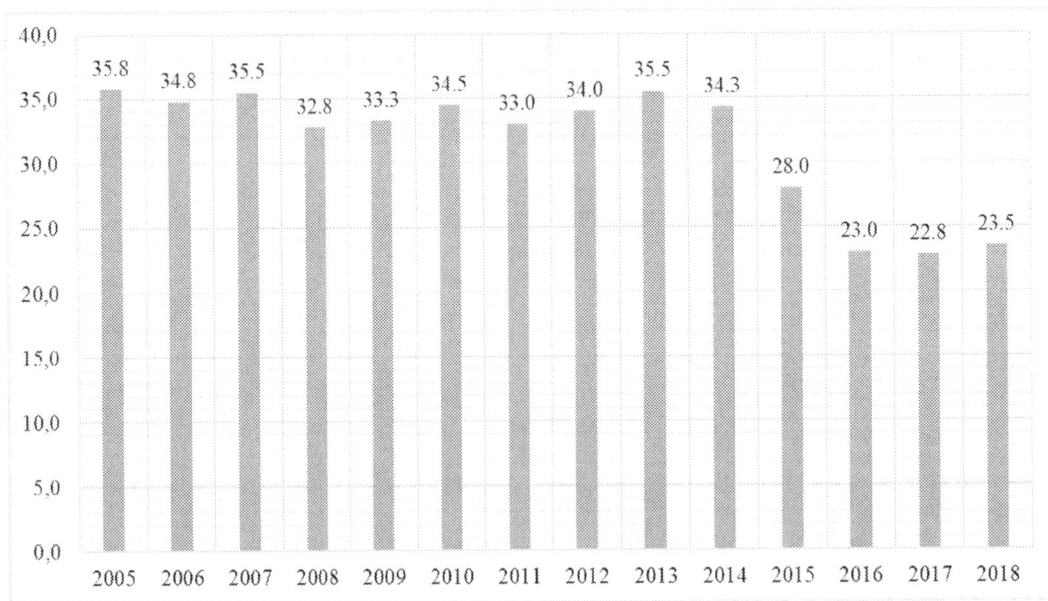

agriculture, on the risks and threats to food security of Russia in the medium and long term. The objectives of the Doctrine do not have time frames or quantitatively defined thresholds. This does not allow assessing their effectiveness (Yakutin, 2014).

According to the National Security Strategy of the Russian Federation (President of the Russian Federation, 2015), food security is provided by the attainment of food independence of the country.

The Strategy of Scientific and Technological Development of the Russian Federation prioritizes ensurance of food security and food independence of Russia among the key development goals of the country along with improvement of competitiveness of domestic products in the world food markets and reduction of technological risks in agricultural production.

Russia's Foreign Trade in Agricultural Raw Materials and Food

Since the Food Security Doctrine defines food security as the main criterion of import dependence, it is necessary to measure the share of imports in the total consumption or production of agricultural products. In 2005-2014, the share of imports in the retail trade was 33-36% (with the existing safety threshold that was below 20%). Since 2015, the dependence of Russia's domestic food market on imports has been decreasing (Figure 1).

In the 1990s, liberal trade policy led to the eradication of food independence. During 2000-2010, food imports increased fivefold. In turn, food-exporting countries used the situation to strengthen their positions in Russia's market. Until 2014, Russian farmers had been uncompetitive and continued to be displaced from the domestic market (Mintusov, 2016). Foreign trade in food products has negative trade balance. Since 2014, it has been declining and amounted to $4.7 billion in 2018 (Figure 2). According to Mintusov (2016), in the medium term, Russia's trade balance in agricultural products and food will remain negative.

Figure 2. Foreign trade in food products and agricultural raw materials in Russia in 2000-2018, $ million
Source: Authors' development based on Federal Service for State Statistics of the Russian Federation (2019)

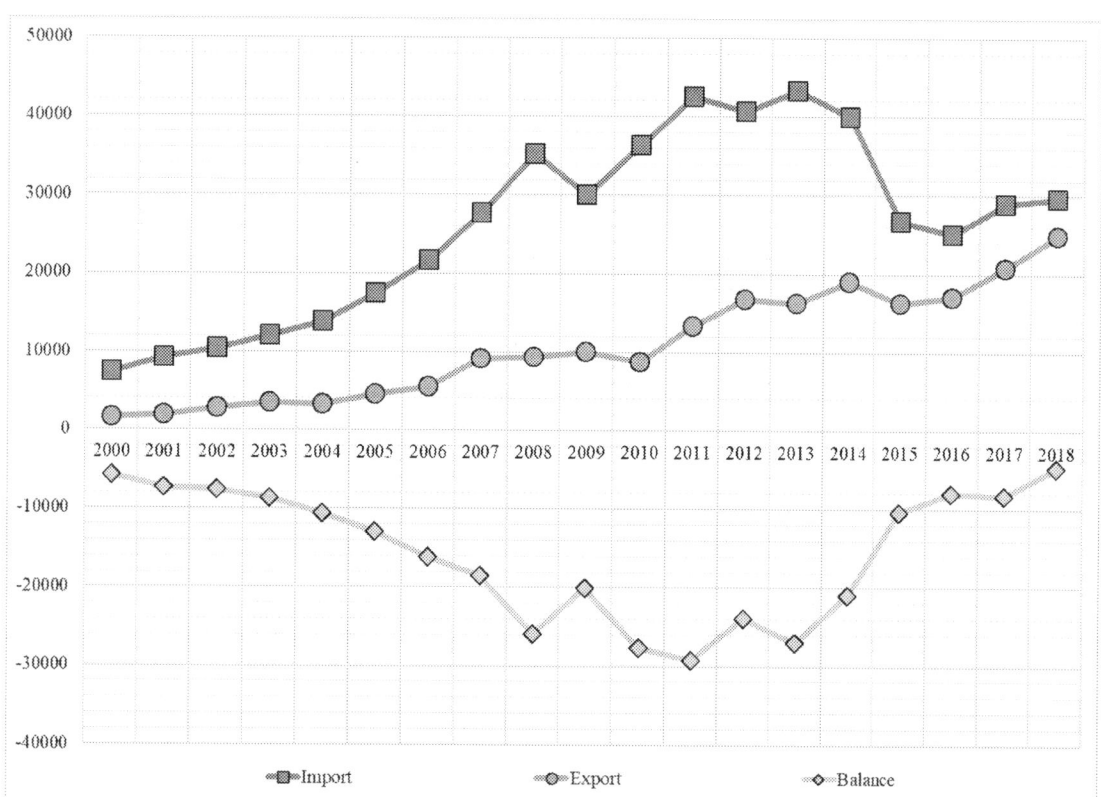

In 2018, the cost of imports of food products and agricultural raw materials reached $29.6 billion (the third biggest in the total volume of commodity imports of Russia). In essence, this is an unrealized reserve of domestic investment in the agricultural sector of the economy, support for foreign producers (Altukhov, 2014).

In recent years, there has been a positive trend in exports of agricultural raw materials and food (Table 3). The significant increase in value observed since 2011 is explained by the growth of grain exports due to high yields and price shifts in the world grain market (Lomakin, 2017).

The main obstacles in solving the problem of import substitution and increasing the share of domestic products in the total food resources of Russia are as follows:

- Lack of large-scale target state support of producers, that corresponds to the fact that production of import-substituting goods does not receive sufficient incentives, and it is also not coordinated and ineffective
- Low awareness of producers about the existing measures of state support, as well as the difficulties of obtaining it
- Difficulties with production lending, contradictory nature of lending to the agricultural sector of the Russian economy
- Weak price support for agricultural enterprises from the government (Lomakin, 2017)

Table 3. Share of exports of food products and agricultural raw materials in the commodity structure of Russia's exports in 1995-2018

Year	Volume, $ billion	Share in aggregated export of Russia, %	Year	Volume, $ billion	Share in aggregated export of Russia, %
1995	1.4	1.8	2007	9.1	2.6
1996	1.8	2.1	2008	9.3	2.0
1997	1.6	1.9	2009	10.0	3.3
1998	1.5	2.1	2010	8.8	2.2
1999	1.0	1.3	2011	13.3	2.6
2000	1.6	1.6	2012	16.8	3.2
2001	1.9	1.9	2013	16.3	3.1
2002	2.8	2.6	2014	19.0	3.8
2003	3.4	2.6	2015	16.2	4.7
2004	3.3	1.8	2016	17.1	6.0
2005	4.5	1.9	2017	20.7	5.8
2006	5.5	1.8	2018	24.9	5.5

Source: Authors' development based on Federal Service for State Statistics of the Russian Federation (2019)

Current State of Food Security in Russia

Today, the problems of ensuring food security are of particular relevance in the context of Russia's membership in the World Trade Organization (WTO), development of the single food market within the Eurasian Economic Union, and economic sanctions imposed against Russia.

In connection with the accession to the WTO, Russia agreed to reduce some of the import tariffs for food on the accession, while the remaining ones – during three years of transition period. For some goods, the duties were to be reduced during seven-eight years. Russia also committed to reduce subsidies to agricultural producers. Russia's food market became opened to the competitors from the EU and the USA (Mintusov, 2016). Russia's membership in the WTO can contribute to the deterioration of Russia's current position in the world food market. Thus, the reduction of customs duties on certain types of food (especially, meat and milk) and the opening of domestic market for foreign competition will reduce food security of Russia. In the medium term, membership in the WTO will allow retaining Russia's dependence on food imports (Mintusov, 2016).

Adaptation to the WTO regulations require more active state intervention to overcome or mitigate negative trends in the agro-industrial complex. The insufficient level of financing and the reduction in the volume of state support are the threats to the decline in the rate of agricultural production. These are the threats of investment attractiveness of the industry. The experience of many economies in transition shows that the implementation of the WTO regulations may leads to the degradation of agricultural and food complexes when food availability on the domestic market is ensured by imports (Lomakin, 2017).

In Russia, the WTO-related risks for agriculture and food security include, first, reduction in the level of state support which is stipulated by the international regulations. Second, food insecurity risks are attributable to those sectors of agricultural production where the reduction in import tariffs is the

Table 4. State support of agricultural production and fishing in Russia in 2000-2017

	2000	2010	2015	2016	2017
Expenditures on agricultural production and fishing, $ billion	2.0	8.6	5.9	4.9	5.9
Share in total expenditures, %	2.8	1.5	1.2	1.1	1.1

Source: Authors' development based on Federal Service for State Statistics of the Russian Federation (2018a)

biggest (as well as the elimination of import quotas). Third, among the threats to domestic food market is the growth of import dependence against the decrease in the level of self-sufficiency (Lomakin, 2017).

The ways to ensure the competitiveness of Russian farmers in the domestic market include the expansion of state support to the level of the EU and the USA and the increase in the level of customs tariff by 1.5 – 2.0 times.

According to the Ministry of Agriculture of the Russian Federation (2018), agro-industrial complex requires an annual support in the amount of $3.0 billion to $3.4 billion to mitigate the effects of the WTO membership (Podkopaev, 2013). Otherwise, Russia will have to spend $5-10 billion annually to eliminate the negative trade balance in food and agricultural products.

The sanctions were imposed by the USA and the EU in relation to major state-owned banks of Russia – development institutions of agricultural sector that provide industrialization and modernization of agricultural production through the resources raised in the interbank market (Lomakin, 2017).

In agriculture, anti-sanctions policy of the government was expressed in the form of import substitution of food products from the countries imposed the sanctions. Russia banned import of food and agricultural products from the EU, the USA, and other developed countries. This has led to a decrease in the diversity of food in the domestic retail network, due to the loss of trade, on the one hand, cheap products for the poor, and on the other hand, high-quality products from the EU countries, which made up the allowance of the middle class (Mintusov, 2016).

The system of restrictive measures restrains the growth of import dependence but solves the problem of low efficiency of agricultural production in Russia. The solution is not much related to the ban on imports but to the development of agro-industrial complex, expansion and strengthening of storage and processing systems, and increased funding for the reorganization and modernization of agricultural production (Lomakin, 2017).

State Support of Agricultural Production

Food Security Doctrine demonstrated the changes in the field of state support of farmers in Russia, particularly, a gradual shift from the extremes of a liberal agricultural strategy (Table 4).

A significant increase in state support is confirmed by the dynamics of the indicator that defines the assessment of support for agricultural producers, calculated in monetary units and as a percentage of gross income of farms. A significant increase in relative state support in Russia was observed in 2010-2014, but the level of support is lower than in the EU (tenfold) and the USA (fourfold) (Table 5).

Developed countries widely subsidize export-oriented agricultural production (Figure 3). In Russia, the volume of support is very low compared to many other WTO member countries. In 2013, Russia allocated about $471 million as subsidies ($7.9 – $9.4 per hectare of arable land). For comparison, the

Table 5. State support of farmers in Russia, the EU, and the USA in 2000-2017

Years	Russia		EU (28)		USA	
	$, billion	% of gross income	$, billion	% of gross income	$, billion	% of gross income
2000	0.4	1.53	87.8	33.19	50.9	22.67
2005	6.4	14.70	127.5	31.14	40.1	15.05
2010	16.6	22.40	104.7	20.11	30.8	8.58
2011	12.5	12.59	110.5	18.31	32.7	8.02
2012	13.8	14.74	111.8	19.46	36.0	8.46
2013	13.6	13.20	122.8	20.23	29.1	6.91
2014	12.5	13.09	106.2	17.72	40.5	9.26
2015	9.5	12.89	93.8	19.00	38.2	9.45
2016	10.5	14.75	100.0	20.75	36.5	9.56
2017	9.9	12.28	93.2	18.32	39.6	9.88

Source: Authors' development based on Organisation for Economic Cooperation and Development [OECD] (2018)

Figure 3. Export subsidies in some of the WTO countries, $ million
Source: Authors' development based on Ushachev (2013)

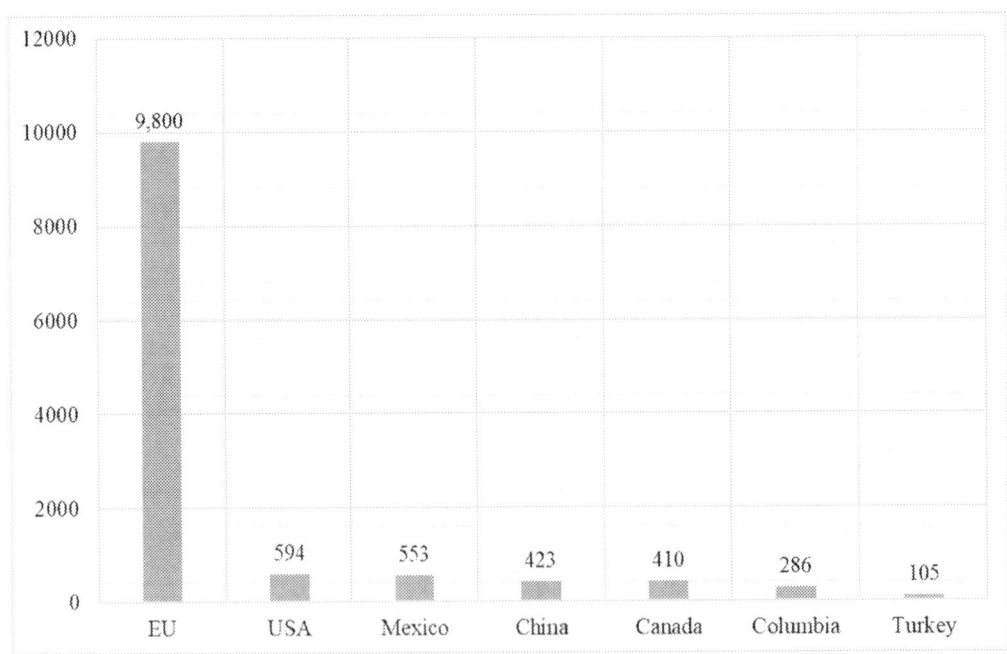

EU allocates $485 per hectare, in particular, France – $514, Germany – $590, Finland – $1,267 (Plotnikov, 2013).

Russia's system of subsidizing interest on loans, compensation for costs of fuels, lubricants, and electricity requires optimization. The funds are allocated mainly to big farms, while small and medium producers lack support. The distribution of funds is not transparent, especially in the sphere of renovation of fixed assets (Lomakin, 2017). One of the critical problems is high interest rate on loans. In Russia, the average interest rate for livestock is 15-16%, while in Germany, the long-term investment credit for the construction of a new meat processing complex can be obtained at 3% per annum for 10-15 years (Samedova, 2013). Loan terms are low and presented up to 3 years, while agricultural production requires a loan term of 5-10 years. Consequently, the level of credit indebtedness of Russian farmers is high (Lomakin, 2017).

The leading food exporting countries provide support to national producers not only in the implementation of export activities but also at the stage of organization of export production. The engineering industry is the priority for credit and credit insurance support for exports in Russia (Lomakin, 2017), while food industry receives insufficient amount of funds (Ivanter, Khazbiev, & Yakovenko, 2014).

Food Security Indicators

In accordance with the Doctrine, to assess the status of food security at the national level, Russia uses the indicator of the share of agricultural products, raw materials and food of domestic production in the total amount of resources (including rolling stocks). Threshold levels are set for eight products and product groups (Table 6). Food security is thus assessed by food independence indicator.

In 2017, Russia reached the threshold level for grain, potatoes, meat and meat products, seed oil, and sugar. For milk and dairy products, the threshold level was not reached, but the dynamic was positive. Thus, according to the criterion established by the Doctrine, both food independence and food security statuses have been improving. However, the share of domestic food in the total volume of commodity resources in the domestic market is the only criterion for assessing food security, as it does not reflect the sustainability of food security in terms of trade restrictions or the cessation of the supply of imported food (Belugin, 2017). Other indicators should be used to assess the stability of food security. For example, the ratio of food production (on average per capita or nationwide) to actual consumption of these products. This indicator is the measure of the level of food self-sufficiency (Charykova & Nesterov, 2014).

According to the results of the Global Food Security Index (GFSI) study in 2018, Russia is ranked 42nd in the overall ranking with 67 points. In terms of economic availability of food, the rank is 37

Table 6. Share of agricultural products, raw materials, and food of domestic production in total resources in Russia, %

Types of agricultural products, raw materials, and food products	2011	2012	2013	2014	2015	2016	2017	Threshold values of the Doctrine
Grain	99.3	98.8	98.4	98.9	99.2	99.2	99.3	95.0
Seed oil	78.0	83.6	81.4	85.0	82.5	83.7	84.8	80.0
Sugar (made from sugarbeet)	62.4	77.9	84.3	81.9	83.3	88.3	94.6	80.0
Potato	95.3	96.8	97.6	97.1	97.1	97.5	97.0	95.0
Milk and dairy products (equivalent to milk)	79.9	78.9	76.5	77.0	79.4	80.3	82.4	90.0
Meat and meat products (equivalent to meat)	73.4	74.8	77.3	81.9	87.2	88.7	90.4	85.0

Source: Authors' development based on Ministry of Agriculture of the Russian Federation (2018)

Figure 4. Russia's scores in the GFSI 2018
Source: Authors' development based on The Economist Intelligence Unit (2018)

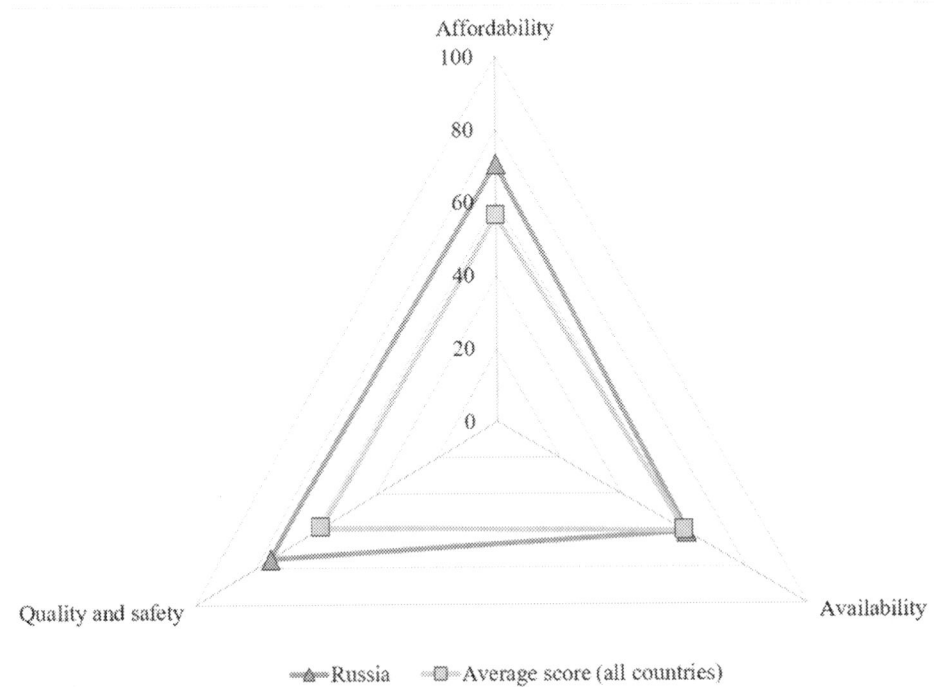

(70.5 points), in terms of availability and sufficiency – 51 (61 points), and in terms of quality and food safety – 25 (75.2 points) (Figure 4).

Russia is scored among the countries where food products are of high quality and safety, the levels of economic availability, affordability, and sufficiency are high. The overall level of food security in Russia remained stable throughout 2012-2018. The quality and safety of food had been improving until 2015 and are now consistently at high level. However, it is necessary to note the negative dynamics of some indicators. Since 2016, there has been a gradual decline in the economic availability of food. In 2013-2015, there was a significant decline in the availability and adequacy of food, which had not reached the level of 2012 despite a certain growth (Table 7).

Russia's strengths in terms of ensuring food security include low proportion of population under global poverty line, food safety net programs, access to financing for farmers, nutritional standards,

Table 7. Russia's GFSI scores, points

Food security indicators	2012	2013	2014	2015	2016	2017	2018
Overall	67.4	66.4	65.4	66.0	66.8	66.2	67.0
Economic availability	68.2	68.5	69.3	72.2	72.0	70.6	70.5
Affordability and sufficiency	64.4	61.9	58.4	56.8	58.8	58.7	61.0
Quality and safety	73.7	73.7	74.5	75.7	75.7	75.7	75.2

Source: Authors' development based on The Economist Intelligence Unit (2018)

food safety, low food loss, sufficiency of supply, low agricultural import tariffs, volatility of agricultural production, urban absorption capacity. Among the weaknesses, there are relatively low GDP per capita, low expenditures on research and development in agriculture, and high corruption (The Economist Intelligence Unit, 2018).

Production of Agricultural Raw Materials

Russia is the biggest country in the world with a territory of 17.12 million km^2. The country is abundant in agricultural land (222.0 million hectares, where 122.7 million hectares of which are arable lands). Water resources (29,013 m^3/capita) add to the establishment of a solid basis of agricultural production (Federal Service for State Statistics of the Russian Federation, 2018b). Russia is the world's biggest producer of sugarbeet (51.9 million tons), third biggest producer of potatoes (21.7 million tons), fourth biggest producer of grain and leguminous crops (135.5 million tons), livestock and poultry for slaughter (10.3 million tons in slaughter weight), and seventh biggest producer of milk (30.2 million tons) (Federal Service for State Statistics of the Russian Federation, 2018a).

In agriculture, however, there are many factors apart from land area and resources that the output. One of the most important is the ability to provide high returns on agricultural land due to the implementation of advanced technologies and innovations (Usenko, 2014). In Russia, agricultural production lags behind the developed countries in labor productivity. There is a lack of equipment and fertilizers; specialization and diversification of export potential are weak (Lomakin, 2017). Agricultural sector is negatively affected by the following aspects:

- WTO obligations and vulnerability of Russian agricultural producers in terms of competition;
- "Oil for food" development model, when Russia exports resources and raw materials and imports food products;
- Economic sanctions;
- Diversities between the territories of Russia in terms of agricultural production and ensurance of food security;
- Monopolization of certain segments of domestic food market under the influence of international trade networks and transnational corporations (Mintusov, 2016).

Domestic producers are slow in their reactions to the changes in demand due to the following reasons:

- High degree of dependence of some sub-sectors of agricultural sector on imports of genetic material, seeds, technological equipment, technologies, veterinary drugs, and plant protection products.
- Low level of cooperation compared to the developed countries, where production chains and networks drive the development of agricultural sector.
- Lack of access to cheap credit resources. In 2017, indebtedness of agricultural enterprises to banks amounted to $44.4 billion, and the amount of overdue debt exceeded $3.6 billion.
- Technological gap between livestock and crop production, chronic problems in inter-sectoral relations and territorial and sectoral division of labor in agriculture, imperfection of the organizational and economic mechanisms in agriculture.
- Low rates of structural and technological modernization of agricultural sector, modernization of production assets in the context of general unfavorable conditions of the industry.

- Undeveloped infrastructure of food market, primarily, wholesale and logistics centers, inability for many agricultural producers to access it.
- Observation of non-equivalence of commodity exchange of agricultural and industrial products, existing inter-sectoral imbalances in the agro-industrial sector of the country (Mintusov, 2016).

However, the potential for agricultural development in Russia is higher than in a number of countries that export food to Russia. In many countries, the expansion of the acreage has been exhausted. As the problem of food insecurity emerges, the role of land resources grows. Russia has an opportunity to grow into one of the global centers for provision of food security, at least in the region of Asia-Pacific (Mintusov, 2016). With climate change and reduced precipitation caused by global warming, Russia's advantages in availability of water resources are gaining importance (The Intergovernmental Panel on Climate Change, 2019). According to Ushachev (2015), as a result of the full use of the potential of agricultural sector, Russia may not to relate on import substitution, but increase the volume of exports of certain products (vegetable oil, sugar, grain, egg, potatoes, and meat) under the most favorable economic conditions. Intensive development of animal husbandry could contribute to the growth of food independence. It is necessary to find the optimal ratio between the volume of feed for livestock and grain exports (Mintusov, 2016).

SOLUTIONS AND RECOMMENDATIONS

There are many ways in which food insecurity may be combated, including adaptation to climate change processes; increasing agricultural production; efficient use of agricultural raw materials; creating an enabling environment for production and trade; reduction of food waste, both at the production stage and at the stage of consumption; and reduction of excessive food consumption (OECD, 2013).

In the course of aggravation of food insecurity problem and depletion of land and natural resources, Russia is expected to play an increasing role. However, having a significant resource potential, Russia has only 2.23% of the world market of food products and agricultural raw materials. The country exports mainly low-value-added products (grains, fish, fats, and oils) and is not a significant food exporter except for grains. If the output of agricultural raw materials and food does not grow, the dependence on imports of basic food products will remain high (Mintusov, 2016).

The prospective ways to promote agricultural production in Russia include the development of rural areas and prevention of desertification; restoration of neglected arable land; expansion of employment in agricultural production; and increasing state support for agricultural sector to the level of developed countries. Major directions of state regulation in the sphere of agriculture should include the unification of methods of providing state support to producers of agricultural goods and food; improvement of the system of subsidies to producers; tighter control over trade and intermediary activities; price regulation; comprehensive development of agricultural insurance; promotion of cooperation in agri-food sector; establishment and development of infrastructure in rural areas; export promotion of excess food; further restriction of imports (Mintusov, 2016).

In the context of Russia's membership in the WTO, economic sanctions against Russia, and economic and political confrontation with the USA, the EU, and some other countries, the most optimal and effective ways to protect domestic food market should include:

- Application of customs and tariff regulation measures to rationalize the ratio of exports and imports of agricultural and fish products, raw materials, and food
- Active use of economic and administrative protection measures with significant imports of agricultural and fish products, raw materials, and food, as well as in cases of dumping and subsidies
- Gradual reduction of dependence of domestic agro-industrial and fisheries complexes on imports of technologies, machinery, equipment, and other resources (Lomakin, 2017)

The increase in the competitiveness of Russian food products in the global market is possible by means of technological and technical modernization of production, primarily in terms of expanding capacities and deepening the processing of raw materials in order to obtain products with higher added value (Lomakin, 2017).

FUTURE RESEARCH DIRECTIONS

In the future, food insecurity problem will become increasingly acute, affecting the socio-economic and political situation in the world. Further research is needed at regional, national, and international levels.

At regional and international levels, the studies should focus on the assessment of the resource base (land, water, climate, and human and financial resources) and the potential of food-supplying countries (Brazil, China, Indonesia, India, Thailand, Mexico, Russia, and others) to ensure world food security in the middle- and long term.

At national level in Russia, future research directions should address:

- Reassessment of objectives and indicators for ensuring food security in the new environment (rising imports, domestic prices for food and fuel and lubricants, taxes, decline in the number of economically active people in rural areas, inflation and depreciation of national currency, effects of economic sanctions, and climate change).
- Development of measures to increase labor productivity and production of high value-added agricultural and food products.
- Development of mechanisms for diversifying food exports and its state support.

CONCLUSION

The fight against hunger and the provision of food to the population have acquired the status of a global problem of mankind after the commodity crises of the 1970s. As a result of theoretical and practical studies, food security is now defined as provision of all people, at all times, with physical, social, and economic access to sufficient, safe, and nutritious food that meets their dietary needs and food preferences for an active and healthy life.

UN Sustainable Development Goals include, among others, ensuring food security, improving nutrition, and the promotion of sustainable agriculture by 2030. Provision of people with food is a fundamental factor of economic, social, and political security of any state. In 2050, population of the world is forecasted to grow up to 9.7 billion people, which will cause the growth in demand for food and the burden on existing natural resources, including land and water. In most of the developed countries, the possibility of expanding agricultural production has been almost exhausted. In this context, the role of Russia, the country with significant agro-climatic, soil-land, and biological resources, in establishing food security is growing.

Russia has great potential to increase output, but labor productivity in agricultural sector is now low. Agricultural products and food produced in Russia are low competitive in the global market due to inefficient agricultural policy of the state. To solve the internal problems of the industry and increase food exports, Russia should address the experience of the world's biggest agricultural producers, such as the USA, the EU, Canada, Brazil, and China. It is necessary to increase the volume and improve the forms of state support, put into circulation unused and abandoned land, improve technological background of the industry, and establish the production of products with high added value. The implementation of these measures will allow Russia becoming a key actor in establishing food security in the world.

REFERENCES

Altukhov, A. (2014). Russia Needs a New Agricultural Policy. *The Economist, 8*, 28-39.

Belugin, A. (2017). *Food Security of the Russian Federation and Its Measurement in Modern Conditions*. Moscow: Moscow State University.

Botkin, O., Sutygina, A., & Sutygin, P. (2016). National Issues of Food Security Assessment. *Bulletin of Udmurt University, 26*(4), 20–27.

Charykova, O., & Nesterov, M. (2014). Theoretical Aspects of Substantiation of Rational Level of Self-Sufficiency of the Regions with Food. In Encyclopedia of Russian Villages (pp. 168-171). Academic Press.

CIS Council of Heads of Government. (2010). *Decision from November 19, 2010, "On the Concept of Improvement of Food Security of CIS Member States."* Retrieved from https://www.fsvps.ru/fsvps-docs/ru/news/files/3143/concept.pdf

Federal Service for State Statistics of the Russian Federation. (2018a). *Statistics Yearbook of Russia. Statistics Digest*. Moscow: Federal Service for State Statistics of the Russian Federation.

Federal Service for State Statistics of the Russian Federation. (2018b). *Russia and Countries of the World. Statistics Digest*. Moscow: Federal Service for State Statistics of the Russian Federation.

Federal Service for State Statistics of the Russian Federation. (2019). *Russia's Foreign Trade in 2018*. Retrieved from http://www.gks.ru/

Food and Agriculture Organization of the United Nations. (1981). *FAO: Its Origins, Formation and Evolution, 1945-1981*. Retrieved from http://www.fao.org/docrep/009/p4228e/P4228E04.htm

Food and Agriculture Organization of the United Nations. (1996). *Rome Declaration on World Food Security*. Retrieved from http://www.fao.org/3/w3613e/w3613e00.htm

Food and Agriculture Organization of the United Nations. (2006). *Food Security. Policy Brief*. Retrieved from http://www.fao.org/forestry/13128-0e6f36f27e0091055bec28ebe830f46b3.pdf

Food and Agriculture Organization of the United Nations. (2009). *Declaration of the World Summit on Food Security.* Retrieved from http://www.fao.org/tempref/docrep/fao/Meeting/018/k6050e.pdf

Food and Agriculture Organization of the United Nations. (2013). *Food Systems for Better Nutrition.* Rome: Food and Agriculture Organization of the United Nations.

Food and Agriculture Organization of the United Nations. (2014). *Second International Conference on Nutrition. Rome Declaration on Nutrition.* Rome: Food and Agriculture Organization of the United Nations.

Food and Agriculture Organization of the United Nations. (2015a). *Thirty-Ninth Session of the European Commission on Agriculture.* Retrieved from http://www.fao.org/europe/commissions/eca/eca-39/en/

Food and Agriculture Organization of the United Nations. (2015b). *Regional Overview of Food Security: Europe and Central Asia. Focus on Healthy and Balanced Nutrition.* Rome: Food and Agriculture Organization of the United Nations.

Food and Agriculture Organization of the United Nations. (2016). *Regional Food Security Review: Europe and Central Asia. Change the Status of Food Security.* Rome: Food and Agriculture Organization of the United Nations.

Food and Agriculture Organization of the United Nations. (2018a). *The Situation of Food Security and Nutrition in Europe and Central Asia.* Rome: Food and Agriculture Organization of the United Nations.

Food and Agriculture Organization of the United Nations. (2018b). *The State of Food Security and Nutrition in the World 2018. Building Climate Resilience for Food Security and Nutrition.* Rome: Food and Agriculture Organization of the United Nations.

Ivanter, A., Khazbiev, A., & Yakovenko, D. (2014). *Time to Change Stereotypes.* Retrieved from https://expert.ru/expert/2014/49/pora-menyat-stereotipyi/

Lomakin, P. (2017). *Ensuring Food Security in Russia: Domestic and International Aspects.* Moscow: Moscow State Institute of International Relations.

Ministry of Agriculture of the Russian Federation. (2018). *National Report "On the Progress and Results of the Implementation of the State Program for the Development of Agriculture and Regulation of Markets of Agricultural Products, Raw Materials, and Food for 2013-2020."* Retrieved from http://mcx.ru/upload/iblock/ec8/ec8f3b2c7fa3b4642f76d3fbda07804b.pdf

Mintusov, V. (2016). *Directions of Improvement of Regulation of Food Imports in the Russian Federation.* Moscow: State University of Management.

Organisation for Economic Cooperation and Development. (2013). Better Policies for Development. In *Focus: Policy Coherence for Development and Global Food Security.* Paris: OECD.

Organisation for Economic Cooperation and Development. (2018). *Monitoring and Evaluation: Reference Tables.* Retrieved from https://stats.oecd.org/viewhtml.aspx?datasetcode=MON2018_REFERENCE_TABLE&lang=en#

Pena, K. (2008). Opening the Door to Food Sovereignty in Ecuador. *Food First News & Views, 111*(30), 1–4.

Plotnikov, V. (2013). If Tens of Billions Rubles Are Not Sent Now – Hundreds Will Not Save Tomorrow. *Krestyanskij Dvor, 21*(6), 8-9.

Podkopaev, O. (2013). State Support of the Agricultural Sector of the Economy in the Conditions of Russia's Membership in the WTO: The Issue of Food Security of the Country. *Successes of Modern Natural Science, 3*, 156–157.

President of the Russian Federation. (2010). *Decree #120 from January 30, 2010, "On the Approval of Food Security Doctrine of the Russian Federation."* Retrieved from http://www.gks.ru/free_doc/new_site/import-zam/ukaz120-2010.pdf

President of the Russian Federation. (2015). *Decree #683 from December 31, 2015, "On National Security Strategy of the Russian Federation."* Retrieved from http://www.consultant.ru/document/cons_doc_LAW_191669/

Reinert, S. (2015). *How Rich Countries Have Become Rich and Why Poor Countries Remain Poor.* Moscow: Higher School of Economics.

Revenko, L. (2003). *World Food Market in the Era of the "Gene" Revolution.* Moscow: Economy.

Revenko, L. (2015). Parameters and Risks of Food Security. *International Processes, 41*(13), 6–20.

Samedova, E. (2013). *Meat Embargo: Help or Bad Turn for Russian Farmers?* Retrieved from http://dw.de/p/17emB

Shields, D. (2015). *Farm Safety Net Programs: Background and Issues.* Washington, DC: Congressional Research Service.

The Economist Intelligence Unit. (2018). *Global Food Security Index.* Retrieved from https://foodsecurityindex.eiu.com/Home/DownloadIndex

The Intergovernmental Panel on Climate Change. (2019). *Final Report. Summary for Politics.* Retrieved from http://www.ipcc.ch/pdf/assessment-report/ar5/syr/AR5_SYR_FINAL_SPM_ru.pdf

The White House. (2016). *Historical Tables: Budget of the U.S. Government. Fiscal Year 2016.* Retrieved from https://www.whitehouse.gov/omb/historical-tables/

Unified Interdepartmental System of Information and Statistics. (2019). *The Share of Imported Food Products in the Commodity Resources of Retail Trade in Food Products.* Retrieved from https://fedstat.ru/indicator/37164

United Nations. (2000). *United Nations Millennium Declaration.* Retrieved from https://www.un.org/millennium/declaration/ares552e.pdf

United States Department of Agriculture. (2010). *National Export Initiative: Importance of U.S. Agricultural Exports.* Retrieved from https://webarchive.library.unt.edu/web/20130216190207/http://@fas.usda.gov/info/NEI/NEInewrev.pdf

United States Government. (1985). *99-198 Food Security Act of 1985.* Retrieved from http://legcounsel.house.gov/Comps/99-198%20-%20Food%20Security%20Act%20Of%201985.pdf

Usenko, L. (2014). Food Security of Russia: Challenges and Implementation Mechanisms. *Proceedings of VEO*, *187*, 198–203.

Ushachev, I. (2013). *Food Security of Russia in the Framework of the Global Partnership*. Moscow: All-Russian Research Institute of Agrarian Economics and Social Development of Rural Territories.

Ushachev, I. (2014). *Scientific Support of the State Program of Development of Agriculture and Regulation of Markets of Agricultural Products, Raw Materials, and Food for 2013-2020 Report at the General Meeting of the Russian Agricultural Academy*. Retrieved from http://www.vniiesh.ru/news/9671.html

Ushachev, I. (2015). Strategic Approaches to the Development of Agriculture in Russia in the Context of Interstate Integration. *Proceedings of the International Scientific Conference "Agrarian Sector of Russia in the Conditions of International Sanctions: Challenges and Responses."* Moscow: Russian State Agrarian University.

Wittman, H., Desmarais, A., & Wiebe, N. (2010). *Food Sovereignty: Reconnecting Food, Nature and Community. Halifax: Fernwood Publishing*. Oakland: FoodFirst Books.

World Trade Organization. (2019). *World Trade Statistical Review 2018*. Geneva: World Trade Organization.

Yakutin, Y. (2014). Food Security – A Strategic Component of National Security of Russia. *Proceedings of VEO*, *187*, 166–167.

ADDITIONAL READING

Bouet, A., & Laborde, D. (2017). *Building Food Security through International Trade Agreements*. Retrieved from http://www.ifpri.org/blog/building-food-security-through-international-trade-agreements

Frumkin, B. (2015). Russian Agricultural Sector in the "War of Sanctions." *Issues of Economics*, *12*, 147–153.

Gavrilyuk, O., Gaidaenko-Sher, I., & Merkulova, T. (2017). *Agrarian Legislation of Foreign Countries and Russia*. Moscow: Institute of Legislation and Comparative Law under the Government of the Russian Federation.

International Food Policy Research Institute. (2018). *Global Food Policy Report 2018*. Washington, DC: IFPRI.

Makarov, I. (2013). *Economic Mechanisms of Providing the World Economy with Water and Food in the Conditions of Global Climate Change*. Moscow: Higher School of Economics.

United Nations. (2015). *Transforming Our World: The 2030 Agenda for Sustainable Development*. Retrieved from https://sustainabledevelopment.un.org/post2015/transformingourworld/publication

United Nations. (2018). *Implementation of the United Nations Decade of Action on Nutrition 2016-2025*. Retrieved from https://www.unscn.org/uploads/web/news/document/MW088-FoodSec-UNGA-Report-en.pdf

KEY TERMS AND DEFINITIONS

Food Access: An access by individuals to adequate resources (entitlements) for acquiring appropriate foods for a nutritious diet. Entitlements are defined as the set of all commodity bundles over which a person can establish command given the legal, political, economic and social arrangements of the community in which they live (including traditional rights such as access to common resources).

Food Availability: An availability of sufficient quantities of food of appropriate quality, supplied through domestic production or imports (including food aid).

Food Independence: A sustainable domestic food production in the volumes not less than the established threshold levels of its share in the commodity resources of the domestic market of relevant products.

Food Safety: An availability of safe and quality food products on the domestic market.

Food Security: A situation where all people are provided with a permanent physical, social and economic access to a sufficient amount of safe and nutritious food that satisfies their nutritional needs and taste preferences for maintaining an active and healthy lifestyle.

Food Self-Sufficiency: A state of the economy which ensures food independence of a country, guarantees physical and economic accessibility for every citizen of the country of food products that meet the requirements of technical regulations, in amounts not less than rational food consumption standards necessary for active and healthy lifestyle.

Stability: To be food secure, a population, household or individual must have access to adequate food at all times. They should not risk losing access to food as a consequence of sudden shocks (e.g. an economic or climatic crisis) or cyclical events (e.g. seasonal food insecurity). The concept of stability can, therefore, refer to both the availability and access dimensions of food security.

Utilization: A utilization of food through adequate diet, clean water, sanitation, and health care to reach a state of nutritional well-being where all physiological needs are met. This brings out the importance of non-food inputs in food security.

Chapter 18
Integrative Associations and Food Security:
Case of China–Russia Interregional Cooperation

Alexander Voronenko
Khabarovsk State University of Economics and Law, Russia

Sergei Greizik
Khabarovsk State University of Economics and Law, Russia

Mikhail Tomilov
https://orcid.org/0000-0002-5048-984X
Economic Research Institute, Far Eastern Branch, Russian Academy of Science, Russia

ABSTRACT

This chapter presents the analysis of agricultural regulations in the frames of the World Trade Organization (WTO) along with an overview of control measures in the sphere of food security in bilateral and multilateral trade unions. The main attention is given to the associations in the Asia-Pacific Region (APR). The chapter concludes with an overview of interregional cooperation between Russia and China in the sphere of agriculture and analysis of its impact on food security of the region. Recommendations for improving and establishing food security are made and future research directions are discussed.

INTRODUCTION

Despite the fact that trade in agricultural products has rather modest share in the world's global trade, agriculture has been one of the most difficult and controversial areas to negotiate trade agreements. Most of the countries implement high tariffs, as well as non-tariff barriers to protect domestic markets from agricultural imports and in such a way ensure food security. The term "food security" was first introduced in 1974 at the Rome Conference on Food Security held by the Food and Agriculture Organization of the

DOI: 10.4018/978-1-7998-1042-1.ch018

United Nations (FAO) after a sharp rise in world prices for grain. In 1996, the definition of food security was included to the Rome Declaration on World Food Security: food security exists when all people, at all times, have physical, social and economic access to sufficient, safe and nutritious food to meet their dietary needs and food preferences for an active and healthy life (Food and Agriculture Organization of the United Nations [FAO], 1996).

In the academic community, however, the understanding of food security has not been unified. Some scholars define food security as a state of food resources in which food needs are met mainly by domestic production in amounts sufficient for normal life of the population (Uskova, 2014). The authors consider this definition as more appropriate. This is proved by strict tariff and non-tariff policy of many countries aimed at supporting the competitiveness of domestic agricultural producers. According to international standards, when a share of imports in domestic consumption of agricultural products exceeds 35-40%, a country is not secured against the interruptions in supply of foreign food products or volatilities in food prices in the global market (Sukhomirov, 2012).

The measures undertaken by governments to support domestic agriculture have a huge impact on food security. Participation in various integration and trade unions may limit sovereign rights to apply agricultural regulatory measures, as well as tariff and non-tariff policies. The advantages of trade liberalization include increased availability of food and its diversity through lower prices, increased incomes of food-producing countries, establishment of stable and uninterrupted supply of food products, technological progress in agriculture through the exchange of technology and diffusion of innovation effect (Martin & Laborde Debucquet, 2018). Negative factors include increased risk of purchasing low-quality products and adverse impact on public health, as well as environmental problems, including the expansion of land use, fertilizers, and energy (Martin & Laborde Debucquet, 2018). Increase in food prices in export-oriented countries due to increased demand for products may also have a negative effect on food security in food-importing states. Liberalization may also bring an increase in food import and a decrease in the competitiveness of national agricultural producers (with possible withdrawal from the market in «non-agricultural» countries), as well as limiting national policies in the regulation of agriculture.

BACKGROUND

Globally, food security studies are conducted by the FAO, International Food Policy Research Institute (IFPRI), and The Economist Intelligence Unit. These organizations collect data on agricultural production and trade, calculate indexes of food security for different countries, describe the situation in this sphere in different regions, and develop measures to address emerging food insecurity problems.

Along with this, IFPRI scholars study the impact of free trade, protectionism, and integration processes on food security across the continents and countries (Martin & Laborde Debucquet, 2018; Wang, 2015; Qi et al., 2017; Grishkova & Poluhin, 2014; Uskova, 2014). So far, however, few studies have actually paid enough attention to the impact of specific measures implemented within various integration and trade unions on food security in member countries. In particular, food insecurity problem has been poorly addressed in China-Russia interregional trade and economic cooperation.

In this chapter, the authors conducted an analysis of agricultural trade regulations within the World Trade Organization (WTO) and made an overview of food security control measures in bilateral and multilateral trade unions, specifically, in Asia Pacific Region (APR). The chapter is concluded by an

Figure 1. Major (a) exporters and (b) importers of agricultural products in 2017, $ billion
Source: Authors' development based on World Trade Organization [WTO] (n.d.)

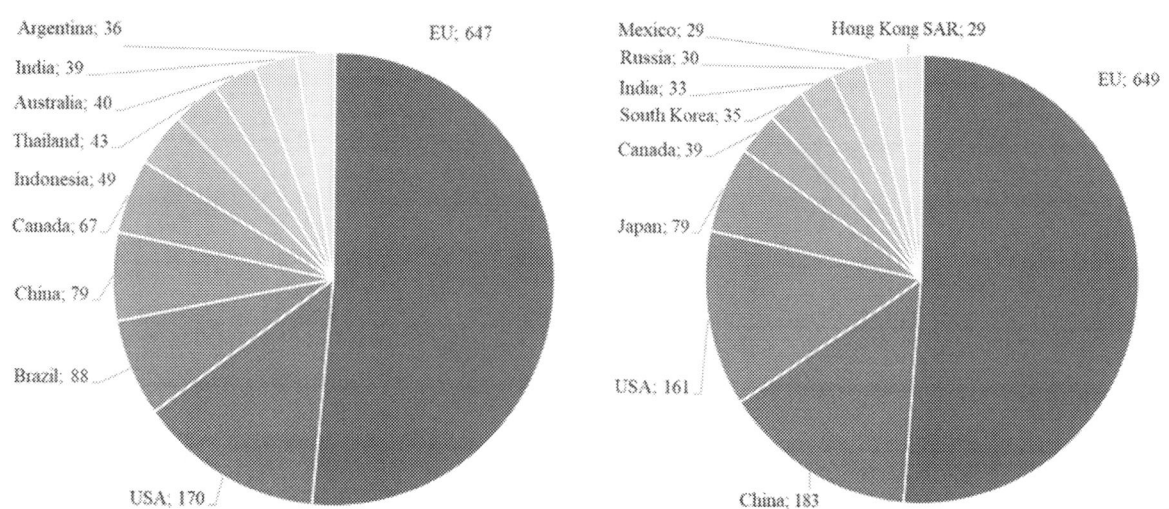

overview of interregional agricultural cooperation between Russia and China and its impact on food security in the two countries.

MAIN FOCUS OF THE CHAPTER

World Agricultural Market: An Overview

The European Union (EU) is the world's biggest supplier and consumer of agricultural products. Other major exporters and importers are APR countries (six out of ten major exporters, seven out of ten major importers) (Figure 1).

Ten top food suppliers occupy 73% of world agricultural export. The share of agricultural trade in gross domestic product (GDP) of EU and APR countries is rather modest (Table 1), but still these states are very active in regulating agricultural markets.

Paradoxically, in some of the major food exporting countries, the share of undernourished people in total population is high. In 2015-2017, about 200 million undernourished people were located in India (1st place in the world), 150 million people – in China (2nd place), over 20 million people – in Indonesia (7th place) (FAO, 2018). In these countries, food security problem is among the most serious ones. The highest average and maximum import tariffs are imposed on agricultural products (Table 2).

In order to support competitiveness of domestic agricultural producers in the domestic markets, the governments impose high tariff barriers. The WTO is working to liberalize agricultural trade and prevent unjustified protectionism in this area.

Table 1. Major exporters of agricultural products in 2017

Country	Agricultural export, $ billion	Share in aggregated national export, %	Share in GDP, %
EU	647	11.0	3.7
USA	170	11.0	0.9
Brazil	88	40.4	4.3
China	79	3.5	0.7
Canada	67	15.9	4.1
Indonesia	49	29.1	4.8
Thailand	43	18.2	9.4
Australia	40	17.3	2.9
India	39	13.1	1.5
Argentina	36	61.5	5.6

Source: Authors' development based on WTO (2018); United Nations Comtrade Database (n.d.)

Table 2. Tariffs on agricultural and non-agricultural products in selected countries in 2017, %

Country	Average		Maximum	
	agricultural products	non-agricultural products	agricultural products	non-agricultural products
EU	10.8	4.2	189	26
China	15.6	8.8	65	50
USA	5.3	3.1	350	56
Japan	13.3	2.5	736	262
Canada	15.7	2.1	484	25
Republic of Korea	56.9	6.8	887	59
India	32.8	10.7	150	125
Russia	10.2	6.2	230	64
Mexico	13.5	5.8	75	50

Source: WTO (2018)

WTO Regulations in Agriculture

Agriculture was excluded out of the WTO trade agreements during the Uruguay round (1986-1994). The main reason was a common opinion that agriculture could not remain competitive under free market regime. Within the WTO, agriculture is now regulated by several agreements including the Agreement on Agriculture, the Agreement on the Application of Sanitary and Phytosanitary Measures, and the Agreement on Subsidies and Countervailing Measures.

The Agreement on Agriculture is based on the following approaches:

- Expanding access to the domestic markets of participating countries (liberalization of tariff regulation)
- Reducing domestic support for national agricultural producers
- Increasing export competition by the reduction of domestic export support

WTO classifies the agriculture-related national regulations across Amber, Green, and Blue boxes.

Green Box contains the measures which are classified as those having no distorting effect or having little distorting effect on trade (WTO, 1995). If the criteria are met, support measures are exempted from the reduction obligation. In addition, such measures are not limited in scope. Green Box measures have to be subsidizing from the budget only, not at the expense of consumers. Also, they cannot be used to support prices.

Blue Box includes the programs aimed at self-limitation of agriculture production. In this case, state budget payments should be related to a fixed number of livestock, or tied to fixed areas and crops, or payments should be related to 85% or less of the basic production level. Such programs are not restricted in volume but are considered to have a distorting effect on trade.

The measures that do not fall under Green Box or Blue Box are the subjects to reduction and are limited in scope for each the WTO member in accordance with its obligations. Such measures are referred to Amber Box and recognized as those having a distorting effect on trade. Amber Box measures are differentiated between product-specific and product-non-specific support. Product-specific support is a support for production, sale, and transportation of particular agricultural products. Such measures include market price maintenance and subsidies for certain types of products. Product-non-specific support is a support for agricultural production, which cannot be distributed between specific goods. Such measures include preferences for electricity, fuel and lubricants, preferential loans, and state investment in agriculture production. Considering the level of support, WTO members make a commitment not to exceed the established indicators of Aggregated Measure of Support (AMS) and to reduce them step by step.

Along with the boxes, de minimis rule is applied which is a threshold level of financing of Amber Box measures and is a kind of quantitative criterion of distorting effect on trade. De minimis is set at 5% or 10% of total agricultural production for developed and developing countries, respectively. For example, a country can use support for specific goods at a level not exceeding 5% of their gross output, and at the same time implement subsidy programs related to product-non-specific support at a level below or equal to 5% of total agricultural production. It is assumed that 5-10% level of support has minimal distorting effects on trade, therefore de minimis is out of reduction, and such subsidies are excluded from the calculation of countries support volumes obligations and current level of AMS.

Thus, WTO has developed and implemented the system that allows not only taking into account various support measures but also differentiating them depending on their effect on trade. However, this system is often criticized (Bouet & Laborde, 2017). For instance, the WTO still favors some measures that distort trade and damage developing countries (export subsidies, insurance, grants, etc.). Moreover, the WTO has limited capacity to influence food price volatility, as well as to accurately monitor country measures that distort agricultural trade. In this regard, countries increasingly prefer to conclude separate bilateral and multilateral trade agreements in the attempts to solve food security problems in a narrower format.

Regulation of Agriculture in the Framework of Multilateral and Bilateral Integrative Associations

European Union

In agriculture, support measures have been developing under different bilateral and multilateral integration and trade agreements. One of the notable examples of such practice is the EU, where about 70% of agricultural export is provided by six member countries (Figure 2).

In the EU, the development of agricultural sector is conducted under the Common Agricultural Policy (CAP) which key objectives are to increase agricultural production, ensure farmers' income and protection from foreign competitors, stabilize agriculture markets, supply the market with food products at affordable prices, and maintain environmental and natural balance.

Within Green Box, the EU provides domestic food aid, non-production assistance to local agricultural producers, compensates losses after natural disasters, subsidizes retired farmers, funds environment protection programs, regional help programs and research and development programs in agriculture. Within Blue Box, subsidy programs for fixed livestock population and fixed acreage are implemented. Within Amber Box, support measures are allocated to support the production of soft wheat, barley, sugar, butter, and milk powder. Basic product-non-specific measures include subsidies for the collection of fruits before their ripening and refusal to harvest, as well as subsidizing interest rates and providing crops' insurance.

Figure 2. Shares of selected member countries in aggregated agricultural export of the EU in 2017, %
Source: Authors' development based on WTO (n.d.)

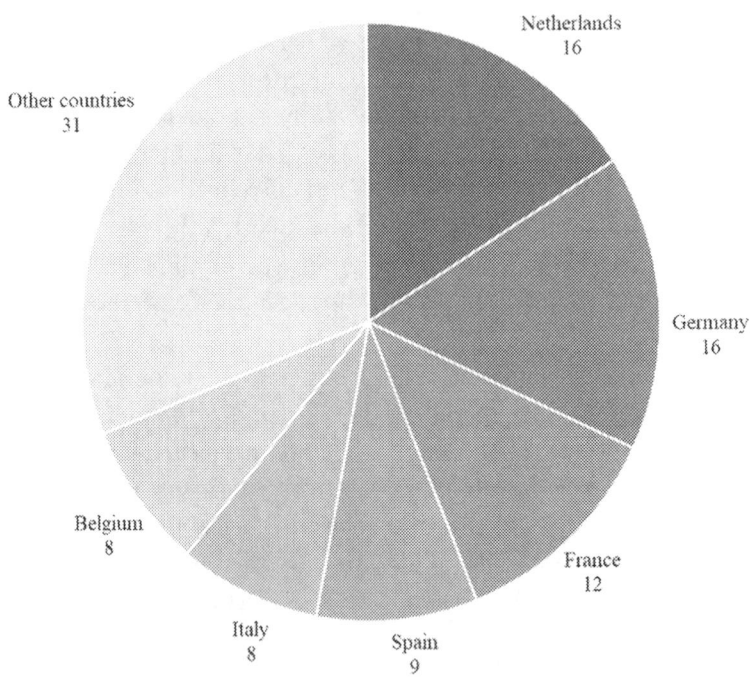

Association of Southeast Asian Nations

In contrast to the EU, the countries of ASEAN (Association of Southeast Asian Nations) almost do not implement coordinated agricultural policy. Basic measures for stimulation of agricultural trade are the elimination and reduction of import tariffs within ASEAN Free Trade Area (AFTA) framework.

The main instrument of trade liberalization is the Common Effective Preferential Tariff (CEPT). CEPT categorizes the products into the four groups:

- Included goods, which are the priority subjects for liberalization (fifteen groups, including agricultural products, such as vegetable oil, rubber, leather products, etc.)
- Temporary exemptions or goods temporarily exempted from liberalization (special transition period)
- Sensitive products, for example, agricultural raw materials, which have a special status and are the objects for negotiations
- General exemptions, which have strategic importance for member countries and thus excluded from liberalization

The AFTA does not include the paragraphs related to the implementation of compensatory and anti-dumping duties. That allows ASEAN members applying such measures to protect domestic markets from imports. In addition to the elimination of tariffs in the mutual trade, the AFTA also covers the cancellation of quantitative restrictions, harmonization of national standards, mutual recognition of test results and quality certificates, as well as the possibility of other non-tariff barriers elimination (Eurasian Economic Commission, 2015). However, the agreement does not affect regulation measures of domestic and export support for farmers.

North American Free Trade Agreement

Within North American Free Trade Agreement (NAFTA) framework, the USA, Canada, and Mexico signed an additional trilateral agreement, which covered the elimination of all quotas, quantitative restrictions, and half of the tariffs. In some cases, quotas were replaced by tariff quotas. In other words, duty-free import was limited by quota, and over-quota import became liable to customs duties. Remaining tariffs and tariff quotas were supposed to be eliminated within the adaptation period. The USA provided the longest adaptation period to domestic producers of sugar, peanuts, orange juice, and some kinds of fruits and vegetables. Mexico provided a transition period to the producers of cereals, including wheat, beans, and milk powder. NAFTA established the trilateral commission to monitor the implementation of the Agreement. At the same time, NAFTA participants reserved the right to apply agricultural production and export support measures, including export subsidies (Schott, Kotschwar, & Muir, 2013).

Agriculture within New APR Integration Unions

In the region of Asia Pacific, three multilateral trade agreements are implemented: Comprehensive and Progressing Agreement for Trans-Pacific Partnership (CPTPP), Regional Comprehensive Economic Partnership (RCEP) (under negotiation), and Free Trade Area of Asia-Pacific (FTAAP) (under negotiation).

RCEP unites the countries in which agriculture is one of the most sensitive areas (the Republic of Korea, India, Indonesia, Philippines, and Thailand). Most of the RCEP members use Amber Box measures which significantly exceed de minimis, as well as impose high tariffs and non-tariff barriers in agriculture.

According to the IFPRI Global Hunger Index (GHI), Malaysia, Vietnam, and Thailand have moderate risk of undernourishment, while in the Philippines, Indonesia, Cambodia, Myanmar, Laos, and India, the level of risk is rated as serious (International Food Policy Research Institute [IFPRI], 2018). Thus, over a half of RCEP member countries experience food security problems. Therefore, liberalization of agricultural trade is not the core subject of the agreement. Countries reserve the right to impose additional import tariffs in order to ensure food security.

APEC established a number of institutions which affect food security, including Agriculture Technical Cooperation Working Group, High-Level Policy Dialogue on Agricultural Biotechnology, Sub-committee on Standards and Conformance, Ocean and Fisheries Working Group, APEC Food System, and APEC Policy Partnership on Food Security. However, their activities are mainly limited to the exchange of experience in the agricultural sector, adoption of non-mandatory recommendations for APEC countries in the sphere of food security, as well as the development of emergency response measures.

Within APEC, the main areas of cooperation in the field of food security are agricultural productivity and economic growth, international trade in food and agricultural products, development and adaptation of new agricultural technologies. The main objectives include sustainable development of the agricultural sector and support for investment, trade, and markets (Antonova, 2012). This work is conducted through the assistance to developing countries, exchange of experience, and joint research, etc. The creation of the FTAAP is hampered by the different positions of the key players in Asia Pacific (the USA, Japan, China, Russia, and the Republic of Korea) in a number of areas, including agriculture, as well as the reluctance to assume any mandatory conditions.

The CPTPP is the only valid agreement out of the three considered. According to the GHI, all the CPTPP members (except Malaysia and Vietnam) do not have a risk of population's undernourishment (IFPRI, 2018). Canada, Australia, New Zealand, Malaysia, Vietnam, Peru, and Chile are the members of Cairns group, an association of agricultural exporters which aims at gradual elimination of support measures that distort agricultural trade.

Canada's position is to eliminate all export subsidies. At the same time, Canada itself does not always committed to the principles of liberalization. In its agricultural policy, Canada increases Amber Box measures. The country applies high tariffs on agricultural products, as well as runs numerous domestic programs to support farmers (dairy and pork producers) (WTO, 2015b).

Australia and New Zealand do not use export subsidies; their tariffs are among the lowest in the world. At the same time, various subsidies are less than 3% of the total farmers' income (average index for developed countries is 19%). Nearly 85% of support measures in Australia and New Zealand refer to Green Box, 15% – to Amber Box, Blue Box measures are not implemented. Most of Amber Box measures are directed on drought control (WTO, 2015e; WTO, 2015a).

Vietnam at its accession to the WTO committed to abandon export subsidies. Nevertheless, there were facts of establishing minimum export prices and the suspension of signing new rice export contracts in 2007-2013. Vietnam applies high tariffs on eggs, sugar, rice, and tobacco. Amber Box measures are used to support the production of rice, sugar (exceeds 10% of de minimis), cotton, and pork. The government provides farmers with preferential loans and subsidies on the usage of irrigation systems (WTO, 2013d).

Malaysia keeps relatively low import tariffs on agricultural products, but actively employs licensing. Special rules apply to rice import. Malaysia does not use export subsidies, but implements tax preferences for particular agricultural products (palm oil) and reduces export duties and licensing barriers. Amber Box measures are used to support the production of rice, sugar, wheat flour, and vegetable oil, as well as to provide farmers with preferential loans and subsidies for the purchase of fertilizers and the construction of irrigation systems (WTO, 2014).

Chile and Peru impose tariffs on agricultural products at a rate of 4-6%. The so-called "system of prices for particular groups of goods" is used. When prices on particular products go below the established threshold, import tariffs are increased, and vice versa. This system is used only for the most sensitive goods. Chile and Peru do not employ Amber Box product-specific support. The most popular product-non-specific measures include preferential lending and taxation, as well as various types of insurance (WTO, 2015c; WTO, 2013c).

Japan is one of the leaders in the area of governmental support for agricultural producers. The country uses import quotas and tariffs to protect national production and employs a complicated licensing and certification system for imports. About 65% of government support measures refer to Green Box (insurance programs, infrastructure construction, compensation for losses from natural disasters), 30% – to Amber Box (support for the production of rice, pork, beef, tobacco, sugar, and dairy products), and 5% – to Blue Box. In addition, Japan uses export credits, guarantees, and insurance (WTO, 2015d).

Mexican government plays significant role in the development of agricultural sector. Mexico, as well as Japan, has high import tariffs on agricultural goods. In addition, Mexico provides export subsidies to wheat and maize producers, runs domestic support programs for farmers, including preferential credits, insurance, subsidies for electricity, and infrastructure construction (WTO, 2013b).

Singapore has one of the most liberal tariff regimes for agricultural goods. This is due to the fact that 90% of food demand is provided by imports. Singapore does not use export subsidies, credits, and insurance to promote agricultural products to international markets. Green Box measures are the most used. Amber Box measures are used to control the import of rice in order to maintain the price (Eurasian Economic Commission, 2015).

Brunei also imports most of its agricultural goods. In this regard, it has low import tariffs, but there are so-called "Halal" requirements for products (WTO, 2013a).

The US government (the USA withdraw the TPP in 2017) plays significant role in the development of agricultural sector. It spends 30% more capital per unit of agricultural production than in any other industry (Neschadin, 2008). However, about 90% of support measures are within Green Box (the most expensive item is the program of providing financial assistance to the population with low incomes for the purchase of food products), 10% – Amber Box (price support, preferential loans and insurance, the construction of infrastructure, etc.). At the same time, support measures for the production of dairy products, sugar, wheat, cotton, mustard, buckwheat, canola, and sunflower exceed de minimis (WTO, 2012). The promotion of national agricultural production to international markets is a separate direction in the support system in the USA. It includes the program of external market development, the program of export activities expansion, the program for developing markets, the program of export products' samples distribution, and the program of compliance with technical standards for the export of certain crops.

State support for agriculture and its regulation within various regional and bilateral trade agreements is carried out in accordance with the WTO rules. In contrast, CPTPP participants agreed to take measures that had not yet been implemented by the WTO (Hufbauer & Cimino-Isaacs, 2015). CPTPP agreement consists of 30 chapters covering a large number of different topics. Agricultural trade liberalization is described in chapter 2 (market access for goods), 3 (rules of origin), 7 (sanitary and phytosanitary measures), 8 (technical barriers to trade), and 18 (intellectual property) (New Zealand Foreign Affairs and

Trade, 2018). The main instrument of trade liberalization within CPTPP is the elimination (significant reduction) of import tariffs and the rising of quotas for duty-free goods. In comparison with other agreements, CPTPP participants managed to agree on the inclusion of "sensitive" goods (rice and pork for Japan, dairy products for Canada, etc.). However, new tariff rules will be imposed step by step during a transition period from two to twenty years (Cimino-Isaacs & Schott, 2016).

Along with tariffs, CPTPP regulates non-tariff barriers. In this regard, several chapters are related to sanitary and phytosanitary regulation. These chapters are ahead of the WTO Agreement on the Application of Sanitary and Phytosanitary Measures in many areas, including risk assessment and management, transparency of regulation, border and laboratory control, as well as the rapid resolution of export-import issues (McMinimy, 2016). It is also planned to create a special Committee on Sanitary and Phytosanitary Control for mobile solutions to emerging problems within CPTPP framework. In addition, CPTPP participants adopted a simplified approach to potentially harmful products based on risk management principle. It assumes that it is possible to produce and sell products if its harm to humans has not yet been proven (Office of the U.S. Trade Representative, 2016). In this regard, it is planned to establish a group on biotechnology in agriculture, which will help to exchange information on CPTPP members' legislative regulation of the biotechnology usage in agriculture.

Regulation of agricultural export within CPTPP also has several innovations in comparison with the WTO and other regional and bilateral agreements. Countries agreed to eliminate export subsidies for agriculture goods distributed within the borders of the partnership. In addition, CPTPP participants committed not to impose export restrictions more than for 6 months, in each case consulting on the issue with an importer. CPTPP members also agreed to continue consultations on reducing export credits, guarantees, and insurance, as well as to promote proposals to limit government support for state agricultural corporations. The countries will work together to promote these initiatives within the WTO framework.

Such measures may reduce distortions in agricultural trade. At the same time, export growth may lead to sharp increase of prices on main agricultural products in the developing countries which are most vulnerable to food insecurity. Some measures (such as biotechnology, intellectual property) can cause a decrease in the quality of agricultural products, and consequently, some health problems among population. Quality and ecological safety can become a victim of economic growth. Certain conclusions can be made only at least five-ten years after the implementation of the CPTPP.

Food Security in China

Trade between China and Russia is one of the examples of the rapidly growing agricultural trade between two countries. In 2013-2016, agricultural trade turnover increased by 30% (United Nations Conference on Trade and Development, 2019). Food security concerns are important for both sides.

In China, food security problem was first addressed in economic policy in December 1974. At that time, food shortage emerged as a result of the upturn of world food prices. Later, the term "food security" was used in Basic Agrarian Policy for the 1980s document, which announced the objective to increase the level of food self-sufficiency and comply it with the program of food market liberalization (Gao, 2017).

Demands for food in China's domestic market is increasing. in the 1970s, Chinese people periodically encountered hunger, many Chinese people suffered from chronic malnutrition. Due to the economic growth of China's in recent years, purchasing power of Chinese people along with the diversity and quality of food basket have been increasing steadily. Consumption of bread and bakery products, noodles, flour confectionery, vegetable oils, and meat and meat products has increased. China's policy on providing the population with essential food is rather efficient. The transition from chronic food shortages to established food security is a result of long-term planning, consistent policies to stimulate food production, and pursuit of advanced technologies (Bobrova, 2013). The country has managed to become one of the largest agricultural producers in the world (Osipova, 2016), a net exporter of grain, and thus demonstrated an example of an effective solution of food security problem (Chun, 2015).

The five factors influence the current state of food security in China, namely, depletion of natural resources, climate change, reduction of arable land, increase in grain imports, and demographic factor. The most sensitive to China's economy and the living conditions of the people is the depletion of natural and strategically important resources that cannot be substituted or imported, such as water, arable land, and forests (Kravchenko & Sergeeva, 2013). There are over 100 million hectares of arable land in China, but the acreage tends to decrease due to the intensive use of land, environmental problems, and climate change.

In this situation, China has been reorienting its agricultural strategy from reliance on domestic production to organization of agricultural enterprises abroad, as well as to increase food imports. One of prospective options of diversification of food imports and ensuring food security is the collaboration with Russia, specifically, with the regions of Russia's Far East.

Agricultural Production in the Russia's Far East

In Russia's Far East, the acreage of agricultural land is about 8.0 million hectares (only 1.3% of the total area of the region). Across Russia, the average share of agricultural lands in total territory of an administrative entity is about 13%. Out of 8 million hectares, only 2.5 million hectares are arable. In general, the level of agricultural development of the Far East is extremely low. This vast, sparsely populated area of 6 million km^2. In 2008, the sown area per capita was 0.2 ha (Russia's average was 0.54 ha). Most of the arable lands are concentrated in the southern part of the district (14:1 ratio against the northern part) (Romanov & Stepanko, 2018). Only southern territories (Khabarovsk and Primorsky krais, Amur Oblast, and Jewish Autonomous Region) are considered a promising area for agricultural production. The sum of active temperatures (the sum of average daily temperatures above 10°C during the year) is above 2,000°C (relatively comfortable for temperate crops) is observed in Zeya-Bureya (Amur Oblast) and Khanka (Primorsky Krai) plains, Jewish Autonomous Region, and southern parts of Khabarovsk Krai (Romanov & Stepanko, 2018). The largest sum of active temperatures is in Primorsky Krai. In this regard, it is possible to cultivate and obtain high yields of both early and late heat-loving crops, such as soybeans and corn. These crops are adapted to a monsoon climate when the main amount of precipitation falls in the second half of summer and autumn. Perennial crops can be successfully cultivated too, including fodder grasses, which not only provide complete feed for livestock but also enrich the soil with organic matter and restore the fertility. The availability of surface waters is the highest among the regions of the country due to a dense network of rivers and lakes (Potenko & Emelyanov, 2018).

Until the 1990s, agricultural sector in Russia's Far East largely met the needs of local population in food (milk – 53.5%, meat – 59.6%, eggs – 98.2%, potatoes – 100%, vegetables – 49.8%). Grain production (including fodder) reached 163 kg per capita, while the average per capita consumption was 114 kg (Romanov & Stepanko, 2018). After the collapse of the USSR, production volumes reduced dramatically. In the beginning of the 1990s, a great part of agricultural products was produced by state and collective farms (average size – six thousand hectares of agricultural land (of which three thousand hectares of arable land), 1.4 thousand heads of cattle, 4.3 thousand pigs, 24 thousand heads of poultry). Over 97%

of large farms were profitable. In most of the farms, the level of profitability was about 20-30%, in some of them – above 40%.

Agrarian reforms of the 1990s resulted in significant structural changes in the size and property of farms. Most of the large farms were divided into several smaller units or completely destroyed. At the beginning of 1999, only 15% of collective farms retained the form of a collective or state farm. Other farms were transformed into partnerships, joint-stock companies, production agricultural cooperatives, or private enterprises (Romanov & Stepanko, 2018). Agricultural sector was not ready for introduction of free-market regime and withdrawal of state support. Since 1920s, there have been no private property in Russia's agriculture which actually predetermined the collapse of the agricultural sector amid the market reforms.

Today, there is observed a recovery. In 2000-2015, the production of grain and soybeans in the Far East almost doubled, while the production of meat (livestock and poultry in slaughter weight) increased by 50%. In 2018, the Far East became self-sufficient in several types of food products and even started exporting grain. The potential of agricultural exports is largely due to the existing competitive advantages of the Far East over some other regions of Russia. First, transport access to the markets of APR countries (China, Republic of Korea, and Japan). Economic growth in APR is one of the major drivers for Russia's export of grain crops, soybeans, potatoes, and compound feedstuff. Second factor is natural resource potential, in particular, the availability of water and land resources.

There is observed the rising interest of investors in the development of grain and legumes production in the Far East. On the one hand, this interest is due to the increase in domestic demand for grain and legumes. Steady growth of meat production in Russia has been increasing the demand for feed and enlarging domestic market of cereals and legumes. On the other hand, interest in grains and legumes production is due to the increased demand from the countries of East Asia. China has become the main market for grain products (99% of exports of legumes from Russia's Far East goes to China). China accounts for about half of the world's supply of pork, which is undoubtedly the main driver of price increases, primarily for soybean, the main feedstuff. Considering the high nutritional value of soya and growing demand for it in the world market, this crop should be considered as one of the most promising export products of the Far East.

Given the high interest of China in ensuring its food security, this situation creates the opportunities for close collaboration between China and Russia's Far East with an aim to diversify food supplies for China. The need to diversify import of agricultural products for China is also stimulated by the ongoing trade tensions with the USA.

China-Russia Interregional Cooperation in Agriculture

Cooperation of China and Russia in the agricultural sector began in the last years of the Soviet era with the liberalization of foreign policy and border regime in the USSR and received a new impetus after the establishment of the Russian Federation. The region of the Far East has emerged as a platform for geopolitical and economic interaction with the APR. However, the real development of this process is ambiguous, accompanied by certain achievements, problems, and difficulties (Chernolutskaya, 2017).

During the 1990s, agricultural sector in the Far East suffered the lack of labor and capital for development (Duncan & Ruetschle, 2002). At the same time, north-eastern province of China (Heilongjiang, Jilin, and Liaoning) were developing (Wishnick, 2005). In China, agrarian reforms of the 1980s increased the capitalization of agricultural production. However, small peasant farms still dominated

Integrative Associations and Food Security

in the North East of China. The land was owned by groups of rural people and distributed among the rural dwellers on an equal basis. The average acreage per capita, however, was very low which did not allow the development of large-scale agricultural production (Zhou, 2015). Environmental degradation, urbanization, water scarcity, and volatile food prices have led to the dramatic reduction of the possibilities of agricultural production in the North East of China. Furthermore, China's accession to the WTO increased the competition from abroad and resulted in the degradation of the production of soybeans, one of the main crops. These facts forced many peasants from northern China who "wanted to improve their economic conditions in the overheated and inflationary Chinese economy" (Moltz, 2002) to move to Russia's Far East (Wishnick, 2005).

Russian legislation of that time in the sphere of land lease was rather contradictory in relation to the foreigners. The new-comers often did not have proper immigration documents. For these reasons, renting a land in Russia by Chinese farmers was often semi-legal or hidden. Chinese farmers often used a mechanism of registration of leased land by Russian citizens. With the adoption of the Federal Law "On the Turnover of Agricultural Land" in 2002, the lease of land to the foreigners in Russia received a legal basis. Land rent increased both in size and in terms. The lease terms increased from one year to five, twenty, or even up to 49 years (the longest term allowed). Land sublease emerged, but that fact was hidden from official bodies in order to avoid tax payments or disguise violations of the law (Zhou, 2015).

In the early 2010s, Chinese farmers cultivated above 850 thousand hectares of agricultural land in the Far East, including 426 thousand hectares in Khabarovsk Krai and the Jewish Autonomous Region. Real figures are even higher since there has been no statistics regarding this issue (Lee, 2013).

In the 2000s, large Chinese companies emerged their operations in the Far East by means of a broader access to financial resources and often hidden or overt government support from both the Chinese and Russian sides. For example, in 2004, "Baoquanling Land Reclamation Far East Agricultural Development Company", one of the state-owned Chinese companies, established two subsidiaries in the Jewish Autonomous Region. In the early 2010s, it operated 51 soybeans and corn farms in Russia with a total acreage of over 10,000 hectares leased from the local administration (Chernolutskaya, 2017).

The adoption of the Federal Law "On the Turnover of Agricultural Land" allowed the acquisition of land by joint-stock companies with a Russian majority. On this basis, Chinese investors entered into a structural symbiosis with Russian organizations forming a number of joint companies. An example of such cooperation is "Armada" company, one of the biggest joint Russia-China agricultural enterprises in Primorsky Krai. It was founded in 2004 and engaged in animal husbandry and arable farming. In 2015, it already used 50,000 hectares of land in the Far East. Its Chinese co-owner Lee Dimin is also the chairman of a larger-scale structure – Dongning Huaxin Industry and Trade Group, which has a wide range of investments in Russia (trade, transport, construction, and real estate), including agriculture. In 2010, "Armada" initiated the establishment of China-Russia zone of agricultural economic cooperation, in 2012, – the establishment of Heilongjiang-Russian Agricultural Cooperative Association, which currently dominates the agriculture of the Far East. It consists of almost 100 corporate members, who have received a total of 380 thousand hectares of land in the Far East and Siberia, each of them of at least 500 hectares (Zhou, 2015). The principles of the organization of this association are known as "Huaxing Model" consisting of many levels and structures. It includes cross-border trading enterprises, the largest state-owned agricultural holding, Beidahuang Group, small and medium-sized business cooperatives, sponsors, each of which has its own responsibilities – from land leasing, access to markets, logistics, and customs clearance to financial support and management of specific manufacturing operations. The group also implements contract farming. The focus of the Group is the cultivation of cereals and legumes, as well as dairy and beef cattle. In addition, it operates three grain processing centers and oil and feed mills. The management of the Group states that it pursues a policy of attracting 60% of local labor resources, and the total number of created jobs exceeds 600 (Zhou, 2015).

Along with this, it is necessary to recognize that the presence of Chinese agricultural companies in Siberia and the Far East of has so far inevitably been accompanied by protest moods, informational actions, and campaigns from the local population. The reason is the negative experience of the late 1990s – early 2000s and, as a result, concern about the environmental component. An example is an agricultural investment project in Transbaikalia region, which is tentatively estimated at 2 billion yuan. However, the serious fears caused by the possible violation of agrotechnical standards adopted in Russia led to the intervention of the federal authorities. As a result, the project was terminated.

In general, Russian authorities have not paid much attention to large agricultural holdings and concerns. The above described period was characterized by high activity of Chinese companies that independently overcome certain difficulties on their own and solving all problems at the local level. State participation in the development of cooperation, if it was present at all, was only from Chinese side. From Russian side, only regional authorities demonstrated some kind of an implicit support. Export of agricultural products to China was carried out mainly by Chinese companies with a large number of barriers, which made it difficult to achieve a substantial level of supply of agricultural products to China. However, due to a number of events on the political map of the world, global food market is transforming. It creates new opportunities for such countries as Russia to increase the share in the market.

Prerequisites for Further Intensive Interregional China-Russia Cooperation in Agricultural Production in the Far East

The conditions for intensive development of cooperation in agriculture in the Far East with an aim to increase exports to China and APR are being actively. The following factors are considered as the most important ones:

- In 2013, Russia's President Putin announced the development of Siberia and the Far East of Russia as "a national priority for the XXI century" (President of the Russian Federation, 2013).
- The tensions between Russia and some developed countries, including the introduction of various trade and economic restrictions from both sides.
- China-USA trade conflict which may affect food security of China. As one of the consequences, China announced its readiness to increase the openness of domestic market for food imports from the countries other than the USA, as well as to expand international cooperation in agriculture.
- Focus on the cooperation in agriculture at the Second Meeting of the Intergovernmental Russia-China Commission on Investment Cooperation (August 2018).
- Memorandum of Understanding between Russia's Ministry of Economic Development and China's Ministry of Commerce on the promotion of bilateral trade.
- Memorandum of Understanding between Russia's Ministry for the Development of the Far East and China's State Committee on Development and Reform on strengthening Russia-China regional, industrial, and investment cooperation in the Far East.
- Agreement on cooperation between Russia's Direct Investment Fund, Vnesheconombank, and China's Eurasian Fund for Economic Cooperation in the field of investment.

- Protocols between Russia's Federal Service for Veterinary and Phytosanitary Surveillance and China's Main State Administration for Quality Control, Inspection, and Quarantine on phytosanitary requirements for wheat, corn, rice, soybeans, and rapeseed exported from Russia to China.

The latter two documents mark a significant step in the development of the relations in the sphere of agriculture and lay a basis for practical implementation of the identified priorities.

According to Gao Feng, official representative of China's Ministry of Commerce, trade in food and agricultural products is one of the main sources of growth of China-Russia bilateral trade: "China is ready to continue strengthening cooperation in agricultural and food trade with Russia by further improvement of trade facilitation, supporting enterprises in their participation in specialized exhibitions and fairs, and actively developing cross-border e-commerce and other measures" (RIA Information Agency, 2018).

In 2018, Russia's Federal Service for Veterinary and Phytosanitary Surveillance (Kulikov, 2018) after the negotiations with China's General Administration of Customs announced that China might open the market for Russian cereals (wheat, soybeans, corn, rice, and rapeseed) when Russia would provide the information required for conducting phytosanitary risk analysis and on-site inspections. This is one of the ways how China sees the opportunities to diversify agricultural imports amid the increasing trade tensions with the USA. Moreover, China is interested in establishing reliable partnerships in the sphere of food supply to ensure its food security.

Meanwhile, there are particular factors that constrain the development of bilateral cooperation in agriculture in the Far East:

- Non-tariff barriers. One of the main obstacles to the development of agricultural exports from Russia to China are trade barriers imposed by China. There are restrictions on the Russia's side too, however, lower ones compared to those applied in China. In China, the major barriers are the compliance with national standards, including sanitary and hygienic, certificates of quality, specific requirements to packaging and labeling of goods, complicated customs formalities, quotas, and licensing, among others (Bokarev, 2016). The barriers are established as trade protection measures in order to prevent imports of those goods that might harm the competitiveness of local producers. In Russia, non-tariff barriers include a set of various organizational decisions of the authorities at various levels, prohibitions and restrictions on import and export of certain agricultural products, as well as the establishment of barriers for non-residents' agricultural businesses (Gavrilyuk, Gaidaenko-Sher, & Merkulova, 2017).
- Underdeveloped infrastructure. Infrastructure of agricultural sector in the Far East has suffered significantly since the collapse of the USSR. Many rural areas are poorly connected with urban and economic centers; the road network is not properly developed or even missed in some territories. Part of the farmland is located outside human settlements and even at a considerable distance from large urban settlements (Pechatnova, 2014; Dobrinsky, 2017). Moreover, there is insufficient number of cross-border crossings (Podberezkina, 2015).
- Labor resources. There is a shortage of specialists and workers at all levels, which significantly slows down the development of Russian agricultural science and impedes the intensive development of the agro-industrial complex. This problem is directly related to the underdevelopment of rural infrastructure (Novikova, 2015).
- Environment. Many experts in Russia are rather skeptical about the possibility of rapid development of agriculture in Far East in cooperation with Chinese companies due to environmental

concerns. There are many examples of negative impact on the environment made by agricultural activities of Chinese companies in Russia in the 1990s and the beginning of the 2000s (Bokarev, 2016). China itself faces environmental problems, including depletion of land resources and lowering fertility of land, many of which have been caused by improper farming practices (Sozinova & Kolpakova, 2015). In the Russian society, the fears are caused by both the intensive use of pesticides by Chinese farmers and non-compliance with the rules of land management.

- Legal issues. Russia and China have not entirely harmonized the legislation in the sphere of bilateral agricultural trade. The progress has been achieved, but some of required technical regulations are still missing. There are no comprehensive program-target documents aimed at the development of conditions for the implementation of bilateral projects (Gavrilyuk et al., 2017).
- Investment risks, which actually result from the problems emphasized above. They include substantial volume of initial investments required to set up a business, low profitability of the invested capital and long payback period, seasonality of production, and uneven receipt and expenditure of funds (Kovarda & Bezuglaya, 2013).

SOLUTIONS AND RECOMMENDATIONS

An increase in agricultural exports to China is hardly possible without support from the administrative bodies in both countries, as well as involvement of large Russian and Chinese agricultural corporations, which are underrepresented in the Russia's Far East. Governmental authorities, as well as the management of large agricultural holdings, should be focused on the elimination of non-tariff barriers. China has already started a move towards business consolidation, which will determine the vector of agricultural development in the region in the near future. An illustrative example is the establishment of the Park of Modern Agriculture and Economy of Heilongjiang Province in Primorsky Krai (Huaxin Corporation). China is now working on launching similar parks in Khabarovsk Krai, Amur Oblast, and Jewish Autonomous Region.

One of the key factors that could increase the volume of agricultural exports to China is the development of transport infrastructure, storage and processing facilities, and cross-border overpasses. At first stage, China-Russia collaboration in the sphere of agriculture in the Far East should result in the involvement of unused land into commercial circulation. In the Far East, about 51% of agricultural land is arable land, the highest share being in Primorsky and Khabarovsk krais and Amur and Sakhalin oblasts. There are regional programs which stimulate the introduction of land into circulation. However, up to 50% of arable land is not used. In 2016, only 62% of arable land was used in Primorsky Krai (in comparison, 80% in Amur oblast).

The process of development of agricultural production should be accompanied by an increase in control over land turnover. In the Jewish Autonomous Region, up to 80% of agricultural land is obtained by Chinese farmers in various illegal ways. In order to solve this problem, it is necessary to develop specific simple administrative and legal mechanisms of rent and create an information system which would accumulate information on agricultural land turnover (owner, possessor, structure of farmland (arable land, deposits, fodder lands, etc.), lands in circulation, soil fertility, and other parameters).

Joint development of farming technologies and crop productivity enhancement are another potential areas of collaboration. In 2010-2015, in the Far East, the average productivity of grain crops was about 1.2 tonnes per hectare, while in Northeast China – 2.7-4.6 tonnes per hectare.

A joint Russia-China research center established in 2018 by the Agricultural Research Center of the Academy of Agricultural Sciences of Jilin Province and Far Eastern Scientific Research Institute of Agriculture may serve as one of the examples of cooperation in the sphere of development of new varieties of plants resistant to natural and climatic conditions of the Far East.

An attention should be paid to the improvement of the system of state support of agricultural sector. To compete with the US farmers, who enjoys governmental subsidies up to $100-150 per hectare, a clear, production-oriented support policy is needed. It is important to improve the legislation in agrarian sphere and elaborate specific production-oriented program documents aimed at the development of the agricultural sector.

In the view of the negative opinion among the society regarding agricultural practices of Chinese farmers, it is worth paying increased attention to positive information campaign.

Both Russian and foreign investors are more interested in crop production as the most profitable industry. Soya is the only highly profitable crop (profitability up to 40%) grown in the southern part of the Far East. However, without proper crop rotation, its constant cultivation may lead to soil depletion. In order to minimize the risk of land depletion and pollution, it is important to make clear requirements in the sphere of soil protection and reclamation.

FUTURE RESEARCH DIRECTIONS

Given the importance of establishing food security for China, as well as the intention of the Russian authorities to revitalize the economy of the Far East, future studies in this area should be carried out on the following directions:

- Assessment of the state of agricultural lands, identification of positive and negative aspects of bilateral cooperation in the field of agriculture, assessment of potential threats to food security as a result of bilateral collaboration in the sphere of agriculture.
- Proposal, evaluation, and analysis of specific joint investment projects in the agricultural sector in Russia's Far East.
- Legal aspects of the implementation of bilateral cooperation, analysis of legal barriers and suggestions for their elimination, studies on the harmonization of legislation in agriculture.
- Tracking global trends in the agricultural sector and studying the possibility of their use in regional cooperation.

CONCLUSION

Recent political challenges, including trade "wars" and tensions, force the countries to increase domestic food production and diversify food imports in order to ensure food security. The escalation of the conflict between China and the USA, as well as the presence of smaller conflicts, forces China to seek new partners to ensure its food security.

The potential of Russia's Far East in the field of food production is significant. Russia is actively working on the attraction of investors to the Far East from China and other countries, in particular, Japan

(Russia-Japan joint venture JGC Evergreen, a resident of the special economic zone "Khabarovsk") and the Republic of Korea.

In "May Decrees", Russia's President Putin set a goal to increase the volume of agricultural export up to $45 billion by 2024 (President of the Russian Federation, 2018). This goal is also included in the "International Cooperation and Export" national project (Government of the Russian Federation, 2019). China is intended to become the major direction for agricultural products from Russia.

Russian government has already paid attention to the development of the agricultural sector in the Far East. It is necessary to harmonize the interest of the Russian authorities in the comprehensive development of the Far East and the needs of Chinese partners in ensuring food security. Given the high rates of agricultural production in the Far East during the Soviet period, as well as the presence of significant scientific and technical progress in this area over the past three decades, a significant agricultural potential of the region with an export orientation may be realized in the future.

REFERENCES

Antonova, N. (2012). Collaboration of APEC Member Countries in the Sphere of Food Security. *Spatial Economics*, *2*, 146–151. doi:10.14530e.2012.2.146-151

Bobrova, L. (2013). Food Security of China as a Reflection of the Economic Growth. *Approbation*, *6*(9), 41–43.

Bokarev, D. (2016). *Russian-Chinese Agricultural Cooperation*. Retrieved from https://journal-neo.org/2016/04/08/russian-chinese-agricultural-cooperation/

Bouet, A., & Laborde, D. (2017). *Building Food Security through International Trade Agreements*. IFPRI. Retrieved from http://www.ifpri.org/blog/building-food-security-through-international-trade-agreements

Chernolutskaya, E. (2017). English-Language Historiography about Chinese Participation in the Agricultural Sector of the South of the Russian Far East during 1990-2010. *Regional Problems*, *20*(3), 50–57.

Chun, Y. (2015). Security at the Tip of the Tongue. *China Journal (Canberra, A.C.T.)*, *10*, 25–26.

Cimino-Isaacs, C., & Schott, J. (2016). *Trans-Pacific Partnership: An Assessment*. Washington, DC: Peterson Institute for International Economics.

Dobrinsky, V. (2017). Analysis of Placing Customs and Logistics Infrastructure in the Russian Federation and the Cargo Volumes Passing through Its Facilities. *Transport Business in Russia*, *1*, 10–13.

Duncan, J., & Ruetschle, M. (2002). Agrarian Reform and Agricultural Productivity in the Russian Far East. In J. Thornton & C. Ziegler (Eds.), *Russia's Far East: A Region at Risk* (pp. 193–220). Seattle, WA: University of Washington Press.

Eurasian Economic Commission. (2015). *Analysis of the Barriers for Access of Agricultural Goods to the Market of the Countries of Southeast Asia*. Retrieved from http://www.eurasiancommission.org/ru/act/prom_i_agroprom/dep_agroprom/monitoring/Documents/Барьеры%20в%20Юго-Восточной%20Азии.pdf

Food and Agriculture Organization of the United Nations. (1996). *Rome Declaration on World Food Security*. Retrieved from http://www.fao.org/3/w3613e/w3613e00.htm

Food and Agriculture Organization of the United Nations. (2018). *World Food and Agriculture. Statistical Pocketbook 2018*. Rome: Food and Agriculture Organization of the United Nations.

Gao, Y. (2017). Policy of Ensuring Food Security of China in the 1990s. *Nauchnyy Dialog*, *10*(10), 201–207. doi:10.24224/2227-1295-2017-10-201-207

Gavrilyuk, O., Gaidaenko-Sher, I., & Merkulova, T. (2017). *Agrarian Legislation of Foreign Countries and Russia*. Moscow: Institute of Legislation and Comparative Law under the Government of the Russian Federation.

Government of the Russian Federation. (2019). *National Project "International Cooperation and Export."* Retrieved from http://government.ru/rugovclassifier/866/events/

Grishkova, Y., & Poluhin, I. (2014). Interaction Problems of Customs and Logistics and Transportation. *Reshetnikov's Reading*, *18*, 433–434.

Hufbauer, G. C., & Cimino-Isaacs, C. (2015). How Will TPP and TTIP Change the WTO System? *Journal of International Economic Law*, *18*(3), 679–696. doi:10.1093/jiel/jgv036

International Food Policy Research Institute. (2018). *2018 Global Food Policy Report*. Washington, DC: International Food Policy Research Institute.

Kovarda, V., & Bezuglaya, Y. (2013). Influence of Infrastructure on the Development of the Agro-Industrial Complex in Russia. *Young Scientist*, *55*(8), 195–198.

Kravchenko, A., & Sergeeva, O. (2013). China Policy in the Area of Food Security: Modernization of Agriculture. *Pacific Rim: Economics, Politics. Law*, *16*(3-4), 57–65.

Kulikov, S. (2018). *Dispatching to the East. China Intends to Increase Imports of Russian Grain*. Retrieved from https://rg.ru/2018/07/23/kitaj-gotov-uvelichit-import-zerna-iz-rossii.html

Lee, R. (2013). *The Russian Far East and China: Thoughts on Cross-Border Integration*. Philadelphia, PA: Foreign Policy Research Institute.

Martin, W., & Laborde Debucquet, D. (2018). Trade: The Free Flow of Goods and Food Security and Nutrition. In *International Food Policy Research Institute, 2018 Global Food Policy Report* (pp. 20–29). Washington, DC: International Food Policy Research Institute.

McMinimy, M. (2016). *TPP: American Agriculture and the Trans-Pacific Partnership (TPP) Agreement*. Washington, DC: Congressional Research Service.

Moltz, J. (2002). Russo-Chinese Normalization from an International Perspective. Coping with the Pressures of Change. In T. Akaha (Ed.), *Politics and Economics in the Russian Far East* (pp. 187–197). New York, NY: Routledge.

Neschadin, A. (2008). Experience of Governmental Regulation and Support of Agriculture Abroad. *Society and Economy*, *8*, 132–150.

New Zealand Foreign Affairs and Trade. (2018). *Comprehensive and Progressive Agreement for Trans-Pacific Partnership Text and Resources.* Retrieved from https://www.mfat.govt.nz/en/trade/free-trade-agreements/free-trade-agreements-in-force/cptpp/comprehensive-and-progressive-agreement-for-trans-pacific-partnership-text

Novikova, V. (2015). Problems of Development of Agriculture in Russia. In *Proceedings of the International Conference "Relevant Issues of Economics, Management, and Finance in Modern Conditions."* Saint Petersburg: Innovation Center for the Development of Education and Science.

Office of the U.S. Trade Representative. (2016). *How TPP Benefits U.S. Agriculture.* Retrieved from https://ustr.gov/sites/default/files/TPP-Benefits-for-US-Agriculture-Fact-Sheet.pdf

Osipova, A. (2016). *China's Policy in the Sphere of Food Security: Environmental Aspect.* Retrieved from https://scienceforum.ru/2016/article/2016026522

Pechatnova, A. (2014). Innovative Development of Agriculture: Problems and Prospects. *Young Scientist, 4*, 427–429.

Podberezkina, O. (2015). Transport Corridors in Russian Integration Projects, the Case of the Eurasian Economic Union. *MGIMO Review of International Relations, 40*(1), 57–65.

Potenko, T., & Emelyanov, A. (2018). Export Potential of Agriculture in the Far East of Russia. *Agrarian Bulletin of the Far East, 45*(1), 125–133.

President of the Russian Federation. (2013). *Presidential Address to the Federal Assembly.* Retrieved from http://en.kremlin.ru/events/president/news/19825

President of the Russian Federation. (2018). *The President Signed Executive Order on National Goals and Strategic Objectives of the Russian Federation through to 2024.* Retrieved from http://en.kremlin.ru/events/president/news/57425

Qi, J., Qian, B., Shang, J., Huffman, T., Liu, J., Pattey, E., ... Wang, J. (2017). Assessing the Options to Improve Regional Wheat Yield in Eastern Canada Using the CSM-CERES-Wheat Model. *Agronomy Journal, 109*(2). doi:10.2134/agronj2016.06.0364

RIA Information Agency. (2018). *Russian Agricultural Imports to China Grew by 48% over Three Quarters.* Retrieved from https://ria.ru/20181123/1533386548.html

Romanov, M., & Stepanko, A. (2018). Dynamics of Territorial Structures of Agriculture of the Far East of Russia. *Agrarian Bulletin of the Far East, 45*(1), 133–143.

Schott, J., Kotschwar, B., & Muir, J. (2013). *Understanding the Trans-Pacific Partnership.* Washington, DC: Peterson Institute for International Economics.

Sozinova, S., & Kolpakova, T. (2015). Trends in the Development of Cooperation between Russia and China in the Sphere of Agriculture. *Russia and China: Problems of Strategic Cooperation. Proceedings of the Eastern Center, 16*(2), 70–81.

Sukhomirov, G. (2012). World Food Crisis and Food Security of the Far Eastern Federal District. *Spatial Economics, 2*(30), 158–160. doi:10.14530e.2012.2.158-160

United Nations Comtrade Database. (n.d.). *International Trade Statistics Database*. Retrieved from http://comtrade.un.org

United Nations Conference on Trade and Development. (2019). *Statistics Database* [Data file]. Retrieved from http://unctad.org/en/Pages/Statistics.aspx

Uskova, R. (2014). *Regional Food Security*. Vologda: Institute for Social and Economic Development of Territories of the Russian Academy of Science.

Wang, X., Cai, D., Grant, C., Hoogmoed, W. B., & Oenema, O. (2015). Factors Controlling Regional Grain Yield in China over the Last 20 Years. *Agronomy for Sustainable Development*, *35*(3), 1127–1138. doi:10.100713593-015-0288-z

Wishnick, E. (2005). Migration and Economic Security: Chinese Labour Migrants in the Russian Far East. *Crossing National Borders: Human Migration Issues in North Asia*, 68-92.

World Trade Organization. (1995). *Agreement on Agriculture*. Retrieved from https://www.wto.org/english/docs_e/legal_e/14-ag_01_e.htm

World Trade Organization. (2012). *Trade Policy Review: United States of America*. Retrieved from https://www.wto.org/english/tratop_e/tpr_e/tp375_e.htm

World Trade Organization. (2013a). *Trade Policy Review: Brunei Darussalam*. Retrieved from https://www.wto.org/english/tratop_e/tpr_e/s309_e.pdf

World Trade Organization. (2013b). *Trade Policy Review: Mexico*. Retrieved from https://www.wto.org/english/tratop_e/tpr_e/s279_e.pdf

World Trade Organization. (2013c). *Trade Policy Review: Peru*. Retrieved from https://www.wto.org/english/tratop_e/tpr_e/s289_e.pdf

World Trade Organization. (2013d). *Trade Policy Review: Vietnam*. Retrieved from https://www.wto.org/english/tratop_e/tpr_e/s287_e.pdf

World Trade Organization. (2014). *Trade Policy Review: Malaysia*. Retrieved from https://www.wto.org/english/tratop_e/tpr_e/s292_e.pdf

World Trade Organization. (2015a). *Trade Policy Review: Australia*. Retrieved from https://www.wto.org/english/tratop_e/tpr_e/s312_e.pdf

World Trade Organization. (2015b). *Trade Policy Review: Canada*. Retrieved from https://www.wto.org/english/tratop_e/tpr_e/s314_e.pdf

World Trade Organization. (2015c). *Trade Policy Review: Chile*. Retrieved from https://www.wto.org/english/tratop_e/tpr_e/s315_e.pdf

World Trade Organization. (2015d). *Trade Policy Review: Japan*. Retrieved from https://www.wto.org/english/tratop_e/tpr_e/s310_e.pdf

World Trade Organization. (2015e). *Trade Policy Review: New Zealand*. Retrieved from https://www.wto.org/english/tratop_e/tpr_e/s316_e.pdf

World Trade Organization. (2018). *World Tariff Profiles 2018*. Retrieved from https://www.wto.org/ENGLISH/res_e/publications_e/world_tariff_profiles18_e.htm

World Trade Organization. (n.d.). *Statistics Database*. Retrieved from http://stat.wto.org/Home/WSDBHome.aspx?Language=E

Zhou, J. (2015). *Chinese Agrarian Capitalism in the Russian Far East*. Retrieved from https://www.tni.org/files/download/bicas_working_paper_13_zhou.pdf

ADDITIONAL READING

Li, M. (2018). Barriers in the Cooperation of China and Russia in the Agrarian Sector. *Economics: Yesterday. Today and Tomorrow, 8*, 313–323.

Minina, E. (2017). Legal Support of Agricultural Development in the Far East. *Russian Law Magazine, 6*(6), 134–144. doi:10.12737/article_59240612b42d52.36179719

Reimer, V., & Ulezko, A. (2016). Conceptual Approach to the Policy Design of Innovation Development of Agro-Industrial Sector of the Far East. *Agricultural Economics of Russia, 1*, 20–26.

Reimer, V., Ulezko, A., & Tyutyunikov, A. (2016). *Innovation-Orientated Development of Agro-Industrial Sector of the Far East*. Voronezh: Voronezh State Agrarian University.

Shelepa, A., & Sukhomirov, G. (2012). Forecast of Development of Agriculture in the Far East till 2015-2020. *Spatial Economics, 4*, 155–165.

KEY TERMS AND DEFINITIONS

Arable Land: A land capable of being plowed and used to grow crops.

Collective Farm: A type of agricultural production in which multiple farmers run their holdings as a joint enterprise. The process by which farmland is aggregated is called collectivization.

Comprehensive and Progressive Agreement for Trans-Pacific Partnership (CPTPP): A trade agreement between Australia, Brunei, Canada, Chile, Japan, Malaysia, Mexico, New Zealand, Peru, Singapore, and Vietnam.

Comprehensive Regional Economic Partnership (CREP): A proposed free trade agreement between Australia, Brunei, Vietnam, India, Indonesia, Cambodia, China, Laos, Malaysia, Myanmar, New Zealand, Republic of Korea, Singapore, Thailand, Philippines, and Japan.

Free Trade Zone of Asia-Pacific (FTZAP): A proposed free trade agreement between Australia, Brunei, Vietnam, Hong Kong SAR, Indonesia, Canada, China, Malaysia, Mexico, New Zealand, Papua New Guinea, Peru, Russia, Republic of Korea, Singapore, USA, Taiwan, Thailand, Philippines, Chile, and Japan.

Northeast Asia: A sub-region of Asia, which consists of the northeastern landmass and islands, bordering the Pacific Ocean. In the chapter, the authors refer this term to China, Japan, and the Republic of Korea as major actors in the sub-region.

Integrative Associations and Food Security

Russia's Far East: The eastern territories of Russia between Lake Baikal in Eastern Siberia and the Pacific Ocean. The region includes the Republic of Sakha (Yakutia), Amur Oblast, Jewish Autonomous Region, Kamchatka Krai, Magadan Oblast, Primorsky Krai, Sakhalin Oblast, Khabarovsk Krai, and Chukchi Autonomous Region. These territories compose the Far Eastern Federal District.

Chapter 19
Water, Food Security, and Trade in Sub-Saharan Africa

Sinmi Abosede
Pan Atlantic University, Nigeria

ABSTRACT

Water is essential for food production and it plays an important role in helping countries achieve food security. The effect of climate change poses significant threats to agricultural productivity in Sub-Saharan Africa, where 95% of agriculture is rain-fed. Changes in weather patterns in the form of prolonged drought and severe flooding, in addition to poor water and land agricultural management practices, has resulted in a significant decline in crop and pasture production in several African countries. The agricultural sector in the region faces the challenge of using the existing scarce water resources in a more efficient way. Most of the countries have failed to achieve food self-sufficiency and rely on imports to meet the demand for food. Agricultural trade can play a significant role in helping countries in Africa achieve food security by increasing availability and access to food in countries that are experiencing food insecurity.

INTRODUCTION

The Sustainable Development Goal (SDG) 2 aims to end hunger, achieve food security and improved nutrition, and promote sustainable agriculture. It calls for the universal access to food while increasing productivity and income of small-scale farmers. Achieving food security will mean that consumers have access to the resources they need to purchase food. In addition, achieving food security depends on a healthy and sustainable food system. Water is essential for food production and rain-fed and irrigated agriculture play a key role in achieving food security. 815 million people are currently undernourished (Food and Agriculture Organization of the United Nations [FAO], International Fund for Agricultural Development [IFAD], United Nations International Children's Emergency Fund [UNICEF], World Food Programme [WFP], & World Health Organization [WHO], 2018) and the global population is projected to increase by 30% to 9.3 billion people by 2050 (United Nations [UN], 2017). To achieve SDG goal 2, these people will have to be fed in a sustainable manner utilizing scarce water and resources. To feed

DOI: 10.4018/978-1-7998-1042-1.ch019

these additional people, it has been estimated that food production will have increased by about 60% by 2050, utilizing the existing resources of land and water (Breene, 2016). In addition to population growth, as a result of increased incomes and changing diet patterns, consumption patterns are switching to increased consumption of beef, dairy, and other livestock products, which require substantial amounts of water to produce. In spite of this, it has also been predicted that approximately 440 million people will still suffer from hunger by 2030.

Achieving the required increase in food production is a daunting task and it is of utmost importance that changes are made across the entire agriculture production chain, to ensure optimum use of the global water resources to meet the growing demand for food. About 80% of the global agriculture is rain-fed. However, to meet the growing demand for food over the years, irrigation has been adopted by many countries as a strategy to increase crop yield per unit area of land. The increased adoption of irrigated agriculture to boost food supplies has placed severe strain on available water resources and agriculture is now responsible for about 71% of the global freshwater withdrawals (FAO, 2014).

Studies on climate projections suggest that that global freshwater resources are vulnerable and have the potential to be strongly impacted by climate change (Bates, Kundzewicz, Wu, & Palutikof, 2008; Parry, Canziani, Palutikof, van der Linden, & Hanson, 2007). In sub-Saharan Africa, climate change poses significant threats to agriculture systems and can impact agriculture productivity, which is severely affected by drought and flooding conditions. The Gross Domestic Product (GDP) of some countries in the region is heavily dependent on agricultural production and it has been predicted that climate change may reduce potential crop yields in some sub-Saharan Africa by 50% leading them to spend 5-10% of their GDP to adapt to the effects of climate change (Intergovernmental Panel on Climate Change [IPCC], 2007). Changes in weather patterns in the form of prolonged drought and severe flooding has resulted in significant declines in crop and pasture production in several countries. For example, Malawi loses approximately 1.7% of its GDP to drought-related events and flooding incidents (Pauw, Thurlow, & Van Seventer, 2010), while it has been estimated that Kenya loses 2.4% of its GDP to drought and flooding (Sanctuary, Tropp, & Haller, 2005). Changes in the quantity and intensity of rain and changes in rainfall patterns make it difficult for farmers to plan their operations, may reduce the cropping season, and lead to low germination, reduced yield, and crop failure. In addition, the increased frequency of storm events damages farmland, crops, and livestock.

The agricultural industry in Africa is faced with the challenge of using the existing and scarce water resources in a more efficient way. Population growth in sub-Saharan African countries has resulted in an increase demand for food. However, most of the countries have experienced low agriculture productivity in recent years, which has led to a change from being net agricultural exporters of food to net agricultural importers. Agriculture international trade can play a significant role in helping countries in Africa achieve food security by increasing availability and access to food in countries that are experiencing food insecurity. It can also help to solve the problem of instability of food prices as a result of severe weather events and increase diversity and range of food available in a particular region.

This chapter will focus on the relationship between water, food security, and agricultural trade in Africa. The work will be conducted as a desktop survey reviewing literature from various academic publications and national and international documents.

BACKGROUND

There have been many definitions of food security and the concept has evolved over the years. In the 1970s, the concept of food security was seen as a function of supply and the main focus of policymakers was to ensure availability of food through the production of food and price stability. In 1975, at the World Food Conference, food security was defined as the availability at all times of adequate world food supplies, to sustain a steady expansion of food consumption and to offset fluctuations in production and prices (UN, 1974). In the 1980s, with the Green Revolution, food production increased, however, poverty levels did not decrease and it became clear that the purchasing power of households was also an important factor in determining food security. In 1983, Food and Agriculture Organization of the United Nations (FAO) gave the following description of food security – "the ultimate objective of world food security should be to ensure that all people of all times have both physical and economic access to the basic food they need. Food security should have three specific aims, namely ensuring production of adequate food supplies; maximizing stability in the flow of supplies; and securing access to available supplies on the part of those who need them" (FAO, 1983).

Currently, the most widely accepted definition of food security is that of the World Food Summit which states that "food security exists when all people, at all times, have physical, social and economic access to sufficient, safe and nutritious food, which meets their dietary needs and food preferences for an active and healthy life" (FAO, 1996). This definition has led to the identification of four dimensions of food security, which are food accessibility, food availability, food utilization, and stability (FAO, 2015b). Food availability is the supply of the sufficient quantities of food, of the appropriate quality through domestic agricultural, production, existing stocks, importation, and food aid. Food access deals with the demand side of food requires individuals to have the necessary resources to acquire food, such as means to physically access the available food and financial resources to purchase food. Utilization of food requires individuals to make good use of the food they access (Barrett, 2010), this will be achieved through adequate diet, clean water, sanitation, and health care, which will ensure all their nutritional and physiological needs are met (FAO, 2016a). Stability involves the ability to withstand risks and uncertainty relating to the availability of and access to food (FAO, 2015b). Possible risks that can affect stability, include the effects of rainfall variability on crop production, economic crisis, and seasonal and regional food scarcity. The four dimensions are interrelated as food availability is essential but not sufficient to guarantee access to food, which is in turn not sufficient for the effective utilization of food. Individuals will not be able to effectively utilize food if they do not have adequate protection from sudden shocks affecting availability of and access to food.

The population of Sub-Saharan Africa was approximately 1.1 billion in 2017 (World Bank, n.d.) accounting for 14% of the global population. The percentage of undernourished people in the region increased from 20.8% in 2015 to 22.7% in 2016. With 224 million undernourished people in 2016, it has the highest rate of undernourishment in the world with 25% of the population experiencing food insecurity (FAO, 2017). The prevalence of undernourishment is often used as a measure of food security. It measures the share of the population who have a caloric intake which is insufficient to meet the minimum energy requirements for an active and healthy life (Kidane, Maetz, & Dardel, 2006). This prevalence of undernourishment for sub-Saharan Africa is 23.3% in comparison to a global average of 10.9% (FAO, 2016a). However, there are limitations to this indicator, as it does not give an indication of the depth of hunger of the undernourished. Another indicator, depth of the food deficit, is used to estimate the intensity of undernourishment. This provides a measure of the number of calories the average individual, would need in order to balance their caloric intake with energy requirements (Roser & Ritchie, 2013). The average depth of food deficit in sub-Saharan Africa was 130 kcal/person/day in 2016 in comparison to a global average of 88 kcal/person/day.

Slow progress towards food security in sub-Saharan Africa can be attributed to low productivity of agricultural resources, high population growth rates, political instability, and civil strife (Organization for Economic Cooperation and Development [OECD[, 2016). Agriculture plays a crucial role in the overall growth and development of region (Ndamani & Watanabe, 2014), contributing about 16% of total GDP of the region and employing about 55% of total workforce (World Bank, n.d.). Approximately 80% of the farms in sub-Saharan Africa are smallholder farms (Alliance for a Green Revolution in Africa [AGRA], 2014) owned predominantly by farmers living in rural communities. Agriculture thus plays an important role in reducing poverty levels and achieving food security, since increases in agriculture output help to reduce food prices. It has been estimated that a 1% increase in crop productivity will reduce the number of poor people in Africa by 0.72% (Thirtle, Lin, & Piesse, 2003).

Water is essential to all life and is crucial for economic development and any poverty alleviation initiatives. It is essential in achieving food security and sustaining the livelihoods of rural communities in developing countries. Sub-Saharan Africa has 3,884 km^3 of total renewable resources, representing about 9.1% of the global freshwater resources and an average water availability per person of 3,967 m^3/person, well below the global average of 5,921 m^3/person (World Bank, 2014). However, within the different sub-regions in sub-Saharan Africa, the availability of freshwater resources are unevenly distributed, with Central and Western Africa having the largest endowments at 51% and 23%, respectively (United Nations Environment Programme [UNEP], 2010). There are also wide variations in the per capita availability of freshwater in the region, ranging from 98 m^3/person in Mauritania to 87,433 m^3/person in Gabon. Water scarcity has been described as the lack of access to adequate quantities of safe water for human and environment uses (White, 2012) and lack of sufficient water resources to meet the water needs within a region. The Falkenmark Water Stress Indicator is a widely used indicator to measure water scarcity. It measures water scarcity as the amount of renewable freshwater that is available per person per year and proposes a threshold of 1,700 m^3 per person per year to identify the regions that suffer from water stress. There are two dimensions to water scarcity – economic and physical scarcity. Economic scarcity occurs when there is a lack of investment in water infrastructure to satisfy the demand for water and physical scarcity occurs when there is a lack of water availability to meet the demand (Molden, 2007). Only 58% of the population in sub-Saharan Africa have access to at least a basic source of improved water, in comparison to a global access level of 89% (World Health Organization [WHO], 2017). Majority of the countries in the region suffer from economic scarcity as there has been a lack of adequate investment in the management of water resources in some areas, resulting in a lack of availability and access to safe water to be used for domestic and agricultural production. As a result, water scarcity presents a challenge to meeting food security in the region.

In sub-Saharan Africa, the effects of climate change and rainfall variability has negative impacts on food availability, affecting crop yields, fish stock, and animal health (FAO, 2016b). In the region, rain-fed agriculture accounts for about 95% of food production (Abrams, 2018), thus making agricultural production limited by the availability of water. Rainfall variability during a growing season generally translates into variability in crop production. Lack of rain or too much rain can lead to famine and increase in poverty levels, as a result of a decrease in food availability. A reduction in the supply of food leads to an increase in the price of food, which decreases the purchasing power of households, limiting access to food. In addition to rainfall variability, there has been an increase in the intensity and frequency of natural disasters in the region, as climate change and various man-made intervention has led to recurrent drought and floods and left many areas vulnerable to desertification, deforestation, and land degradation. This has had significant socioeconomic and environmental impacts on the continent and the situation will worsen in the absence of new investments and successful adaptation (Buhaug, Benjaminsen, Sjaastad, & Theisen, 2015).

In order to meet increased food production, it will be necessary to increase the yields of the land already being cultivated by irrigation. Irrigation has the capacity to boost agricultural production in sub-Saharan Africa by at least 50% (You et al., 2011), as irrigated agriculture produces a crop yield that is three times higher than of rain-fed agriculture (Sasson, 2012), yet for sub-Saharan Africa, the percentage of arable land that is irrigated is approximately 4%, compared to 28% in North Africa (You et al., 2011) and 41% in South Asia (New Partnership for Africa's Development [NEPAD], 2002). It will be necessary to improve irrigation and rain-fed water management practices, to boost agricultural production. Most of the soil used for food production in the region suffer from low fertility, with low concentrations of phosphorus, sulfur, magnesium, zinc, and soil organic matter (Donovan & Casey, 1998). This is compounded by the effect of land degradation in some areas, resulting from extensive farming, deforestation, and overgrazing by livestock. It is essential that farmers in the region adopt sustainable land management practices to improve agricultural productivity and food security.

In spite of the challenges facing the agricultural sector in sub-Saharan Africa, only about 8% of the total arable land is used for agricultural production (Bruinsma, 2003). Substantial investment in the management of land and water resources will need to be made by countries in the region to reach the levels of agricultural production required to meet the various targets for poverty alleviation, food production, and economic recovery (NEPAD, 2002). In the absence of the necessary investment need to improve agricultural productivity, international trade can play a significant role in helping countries in Africa achieve food security, by increasing availability and access to food in countries that are experiencing food insecurity.

MAIN FOCUS OF THE CHAPTER

Agricultural Trade and Food Security

Agricultural trade has helped many countries meet growing food demand in areas faced with these challenges. It impacts each of the four dimensions of food security, through its effect on availability, incomes, prices, stability of supply and through the provision of variety and quality of food products (Brooks & Matthews, 2015). Trade of agricultural products helps to increase food availability and improve access by increasing the incomes of the farmers, which they use to procure food products. It also helps to improve food utilization by increasing the variety and quality of food consumed leading to better nutritional benefits for the consumers. Trade can contribute to the stability of food supply because domestic food production is more likely to be affected by adverse shocks than food obtained from international markets.

Trade can be used as an adaptation mechanism to limit the impact of climate change. It can help countries reduce the impact of rainfall variability and provide an important mechanism to offset reductions in agricultural productivity induced by climate change, thereby improving availability and access to food (Muller & Bellmann, 2016). It can help to mitigate the effects of price instability arising from extreme weather conditions and it allows countries to enjoy greater diversity in their food options. Studies have shown that climate change will result in a decline in global agricultural productivity, leading to an increase in price of agricultural products, which will subsequently result in an increase in international agricultural trade (Ahammad et al., 2015; Von Lampe et al., 2014). A World Bank Study found that by 2080, agricultural trade will become more significant and its share of global agricultural production will increase between 0.4% and 1.2%, as the effect of climate change becomes more pronounced (Havlík et al., 2015).

Rainfall variability induced by climate change can alter the net trade status of countries. Regions that were once food self-sufficient or net exporters may increase imports to meet food demand and become net importers, while regions that gain competitive advantage from the effects of climate change may increase their exports. Elbehri, Elliott, and Wheeler (2015) suggest that the impact of climate change on future food supply will result in increased levels of trade from mid to high latitude regions to low latitude regions, which may suffer from low productivity and decrease in export levels. At high latitudes, water availability is predicted to increase and decrease in low latitudes leading to a knock-on effect on agricultural productivity (Calzadilla, Rehdanz, & Tol, 2011).

Countries have followed two broad options in an attempt to achieve food security – food self-sufficiency and food self-reliance (FAO, 2003). Food self-sufficiency aims to meet the demand for food by supplying food from its own domestic production, while food self-reliance focuses on food availability through domestic food production and importation of food products. Supporters of food self-sufficiency argue that depending on international markets to meet food demand is risky, as it exposes countries to the risk of price volatility and possible disruption of supplies. Following the 2007-2008 global food crisis, many countries expressed increased interest in implementing policies to increase their levels of food self-sufficiency, to protect themselves from any future challenges arising from high and volatile food prices (Clapp, 2015). Supporters of food self-reliance argue that improved food security and efficiency gains can be achieved if the country focuses on producing according to its comparative advantage and sources other food products from the international markets (FAO, 2003).

Agricultural productivity in Sub-Sahara Africa is relatively low in comparison to other regions. The net production value has increased by about 329% from 1961 to 2016, however, the net per capita production has decreased by about 10% during the same period (FAO, 2016a), implying that food production in sub-Saharan Africa did not keep pace with the increase in population over the period. The performance of the food production sector in sub-Saharan Africa is important because domestic production is the main path for ensuring access to affordable food (Demeke, Di Marcantonio, & Morales-Opazo, 2013). However, over the last four decades, most countries have lost their capacity to achieve food self-sufficiency because of the various water and land management challenges and the impact of climate change facing the agricultural sector in the region. In addition to these challenges, poor infrastructure, lack of productive technologies, lack of access to inputs, and weak institutions have contributed to the poor performance of the agricultural sector in the sub-Saharan countries (FAO, 2018b). As a result, the region depends on global markets to meet growing food demand. As a result of a failure to meet food demand through domestic production, most of the countries in sub-Sahara have increased their spending on food imports in recent years. In the 1960-1980s, countries in the region were about 95% self-sufficient in cereals, with the main exports being groundnut, palm kernels, vegetable oil, coffee, cocoa, sugar, and cassava (Kidane et al., 2006), however, in the last four decades, countries in the region have changed from being net exporters of food to being net importers of food (FAO, 2016a). This increase in demand for food can be attributed predominantly to population growth on the continent and growing demand for cereals, dairy products, and protein. FAOSTAT data (FAO, 2016a) shows that in the last four decades, rice and wheat imports have increased by 2,540% and 3,770% respectively. Import of maize which was a major export commodity in the 1960s has increased by 2,751%, while import of other traditional export products like sugar and vegetable oil have also increased by 1,551% and 1,470%, respectively. Milk imports have grown by 7,433%, while import of chicken has increased by 46,745% in last 40 years. Cocoa exports has performed particularly well among the export products and has increased by 183% and imports have decreased by 99% in the last 40 years. However, while increased spending on imported food, has increased food availability in the region, it has not translated to a reduction in food insecurity over the period. This can be attributed to the fact that poor households predominantly those in rural areas, do not have access to imported food products, as they do not have the disposable income to pay for imported food to improve their food security status. In a world where adequate food is globally available, trade policies should make provision for countries facing food insecurity by providing them with the necessary volume of food required to feed their populations (Boussard, Daviron, Gerard, & Voituriez, 2006). In addition, the purchasing power of poor people in rural communities in Sub-Saharan Africa can be increased by helping their products and services gain access to global markets through trade liberation, thus improving their overall welfare (Muller & Bellmann, 2016).

Regional trade within sub-Saharan Africa can also help to improve food security within the continent, however, this opportunity is yet to be fully exploited, as African farmers face more trade barriers and obstacles in accessing other African countries than suppliers from the rest of the world (Brenton, 2012), just 5% of Africa's imports come from other African countries. The countries are unable to derive maximum benefit from the synergies and complementarities of their economies and benefit from economies of scale which increased regional trade will have provided (African Union [AU], 2012). As a result of limited regional trade with Africa, the growing demand for food in the continent is being met by suppliers from the rest of the world. There is uneven distribution of agricultural products across the different countries, with some countries having food surplus, while some suffer from food deficiency. Countries within the region can use regional trade to improve food security, connecting farmers to consumers in other countries, thus mitigating against the effect of food shortages, food price increases and high import bills. Ease of trade across borders will create economies of scale in food production, increase market size for farmers, and decrease the exposure of poor rural families to price instability, arising from climate-induced food shortages (World Bank, 2015). Regional trade has the potential to improve food security, especially in countries that find it difficult to access international markets (Brooks & Matthews, 2015). Accessing food from regional markets can help to improve access to food and stability, as it creates access to more affordable food, less prone to volatility and uncertainty. The low level of intra-regional trade amongst African countries means that products that could have been procured within the region at competitive prices are procured out the region. This has exposed countries in the region to price increases, volatility, and supply uncertainty, resulting from food crises in other parts of the world, this was particularly evident during the 2007-2008 food crisis. The countries in the region can reduce their exposure to risks by trading more with each other. The main advantage of intra-regional trade is to connect areas with food surplus with areas with food deficit. For example, South Africa, Northern Mozambique, Southern Tanzania, and Eastern Uganda produce maize in surplus quantities, local farmers from these countries can supply markets in Southern Mozambique, Malawi, and Kenya, who have a major deficit of maize (Dorosh, Dradri, & Haggblade, 2009). Intra-regional trade can also help to reduce food price volatility in the countries affected by rainfall variability. Increase in imports of stable food can help to protect consumers from price shocks (Haggblade, Me-Nsope, & Staatz, 2017). To protect consumers from price volatility, governments have used export restrictions to prevent price increases, however, this has been shown to have an opposite effect, as it increases prices and volatility (FAO, 2015a). According to Brooks and Matthews (2015), it is more efficient to rely on regional stocks than on national stocks without cooperation with other countries. Other impediments to the increase of intra-regional trade in sub-Saharan Africa include poor road infrastructure, roadblocks, unnecessary permits, and licenses and costly document requirements (Brooks & Matthews, 2015). Intra-regional trade between countries in the region will improve if barriers to trade are lifted, as this will create larger markets for farmers, leading to increased regional sourcing lower imports from outside the region and capture economies of scale, all of which will contribute to price reduction (FAO, 2018a).

Virtual Water Trade

International trade can help to mitigate the effects of water scarcity in a region if a water-intensive product is traded from an area with abundant water resources to an area with scarce water resources. Virtual water has been described as the volume of water used in production of goods or services (Allan, 1993). International trade moves the 'virtual water' from comparatively advantaged regions, rich in water resources to comparatively disadvantaged regions experiencing physical scarcity (Allan, 1998). A country experiencing water scarcity can import water-intensive products and export water-intensive products, this will be an attractive option instead of using scarce water resources to produce the same products domestically (Hoekstra & Hung, 2002). As a result, a country rich in water resources can gain by producing water-intensive products for export, thus contributing to global water use efficiency, thus relieving the pressure of achieving food security on countries experiencing water scarcity (Global Water Partnership [GWP], 2017). Global water use efficiency is achieved because virtual water trade saves water for the countries importing food products and global water savings are achieved from the net water productivity between countries exporting and countries importing.

For example, Jordan imports about 5 km^3 to 7 km^3 of virtual water annually from imported agricultural products and withdraws only 1 km^3 of water from its available water resources (Hoekstra, 2010). Countries in the Middle East/North Africa (MENA) region have used virtual water trade to achieve food security and have moved from food self-sufficiency to food self-reliance, through a substantial increase in food imports, especially grains, to increase food availability. These countries have chosen to focus the use of their limited water resources on urban supply and industrial use instead of on the production of food.

The negative side of the virtual water trade is that as countries produce more food for export, they tend to extract more groundwater for irrigation. As a result, though the virtual water trade is helping some countries to solve their food and water needs, it is putting excessive pressure on non-renewable groundwater. This excessive abstraction of groundwater for irrigation is resulting in the rapid depletion of groundwater in key food-producing regions around the world, with India, China, and the USA, being the worst offenders (Dalin, Wada, Kastner, & Puma, 2017), this exerts negative pressure on the environment and can lead to environmental issues such as land subsidence and seawater intrusion (Konikow & Kendy, 2005). The growing use of irrigation water for the production of food exports has also led to competition and conflict with local users in many regions. In some regions, large companies have gained access to water resources without the consent of local community.

Despite these challenges, trade in virtual water can be used in Africa as a means to achieve water and food security. Countries experiencing water scarcity in Africa can import water-intensive commodities such as cereals from countries with water surplus and use their limited water resources to produce and export high-value crops like fruits, vegetables, and flowers (Diouf, 2003). Konar and Caylor (2013) estimated that the total virtual water traded from the rest of the world to Africa is 61.67 km^3, total virtual water traded from Africa to the rest of the world is 1.8 km^3 and the total virtual water traded within Africa is 3.59 km^3.

SOLUTIONS AND RECOMMENDATIONS

Agriculture trade can play an important role in improving food security in many countries in sub-Saharan Africa. As food is moved from areas of surplus to deficit areas, trade can improve food availability and food access and help address shortfalls caused by water-related events such as droughts and floods.

Countries experiencing food insecurity will benefit from reforms in trade policy from regional and WTO negotiations and bilateral agreements favoring tariff reductions for agricultural products, thereby leading to an increase in imports and reduction in price of imported goods (Chikhuri, 2013). Studies have shown there is a positive economic effect of agricultural trade liberalization for developing countries, as this is expected to compensate for production shortages or changes in production patterns resulting from water scarcity induced by climate change (Anderson, Martin, & Van Der Mensbrugghe, 2005; Francois, Van Meijl, & Van Tongeren, 2005). Wiebe et al. (2015) find that agricultural price increases due to rainfall variability are more significant when trade is restricted across region, in comparison to when all tariffs and export subsidies on agricultural products are lifted. Another study suggests that severe climate change could increase number of malnourished by about 55% if there are restrictions in trade. When the restrictions are lifted for the same scenario, this value reduces by about 25%, as poor households can access imported food products at a lower cost (Baldos & Hertel, 2015).

Water-related events such as drought and flooding can lead to an increase in the price of food, thereby weakening the diversity of food consumed, as a result of a reduction in the purchasing power of consumers from poor households. A study on the impact of food price changes on food security in Ethiopia found that increases in the price of cereals, resulting in households reducing their number of daily meals and switching to less preferred foods (Matz, Kalkuhl, & Abegaz, 2015). Agricultural trade can also help to improve food utilization by strengthening the diversity and safety of foods available, contributing to better nutrition and overall well-being of consumers (FAO, 2018a).

FUTURE RESEARCH DIRECTIONS

While agricultural trade can help to improve food security in countries in sub-Saharan Africa, it can result in the over-reliance of importation to meet the food demand of the countries. This exposes the countries to price volatility arising from food crises in the international markets. As agriculture makes the largest contribution to the GDP of countries in the region, a degree of self-sufficiency is still important, as the purchasing power required to buy imported goods is quite low. These countries may be wary of relying on international markets to meet their food demand. There is a need for research to study means by which countries in the region can promote policies that will build resilience in domestic agricultural production. This will most likely be achieved by investing in the necessary infrastructure and providing the appropriate fiscal environment to boost agricultural productivity of foods that the countries have comparative advantage. This will help to reduce poverty by improving the productivity of small-scale farmers, who can increase their purchasing power by benefiting from the gains of an increase in agricultural production.

CONCLUSION

Sub-Saharan Africa has the highest number of undernourished people in the world. Agriculture plays a crucial role in increasing food security on the continent. However, agricultural productivity is impacted by the effects of rainfall variability, poor infrastructure, lack of productive technologies, lack of access to inputs and weak institutions. As a result of these factors, most sub-Saharan countries have become net importers of food to meet a growing food demand resulting from population growth and change in dietary patterns. Agricultural regional and international trade can play a significant role in helping countries in Africa achieve food security by improving availability and access to food and providing greater diversity.

REFERENCES

Abrams, L. (2018). *Unlocking the Potential of Enhanced Rainfed Agriculture*. Stockholm: Stockholm International Water Institute.

African Union. (2012). *Synthesis Paper on Boosting Intra-African Trade and Fast Tracking the Continental Free Trade Area*. Addis Ababa: African Union.

Ahammad, H., Heyhoe, E., Nelson, G., Sands, R., Fujimori, S., Hasegawa, T., & Tabeau, A. A. (2015). The Role of International Trade under a Changing Climate: Insights from Global Economic Modelling. In A. Elbehri (Ed.), *Climate Change and Food Systems: Global Assessments and Implications for Food Security* (pp. 293–312). Rome: Food and Agriculture Organization of the United Nations.

Allan, J. A. (1993). *Fortunately There Are Substitutes for Water: Otherwise Our Hydropolitical Futures Would Be Impossible*. London: Overseas Development Administration.

Allan, J. A. (1998). Virtual Water: A Strategic Resource Global Solutions to Regional Deficits. *Ground Water*, *36*(4), 545–546. doi:10.1111/j.1745-6584.1998.tb02825.x

Alliance for a Green Revolution in Africa. (2014). *Africa Agriculture Status Report 2014: Climate Change and Smallholder Agriculture in Sub-Saharan Africa*. Nairobi: Alliance for a Green Revolution in Africa.

Anderson, K., Martin, W., & Van Der Mensbrugghe, D. (2005). *Would Multilateral Trade Reform Benefit Sub-Saharan Africans?* Adelaide: Centre for International Economic Studies. doi:10.1596/1813-9450-3616

Baldos, U. L. C., & Hertel, T. W. (2015). The Role of International Trade in Managing Food Security Risks from Climate Change. *Food Security*, *7*(2), 275–290. doi:10.100712571-015-0435-z

Barrett, C. B. (2010). Measuring Food Insecurity. *Science*, *327*(5967), 825–828. doi:10.1126cience.1182768 PMID:20150491

Bates, B. C., Kundzewicz, Z. W., Wu, S., & Palutikof, J. (2008). *Climate Change and Water. Technical Paper of the Intergovernmental Panel on Climate Change*. Geneva: Intergovernmental Panel on Climate Change.

Boussard, J. M., Daviron, B., Gerard, F., & Voituriez, T. (2006). *Food Security and Agricultural Development in Sub-Saharan Africa*. Rome: Food and Agriculture Organization of the United Nations.

Breene, K. (2016). *Food Security and Why It Matters*. WE Forum. Retrieved from https://www.weforum.org/agenda/2016/01/food-security-and-why-it-matters/

Brenton, P. (2012). *Africa Can Help Feed Africa: Removing Barriers to Regional Trade in Food Staples*. Washington, DC: World Bank.

Brooks, J., & Matthews, A. (2015). *Trade Dimensions of Food Security*. Paris: OECD Publishing.

Bruinsma, J. (Ed.). (2003). *World Agriculture: Towards 2015/2030*. London: Earthscan Publications Ltd.

Buhaug, H., Benjaminsen, T. A., Sjaastad, E., & Theisen, O. M. (2015). Climate Variability, Food Production Shocks, and Violent Conflict in Sub-Saharan Africa. *Environmental Research Letters*, *10*(12), 125015. doi:10.1088/1748-9326/10/12/125015

Calzadilla, A., Rehdanz, K., & Tol, R. S. (2011). Trade Liberalization and Climate Change: A Computable General Equilibrium Analysis of the Impacts on Global Agriculture. *Water (Basel)*, *3*(2), 526–550. doi:10.3390/w3020526

Chikhuri, K. (2013). Impact of Alternative Agricultural Trade Liberalization Strategies on Food Security in the Sub-Saharan Africa Region. *International Journal of Social Economics*, *40*(3), 188–206. doi:10.1108/03068291311291491

Clapp, J. (2015). *Food Self-Sufficiency and International Trade: A False Dichotomy?* Rome: Food and Agriculture Organization of the United Nations.

Dalin, C., Wada, Y., Kastner, T., & Puma, M. J. (2017). Groundwater Depletion Embedded in International Food Trade. *Nature*, *543*(7647), 700–704. doi:10.1038/nature21403 PMID:28358074

Demeke, M., Di Marcantonio, F., & Morales-Opazo, C. (2013). Understanding the Performance of Food Production in Sub-Saharan Africa and Its Implications for Food Security. *Journal of Development and Agricultural Economics*, *5*(11), 425–443. doi:10.5897/JDAE2013.0457

Diouf, J. (2003). *Agriculture, Food Security and Water: Towards a Blue Revolution*. Retrieved from http://oecdobserver.org/news/fullstory.php/aid/942/Agriculture,_food_security_and_water_:_Towards_a_blue_revolution.html

Donovan, G., & Casey, F. (1998). *Improving Soil Fertility in Sub-Saharan Africa*. Washington, DC: World Bank. doi:10.1596/0-8213-4236-3

Dorosh, P. A., Dradri, S., & Haggblade, S. (2009). Regional Trade, Government Policy and Food Security: Recent Evidence from Zambia. *Food Policy*, *34*(4), 350–366. doi:10.1016/j.foodpol.2009.02.001

Elbehri, A., Elliott, J., & Wheeler, T. (2015). Climate Change, Food Security and Trade: An Overview of Global Assessments and Policy Insights. In A. Elbehri (Ed.), *Climate Change and Food Systems: Global Assessments and Implications for Food Security and Trade* (pp. 1–27). Rome: Food Agricultural Organisation of the United Nations.

Food and Agriculture Organization of the United Nations. (1983). *Director-General's Report on World Food Security: A Reappraisal of the Concepts and Approaches: Item IV of the Provisional Agenda*. Rome: Food and Agriculture Organization of the United Nations.

Food and Agriculture Organization of the United Nations. (1996). *Rome Declaration on World Food Security*. Retrieved from http://www.fao.org/3/w3613e/w3613e00.htm

Food and Agriculture Organization of the United Nations. (2003). *Trade Reforms and Food Security: Conceptualizing the Linkages*. Rome: Food and Agriculture Organization of the United Nations.

Food and Agriculture Organization of the United Nations. (2014). *Water Withdrawal and Pressure on Water Resources*. Retrieved from http://www.fao.org/nr/water/aquastat/didyouknow/index2.stm

Food and Agriculture Organization of the United Nations. (2015a). *The State of Agricultural Commodity Markets. Trade and Food Security: Achieving a Better Balance Between National Priorities and the Collective Good*. Rome: Food and Agriculture Organization of the United Nations.

Food and Agriculture Organization of the United Nations. (2015b). *Towards a Water and Food Secure Future. Critical Perspectives for Policy*. Rome: Food and Agriculture Organization of the United Nations.

Food and Agriculture Organization of the United Nations. (2016a). *FAOSTAT*. Retrieved from http://www.fao.org/faostat/en/#data/QV

Food and Agriculture Organization of the United Nations. (2016b). *The State of Food and Agriculture. Climate Change, Agriculture and Food Security*. Rome: Food and Agriculture Organization of the United Nations.

Food and Agriculture Organization of the United Nations. (2017). *Africa. Regional Overview of Food Security and Nutrition. The Food Security and Nutrition-Conflict Nexus: Building Resilience for Food Security, Nutrition and Peace*. Accra: Food and Agriculture Organization of the United Nations.

Food and Agriculture Organization of the United Nations. International Fund for Agricultural Development, United Nations International Children's Emergency Fund, World Food Programme, & World Health Organization. (2018). The State of Food Security and Nutrition in the World 2018. Building Climate Resilience for Food Security and Nutrition. Rome: Food and Agriculture Organization of the United Nations.

Food and Agriculture Organization of the United Nations. (2018a). *Africa. Regional Overview of Food Security and Nutrition. Addressing the Threat from Climate Variability and Extremes for Food Security and Nutrition*. Accra: Food and Agriculture Organization of the United Nations.

Food and Agriculture Organization of the United Nations. (2018b). *The State of Agricultural Commodity Markets. Agricultural Trade, Climate Change and Food Security*. Rome: Food and Agriculture Organization of the United Nations.

Francois, J., Van Meijl, H., & Van Tongeren, F. (2005). Trade Liberalization in the Doha Development Round. *Economic Policy*, 42(20), 350–391. doi:10.1111/j.1468-0327.2005.00141.x

Global Water Partnership. (2017). *Virtual Water (C8.04)*. Retrieved from https://www.gwp.org/en/learn/iwrm-toolbox/Management-Instruments/Promoting_Social_Change/Virtual_water/

Haggblade, S., Me-Nsope, N. M., & Staatz, J. M. (2017). Food Security Implications of Staple Food Substitution in Sahelian West Africa. *Food Policy*, 71, 27–38. doi:10.1016/j.foodpol.2017.06.003

Havlík, P., Valin, H., Gusti, M., Schmid, E., Forsell, N., & Herrero, M. ... Obersteiner, M. (2015). *Climate Change Impacts and Mitigation in the Developing World: An Integrated Assessment of the Agriculture and Forestry Sectors*. Washington, DC: World Bank Group.

Hoekstra, A. Y. (2010). *The Relation between International Trade and Freshwater Scarcity*. Geneva: World Trade Organization.

Hoekstra, A. Y., & Hung, P. (2002). *Virtual Water Trade: A Quantification of Virtual Water Flows between Nations in Relation to International Crop Trade*. Delft: IHE Delft Institute for Water Education.

Intergovernmental Panel on Climate Change. (2007). *AR4 Climate Change 2007: The Physical Science Basis*. Cambridge: Cambridge University Press.

Kidane, W., Maetz, M., & Dardel, P. (2006). *Food Security and Agricultural Development in Sub-Saharan Africa. Building a Case for More Public Support*. Rome: Food and Agriculture Organization of the United Nations.

Konar, M., & Caylor, K. K. (2013). Virtual Water Trade and Development in Africa. *Hydrology and Earth System Sciences*, *17*(10), 3969–3982. doi:10.5194/hess-17-3969-2013

Konikow, L. F., & Kendy, E. (2005). Groundwater Depletion: A Global Problem. *Hydrogeology Journal*, *13*(1), 317–320. doi:10.100710040-004-0411-8

Matz, J. A., Kalkuhl, M., & Abegaz, G. A. (2015). The Short-Term Impact of Price Shocks on Food Security – Evidence from Urban and Rural Ethiopia. *Food Security*, *7*(3), 657–679. doi:10.100712571-015-0467-4

Molden, D. (2007). *Water for Food, Water for Life: A Comprehensive Assessment of Water Management in Agriculture*. London: Earthscan.

Muller, M., & Bellmann, C. (2016). *Trade and Water: How Might Trade Policy Contribute to Sustainable Water Management?* Geneva: International Centre for Trade and Sustainable Development.

Ndamani, F., & Watanabe, T. (2014). Rainfall Variability and Crop Production in Northern Ghana: The Case of Lawra District. *Proceedings of the 9th International Symposium on Social Management Systems SSMS 2013*. Sydney: Society for Social Management Systems.

New Partnership for Africa's Development. (2002). *Comprehensive Africa Agriculture Development Programme*. Midrand: New Partnership for Africa's Development.

Organization for Economic Cooperation and Development. (2016). *Agriculture in Sub-Saharan Africa: Prospects and Challenges for the Next Decade*. Paris: OECD Publishing.

Parry, M., Canziani, O., Palutikof, J., van der Linden, P., & Hanson, C. (Eds.). (2007). *Climate Change 2007: Impacts, Adaptation and Vulnerability*. Cambridge: Cambridge University Press.

Pauw, K., Thurlow, J., & Van Seventer, D. (2010). *Droughts and Floods in Malawi: Assessing the Economywide Effects*. Washington, DC: International Food Policy Research Institute.

Roser, M., & Ritchie, H. (2013). *Hunger and Undernourishment.* Retrieved from https://ourworldindata.org/hunger-and-undernourishment#depth-of-the-food-deficit

Sanctuary, M., Tropp, H., & Haller, L. (2005). *Making Water a Part of Economic Development: The Economic Benefits of Improved Water Management and Services.* Stockholm: Stockholm International Water Institute.

Sasson, A. (2012). Food Security for Africa: An Urgent Global Challenge. *Agriculture & Food Security, 1*(2). doi:10.1186/2048-7010-1-2

Thirtle, C., Lin, L., & Piesse, J. (2003). The Impact of Research-Led Agricultural Productivity Growth on Poverty Reduction in Africa, Asia and Latin America. *World Development, 31*(12), 1959–1975. doi:10.1016/j.worlddev.2003.07.001

United Nations. (1974). *Communication from the Commission to the Council. SEC (74) 4955 Final.* New York, NY: United Nations.

United Nations. (2017). *World Population Prospects: The 2017 Revision.* New York, NY: United Nations.

United Nations Environment Programme. (2010). *Africa Water Atlas.* Nairobi: United Nations Environment Programme.

Von Lampe, M., Willenbockel, D., Ahammad, H., Blanc, E., Cai, Y., Calvin, K., ... van Meijl, H. (2014). Why Do Global Long-Term Scenarios for Agriculture Differ? An Overview of the AgMIP Global Economic Model Intercomparison. *Agricultural Economics, 45*(1), 3–20. doi:10.1111/agec.12086

White, C. (2012). *Understanding Water Scarcity: Definitions and Measurements.* Retrieved from http://www.globalwaterforum.org/2012/05/07/understanding-water-scarcity-definitions-and-measurements/

Wiebe, K., Lotze-Campen, H., Sands, R., Tabeau, A., van der Mensbrugghe, D., Biewald, A., ... Willenbockel, D. (2015). Climate Change Impacts on Agriculture in 2050 under a Range of Plausible Socioeconomic and Emissions Scenarios. *Environmental Research Letters, 10*(8), 085010. doi:10.1088/1748-9326/10/8/085010

World Bank. (2014). *Renewable Internal Freshwater Resources.* Retrieved from https://data.worldbank.org/indicator/ER.H2O.INTR.K3

World Bank. (2015). *Improving Food Security in West Africa: Removing Obstacles to Regional Trade Markets.*

World Bank. (n.d.). *World Development Indicators.* Retrieved from https://databank.worldbank.org/reports.aspx?source=2&type=metadata&series=SP.POP.TOTL

World Health Organization. (2017). *Progress on Drinking Water, Sanitation and Hygiene.* Geneva: World Health Organization.

You, L., Ringler, C., Wood-Sichra, U., Robertson, R., Wood, S., Zhu, T., ... Sun, Y. (2011). What Is the Irrigation Potential for Africa? A Combined Biophysical and Socioeconomic Approach. *Food Policy, 36*(6), 770–782. doi:10.1016/j.foodpol.2011.09.001

ADDITIONAL READING

Aksoy, M. A., & Beghin, J. C. (2004). *Global Agricultural Trade and Developing Countries.* Washington, DC: World Bank.

Badiane, O., Odjo, S. P., & Collins, J. (Eds.). (2018). *Africa Agriculture Trade Monitor Report 2018.* Washington, DC: International Food Policy Research Institute.

Besada, H., & Werner, K. (2015). An Assessment of the Effects of Africa's Water Crisis on Food Security and Management. *International Journal of Water Resources Development, 31*(1), 120–133. doi:10.1080/07900627.2014.905124

Chauvin, N. D., Mulangu, F., & Porto, G. (2012). *Food Production and Consumption Trends in Sub-Saharan Africa: Prospects for the Transformation of the Agricultural Sector.* New York, NY: United Nations Development Programme.

Dorodnykh, E. (2017). *Economic and Social Impacts of Food Self-Reliance in the Caribbean.* Cham: Palgrave Macmillan. doi:10.1007/978-3-319-50188-8

Mekonnen, M. M., & Hoekstra, A. Y. (2016). Four Billion People Facing Severe Water Scarcity. *Science Advances, 2*(2), e1500323. doi:10.1126ciadv.1500323 PMID:26933676

Organization for Economic Co-operation and Development, & Food and Agriculture Organization of the United Nations. (2018). *OECD-FAO Agricultural Outlook 2018-2027.* Paris: OECD Publishing; Rome: Food and Agriculture Organization of the United Nations.

Sadoff, C. W., Hall, J. W., Grey, D., Aerts, J. C. J. H., Ait-Kadi, M., & Brown, C. ... Wiberg, D. (2015). Securing Water, Sustaining Growth: Report of the GWP/OECD Task Force on Water Security and Sustainable Growth. Oxford: University of Oxford.

United Nations Development Programme. (2006). *Human Development Report 2006. Beyond Scarcity: Power, Poverty and Global Water Crisis.* London: Palgrave Macmillan UK.

Woodhouse, P., & Ganho, A. S. (2011). Is Water the Hidden Agenda of Agricultural Land Acquisition in Sub-Saharan Africa? *Proceedings of the International Conference on Global Land Grabbing.* Brighton: University of Sussex.

KEY TERMS AND DEFINITIONS

Falkenmark Water Stress Indicator: An indicator used to measure water scarcity. It measures water scarcity as the amount of renewable freshwater that is available per person per year and proposes a threshold of 1,700 m^3 per person per year to identify the regions that suffer from water stress.

Food Access: A requirement of individuals to have the necessary resources to acquire food, such as means to physically access the available food and financial resources to purchase food.

Food Availability: A supply of the sufficient quantities of food, of the appropriate quality through domestic agricultural, production, existing stocks, importation, and food aid.

Food Security: An availability at all times of adequate world food supplies, to sustain a steady expansion of food consumption and to offset fluctuations in production and prices.

Food Self-Reliance: An ability of a country to ensure food availability through domestic food production and importation of food products.

Food Self-Sufficiency: An ability of a country to meet the demand for food by supplying food from its own domestic production.

Food Utilization: The ability of individuals to make good use of the food they access. This will be achieved through adequate diet, clean water, sanitation, and health care, which will ensure all their nutritional and physiological needs are met.

Stability: The ability of a country to withstand risks and uncertainty relating to the availability of and access to food.

Virtual Water: A volume of water used in production of goods or services.

Water Scarcity: A lack of access to adequate quantities of safe water for human and environment uses and lack of sufficient water resources to meet the water needs within a region.

Chapter 20
Prospects, Challenges, and Policy Directions for Food Security in India–Africa Agricultural Trade

Ishita Ghosh
Symbiosis International University (Deemed), India

Sukalpa Chakrabarti
https://orcid.org/0000-0003-2841-2771
Symbiosis International University (Deemed), India

Ishita Ghoshal
https://orcid.org/0000-0002-5993-6431
Fergusson College (Autonomous), India

ABSTRACT

This chapter focusses on the agricultural and investment potential between Sub-Saharan Africa and India, in order to combat food insecurity. There is much scope for meaningful collaboration with governments and public-private partnerships, which could be instrumental in reducing hunger and poverty and managing the adverse effects of climate change. Increasing inclusivity, devising sustainable landholding policies, incentivizing exporters, knowledge sharing in terms of technology and expertise will also boost employability, production and trade potential. Moreover, effective financial and technical cooperation between India and the African countries may be the key to achieving the desired synergies that will bring about positive changes towards ensuring food security.

DOI: 10.4018/978-1-7998-1042-1.ch020

INTRODUCTION

It is a well-documented fact that hunger is on the rise in large regions of the world. The Food and Agriculture Organization of the United Nations (FAO) has found that those facing chronic food deprivation has increased to nearly 821 million in 2017 from around 804 million in 2016, and is still rising. The proportion of malnourished population in the world was projected to have reached a worrisome 10.9% in 2017. The perpetual economic and climatic precariousness across the globe has ended up in grave challenges with respect to food security. Some of the worst affected parts of the world include regions of South America and Africa.

Africa is still affected by the highest prevalence of undernourishment (PoU). Approximately, 21% of the population in the continent have been victims of food insecurity. The condition is also worsening in South America with PoU rising from 4.7% in 2014 to 5.0% in 2017. Although the projected PoU for Asia has slowed down, it is unlikely that without extraordinary measures, the Sustainable Development Goals (SDG) target of wiping out food insecurity in any form will be achieved by 2030, especially in Asia and Africa. Complex socio-economic situations that involve chronic malnourishment cannot be addressed by a single stakeholder alone. Since the underlying causes lay in a variety of economic sectors such as agriculture, infrastructure, social security, international trade and markets, it will require very strong local, national, and global level actions such that policies and programs are beneficial for all.

Africa is the most food-insecure region of the world, constantly battling against acute and chronic hunger. Inadequate infrastructure, lack of investments, and extreme poverty have propagated the malnutrition issue to frightful levels. However, the dependency on agriculture still remains at the heart of the problem. While nearly two-thirds of Africa's population make a living through agriculture, the sector contributes less than one-third to the continent's GDP. Although its importance in the economy differs widely across African countries, agriculture remains a prominent sector for most of them. Additionally, a projected 38% of Africa's working youth are presently employed in agriculture (Bojang & Ndeso-Atanga, 2013; Organization for Economic Cooperation and Development [OECD], 2016).

The natural resources of Africa are greatly underutilized, although the continent produces a large variety of staple food crops and some traditional crops that are high in global demand (coffee, cocoa, tobacco, palm oil, sugar, etc.) (Achancho, 2013; PricewaterhouseCoopers, 2015). In spite of natural endowments, large parts of the continent face acute poverty and hence, hunger. While there are large tracts of agricultural land at their disposal, the levels of mechanization, rainwater harvesting, use of fertilizers, storage and distribution networks and such like facilities still remain the lowest in the world. Given the urgent need for cross border investments, collaborations, and cooperation, the African case needs no special argument in support of these. Also, it has been noted that India and Pakistan have the highest import duties for agricultural goods from most parts of Africa, barring sub-Saharan Africa (SSA). This has raised a lot of questions in terms of the level of cooperation that exists for potentially addressing food security (Stroh de Martinez, Feddersen, & Speicher, 2016; The World Bank, n.d.).

The United Nations Sustainable Development Goals 2 (SDG 2) focuses on the target to end hunger and achieve food security and improved nutrition by 2030. Given that the combined population of Africa and India constitute about one-third of that of the world's, it is obvious that this would comprise the most impactful zone for the SDG 2. Both, the country and the continent of Africa have experienced promising growth and have been attractive destinations for trade and investments. It is projected that Africa's GDP may touch $2.6 trillion by 2020 (Leke & Barton, 2016; Campbell, 2019; Leke, Lund, Roxburgh, & van Wamelen, 2010; United Nations Conference on Trade and Development [UNCTAD], 2018b; United Nations Office for South-South Cooperation, 2019). Also, eleven of the world's fastest-growing economies are in Africa. Presently, India and Africa together, have manpower of almost 2.2 billion and a combined GDP estimate of more than $3 trillion (PricewaterhouseCoopers, 2016; OECD, 2018). The agricultural sector in Africa has great potential to contribute to this growth, with the continent having almost 60% of uncultivated land in the world and currently producing only 10% of the global output (Jones, 2017; Leke et al., 2010; The Economist, 2018; Schwartzstein, 2016). Northern developed economies, and India and China aside, Africa's agricultural development is currently being wooed by states like Brazil, Indonesia, Malaysia, Turkey, Republic of Korea, and the Gulf countries. This explains both South-South and Triangular cooperation being operational in Africa (UNCTAD, 2017; International Fund for Agricultural Development [IFAD], 2017; Pigato, 2009). However, the true impact of these programs for Africa would depend on how the countries in the region are able to strategically engage with the external agencies in ensuring their national interests for growth and development.

Apart from the fact that Africa offers a lucrative food and agricultural market, India's cooperation with Africa in the agricultural sector also comes under the policy ambit of India's commitment to South-South cooperation (SSC). The India-UN Fund portfolio is a great example of South-South and triangular cooperation with development projects and partnerships in Africa, Asia-Pacific, the Caribbean, Europe, and Latin America, ranging across varied thematic areas including agriculture. India's South-South cooperation program which includes overseas aid has shown a significant increase over the years. Africa is targeting to raise its agricultural output from $280 billion in 2010 to $880 billion in 2030. India has partnered with Africa to help realize the latter's development agenda through a multipronged approach that involves extension of credit facilities, bilateral and trilateral investments and partnerships in agriculture and allied sector involving public sector, private sector, and civil society in knowledge and technology transfer, capacity building, and skills development. The Indo Africa Forum Summit (IAFS) is a program fully sponsored by the Ministry of External Affairs (MEA) with a view to develop Indo-Africa cooperation by helping African countries to develop their own potential for development in human resource and agriculture. In a recent development, the MEA has signed a MoU in April 2019 with the National Bank for Agriculture and Rural Development Consultancy Service (NABCONS) for setting up the India-Africa Institute of Agriculture and Rural Development (IAIARD) in Malawi in Southern Africa. IAIARD will be a Pan-African Institute wherein participants from across all African countries, will receive training in order to enhance capacity in the areas of agro-financing and entrepreneurship development for African countries. All of these efforts bear testimony to India's SSC outreach towards Africa. The Indo-Japan Asia-Africa Growth Corridor (AAGC) is a recent endeavor in triangular cooperation partnership to promote development, connectivity, and cooperation between Africa and Asia. The African countries have welcomed India's commitment to engage with the region to work together towards realization of the common goals of food security, improved living standards, and economic growth (IFAD, 2017; Ayyappan, 2015; Modi, Desai, & Venkatachalam, 2019; Kumar & Mazumdar, 2017).

A strong contender to India's SSC outreach to Africa is China. Agriculture and rural development has been a priority for China-Africa cooperation efforts that involve a multitude of projects spread across numerous countries. Apart from a China-Africa Development Fund (2007), China-Africa Cooperation on Agricultural Matters include the strengthening of the exchanges, and cooperation in farming, animal husbandry, irrigation, fisheries, agricultural machinery, processing of agricultural produce, sanitary and phytosanitary measures, food safety and epidemic control, and actively explores new forms and ways of agricultural cooperation. China trains around 10,000 African officials each year through its various programs of which agriculture and development occupy a significant position. Such programs also cater to building relationships with a large pool of African officials and is often interpreted by scholars as the exertion of soft power through development cooperation. Interestingly, there also exist commercial links to these training courses. Several Chinese state-owned companies (such as China State Farm and Agribusiness Corporation) have also adopted an active role in design and implementation of agricultural projects in Africa, which also caters to employment opportunities (UNCTAD, 2009; Kumar & Mazumdar, 2017).

While the scale of Chinese investments in Africa are higher than that of India, the former has been criticized by African countries for lack of transparency and exploitation. India's non-imposing development partnership model on the other hand truly acknowledges Africa as a collaborative partner and that a strong agriculture base is critical for ensuring sustainable economic growth in both Africa and India. This has been well-received by the African leaders.

India is the second-highest Asian investor in Africa, next to China. India and African countries have a large list of agricultural commodities that they trade in. This makes a strong case for strengthening the cooperation between these countries. Additionally, there may be a direct contribution to alleviating each other's food insecurity issues.

BACKGROUND

Africa's natural endowments, if utilized optimally, can feed its own and the world. The continent has 60% of the world's uncultivated arable land. In May 2016, the African Development Bank (ADB) launched the Feed Africa: Strategy for Agricultural Transformation in Africa, 2016-2025 as a part of the high-5 priority projects. The key objective of this initiative is to augment agricultural yield through better infrastructure while scaling up the sector as a viable business model with the help of public-private partnerships. The other objective is to transform the continent to a net exporter of agricultural commodities to be able to substitute or replace imports. Africa's agribusiness is potentially valued at more than $110 billion annually by 2025 (Jones, 2017; Mukasa, Woldemichael, Salami, & Simpasa, 2017; African Development Bank [ADB], 2019).

It has been recorded that while some countries in Africa have a large proportion of arable land (202 million ha of total cultivable land), most farms have less than two hectares of land. This holding size makes large-scale farming difficult. The inputs are minimal and the yield is low, sometimes just enough to feed the farmer's family and maybe the farmhands (The Economist, 2018; Grain, 2014). Such farms are also at the mercy of climatic and environmental changes with no recourse to constant water supply. The consumption of fertilizer for all the farms in the continent is approximately 3% of the global requirements. There is significant scope of using fertilizers. Also, the cost of procuring fertilizers is sometimes ten times more than in other developing or underdeveloped nations. This hurdle has to be resolved urgently. Grossly inadequate mechanization makes it impossible to keep up with desired levels productivity and value addition. This, in turn, affects the structural transformation process and economic diversification. The lost opportunities post-harvest is annually to the tune of $4 billion approximately. This situation needs immediate rectification (Munang, 2015).

While food crops have a double-digit share in the world production measures, cash crops such as coffee, tea, cocoa, and sugar account for 50% of the African agricultural exports. Although cash crops bring better revenue, the production of food crops also need to be stepped up in order to ensure food security (ADB, 2019; Food and Agriculture Organization of the United Nations [FAO], 2013, 2003).

Over the last decade, many countries have realized the potential of investing in the African countries. Till early 2017, the accumulative foreign investment into the African agro sector has been to the tune of aggregated to $9,391 million. India has been one of the forerunners in this case competing neck and neck with China. However, India does not participate as much as China does in agro-based processing. This requires intense investment and money which India has not been able to commit to, unlike China (Livemint, 2017).

Developed but food insecure nations find it easier to import agricultural products by investing in foreign lands. This way, there is more access and control over the arable lands in developing countries. Secondly, the demand for alternative fuels (agrofuels) is on the rise (Ben-Iwo, Manovic, & Longhurst, 2016; FAO, 2014). The International Energy Agency projected that to meet the demand for biofuels by 2030 will require 35 million to 54 million hectares of land (2.5% to 3.8% of available arable land). This demand is likely to be met through oilseeds in the coming decades. Thirdly, the soaring food prices coupled with an abundant availability of arable land make it very lucrative to mechanize large tracts of crop fields and increase yield with new technology while generating employment and other economic opportunities. The one concern that remains is sustainability of the methods employed so that the foreign investor, the local farmers and consumers, as well as the environment continues to prosper with minimum negative externalities, as far as possible (Cotula, Dyer, & Vermeulen, 2008).

Fingar et al. (2017) indicated that although investments have taken place across the various agro sub-sectors in Africa, crop production has appealed the most to investors followed closely by agro value-added activities for natural endowments. The top-ten investing countries in Africa constituted 77% share of the cumulative investments of $10.3 billion with most of the investments coming from European countries like the UK, the Netherlands, Switzerland, and France. The countries that received the highest investments (cumulatively during January 2003 – February 2017) are Nigeria, Egypt, Cote d'Ivoire, South Africa, Zambia, Ghana, Ethiopia, Angola, Tanzania, and Mozambique – together constituting almost 78% of the inward investments into processed agricultural and allied industries in the continent. Agricultural exports from Africa aggregate to $40.8 billion in 2015, accounting for 10.9% of the total exports of the continent. Among the agricultural items exported, cocoa and its preparations constituted approximately 20.8% of the total exports. The exports of edible fruits and nuts valuing around $8.1 billion formed 19.8% share of the aggregate agricultural exports by the continent. Other significant agricultural products exported by Africa included edible vegetables (10.4% of the total exports) and small proportions of coffee, spices, and tobacco. The agricultural imports accounted for nearly 9.6% of the total imports by Africa in 2015. Cereals were the leading import item and its share in the aggregate agricultural imports was 35.5% ($17.1 billion). Animal, vegetable fats and oils formed the second-largest agricultural product imported followed by sugar and sugar confectionery (Fingar et al., 2017; UNCTAD, 2018a; Kumar & Mazumdar, 2017).

India largely depends on its agricultural sector to feed a population of over 1.3 billion and has been able to address food insecurity with better success than some of the other Asian or African countries have been able to so far. Historically, the country has also had political and socio-economic ties with various African countries and both regions have food security listed as a top priority as part of the SDG goals. India has knowledge advantage over the African nations in being able to enhance the agro value chain and gradually become net exporters of agricultural goods (Ghose, 2014; Mahapatra, Rattani, Sengupta, Pandey, & Goswami, 2017). According to the 2018-19 projections, India ranks second worldwide in farm outputs. As per Indian Economic Survey Report (2018), the agricultural sector has employed 50% of the Indian workforce and contributed to about 17-18% to country's GDP. India is also globally the top-ranking nation in terms of the highest net cropped area followed by the USA and China. The economic contribution of agriculture to India's GDP is steadily declining with the country's broad-based economic growth. However, agriculture is demographically the broadest economic sector and plays a significant role in the overall socio-economic fabric of India. India has effectively executed Green Revolution, contract farming, drip irrigation, and created agricultural markets. The Green Revolution executed in 1966 was a targeted action plan involving use of high yielding varieties of seeds, advanced farming and irrigation techniques, and financing of agrochemicals. In supplement to this, GOI adopted the Per Drop More Crop scheme that seeks to promote efficient water conveyance through use of devices like drips, sprinklers, rain guns, and building of micro irrigation structures such as tube wells and dug wells. The economy has also slowly but successfully introduced robotics and precision farming. India's R&D expertise in agro, coupled with institutional and market support in building agribusiness could make India a strong knowledge partner for Africa (Ministry of Finance of India, 2018; Sunder, 2018; India Brand Equity Foundation, 2019; Ministry of Finance of India, 2019).

With increasing rural-urban migrations and a rising per capita income, there has been a rise in the demand for protein consumption in India. Traditionally, a large part of the population prefers and can afford vegetarian/vegan sources of protein. Pulses are the key source of protein for the poor while various cereals and pulses are used for feed to livestock. As the demand for pulses is going up, domestic supply has recurrently been falling short due to rising food price inflation. The 2016-2017 production of pulses has been recorded at 17.8 million tons, and India's consumption has touched at 22.2 million tons, making it imperative to import pulses to the tune of around 4-5 million tons This has put a lot of pressure on the yield of pulses and cereals for domestic consumption. However, the lack of water supply and adequate agrarian land that should be irrigated to match the additional demand surge forces India to import such produce (especially, pulses and oilseeds). India aims to increase its production of oilseeds and pulses and be able to meet domestic demand through domestic production. To be able to achieve this, the country will require an additional 7-8 million ha of arable land. Unless India can solve its problems of fragmented land holding, rising labor costs, lack of mechanization, and shortage of farmhands, there seems to be no recourse for India but to import these produces. In such circumstances, India has turned to Africa for pulses import and other cash crops (The Economic Times, 2018; Ministry of Finance of India, 2018).

Given this backdrop, this chapter attempts to examine the prospects, challenges and policy directions that can mutually help India and its African trading partners address the problems of food security. To address these concerns, the authors have analyzed export and import data of agricultural trade between India and Africa available from various repositories and databases, using statistical tools and other ratios that throw light on comparative advantages, trade potentials, and margins. The data analysis is then used for highlighting the potential for future agricultural trade between India and Sub-Saharan Africa to facilitate food security. Combining the findings with the various channels for cooperation, the solutions are provided in the form of policy recommendations for the cooperation efforts to succeed.

MAIN FOCUS OF THE CHAPTER

India-Africa Cooperation Efforts

India is well aware as to how it can help in playing a crucial role in the rampant African situation of food insecurity. As of now, India has extended support for the development of the cotton sector in the Cotton

Four (C4) countries (i.e. Benin, Burkina Faso, Chad, and Mali) and also in Nigeria, Uganda, and Malawi where India is providing technical assistance, support, and cooperation. The Energy Research Institute (TERI), New Delhi has been actively involved in the Indian Technical and Economic Cooperation (ITEC) program offering African students courses on applications of biotechnology and its regulation. There are also some key Africa-India initiatives undertaken at multilateral levels, particularly in the domain of South-South cooperation. ICRISAT, a CGIAR (Consultative Group on International Agricultural Research) Centre, conducts agricultural research for the development in Asia and Sub-Saharan Africa with a wide array of partners throughout the world. A lot of initiatives have been undertaken under the aegis of the India-Africa Forum Summit.

The EXIM Bank of India has offered Lines of Credit (LOC) to permit Indian exporters to explore new territories and develop businesses in existing and new export markets by way of minimum risks involved. The bank has also been offering LOCs to overseas financial institutions, regional development banks, sovereign governments, and other entities overseas, to enable buyers in those countries to import developmental and infrastructure projects, equipment, goods and services from India, on deferred credit terms. As on March 31, 2017, Africa's share in the total value of Exim Bank's LOC program stood at $7.51 billion, which constituted 47.9% of the total LOC portfolio – of which more than $1.65 billion has been to the agricultural sector alone. These have been provided to as many as 25 African countries for projects as varied as acquisition of tractors, harvesters, agricultural processing equipment; farm mechanization; setting up plantation projects and processing plants; development of sugar industry; procurement of design, supply, installation and commissioning of fuel storage facilities, irrigation network, commissioning of sugar processing facility; rice self-sufficiency program; including setting up of the agri-related institutions like the Mahatma Gandhi Institute of Technology and Biotechnology Park in Cote d'Ivoire. More recently, the bank has extended two LOCs of $30 million and $150 million to the Government of the Republic of Ghana on April 8, 2019, one LOC worth $38 million to the Government of the Republic of Mozambique on March 22, 2019, and three LOCs aggregating to $83.11 million to the Government of the Democratic Republic of the Congo on March 18, 2019.

Given Africa's requirements in agricultural production, India has ventured into important investments in the region, reinforcing each other's growing importance of the India-Africa partnership, and the use of South-South and Triangular Cooperation (SSTC) as an instrument to promote food security and capacity building.

The third India-Africa Forum Summit held in New Delhi, India in October 2015 – the largest of its kind – and the subsequent 52nd annual conference of the ADB held in Gandhinagar, India in May 2017, lay emphasis on India's pledge to firming up the demographic, economic, and government commitment with the African trading partners. India's development pact with Africa in the field of agriculture has been proposed through four routes:

- Bilateral Cooperation. These include Government of India-sponsored activities, such as grant disbursement, and the extension of concessional LoCs and foreign direct investments channel finances and technologies for the development of agriculture and allied sectors in Africa. India has been actively involved with several African nations, sharing knowledge, technology and experiences, facilitating training programs, farming equipment procurement and techniques, soil quality assessment and enrichment, building institutional capacity, appropriate storage and processing technologies, and providing concessional credit in the agriculture and allied sectors.

- Trilateral Partnerships. India has partnered with Africa through various trilateral cooperation initiatives to ensure food security in the region. Some such interventions include the India, Brazil, and South Africa (IBSA) Fund, the United States Agency for International Development (USAID), and the United Kingdom's Department for International Development (DFID)-funded Supporting India's Trade Preference for Africa (SITA) program. Through these programs, the Government of India (GoI) is focusing on capacity building through direct training of farmers for better crop yield, crop and seed quality improvement, enhancement of agri-business, providing access to information on market access requirements, and prevalent duties.
- Private Sector Participation. The GoI is keen on promoting private sector partnerships in Africa to enable FDIs led by the private sector through Business to Business (B2B) projects and sharing of best practices, on all levels, from MSMEs to financial institutions. This would help not just to facilitate greater investments but also to address labor and technology issues that act as hindrances to development. In order to attract FDIs, several African countries have crafted favorable domestic policies through the establishment of financial and regulatory frameworks and ensuring ease of doing business. Indian businesses have been found to be responding positively to these measures and the Public-Private Partnership (PPP) model appears to work well.
- Partnering with Civil Society. Both the governments of India and Africa have also been actively promoting people-to-people (P2P) connect which is further facilitated through civil society partnerships. Indian NGOs have been sharing development experiences in sectors such as agriculture and food security. The Self Employed Women's Association (SEWA), for example, has been doing substantial work on female empowerment and self-reliance in rural India and has been sharing the knowledge and working model with their counterparts in Africa. This helps in enhancing women-to-women (W2W) cooperation in the area of agriculture and the allied sectors. Similarly, the highly successful Amul Milk Cooperative business model is being shared with Tanzania under the program, Milk I-T (India-Tanzania).

It makes practical sense for India to invest in agriculture and collaborate with African countries because of the former's advantages in technology and knowledge and the latter's abundance of land and other natural resources. In fact, India and Africa have made significant advances to this end. African countries definitely gain in the form of large scale employment creation, income generation, and augmentation of value chains through agro-processing.

India-Africa Agricultural Trade Data Analysis

The following section delves into a cursory understanding of the export and import data for agricultural trade between India and Africa in recent years and relies on information for select products traded between India and the African nations. The goods that are traded in mostly between the said set of economies have been selected based on the datasets provided by the Agricultural and Processed Food Products Export Development Authority (APEDA), under the Ministry of Commerce and Industry of India. The ten highest trading agricultural products (in terms of value), both exports and imports to/from Africa from India for the year 2017-18 are selected.

Table 1 depicts the top exports from India to Central Africa, East Africa, North Africa, other South African Countries, and West Africa in 2017-2018. Data for basmati rice was not available.

Table 1. India's exports to Africa (Central Africa, East Africa, North Africa, other South African countries, and West Africa) in 2017-2018

Product	Export volume, $ million
Rice parboiled	1,068.078
Boneless meat of bovine animals, frozen	386.120
Broken rice	336.092
Other rice semi milled or wholly milled rice, whether or not polished or glazed	157.617
Basmati rice	120.873
Sweet biscuits	82.593
Other spirit of un-denatured ethyl alcohol	79.762
Chickpeas (Garbanzos), dried, shelled, whether or not skinned/split	44.358
Other sugar confectionery, not containing cocoa	39.330
Glucose and glucose syrup, not containing fructose or containing in dry state less than 20% by weight of fructose, liquid	28.439
Butter	26.157

Source: Authors' development based on Agricultural and Processed Food Products Export Development Authority [APEDA] (n.d.)

Table 2 depicts the top imports from Central Africa, East Africa, North Africa, other South African Countries, and West Africa to India in 2017-2018. Data for clover imports was not available.

The data is reported in terms of the value of exports and imports. The African economies were selected based on the value of trade to and from India that each of them are engaged with. The study focuses on exports and imports to and from Botswana, Ethiopia, Ghana, Mauritius, Nigeria, South Africa, and Zimbabwe. The study also looks at overall exports and imports to entire sub-Saharan Africa from India to identify individual country trends as well as a holistic trend.

Table 2. India's imports from Africa (Central Africa, East Africa, North Africa, other South African countries, and West Africa) to India in 2017-2018

Product	Import volume, $ million
Pigeon peas (Cajanuscajan)	86.747
Chickpeas (Garbanzos), dried, shelled, whether or not skinned/split	51.289
Beans (Vigna Spp., Pjaseolus Spp.) of the species Vigna Mungo (L.) Hepper or VignaRadiata (L) Wilczek	33.607
Cocoa beans, whole/broken, raw/roasted	27.860
Oranges, fresh/dried	20.114
Clover (Trifolium Spp) seed, for sowing (no data available)	17.561
Kidney beans including white pea beans dried and shld	15.516
Cocoa paste, not defatted	15.072
Cowpeas (Vignaunguiculata)	7.847
Dates soft (Khayzur or wet dates)	4.895

Source: Authors' development based on APEDA (n.d.)

Prospects, Challenges, and Policy Directions for Food Security in India-Africa Agricultural Trade

The database from APEDA reports products in the 8-digit HS Code whereas UN Comtrade reports data in the 6-digit HS Code format. Hence, while mapping the products from Table 1 and Table 2 with the list of products available from UN Comtrade, the closest available product or product group has been selected.

Besides the products listed above, the study also looks at trade trends for pulses, cereals, oilseeds, and cotton for the selected regions given that these products are also highly traded between India and African countries.

Results and Discussion

In the figures below, the trends of exports and imports for the individual African countries with reference to India are presented. The trends for exports from India are presented first followed by the trends for imports. Finally, export and import trends with entire sub-Saharan Africa is reported. Figure 1 shows the value of export of rice, boneless meat of bovine animals, broken rice and other items from India to selected African economies.

Figure 1. Import of rice, boneless meat of bovine animals, broken rice, and other rice from India to African countries
Note: (a) import of rice from India; (b) import of frozen boneless meat of bovine animals from India; (c) import of broken rice from India; (d) import of other rice from India
Source: Authors' development based on United Nations (n.d.)

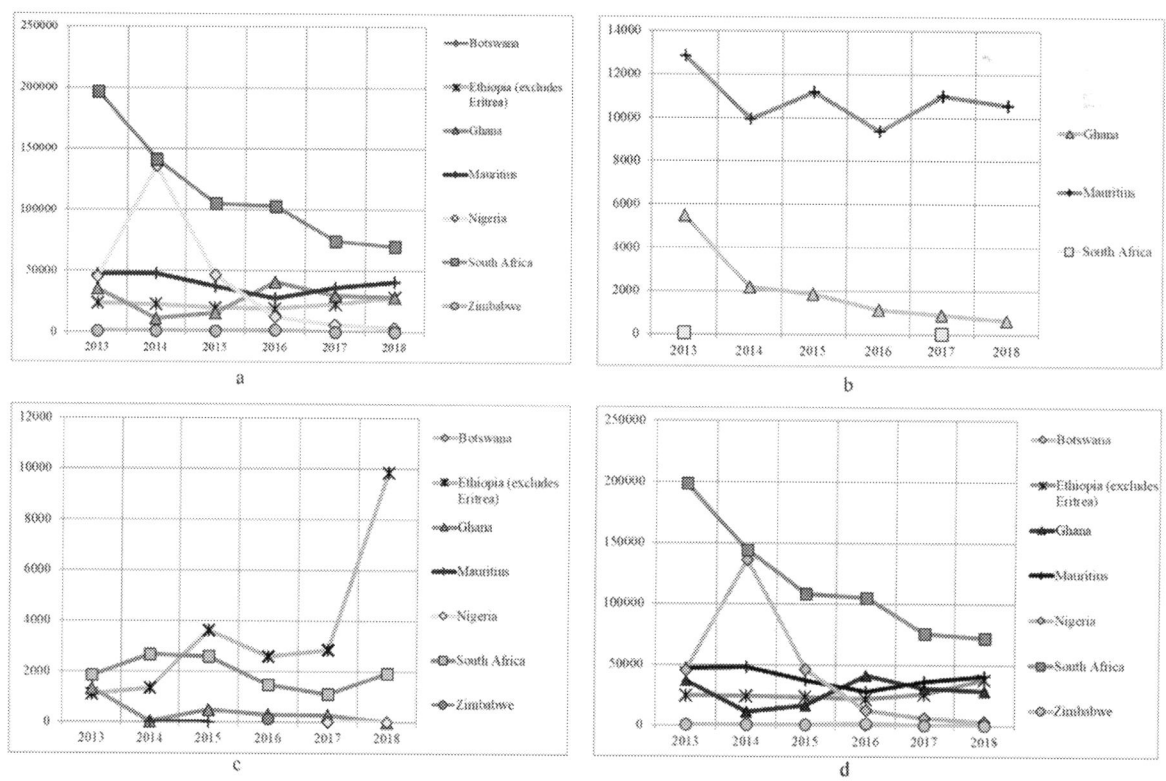

Panel (a) shows that South Africa has consistently been the largest importer of rice from India but the value of imports has reduced over time. Nigeria had a steep level of import from India in 2014 which has consistently reduced since then. Mauritius, Ethiopia, Ghana, and Zimbabwe have had more or less similar levels of imports showcasing slight ups and downs through the years. Mauritius has been a stronger importer than Ethiopia which again is followed by Zimbabwe for all the years. Botswana has seen a much smaller amount of rice imports from India compared to its fellow economies in Africa.

Panel (b) shows the value of exports of frozen boneless meat of bovine animals from India to the selected African economies. The highest importers are Mauritius and Ghana. The imports of Mauritius have been significantly higher than that of Ghana. The exports to Ghana from India have fallen over the years. South Africa is seen to have imported from India only in 2013 and 2017 in small amounts. The rest of the countries did not report any import from India.

Panel (c) shows the value of exports of broken rice from India to the selected African economies. Ethiopia, South Africa, and Ghana are found to be the heaviest importers of broken rice from India. Ethiopia's import value has significantly increased over the years. South Africa's imports decreased after 2015 but took off once again in 2018. Ghana's imports, on the other hand, have experienced a steady but small fall. Data for Botswana, Mauritius, Nigeria, and Zimbabwe were extremely sparse.

Panel (d) shows the value of exports of other rice products from India to the selected African economies. South Africa's imports have been significantly higher though decreasing consistently with time. Nigeria's imports peaked in 2014 but have decreased ever since significantly. Mauritius, Ethiopia, and Ghana reported significant and steady amounts of imports from India with Ghana's imports having increased over time. Botswana and Zimbabwe reported much smaller values of imports with a decreasing trend over the years.

Figure 2 shows the value of exports of sweet biscuits, un-denatured ethyl alcohol, chickpea, and sugar confectionery without cocoa from India to selected African economies.

Panel (a) shows that Ghana and South Africa are found to be the strongest importers of sweet biscuits with Ghana's trend being mildly downwards and South Africa's being upward. Mauritius, Nigeria, Ethiopia, and Botswana report moderately steady imports (Botswana being the smallest importer) but much lesser in volume in comparison to Ghana and South Africa. Zimbabwe reports very small values of imports from 2014 to 2016.

Panel (b) shows the value of exports of undenatured ethyl alcohol from India to selected African economies. Ghana's imports have steadily gone down over the years having slightly peaked in 2016. Nigeria has exhibited quite a high value of imports with a peak in 2015. Imports for South Africa increased in 2014 from 2013 but since 2016, the value has been substantially low. Mauritius reported imports only in 2015 and 2016 and Zimbabwe only a very small value in 2018. Data for Botswana and Ethiopia was not reported.

Panel (c) shows the value of exports of chickpea from India to the selected African economies. Mauritius, Nigeria and South Africa show a positive trend in the imports with South Africa's imports being higher than the other two for all the years till 2017. Mauritius reports a very steep rise in imports in 2018 overtaking South Africa's imports significantly. All three countries report a significant climb in imports in 2018 with respect to the previous years. Ghana reports very low values of imports with a steep fall in 2015 followed by an even steeper recovery in 2016 only to fall in value drastically in 2018. Data for Botswana, Ethiopia, and Zimbabwe was found only for a year or two.

Panel (d) shows the value of exports of sugar confectionery without cocoa from India to the selected African economies. Nigeria and Ghana are the heaviest importers both with positive trends with sig-

Figure 2. Import of sweet biscuits, undenatured ethyl alcohol, chickpea, and sugar confectionery without cocoa from India to African countries
Note: (a) import of sweet biscuits from India; (b) import of undenaturated ethyl alcohol from India; (c) import of chickpea from India; (d) import of confectionary sugar without cocoa from India
Source: Authors' development based on United Nations (n.d.)

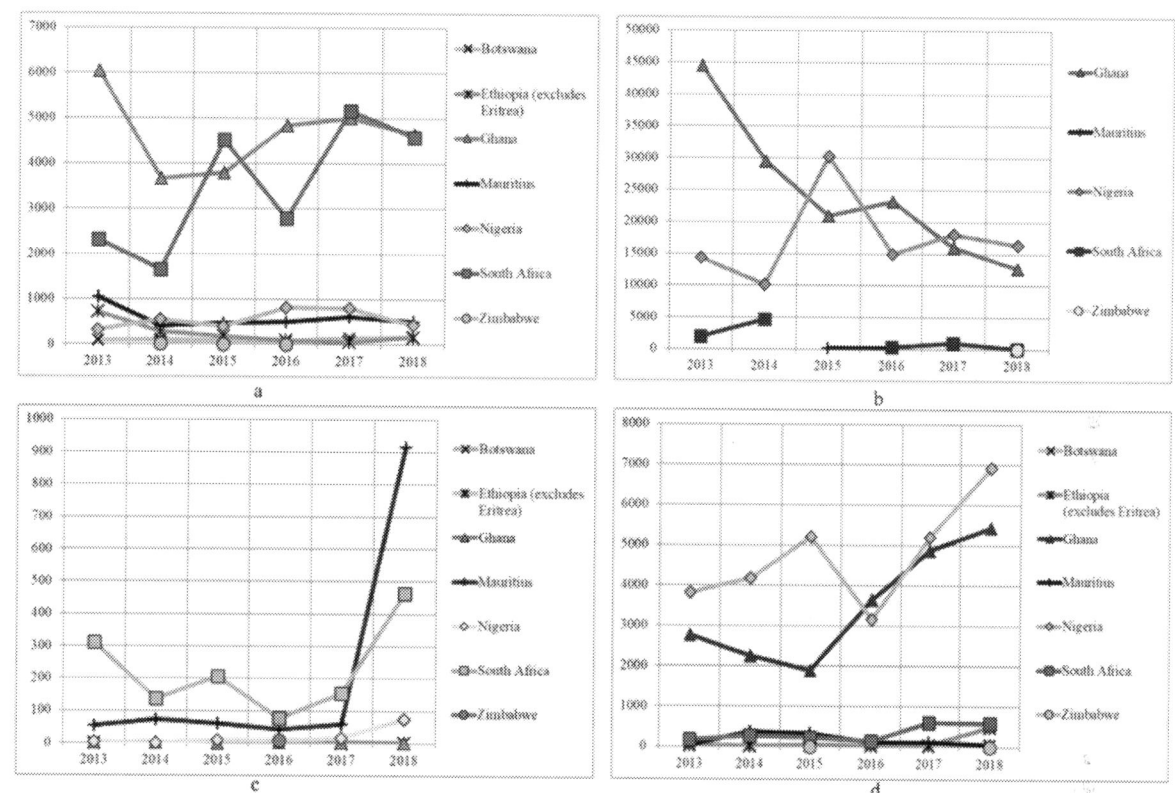

nificant dips in 2016 and 2015, respectively. South Africa and Ethiopia show moderately positive trends with Mauritius taking a massive leap in 2018 and Mauritius depicts a slightly negative trend. Botswana reports ups and downs but the value of imports is much lesser than any of the other countries. Zimbabwe reports very small imports in 2015 and 2018.

Figure 3 shows the value of exports of glucose, butter, and oilseeds from India to selected African economies.

In Panel (a), glucose import data is reported only for 2017 and 2018. Nigeria and Ethiopia exhibit strong dips while Ghana, Mauritius, and South Africa show small improvements in the value of imports. Zimbabwe reports a much smaller value of imports with a dip in 2018.

Panel (b) shows the value of exports of butter from India to selected African economies. Mauritius and Nigeria show slight upward trends in 2016 and 2017. South Africa reports a steep decline in 2018 from 2015 with no data being reported in between. Data for the rest of the years and the rest of the countries were not available.

Panel (c) shows the value of exports of oilseeds from India to the selected African economies. South Africa has been the strongest importer from India with a recent dip in 2018 but still remaining the largest importer. Nigeria exhibits a mild negative trend but remains to be a larger importer than the rest of

Figure 3. Import of glucose, butter, and oilseeds from India to African countries
Note: (a) import of glucose from India; (b) import of butter from India; (c) import of oilseed from India
Source: Authors' development based on United Nations (n.d.)

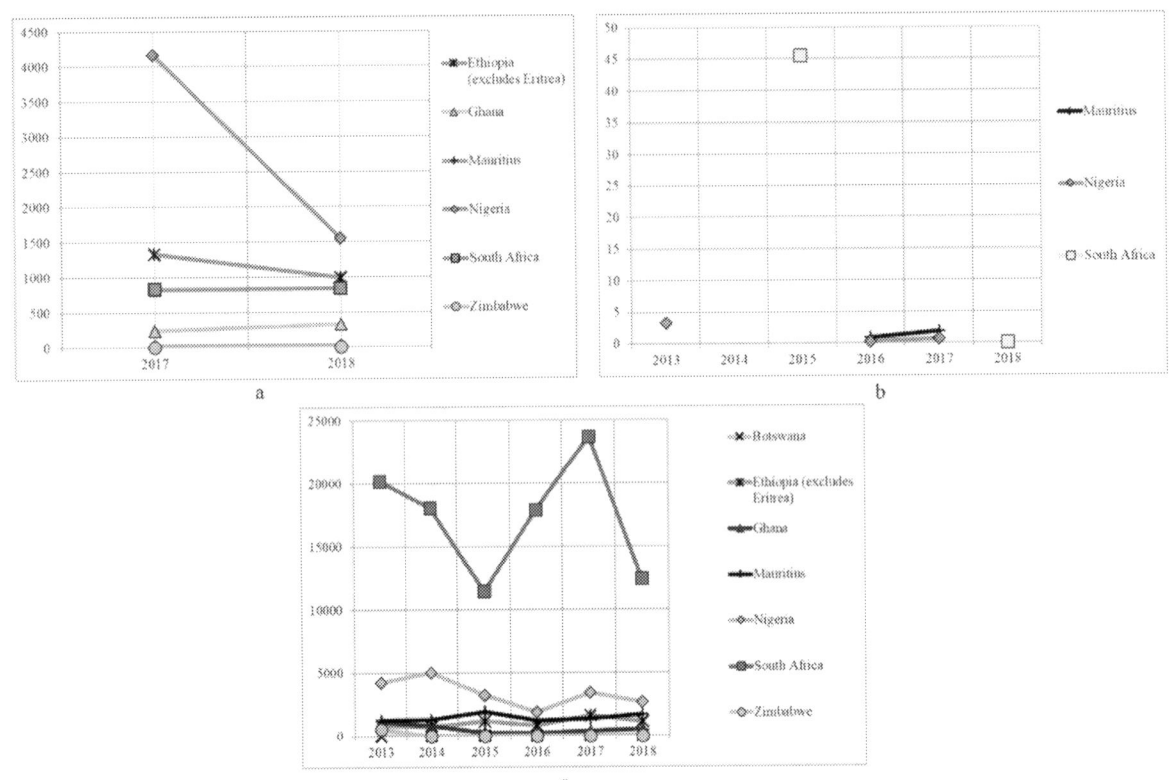

the countries. Mauritius depicts a mild positive trend whereas Ethiopia shows a dip in 2018 despite a general positive trend. Ghana reported a negative trend till 2016 having picked up imports since then. Botswana and Zimbabwe report very small values of imports and has experienced several ups and downs.

Figure 4 shows the value of exports of cereals, pulses, cotton, and all the chosen products from India to selected African economies.

In panel (a) for import of cereals, South Africa reports a negative trend but still has the highest value of imports among the other countries under consideration. Nigeria reports a massive peak in 2014 but thereafter the imports have reduced significantly. Ethiopia, Mauritius, and Ghana have experienced some ups and downs but the imports have remained similar by and large. Botswana and Zimbabwe report comparatively much smaller values of imports and have experience a negative trend with certain peaks and dips in the interim years.

Panel (b) shows the value of exports of pulses from India to selected African economies. Data for the same was not available for most of the countries and the years under consideration. South Africa reports a peak in 2014 but slides down in the subsequent year. Mauritius experienced a recovery in 2017 from a steep fall in 2015 earlier.

Panel (c) shows the value of exports of cotton from India to the selected African economies. Ethiopia exhibits a strong peak in 2015 but slides down significantly thereafter till 2018 whereas Mauritius shows a steep positive trend with an extremely significant leap in 2017. South Africa reports a positive trend with

Prospects, Challenges, and Policy Directions for Food Security in India-Africa Agricultural Trade

Figure 4. Import of cereals, pulses, cotton, and all the chosen products from India to African countries
Note: (a) import of cereal from India; (b) import of pulses from India; (c) import of cotton from India; (d) import of selected crops from India to Sub-Saharan Africa
Source: Authors' development based on United Nations (n.d.)

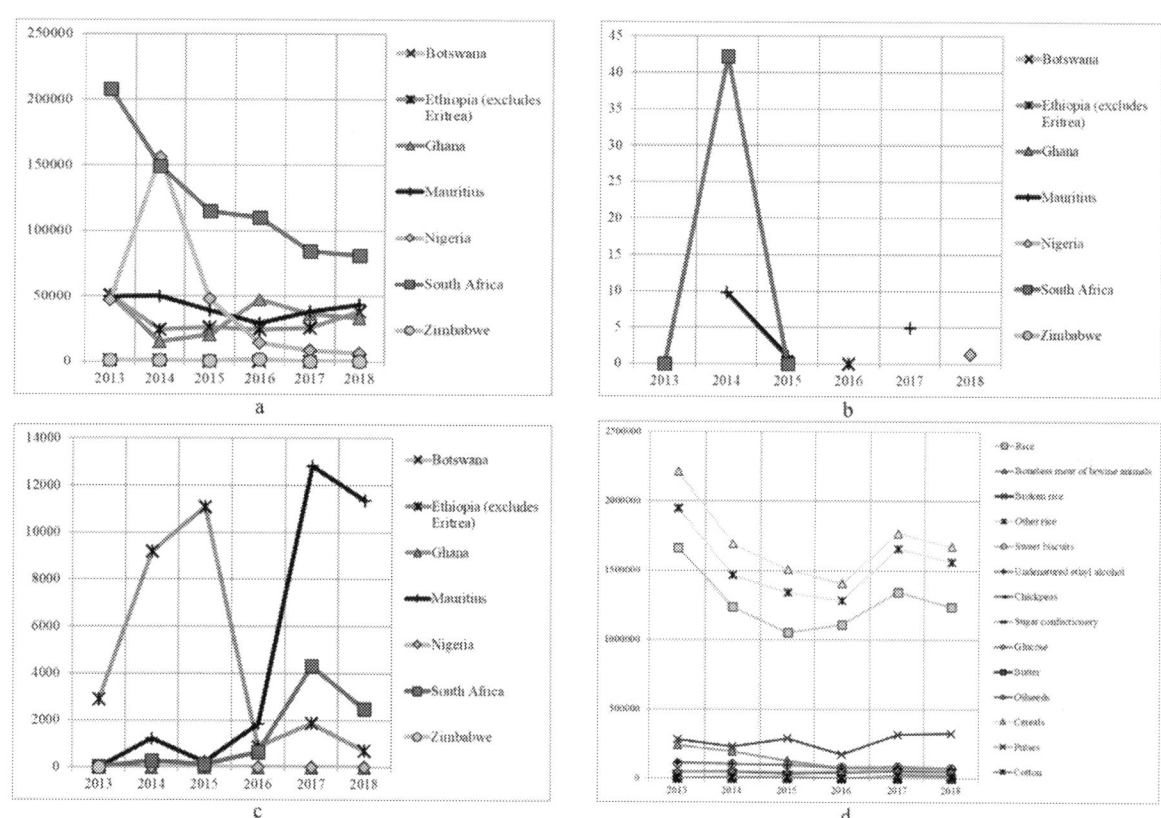

a much smaller value of imports till 2016 in comparison to the two countries mentioned earlier, reaching a peak in 2017 and sliding down a little in 2018 but remains higher than that of Ethiopia. Ghana and Nigeria report extremely small values of imports. Data for Botswana and Zimbabwe was not available.

Panel (d) shows the value of exports of all the selected products from India to sub-Saharan Africa as a region. Among all the products, cereals, rice, and other rice products have the highest value of exports with all of them dipping till 2015, peaking in 2017, and again going down in 2018. Broken rice is found to have a comparatively more consistent movement over the years but the exports in value are significantly lesser than the three products mentioned above. All the other products report much lesser values of exports from India to SSA. Bovine meat, undenatured ethyl alcohol, and sweet biscuits show a negative trend. Chickpeas, sugar confectionery, butter, oilseeds, cereals, and cotton show several ups and downs to various extents.

Next, export data from all the selected countries to India is analyzed. The data for imports to India was found to be extremely sparse.

Figure 5 shows the value of exports of pigeon peas, chickpea, beans, and cocoa beans to India from selected African economies.

Figure 5. Export of pigeon peas, chickpea, beans, and cocoa beans from India to African countries
Note: (a) export of pigeon pea to India; (b) export of chickpea to India; (c) export of beans to India; (d) export of cocoa beans to India
Source: Authors' development based on United Nations (n.d.)

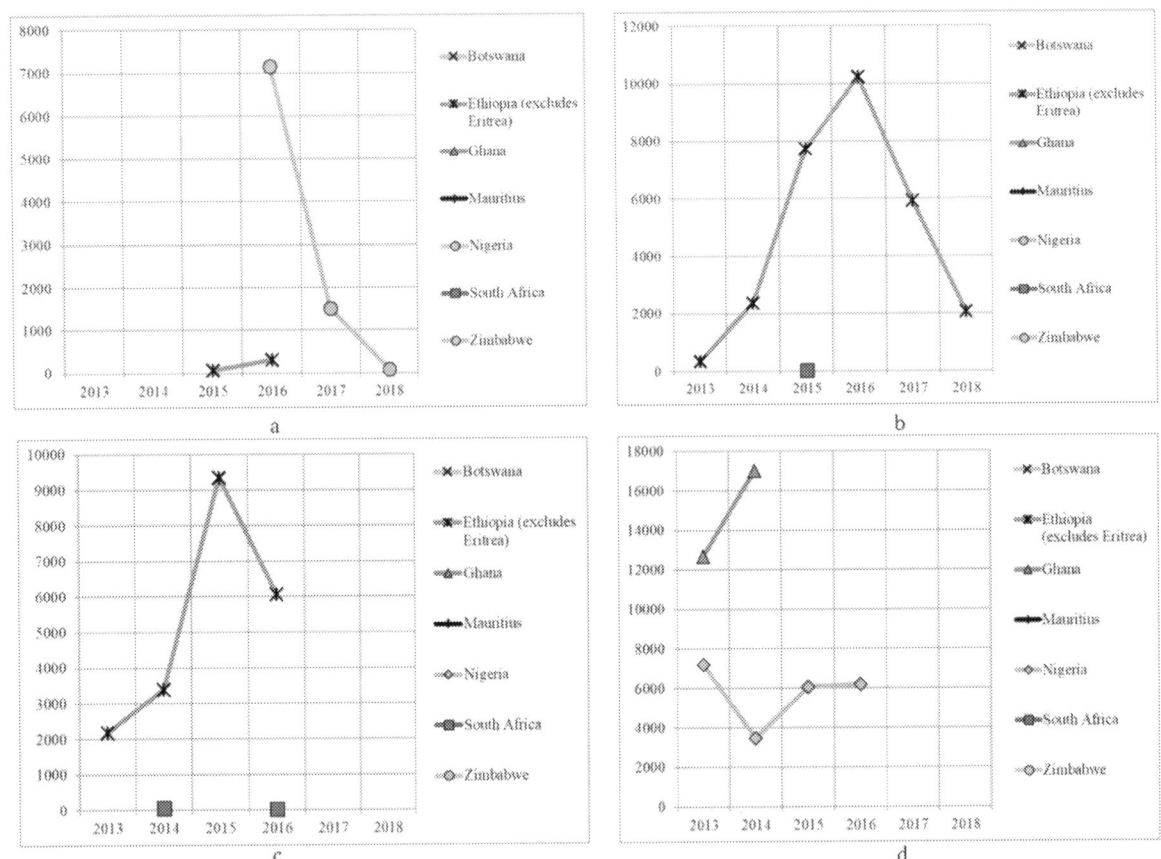

Panel (a) shows Nigeria exhibiting a steep decline in export of pigeon peas between 2016 and 2018 and Ethiopia shows a strong recovery from 2015 to 2016. Rest of the data was not available.

Panel (b) shows the value of imports of chickpeas to India from selected African economies. Ethiopia shows a positive trend till 2016 and the exports to India reduce drastically thereafter.

Panel (c) shows the value of imports of beans to India from the selected African economies. Ethiopia shows a positive trend till 2015 and the exports to India reduce drastically in 2016. South Africa reports a very small value of exports in 2014 which dips in 2016. Rest of the data was not available.

Panel (d) shows the value of imports of cocoa beans to India from the selected African economies. Ghana reports increased exports to India between 2013 and 2014. Nigeria experiences a dip in 2014 with a recovery till 2016. Rest of the data was not available.

Figure 6 shows the value of imports of oranges, kidney beans, cocoa paste, and cowpeas to India from selected African economies.

In panel (a), South Africa reports a positive trend with peaks in 2015 and 2017 and dips in 2016 and 2018 for export of oranges. Rest of the data was not available.

Prospects, Challenges, and Policy Directions for Food Security in India-Africa Agricultural Trade

Figure 6. Export of oranges, kidney beans, cocoa paste, and cowpeas from India to African countries
Note: (a) export of oranges to India; (b) export of kidney beans to India; (c) export of cocoa paste to India; (d) export of cowpeas to India
Source: Authors' development based on United Nations (n.d.)

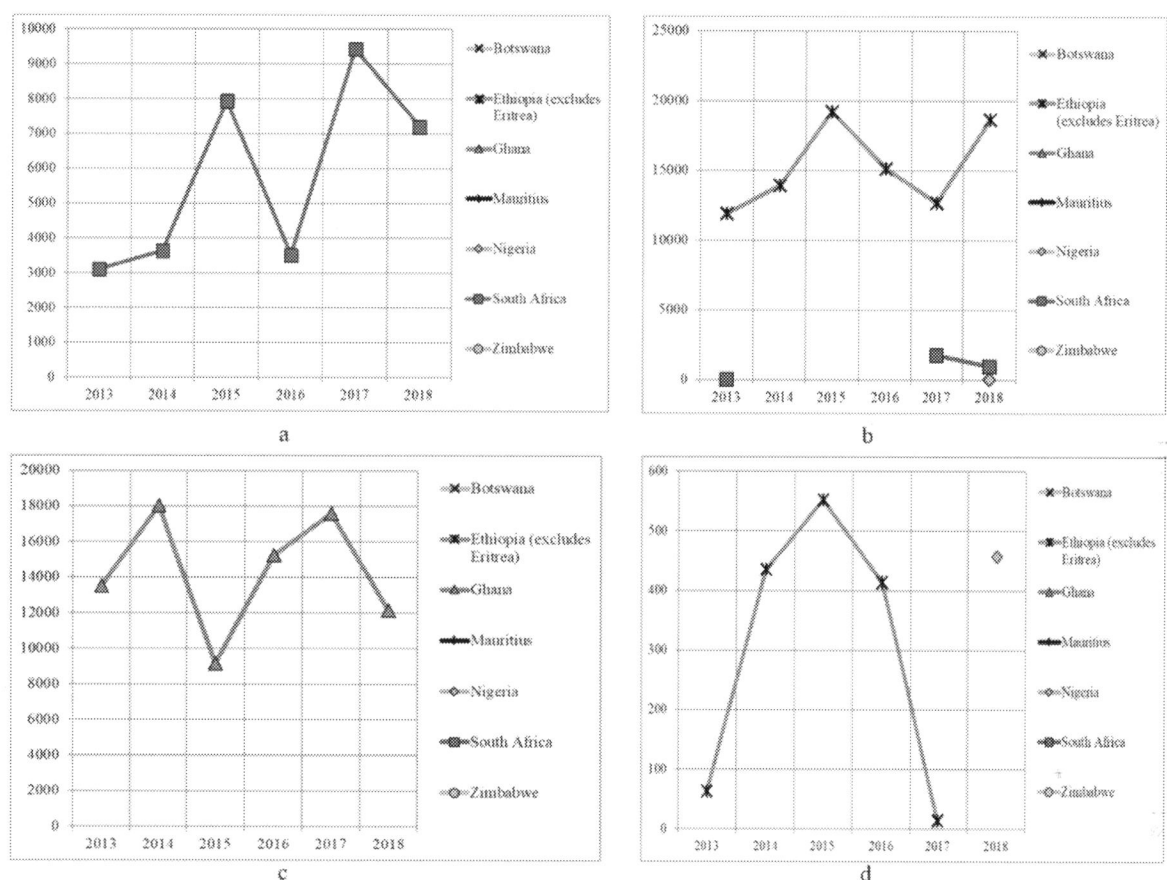

Panel (b) shows the value of imports of kidney beans to India from selected African economies. Ethiopia reports a positive trend with peak in 2015 and dip in 2017 but increased again in 2018. South Africa reports a reduction in exports between 2017 and 2018. Rest of the data was not available.

Panel (c) shows the value of imports of cocoa paste to India from the selected African economies. Ghana reports the exports peaking in 2014 and 2017 but decreasing in 2018 to reach a value of exports lesser than that in 2013. Rest of the data was not available.

Panel (d) shows the value of imports of cowpeas to India from selected African economies. Ethiopia exhibits a rise till 2015 and thereafter a consistent fall in the exports to India. Rest of the data was not available.

Figure 7 shows the value of imports of oilseeds, cereals, cotton, and all the selected products to India from selected African economies.

Panel (a) shows that Ethiopia reports a steep upward movement in 2016 and the exports rose further significantly in 2018. Ghana experiences a positive trend by and large with a dip in 2017. Nigeria reported a negative trend with ups and downs in between the years. South Africa reports a much smaller

Prospects, Challenges, and Policy Directions for Food Security in India-Africa Agricultural Trade

Figure 7. Export of oilseeds, cereals, cotton, and all the selected products from India to African countries
Note: (a) export of oilseeds to India; (b) export of cereals to India; (c) export of cotton to India; (d) export of selected crops to India from Sub-Saharan Africa
Source: Authors' development based on United Nations (n.d.)

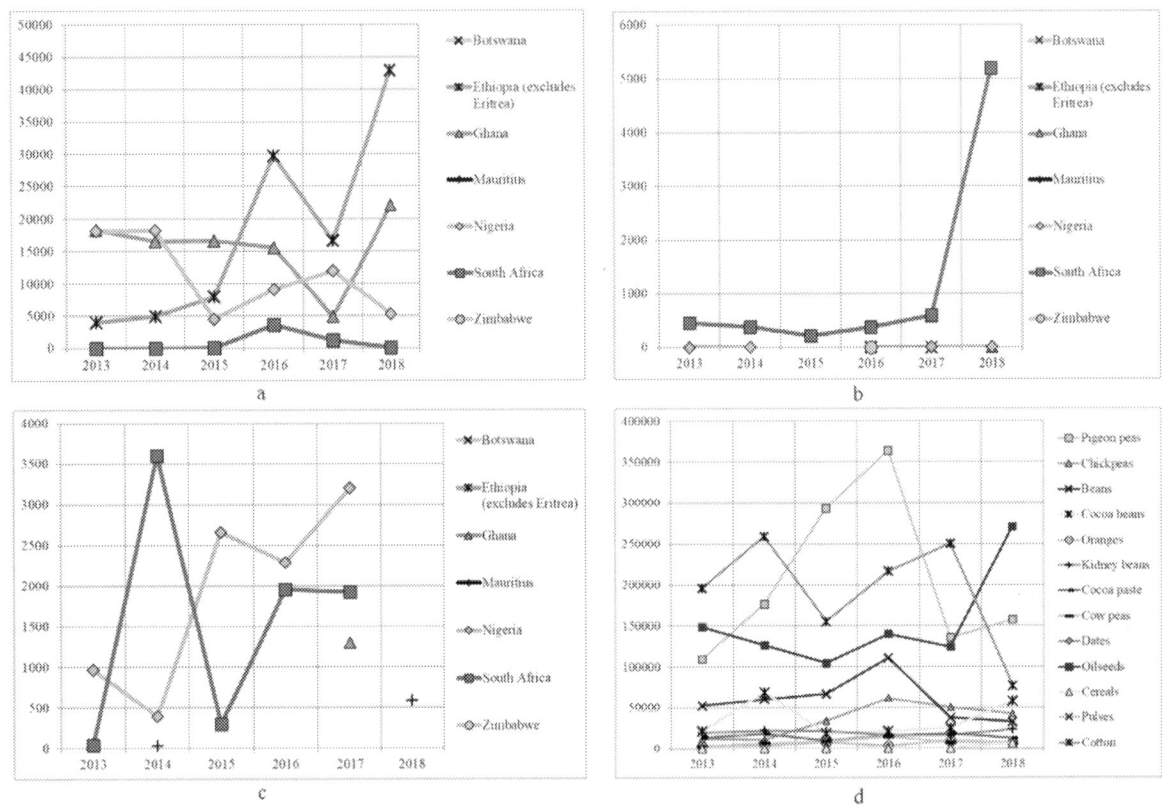

amount of export to India which peaked in 2016 but falling significantly thereafter. Rest of the data was not available.

Panel (b) shows the value of imports of cereals to India from the selected African economies. South Africa reports a steady export till 2017 and in 2018 takes a massive upward leap. Rest of the data was extremely sparse or not available.

Panel (c) shows the value of imports of cotton to India from selected African economies. South Africa reports a significant peak in the exports in 2014, dipping and steadying after 2016. Nigeria exhibits a positive trend with slight ups and downs. Rest of the data was extremely sparse or not available.

Panel (d) shows the value of imports of all the selected products to India from the sub-Saharan African region. Pigeon peas, cotton, and oilseeds are the most heavily imported products from sub-Saharan Africa to India. Pigeon peas show an increasing trend till 2016 taking a fall in 2017 and partially recovering in 2018. Cotton exhibits a negative trend with peaks in 2014 and 2017. Oilseeds, on the other hand, had a steady trend till 2017 and in 2018 experienced a very steep rise in exports to India. Beans, chickpeas, and cocoa beans are next in line with import values being significantly higher than the rest of the products but significantly lesser than the three products mentioned above. Beans and chickpeas have a negative trend but cocoa beans exhibit a mild positive trend. Oranges, kidney beans, cowpeas, and cereals are found to have undergone ups and downs in various years.

The above data clearly shows that there is immense potential to engage in agricultural trade between India and Sub-Saharan Africa.

SOLUTIONS AND RECOMMENDATIONS

Based on the above discussion, the following key policy areas have been identified for furthering the agricultural collaboration between India and Africa:

- Reduction of hunger and poverty
- Conservation of resources
- Reduction in negative impact on climate
- Inclusivity (gender and marginalized groups)

Correspondingly, the key action areas would be as follows.

- There should be concerted efforts at creation of virtual platforms for open sharing of knowledge and expertise to provide solutions to the unique environmental and cultural challenges. Virtual platforms can be used to impart knowledge via MOOCs (massive open online courses) under the GoI SWAYAM (Study Webs of Active – Learning for Young Aspiring Minds) initiative.
- India should increase the number of fellowships offered to African researchers to receive training in India for sustainable agricultural practices and technology.
- Private sector should be provided incentives to invest more in public-private partnerships to support smallholder farmers, enhance the high-quality seed production capability and provide for setting up of agro-business incubators to provide impetus to entrepreneurship and jobs for the youth. This would go a long way in professionalizing small farm holdings.
- The focus should be on promotion of such agricultural practices that help increase rural income and food security while being environmentally sustainable. Some of such sectors are food processing and storage, soil and water conservation, pest management, and agroforestry. The modern technologies should be well integrated with indigenous knowledge to not just enhance crop production but also reduce the post-harvest losses.
- India's development outreach to Africa should provide targeted support for specific agricultural commodities to address global market access challenges while also helping the local producers gain fair market access.
- The focus should also be on decreasing the vulnerability for agro-producers through capacity building and increasing the opportunity for integration with regional value chains.
- At the policy-making level, lessons should be incorporated from India's successful land reforms measures to help draft a development-oriented land policy that is inclusive by design.
- It is critical to acknowledge the significant role women play in regard to food security. Special programs should be designed for women to gain access to funds and land.
- India should continue to promote South-South and Trilateral cooperation and coordination to help Africa achieve its food security goals while also bringing about infrastructure development. The regional organizations, local and state government bodies, grass-root level representation from the farming community, marginalized groups, academics/researchers working in this field, NGOs and

advocacy groups must also be engaged in enhanced dialogue to bring about policy coherence and successful implementation through effective local action. There should be a seamless interaction between financial, technical, and educational cooperation between India and Africa to achieve the desired synergy in sustainable agriculture that would enable the goal of food security.

FUTURE RESEARCH DIRECTIONS

Based on the above policy recommendations, the directions for future research are as follows. The authors are of the view that the research should be interdisciplinary and comparative in nature to account for different variables that affect the sector.

- Mapping of employment and entrepreneurship opportunities within the modernized agriculture system
- Integration of the modernized agri-system into agribusiness value chains
- Gender dynamics in food security
- Impact of the rise in demand for green energy on the agricultural sector in Africa
- Social protection programs, especially in conflict and disease affected zones to ensure to address hunger and poverty
- India-Africa cooperation in leveraging blue economy to address food security

CONCLUSION

It is evident that agrarian sustainability is essentially related to food security and hence, fundamental to the advancement and socioeconomic progress of both Africa and India. Given the established and mutual aims to be food secure and the available opportunities to be trading partners in the agricultural sector, it positively benefits India and Africa to partner and progress together. However, India has a different economic maturity and growth dynamic when compared to Africa but both have complementary characteristics that will ensure that investments towards high-impact priority will bring immediate and sustainable returns.

Seven of the twenty fastest-growing economies hail from Africa. Yet, the continent cannot feed a population that is almost close to India's. Africa also has 44% of arable land, second to Asia. Agriculture added to nearly 32% of Africa's GDP and provided employment to approximately 65% of the labor force in 2015. Additionally, a projected 38% of Africa's working youth are currently employed in agriculture. It is also projected that about 60% of the world's accessible and unexploited cropland is in Sub-Saharan Africa. However, only 5-7% of the continent's cultivated land is irrigated, which leaves farmers exposed to the vagaries of nature. In spite of its inherent strengths, Africa is left with no choice but to import a substantial portion of its food requirements (ADB, 2019; FAO, 2013).

As opposed to the above, India has been actively working towards establishing food security through the collaborations abroad by way of leasing or acquiring land, knowledge sharing, providing know-how to augment agricultural output, and investments. Given the African case, each of these forms of cooperation is vital. There is fragmented and small landholding in Africa with a prevalence of subsistence farming. Investment for the development of quality inputs, markets for produce, good soils and soil management techniques, innovative financing tools, and other resources needed for sustained agricultural production become critical. Additionally, most African farmers use traditional methods of farming which restricts their productivity, while the deficiency of proper irrigation renders them extremely susceptible to weather shocks. This condition is further aggravated by insufficient inputs, lack of efficient markets and the necessary technology to augment productivity to a level where over and above personal consumption, there could also be surpluses for trading.

India and Africa have had longstanding historical ties for various partnerships and collaboration. Among the various reasons for cooperation, food security has been a perpetual and common concern. While agricultural output has increased manifolds in India in the last three decades, pulses are still being imported from Africa. India has major edge in helping Africa overcome its agri-based issues with a win-win situation for both. In spite of this realization, the sector in Africa has been dependent on government subsidy and donor funding. The importance of agriculture being the key driver in African economy has only recently dawned upon them.

As per ADB projections, the total cost for agricultural transformation for the priority commodities and agro-ecological zones in the ADB's strategy is between $315 billion and $400 billion over ten years, which amounts to $32-40 billion per year. Currently, the financing of the sector come from the autonomous and non-sovereign investments from multilateral and bilateral development partners. There is around $9 billion per year of investments into African agriculture leaving a shortage of $23 billion to $31 billion per annum to be mobilized in order to drive growth and development. The financial institutions and banks need to make long term finance available for investing in agricultural infrastructure and ensuring enhancement in technology. This will ensure that agricultural finance is not restricted to only a few stages of the value chain but all across. Procurement of seeds, tilling land with machinery, maintaining land, protecting land from insecurities, mechanical harvesting, storage in warehouses, processing and selling to the market require equal attention from financing in order to maintain the quality and labeling standards (ADB, 2019; FAO, 2013).

Agricultural investments take long to show effect. So do the assessments of funding requirements. Also, many potential exporters do not have the necessary know-how of conducting businesses of this kind and hence, cannot accurately estimate resource requirements and risks.

It has been suggested by experts that India should evaluate the exact sectoral needs in Arica based on their local circumstances and resources. The typical challenges that Africa faces cannot be generalized to the Indian context or understanding. Both, hard and soft infrastructure pertaining to agriculture is a major deficit in Africa. Connecting roads, transportation network, power transmission, communication channels, irrigation canals, distribution, storage, governance, institutions, and rule of law are not up to the mark for a smooth functioning of and output from the large availability of natural resources. Such infrastructure need very high ticket investments – the risks to which are very high, given that agro yield is heavily dependent on the weather as well. The local political scenario can also be a major hindrance to the effectiveness of the incoming investments. This is why not many domestic and private establishments are willing to invest in the sector. This leaves the various African Governments with very little choice but to constantly ask for international loans. In this context, India can be a major player, providing the essential monetary investments and know-how. It is proposed that the African economies could benefit from the development assistance from India to build agri-infrastructure. The agricultural land sustained by such collaborations could be leased to Indian firms for cultivation, too, thereby easing out some of the risks that Indian investors may face. The lease rentals could serve as cash-flow to service the debt. At the same time, Indian investors would require funds to pay upfront lease rentals, besides sourcing of agri-inputs and implements. The funding requirements could be met by Exim Bank's Overseas Investment Finance program and suchlike. It is also proposed that in order to support the Indian investments in African agriculture sector, a dedicated India-Africa Agriculture Development Fund is made functional with a suitable institutional machinery to manage this fund. The Exim Bank of India could ensure this aspect of the partnership while the funds could be consumed for medium and long term foreign currency finance to Indian enterprises planning to invest in the African agriculture and allied sector. The initial amount has been proposed to be around the tune of $5 billion wherein some of the Indian forex can be included. India's current forex reserves stood at approximately $420 million on May 10, 2019. The suggested amount of $5 billion is a smaller amount than 1.5% of the current forex reserves, and would not affect its position. This initiative will also serve as a channel for stimulating investments in African agriculture and to meet India's import requirements of pulses and oilseeds, besides creating remarkable goodwill through cooperation, collaboration, and partnerships (Kumar & Mazumdar, 2017; Livemint, 2019; Reserve Bank of India, n.d.).

India can strengthen its role as a development partner for Africa, in the critical area of food security, through the strategies put forward in the paper and thereby augment its existing goodwill in South-South cooperation efforts.

REFERENCES

Achancho, V. (2013). Review and Analysis of National Investment Strategies and Agricultural Policies in Central Africa: The Case of Cameroun. In A. Elbehri (Ed.), *Rebuilding West Africa's Food Potential* (pp. 117–149). Rome: Food and Agriculture Organization of the United Nations.

African Development Bank Group. (2019). *Feed Africa*. Abidjan: African Development Bank Group.

Agricultural and Processed Food Products Export Development Authority. (n.d.). *Trade Information.* Retrieved from https://apeda.gov.in/apedawebsite/#

Ayyappan, S. (2015). *India-Africa Cooperation in Agricultural Sector for Food Security.* Retrieved from https://www.mea.gov.in/in-focus-article.htm?25950/IndiaAfrica+Cooperation+in+Agricultural+Sector+for+Food+Security

Ben-Iwo, J., Manovic, V., & Longhurst, P. (2016). Biomass Resources and Biofuels Potential for the Production of Transportation Fuels in Nigeria. *Renewable & Sustainable Energy Reviews, 63,* 172–192. doi:10.1016/j.rser.2016.05.050

Bojang, F., & Ndeso-Atanga, A. (2013). *African Youth in Agriculture, Natural Resources and Rural Development.* Accra: FAO Regional Office for Africa.

Campbell, J. (2019). *Unpacking Africa's 2019 GDP Growth Prospects.* Retrieved from https://www.cfr.org/blog/unpacking-africas-2019-gdp-growth-prospects

Cotula, L., Dyer, N., & Vermeulen, S. (2008). *Fuelling Exclusion? The Biofuels Boom and Poor People's Access to Land. Rome: Food and Agricultural Organization of the United Nations and International Fund for Agricultural Development.* London: International Institute for Environment and Development.

Fingar, C., Loewendahl, H., Ewing, G., McMillan, C., Reynolds, J., & Rodriguez, E. ... Whitten, J. (2017). *The FDI Report 2017: Global Greenfield Investment Trends.* London: The Financial Times.

Food and Agriculture Organization of the United Nations. (2003). *WTO Agreement on Agriculture: The Implementation Experience. Developing Country Case Studies.* Rome: Food and Agriculture Organization of the United Nations.

Food and Agriculture Organization of the United Nations. (2013). *FAO Statistical Yearbook 2013. World Food and Agriculture.* Rome: Food and Agriculture Organization of the United Nations.

Food and Agriculture Organization of the United Nations. (2014). *Biodiversity for Food and Agriculture: Contributing to Food Security and Sustainability in a Changing World.* Rome: Food and Agriculture Organization of the United Nations.

Ghose, B. (2014). Promoting Agricultural Research and Development to Strengthen Food Security in South Asia. *International Journal of Agronomy.* doi:10.1155/2014/589809

Grain. (2014). *Growing Corporate Hold on Farmland Risky for World Food Security.* Retrieved from https://ourworld.unu.edu/en/growing-corporate-hold-on-farmland-risky-for-world-food-security

India Brand Equity Foundation. (2019). *Indian Economy.* Retrieved from https://www.ibef.org/economy.aspx

India Times. (2018). *India Imports 50.8 Lakh Ton Pulses for Rs 17,280 Crore in April-December.* Retrieved from https://economictimes.indiatimes.com/news/economy/agriculture/india-imports-50-8-lakh-ton-pulses-for-rs-17280-crore-in-april-december/articleshow/62808888.cms

International Fund for Agricultural Development. (2017). *South-South and Triangular Cooperation (SSTC). Highlights from IFAD's Portfolio.* Rome: International Fund for Agricultural Development.

Jones, C. (2017). *Agricultural Opportunities in Africa: Crop Farming in Ethiopia, Nigeria and Tanzania.* London: Deloitte.

Kumar, A., & Mazumdar, R. (2017). *Feed Africa: Achieving Progress through Partnership.* Mumbai: Export-Import Bank of India.

Leke, A., & Barton, D. (2016). *3 Reasons Things Are Looking up for African Economies.* Retrieved from https://www.weforum.org/agenda/2016/05/what-s-the-future-of-economic-growth-in-africa/

Leke, A., Lund, S., Roxburgh, C., & van Wamelen, A. (2010). *What's Driving Africa's Growth.* Retrieved from https://www.mckinsey.com/featured-insights/middle-east-and-africa/whats-driving-africas-growth

Livemint. (2017). *How Does Indian Investment in Africa Compare with China's?* Retrieved from https://www.livemint.com/Politics/Xd5t4vKx2dRCoPwnScENjO/How-does-Indian-investment-in-Africa-compare-with-Chinas.html

Livemint. (2019). *India's Forex Reserves Up by $1.3 Billion to over $420 Billion.* Retrieved from https://www.livemint.com/market/stock-market-news/india-s-forex-reserves-up-by-1-3-billion-to-over-420-billion-1558155104850.html

Mahapatra, R., Rattani, V., Sengupta, R., Pandey, K., & Goswami, S. (2017). *How to Make Africa Food Self-Sufficient, Again?* Retrieved from https://www.downtoearth.org.in/coverage/agriculture/why-farmers-now-dread-a-normal-monsoon-58206

Ministry of Finance of India. (2018). *Economic Survey 2018.* Retrieved from http://mofapp.nic.in:8080/Economicsurvey/#

Ministry of Finance of India. (2019). *Key Highlights of Economic Survey 2018-19.* Retrieved from http://pib.nic.in/newsite/PrintRelease.aspx?relid=191213

Modi, R., Desai, D. D., & Venkatachalam, M. (2019). *South-South Cooperation: India-Africa Partnerships in Food Security and Capacity-Building.* Mumbai: Observer Research Foundation.

Mukasa, A. N., Woldemichael, A. D., Salami, A. O., & Simpasa, A. M. (2017). Africa's Agricultural Transformation: Identifying Priority Areas and Overcoming Challenges. *Africa Economic Brief, 8*(3), 1–16.

Munang, R. (2015). *Winning Africa's Future: Food Security for All.* Retrieved from https://intpolicydigest.org/2015/07/23/winning-africa-s-future-food-security-for-all/

Organization for Economic Cooperation and Development. (2016). *Agriculture in Sub-Saharan Africa: Prospects and Challenges for the Next Decade.* Paris: OECD Publishing.

Organization for Economic Cooperation and Development. (2018). *Economic Outlook for Southeast Asia, China and India 2018.* Paris: OECD Publishing.

Pigato, M. (2009). *Strengthening China's and India's Trade and Investment Ties to the Middle East and North Africa.* Washington, DC: The World Bank.

PricewaterhouseCoopers. (2015). *Food Security in Africa. Water on Oil.* Retrieved from https://www.pwc.com/gx/en/issues/high-growth-markets/assets/food-security-in-africa.pdf

PricewaterhouseCoopers. (2016). *India-Africa Partnership in Agriculture: Current and Future Prospects.* Retrieved from https://www.pwc.in/assets/pdfs/publications/2016/india-africa-partnership-in-agriculture-current-and-future-prospects.pdf

Reserve Bank of India. (n.d.). *Weekly Statistical Supplement.* Retrieved from https://www.rbi.org.in/scripts/WSSViewDetail.aspx?TYPE=Section&PARAM1=1

Schwartzstein, P. (2016). *African Farmers Say They Can Feed the World, and We Might Soon Need Them to.* Retrieved from https://qz.com/africa/736626/african-farmers-say-they-can-feed-the-world-and-we-might-soon-need-them-to/

Stroh de Martinez, C., Feddersen, M., & Speicher, A. (2016). *Food Security in Sub-Saharan Africa: A Fresh Look on Agricultural Mechanisation. How Adapted Financial Solutions Can Make a Difference.* Bonn: German Development Institute.

Sunder, S. (2018). *India Economic Survey 2018: Farmers Gain as Agriculture Mechanisation Speeds up, but More R&D Needed.* Retrieved from https://www.financialexpress.com/Budget/India-Economic-Survey-2018-For-Farmers-Agriculture-Gdp-Msp/1034266/

The Economist. (2018). *Africa Has Plenty of Land. Why Is It so Hard to Make a Living from It?* Retrieved from https://www.economist.com/middle-east-and-africa/2018/04/28/africa-has-plenty-of-land-why-is-it-so-hard-to-make-a-living-from-it

The World Bank. (n.d.). *Liberalizing Trade in Agriculture: Africa and the New WTO Development Agenda.* Washington, DC: The World Bank.

United Nations. (n.d.). *UN Comtrade Database.* Retrieved from https://comtrade.un.org/

United Nations Conference on Trade and Development. (2009). *The Role of South-South and Triangular Cooperation for Sustainable Agriculture Development and Food Security in Developing Countries.* Geneva: United Nations.

United Nations Conference on Trade and Development. (2017). *World Investment Report 2017. Investment and the Digital Economy.* Geneva: United Nations.

United Nations Conference on Trade and Development. (2018a). *Investment and New Industrial Policies: Key Messages and Overview.* Geneva: United Nations.

United Nations Conference on Trade and Development. (2018b). *Economic Development in Africa. Report 2018. Migration for Structural Transformation.* New York, NY: United Nations.

United Nations Office for South-South Cooperation. (2019). *India-UN Development Partnership Fund.* New York, NY: UNOSSC.

ADDITIONAL READING

Bhayani, R. (2017). *Pulses Farmers' Income Fell 16% in 2016-17: Crisil Study.* Retrieved from https://www.business-standard.com/article/Economy-Policy/Pulses-Farmers-Income-Fell-16-In-2016-17-Crisil-Study-117090600876_1.Html

Food and Agriculture Organization of the United Nations. (2018). The State of Food Security and Nutrition in the World 2018. Building Climate Resilience for Food Security and Nutrition. Rome: Food and Agriculture Organization of the United Nations.

The World Bank. (2013). *Africa's Agriculture and Agribusiness Markets Set to Top US$ One Trillion in 2030.* Retrieved from https://www.worldbank.org/en/news/feature/2013/03/04/africa-agribusiness-report

KEY TERMS AND DEFINITIONS

Agricultural Investment: The expenditures on agriculture including short-term costs as well as long-term investments. Investment in agriculture includes government expenditures directed to agricultural infrastructure, research and development, and education and training.

Agricultural Trade: A reliance on exchanging agricultural and food commodities to supplement and complement domestic production.

Food and Agriculture Organization: A specialized agency of the United Nations that leads international efforts to defeat hunger. The goal of the organization is to achieve food security for all and make sure that people have regular access to enough high-quality food to lead active healthy lives. With over 194 member states, it works in over 130 countries worldwide.

Food Security: A situation when all people, at all times, have physical, social, and economic access to sufficient, safe, and nutritious food that meets their food preferences and dietary needs for an active and healthy life.

India-Africa Institute of Agriculture and Rural Development: A Pan-African Institute wherein trainees not only from Malawi but also from other African countries, will receive training to develop their human resources and build their capacity. The Institute will develop training programs in the areas of micro-financing and agro-financing, among others. The entire expenditure on faculty from India, travel, logistics, and training course expenses for students from other African countries will be borne by the Government of India for an initial period of three years.

Indo-Japan Asia-Africa Growth Corridor (AAGC): An economic cooperation agreement between the governments of India, Japan, and multiple African countries. It aims for Indo-Japanese collaboration to develop quality infrastructure in Africa, complemented by digital connectivity, which would undertake the realization of the idea of creating free and open Indo-Pacific Region. The AAGC will give priority to development projects in health and pharmaceuticals, agriculture and agro-processing, disaster management and skill enhancement. The connectivity aspects of the AAGC will be supplemented with quality infrastructure.

National Bank for Agriculture and Rural Development Consultancy Service (NABCONS): A wholly owned subsidiary promoted by National Bank for Agriculture and Rural Development (NABARD) and is engaged in providing consultancy in all spheres of agriculture, rural development, and allied areas. NABCONS leverages on the core competence of the NABARD in the areas of agricultural and rural development, especially multidisciplinary projects, banking, institutional development, infrastructure, training, etc., internalized for more than two decades.

South-South and Triangular Cooperation: South-South cooperation is a broad framework of collaboration among countries of the South in the political, economic, social, cultural, environmental and technical domains. Involving two or more developing countries, it can take place on a bilateral, regional, intraregional or interregional basis. Triangular cooperation is collaboration in which traditional donor countries and multilateral organizations facilitate South-South initiatives through the provision of funding, training, management and technological systems as well as other forms of support.

State of Food Insecurity (SOFI): The State of Food Security and Nutrition in the World is an annual flagship report jointly prepared by FAO, IFAD, UNICEF, WFP and WHO to inform on progress towards ending hunger, achieving food security and improving nutrition and to provide in-depth analysis on key challenges for achieving this goal in the context of the 2030 Agenda for Sustainable Development. The report targets a wide audience, including policy-makers, international organizations, academic institutions, and the general public.

Sub-Saharan Africa: Geographically, the area of the continent of Africa that lies south of the Sahara. According to the United Nations, it consists of all African countries that are fully or partially located south of the Sahara. The UN Development Program lists 46 of Africa's 54 countries as sub-Saharan, excluding Algeria, Djibouti, Egypt, Libya, Morocco, Somalia, Sudan, and Tunisia.

Sustainable Development Goals (SDG): A universal call through UNDP to action to end poverty, protect the planet, and ensure that all people enjoy peace and prosperity.

Chapter 21
The Role of Agricultural Production and Trade Integration in Sustainable Rural Development:
Evidence From Ethiopia

Henrietta Nagy
Szent Istvan University, Hungary

György Iván Neszmélyi
Budapest Business School, University of Applied Sciences, Hungary

Ahmed Abduletif Abdulkadr
Szent Istvan University, Hungary

ABSTRACT

Ethiopia is the second-most populous country in Africa with rainfed agriculture as a backbone of its economy. Most of the population, 79.3%, are rural residents. Sustainable rural development can be achieved if great attention is given to the labor-intensive sector of the country, agriculture, by improving the level of productivity through research-based information and technologies, increasing the supply of industrial and export crops, and ensuring the rehabilitation and conservation of natural resource bases with special consideration packages. The improvement in agricultural productivity alone cannot bring sustainable development unless supported by appropriate domestic and international trade. The main objective of this study is to identify and examine key determinants that influence agricultural productivity to assure food security, as well as to analyze domestic and foreign trade in agricultural products in Ethiopia.

DOI: 10.4018/978-1-7998-1042-1.ch021

INTRODUCTION

Agriculture remains to be the key for all development aspects in developing countries. It plays a significant role as source of employment, gross domestic product (GDP), and foreign earnings through export. With the increase of the world's population, the need for food is expected to rise. It requires the productivity of agricultural sector to grow. The issue of poverty reduction has been a central point in many countries in order to assure the sustainability of the livelihood of smallholder farmers while agriculture remains a solution and the focus of trade and sustainable development nexus. Brooks (2012) points out that smallholder development is the key to reduce poverty, while Le Goff and Singh (2014) indicate that trade has a positive impact on poverty reduction. According to Markelova and Mwangi (2010), sustainable agricultural growth needs the application of different approaches to the smallholders, as well as requires access to market where the smallholders can sell and make profit out of it which in turn will positively affect the investment of smallholders on enhancing agricultural productivity.

Agriculture is the mainstay of most of the Ethiopian community (79.3% of which is rural involved in agricultural production one way or another), main source of food and cash. The increase in the productivity of agricultural production has a significant impact on the welfare of rural people (Urquia Grande & Rubio-Alcocer, 2015; Abro, Alemu, & Hanjra, 2014). Most of the households get the food they consume and cash they need from agricultural sector to cover other needs to survive. Agriculture is the main potential industry for the overall development of the nation where many other economic activities depend on agriculture, including the processing industry and foreign trade. Most of the commodity exports is generated by smallholders.

The main aim of this study is to critically understand the challenges of agricultural productivity and market, and the role the market plays in enhancing the productivity and sustaining the livelihood of rural farmers which is very important for the overall development efforts of any country. Especially in least developed and developing countries where the share of agriculture in the economy is higher compared to developed states, sustainable agricultural intensification and improving the market system play a prominent role in achieving the goals of becoming a middle-income level country and beyond.

This study attempts to identify and examine key determinants that influence agricultural productivity to assure food security, analyze trend of main agricultural products, domestic and export trade status of the country. A mixed (qualitative and quantitative) research approach is carried out basically to take the tenets of the two approaches. Besides, explanatory research design is employed to describe the analysis that in turn employs document analysis, graphs, charts, and tables to analyze the secondary data obtained from National Bank and Central Statistical Agency of Ethiopia.

BACKGROUND

Agricultural products are the main sources of food, feed, and industrial inputs. In the developed world, where food demand is increasing in a lesser extent compared to developing countries, the demand for agricultural products such as maize, vegetable oils, and sugar cane has increased for biofuel usage. On the other hand, the increase in income in emerging economies has changed the need for agricultural food consumption. According to Food and Agriculture Organization of the United Nations [FAO] (2018a), the global total consumption of meat and fish is expected to increase by 15% by 2027 while global demand for food, feed, and fiber is expected to grow by 70% by 2050 (FAO, 2009). This is an indication that agricultural production needs to be improved to cope up with the global demand.

Most of sub-Saharan African countries are net importers of food due to low productivity of agricultural sector amid the fast growth of population (FAO, 2017). FAO (2018a) also highlighted that the food consumption would continue to rise due to the population growth.

Agriculture in Ethiopia accounts for 41% of the GDP, 80% of the employment, and most foreign exchange earnings although. In 2017-2018, its contribution to GDP decreased down to 36.3% but still remains essential (Gebreegziabher, Stage, Mekonnen, & Alemu, 2011; National Bank of Ethiopia [NBE], 2018). Such a decrease may be considered as an indication of economic progress and diversification of the economy (Byerlee, de Janvry, & Sadoulet, 2009; Brooks, 2012). Similarly, Urquia Grande, Cano, Estebanez, and Chamizo-Gonzalez (2018) stated that the food and agriculture sector offered key solutions for development and was central for hunger and poverty eradication. Likewise, many scholars indicate that the growth of agricultural productivity plays an important role in reducing poverty as agriculture is the main source of income to both rural and urban Ethiopians (Abro et al., 2014; Bezabih, Di Falco, & Mekonnen, 2014; Robinson, Strzepek, & Cervigni, 2013; Wagesho, Goel, & Jain, 2013).

Bali Swain and Varghese (2013) stated that the government of Ethiopia paid a great attention towards the development of entrepreneurial groups in extensive farming. In line to this, a growth strategy was introduced denominated Agricultural Development Led Industrialization (ADLI) giving priority to the agricultural sector (Shiferaw & Bedi, 2013) and a huge agricultural investment was made to significantly improve the agricultural productivity by the Ministry of Finance and Economic Development of Ethiopia (2012). The growth of the sector is not as expected in improving the livelihood of the rural community.

Under the Sustainable Development Goals (SDG) framework, much research has been done regarding economics in rural areas, linking advances in agriculture with nutrition and health, with improvement being sought in all these areas to alleviate poverty (Cervantes-Godoy & Dewbre, 2010; Christiaensen, Demery, & Kuhl, 2011). Hence, best practices of agricultural production guarantee a virtuous cycle of improving crop quality and environmental protection, as well as nutrition and health of the most vulnerable people. They are the key activities required for sustainability (Urquia Grande & Rubio-Alcocer, 2015).

According to World Bank (2019), Ethiopian poverty line declined down to 27.3% in 2015. International Fund for Agricultural Development [IFAD] (2010) also indicated that the poverty affected the rural community compared to urban population. The impact on poverty reduction depends on labor intensity in various sectors (Loayza & Raddatz, 2016). This implies that the sector where labor intensity is high has better potential to reduce poverty. In case of Ethiopia, agriculture is the most labor-intense sector which has been given due attention so far though the impact in reducing poverty has been not that significant.

According to the Coalition of European Lobbies for Eastern African Pastoralism [CELEP] (2017), 20% of the national export, 90% of live animal export, and 80% of annual milk supply in Ethiopia resulted from the pastoralists. Kristjanson et al. (2014), Walugembe et al. (2016), and Deere, Oduro, Swaminathan, and Doss (2013) indicate that smallholders give a great deal of attention if the system becomes more productive and profit-oriented. Market access is one of the main reasons of weak agricultural growth. Underdeveloped market chains force pastoralists to look for some other sources of income like charcoal production.

Productivity growth in agriculture helps reducing poverty (Cervantes-Godoy & Dewbre, 2010; Christiaensen et al., 2010). However, increasing agricultural productivity alone is not enough to improve the livelihood and assure the food security of the poor people in Ethiopian rural community. Though productivity can be achieved through the implementation of different packages, the level of income generated cannot raise without proper market channel and product prices if sustainable livelihood is to be achieved.

Agricultural trade plays an important role both in motivating production and consumption and hasten the pace of development by increasing demand for money and selection of crop types which have market values in different production seasons. Farmers who produce extra products should have an access or a link to output market where they can sell their products. The importance of trade has been taken seriously as a main source of sustainable development enabling developing countries to devise export-oriented approaches (Berg & Krueger, 2003). This is supported by Greenville, Kawasaki, and Jouanjean (2019) stating that the integration of agriculture and food sector and international market is increasing.

According to FAO (2018b), trade in agricultural products exhibited an annual growth rate of over 6%. Though trade is a key element to economic growth, because of domestic supply-side constraints in least developed countries, it is difficult to benefit from it in a proper manner. In Ethiopia, the main challenges are high transactions costs, lack of finance and credit resources, low level of control, underdeveloped transport infrastructure, and lack of storage facilities (Hailu, Sala, & Seyoum, 2016). An improvement of 10% in the transport and trade-related infrastructure quality index has a potential to increase the export of developing countries by 30% while reduction of tariffs by 10% would increase trade value by about 3.7% (Hallaert, Cavazos-Cepeda, & Kang, 2011). According the Standing Committee for Economic and Commercial Cooperation of the Organization of Islamic Cooperation COMCEC (2017), efficient agricultural market system is needed to produce and compete in the international market.

MAIN FOCUS OF THE CHAPTER

Ethiopia: Fast Facts

Ethiopia is one of the developing countries located in the horn of Africa boarded by Eritrean in the north, Djibouti in the east, Kenya in the west, as well as Somalia, Sudan, and South Sudan. Ethiopia covers an area of 1.14 million km^2, its population is 102 million people (79.3% of which is rural community practicing agriculture as a main livelihood) (World Bank, 2019).

The economy of Ethiopia is heavily dependent on rain-fed agriculture (Arayaa, Keesstra, & Stroosnijder, 2010; Awulachew, Erkossa, & Namara, 2010; FAO, 2012; Hadgu, Fantaye, Mamo, & Kassa, 2013). Irrigation covers only 5% of the cultivated land in the country (Awulachew et al., 2010).

Agricultural sector of Ethiopia is a part of the agro-industrial complex which includes crop production and animal husbandry. Crop production practice in Ethiopia includes production of grain crops, vegetable crops, root crops, fruit crops, permanent crops, and enset trees. Grain crops include cereals which are the most important field crops and the main element in the diet of most Ethiopians, pulses, and oilseeds. Animal husbandry (livestock production) includes cattle breeding, sheep breeding, goat breeding, horse breeding, beekeeping, and poultry production. In the 2010s, crop production accounted nearly 70% of the real value of agricultural output. It is often reported that Ethiopia has the largest livestock population in Africa. According to the Central Statistical Authority [CSA] (2018), there are 60.39 million heads of cattle, 31.3 million sheep, 32.74 million goats, 60.04 million of total poultry population (include cocks, cockerels, pullets, laying hens, non-laying hens, and hicks), 11.32 million equines (donkeys, horses, and mules), 6.52 million beehives, and 1.42 million camels.

Table 1. Contributions to GDP per sectors in Ethiopia in 2011-2017

Sectors	2011-2012	2012-2013	2013-2014	2014-2015	2015-2016	2016-2017
Agriculture	43.1	42.0	40.2	38.7	36.7	36.3
Industry	11.5	13.0	13.8	15.0	16.7	25.6
Services	45.1	45.5	46.6	45.0	47.3	39.3

Source: NBE, 2018

Agriculture and Its Role in the Ethiopian Economy

In 2017, economic growth in Ethiopia was higher compared to that in most African countries. In terms of the growth rates, Ethiopia even overtook Kenya, the largest economy in East Africa. According to the NBE (2018), Ethiopia registered annual economic growth up to approximately 11% in 2016-2017 showing relative improvement from 9.9% average annual growth registered in 2012-2013. The contribution of agricultural sector declined from 43.1% in 2011-2012 to 36.3% in 2016-2017 while both industry and service sectors demonstrated relative improvement (Table 1).

The national development policy (Agricultural Development Led Industrialization) goals are still behind the three actors. The main aim of the policy is to develop the economy based on agricultural sector and growth of industrial production and service sector. According to the NBE (2018), crop production comprised the lion's share of agricultural sector (65.3%), followed by animal farming and hunting (25.3%) and forestry (8.9%). Most importantly, service sector demonstrated a significant improvement compared to its past contribution which was below that of agriculture. Yet, the majority of Ethiopian population (79.3%) still live in rural areas practicing agriculture as main livelihood. Sectoral contribution of agriculture has to improve with an improvement of the agricultural inputs, access to market, and other challenges the agriculture has been facing.

Compared to 2010-2011 fiscal year, in 2016-2017, the percentage contribution of agricultural sector to the GDP decreased while that of other sectors (industry and service) increased (Table 2). But though the percentage share relatively decreased, its absolute growth was better except for the year 2015-2016 due to El-Nino effect. In addition, the percentage contribution of agricultural sector decreased from 25.3% in 2011-2012 down to 22.9% in 2016-2017 and demonstrated its highest and lowest contribution in 2013-2014 and 2015-2016, accordingly.

According to the NBE (2018), the contribution of agriculture to GDP in terms of money has been increasing from year to year and doubled in 2016-2017 compared to the preceding year from $9.8 billion to $20.5 billion.

Table 2. Growth of agricultural production and its contribution to overall economic growth in Ethiopia in 2010-2017

Parameters	2010-2011	2011-2012	2012-2013	2013-2014	2014-2015	2015-2016	2016-2017
Absolute growth	-	4.9	7.1	5.4	6.4	2.3	6.7
Percentage age contribution to the overall growth	-	25.3	31.2	22.3	24.0	11.3	22.9

Source: NBE, 2018

The share of forestry in agricultural GDP was 8.9% in 2016-2017, while the share of fishing was only 0.2%. The average growth of crop increased except for 2015-2016 due to El-Nino which was down to 3.4% but showed an improvement in 2016-2017. Concerning animal farming and hunting, though its growth was lower than crop production, it was highly affected by El-Nino in 2015-2016. This livestock sector has been growing low and contributing less with an average of 21.4% of the agricultural GDP over the last five years which shows the need of attention as the country is one the livestock richest countries in the world.

The total land area covered by grain crops i.e. cereals, pulses, and oilseeds, has slightly increased. From that, a total volume of about 306.13 billion quintals of grains (30.61 million metric tons) was obtained from private peasant holdings.

Within the category of grain crops, cereals are the major food crops both in terms of the area they are planted and volume of production obtained. They are produced in larger volume compared with other crops because they are the principal staple crops. Cereals are grown in all the regions with varying quantity.

The area cultivated for fruit and vegetable crops has shown an increment over the years though the cultivated land for root crops is much higher than fruits and vegetables. Fruit crops grown by the private peasant holders cover only a small token area and production in the country. The number of holders practicing fruit farming is lower than that of grains or cereals. Accordingly, the volume of production of the three crops (fruit, vegetables, and root crop) has increased over the years. Ethiopian farmers also produce stimulant crops such as chat and coffee and both crops are used for both internal consumption and export. Today, the total size of the country's coffee plantation is 725 thousand hectares with annual production of coffee above 400 thousand tons. The total yield obtained for main (meher) season has been increasing though the size of coffee production is above that of chat. These gap in production amount is due to the fact that the area cultivated for coffee is larger than the area covered by chat. The area cultivated for coffee crop has been increasing over the years while the increment of the cultivated area for chat crop has remained small over the last three years. These two stimulant crops are the main sources of foreign currency in Ethiopia. The average yield for a hectare of stimulant crops such as coffee and chat decreased from 10.5 quintals per hectare in 2011-2012 down to 8.98 quintals per hectare in 2017-2018 for chat crop. Per hectare yield for coffee decreased from 7.3 quintals in 2011-2012 to 6.19 quintals in 2017-2018. Considering the number of farmers engaged, area cultivated, and production obtained, those engaged in growing and producing stimulant crops such as coffee and chat are greater than those growing fruits since they earn a considerable amount of cash for the holders.

Agricultural Trade in Ethiopia

Domestic Trade

Ethiopia, the second most populous country in Africa with over 102 million people, has a big domestic markets. But access to domestic market in the country is one of the lowest in the world ranking (124 out of 132 countries) (Lawrence, Hanouz, Doherty, & Moavenzadeh, 2010). Most of grain crops produced in Ethiopia are used for household consumption. The permanent crops include stimulant crops such as coffee and chat. The main distributing channels of agricultural products are open marketplaces that are always available, supermarkets, consumer cooperatives, private commercial, kebele shops, Et-fruit shops, flour mills, regular shops, fruit and vegetable grocery shops, cereal shops, shops, and gulit (micro sell-

Table 3. Crop productions and percentage utilization for household consumption and sale in Ethiopia in 2016-2017

	Grain	Cereals	Pulses	Oilseeds	Vegetables	Root Crops	Permanent Crops	Enset
Household consumption	63.3	67.72	59.33	38.91	78.46	74.17	54.84	82.76
Sale	21.69	17.08	25.42	48.33	19.36	17.54	41.61	12.01

Source: CSA, 2018

ers). Higher percentage of crop products are utilized for household consumption ranging from 82.76% of enset, 67.3% of cereals, 59.33% of pulses, and to 38.91% of oilseeds (Table 3).

Ethiopian livestock supply chain represents four levels based on the number of live animals and traders per market day. Bush markets are markets where live animals are sold on weekly basis between livestock owners (mostly pastoralists) and small-scale buyers for breeding purposes. Another level is the primary market where the sells size does not overdo 500 live animals on weekly basis. In that market, the buyers are the accumulators and medium scale actors while pastoralists and small-scale traders are the sellers. When the weekly supply size of live animals lies between 501 and 1,000 on weekly market and sellers are those medium scale traders while the buyers are the big traders, this level is called secondary market. Finally, terminal markets are those markets that accommodate more than 1,000 live animals in a big city. These are the markets where big traders are the sellers and butchers and consumers are main buyers.

According to Negassa, Rashid, and Gebremedhin (2011), the livestock supply chain of Ethiopian market is too long due to the number of intermediaries which do not have any value-added activities other than cost of transactions. The involvement of intermediaries such as trackers, brokers, collectors, agents, small, medium and big scale buyers, butchers, etc. increased the cost of live animals and made life difficult to the end users. In the meantime, besides the transportation costs, most of the live animals move long distance which forces them to lose weight and decrease their quality of both meat and hide. The other important factor in Ethiopian livestock market system is the distance and scarcity of well-structured market centers. Majority of the pastoralists travel long distance, often more than a day, to reach the market center and are forced to sell with whatever price they got, as domestic livestock market system is based on bargaining. All these factors led the main actors (producers) of livestock benefit less and force them to informal cross border trade.

Most of the livestock trades in the Horn of Africa are informal and unrecorded by respective government authorities and pastoralists are not able to profit from formal markets (Njuki, Kaaria, Chamunorwa, & Chiuri, 2011; Johnson et al., 2015). Food Security and Nutrition Working Group [FSNWG] (2018) indicated that about 355,151 head of livestock were traded across the borders of horn of Africa between October and December of 2011. This is due to the location where the pastoralists are located which is around borders. Besides, the limitations of the government towards devising strong policies to keep informal trade out of track. Though Ethiopia is implementing free market economy aiming at private sector to play an important role in the economic development, limitation/unavailability of strong private sector to help smallholder producers integrate and link to broader commercial value chains plays a key role for the low development of the sector. Though cross-border livestock trade helps about 17 million people from the horn of Africa (Famine Early Warning Systems Network [FEWSNET], 2010) including pastoralists, it hinders the sectorial, regional, and national development. It also only helps trader and others who act as intermediaries than the real smallholder producers.

Export Trade

These days, many countries are interlinked and interconnected via trade as the involvement in the international trade plays an immense role in the economic growth of a nation. The involvement in trade integration benefits Ethiopia to search for alternatives to transform the current traditional agriculture into modern economy. International trade can also play an important role in Ethiopia's economic growth by creating an opportunity to export agricultural commodities which will motivate the domestic producers to produce a high-quality product in excess amount. In this regard, an intervention is required. Commercialization of livestock production requires interventions other than technical solution to feeding, breeding, and animal health problems of the Ethiopian smallholders.

Coffee, oilseeds, pulses, fruits and vegetables, chat, leather and leather products, meat and meat products, and live animal are the major agricultural export items in Ethiopia. Among export destinations, the countries of Asia take the first place with 37.7% of the exports followed by Europe (32.4%), Africa (21.5%), and America (7.5%) (NBE, 2018). Coffee gives the country a major amount of export earnings over the years and reached 30.45% of the total earnings of the country (Figure 1). The contribution of oilseeds decreased from 16.9% in 2014-2015 to 12.1% in 2016-2017. The contribution of pulses has been increasing while the contribution of fruits and vegetables has been increasing slightly.

Total earnings are $883.2 million from coffee, $351.0 million from oilseeds, $279.1 million from pulses, and $56.1 million from vegetables and fruits (Figure 2).

Among export items, in 2016-2017, Ethiopian livestock and livestock products are imported by China, Saudi Arabia, United Arab Emirates, Republic of Korea, India, Pakistan, and Yemen. Out of the countries of Asia, China (including Hong Kong and Taiwan), Republic of Korea, India, and Pakistan are the main

Figure 1. Percentage share of export earnings of crop products in Ethiopia
Source: NBE, 2018

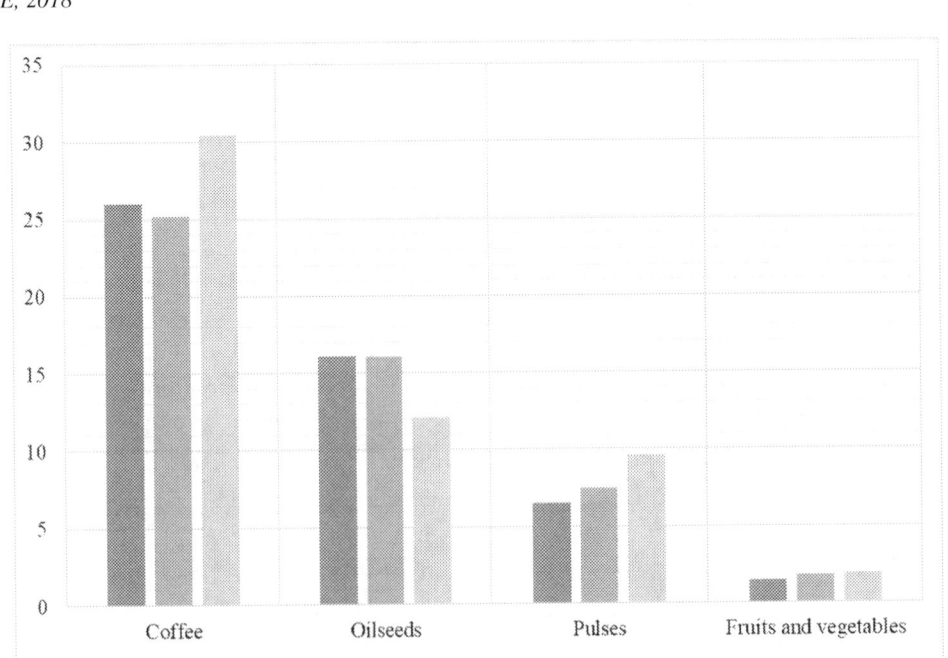

Figure 2. Export of major agricultural products from Ethiopia, $ million
Source: NBE, 2018

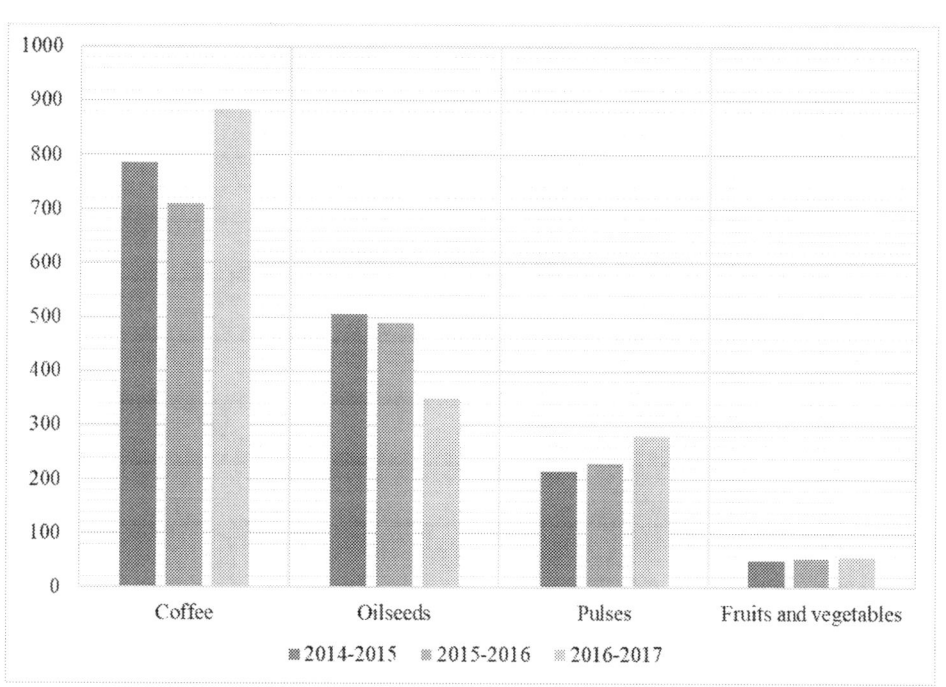

importers of Ethiopian leather and leather product, while Saudi Arabia is one of the major destinations of Ethiopian meat and meat products and live animals. United Arab Emirates import meat and meat products, Yemen imports live animals. Among the countries of Europe, Italy, the United Kingdom, and France import leather and leather products. Live animals are exported from Ethiopia to Somali and Djibouti, while Kenya imports leather and leather products. Leather and leather products are also exported to the USA and Canada (NBE, 2018). The export volume of leather and leather products decreased from $120.45 million in 2012-2013 to $114.0 million in 2016-2017, the volume of live animals fell down to $67.6 million in 2016-2017 from $166.6 million in 2012-2013 (Figure 3). The volume of meat and meat products increased by $24.4 million in 2016-2017 compared to that in 2012-2013.

When the share in the total national export is considered, it is relatively low compared to the amount of livestock population in the country. This indicates the need to be urgency in enhancing livestock type and productivity mechanism to fit the global market. Global price for live animals declined by almost 26% in 2016-2017 compared to 2012-2013 while it remains almost the same for the other two major livestock products exported by Ethiopia (Figure 4).

The volume of live animals exported decreased dramatically by 53.6% while leather and leather products and meat and meat products did not demonstrate essential change over the same period (Figure 5). The decline might be due to the decline in the volume of live animals exported and fall of the global price. The decline in volume is due to the weak value chain of live animals in the country. The main contributors of the weak value chain live animals were the lack of market information, lack of awareness of the main actors on livestock productivity enhancement, and traditional slaughtering of animals.

One of the main goals of agricultural and rural development policy in Ethiopia is to promote market-oriented economy. To do so, agricultural products should be improved in terms of quality and quantity.

Figure 3. Volume of major livestock export items in Ethiopia in 2012-2017, $ million
Source: NBE, 2018

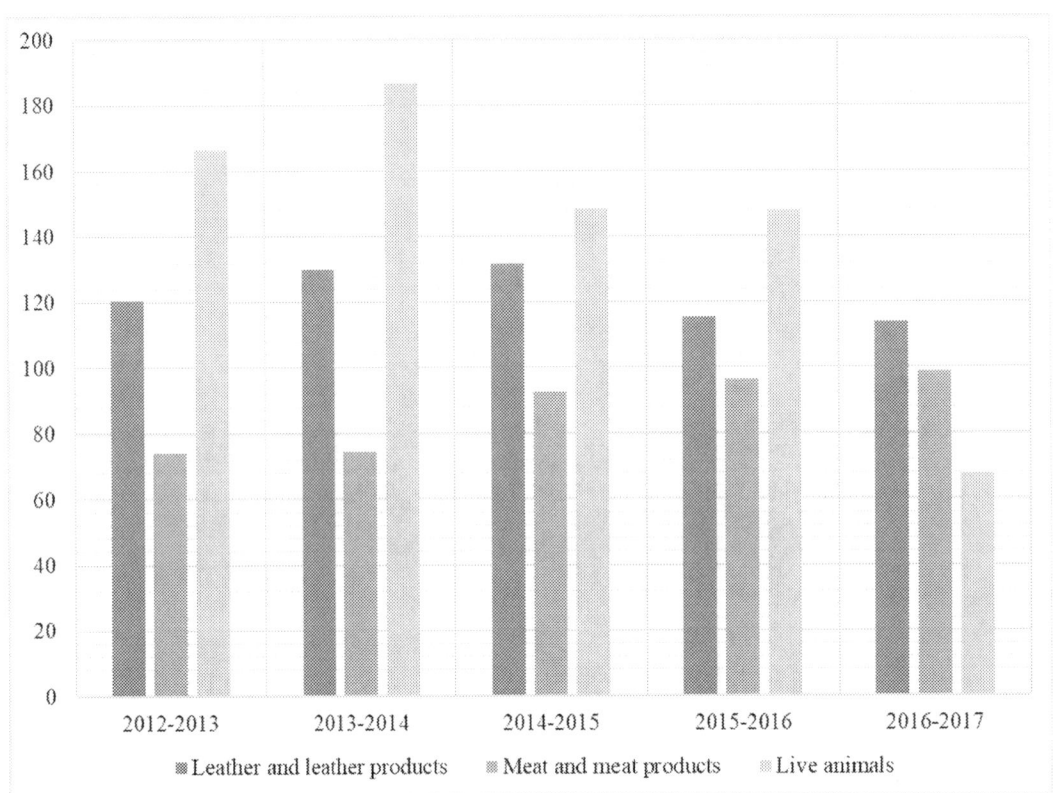

The main distributing channels of processed food are supermarkets, consumer cooperatives, private commercial, kebele shops, Et-fruit shops, flour mills, regular shops, fruit and vegetable grocery shops, cereal shops, baltena shops, gulit (micro sellers), and kiosks.

Agriculture and Trade for Sustainable Development

Challenges of rural areas, especially the low level of livelihood, should be improved in a sustainable way if sustainable development is to be assured. To achieve sustainability of rural development its worth focusing on the labor-intensive sector of a nation. In Ethiopia, most the population live in rural areas leading rainfed agriculture-based livelihood. It is, therefore, a priority to identify the key challenges of the sector and mitigate the problems in a sustainable way. The current rural development policy intends to sustain the development by increasing the productivity of smallholders through research-based information and technologies, increasing supply of industrial and export crops and ensuring the rehabilitation and conservation of natural resource bases with special consideration packages. Yet, the agriculture sector where the rural development can be done duffers due to low resource utilization, low technology and farming techniques (still majority of smallholders use wooden plow by oxen), low level of technological adaptation, over-reliance on fertilizers and underutilized techniques for soil and water conservation. According to the agriculture and rural development policy, the main challenges the sector

Figure 4. Unit value of major livestock export items in Ethiopia in 2012-2017, $/kg
Source: NBE, 2018

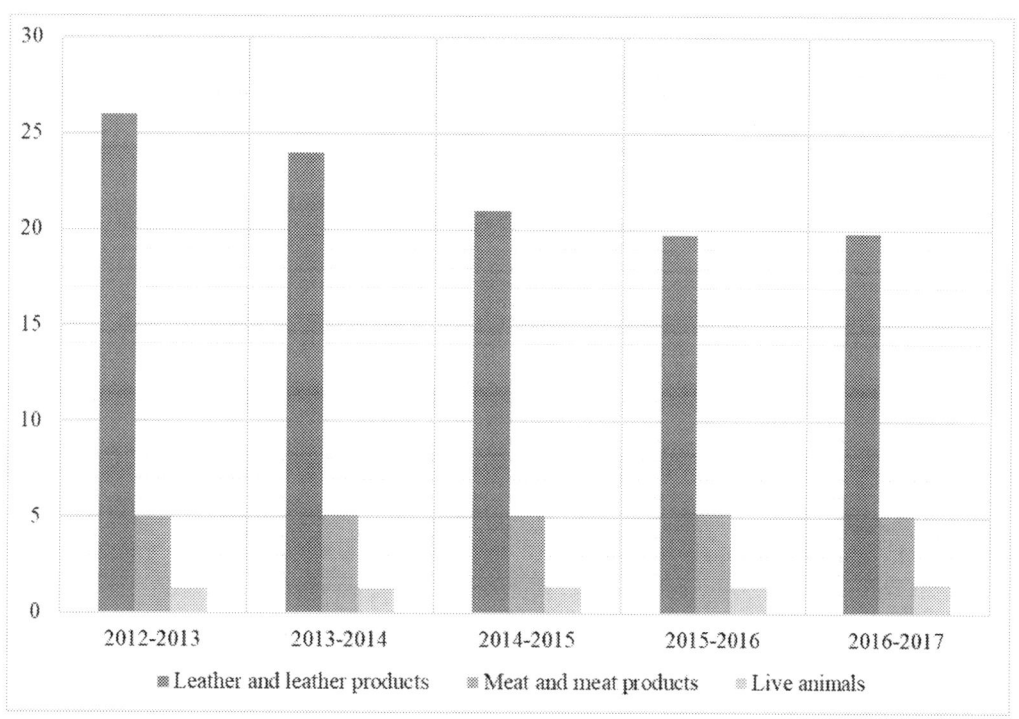

Figure 5. Volume of major livestock export items in Ethiopia in 2012-2017, millions kg
Source: NBE, 2018

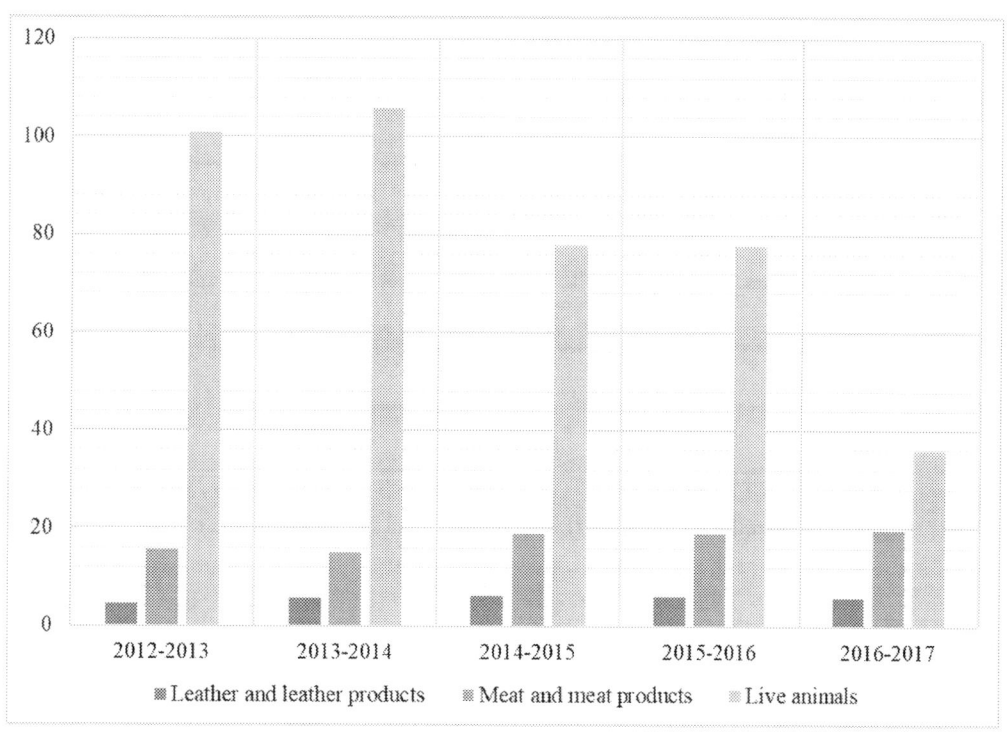

face are the small size and diminishing land due to large family size, unemployment, rapid population growth, soil infertility, on field and post-harvest crop pests, unpredicted patterns of rainfall, input scarcity and outdated technologies, shortage of capital, reduced market access, lack of market information, outbreak of animal diseases and shortage of animal feeds.

The factors of production where the rural development policy focused on are mainly labor and land with main objectives of ensuring rapid economic growth, enhancing the benefit of the people, eliminating food aid dependency and promoting market-oriented economy. As details of the policy indicate, the policy emphasizes making best use of human resource by promoting labor-intensive technology and enhancing the productive capacity of labor, and proper utilization of land. The UN Sustainable Development Goals appreciate the integration of the practice of integrated approaches of different rural development policies such as building resilient infrastructure, promote inclusive and sustainable industrialization regional and transborder and foster innovation. Similarly, Ethiopian government has focused on integration of development of education, health, road infrastructure, trade, finance, and industry, which are all set as the directions to achieve the goals.

Sustainable rural development, therefore, is to be achieved by increasing the level of productivity through proper implementation of agricultural inputs such as technology that meets the level of understanding of the rural, fertilizers which can enhance the efficiency of the soil type, introducing improved seeds that are practically approved by the rural community. With all these inputs, agricultural productivity can be increased. But the improvement in agricultural productivity alone cannot bring sustainable development unless supported by appropriate domestic and international trade. Trade is the key for development of rural livelihood in a way to helps boost the GDP and overall wealth which in turn will advantage the rural. Yet, agricultural and rural development promoters continued to underrate the importance of markets and trade for development, focusing mostly on agricultural production, its management, and organization (Engel, 2018).

As it has been shown earlier, the amount of foreign earnings from agricultural products is not to the level which satisfies the implementation of development activities. Foreign earnings enable exporters to easily access the required amount of foreign currency so that they can import that equipment that enhance the development of the country. The exchange rate between Ethiopian national currency (ETB) with USD has been tripled compared to two decades ago diminishing the buying power of the local currency. Ethiopia has been suffering from drought of foreign currency in the past ten year especially the last five years. In the year 2018-2019, following the leadership change in the country, over $13 billion has been obtained and high illegal foreign currency movements and black market were regulated. This helped the country to just settle for a year. Currently available foreign currency can only allow investing in imports of gas and medicine for short period of time.

This implies that the government should focus on the increasing the productivity of crop and livestock products in size and quality. Most importantly, access to market should be improved and liberation of trade should be exercised to motivate the low-cost producers. Higher economic growth can be achieved by the implementation of trade liberalization policy as trade liberalization improves export of a country (Ivolga, 2016). One of the main reasons of the advantage of liberalization of trade for the economic growth of a developing country is the nature of the backbone the economy of developing country, labor-intensive sector. Agricultural sector has been the base of the Ethiopian economy. It is still aggregates 79.3% of employment in Ethiopia with low level of per capita income and limited access to market. The liberalization of trade is therefore expected to allow the smallholder producers increasing their productivity and reaching out foreign markets which will enable them to gain higher income. The more income generated, the more investment in agricultural improvement can be realized thus improving the livelihood of rural dwellers beyond securing their basic food needs. Inline to this, the government should monitor the illegal cross border market which greatly affects the national revenue.

SOLUTIONS AND RECOMMENDATIONS

In order to improve the livelihood of the Ethiopians and enhance the growth level of the country in a sustainable way, the government should revise and devise policies with center point of smallholder farmers as most Ethiopian farmers are smallholders. The main challenges hindering the development of agricultural sector should be tackled to stimulate production level by improving transportation infrastructure, storage facility, providing and subsidizing agricultural inputs, access to financial credit center, access to market information, and appropriate market system. The livestock types are mostly indigenous where the output is low. To meet the growing global demand genetically modified livestock types should be introduced and agricultural extension workers as to be well trained in order to transfer and convince the importance. The government should also have to devise the mechanisms to control the illegal trade across the borders which has been observed over the years affecting the benefit of the farmer and reducing the revenue the country could have collected.

FUTURE RESEARCH DIRECTIONS

This research is only based on secondary information. There are few studies related to the perception of smallholder farmers towards the use of genetically modified livestock types especially in pastoral areas where the livelihood of the pastoralists is mainly nomadic livestock rearing.

CONCLUSION

The main challenges in rural areas, where agriculture is the main activity, are the scarcity of land due to the increase in population, unemployment, and rapid population growth, soil infertility, on-field and post-harvest crop pests, unpredicted patterns of rainfall, input scarcity, outdated technologies, shortage of capital, reduced market access, lack of market information, outbreak of animal diseases, and shortage of animal feeds. Many countries of sub-Saharan African, including Ethiopia, are net importers of agricultural products and food although the agricultural productivity in Ethiopia has been increasing despite the decrease in the contribution of agricultural sector to GDP. However, there is still a need to work towards increasing the level of productivity in rural areas of the country. The population size has also been increasing which leads to decrease in farmland. Besides, the agricultural technology adaptation status is minimal. The production of major export crops such as coffee and chat production need to be improved in parallel to the production of grain used for domestic consumption. The share of livestock products in the total volume of national export is relatively low compared to the amount of livestock population. That is why the attention is required to improve the quality and quantity of livestock and livestock products to meet the global demand. Increasing the productivity alone cannot ensure sustainable livelihood of rural farmers. Trade is the key for the development of rural livelihood in a way to help boosting the GDP and overall wealth which in turn will advantage rural people. To assure sustainable livelihood of rural community, it is necessary to integrate the agricultural productivity with increased access to market with infrastructures such as road, information, and agricultural extension services. It is worth developing the qualification of rural people involved in agricultural production since high quality of human resource may contribute to establishing sustainable development. Despite the fact that most of agricultural activities in rural territories in Ethiopia are labor-intensive, farmers should use the latest technologies and the most efficient fertilizers. Liberalization of trade can also play an important role in motivating the smallholders produce and aim at reaching broader market destinations domestically and internationally. This will enable rural farmers to increase their income which can be then invested in the development of agricultural production and infrastructure.

REFERENCES

Abro, Z. A., Alemu, B. A., & Hanjra, M. A. (2014). Policies for Agricultural Productivity Growth and Poverty Reduction in Rural Ethiopia. *World Development*, *59*(C), 461–474. doi:10.1016/j.worlddev.2014.01.033

Arayaa, A., Keesstra, S. D., & Stroosnijder, L. (2010). A New Agro-Climatic Classification for Crop Suitability Zoning in Northern Semi-Arid Ethiopia. *Agricultural and Forest Meteorology*, *150*(7-8), 1057–1064. doi:10.1016/j.agrformet.2010.04.003

Awulachew, S. B., Erkossa, T., & Namara, R. (2010). *Irrigation Potential in Ethiopia: Constraints and Opportunities for Enhancing the System*. Colombo: International Water Management Institute.

Bali Swain, R., & Varghese, A. (2013). Delivery Mechanisms and Impact of Microfinance Training in Indian Self-Help Groups. *Journal of International Development*, *25*(1), 11–21. doi:10.1002/jid.1817

Berg, A., & Krueger, A. (2003). *Trade, Growth, and Poverty: A Selective Survey*. Washington, DC: International Monetary Fund.

Bezabih, M., Di Falco, S., & Mekonnen, A. (2014). *Is It the Climate or the Weather? Differential Economic Impacts of Climatic Factors in Ethiopia*. Leeds: Centre for Climate Change Economics and Policy.

Brooks, J. (2012). *A Strategic Framework for Strengthening Rural Incomes in Developing Countries*. Paris: Organization for Economic Cooperation and Development. doi:10.1079/9781780641058.0023

Byerlee, D., de Janvry, A., & Sadoulet, E. (2009). Agriculture for Development: Toward a New Paradigm. *Annual Review of Resource Economics*, *1*(1), 15–35. doi:10.1146/annurev.resource.050708.144239

Central Statistical Authority. (2018). Agricultural Sample Survey.: Vol. II. *Report on Livestock and Livestock Characteristics (Private Peasant Holdings)*. Addis Ababa: Central Statistical Authority.

Cervantes-Godoy, D., & Dewbre, J. (2010). *Economic Importance of Agriculture for Poverty Reduction*. Paris: Organization for Economic Cooperation and Development.

Christiaensen, L., Demery, L., & Kuhl, J. (2010). *The (Evolving) Role of Agriculture in Poverty Reduction: An Empirical Perspective*. Helsinki: World Institute for Development Economics Research.

Coalition of European Lobbies for Eastern African Pastoralism. (2017). *Recognising the Role and Value of Pastoralism and Pastoralists.* Retrieved from http://www.celep.info/wp-content/uploads/2017/05/Policybrief-CELEP-May-2017-Value-of-pastoralism.pdf

Deere, C. D., Oduro, A. D., Swaminathan, H., & Doss, C. (2013). Property Rights and the Gender Distribution of Wealth in Ecuador, Ghana and India. *The Journal of Economic Inequality, 11*(2), 249–265. doi:10.100710888-013-9241-z

Engel, P. (2018). *Aligning Agricultural and Rural Development and Trade Policies to Improve Sustainable Development Impact.* Maastricht: European Centre for Development Policy Management.

Famine Early Warning Systems Network. (2010). *Cross-Border Livestock Trade Assessment Report: Impacts of Lifting the Livestock Import Ban on Food Security in Somalia, Ethiopia, and the Djibouti Borderland.* Washington, DC: Famine Early Warning Systems Network.

Food and Agriculture Organization of the United Nations. (2009). *How to Feed the World in 2050.* Rome: Food and Agriculture Organization of the United Nations.

Food and Agriculture Organization of the United Nations. (2012). *Initiative on Soaring Food Prices: Ethiopia.* Rome: Food and Agriculture Organization of the United Nations.

Food and Agriculture Organization of the United Nations. (2017). *The Future of Food and Agriculture: Trends and Challenges.* Rome: Food and Agriculture Organization of the United Nations.

Food and Agriculture Organization of the United Nations. (2018a). OECD-FAO Agricultural Outlook 2018-2027. Rome: Food and Agriculture Organization of the United Nations; Organization for Economic Cooperation and Development.

Food and Agriculture Organization of the United Nations. (2018b). *The State of Agricultural Commodity Markets: Agricultural Trade, Climate Change and Food Security.* Rome: Food and Agriculture Organization of the United Nations.

Food Security and Nutrition Working Group. (2018). *East Africa Crossborder Trade Bulletin.* Nairobi: Food Security and Nutrition Working Group.

Gebreegziabher, Z., Stage, J., Mekonnen, A., & Alemu, A. (2011). Climate Change and the Ethiopian Economy: A Computable General Equilibrium Analysis. *Environment and Development Economics, 21*(2), 1–21.

Greenville, J., Kawasaki, K., & Jouanjean, M. (2019). *Dynamic Changes and Effects of Agro-Food GVCS.* Paris: Organization for Economic Cooperation and Development.

Hadgu, G., Fantaye, K.T., Mamo, G., & Kassa, B. (2013). Trend and Variability of Rainfall in Tigray, Northern Ethiopia: Analysis of Meteorological Data and Farmers' Perception. *Academia Journal of Agricultural Research, 6*(1), 88-100.

Hailu, T., Sala, E., & Seyoum, W. (2016). Challenges and Prospects of Agricultural Marketing in Konta Special District, Southern Ethiopia. *Journal of Marketing and Consumer Research, 28*, 1–7.

Hallaert, J.-J., Cavazos-Cepeda, R. H., & Kang, G. (2011). *Estimating the Constraints to Trade of Developing Countries*. Paris: Organization for Economic Cooperation and Development.

International Fund for Agricultural Development. (2010). *Rural Poverty Report: New Realities, New Challenges, New Opportunities for Tomorrow's Generation*. Rome: International Fund for Agricultural Development.

Ivolga, A. (2016). Sustainable Rural Development in the Conditions of Trade Integration: From Challenges to Opportunities. In V. Erokhin (Ed.), *Global Perspectives on Trade Integration and Economies in Transition* (pp. 262–280). Hershey, PA: IGI Global. doi:10.4018/978-1-5225-0451-1.ch013

Johnson, N., Njuki, J., Waithanji, E., Nhambeto, M., Rogers, M., & Kruger, E. H. (2015). The Gendered Impacts of Agricultural Asset Transfer Projects: Lessons from the Manica Smallholder Dairy Development Program. *Gender, Technology and Development*, *19*(2), 145–180. doi:10.1177/0971852415578041

Kristjanson, P., Waters-Bayer, A., Johnson, N., Tipilda, A., Njuki, J., & Baltenweck, I. ... MacMillan, S. (2014). Livestock and Women's Livelihoods. In A. Quisumbing, R. Meinzen-Dick, T. Raney, A. Croppenstedt, J. Behrman, & A. Peterman (Eds.), Gender in Agriculture: Closing the Knowledge Gap (pp. 209-233). Amsterdam: Springer Netherlands.

Lawrence, R. Z., Hanouz, M. D., Doherty, S., & Moavenzadeh, J. (Eds.). (2010). *The Global Enabling Trade Report 2010*. Geneva: World Economic Forum.

Le Goff, M., & Singh, R. J. (2014). *Does Trade Reduce Poverty? A View from Africa*. Washington, DC: The World Bank.

Loayza, N., & Raddatz, C. (2016). *The Composition of Growth Matters for Poverty Alleviation*. Washington, DC: The World Bank.

Markelova, H., & Mwangi, E. (2010). Collective Action for Smallholder Market Access: Evidence and Implications for Africa. *The Review of Policy Research*, *27*(5), 621–640. doi:10.1111/j.1541-1338.2010.00462.x

Ministry of Finance and Economic Development of Ethiopia. (2012). *Ethiopia's Progress towards Eradicating Poverty: An Interim Report on Poverty Analysis Study (2010/11)*. Addis Ababa: Development Planning and Research Directorate, Ministry of Finance and Economic Development of Ethiopia.

National Bank of Ethiopia. (2018). *National Bank of Ethiopia Annual Report*. Addis Ababa: National Bank of Ethiopia.

Negassa, A., Rashid, S., & Gebremedhin, B. (2011). *Livestock Production and Marketing*. Addis Ababa: International Food Policy Research Institute.

Njuki, J., Kaaria, S., Chamunorwa, A., & Chiuri, W. (2011). Linking Smallholder Farmers to Markets, Gender and Intra-Household Dynamics: Does the Choice of Commodity Matter? *European Journal of Development Research*, *23*(3), 426–443. doi:10.1057/ejdr.2011.8

Robinson, S., Strzepek, K., & Cervigni, R. (2013). *The Cost of Adapting to Climate Change in Ethiopia: Sector-Wise and Macro-Economic Estimates*. Addis Ababa: Ethiopia Strategy Support Program.

Shiferaw, A., & Bedi, W. (2013). The Dynamics of Job Creation and Job Destruction in an African Economy: Evidence from Ethiopia. *Journal of African Economies, 22*(5), 651–692. doi:10.1093/jae/ejt006

Standing Committee for Economic and Commercial Cooperation of the Organization of Islamic Cooperation. (2017). *Improving Agricultural Market Performance : Creation and Development of Market Institutions*. Ankara: COMCEC Coordination Office.

The World Bank. (2019). *World Bank Open Data*. Retrieved from https://data.worldbank.org

Urquia Grande, E., Cano, E., Estebanez, R. P., & Chamizo-Gonzalez, J. (2018). Agriculture, Nutrition and Economics through Training: A Virtuous Cycle in Rural Ethiopia. *Land Use Policy, 79*, 707–716. doi:10.1016/j.landusepol.2018.09.005

Urquia Grande, E., & Rubio-Alcocer, A. (2015). Agricultural Infrastructure Donation Performance: Empirical Evidence in Rural Ethiopia. *Agricultural Water Management, 158*, 245–254. doi:10.1016/j.agwat.2015.04.020

Wagesho, N., Goel, N., & Jain, M. (2013). Temporal and Spatial Variability of Annual and Seasonal Rainfall over Ethiopia. *Hydrological Sciences Journal, 58*(2), 354–373. doi:10.1080/02626667.2012.754543

Walugembe, M., Tebug, S., Tapio, M., Missohou, A., Juga, J., Marshall, K., & Rothschild, M. F. (2016). *Gendered Intra-Household Contributions to Low-Input Dairy in Senegal*. Ames, IA: Iowa State University. doi:10.31274/ans_air-180814-208

ADDITIONAL READING

Badiane, O., Makombe, T., & Bahiigwa, G. (Eds.). (2014). *Promoting Agricultural Trade to Enhance Resilience in Africa*. Washington, DC: International Food Policy Research Institute.

Chanie, A. M., Pei, K. Y., Zhang, L., & Zhong, C. B. (2018). Rural Development Policy: What Does Ethiopia Need to Ascertain from China Rural Development Policy to Eradicate Rural Poverty? *American Journal of Rural Development, 6*(3), 79–93. doi:10.12691/ajrd-6-3-3

Khairo, S. A., Battese, G. E., & Mullen, J. D. (2005). Agriculture, Food Insecurity and Agricultural Policy in Ethiopia. *Outlook on Agriculture, 34*(2), 77–82. doi:10.5367/0000000054224300

Ministry of Agriculture and Rural Development of Ethiopia. (2010). *Ethiopia's Agricultural Sector Policy and Investment Framework (PIF) 2010-2020*. Retrieved from https://www.unisdr.org/files/28796_ethiopiaagriculturepif%5B30%5D.pdf

Sisaye, S. (1985). Education and Rural Development in Ethiopia. *Agricultural Administration, 20*(4), 237–255. doi:10.1016/0309-586X(85)90015-9

Welteji, D. (2018). A Critical Review of Rural Development Policy of Ethiopia: Access, Utilization and Coverage. *Agriculture & Food Security, 7*(1), 55. doi:10.118640066-018-0208-y

KEY TERMS AND DEFINITIONS

Domestic Trade: An exchange of internally produced goods and services within a country.

Food Security: A situation that exists when all people, always, have physical, social and economic access to enough, safe and nutritious food that meets their dietary needs and food preferences for an active and healthy life.

Holder: A person who exercises management control over the operation of an agricultural holding and makes the major decision regarding the utilization of the available resources.

Meher (Main) Season: Any temporary crop harvested between the months of September and February are considered as meher season crop.

Quintal: A measure which is traditionally used in Ethiopia; one quintal is equivalent to 0.1 metric ton.

Smallholder: A person who have less than one hectare of agricultural holding.

Sustainable Agricultural Intensification: An activity involved in increasing the agricultural productivity without damaging the environment.

Sustainable Development: A development that meets the needs of the present without compromising the future.

Chapter 22
International Competitiveness of Niche Agricultural Products:
Case of Honey Production in Serbia

Drago Cvijanović
University of Kragujevac, Serbia

Svetlana Ignjatijević
https://orcid.org/0000-0002-9578-3823
University Business Academy in Novi Sad, Serbia

ABSTRACT

The subject of the study is the analysis of honey production in Fruska Gora. Specifically, the authors determine the possibilities of honey production in the monasteries of Fruska Gora and the possibilities of increasing production and its impact on rural development. The chapter introduces the readers to the sector of honey production and sale in the monasteries of Fruska Gora, as well as to the problems related to the procurement of production material, state of marketing, and engagement of human resources. Since the monasteries represent sacred places, honey produced in such an environment can be considered a unique and special product. The authors reveal the factors which have a restrictive effect on the development of the honey sector in a specified geographical area, as well as explore the significance and role of production growth in the economic development of both the sector and the area.

INTRODUCTION

People live together with nature and therefore it is important to preserve resources and use them in a sustainable manner. In past few years, sensitive eco-system in Serbia has been significantly violated. The decrease of population in active working age has been registered in rural areas, as well as the aging trend, the disturbed educational structure, and the decrease of productivity and efficiency of agricultural production. All those issues mean the challenges for using the comparative natural conditions for pro-

DOI: 10.4018/978-1-7998-1042-1.ch022

duction, use of available production capacities, and intensifying agricultural production for the needs of domestic market and export.

Serbia disposes favorable natural conditions, relatively unpolluted resources, moderate continental climate, and abundance of honey plants. Honey production is constantly growing. Nevertheless, the production of honey in rural area of Vojvodina, especially in the monasteries in Fruska Gora, is at the pre-transitional level. Development of this market segment would contribute to the improvement of overall supply and export of honey, as well as the development of rural territories.

In this chapter, the authors studied Fruska Gora monastery honey as one of the segments of honey market in Serbia. Both global and Serbian markets of honey have been increasing constantly. Consumption patterns and preferences have been changing, that is why the understanding of this market is of exceptional importance. In the Background section, the authors introduce a reader to a structure and dynamics of honey production at the monasteries of Fruska Gora, and a role of beekeeping in rural development of Vojvodina and Serbia. The authors demonstrate that honey manufactured in Fruska Gora monasteries can satisfy the specific needs of consumers (quality, psychography, etc.) which actually confirms the results of many prior studies on honey consumption (Ćirić, Ignjatijević, & Cvijanović, 2015; Ignjatijević et al., 2019). Moreover, Fruska Gora monastery honey has a potential for branding and distinguishability on the domestic and international markets. The authors reveal the factors which restrict the development of honey sector in a limited geographical area of Fruska Gora. They also demonstrate a significance and role of increasing production in a market niche to the economic development of both honey sector and the area. This chapter discusses the findings of many previous studies which analyzed honey production issues internationally. In the Solutions and Recommendations section, the authors summarize the recommendations and discuss them across the findings of many scholars of different economic, political, cultural, and religious beliefs. Issues of honey production in monasteries have not been explored to date, whether it was about the context of rural development or creating a niche market strategy. Therefore, the conclusions of this research can be of particular interest to the scholars in the region of South-East Europe. In particular, it should be emphasized that the research results represent a starting point for further development of this market segment and creating a strategy for improvement of the competitiveness of Serbian honey in the domestic and international markets.

BACKGROUND

Most of the studies published internationally deal with the competitiveness of national economies in general and beekeeping sector in particular. There have been few studies related to the analysis of the competitiveness of niche agricultural products, especially the monastery honey as the market niche of honey sector. In Serbia, few authors have explored the significance of monasteries in the spiritual and cultural development of Serbian identity, and when it comes to their empowerment or contribution to the economic development, a number of studies is almost negligible; in fact, there is only few similar studies in the world. As Della Fave and Hillery (1980) stated, it was important that a monastery was economically self-sustainable, with the fact that the economic empire was subordinated to the religious one. Regarding that monasteries exist as the longest intra-national living communities they should offer some valuable traces (monastery products are certainly the examples of such goods).

Klimova (2011) pointed out that the monasteries were "the social world foundations" and helped in their empowerment by the transfer of religious knowledge and institutional refinements. Exactly, monasteries support a community and establish a reciprocal connection in which protagonists supplement and feed each other. Naidenova (2013) asserted that the monks' purpose was not to fight against an external enemy, i.e. the success was not in material things or the successful production of goods and services. Although it is possible to find many discussions about the economy in monasteries. Thus, Paganopoulos (2009, p. 363) stated that economy represented an internal and external dichotomy: an economy within the spiritual self ("economy of passions") and the monastery ("law of the house"), expressed in traditional practices, such as prayer, confession, psalmody, and painting. According to Paganopoulos (2009), nature grants the monasteries such gifts as honey, fruits, fish, bread, and wine, and in return, the monks keep the natural environment, which includes the way of life in monasteries, but the sustainable use of resources as well.

Erşan and Çiftçi (2016) analyzed one of the most attractive wine roads in which the monastery St. Ioannis Theologos in Turkey's Europe side is located. Rich historical and cultural heritage and a tradition of a winery road in combination with the unique natural environment makes the location ideal for cultural and natural tourism. According to Erşan and Çiftçi (2016), it is necessary to create the programs to inform the community about the use of the monastery potential to improve the life quality of urban and rural population. This concept has been applicable in the area of Fruska Gora National Park.

Although population of a rural area of Serbia in most cases have been employed in their households, the development of beekeeping would also contribute to mobility and participation in private business. Rural people place their folk arts and crafts' products, primarily handicraft, embroidery, fabric, souvenirs, and traditional food (for example honey) through the personal sales channels. People in rural areas, however, face problems: low level of investments in the additional education in regard to the need of lifelong education, poor incidence of educational institutions, especially in rural areas, dysfunctionality of the educational system with necessary skills on the job market, lack of activities related to education and training of adults, especially in rural areas, lack of measures that would stimulate self-employment, low level of workforce mobility, among others. (Paraušić & Cvijanović, 2007; Mihailović, Cvijanović, Milojević, & Filipović, 2014). Rural areas in Serbia have a potential for production of healthy food, as well as vacation and consumption of local food products. Rural development consolidates the demographic renaissance, the use of resources for the healthy food or organic food production, as well as the development of non-agricultural activities. Strengthening the connection between agriculture and tourism is important, therefore this integral connection represents a base for food consumption and the employment of workers in a rural area. In Serbia, rural development is directly related to the improvement of agricultural production and development of agri-tourism. In the context of the development of rural tourism, monastery tourism plays a special role and thereby it has a direct effect on the improvement of life standard and creates a barrage for further migration of population from villages to towns, food supply, especially honey, cheese, wine, and rakija, i.e. niche agricultural products.

Pocol and Popa (2012) stated that beekeepers were more interested in the stationary, conventional production. The reasons for that are high transport costs and the risk in moving the hives, very expensive periodic inspections, bureaucracy and the difficulty of selling. Pocol and Popa (2012, p. 243) concluded that "ecologic beekeeping is justified in terms of profitability only when it comes to high productions". Popescu (2010) emphasized that beekeeping was a "good job" that could contribute to a positive balance of payments. Tsafack Matsop, Muluh Achu, Kamajou, Ingram, and Vabi Boboh (2011, p. 3) pointed out that "there was a significant difference in output and in net benefit between traditional and semi-modern bee farms," and suggested that "beekeepers should adopt the semi-modern (Kenyan) hives." Pocol,

Bârsan, and Popa (2012) stated that a purpose of a social company in the field of beekeeping was to use agricultural resources and to integrate vulnerable groups from the region on the labor market.

Tesser and Cavicchioli (2014) emphasized the significance of being well aware of the prices, incomes, expenses, and profit in beekeeping. They said that the greatest contribution of their study was exactly in the judgment that it represented a first step in studying beekeeping in Italy and a base for further discussions.

If the research is directed to the competitiveness of honey sector, it is necessary to start from the specialization in the national domain of crucial issues, such as resources, potentials and possibilities, supply, demand, quantities, etc. Kruijssen, Keizer, and Giuliani (2007) pointed out to the market role in the agrobiodiversity conservation through product diversification and increasing competitiveness in niche and novelty markets. The authors said that a market approach would contribute to a better management of agrobiodiversity and the improvement of life conditions. Šerić and Uglešić (2014) emphasized a role in tracking and segmentation, as well as a task of marketing to follow trends on the domestic and international markets, in order to find the profitable niches, distinguishability of products, and satisfaction of consumers. The market research, aiming to find a recommendable market niche, will contribute to the consumers' loyalty, while the adjustment of a marketing strategy will have an effect on the increase of a selected market niche sale.

Previous researches have referred to the condition of national resources; the analysis of infrastructural and other conditions for honey and honey products production. Natural conditions are essential for the competitiveness in the national framework (Grubić, 2008), while the modernization of production is essential for international competitiveness (Bekić, Jeločnik, & Subić, 2013), the strategy of performance (Marinković & Nedić, 2010), and organization of the production process improvement (Pocol et al., 2012; Popa, Mărghitaş, & Pocol, 2011; Pocol, Ignjatijević, & Cavicchioli, 2017).

In beekeeping sector, the factors of development can be divided into developmental, i.e. positive, and limited ones. Factors that affect the development of beekeeping in a positive way include favorable geographic position for production, low labor costs, and cheap raw materials. Bad models of the manufacturers' organization, lack of strategic partnerships with purchasers of honey, or exporters and distributors should be mentioned as the key limiting factors of competitiveness on the domestic and international markets. Lack of financial resources limits the development of the market niches and the research activities and development, as it is the case in Malesia (Bohari, Hin, & Fuad, 2013). Due to the stated constraints, there is a justified fear that honey sector will continue existing as an extensive agricultural activity (Cvijanović, Mihajlović, & Cvijanović, 2012; Ignjatijević & Milojević, 2011). Active measures of state support are inevitable especially in the production of ecologically acceptable products, such as honey, as well as the engagement of science in finding an adequate organizational solution. Beekeeping has the potential to develop in the context of a total export contribution, in the context of strengthening an agricultural activity (Ignjatijević, Milojević, & Andžić, 2018). The incentive measures of the state will encourage manufacturers to find and develop their market niche (Kravchenko, Volchyonkova, & Esina, 2014).

Popescu (2010) accentuated that the promotion of quality of honey could stimulate export, while Popa and Pocol (2011) pointed out to significance of the beekeeping exploitation modernization, cooperation with other companies from the beekeeping sector, and export strategy of the beekeeping production. The authors concluded that the most important factors were the modernization of beekeeping exploitation, cooperation with other companies in the sector of beekeeping, which belonged to other beekeeping organizations, strategies for export of the beekeeping products. They concluded that the efforts should follow the modernization of apiaries and establishing a partnership with domestic and foreign companies. Finally, Popa and Pocol (2011, p. 194) emphasized that "entrepreneurs should identify opportunities to add value to products (branding, certification) and improve their marketing strategies in order to meet consumers' requirements and identify the most profitable markets. These entrepreneurial strategies represent a solution for the development of beekeeping sector, generating the increase of the sector competitiveness and strong value chain linkages through collaboration and strategic alliances".

The study of Ignjatijević and Cvijanović (2018) is a result of overall research of the Serbian agri-business sector. The authors deal with the issues of global competitiveness of Serbian agri-business sector, emphasizing sectors and products – the carriers of export and the improvement of international competitiveness.

In the study related to the competitiveness of honey sector, Ignjatijević, Ćirić, and Carić (2015) stated that the limiting factors of the honey production competitiveness were the lack of managerial and marketing skills. Honey manufacturers are focused on making profit in short term, as it is opposing to the long-term market positioning strategies. Furthermore, the manufacturers of honey identify marketing with promotion, while investing in marketing research, development and innovation, design of packaging, and creating a recognizable brand were neglected. Likewise, sales channels have not worked out enough, since a direct sale of honey has been dominated, directly from manufacturers. This all makes the limiting competitiveness factors on the niche level, while exactly the differentiation of Fruska Gora honey quality leads to the successful international competitiveness.

MAIN FOCUS OF THE CHAPTER

Most of the territory of Fruska Gora National Park is located in AP Vojvodina, the Republic of Serbia, while a smaller part of the territory belongs to Vukovar-Srem Country, the Republic of Croatia. The mountain has a specific position, different micro-climatic conditions, numerous endangered animal species, and protected plant species (Vujičić et al., 2011). Owing to the abundance of honey plants, water resources, as well as the coverage of large areas under meadows and forests, bees are provided with quality grazing. The most nectariferous, and at the same time the most widespread woody species on Fruska Gora mountain is linden, which provides the production of Fruska Gora linden honey, honey with a geographical origin (the only monofloral honey in Serbia with over than 60% of pollen). Significance of the geographic position for the production of Fruska Gora linden honey was presented in an Elaborate on protection of the designation of geographical origin for the production of Fruska Gora honey (Mitrovic, 2011). A special value and a pearl of Fruska Gora represents 16 Serbian orthodox monasteries, famous by their specific architecture, rich repositories, libraries, and frescoes (Fruska Gora National Park, 2018). In this chapter, the emphasis is made on the analysis of honey production in Fruska Gora National Park with the special focus on the production potential of the monasteries. In terms of a reduced number of bees owing to diseases, poisoning with pesticides, climate changes, the increase in production of honey in the Fruska Gora National Park, by increasing the monastery production, would ensure a top quality and absolute healthy products, together with the preservation and improvement of environment (Prodanović, Bošković, & Ignjatijević, 2016; Gibbs and Muirhead, 1998).

In Fruska Gora National Park, in the area of 50 km in length and 10 km in width, there are located sixteen active orthodox monasteries out of formerly thirty monasteries, namely, Krušedol, Petkovica, Rakovac, Velika Remeta, Divša, Novo Hopovo, Staro Hopovo, Jazak, Mala Remeta, Grgeteg, Beočin, Privina Glava, Šišatovac, Kuveždin, Fenek, and Vrdnik-Ravanica. In 1990, the unique cultural-historical entirety was proclaimed a cultural asset of exceptional significance for the Republic of Serbia and was proposed for the enrolment in the UNESCO Heritage List (Fruska Gora National Park, 2018). Every year, Fruska Gora National Park provides individual manufacturers with a possibility to locate their beehives in the park. These are specifically defined and arranged locations for which the Park guarantees keeping and surveillance. Besides Fruska Gora, Tara national park and Ovcar-Kablar gorge are also favorable for beekeeping (Ignjatijević et al., 2015).

As Antonić (1998) stated, in medieval age, Emperor Dusan encouraged beekeeping in the monasteries. The Emperor personally named ten beekeepers to deal with beekeeping, but there was no inscription on the methods of production. However, due to the extraordinary natural conditions and professionalism of beekeepers, honey became a subject of intensive trade, while the tradition has been held till now. Nowadays, honey production is not that active as it was several decades or even centuries ago. Economic independence has been one of the basic principles of monastic communities from the earliest times. Today, production and distribution of wine, rakija, and food products is only a tiny trail of the former strength of monastic communities. Tradition of production has not been abandoned to this day, people gladly purchase the goods manufactured by monks (Dragović, 2013). Due to the space restraints, the production in monasteries would significantly increase the usability of resources. It is important to point out the increasing interest and the increase in diversity of consumers' demands: besides high quality, healthy products, clean and improved production, one can also mention vacation and tourism, education and other forms of rural area exploitation, as well as beekeeping. Beekeeping would contribute to keeping biodiversity and improvement of ecosystem, and, on the other hand, improvement of economic well-being of the farmers, what furthermore makes a base for the sustainable rural development.

Production of high quality products, a branded product, with a geographical indication, may contribute to the distinguishability of the area (Ignjatijević, Ćirić, & Carić, 2013), increase in incomes and life standard of manufacturers, engagement of all community members, development of teamwork, and finally stopping out-migration to urban areas or other countries (Ignjatijević et al., 2018).

This chapter analyses honey production in Fruska Gora monasteries, one segment of Serbian honey market. Data obtained during a survey of the producers demonstrate the production potential of the market niche in Fruska Gora monasteries in the period from 2008 to 2018, aiming to study the market niche competitiveness factors of Serbian honey sector. For that purpose, there were determined the potential and possibilities for the production of honey and honey products, opportunities for increasing production, and impact on the rural development of an analyzed area. Selected data offer the directions for the preservation and improvement of production and point out to elements of the strategy for the improvement of competitiveness of the analyzed market niche. The expected contribution of the study is the elaboration of measures for the improvement of the competitiveness of both market niche and entire honey sector. In terms of the market adjustment, it is important to manage the factors of competitiveness, from purchasing habits (Kotler, 2003), perception of products, price, and distribution to promotion (Šerić, 2003). That is why the strategies for improvement of competitiveness of the market niche should be based on the decrease in market losses, development of new forms of sales, production innovations, and customers' relations.

The authors employed the following methods: an inductive and deductive method (following trends and changes in the production volume in a sample); causality induction method (analysis of the causes for changes in production volume); classification method (sorting the research conclusions); comparative method (analysis of a niche specificity in regard to Serbian honey sector); descriptive statistics (analysis of the selected survey data). At the pre-analysis stage, there was determined which monasteries made a

Table 1. Production of honey in Beočin monastery

	Number of beehives	Yield per beehive	Quantity, kg	Accompanying products	Types of honey
2018	28	<6	300	Propolis	Acacia, meadow, and linden honey
2017	25	<6	150		Acacia, meadow, and linden honey
2016	25	6-10	200		Acacia, meadow, and linden honey
2015	20	6-10	140		Meadow and linden honey
2014	20	6-10	160		Meadow honey
2013	20	<6	120		Acacia and meadow honey

Source: Authors' development

market niche. Then, there were determined the criteria manufacturers had to meet to be comprised with a sample (monasteries which manufactured honey with an expressed market share). Since the study encroached in the field of some monasteries' business secrets, not all manufacturers of the market niche were considered. The study was carried out on the territory of Fruska Gora from October 2018 to January 2019. All 16 Serbian orthodox monasteries were comprised. The survey questions were classified in three groups. The first group referred to the production of honey, second group – to sales and procurement of the production facilities, and third group – to research and development, the accompanying activities besides production. During the interviews, the answers were recorded. Data processing was completed in SPSS software. Descriptive statistics was implemented for in detailed analysis.

Beočin Monastery

In Beočin monastery (Table 1), the beekeeper keeps the hives in a rented monastery estate (the beekeeper has twenty hives and is engaged amateurish in production for ten years). The beekeeper used neither the state incentives (subsidies) for production, nor had some tax reliefs. As he was completely relied on his own funds, there is no possibility for the research and development in production.

The study showed that first bee hives were set in 2013, and until today, a number of hives have grown. A yield of honey has also changed, according to weather conditions and experience and knowledge of the beekeeper. Besides the production of honey, the beekeeper has also been engaged in the production of propolis, while most represented had been acacia, meadow, and linden honey. Considering the manufactured amount of honey on the monastery estate, 10-50% of the manufactured amount is donated to the monastery (depending on a yield in a specific year), along with the entire amount of the manufactured propolis.

Rakovac Monastery

In Rakovac monastery, similar to Beočin monastery, the monks, do not deal with direct manufacture of honey but have leased the monastery estate to the beekeeper's family since 2008. The beekeeper's family has been involved in the state support system, in form of subsidies. Since 2013, production and a number of hives have significantly decreased (Table 2). Besides unfavorable climatic conditions, the reasons of that included insufficient financial resources. Experience and knowledge in the field of beekeeping are currently crucial for the sustainability of family business.

Table 2. Production of honey in Rakovac monastery

	Number of beehives	Yield per beehive	Quantity, kg	Accompanying products	Types of honey
2018	24	>20	700	Propolis, pollen	Oilseed rape, fruit, acacia, and linden honey
2017	24	11-15	400		Oilseed rape, fruit, acacia, and linden honey
2016	25	11-15	350		Oilseed rape, fruit, acacia, and linden honey
2015	25	11-15	330		Oilseed rape, fruit, acacia, and linden honey
2014	50	6-10	550		Fruit, linden, and meadow honey
2013	50	6-10	350		Fruit, linden, and meadow honey
2012	55	>20	1,100		Acacia, linden, fruit, and meadow honey
2011	55	>20	1,200		Acacia, linden, fruit, and meadow honey
2010	155	16-20	2,000		Acacia, linden, fruit, and meadow honey
2009	155	16-20	2,000		Acacia, linden, fruit, and meadow honey
2008	155	16-20	2,000		Acacia, linden, fruit, and meadow honey

Source: Authors' development

In 2018, the pasture was very rich and the yield reached a record 30 kg of honey per beehive. In 2000-2004, the beekeepers invested around €20,000 in the equipment necessary for the production of honey, such as beehives and the accessories for beekeeping, as well as for the equipment inevitable for extracting honey and preparation for packaging and further sales. In addition to honey, the beekeepers produce pollen, propolis, and honey-based preparations, such as various balms, creams, waxes, rakijas, etc.

Vrdnik Monastery (Nova Ravanica)

Nova Ravanica monastery (Table 3) deals with the production of domestic products, but unfortunately, honey and honey products are not among them. However, the monastery leases the monastery land to other beekeepers for placing their hives. The monastery has an extraordinary potential for beekeeping and therefore has been visited by many beekeepers.

The monastery estate has a huge potential for beekeeping due to the untouched nature and non-sprayed plants. The estate is suitable for acacia, meadow, and linden pasture, and therefore these three types of honey are mostly represented. Besides honey, there is a production of beeswax that is used for the production of candles. A price of acacia honey ranges between €5 and €6, linden and meadow honey – €5-8. Although there are favorable natural conditions for production, yields are not high. This situation is a result of numerous set beehives, owing to the presence of many bees in a radius of five km. In this situation, bees come into conflict which can affect a yield per a hive. The research shows the presence of the significant oscillations in the production volume, as well as that the years 2011-2012 were favorable for production, and also that the year 2008 was record-breaking both in a number of hives and yield.

Jazak Monastery

Jazak monastery (Table 4) deals with the production of honey since 2008. Local residents of neighboring Spa Vrdnik have donated beehives and a part of equipment to the monastery, in order monks could

Table 3. Production of honey in Nova Ravanica monastery

	Number of beehives	Yield per beehive	Quantity, kg	Accompanying products	Types honey
2018	500	11-15	7,000	Wax	Acacia, linden, and meadow honey
2017	300	6-10	2,500		
2016			2,500		
2015	150		1,200		
2014	100		700		
2013			600		Meadow and linden honey
2012	300	11-15	4,500		Acacia, linden, and meadow honey
2011	200		2,000		
2010	100	6-10	800		
2009			800		
2008			700		

Source: Authors' development

manufacture honey and beeswax by themselves. In total, thirty-five hives and some of the beekeeping equipment were granted to the monastery.

The study shows that a yield per hive depends mainly on weather conditions, as well as the knowledge of beekeepers – monks. From the beginning of beekeeping, a yield ranges from 10 kg to 15 kg per hive, which belongs to a solid yield, except in those years when weather conditions were hostile to bees. The area in which the monastery is located abounds in the acacia and linden plants, and as the result, the acacia and linden honey is manufactured.

Table 4. Production of honey in Jazak monastery

	Number of beehives	Yield per beehive	Quantity, kg	Types of honey
2018	35	11-15	400	Acacia and linden honey
2017		6-10	250	
2016			250	
2015			230	
2014			230	
2013		11-15	200	Linden honey
2012			550	Acacia and linden honey
2011			380	
2010			350	
2009			350	
2008				

Source: Authors' development

Table 5. Production of honey in Mala Remeta monastery

	Number of beehives	Yield per beehive	Quantity, kg	Types of manufactured honey
2018	21	11-15	300	Acacia, linden, and meadow honey
2017			70	
2016	7		80	
2015			75	
2014	5	6-10	50	
2013			40	Linden and meadow honey

Source: Authors' development

Table 6. Production of honey in Staro Hopovo monastery

	Number of beehives	Yield per beehive	Quantity, kg	Types of honey
2018	10	11-15	115	Acacia and linden honey
2017		6-10	90	
2016			95	
2015			70	
2014				

Source: Authors' development

Mala Remeta Monastery

The female Mala Remeta monastery (Table 5) started the production of honey in 2013. Local residents of neighboring settlements have donated beehives to the monastery in order to start beekeeping and honey production.

Nuns started to deal with beekeeping with only five hives, of which a yield of approximately 10 kg per hive was obtained. Since 2015, as the manufactured quantity has increased, more honey has remained for the nuns' own needs and gifts. The monastery is not included in the state support system in a form of subsidies. Due to the limited financial resources, there is no possibility for engaging the external human resources who would deal with marketing, management, promotions, research, and development of sales.

Staro Hopovo Monastery

Beekeeping in Staro Hopovo monastery was initiated with ten hives, which were donated to the monastery in order the monks could start with the production of honey. In the first years of production, when the monastery did not sell honey, yields were low (around 7 kg per hive) (Table 6). Forests that surround the monastery estate are rich in acacia and linden trees, and therefore monks produce and sell acacia and linden honey.

Sales of honey in Staro Hopovo monastery started in 2006 when honey was placed for the price of 8-9 €/kg depending on a type. This price has been kept until today. The price of honey for customers from abroad is 10 €/kg, while the customers are mainly tourists visiting Fruska Gora and the monasteries.

Novo Hopovo Monastery

Residents of Novo Hopovo monastery have been engaged in the production of honey since 2015. The production started with only five hives donated to the monastery. Besides honey, the monks have started to produce pollen and propolis (Table 7).

The monastery is engaged in the production of acacia, forest, and meadow honey, according to the environment. Every year, a part of the manufactured amount of honey is allocated for own needs, sale, and donation. The distribution price of acacia, meadow, and forest honey is €8 €. In past two years, the demand of foreign tourists for honey has increased, and they were charged €10 regardless the type of honey.

Krušedol Monastery

Krušedol monastery (Table 8) has been dealing with the production of honey since 2014. Few years earlier, the equipment and hives were donated to the monastery, in order to start the production of honey. The monastery residents have neither financial resources nor personnel, and therefore is uncertain if the production of honey continues in this monastery (production capacity remained at the level of the initial five hives).

Extremely high yield of honey in past years has instigated the monks to do in-depth analysis of production. Still, due to the lack of funds, there is no space for the engagement of the experts in marketing, management, and research. Honey is marked by the simple stickers with a monastery inscription and a type of honey.

Table 7. Production of honey in Novo Hopovo monastery

	Number of beehives	Yield per beehive	Quantity, kg	Accompanying products	Types of honey
2018	8	11-15	110	Propolis and pollen	Acacia, forest, and meadow honey
2017		6-10	70		
2016					
2015	5		30		

Source: Authors' development

Table 8. Production of honey in Krušedol monastery

	Number of beehives	Yield per beehive	Quantity, kg	Types of honey
2018	5	11-15	50	Fruit, meadow, and linden honey
2017		6-10	40	
2016				
2015			35	
2014			25	

Source: Authors' development

Velika Remeta Monastery

Velika Remeta monastery has started the production of honey long time ago. The study covers the period 2008-2018 (Table 9). The specifics of this monastery is that the production of honey is made by the stationary and moving hives. Accordingly, the acacia and linden honey are mostly manufactured. Nevertheless, the hives are moved to sunflower fields which allows producing sunflower honey. In addition to these three types of honey, the monastery is engaged in the production of propolis.

Monks and nuns take part in the honey production process; however, the monastery has no sufficient resources in order to deal with marketing, management, and research. From the beginning of the observed period, a number of beehives in Velika Remeta monastery has remained homogenous, until 2016, when a number of hives decreased due to the diseases of bees.

Monasteries in Which Honey Is Not Produced

Šišatovac monastery is still in the process of reconstruction and is not in the position to deal with beekeeping. Kuveždin monastery is also in the process of reconstruction and is not engaged in the production of honey. In Grgeteg monastery, the monks are engaged in the production of honey, yet it is a question of internal activities and not available to the public. Petkovica monastery has a positive attitude towards the production of honey, but the position of this monastery is not favorable for breeding bees. Djipsa monastery due to its remoteness from the settlements has always been poor, demolished, and robbed many times, as well as burdened by the problems of survival. Therefore, the residents of Djipsa monastery are not engaged in the production of honey. In Privina Glava monastery, the monks do n0t deal with beekeeping because the location of the monastery is not suitable for beekeeping. The situation is similar in Fenek monastery – the monks are not engaged in the production of honey and honey products since the location of the monastery is not favorable for breeding bees.

Table 9. Production of honey in Velika Remeta monastery

	Number of beehives	Yield per beehive	Quantity, kg	Accompanying products	Types of honey
2018	55	11-15	1,000	Propolis	Acacia and linden honey
2017			600		
2016					
2015	105	6-10	900		
2014					
2013			700		
2012		11-15	2,000		
2011		6-10	900		
2010					
2009			850		
2008					

Source: Authors' development

International Competitiveness of Niche Agricultural Products

Table 10. Aggregated indicators of honey production in Fruska Gora monasteries

	2018	2017	2016	2015	2014	2013	2012	2011	2010	2009	2008	average
Number of bee hives	686	469	470	362	330	315	490	395	395	395	395	427.45
Yield per hive	13.6	9.4	9.7	9.1	8.6	9.0	16.0	14.8	11.8	11.8	11.8	11.40
Quantity, kg	9,975	4,170	4,185	3,010	2,685	2,010	8,150	4,480	4,050	4,000	3,900	4,601.36
Sale, kg	2,750	2,555	2,250	1,540	1,070	870	4,600	2,550	2,550	2,150	2,150	2,275.91
Number of beehives in Vojvodina, thousands	209	165	149	143	123	150	124	111	97	92	91	132.18
Share of honey production in monasteries in regard to Vojvodina, %	0.33	0.28	0.32	0.25	0.27	0.21	0.40	0.36	0.41	0.43	0.43	0.32
Production of honey in Vojvodina, tons	2,440	2,093	1,652	2,415	1,396	2,532	-	1,134	981	911	921	1,647.50
Production of honey in Serbia, in tons	11,427	7,014	5,761	12,263	4,383	8,554	6,983	6,963	7,281	7,356	4,161	7,467.82

Source: Authors' development

Production and Sale of Honey in Fruska Gora Monasteries

The study allowed to calculate the overall amount of beehives, average yield per a hive, and a total volume of honey production in the monasteries of Fruska Gora (Table 10). The observed period is 2008-2018, although, in some monasteries, the production has started only few years ago.

An average number of beehives in Vojvodina was 132.18 thousand, in the market niche in total – 427.45 thousand. The production of honey in Vojvodina was 1647.50 tons, 0.32% of which was manufactured in the market niche. The study demonstrates that the market niche has a small volume of production that surely should not be understood as a static category, but as the potential to be developed.

Linden, meadow, acacia, and forest pastures are the most represented pastures in the monasteries of Fruska Gora. A yield per hive depends on the experience and knowledge of a beekeeper. Out of nine monasteries engaged in the production of honey, five monasteries in the beginning of beekeeping process what can be seen in a yield per hive. In eight monasteries, out of nine observed ones, beekeeping is conducted stationary due to low financial investments. Applying the stationary method of beekeeping, the monastery estates can receive a limited number of beehives. The researches of Fruska Gora National Park potential show that this area can receive three times more bee communities without being jeopardized. Accordingly, there is a potential for the increase in top-quality honey production. Thereby potentials for increasing the production of top-quality honey represent a factor that affects positively the increase in competitiveness.

The monasteries do not dispose human resources and therefore miss marketing, management, and research activities. Individual and average quantities of sold honey do not represent a real state of sales, because they also include donated honey. There are no records which separate donated honey from the sold one. No matter if honey is donated or sold, an important data is that this is the correct quantity of honey manufactured in the monastery since every gift can be good recommendation for further sale.

Figure 1. Price of honey in Fruska Gora monasteries in (a) RSD/kg and (b) €/kg
Source: Authors' development

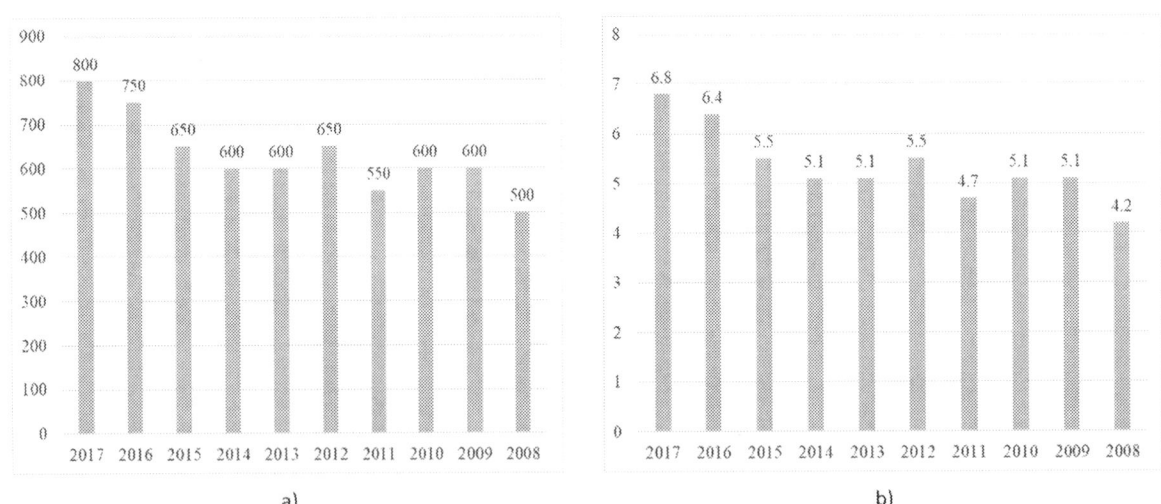

The prices of honey differ from one monastery to another. Research of a price of honey shows that an average price ranges from RSD 500 to RSD 800 per kilogram (Figure 1). The price varies depending on a type of honey, but also on the experience in production, as well as on quality.

Additionally, there are other accompanying factors that affect a price of honey, including packaging (glass or plastic), label (print quality), design, size of packaging, among others. Price is a factor which has a positive effect on supply, while packaging and distribution are limiting factors of competitiveness on the local and international markets.

There are the monasteries which manufacture honey products besides honey (Figure 2). The most popular ones are propolis, pollen, and beeswax. Nevertheless, no precise data on the qualities of the manufactured accompanying products are available.

SOLUTIONS AND RECOMMENDATIONS

The authors, who were engaged in the analysis of beekeeping, demonstrated the interest in studying the competitiveness of honey production sector in Serbia. According to the obtained results, it is possible to define particular measures for improvement of current state of this sector. The obtained results should be compared with the results of similar studies carried out in developing countries, developed ex-communist countries, and transition economies. The obtained results can equally attract scientists, businessmen, and decision-makers of agri-food sector developmental strategy. Regardless whether they were engaged in scientific analysis, consulting, educational, or entrepreneurship, they all were invited to contribute to the overcoming of the situation described as "Serbia has a potential, but it is used insufficiently".

Previous studies of honey production, especially production in monasteries worldwide and in Serbia, are insufficient. Many scholars have studied honey and beekeeping from various points of view, but in Serbia, in-depth economic studies of beekeeping production are missing. The authors show that beekeeping has the insufficiently used potential when it comes to the monastery production of honey.

Figure 2. Number of monasteries in which the accompanying bee products are manufactured
Source: Authors' development

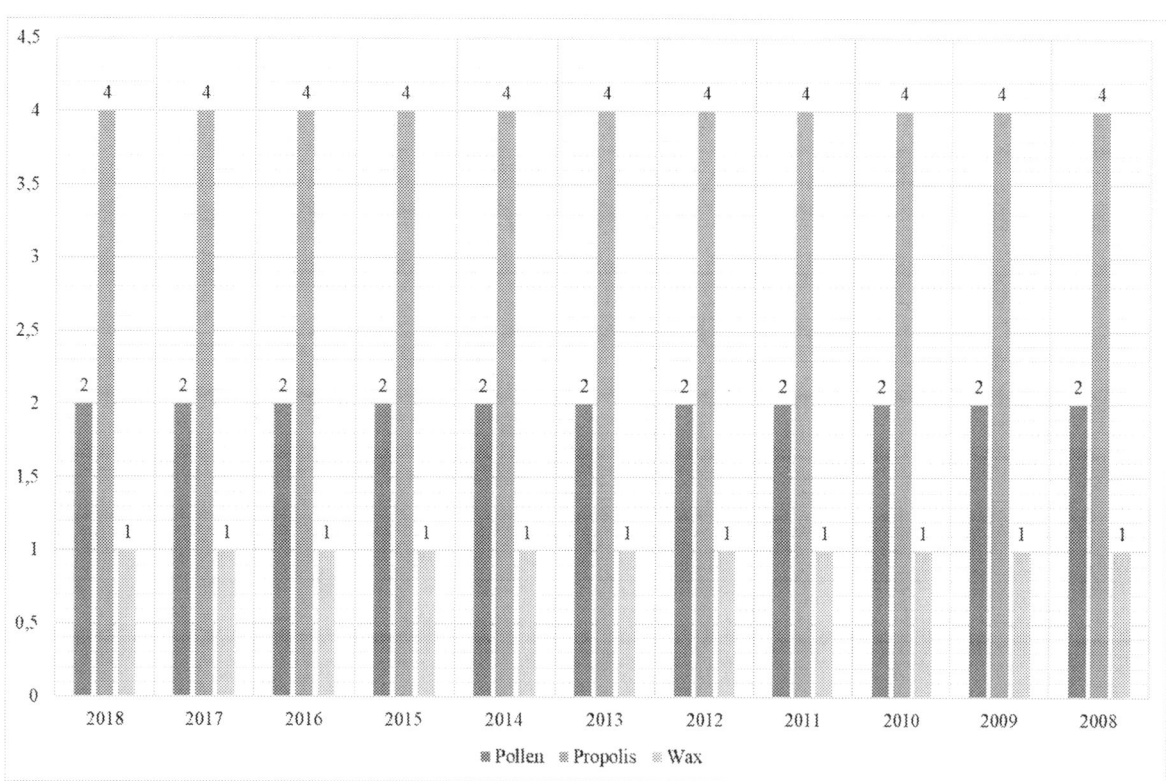

If the one starts from the fact that linden honey from Fruska Gora is the only honey with a protected geographic origin in Vojvodina, then there can be ascertained the existence of production potential. The research shows that the production of honey in monasteries has increased in past few years, the niche has strengthened. This finding is in accordance with the previous conclusions on increasing the total honey production in Serbia (Ignjatijević et al., 2015), as well as with Pocol's (2012) conclusion on the production of honey issues in the EU. Discussing beekeeping in the context of rural development in Serbia, the obtained conclusions were similar to the conclusions of Tesser and Cavicchioli (2014) about Italy. They emphasized that data could fulfill an information vacuum in Serbia. Research of the market niche potentials of honey sector has not been stimulated by the state. It is necessary to overcome such situation by using a set of measures which is in accordance with Popescu's (2010) conclusion. demonstrated the increase in organic honey consumption and emphasized the inclination towards the consumption of top-quality certified honey or honey with a geographical indication. Ćirić et al. (2015) found a great interest in buying honey in the monasteries, which was also confirmed by this study. The consumers increasingly care about the quality of honey they buy. It is important for consumers to be informed on the quality of honey and its clear differentiation from a forgery on the market (Dugalić-Vrndić, Kečkeš, & Mladenović, 2011). When it comes to honey manufactured in the monasteries, quality is not called into question due to its manufacture in a "holy place".

The issue of increasing the production of honey in the monasteries, in order to empower them and the indirect rural development, was not a backbone in the previous studies except the assertion of Della

Fave and Hillery (1980) that the monasteries should be independent. The conducted research show that the modernization of production, education (Popa et al., 2011), and the development of entrepreneurial behavior are necessary for increasing the manufactured amount of honey. Previous studies of Ignjatijević et al. (2015) demonstrated that beekeeping in Serbia represented a supplemental activity without serious approach, as Ignjatijević and Cvijanović (2018) also pointed out. Within the strategy of rural development in Serbia, it is inevitable to create a strategy of beekeeping development, which will comprise an incentive in honey production in monasteries, i.e. the modernization and cooperation at all levels are necessary (Popa & Pocol, 2011). If the beekeepers' network expands, it will use agricultural resources in an effective manner, employ people that have stayed out of work in the process of transition, as well as support rural development.

FUTURE RESEARCH DIRECTIONS

In Serbia, the scholars have analyzed the volume of production with a focus on the total number of beehives, total quantity of honey, and yield per hive. However, in-detail research has not been conducted in Serbia. The authors' desire is that the readers, scientists, and developmental managers and strategists become interested in the in-depth analysis of beekeeping in the context of rural development. Future research should be directed at the measurement of a level of economic power of honey manufacturers, definition of the criteria, and their classification in small, medium, and big manufacturers. Finally, it would be useful to perceive the level of investments in beekeeping. It would be significant to conduct the research on the possibilities of connection of medical and aromatic sector with beekeeping, pharmaceutics, cosmetics, and production of auxiliary remedies and cosmetic products. A wider study of beekeeping through all educational levels and from different points of view within the different scientific disciplines is needed.

CONCLUSION

This study demonstrated the increase in the volume of manufactured honey in the monasteries of Fruska Gora. It can be concluded that a realized total amount of honey was high, with small share in the total honey supply of Vojvodina, by realizing a total number of hives on the monastery estates. Results of the research show that the transition from the stationary beekeeping to moving beehives affects the increase in a number of hives, yields, and total production. Realized yield per hive is satisfying, with the growing tendency, particularly, taking into account bad weather conditions in past few years and the fact that most of the monasteries are in the beginning of beekeeping activities. It can be concluded that the sales volume ranges from 40% to 60%, by observing the quantities of sold honey. In the observed quantity of sold honey, there is also honey that the monks donated. If it is well known that the preferences affect primarily the consumption of honey and that consumers buy honey in monasteries due to the high confidence in quality, it is clear that the improvement of production may contribute to the development of market niche and entire beekeeping sector.

The monasteries have a potential to expand production capacities in terms of the products assortment. Most of the monasteries manufacture only honey, while such products as beeswax, pollen, royal jelly, bee poison, and propolis are neglected. Production of the accompanying honey products may have an effect on the development of demand for the niche products, so there should be developed a need for such products as medicinal creams and beverages.

The limiting factors of the competitiveness of market niche are the lack of professionals in the spheres of sales and marketing. The analyzed niche does not dispose with such resources, the activities in the field of marketing, management, market research, introduction of innovation in production are missed. The growth of production and sale of honey not happen by themselves, but should be supported by the experts.

The monastery honey and honey products from Fruska Gora as the main products of the market niche of honey sector in Serbia have the branding potential, obtaining a geographical indication of origin, and in this way increase the market recognition. Consumers are aware that the monasteries are holy places, pure, moral, and therefore honey manufactured in such environment has all characteristics of a unique and special product. If manufacturers from Fruska Gora monasteries join, consolidated supply will make branding of honey easier, i.e. linden honey brand from Fruska Gora will be improved, and in that way, the competitiveness on domestic and international markets will grow.

Profitability of the analyzed market niche depends on its size, but also the engagement of its members. The monks and the beekeepers who manufacture honey in the monastery estates must pay attention to the problems, needs, and habits of consumers in order to personalize their products and offer a higher value. Communication and adjusting a marketing strategy will become easier by a clear definition of consumer's profile. Regarding the expressed inclination for purchasing the Fruska Gora monastery honey, consumers should be offered honey products, which will be realized by higher sales and the increase in business. Within the marketing planning, a starting point is on following trends on the domestic and international markets. Regarding the honey quality, the improvement of products distinguishability, redefining the performance strategy is a precondition for the growth of a selected market niche sale. Experiences show that the strategic approaches in market positioning are recommendable, with activities on strengthening a position of Fruska Gora linden honey brand and those activities which contribute to buyers' loyalty. In the context of research results, there can be concluded that the market niche has a potential that can preserve or even increase the market share by modification of production, assortment, promotion, and distribution. Aiming to increase the niche and entire honey sector export, it is necessary to provide the sustainable increase in production, increase productivity and business efficiency. It is necessary to work on the products diversification in the future, finally aiming to improve the honey sector competitiveness on the domestic and international markets.

The concept of rural development in Serbia should give its contribution to reducing pressure on urban areas and the balanced development of Serbia. Development of other contents in the field of agricultural production, industry, services, tourism, and handicrafts should be provided by the state intervention. Conclusion of the research is that the geographic, demographic, economic, social, and ecological factors are very important for the development of agritourism. Multifunctional agritourism development will contribute to the development of agricultural production, infrastructure, quality improvement and safety of products, by employing population, preventing the migration of population, and the preservation of cultural heritage.

ACKNOWLEDGMENT

The paper is part of the research at the project III-46006 "Sustainable agriculture and rural development in terms of the Republic of Serbia strategic goals realization within the Danube region", financed by the Ministry of Education, Science and Technological Development of the Republic of Serbia.

REFERENCES

Antonić, M. (1998). Razvoj pčelarstva u srednjevekovnoj Srbiji. *Pčelarstvo, 1*(1), 3–6.

Bekić, B., Jeločnik, M., & Subić, J. (2013). Honey and Honey Products. *Works with XXVI Consulting Agronomists, Veterinarians. Technologists and Economists, 18*, 3–4.

Bohari, A. M., Hin, C. W., & Fuad, N. (2013). The Competitiveness of Halal Food Industry in Malaysia: A SWOT-ICT Analysis. *Geografia Online. Malaysian Journal of Society and Space, 9*(1), 1–9.

Ćirić, M., Ignjatijević, S., & Cvijanović, D. (2015). Research of Honey Consumers' Behavior in Province of Vojvodina. *Economics of Agriculture, 62*(3), 627–644.

Cvijanović, D., Mihajlović, B., & Cvijanović, G. (2012). General Trends on Agricultural-Food Products Market in Serbia. *Proceedings of the Third International Scientific Symposium Agrosym 2012*. Jahorina: University of East Sarajevo.

Della Fave, L. R., & Hillery, G. A. Jr. (1980). Status Inequality in a Religious Community: The Case of a Trappist Monastery. *Social Forces, 59*(1), 62–84. doi:10.1093f/59.1.62

Dragović, R. (2013). *Uspešne ekonomije srpskih manastira*. Novosti. Retrieved from http://www.novosti.rs/vesti/naslovna/reportaze/aktuelno.293.html:449678-Uspesne-ekonomije-srpskih-manastira

Dugalić-Vrndić, N., Kečkeš, J., & Mladenović, M. (2011). The Authenticity of Honey in Relation to Quality Parameters. *Biotechnology in Animal Husbandry, 27*(4), 1771–1778. doi:10.2298/BAH1104771D

Erşan, Ş., & Çiftçi, A. (2016). Preservation Proposals for the Cultural Landscapes in Context of St. Ioannis Theologos Monastery and the Surrounding Vineyards. *Proceedings of TCL 2016 Conference "Tourism and Cultural Landscapes: Towards a Sustainable Approach"*. Budapest: Foundation for Information Society.

Fruska Gora National Park. (2018). *Fruškogorski manastiri*. Retrieved from http://www.strazilovo.org/page/np/manastiri.html

Gibbs, D., & Muirhead, I. (1998). *The Economic Value and Environmental Impact of the Australian Beekeeping Industry*. Canberra: The Australian Honeybee Industry Council.

Grubić, R. (2008). Pčelarstvo u Zrenjaninu. *Rad muzeja Vojvodine, 50*, 273-283.

Ignjatijević, S., Ćirić, M., & Carić, M. (2013). International Trade Structure of Countries from the Danube Region: Comparative Advantage Analysis of Export. *Ekonomicky Casopis, 61*(3), 251–269.

Ignjatijević, S., Ćirić, M., & Čavlin, M. (2015). Analysis of Honey Production in Serbia Aimed at Improving the International Competitiveness. *Custos e Agronegocio, 11*(2), 194–213.

Ignjatijević, S., & Cvijanović, D. (2018). *Exploring the Global Competitiveness of Agri-Food Sectors and Serbia's Dominant Presence: Emerging Research and Opportunities*. Hershey, PA: IGI Global. doi:10.4018/978-1-5225-2762-6

Ignjatijević, S., & Milojević, I. (2011). Kvalitet poljoprivredno-prehrambenih proizvoda kao faktor konkurentnosti na međunarodnom tržištu. *Proceedings of the 14 ICDQM International Conference*. Belgrade: Istraživački centar za upravljanje kvalitetom i pouzdanošću.

Ignjatijević, S., Milojević, I., & Andžić, R. (2018). Economic Analysis of Exporting Serbian Honey. *The International Food and Agribusiness Management Review*, *21*(7), 929–944. doi:10.22434/IFAMR2017.0050

Ignjatijević, S., Prodanović, R., Bošković, J., Puvača, N., Tomaš Simin, M., Peulić, T., & Đuragić, O. (2019). Comparative Analysis of Honey Consumption in Romania, Italy and Serbia. *Food & Feed Research*, *46*(1), 125–136. doi:10.5937/FFR1901125I

Klimova, J. (2011). Pilgrimages of Russian Orthodox Christians to the Greek Orthodox Monastery in Arizona. *Tourism: An International Interdisciplinary Journal*, *59*(3), 305–318.

Kotler, P. (2003). *Marketing Management*. Upper Saddle River, NJ: Pearson Education International.

Kravchenko, T., Volchyonkova, A., & Esina, Y. (2014). Small Business Development as a Factor of Higher Competitiveness of Agricultural Production. *Vestnik OrelGAU*, *49*(4), 44–50.

Kruijssen, F., Keizer, M., & Giuliani, A. (2007). *Collective Action for Small-Scale Producers of Agricultural Biodiversity Products*. Washington, DC: Systemwide Program on Collective Action and Property Rights.

Marinković, S., & Nedić, N. (2010). *Analysis of Production and Competitiveness on Small Beekeeping Farms in Selected Districts of Serbia*. Budapest: Agroinform Publishing House. doi:10.19041/Apstract/2010/3-4/10

Mihailović, B., Cvijanović, D., Milojević, I., & Filipović, M. (2014). The Role of Irrigation in Development of Agriculture in Srem District. *Economics of Agriculture*, *61*(4), 989–1004.

Mitrovic, I. (2011). *Beekeepers Association "Jovan Zivanovic". Study on Protection Geographical Indications Product Fruškogorski Med*. Retrieved from http://www.pcelarins.org.rs/ns/fruskogorski-lipov-med/

Naidenova, L. (2013). Life Within Monastery Walls (A Case Study of the Monastery at Solovki). *Russian Studies in History*, *52*(1), 66–88. doi:10.2753/RSH1061-1983520103

Paganopoulos, M. (2009). The Concept of "Athonian Economy" in the Monastery of Vatopaidi. *Journal of Cultural Economics*, *2*(3), 363–378. doi:10.1080/17530350903345595

Paraušić, V., & Cvijanović, D. (2007). Serbian Agriculture: Programmes of Credit Support by the State and Commercial Banks between 2004-2007. *Economic Annals*, *52*(174-175), 186–207. doi:10.2298/EKA0775186P

Pocol, C. B. (2012). Consumer Preferences for Deferent Honey Varieties in the North West Region of Romania. *Agronomy Series of Scientific Research*, *55*(2), 263–266.

Pocol, C.B., Bârsan, A., & Popa, A. (2012). A Model of Social Entrepreneurship Developed in Barcău Valley, Sălaj County. *Analele Universității din Oradea, Fascicula: Ecotoxicologie, Zootehnie și Tehnologii de Industrie Alimentară*, 183-190.

Pocol, C. B., Ignjatijević, S., & Cavicchioli, D. (2017). Production and Trade of Honey in Selected European Countries: Serbia, Romania and Italy. In V. A. A. De Toledo (Ed.), *Honey Analysis* (pp. 1–20). London: IntechOpen Limited. doi:10.5772/66590

Pocol, C. B., & Popa, A. A. (2012). Types of Beekeeping Practiced in the North West Region of Romania-Advantages and Disadvantages. *Bulletin of University of Agricultural Sciences and Veterinary Medicine Cluj-Napoca. Horticulture*, *69*(2), 239–243.

Popa, A.A., Mărghitaş, L.A., & Pocol, C.B. (2011). Economic and Socio-Demographic Factors that Influence Beekeepers' Entrepreneurial Behavior. *Agronomy Series of Scientific Research*, *54*(2).

Popa, A. A., & Pocol, C. B. (2011). A Complex Model of Factors that Influence Entrepreneurship in the Beekeeping Sector. *Bulletin of University of Agricultural Sciences and Veterinary Medicine Cluj-Napoca. Horticulture*, *68*(2), 188–195.

Popescu, A. (2010). Study on the Economic Efficiency of Romania's Honey Foreign Trade. *Scientific Papers Series D, Zootechnics. Faculty of Animal Science Bucharest*, *53*, 176–182.

Prodanović, R., Bošković, J., & Ignjatijević, S. (2016). Organic Honey Production in Function of Enviromental Protection. *Ecologica*, *82*(23), 315–321.

Šerić, N. (2003). Importance of Remodeling of Marketing Strategies for the Market in the Countries in Transition. *Proceedings of the 5th International Conference "Enterprise in Transition"*. Split: University of Split.

Šerić, N., & Uglešić, D. (2014). The Marketing Strategies for Market Niches during Recession. *Proceedings of the 3rd REDETE 2014 Conference "Economic Development and Entrepreneurship in Transition Economies: Challenges in the Business Environment, Barriers and Challenges for Economic and Business Development"*. Banja Luka: University of Banja Luka.

Tesser, F., & Cavicchioli, D. (2014). Economic Aspects of Beekeeping and Honey Productions in Italy and in Lombardy Region. *Proceedings of the Seminarion sugli insetti utili – sala dei Cavalieri, Castelo Visconteo di Sant`Angelo Lodigiano*. Lodi: Lombardo Museum of the History of Agriculture.

Tsafack Matsop, A. S., Muluh Achu, G., Kamajou, F., Ingram, V., & Vabi Boboh, M. (2011). Comparative Study of the Profitability of Two Types of Bee Farming in the North West Cameroon. *Tropicultura*, *29*, 3–7.

Vujičić, M. D., Vasiljević, D. A., Marković, S. B., Hose, T. A., Lukić, T., Hadžić, O., & Janićević, S. (2011). Preliminary Geosite Assessment Model (GAM) and Its Application on Fruška Gora Mountain, Potential Geotourism Destination of Serbia. *Acta Geographica Slovenica*, *51*(2), 361–376. doi:10.3986/AGS51303

ADDITIONAL READING

Bojnec, Š., & Fertő, I. (2009). Agro-Food Trade Competitiveness of Central European and Balkan Countries. *Food Policy*, *34*(5), 417–425. doi:10.1016/j.foodpol.2009.01.003

Brester, G. W. (1999). Vertical Integration of Production Agriculture into Value-Added Niche Markets: The Case of Wheat Montana Farms & Bakery. *Review of Agricultural Economics*, *21*(1), 276–285.

Ingram, J., Maye, D., Kirwan, J., Curry, N., & Kubinakova, K. (2015). Interactions between Niche and Regime: An Analysis of Learning and Innovation Networks for Sustainable Agriculture across Europe. *Journal of Agricultural Education and Extension, 21*(1), 55–71. doi:10.1080/1389224X.2014.991114

Irungu, K. R. G., Mbugua, D., & Muia, J. (2015). Information and Communication Technologies (ICTs) Attract Youth into Profitable Agriculture in Kenya. *East African Agricultural and Forestry Journal, 81*(1), 24–33. doi:10.1080/00128325.2015.1040645

Lebel, L., Mungkung, R., Gheewala, S. H., & Lebel, P. (2010). Innovation Cycles, Niches and Sustainability in the Shrimp Aquaculture Industry in Thailand. *Environmental Science & Policy, 13*(4), 291–302. doi:10.1016/j.envsci.2010.03.005

Loureiro, M. L., & Hine, S. (2002). Discovering Niche Markets: A Comparison of Consumer Willingness to Pay for Local (Colorado Grown), Organic, and GMO-Free Products. *Journal of Agricultural and Applied Economics, 34*(3), 477–487. doi:10.1017/S1074070800009251

Puška, A. (2013). Competitive Marketing Strategy. *Anali Poslovne Ekonomije, 8*(1), 25–44.

Shiferaw, B., Hellin, J., & Muricho, G. (2016). Markets Access and Agriculture Productivity Growth in Developing Countries: Challenges and Opportunities for Producer Organizations. In J. Bijman, R. Muradian, & J. Schuurman (Eds.), *Economic Democratization, Inclusiveness and Social Capital* (pp. 103–124). Cheltenham: Edward Elgar Publishing.

Van Duren, E., Martin, L., & Westgren, R. (1991). Assessing the Competitiveness of Canada's Agrifood Industry. *Canadian Journal of Agricultural Economics, 39*(4), 727–738. doi:10.1111/j.1744-7976.1991.tb03630.x

KEY TERMS AND DEFINITIONS

Competitiveness: An ability of a firm or a nation to offer products and services that meet the quality standards of the local and world markets at prices that are competitive and provide adequate returns on the resources employed or consumed in producing them.

Customer Profile: A description of a customer or set of customers that includes demographic, geographic, and psychographic characteristics, as well as buying patterns, creditworthiness, and purchase history.

Geographical Indication (GI): A name or sign used on products which corresponds to a specific geographical location or origin (e.g. a town, region, or country). The use of a geographical indication, as a type of indication of source, acts as a certification that the product possesses certain qualities, is made according to traditional methods, or enjoys a certain reputation, due to its geographical origin.

Market Niche: A small but profitable segment of a market suitable for focused attention by a marketer.

Orthodox Monastery: Monasteries have always been the strongholds of Orthodoxy and bastions of spirituality. Monasteries were built over the course of centuries. As a result, in them emerged an unrepeatable beauty and combination of styles of architectural ensembles, many of which have been entered into UNESCO's World Heritage List, as their beauty astounds even outsiders.

Product Differentiation: A marketing process that strives to identify and communicate the unique benefits or qualities of a product compared to its competitors.

Rural Area: A sparsely populated area outside of the limits of a city or town or a designated commercial, industrial, or residential center. Rural areas are characterized by farms, vegetation, and open spaces.

Segmentation: A subdivision of a population into segments with similar characteristics, such as age, education, income.

Chapter 23
Agricultural Cooperatives for Sustainable Development of Rural Territories and Food Security:
Morocco's Experience

Maria Fedorova
Omsk State Technical University, Russia

Ismail Taaricht
Cadi Ayyad University, Morocco

ABSTRACT

This chapter deals with the elaboration of a conceptual framework for agricultural cooperatives in Morocco: sustainable development of rural territories. The farming cooperative associations form an effective means for the advancement of the agricultural sector, being one of the elements of agricultural policy, which play an important role in the development of agricultural production, both plant and animal, as well as in the development process in Morocco, especially for rural development, and through it, rural income of the farmers and their social statuses. In this chapter, the authors have taken the Moroccan agriculture cooperatives as a case of cooperative longevity and survival in order to observe the evolution and processes of adaptation to the distinct economic, social, and environmental demands of a broad range of member-owners. The demands of the farming community, members, and society have resulted in social and environmental factors being as much a priority as economic aspects.

DOI: 10.4018/978-1-7998-1042-1.ch023

INTRODUCTION

Cooperative movement dates back to the end of the XIX century – the beginning of the XX century and covers over 700 million cooperatives worldwide. The homeland of cooperation is considered to be the UK. As early as 1761, sixteen weavers from East Ayrshire, Fenwick village created the Fenwick Weavers' Society, the first cooperative organization of the industrial age. In 1795, in the English village of Hull, local residents rebelled against high prices set by local millers and also cooperated to establish a fair price for the products which resulted in forming a consumer cooperative in Ayrshire in 1796.

In the mid-1990s, almost half of the world's population was provided with products and services of cooperative enterprises. According to the International Labor Organization (2007), in 47 countries, there were 330,000 agricultural cooperatives with a total number of individual members of 180 million people. Currently, almost 800 million people are the members of cooperatives, which is four times more than fifty years ago. Cooperatives provide about 100 million jobs worldwide, as well as various services to half of the world's population. International Co-operative Alliance established in 1895 to promote the cooperative model brings together 315 organizations from 110 countries (International Co-operative Alliance [ICA], n.d.).

In the EU, over 50% of agricultural products are grown, processed, and sold through cooperative marketing systems. European Community of Consumer Cooperatives (Euro Coop) brings together 2.5 million consumer cooperatives with a membership of more than 21 million people and 359 thousand employees. In the EU, agricultural cooperatives perform a large part of farming from almost full coverage (the Netherlands, Denmark, and Ireland) to 80% of agricultural organizations (France and Germany). In other EU countries, these figures are lower. In the countries with a high level of cooperation development, the total number of members in cooperative organizations significantly exceeds the number of farms since each farmer is usually a member of several cooperative societies.

Nowadays, the turnover of cooperative enterprises is $2.2 trillion. In Japan, over 90% of all farmers are the members of cooperatives. In Canada, 40% of the population are the members of at least one cooperative. In New Zealand, cooperatives are responsible for 95% of the dairy market. Romanian cooperatives keep the country's best resort facilities. Agricultural cooperatives in France have the second-largest system in the world of banking and credit institutions – Credit Agricole.

Based on the described background, this chapter considers conceptual framework for agricultural cooperatives in Morocco aimed at sustainable development of rural territories and gives some examples of such organizations.

BACKGROUND

Basic Definitions

Let us now consider some terms and definitions important for the sphere of agricultural cooperatives operation. According to the International Cooperative Alliance, a cooperative is an autonomous association of women and men, who unite voluntarily to meet their common economic, social and cultural needs and aspirations through a jointly-owned and democratically-controlled enterprise (ICA, n.d.).

A member of a cooperative is a person or a company who meets all the requirements of the current law and the organization's charter itself. Usually, cooperative member is an owner or a co-owner of a cooperative who economically contributes his capital through the purchase of a share. If everything is done according to the accepted procedure, the new member of the organization gets the right to vote.

Subsidiary liability of cooperative members is also an important term. In this case, the authors mean additional obligations that are not related to the standard list of requirements for a participant when making a contribution. Such additional responsibility may be relevant, for example, in a situation where creditors presented legal requirements to the cooperative, but the organization is not able to fulfill them in a timely manner. It is worth paying attention once again to the fact that both the size and the degree of subsidiary responsibility are determined by the charter of the structure and the legislation of this or that country.

An employee in such organizations should be understood as a person who is not a member of the organization and is employed in a certain type of activity through an employment contract.

It is also important to understand who an agricultural producer is. He is person who is engaged in the production of any product. The percentage of agricultural products from this category should be more than 50% of the total volume of products manufactured by a particular company.

A share contribution is a contribution made by a member of a cooperative to a mutual fund of an organization. This can be finance, land or any property, as well as property rights that have monetary value. There are both basic and additional shares.

Cooperative payments are payments to the participants of the organization according to the contribution and labor activity of each of them.

Johnson (2013) lists some other terms used in agricultural cooperatives studies, among which the following ones seems the most useful: agricultural co-operative – a co-operative involved in agro-allied activities; credit facilities – loanable funds provided by a financial intermediary used to enhance production activities; group farming – a system of collective agricultural practice by association of people with similar interest.

Types and Functions of Cooperatives and Supporting Organizations

Operation of agricultural cooperatives mainly provides food security in rural areas. Besides, they develop economic and social skills of people. A lot attention is now paid to the development of the territories, ecological production, and incorporating women into economic activities. The analysis of the publications and the activities of agricultural cooperatives in some countries let the authors outline the following main functions of cooperatives:

- Meeting needs and interests of the stakeholders
- Providing stakeholders with economic opportunities
- Supporting stakeholders with methods and technologies
- Providing credits for the development

cooperatives exist in some economic sectors and have corresponding variety of forms. Agricultural production cooperatives have three main forms:

- **Agricultural Artel:** A cooperative aimed at production, selling or processing with personal participation in cooperative activities. The land of the cooperative members is also employed in its

operation. Each member makes a share contribution, that is, donates money, land, or other property for public use to the cooperative.
- **Fishing Artel:** Association of fish farms operating in the same conditions as the agricultural artel.
- **Cooperative Farm:** An association established for tillage or production of livestock products. The difference from the artel is that land is not given to the mutual fund.

In all the cases, a cooperative must have at least 3-10 members (depending on the sector of the economy and the main operations and aims). Besides, generally, the number of employees of a cooperative shall not exceed the number of its members. Thus, the mandatory personal labor participation of the members is guaranteed.

Agricultural consumer cooperatives have other forms:

- **Processing Cooperatives:** Any production, including meat and dairy products.
- **Sales (Trade) Companies:** Not only the sales of products but also its packaging and storage.
- **Service Cooperatives:** Any activity related to the repair, tillage, plant protection, and even legal activity.
- **Supply Cooperatives:** Created for the joint procurement of feed, fertilizer and other goods to save money.
- Horticultural, vegetable gardening, and livestock cooperatives are created to provide services to various industries from sales to processing.

However, it should be mentioned, that the terminology differs throughout countries. For example, it is still difficult to distinguish Russian (Soviet) "kolkhoz", a production agricultural cooperative and a cooperative farm.

Agricultural cooperatives are not necessarily small organizations located in a village with an office in the basement, but also international producers and processors. Cooperatives, for example, are Valio from Finland, Fonterra from New Zealand, DMK from Germany, Dairy Farmers of America from the USA, and Friesland Campina from the Netherlands. All of them are in the TOP-20 of the largest dairy companies in the world, and the cooperative form of organization does not interfere with their development. Some of them have become powerful transnational associations with thousands of employees and billions of dollars of income. For example, the turnover of the world's largest South Korean cooperative NH Nonghyup reaches $63 billion.

In the case of the dairy industry, cooperatives can be successful producers because they are not aimed at making a profit for the organization itself. The main goal of Valio, for example, is the profit of the cooperative members, so the company is trying to keep the prices of raw milk attractive. The New Zealand Fonterra is forced to fight for raw materials which is gradually deprived of its monopoly. The high price of milk does not prevent them from competing and often contributes to modernization and innovation. In addition, the cooperative members themselves know that without successful processing there will be no demand for raw milk.

There are some organizations throughout the world which support agricultural cooperatives, such as the Food and Agriculture Organization of the United Nations (FAO), the International Fund for Agricultural Development (IFAD), and the World Food Programme (WFP), are working closely with agricultural cooperatives. Their main roles in supporting farmers organized in cooperatives are (Food and Agriculture

Organization of the United Nations [FAO], International Fund for Agricultural Development [IFAD], & World Food Programme [WFP], 2012):

- Raising awareness of the role of agricultural cooperatives in reducing poverty and improving food security
- Assisting the development of agricultural cooperatives' capacities
- Supporting the development of enabling environments and better governance frameworks for agricultural cooperatives

MAIN FOCUS OF THE CHAPTER

Challenges Facing the Agricultural Cooperative Society

Among the challenges facing agricultural cooperatives are poor capitalization, corruption, illiteracy, poor inspection, and government interference (Johnson, 2013). Daman (2003) recognized internal and external factors that should be looked into in order to enhance the performance of the agricultural co-operatives.

Internal Factors

Internal factors examine the activities within the organization of the co-operative society. The areas of internal factors are focused on the inclusion of the following:

- Trained professional and motivated staff
- Dedication and selfless leadership
- Means of encouraging member's involvement and participation
- Comprehensive programs for member's education and information
- Provision for reasonable coverage of risk for loss of deposits
- Value-added activities through the use of advanced technologies

Onyima and Okoro (2009) underline the role of the cooperative member, claiming that they are the foundation of the co-operative, their support through patronage and capital investment keeps it economically healthy and their changing requirements shape the co-operatives' future. They also stated that the most important obligation of co-operative members is participation in its managing which in practice means to be kept informed about the co-operative activities, attend co-operative meetings and take their turns at the committee.

External Factors

External factors refer to the actions and decision of the agents, organizations, groups, and institution other than the co-operative society which have influence on the performance of the co-operative society (Johnson, 2013). Among the external actors, the following ones should be mentioned:

- Government support

- Market reforms
- Agriculture growth rate
- Availability of basic infrastructure
- Regulatory and development agencies and institutions

Agricultural Cooperatives in Various Countries

With the purpose to understand the overall situation with agricultural cooperatives, the authors have analyzed a number of papers devoted to rural cooperatives in various countries.

Russia

In Russia, the cooperative movement has been developing slowly. Consumer agricultural cooperatives can only be non-commercial. In this form, however, consumer cooperatives greatly helped agriculture in the late 1980s. In 1990, in the Russian Soviet Republic, consumer cooperation served 40% of the population (a quarter of retail turnover, as well as 50% of potato production and around 30% of vegetables and bread baking) and 30 million rural residents were the members of cooperatives. Then, during the 1990s, the value of cooperation in agriculture decreased rapidly.

Italy

In Italy, cooperatives have been developed both quantitatively and qualitatively. Due to the lack of common actions and objectives, their heterogeneity, and different level of economic development of northern and southern parts of the country, two opposing trends have emerged, namely, socialist, or "red" cooperatives, and Catholic, or "white" cooperatives. The predominant type was still large consumer cooperatives in urban centers, although they also operated in rural areas. In 1913, Istituto Nazionale di Credito della Cooperazione was established with the aim of funding cooperatives. Cooperatives were provided with an access to initial credit to carry out their investment activities. At the beginning of the First World War, Italian cooperatives had already acquired the features of a mass movement, even with insufficient organization. Despite the fact that many cooperatives disappeared during the war, this form of association was stronger than before, as it proved to be effective in adverse conditions. Cooperatives have become a key element in the economic restructuring of the country. In less than two years, the number of production cooperatives doubled, consumption rates remained unchanged. The loans were used to stimulate export of agricultural products and construction of agro-industrial processing plants.

Denmark

In Denmark, cooperation developed in the last two decades of the XIX century and was closely related to processing in the field of livestock. The first dairy cooperative center in Denmark was founded by the initiative of local farmers in 1882. Such cooperatives spread throughout the country. By 1888, 244 cooperatives had been established, and after a while one-third of Danish livestock farms delivered their milk to the cooperative center. Soon after, dairy cooperatives were able to compete in the butter market by moving to processing (Just, 1990; Serdyukova & Nikolaeva, 2017).

Nigeria

In the effort to improve the agricultural sector in Nigeria, the government embarked on various programs some of which were listed by Iwuchukwu and Igbokwe (2012): National Economic Empowerment and Development Strategy (NEEDS) – 1999, National Special Program of Food Security (NSPFS) – 2002, and the Root and Tuber Expansion Program (RTEP) – 2003. However, the growth rate of Nigeria's population of about 144 million at 3.2% per year, which is predicted to be doubled in less than 25 years, is a challenge in a country where more than 90% of the agricultural output is accounted for by small-scale farmers.

Japan

Japan is often considered as a country where cooperation has actually become a cult and is supported at all levels, mostly, by the state. Farmers cooperatives sell about 90% of all agricultural products (almost 100% of grain, 95% of potatoes, vegetables, fruit, and milk, 85-90% of pork, eggs, and poultry). The country has created a centralized system that protects producers from monopolistic capital. At the same time, the volume of farms that are not members of cooperatives is not more than one hectare.

One of the most striking examples of Japanese cooperation is ZEN-NOH or the National Federation of Agricultural Cooperatives of Japan. Founded in 1972 in Tokyo, the Federation is one of the most influential structures in the country. The organization has about 1,200 cooperative unions, serving about 4.78 million members and 4.8 million associate members. These cooperatives cover almost all segments of agriculture – from rice production to the construction of powerful logistics centers and international trading. The cooperative consists of nine logistics centers, markets and retail outlets, and the bank. The number of employees of the Corporation is more than 8000 people around the world (Serdyukova & Nikolaeva, 2017).

USA

The USA has become a major driver for the development of cooperation in times of crisis. Farmers' associations became the most popular during the great depression. It was this time when a powerful business giant cooperative CHS started developing. Currently, there are three thousand farm cooperatives in the USA. There are four types of agricultural associations: sales, transactions, provision of farms, and provision of credit cooperatives. The CHS cooperative is often referred as a state within a state. Established in 1929, it still remains one of the most interesting business structures in the USA. All its members, more than 1,100 cooperatives and 75,000 farmers, as well as 20,000 preferred shareholders, are free to use mineral fertilizers and fuel and have access to technology and insurance of their own risks.

Spain

In Spain, cooperatives cover only 15% of the population employed in agriculture. Most of them are based on regional legislation. At the same time, the associations are small – only one-third of local cooperatives have more than 1,000 members. There are also local cooperatives that serve their members only and large cooperatives that process products and sell them to retailers. Cooperation is characterized by the consolidation and absorption. One of the largest business groups in Spain is Mondragon which combines different types of cooperatives with powerful production subsidiaries. The Mondragon Corporation incorporates 261 organizations, including 104 cooperatives, 125 communities and branches, 8 funds, a foundation of mutual insurance, and 13 risk hedging companies. The company employs 74,000 workers. Corporate offices are located in 41 countries, the sales are carried out in 150 countries (ICA, n.d.).

Agriculture in Morocco

In Morocco, 44.6% of the population in working age is employed in agriculture covering 17.1% of the GDP (in 2010). The share of agricultural products in exports is 25%. In the agricultural production of the country, the weather factor remains a main one, therefore, irrigation is used (1.44 million hectares). Cereals, legumes, and sugar beets occupy 80% of the cultivated land. In terms of the output, the main products are wheat (3.0 million tons), tomatoes (1.2 million tons), and potatoes (1.4 million tons).

Since the 1960, a program for the construction of reservoirs and the development of water resources has allowed to provide drinking water to the population as well as agriculture and other sectors of the economy while maintaining the water resources of the country.

Meanwhile, Morocco is still an agrarian country. Despite the establishment and development of the irrigation system, weather conditions remain a decisive factor in Morocco's agricultural production. There is little use of agricultural machinery on small farms, and the output is consumed mainly by the producers themselves. The average land plot does not exceed five hectares. Many of local farmers do not have their own land and are forced to hire agricultural workers or become sharecroppers. At the other pole is the modern agricultural sector with its large and modern farms producing commercial products. The area of only 1% of agricultural enterprises is equal to or exceeds 50 hectares of land, but they provide 80% of export and 25% of all agricultural output of the country.

In 1969-1982, as the needs of the urban population in bakery products and dairy products increased, the volume of food imports increased by eight times. During the same period, exports of crops, especially citrus, tomatoes, and early vegetables, doubled. Despite the fact that in 1981-1992, the total volume of agricultural production doubled, its share in the GDP has steadily decreased. In 1991-1993, unfavorable weather conditions adversely affected the yield of grain crops. The worst indicators were observed in 1980-1988 when drought raged in Morocco. Four-fifths of the acreage is wheat, barley, legumes, and sugar beets. It was only in the 1980s that the agro-industrial sector turned to large-scale production of chicken meat and eggs. Cattle breeding has become intensive.

Fishing

Morocco's coastal waters are rich in fish resources. The country occupies a leading place in Africa for catching sardines, octopuses, and tuna. The largest fishing ports are Agadir, Tantan, and Safi. Three-quarters of the catch is sardines which are canned and exported. Since the mid-1980s, when the government began to provide significant subsidies for the development of the industry, national fisheries have reached new frontiers. In 1992, the catch in coastal waters amounted to 421 thousand tons, while in the open sea – 125 thousand tons.

Forestry

During the XX century, Morocco lost 70% of its forests. In 1914, the country had 14 million hectares of forest land, by now, its area has reduced to only 4 million hectares. Forests cover about 9% of the territory of the country. Morocco annually destroys 30 thousand hectares of forests and, although new plantations are carried out on the area of 45 thousand hectares, only 40-50% of seedlings are established.

In 1990, the production of wood amounted to 2.1 million m^3. Oak and Atlantic cedar which wood is used for decorative purposes and various crafts grow in the timber forests of the Middle Atlas region and the Rif Mountains. Cork oak forests in Garba area are the sources of commercial cork. The rest of the wood is used for the production of charcoal which is widely used in the Middle Atlas and the Reef Mountains in cooking. In the southern provinces of the country, this activity is closely related to adverse climatic conditions. Due to the nomadic life of the local population, the main agricultural activity is breeding of cattle, especially one-humped camels.

Traditional Agriculture

Morocco's traditional agriculture employs about a half of the country's population in working age which handles 70% of the country's arable land. The main production crops are citrus, olives, cereals, sugar cane, grapes, vegetables, and fruit. In addition, due to the ever-changing amount of precipitation, the area of cultivated land changes from year to year. Naturally, under such conditions, the main consumers of the products are mainly the producers. Besides, there are different farm pests in Morocco which are also infection carriers. Particularly, the raids of the locust called "cotton flea hopper" cause significant damage to agriculture.

Morocco is one of the countries that has paid attention to the agricultural cooperative associations from issuing the agrarian reform law. The laws emphasized the need for collaborative work in agriculture and the establishment of farming cooperative societies which practiced various agricultural economic activities in terms of agricultural production, agricultural equipment, credit and cooperative marketing, farm mechanization services, and others.

From a local point of view, these organizations have managed to consolidate the supply thanks to the economic and social benefits growers derive from belonging to these associations. Cooperatives show greater concern for keeping and satisfying the growers, via market prices, allowing their economic sustainability and maintaining an equity income.

As far as eco-social aspects are concerned, cooperatives have made substantial efforts as a driver adaptation to demand could not have been implemented so swiftly without the presence of these of innovation in the production and commercial sector and the improvements to production and organizations. At the same time, they play a role in the transmission of social responsibility and awareness for efficient use of natural resources to the various generations.

The longevity and survival of the cooperatives are linked to their constant and rapid adaptation to crises in which the sector finds itself or to the challenges it faces.

The Green Morocco Plan

One of the conditions for improving the efficiency of small businesses development is favorable investment climate which is measured by not only preferential tax policy but also the creation of economic conditions through a combination of many factors that ensure profitability of a company.

The Green Morocco Plan agricultural strategy launched in 2008 was designed to make agriculture the main growth engine of the national economy over the next ten to fifteen years with significant benefits in terms of GDP growth, job creation, exports, and poverty mitigation (FAO, IFAD, & WFP, 2012; Saidi & Diouri, 2017).

The Green Morocco Plan aims at developing a pluralistic agriculture that is open to foreign markets, locally diversified and especially sustainable. The strategy concerns a sector which contributes 19% of the GNP, with 15% from agriculture and 4% from agro-industry. This sector employs more than 4 million rural inhabitants and has created approximately 100,000 jobs in agriculture. This sector plays a substantial role in the macroeconomic balance of the country. It also plays an important social role as 80% of the 14 million rural inhabitants depend on revenues from agricultural production. Besides, it is important to remember that the sector is directly responsible for ensurance of food security of 30 million consumers. This reaffirms the critical role that agriculture plays in the economic and social stability of Morocco.

The Moroccan strategy has planned, for the accomplishment of its objectives, the conservation of natural resources in view of ensuring sustainable agriculture through the following steps:

- Integration of "Climatic Changes" dimension in the conception of the Green Morocco Plan;
- Conversion of nearly a million hectares from cereal crops to fruit tree plantations, which will help protect agricultural spaces;
- Experimental use of semi-desert zones to increase the usable agricultural surface area;
- Support for the water conservation irrigation systems (from the current 154,000 ha to 692,000 ha);
- Support for the use of renewable energies in agriculture (solar and wind energy, biofuels) (Sayouti & El Mekki, 2015; Oulhaj, 2013).

The implementation of the Green Morocco Plan necessitates the restructuring of the Ministry of Agriculture and Maritime Fisheries with the objective of reorganizing the resources to align itself with a new wave of changes created by the arrival of private actors; refocusing of regulatory functions; increased transfer of functional operations towards private sector; and establishment of two new entities, the Agency for Agricultural Development (ADA) and the National Office for Food Safety (ONSSA), capable of attracting growth potential and of playing the role of renewal and leadership.

The Green Morocco plan makes Morocco increasingly dependent on the world market either to export its agricultural products (tomatoes and citrus in particular) or to import its needs for cereals. Under this policy, cereal imports increased, without being covered by an adequate food export increase. In order to cope with this deterioration, we believe that grain crops must be encouraged to ensure food security, given their high share in the diet of a growing population. In order to rationalize water consumption and combat degradation of natural resources, ecological and biological practices should be adopted. As for fruit growing, GMP should encourage it on lands that are not suitable for cereals, such as mountains, hills and rugged terrains. In general, Morocco should encourage practices that locally and sustainably ensure the availability of basic foodstuffs for its population. Its food policy should also aim at increasing fish consumption, from 12 kg per capita per year to a global average exceeding 20 kg.

SOLUTIONS AND RECOMMENDATIONS

Taaricht Family Farm, Afourer

The farm with 15-year history of producing Extra Virgin Olive Oil and organic oranges. The aim of the farm is not quantity but quality. The farmers, preserving the family traditions, produce the finest Biological Extra Virgin Olive Oil from both black and green organic olives. The production of olive oil has basically not changed for the last years. The olives get picked before their ripeness. This olive juice is extracted in a traditional way of a cold spin process without use of solvents or refining methods. This preserves all the valuable ingredients of the ripe olives. After washing, the olives are mashed into a paste with a simple mortar and pestle. Extra virgin olive oil is the highest grade of virgin oil derived by cold mechanical extraction without use of solvents or refining methods. This preserves all the valuable ingredients of the ripe olives, which are crucial for the taste, smell, color and also the vitamin content of the oil. Depending on the variety, you need about 5-10 kilograms of olives for one liter of olive oil. On average, 50-70 kg of olives are harvested from an olive tree. A single tree, therefore, produces about 5-10 liters of oil per year.

In addition to producing Extra Virgin Olive Oil, they also produce organic oranges. The oranges are also grown without the usage of any herbicides or pesticides. The oranges of the farm are cultivated on the family's own plantation. Moreover, the magic about these oranges is that in this moment they are still hanging on their trees in the gardens and are only picked.

This family farm is an ecological and sustainable farm at the beginning of its way.

The Taymate Agricultural Cooperative

The cooperative was created in March 2008, comprising 16 members, unemployed young people, and women with technical and vocational skills. Its members have created a legal framework (cooperative) which enables them to exploit their skills in order to develop economic activities and thus improve their living conditions and that of their families, and thus to form a cooperative more competitive insertion in the markets. The aims of the cooperative are the following: participation in the socio-economic development of the region; involvement of marginalized groups (women and young people in situations of poverty and precariousness) in the development process; valorization of local natural resources (olive trees, olive production); creation of income-generating activities for beneficiaries and their Families; workplaces creation.

The Taymate cooperative considers that the realization of a socio-economic development program is imperative, as long as the members of the Taymate cooperative are a reliable tool to be involved as partners and not as mere beneficiaries in sustainable development while considering the social and economic gap existing between the different categories of society. The program in question will be initiated by the financing of projects whose objective is the valorization of local agricultural products.

Agricultural activities, including livestock, constitute the bulk of the economic activity of the rural community of Timoulilt; olives are among the main crops, and as a first experience, members of the Taymate cooperative worked on the transformation and marketing of green and black olives during the 2008-2009 seasons (exploitation of local potentialities).

For green olives, the test was carried out on 400 kg in 2008 and 1,000 kg in 2009 of olives which were processed in a traditional way (crushed with a flat stone) and immersed in clear water for two months to give them a very sweet taste, subsequently coated with rock salt and sold in bulk on the market of Beni-Mellal (town at 18 km from Timoulilt), this action took place in kitchen space of the local Timoulilt association for the development which has been loaned free of charge to the cooperative Taymate.

For black olives, the work was done on 1,200 kg in 2008 and 3,000 kg in 2009, purchased from the cooperative members' families, the black olives were first sorted and treated with rock salt in the house of one of the members, where once a week a group of three to four people comes back to the pile for three months, once the margins are extracted totally their taste becomes soft, then the olives are washed and mixed with a small amount of table oil. Some of the production was sold in bulk to a wholesaler in Beni-Mellal, the remainder is sold for sale in the association's premises (ATD) in 1 kg or 1/2 kg.

Women Agricultural Cooperative Toudarte, Agadir, Souss-Massa-Draa

The Toudarte (which means "life" in Berber) was founded in 2004. It specializes in the production of high-quality argan oil for cosmetic and culinary use. It is located in the center of the argan forest in the region Imsouane at 85 km of Agadir and 85 km from Essaouira. Its objectives include improvement of the socio-economic situation of rural women in Imsouane commune and promotion of their involvement in the sustainable development of their country. Currently, the cooperative is one of the biggest and most successful suppliers of argan oil in the region. The activities include production, packaging, and marketing of argan oil and its derivatives; analyzing international markets for the marketing of the products of the cooperative; production of almond oil, amlou, honey, and hammam products (shampoo, douche gel, and soap).

Employment in the co-operatives provides women with the income, which many have used to fund education for themselves or their children. It has also provided them with a degree of autonomy in a traditionally male-dominated society and has helped many become more aware of their rights. Much of the argan oil produced today is made by a number of women's co-operatives. Co-sponsored by the Social Development Agency with the support of the EU, the Union of Women Cooperatives of the Arganeraie is the largest union of argan oil co-operatives in Morocco. It comprises 22 cooperatives that are found in other parts of the region.

In Toudarte cooperative, women are given opportunities to succeed and to introduce their traditions to the rest of the world through their argan oil. It seemed almost too good to be true that in a developing country, women were forging ahead and providing for their families comfortably and with pride. The main purpose is to improve the socio-economic situation of rural women in Imsouane commune and promote their involvement in the sustainable development of their country. Cooperative Toudarte produces authentic, biological argan oil of the highest quality. More than 100 Berber women participate in the corporation, providing them a steady income, literacy classes, medical care, and an important role in local society. The production of high-quality argan oil provides them an income that is 20% above minimum wages. In addition, the cooperative takes care of literacy classes, medical care, childcare, and interest-free microcredits. Through partnerships in corporations like Cooperative Toudarte, women have paid jobs and they reach a higher social status. Women are encouraged to study and get education. More importantly, women are aware of the power of education. They can encourage their school-age daughters and help them with their homework. Illiteracy is still rife among rural women in Morocco. In a bid to improve both the economic and educational status of Berber women, a percentage of the profits made by these co-operatives is invested in the rural community. Women working in Argan co-operatives are offered free afternoon classes in literacy skills and hygienic practices.

Cooperative Toudarte is not a project just to earn money but also a place where the adherents meet to discuss different daily life matters and benefit from illiteracy classes. Moreover, it is also a place where children have support lessons. It also organizes professional trainings in cooperative management and offers different aids to the adherents in need.

Cooperative Toudarte is very important for the village Akhsmou. The production of argan oil brings new economic activities in the area where there almost no any jobs opportunities. The cooperative has become the central meeting point for approximately thousand people in 10-15 villages in the Imsouane region. The compound of the cooperative has a children playground, a crèche, a small shop, and a mosque for women.

The success of the argan co-operatives has also encouraged other producers of agricultural products to adopt the cooperative model. The establishment of the cooperatives has been aided by support from within Morocco, notably the Foundation Mohamed VI.

Multidisciplinary Social and Cultural Center in Aghenbo Village

The center is located in Atlas Mountains, about 120 km from Azilal city. It is a new one and thanks to Taghlast association gets the funds from Joud foundation. The aim of the project is to stop the migration of people from the area to big cities. Lots of people from Aghenbo village migrate to big cities because people in this area are unemployed young people and women. Although the poverty-stricken areas are spacious and bucolic, some of the land is simply not suited for crop growing.

The Association Taghlast for development, cooperation, and environmental protection decided to build a multidisciplinary social and cultural center in Aghenbo village, in high Atlas Mountains in the center of Morocco.

A lot of services and events will be provided in this center:

- A workshop for women cooperative who make traditional carpets and are engaged in knitting and weaving;
- Library for children;
- Information and internet hall;
- Meeting hall;
- educational and recreation areas for pupils.

This center is expected to give opportunities to people who live in Aghenbo village, especially women. There is a hope that people will stay in their area and will not think to immigrate to big cities.

FUTURE RESEARCH DIRECTIONS

While the authors attempted considering conceptual framework for agricultural cooperatives in various countries with a special focus on Morocco, further research is required to study country-specific experiences of cooperation in agriculture, specifically, with an aim to reveal the effects of cooperative movement on ensuring sustainable development of rural territories, improvement of farming practices, increase of agricultural production, and, ultimately, establishing food security.

CONCLUSION

This chapter examines the contributions of cooperatives towards agricultural development in Morocco. It has been shown that cooperatives boost opportunities for Moroccans by the improvement of their social and economic conditions and by creating jobs and generating income, making people more aware of their rights, the reforestation of argan forests with the support of the women's cooperatives, and the promotion of regional tourism.

Some of the above-described cases were presented at INNOSIB forums – 2017 and 2018 in Omsk, Russia and were considered very useful and of great interest for other countries. The forums are for businesses and social entrepreneurs and are aimed at sharing best practices in social-entrepreneurial sphere.

REFERENCES

Daman, P. (2003). *Rural Women, Food Security and Agricultural Cooperatives*. New Delhi: Rural Development and Management Centre.

Food and Agriculture Organization of the United Nations. International Fund for Agricultural Development, & World Food Programme. (2012). *Agricultural Cooperatives: Paving the Way for Food Security and Rural Development*. Retrieved from http://www.fao.org/3/ap088e/ap088e00.pdf

International Co-operative Alliance. (n.d.). *Statistical Information on the Cooperative Movement*. Retrieved from https://www.ica.coop/en/about-us/international-cooperative-alliance

International Labor Association. (2007). *COOP Fact Sheet #1. Cooperatives and Rural Employment*. Geneva: International Labor Association.

Iwuchukwu, J. C., & Igbokwe, E. M. (2012). Lessons from Agricultural Policies and Programmes in Nigeria. *Journal of Law. Policy and Globalisation*, 5, 11–21.

Johnson, D. S. (2013). *Contributions of Co-operative to Agricultural Development: A Study of Agricultural Co-operatives in Awka North Local Government Area*. Nnamdi Azikiwe University.

Just, F. (1990). Butter, Bacon and Organisational Power in Danish Agriculture. In F. Just (Ed.), *Co-operatives and Farmers' Unions in Western Europe – Collaboration and Tension* (pp. 137–156). Esbjerg: South Jutland University Press.

Onyima, J.K.C., & Okoro, C.N. (2009). *Co-operatives: Elements, Principles and Practices*. Awka: Maxiprint.

Oulhaj, L. (2013). *Evaluation of the Agricultural Strategy of Morocco (Green Morocco Plan) with a Dynamic General Equilibrium Model*. Marseille: Femise Research Programme.

Saidi, A., & Diouri, M. (2017). Food Self-Sufficiency Under the Green-Morocco Plan. *Journal of Experimental Biology and Agricultural Sciences*, 5(Spl-1- SAFSAW), 33–40. doi:10.18006/2017.5(Spl-1-SAFSAW).S33.S40

Sayouti, N. S., & El Mekki, A. A. (2015). Le Plan Maroc Vert et l'autosuffisance alimentaire en produits de base à l'horizon 2020. *Alternatives Rurales*, 3(1), 78–90.

Serdyukova, M., & Nikolaeva, E. (2017). Development of Agricultural Cooperation and Small Forms of Economic Activities in Foreign Countries. *Bulletin of Chelyabinsk State University. Economic Sciences*, 406(10), 147–155.

ADDITIONAL READING

Bailey, K. G. (1988). *The Principles of Co-operation and an Outline of Agricultural Cooperative Development*. London: Food from Britain.

Belo Moreira, M. (1984). *L'economie et la production laitiere au Portugal*. Grenoble: University of Grenoble.

Bjorn, C. (1988). *Co-operation in Denmark*. Copenhagen: Danske Andelsselskaber.

Dedieu, M.-S., & Courleux, F. (2011). *Agricultural Cooperatives: The Reference in Term of Farmer Economic Organization*. Centre for Studies and Strategic Foresight.

Ferreira da Costa, F. (1980). Etude bibliographique de la cooperation au Portugal. *Revue des Estudes Cooperatives*, 1-29.

Henriques, M. A., & Reis, J. (1993). *Heterogeneidad estructural de la agricultura portuguesa y deficit neo-corporativista. Las organizaciones profesionales agrarias en la CEE*. Madrid: MAPA.

International Labor Association. (2002). *ILO Recommendation on the Promotion of Cooperatives*. Geneva: International Labor Association.

Knapp, J. G. (1965). *An Analysis of Agricultural Co-operation in England*. London: Agricultural and Central Co-operative Association.

Moyano, E. (1988). *Sindicalismo y politica agraria en Europa*. Madrid: MAPA.

Moyano, E. (1993). *Accion colectiva y cooperativismo en la agricultura europea*. Madrid: MAPA.

Ogunnaike, O. O., & Ogbari, M. (2007). Analysis of the Effectiveness of Co-operative Society as a Tool for Satisfying Human Needs. *Nigerian Journal of Co-operative Economics and Management*, 1(1), 11–15.

Okechukwu, E. (2006). *Manual for Co-operative Professional. Nkpor*. Optimal Press Ltd.

Tortia, E., Valentinov, V., & Iliopoulos, C. (2013). Agricultural Cooperatives. *Journal of Entrepreneurial and Organizational Diversity*, 2(1), 23–36.

Volobueva, T. (2013). *Development of Small Economic Entities in Agriculture. Orel*. Orel State Agrarian University.

KEY TERMS AND DEFINITIONS

Agricultural Cooperative: A co-operative involved in agro-allied activities.

Cooperative: An autonomous association of women and men, who unite voluntarily to meet their common economic, social and cultural needs and aspirations through a jointly-owned and democratically-controlled enterprise.

Cooperative Payments: The payments to the participants of the organization according to the contribution and labor activity of each of them.

Credit Facilities: Loanable funds provided by a financial intermediary used to enhance production activities.

Group Farming: A system of collective agricultural practice by association of people with similar interest.

Member of a Cooperative: A person or a company who meets all the requirements of the law and the organization's charter itself; an owner or a co-owner of a cooperative who economically contributes his capital through the purchase of a share.

Share Contribution: A contribution made by a member of a cooperative to a mutual fund of an organization (finance, land, property, or property rights that have monetary value).

Chapter 24
Participatory Poverty Assessment Effort in Food Security and Extension Policy:
Evidence From Indonesia

Muhamad Rusliyadi
https://orcid.org/0000-0003-4632-5671
Polytechnic of Agricultural Development Yogyakarta-Magelang, Indonesia

Azaharaini Bin Hj. Mohd. Jamil
University Brunei Darussalam, Brunei

ABSTRACT

The impact study assessment aims to evaluate policies and monitor the achievement of targets and the results of a development program such as DMP. The output obtained is information that is an evaluation of how the policy was planned, initiated, and implemented. Participatory monitoring and evaluation analyze the outcome and impact of the DMP Program. PPA seeks to answer the question of whether or not the policy or program is working properly. A participatory approach may improve the outcomes in the form of a new policy model for the future. The output of the PPA process from this study is the agricultural policy formulated in terms of practical ways of approaching poverty problems from a local perspective. The success of alternative policy options applied by local government such as physical, human resources, and institution development at the grassroots level should be adopted at the national level. It should represent the best example of a case of successful program implementation at the grassroots level which can then be used in formulating national policies and strategies.

DOI: 10.4018/978-1-7998-1042-1.ch024

INTRODUCTION

Poverty reduction is key to the success of a country's development. Encouraging economic development is an effective policy for poverty reduction. One of the policy measures to effectively address the issue of poverty is the adoption of participatory approaches in the planning, monitoring, and evaluation of poverty reduction policies. A participatory approach is used to address weaknesses which could not be effectively addressed previously. Variations regarding the evaluation of poverty reduction policies are dominated by top-down approaches. A less top-down approach to evaluating policies will accommodate the needs and aspirations of the poor.

Participatory Poverty Assessment (PPA) was introduced by the World Bank to assess the poverty in a region from top-level down to village level. People facing poverty were identified and so were the causes of poverty, and recommendations for poverty reduction effort at household level were proposed. The variation or diversity of indicators that exists in the country and villages are taken to be representative of poverty at household level. However, this implies a methodology that is diverse and is capable and flexible enough to respond to the needs of the subject of policies (Asian Development Bank [ADB,] 2001).

PPA is a method of analysis of data aggregation. It seeks to recognize deficiency scope within the public, cultural, financial, and political atmosphere in a particular region. It focuses on the local peoples' perceptions. Different research methods can be contextualized to different degrees. These categorizations with participatory methods aim to standardize the data collection and analysis with the sample in a large household study. The course of classification approaches offers the advantage of inferring qualitative information from quantitative evaluations, which characterizes the comparison of survey and participatory data. Nevertheless, it does not include the perspective of acquiring quantitative data from the PPA.

BACKGROUND

PPA initiates a participatory method that thrives in the viewpoint of Participatory Rural Appraisal (PRA). They become key to mechanisms for advance policy agency by embodiment of the community's participation. PRA has been seen as a family of approaches and methods to enable local people to share and analyze their knowledge of life and conditions, and to plan and to act (Chambers, 1994). It emerged in the early 1990s building on insights and methodological innovations from various study methods. Agro-ecosystem analysis provided the series of diagramming, mapping, scoring, and grading methods of the different natural processes; insight provided by the work of practical and development anthropologists and those of field research in farming systems emphasized villagers' capability in conducting their own personal analysis; and most notably there was the development of Rapid Rural Appraisal (RRA) (Ruggeri Laderchi, 2001).

Conservative poverty estimates, together with equally conservative financial and qualification estimates, have been criticized for leading to external enforcement, and for not taking into account the views of poor people themselves. The participatory approach's purpose is to change this and to enable the people themselves to take part in decisions about what it entails to be poor and the level of poverty (Chambers, 1994, 1997). The implementation of PPA, evolved from PRA, is described as "a rising family unit of approaches and methods to enable local people to distribute, develop and analyze their understanding of living and circumstances, to plan and to effort" (Chambers, 1994). Originally, it was intended for small projects, then the World Bank, to complement to its poverty assessments, scaled up PPA. By 1998, the

World Bank's poverty assessments included a participatory aspect. A widespread multi-country workout (of 23 countries) was similarly incorporated as conditions for the World Development Report, available as Voices of the Poor (Narayan, Chambers, Shah, & Petesch, 2000). According to Cornwall (2000), there are three types of PA: (1) those connected with self-determination and empowerment; (2) those linked with raising the effectiveness of programs; and (3) individuals' emphasis on common learning. The role of participatory movements as defined by the World Bank, particularly in their poverty assessments, has tended to be compliant. It adopts PPA so that the poor will join forces with programs rather than to modify the pattern of the programs themselves (type 2), and the Voices of the Poor emphasizes the third type.

Study and policy points of reference on poverty issues have undergone a shift in conceptualization of the poor from passive recipients of development efforts to those who can contribute to their own livelihood improvements (Chambers, 1996). To facilitate this shift, state institutions should provide the conditions for the poor to do so. The terms 'poor' and 'non-poor' can be perceived as externally imposed categories that do not reflect locally perceived differences, as well as possibly being derogatory. Subjective valuation of wellbeing can provide a divergence and more contextually specific perspective on living conditions. Since poverty is embedded in a socioeconomic system, it is always relative to a socially defined threshold. An individual with a given level of income, food, protection, or clothing may be considered inadequate in one context, but not in another. Thus, the distinction between absolute and relative poverty is important (Lok-Dessallien, 1998).

According to Norton, Bird, Brock, Kakande, and Turk (2001), the purpose of PPA is to improve the strength of public activities aimed at poverty reduction. PPAs are generally taken to be policy research exercises, linked to governmental policy processes, directed at understanding poverty from the perspective of poor people, and their priorities in terms of actions to improve their conditions. PPAs can strengthen poverty assessment processes through broadening stakeholder involvement, thereby increasing general support and legitimacy for anti-poverty strategies, enriching the analysis and understanding of poverty by letting in the perspectives of the poor, and providing a diverse scope of valuable information on a cost-effective, rapid and timely basis, creating new relationships between policymakers, service suppliers and people in poor communities. PPAs may originate from a mixture of dissimilar institutions, including NGOs, donors, and research establishments. They may address different audiences – including policy-makers, politicians, advocates, and activists.

According to Norton et al. (2001), the impact to date of PPAs has been the finding that all stakeholders have had significant changes in knowledge, understanding, and military strengths in one or more of the following areas: the nature and causal factors of poverty; greater solidarity with or sympathy for the poor; increased commitment to consulting the poor; better understanding of and/or increased commitment to participate or 'bottom-up' research, planning, and monitoring; better understanding of local conditions such as recognition of inequities which are important at the household level, particularly gender-based issues; greater willingness to acknowledge sensitive issues such as domestic violence or marginalisation of some social groups; increased need for better social services and programs; and increases in commitment to or demand for participation and grassroots democracy (Table 1).

MAIN FOCUS OF THE CHAPTER

Poverty reduction and fair distribution of economic growth are significantly meaningful to the Indonesian government. Poverty reduction is constantly determined in terms of the social safety nets, as opposed

Table 1. Approaches to PPAs

Approach	First generation PPAs	Second generation PPAs
Focus of PPA process	Generating textual representation of realities of the poor, to contribute to policy recommendations.	Creating new relationships within the policy process – bridging public policy, civil society, people in poor communities, donor agencies.
Means of influencing policy change	PPA influences donor country assessment documents (country poverty assessment). PPA is a 'product' which seeks to influence another product. Policy influence, therefore, is largely contingent on the quality of the policy process associated with the overall document.	PPA influences country policy-making processes: budget process (pro-poor allocation of public resources); sector policy; regulatory function (land reform, informal sector); poverty monitoring system; local government policy and budget processes.
Institutional location	Designed/supervised by donor agency staff. Donor 'publishes' PPA. PPA managed and implemented by NGO or research institute, chosen by the donor according to criteria of technical and logistical competence to deliver PPA report.	Chosen to maximize potential contributions to policy process. Institutional partnerships created to introduce logistical and technical capacity.
Time-frame	Period necessary to complete one round of national fieldwork and generate report.	Process approach allowing for PPA to work with planning and budget formulation processes at national, sectoral and local levels. PPA may or may not be produced. Follow-up with specific studies possible.
Substantive focus	Focus on: poor people's understanding of poverty and deprivation, constraints in accessing public benefits and services, poor people's priorities for public policy.	As in left column, but with increasing emphasis on access to information for action for people in poor communities; governance, accountability, and transparency.
Key skills	For PPA implementation. Social analysis (synthesis of results). Training skills in research field methods. Research design logistical support for policy analysis.	For PPA implementation social analysis (synthesis of results). Training skills in research field methods. Research design. Logistical support for policy analysis for policy mainstreaming. Understanding macro policy, sector policy, and budget formulation processes. Advocacy skills institutional change management

Source: Norton et al. (2001)

to sharpening and focusing on addressing macro weakness, at the structural and sectoral level. Therefore, this requires an approach that is able to consider all of them in the form of PPA. People rarely get information about government plans and in the policy formulation phase. The government should be transparent in every policy. The role of stakeholders is indispensable in policy implementation. This method is robust because it involves many stakeholders in finding alternative solutions to address public problems, especially in poverty reduction policies.

The blueprint of a policy initiation plan provides users with direct impacts on policy strategy. It intends to clarify the problems from the grassroots level. Discussion at the household level facilitates the government in allocation of funds and budgets in accordance with the scheme of policy implementation. The government's current top-down policy looks at civil society as an entity that cannot contribute to the development plan. This is the bridge that the participatory approach is expected to address in a strategic plan based on assessment of threats and problems encountered at the grassroots level.

PPA is the basic method, like a baseline study, to assess the poverty level in the country. The focus in this context is rural areas in Indonesia. In this case, rural areas cannot be separated from the agriculture

Participatory Poverty Assessment Effort in Food Security and Extension Policy

Figure 1. The link of PPA findings on agricultural and food security analysis with poverty assessment
Source: Authors' development based on Turk (2001)

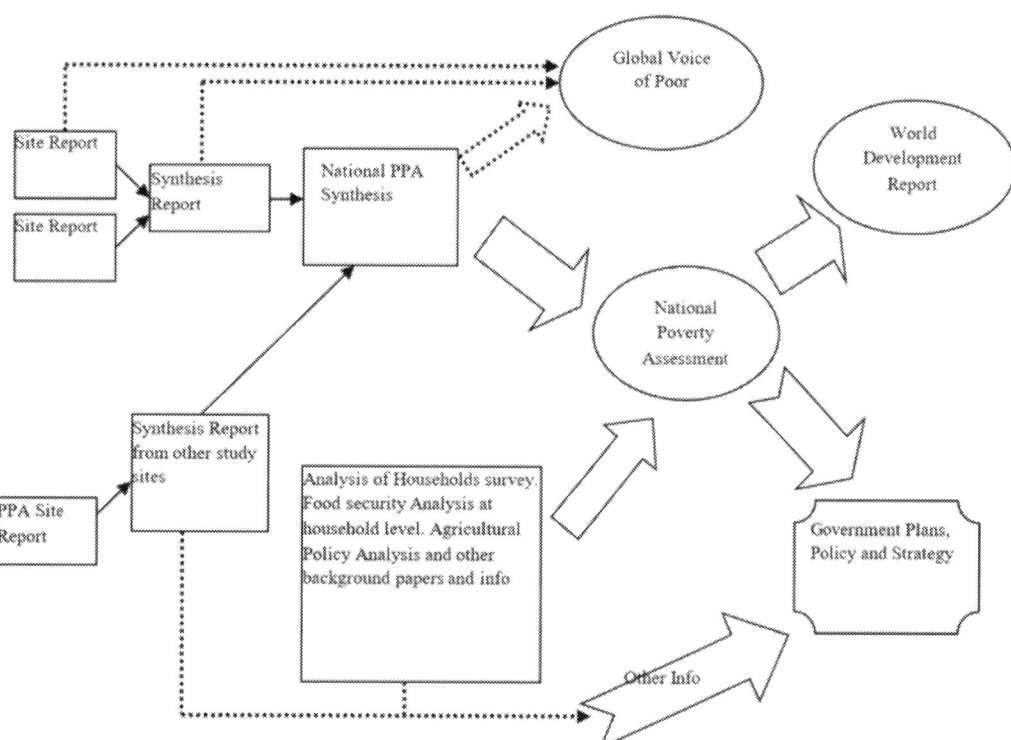

sector. Most rural policies are derived from agricultural policy. The closeness and relationship between agricultural policy and objects in PPA are strongly interrelated. The PPA results from other countries are based on the analytical characterization of livelihoods in agriculture, forestry, and fisheries. The root causes of poverty are expected to be revealed from livelihood analysis. Therefore, the strategic plan becomes more concentrated and comprehensive for the policy focus after the DMP Program.

The four villages in Central Java province in this study are regarded as the main representatives for assessment of the Program by PPA analysis. This study is a pioneer in the area of PPA in Indonesia, and it links to agricultural policy. The PPA concept is used in synthesizing the problem of poverty at village level with the participation approach. Every report shows a naturalistic depiction of poverty commonly found in rural areas.

There is the link of PPA findings in each village on agricultural and food security analysis with poverty assessment (Figure 1). These PPA results can be synthesized for each village in order to draft a proposal for provincial and national poverty assessment reports. The reports can be circulated at district, regency, and province levels for discussion by the local government and agencies to assist them produce a model for development. Hopefully, the synthesis of information using the PPA approach will assist the local government in the formulation of policy actions.

A collection of site 'synthesized reports' of each hamlet/village can become a national synthesized report that can be published internationally as a global national voice of the poor. A site report of each village can also be specified in representing a unique case study from each village. This is a contribution to global reporting on the poor: a valuable document because it contains rich information as a result of the PPA process. Firstly, it highlights specific information in each site report. Second, it provides opportunities to reinforce the finding of sites with varying socioeconomic and geographical environments. Thirdly, it enables synchronization and merging of local-level findings to be turned into national poverty quantitative data for analysis.

Figure 1 shows the results of the research report covering all village sites in terms of the PPA. The analysis of household surveys is mainly told in terms of food security and agricultural policy analysis. The analysis can become a resource for national poverty assessment resources. PPA findings are a national poverty assessment that can attract the attention of national policymakers and be published internationally. The Indonesian study findings can be employed to represent the voices of the poor in global studies. Government involvement in the synthesis of the results of the PPA is indispensable, especially in making constructive comments and giving feedback to improve implementation policy and practice in the alleviation of poverty. National poverty assessment summaries obtained as the outcomes of the PPA and other data can be used in future strategic planning operations by the government.

The relationship between PPA results and agricultural policy can be considered complementary. The PPA result is an official document, while the outcome is the strategic plan involving participation of the community. A part of the strategic plan is a strategic program of agricultural policy on poverty reduction and other research reports. The agriculture policy in the PPA is practical; it examines policy implementation at the grassroots level, especially in how to increase farmers' income and reduce poverty.

The agricultural policy formulated using PPA in terms of practical policy analysis is sourced from local understanding/knowledge. Indonesia has many provinces with a progressive and proactive policy in agriculture. The success of alternative policy options by local government should be enlarged to the national stage.

The local government (province-level) is required to carry out a specific policy to alleviate hunger and poverty. Specific local policies should strong and have comprehensive analysis at grassroots level and solve the problem for sustainable development. The PPA process of collecting data can be adopted as a tool for monitoring and planning based on local practice in pro-poor policies with real objectives. In attempting to reduce poverty, the local policies should take into consideration the strength of local economic conditions in terms of agricultural and potential resources.

PPA is a tool used in the analysis of household- or community-level framework, with data derived from discussions with participants. The DMP is meant for poor rural villages. More than 30% of the people in each village are poor, which is why these villages are suitable for PPA implementation PPA. Contributions of PPA in the DMP Program are based on analysis of the sustainability of existing livelihoods of villagers in order to see the strengths and shortcomings in such areas as human imagination, social capital, natural resources, infrastructure, and institutions at the community level. The PPA process involves consolidation, observation, and intensive discussion of the poor in the rural Program sites about the causes of impoverishment. The conditions of the local community become the baseline in the analysis so that tool used are adjusted from real conditions.

Livelihood Resources Analysis in the Community

The framework of analysis employed in this work is the sustainability of living resources from time to time. Indexes in the analysis of livelihood focus on how the availability of livelihood resources can cause poverty at the community level (Table 2).

Table 2. Changes in livelihood availability and their causes

Resources	1990	Condition	2000	Condition	2010	Condition
Fish	10	• Fewer people use traditional fishing gear • Few market	8	• Many fish from outside and inside village • Use of hand grenades, poison • More markets • Merchants want to buy • Water supply for fishpond decreased	6	• Increased population • High consumption • Use illegal fishing • Competitive fish market
Rice	9	• Fertile soil • Fewer people • Plenty of farmland and rice available • Fewer people	8	• Increased population • Low crop productivity • No maintenance	8	• More people • Infertility of soil floods • Insufficient food consumption
Domestic vegetables and fruits	9	• Newly settled • Not many crops	7	• Additional planting • Increasing population • Few market needs • No processing of product	6	• Vegetable gardening • Expansion of crop areas • More markets • Processing product
Wild vegetables and fruit		• Plenty of forest • Not many people extracting • Far from market • Small population • Household consumption		• Extracting by people from outside • Availability of markets • Clearing for farmland, increasing paddy land		• Logging concession • Availability of market • Clearing for paddy land and plantation • Bush fires
River water	10	• Good quality • Deepwater • Narrow river • Riverbank not yet broken • Fewer floods	8	• Floods damage crops • River became wider and broke riverbanks • Water got shallow planting crops in riverbank	6	• Quality of water got worse • Floods • Shallow river • Many boats • Cutting forest on the riverbanks
Water resources	9	• Easy to get water • Plenty of water resources	7	• Degradation of soil • More difficult to get	6	• Population increased • Construction more develop • Difficult to get • Climate change
Wild animals	10	• Many wild animals in the wild habitat • Household consumptions • Few hunters • No market	8	• Availability of markets • Decreased habitat • More hunters • Markets available • Loss of certain species	4	• Markets increased • Exotic wild animal markets • Illegal hunting
Timber	9	• Plenty of forest • Not many people logging	8	• Availability of markets • Clearing for plantation, settlement	6	• Bush fires • Illegal logging • Clearing for plantation
Total Score	38		31		22	

Note: Scoring is out of 10; the higher the score, the higher the availability of food
Source: Authors' development based on fieldwork primary data

Livelihood analysis shows different results, depending on the local natural resources available in the village. The results of observations in four villages participating in the DPM Program show that farming, livestock, fisheries, and forests have been the key support resources for village livelihood. These sectors have played a significant role in the village in terms of providing primary income as the main livelihood of most villagers. Rice cultivation is common in Madukoro and Selogiri, while in Candirejo and Kedungdowo, dry-land cultivation is common. Dry-land cultivation usually involves planting cassava, maize, bean crops, legume crops, and fruit crops. The yards of villagers are used to grow many vegetables for the daily needs of the family. The yield of vegetables is helpful in reducing expenditure for daily food. Some people raise animals and employ fishing for their livelihoods, which help communities bring in additional income. The products of farming activities are largely used for consumption and only a small proportion is for the purpose of cash.

Agriculture

The most important aspect of the livelihood of rural communities is paddy cultivation. Ownership of paddy land varies from around 0.1 ha to 3 ha. The land area that causes a relatively small income of farmers is very small (0.25 ha). Irrigation in the four villages is only a small part of technical irrigation; the other aspect is non-technical irrigated rice, rain-fed, dependent on the rainy season. With cultivation technology handed down, it is to be expected that the productivity is not high. Thus, farmers receive a yield which is enough for household consumption. The crop is generally enough to meet needs for three to six months. To cover daily need over a full year, farmers usually work in the off-farm sector as laborers, or in livestock or non-timber forest production (NTFS). For example, wild pandan leaves are used as a handcraft in Candirejo, resulting in additional income for villagers.

Every household in almost every village has a garden and yard which is generally wide in area. Various crops are planted in their garden, such as bananas, pineapples, and some vegetables like spinach and water spinach. Agricultural products from the garden usually show that most crops grown are for their own consumption. However, when communities run short of cash, they sell some of the products. In contrast to the farmers with coconut gardens, most households in many villages earn some money by selling palm to coconut buyers. Coconut sales can occur four or five times a year with harvests of young productive plants but only two or three time with older plants.

According to the villagers in Candirejo, "because of land degradation and the limited availability of water, it is very difficult to cultivate paddy and also the lack of fertile land". These conditions are some of the causes of food shortage, which results in poverty being experienced by some villagers. With the increasing population in the village, the land area is reduced and there is a tendency to decrease production, which hinders attainment of food self-sufficiency in the villages. It is necessary for the food supply to come from outside the area for daily consumption and this is the experience of most villages. It can be concluded, then, that land ownership is a key factor in household welfare.

Livestock

Livestock plays a very important role in the livelihood of the villagers. Most villagers are engaged in keeping livestock (cattle raising). Other community members are engaged in raising chickens. Community members are aware that chickens are nutritious food. People can also readily sell chickens at the local market. The cattle production system implemented in the villages is based on traditional practices and therefore productivity tends to be low. Although some activities of the DMP Program had already started in villages at the time of study, there remains a need to improve the Program, especially in rural Madukoro and Candirejo, by introducing, for example, a mix of local and etawa goat farming, which can show a fairly high production yield. Livestock technology should be implemented systematically so that diseases can easily be controlled and detected should there be any problem in the livestock. This can be done with the assistance of extension service officers of animal husbandry in the villages.

Livestock production in the countryside is a direct savings mechanism for villagers. If they need small amounts of cash, such as for buying cooking oil, they can sell chickens to mitigate their financial constraints. This practice occurs in almost all four villages. However, for issues like paying high school fees, sending children to university, and costs of illness or family, people sell goats or sheep, cattle, and timber at their backyards. Livestock health is the major issue for the villagers. It is an even greater concern than people's health from the villagers' perspective.

Fisheries

Fisheries play an important role in people's lives in the village through the DMP Program in Madukoro and Kedungdowo, which have developed means of livelihood around fishing, with the availability of lakes and rivers, and people have utilized these available resources to keep fish farms and ponds. Kedungdowo is surrounded by the large river almost like a lake, so most villagers keep fish cages or nets on almost every bank of the Luk Ulo river. The villagers keep catfish, carp, tilapia, and pomfret. In Madukoro, the river water source is relatively small, but it can be used for small fish farmed in ponds. The villagers in Madukoro usually maintain types of fish such as carp and catfish or fish that are in high demand in the market.

Findings from several group discussions in the four villages indicate that families generated additional income in the range of 20-40% in a year from the contribution of aquaculture. This is a great help economically for villagers to support their livelihood. In small-scale aquaculture, it remains very profitable, but flooding sometimes affects Kedungdowo so that cages installed in the river are washed away by the strong currents due to continuous rain. The concept of small-scale aquaculture is aimed at creating a small pond at the back or side of the house by using a large tarp box shaped like a pool. It is quite efficient because the capital investment required is reasonable in buying a tarp and fingerlings and fish are fed by using food scraps.

Forestry

In general, macro-level forest resources are important factors in socio-economic development, especially around the villages in the PPA site. Timber and non-timber forest products (NTPFPs) provide a significant contribution to national GDP, for domestic use and for overseas exports. This is a very high-value contribution to poverty reduction in the rural population, especially for those who live on a forest's edge. The question is: why does the population remain poor in these areas? One needs to examine the case study in the research site. Local villagers usually obtain resources from the forest, particularly for their own consumption. Most villages in the hills are surrounded by tropical forests and therefore they are rich in forest resources, especially timber. However, government regulation prohibits them from selling forest resources. Local villagers normally use these resources for housing and clear them for farming. However, illegal cutting of timber by outsiders still happens.

Villagers reported that tropical forests are found around villages, as well as NTFPs such as bamboo, rattan, mushrooms, wild animals, wild fruits, wild pandanus leaves, herbs, wood, and fuel. NTFPs are major resources for poor communities that currently do not have paddy land: they collect firewood, wild animals, bamboo, and rattan for sale as daily income. One example is Candirejo, where the majority of the poor collect wild pandanus leaves, which they dry and sell. If villagers can make crafts from the pandanus leaves, the sale value of the product increases and they can sell to the tourist villages nearby. The government's efforts need to be intensified by providing the villagers with extension services and training in making handcrafts from local raw materials. This will help villagers earn higher incomes than before.

Discussion with various groups in the four villages revealed the presence of poaching of wildlife in the forests nearby. The products are sold in the city, where the selling price of wildlife and their products is relatively high, and this is gradually decreasing the number of wild animals in the forest. Commercialization and trade of wildlife are already a matter of concern; this can lead to wildlife in the forests becoming extinct because of extreme hunting by villagers. Hunters, who may also come from the village borders, are difficult to control because they often violate the prohibition notices against hunting in wildlife-protected areas. The villagers proposed supporting local customary practices for the protection of forest resources around the village. Village regulations, which protect wild plants and animals, are also important for sustainable use of natural resources. Effort should be made to strengthen the role and capacity mechanisms of forest management officers at grassroots level. Strict measures and stringent laws to control illegal excessive exploration of forestry resources should be enforced: for example, prohibition of exploiting the forest without official permission from local village leaders.

Tourism Resources

Tourism is actually not a natural resource, but it is a social capital resource. However, there is potential for tourism to become a promising industry nowadays. The authors' opinion is that the main livelihood for two villages in the studied sites is tourism because the villages have many tourist facilities frequently visited by foreign and domestic tourists. Candirejo has become a tourist destination since the initiation of a Jogjakarta NGO because the location of the village is close to the largest temple in the world, Borobudur Temple. The other village is Kedungdowo in Kebumen Regency, which is adjacent to Jembangan tourist village.

Candirejo was a poor rural village before the development of tourism. The shift of this village into a tourist village started in the 2000s by one of the NGOs in Jogjakarta to meet demand for long-term cheap hotel accommodation for foreign tourists. Furthermore, in serving accommodation needs of tourists, this village has made house rental facilities available. The villagers have thus earned a hefty additional monthly income from renting rooms. Further contributions have been made through the government's Department of Tourism and Rural Development program, PNPM (Program National Pemberdayaan Masyarakat Mandiri), in the form of established facilities and the provision of tourist bureau guides for local and foreign tourists. However, only a small part of the community in the village has experienced the positive impacts of tourism. The research village sites are in fact quite distant from the community village center – about 3-5 km from the village municipal office, with a steep uphill road access, such that not all modes of transport can reach it. Focus group discussions in the DMP Program village show that some villagers are able to earn extra income by selling food and products. However, their homes are distant from the village community center, so they cannot use them as guesthouses for tourists. In addition, many rural community members are unable to take advantage of local resources such as wild pandanus leaves that grow wild in the village for handicraft. This is because they do not have the requisite craft skills. Villager efforts have been frequently mentioned in discussion but the village leadership has done little to drive the potential of the village as a producer of crafts using wild pandanus leaves.

Participatory Poverty Assessment Effort in Food Security and Extension Policy

Wealth Ranking

Wealth ranking analysis is used to simplify the classification of welfare. Villagers are divided into groups based on differences among families in the community. The wellbeing criterion is used to classify and categorize the living standards of the people. The number and categories explored in the discussion in each village were aimed at finding the dimensions of poverty and its characteristics (Table 3).

Table 3. Wealth and wellbeing ranking in DMP Program villages

Categories	Kebumen Regency		Magelang Regency	
	Kedungdowo	Selogiri	Candirejo	Madukoro
Rich	• Several cows / Ownership of cattle (3-5) • Large permanent house • Large paddy field (2-5 ha) • Some pond fish (1-3) • Large forestry (2-3 ha) • Have ownership of rice milling units and own barns • Have ownership of car transportation • Enough food for the entire year • Large garden and other small livestock • Inheritance from parents	• Several cows/cattle (2-3) • Several goats (8-10) • Large and permanent house • Large yard (2-3 ha) • Large garden / Forestry (3-5 ha) • Have their own business • Enough food for the whole year • Inheritance from parents • Good health and enough labor	• Several cows (1-3) • Several goats (10-25) • Permanent large house • Large yard (3-5 ha) • Large paddy field (2-3 ha) • Large garden / Forestry (2-3 ha) • Have their own business • Enough food for the whole year • Inheritance from parents	• Several cows (1-3) • Several goats (10-25) • Permanent house • Large yard (3-5 ha) • Several fish ponds (2-5) • Large paddy field (2-3 ha) • Large garden / Forestry (2-3 ha) • Have their own business • Have their own car transportation • Enough food for the whole year • Good health and enough labor
Average	• Some cattle (1-2) • Permanent house • Paddy field (1-2 ha) • Food shortage for 1-3 months a year • Small garden • Small livestock (chickens) • Enough labor forces	• Some goats (10-25) • Paddy field (0.5-1.0 ha) • Some cattle • Permanent house • Enough labor • Little food for 1-4 months a year • Selling some goods in and outside the village • Daily labor	• Some goat (10-25) • Paddy field (0.5-1.0 ha) • Children attend school • Some cattle • Permanent house • Enough labor • Little food for 1-4 months a year • Selling some goods in and outside the village • Daily labor	
Poor	• Not good house • No cattle • Small house • Paddy field less than 0.1-0.2 ha • Stock food for 2-5 months a year • With or without a boat • Poor health	• With or without cows/cattle • With or without paddy field • Stock food for 2-6 months a year • Not enough clothes • Not all children attend school • No money for medicine when ill		
Hungry	• No cow • No paddy field • Woodhouse • Food more than 6 months per year • Elderly or young couples • Really poor health	• No cow • With or lacking paddy field • Woodhouse • Food more than 6-7 months per year • Not enough clothes • No money for children to attend school • No money for drugs when ill • Disabled- and women-headed		• No cow • No paddy field • Woodhouse • Food more than 6 months per year • Elderly • Really poor health

Source: Authors' development based on fieldwork primary data

The primary aspect of the livelihood in every village shows that the key indicator in wellbeing is availability or adequacy of rice and ownership of livestock, especially cattle and goats. Households face poverty when rice and livestock ownership reduce significantly and suddenly. This means that poverty remains a very great challenge to communities that have experienced severe food shortages and lack of livestock ownership. In a focus group discussion, all group members said that the context of poverty for them is ownership of rice for food, and livestock (cattle, goats, and chickens) for savings, and these are used for an emergency situation such as daily food needs and children's schooling. Rice and livestock are crucial. Decline in rice and livestock ownership results in imbalance in their livelihoods and daily lives. Therefore, the core of poverty from the villagers' perspective is the inability to meet basic needs through income improvement and development.

An interesting finding that needs to be addressed by the government is seen in the 'poor' and 'hunger' categories. Some of the families could afford to send their children to school because there are no school fees, but the school is comparatively distant from home. Some villagers were able to send their children to primary school but were unable to send them to middle and high school. On average, every village has a primary school, but middle and high schools are quite far from the village. The daily cost for sending a child to school is relatively high for middle and high school. Although public schools are free, the location of the school is still relatively distant. In addition, poor family awareness on the need to send their children to school in the village is low and some children are engaged in 'child labor' and some drop out of school. The government needs to address the problem other than by adding more school buildings and to emphasize the policy focus on increasing understanding of the importance of schooling for poor families.

A diagram of cause and effect is very useful in characterizing the root problems of poverty at village level. This helps to coordinate the development of action plans and sets priorities in the completion of policy plans. The goal in using causal diagrams is to determine the top priority issues in the research and bottlenecks in analyzing problems. This is done by PPA ranking, looking at any obstacles and problems faced by a representative sample of key informants. The respondents were interviewed to learn the various obstacles and problems in poverty alleviation efforts. This helped to determine the level of priorities in a single diagram (Figure 2).

The main cause of poverty, in general, from several focus group discussions in the villages is lack of access to income resources. Revenue, in general, is obtained by earnings of permanent employees from resources such as land, crops, and livestock or fisheries and forestry. Figure 2 is a general causal diagram of the four study villages. The first cause of poverty is the lack of physical capital due to the effect of rural isolation. Secondly, there is lack of natural capital caused by marginal or low productivity, the environment's decreased stability, climate change and environment shock, and underutilized options. Thirdly, there is lack of financial capital, which is due to lack of saved financial assets, few credit facilities, and few employment options. Fourthly, there is lack of human capital, caused by effects of increasing population, decreases in health quality, low entrepreneurial skills, lack of public sector leadership, and lack of social capital due to overly rapid decentralization, socially inadequate institutions, increasing mortality, violence and corruption, and inadequately focused policies.

Lack of access to financial capital is a major factor in causing poverty. The results show that according to the villagers, poorest people were unable to get financial capital for their businesses. There are some opportunities for opening up businesses in rural areas. However, the access of villagers to credit

Figure 2. Diagram of the causes and effects of poverty in DMP Program villages
Source: Authors' development based on fieldwork

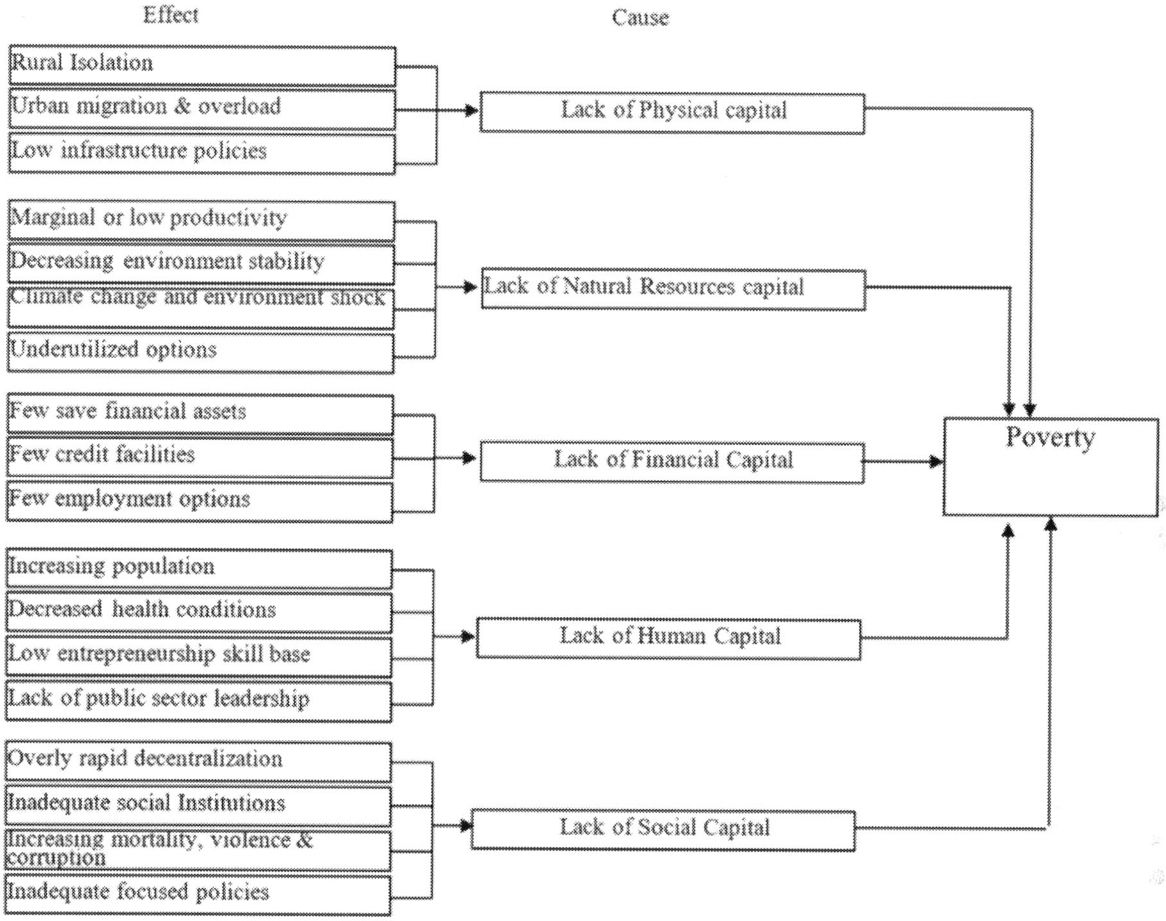

Effect → **Cause**

- Rural Isolation
- Urban migration & overload
- Low infrastructure policies
→ Lack of Physical capital

- Marginal or low productivity
- Decreasing environment stability
- Climate change and environment shock
- Underutilized options
→ Lack of Natural Resources capital

- Few save financial assets
- Few credit facilities
- Few employment options
→ Lack of Financial Capital

- Increasing population
- Decreased health conditions
- Low entrepreneurship skill base
- Lack of public sector leadership
→ Lack of Human Capital

- Overly rapid decentralization
- Inadequate social Institutions
- Increasing mortality, violence & corruption
- Inadequate focused policies
→ Lack of Social Capital

All causes → **Poverty**

facilities from financial institutions is limited. Financial capital with low interest is preferred by villagers. The easy way to get financial capital is by selling livestock. It is a rare case for livestock or saving to be used for starting a business, as such saving is normally used for urgent matters such as paying when a family member is sick or for school fees of children. Therefore, there is a need for alternative capital resources that could be used for opening a business. One solution for this is borrowing from private financial institutions in the village. There is little availability of credit and credit institutions that secure low-interest loans for villagers: unfortunately, many financial institutions have high-interest rates, using, for example, middleman. The DMP Program has established a micro-financial institution (Lembaga Keuangan Desa – LKD) in the village participated in by farmer groups. However, the funds allocated for the LKD are limited and they cannot cover the needs of all the villagers. There remains a lack of capital necessary to start businesses for some farmers. The villagers proposed the establishment of a farmer cooperative in the village in the off-farm sector and on-farm activity, to be managed by farmer groups. The aim of cooperative would be to generate savings and provide loans at low-interest rates to facilitate profit-making for villagers.

Lack of human capital in the village is an important factor that causes poverty. Villagers revealed that they need training to develop the skills necessary to secure capital. The level of motivation to work in the village is so low that concerted efforts should be made to increase the motivation of village heads and their staff members to start enterprises and provide jobs. The solution to some of these problems has been incorporated in the DMP Program, which has proven itself to be helpful to villagers. Health is also one of the important contributing factors in terms of its impact on poverty. Health services in the village are still very basic. For example, one village is served by only one midwife. The Community Health Centre is only provided at the district level. For remote villages like Kedungdowo, it is even more difficult to access health services in case of emergencies.

In regard to social capital, decentralization policies give local governments the authority to manage their own regions effectively. This means the local government needs to create effective policies to be implemented at district and regency level. One example is the policy of the local government that allowed conversion of agricultural land to settlement. In addition, the focus of implementation of pro-poor policy should be geared towards addressing specific problems at the village level. There was a policy designed to integrate the needs of different sectors in terms of implementation. However, the initiated pattern of policy integration should start from the province level right to the village level with the creation of institutions responsible for policy execution.

The village community believes that poverty is caused by different reasons in each village and they are multidimensional. This research revealed that the most important factor in poverty is scarcity of water. This is because, for many years, people have had difficulty in accessing sources of clean and safe water due to the location of their villages in hilly areas. The diagram below describes the condition of Candirejo with respect to lack of water as the main cause of poverty (Figure 3).

Figure 3. Root causes/causal analysis of poverty in the village
Source: Authors' development based on fieldwork primary data

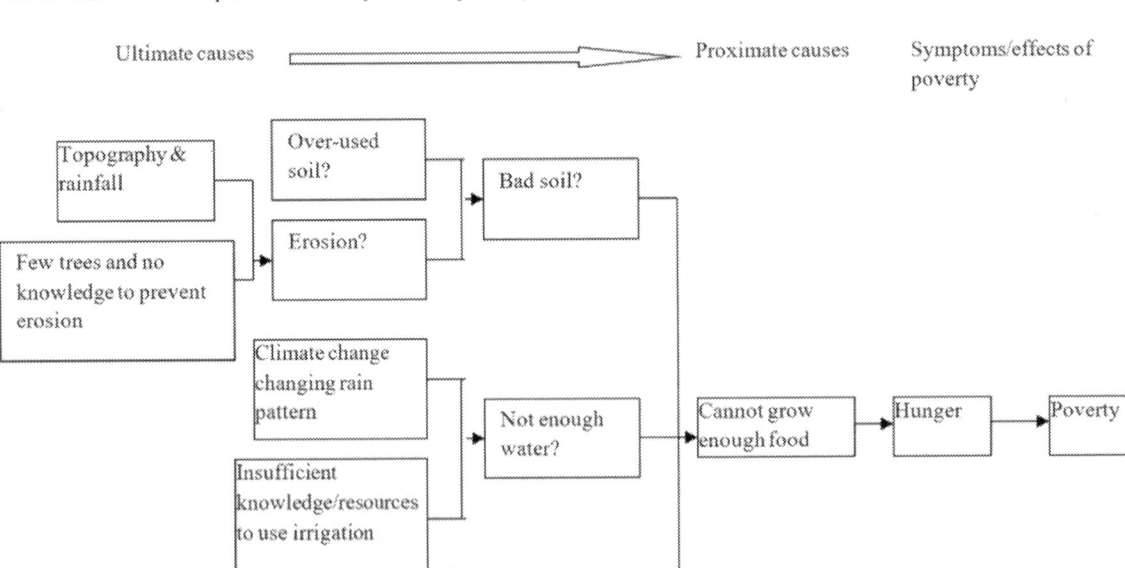

SOLUTIONS AND RECOMMENDATIONS

Active institutions play a very important role in supporting the DMP Program. There were institutions in the village sites before the introduction of the DMP Program, such as the village leader, village councils, farmer groups, rice milling units, a village health station unit, a school, a mosque, a church, private lenders, village shops, and traders. New institutions introduced by the Program are affinity groups, food village teams, rice barns, and micro-financial institutions. In Kedungdowo, the village leader is serving his second elected term. This shows the trust that villagers have for the village head. The relationship of village leader and village council should be harmonious in order to ensure good consultation with the village council, which typically advises and monitors every village activity.

The institutions, organizations, groups, and individuals provide information on the relationships between these entities in the delivery of program activities and these help planners in identifying and characterizing the different factors (impacts, cooperation, linkages, and leadership) involved in different circumstances to bring about successful implementation in rural development. Firstly, in Kedungdowo, the important institution is the RMU. Although it is relatively small, it is useful because it is close by. Its function is like a rice barn or financial institution. It provides villagers with farming inputs and stores of rice stock. Since the launching of the Program in Kedungdowo, village rice barns and micro-financial institution (LKD) have been established, making the function of the RMU less important and its role has been taken over by farmer groups now managing the rice barn and LKD. Low-interest loans are an important factor for poor farmers to be motivated to apply and the loan applications are processed in the regular meeting of farmer groups. The poor families find the rice barn beneficial, especially during periods of food shortages. Secondly, in Selogiri, the Affinity Group (AG) formed as a result of the DMP activities consists of four groups of farmers with different business areas of agriculture, livestock, fisheries, and cassava cracker manufacturers. 50% of funds are channeled to agricultural groups so that many more farmers can become involved. The village food team (FVT), which consists of representatives of each affinity group, holds regular meetings every month attended by the head of the village and the village council. The constraints faced in the implementation of activities and future implementation plans are being worked out together so as to encourage the participation of all levels of the society. Thirdly, in Candirejo, farmer groups are relatively more dynamic than those in other villages. The fundamental difference is absence rice milling units because farmers grind their rice in a neighboring village. Grants are not allocated to the building of rice barns but are used to purchase livestock (local and etawa goats). The allocation of funding is for poor families who are involved as farmer groups. Their tourism resources contribute to revenue. Some villagers earn money through sales to local and foreign tourists. Fourth, in Madukoro, an individual or group has self-motivation initiated by the village leader. High responses of the public to the program makes it easy to introduce and implement. The important success factor is the efforts of the leader driving activities initiated by the DMP Program. The role of the leader is critical to the village's development. The leader should have the ability to mobilize the community to be active in most programs. A key person with great influence can have a significant impact on program success through community participation. Policies targeted at the family level such key people to have a significant impact on rural development.

Several programs have been implemented successfully in one of the villages associated with the Program, various agencies' cooperation gave rise to high impact contribution to the development of the village. In some villages the Program was not implemented with other programs but different times, making them less successful. Programs should be integrated for more significant impact.

FUTURE RESEARCH DIRECTIONS

Analysis on sources of funding shows that villages receive funding mostly from the national budget, implemented through local government. To ensure sustainability in efforts to alleviate poverty central government should continue to consider the local situation in allocation of funds. The local government is currently striving to boost revenues by looking to local and foreign investors to invest in their villages so that the budget can be increased. The local government should be proactive in generating financial sources through local revenue. In this way, the gap between the national and regional budgets can be reduced.

Extension officers play an important role in every activity in the village. Their function is to disseminate knowledge and information on program implementation. However, that information is mostly obtained from friends, relatives, and parents. If extension officers are to play their role, they need to frequently visit villages and come to the problems faced. Extension activities should not be only from the district level but also from regency or province level. Information dissemination is more effective if officers are experienced and come from various sectors, such as agriculture, husbandry, fisheries, and health. Extension officers need to know villagers' problems through close observation. Information should include up-to-date technology and information in terms of program implementation.

CONCLUSION

PPA generally uses several tools to analyze the characteristics of poor families with poverty reduction in mind. PPA is a method of evaluating the characteristics of the poor through observation and analysis of policy implementation. The requirement of this method is the participation of the community and stakeholders to discover alternative solutions to poverty problems at the grassroots level. It is a practical tool of solving poverty problems, such as increasing farmer income and reducing poverty.

The output of the PPA process from this study is the agricultural policy formulated in terms of practical ways of approaching poverty problems from a local perspective. Indonesia has many provinces with a progressive and proactive policy in agriculture. The success of alternative policy options applied by local government at the grassroots level should be adopted at the national level. It should represent the best example of a case of successful program implementation at the grassroots level which can then be used in formulating national policies and strategies.

The livelihood analysis shows different results, depending on the availability of local natural resources in the villages. The four villages in the DPM Program show that the main livelihood of the villagers is based on farming, livestock, fisheries, forests, and tourism. These sectors play a significant role in the village in terms of providing primary income for most villagers. The features of livelihood of the villages can be described as follows. Firstly, the most important aspect of the livelihood of rural communities is paddy cultivation. Ownership of paddy land varies from around 0.1 ha to 3 ha. The land area that generates a relatively small income of farmers is very small (0.25 ha). Secondly, livestock production in the countryside is a direct savings mechanism for villagers. If they need small amount of cash such as for buying cooking oil, they can sell chickens to solve their financial constraints. This practice occurs in almost all four villages. Thirdly, fishery brings additional income for most rural communities. Alternatively, fish can also be used as a source of protein to meet the dietary needs of families. Fishing also plays an important role in the welfare mechanism of the villages by maintaining social security. In

Madukoro, fishing is the main source of livelihood for some people. Fourth, NTFPs are major resources for poor communities that currently do not have paddy land; they can collect firewood or wild animals, bamboo, and rattan for sale as daily income. Fifth, tourism is the main livelihood for two villages in the researched village sites, because they have many tourist facilities and are frequently visited by foreign and domestic tourists. Tourism has been contributing significantly to the local economy by raising income of the villagers.

The availability of food in the villages from time to time is found to be unstable. The results from community discussions reveal that change in the availability of natural resources is one of the causes of poverty in the villages studied. Local natural resources availability is the main source of income of the villagers, either directly or indirectly, for their livelihood.

The wealth ranking analysis of family groups is divided into three categories: rich, average, poor, and hungry. Several indicators are used in the analysis: livestock ownership, possession of a deficit of food stocks, and land ownership of paddy fields, dry land, and gardens. There are differences in the ownership of livestock within the category of 'rich'. This is because each village has different characteristics of ownership of animals, which is determined by the estimation of livestock price. In general, the greater proportion of the population do not have enough food to sustain their living because paddy land ownership is relatively low and rice production is not high enough to meet their daily needs. The 'average' ranking shows that the rice shortage period is about 1-4 months, which usually occurs during August to October. This period represents the time of drought or dry season, so plants are not growing very well. Rice cultivation is only possible in areas where water is available for growth. Nevertheless, the key indicator in wellbeing is the availability and adequacy of rice and ownership of livestock, especially cattle and goats. When the quantity of rice and livestock is reduced significantly, the family is said to be poor. This means that poverty remains a very big challenge to communities who have severe food shortages and the lack of livestock ownership.

The DMP Program has resulted in the establishment of a micro-financial institution (Lembaga Keuangan Desa – LKD) in the village, participated in by farmer groups. However, the funds allocated for the LKD are limited and they cannot cover the needs of all villagers. There remains a lack of capital necessary to start businesses for some farmers. The villagers proposed establishing a farmer cooperative in the village to support 'off-farm' and 'on-farm' activities and to be managed by farmer groups. The aim of the cooperative is to provide the place for savings and loans with low-interest rates and to gain profits for the villagers.

The PPA analysis carried out in this study revealed that the main problem facing most villages is physical development. The government needs to pay more attention to infrastructure development. Infrastructure development such as irrigation and roads should be given high priority in poor villages. The impact of infrastructure development is clear, in that it raises the chances of the poor family fighting against poverty and enhancing food security. The policies on rural development should be comprehensive enough to integrate infrastructure development with community development. The integration between various sectors also plays a crucial role in enhancing infrastructure development.

The community coping mechanisms in time of food shortages seem to be the same for all the villages studied. To survive in the face of food shortages, poor families formed cooperatives in farming, fisheries, livestock, forestry, savings, loans, and small business for their livelihood. This practice is common for villagers in Central Java in a situation where it is still relatively easy for them to have access to food. The next level of coping mechanism for the survival of poor families is to sell household assets such as the television, motorcycle, bicycle, and the like. This is a forced condition when all the opportunities that exist from farming, loans or savings are not available. Villagers said, "most of the poor are dependent on the presence of government assistance". Coping mechanisms in facing food insecurity and poverty are closely interrelated. In facing food insecurity, the coping mechanisms of poor families involve applying specific strategies to meet their food needs, such as borrowing and paying back when conditions return to normal. It was found that, in general, the majority of the villagers in the research sites experienced food shortage for a relatively long period (2-5 months). These villagers had to struggle to meet their needs. The focus group discussion proposed more employment opportunities be provided in the village. This requires a policy of opening more traditional markets to sell agriculture product for villagers. The presence of traditional markets in every district or in any two adjacent villages makes it easier to sell crops and products. In addition, there is a need to establish a cooperative institution, where well-off people could commit themselves to providing cash capital to the poor to enable them to start their business. Monitoring and evaluation of each cooperative by the relevant department needs to be undertaken in order to achieve sustainability.

REFERENCES

Asian Development Bank. (2001). *ADB Annual Report 2001*. Manila: Asian Development Bank.

Chambers, R. (1994). The Origins and Practice of Participatory Rural Appraisal. *World Development*, *22*(7), 953–969. doi:10.1016/0305-750X(94)90141-4

Chambers, R. (1996). *Relaxed and Participatory Appraisal: Notes of Practical Approaches and Methods*. Brighton: Institute of Development Studies.

Chambers, R. (1997). Editorial: Responsible Well-Being – A Personal Agenda for Development. *World Development*, *25*(11), 1743–1754. doi:10.1016/S0305-750X(97)10001-8

Cornwall, A. (2000). *Making A Difference? Gender and Participatory Development*. Brighton: Institute of Development Studies.

Lok-Dessallien, R. (1998). *Review of Poverty Concepts and Indicators*. New York, NY: Social Development and Poverty Elimination Division, United Nations Development Program.

Narayan, D., Chambers, R., Shah, M. K., & Petesch, P. (2000). *Voices of the Poor. Crying Out for Change*. New York, NY: Oxford University Press. doi:10.1596/0-1952-1602-4

Norton, A., Bird, B., Brock, K., Kakande, M., & Turk, C. (2001). *A Rough Guide to PPAs Participatory Poverty Assessment: An Introduction to Theory and Practice*. London: Overseas Development Institute.

Ruggeri Laderchi, C. (2001). *Participatory Methods in the Analysis of Poverty: A Critical Review*. Oxford: University of Oxford.

Turk, C. (2001). *Linking Participatory Poverty Assessments to Policy and Policymaking: Experience from Vietnam*. Hanoi: The World Bank.

ADDITIONAL READING

Palmino-Reganit, M. (2005). *Analysis of Community's Coping Mechanisms in Relation to Floods: A Case Study in Naga City, Philippines*. Enschede: International Institute for Geo-Information Science and Earth Observation.

Patnaik, U. (2008). Theorizing Poverty and Food Security in the Era of Economic Reforms. In G. Lechini (Ed.), *Globalization and the Washington Consensus: Its Influence on Democracy and Development in the South* (pp. 161–200). Buenos Aires: CLACSO.

Pearson, S., Gotsch, C., & Bahri, S. (2003). *Applications of the Policy Analysis Matrix in Indonesian Agriculture*. Ithaca, NY: Cornell University Press.

Russell, T. (1997). Pair Wise Ranking Made Easy. *PLA Notes, 28*, 25–26.

Timmer, P. (1989). Food Price Policy: The Rationale for Government Intervention. *Food Policy, 14*(1), 17–27. doi:10.1016/0306-9192(89)90023-7

Timmer, P. (2008). Agriculture and Pro-Poor Growth: An Asian Perspective. *Asian Journal of Agriculture and Development, 5*(1), 1–29.

KEY TERMS AND DEFINITIONS

AG: Affinity Group
DMP Program: Desa Mandiri Pangan Program (Food Self Sufficiency Village Program).
FVT: The village food team.
GDP: Gross Domestic Product.
LKD: Lembaga Keuangan Desa (Micro-Financial Village Institutional).
NGOs: Non-Governmental Institution.
NTFS: Non-Timber Forest Productions.
NTPFPs: Timber and Non-Timber Forest Products.
PNPM: Program National Pemberdayaan Masyarakat Mandiri (National Program Community Empowerments).
PPA: Participatory Poverty Assessment.
PRA: Participatory Rural Appraisal.
RMU: Rice Milling Unit.

Compilation of References

Abazova, F., & Kulova, A. (2004). Major Opportunities to Overcome Crisis in Agriculture in the Conditions of Market Transformation. *Advances in Current Natural Sciences, 8*, 127–128.

AB-Center. (n.d.). *Expert-Analytical Center of Agribusiness*. Retrieved from https://ab-centre.ru/

Abrams, L. (2018). *Unlocking the Potential of Enhanced Rainfed Agriculture*. Stockholm: Stockholm International Water Institute.

Abro, Z. A., Alemu, B. A., & Hanjra, M. A. (2014). Policies for Agricultural Productivity Growth and Poverty Reduction in Rural Ethiopia. *World Development, 59*(C), 461–474. doi:10.1016/j.worlddev.2014.01.033

Abu-Afife, S., Bushuyeva, N., & Kolomiychuk, S. (2005). The Use of Trypsine Inhibitors at Sclera Reinforcement within Children and Teenagers with Progressive Myopia. *Journal of Ophthalmology, 2*, 11–14.

Achancho, V. (2013). Review and Analysis of National Investment Strategies and Agricultural Policies in Central Africa: The Case of Cameroun. In A. Elbehri (Ed.), *Rebuilding West Africa's Food Potential* (pp. 117–149). Rome: Food and Agriculture Organization of the United Nations.

Adom, P. K., Djahini-Afawoubo, D. M., Mustapha, S. A., Fankem, S. G., & Rifkatu, N. (2018). Does FDI Moderate the Role of Public R&D in Accelerating Agricultural Production in Africa? *African Journal of Economic and Management Studies, 9*(3), 290–304. doi:10.1108/AJEMS-07-2017-0153

Adukov, R. (n.d.). *Agrarian Reforms and Rural Economy Development in Russia*. Retrieved from https://www.adukov.ru/articles/agrarnye_reformy/

African Development Bank Group. (2019). *Feed Africa*. Abidjan: African Development Bank Group.

African Development Bank. (2015). *African Development Report 2015. Growth, Poverty and Inequality Nexus: Overcoming Barriers to Sustainable Development*. Abidjan: African Development Bank Group.

African Union. (2012). *Synthesis Paper on Boosting Intra-African Trade and Fast Tracking the Continental Free Trade Area*. Addis Ababa: African Union.

Agarkov, N. (2001). To Increase Efficiency of Production. *Economics of Agriculture in Russia, 3*, 9–10.

Agricultural and Processed Food Products Export Development Authority. (n.d.). *Trade Information*. Retrieved from https://apeda.gov.in/apedawebsite/#

Agricultural Bulletin. (2019). *Agricultural Production in Russia Decreased by 0.6% in 2018*. Retrieved from https://agrovesti.net/news/indst/proizvodstvo-selkhozproduktsii-v-rf-v-2018-g-snizilos-na-0-6.html

Agricultural Market Information System. (n.d.a). *About AMIS*. Retrieved from http://www.amis-outlook.org/amis-about/en/

Compilation of References

Agricultural Market Information System. (n.d.b). *Market Database.* Retrieved from https://app.amis-outlook.org/#/market-database/supply-and-demand-overview

Agricultural Market Information System. (n.d.c). *Prices and Price Volatility.* Retrieved from http://www.amis-outlook.org/indicators/prices/en/

Agriculture Equipment Market. (2019). *Report.* Retrieved from https://www.farmmachinerysales.com.au/

Agronews. (2013). *How Russia Loses the Market of Organic Products.* Retrieved from http://agronews.ua/node/35624

Ahammad, H., Heyhoe, E., Nelson, G., Sands, R., Fujimori, S., Hasegawa, T., & Tabeau, A. A. (2015). The Role of International Trade under a Changing Climate: Insights from Global Economic Modelling. In A. Elbehri (Ed.), *Climate Change and Food Systems: Global Assessments and Implications for Food Security* (pp. 293–312). Rome: Food and Agriculture Organization of the United Nations.

Aharoni, Y., & Nachum, L. (Eds.). (2000). *Globalization of Services: Some Implications for Theory and Practice.* London: Routledge. doi:10.4324/9780203465363

Akhmadullina, I. (2018). *Enhancement of the Techniques of Primary Seed Breeding and Pea Cultivation for Seed Purpose in the CIS-Urals Region of the Republic of Bashkortostan.* Kazan: Kazan State Agrarian University.

Akhramovich, V., Chubrik, A., & Shymanovich, G. (2015). *AGRICISTRADE Country Report: Belarus.* Retrieved from http://www.agricistrade.eu/document-library

Alakoz, V., Kiselev, V., & Shmelev, G. (1999). *Reasons for Land Reform in Russia.* Moscow: Interdesign.

Alakoz, V., Vasiliev, I., Kiselev, V., & Pulin, A. (2001). *Joint Shared Land Property: Theory, Data, Practice.* Moscow: Construction Formula.

Alam, A., Murthi, M., Yemtsov, R., Murrugarra, E., Dudwick, N., Hamilton, E., & Tiongson, E. (2005). *Growth, Poverty, and Inequality: Eastern Europe and the Former Soviet Union.* Washington, DC: The World Bank. doi:10.1596/978-0-8213-6193-1

Aleksandrov, A. (2001). Management of Agricultural Production Requires Improvement of Legislation. *Economist, 3,* 83–89.

Alekseyeva, Y. (2018). *Russia – Agricultural Equipment.* Retrieved from https://www.export.gov/article?id=Russia-Agricultural-Equipment

Alexandrova, E., Shramko, G., & Knyazeva, T. (2010). New Composition of Mineral Fertilizers for Foliar Feeding of Winter Wheat. *Proceedings of Kuban State Agrarian University, 22*(1), 71–74.

Algazin, D., Vorobiev, D., Zabudsky, A., & Zabudskaya, E. (2016). *Patent #1600896 "Device for Growing Plants."* Retrieved from https://yandex.ru/patents/doc/RU160896U1_20160410

Al-Haq, M. I., Seo, Y., Oshita, S., & Kawagoe, Y. (2002). Disinfection Effects of Electrolyzed Oxidizing Water on Suppressing Fruit Rot of Pear Caused by *Botryosphaeria Berengeriana. Food Research International, 35*(7), 657–664. doi:10.1016/S0963-9969(01)00169-7

Allan, J. A. (1993). *Fortunately There Are Substitutes for Water: Otherwise Our Hydropolitical Futures Would Be Impossible.* London: Overseas Development Administration.

Allan, J. A. (1998). Virtual Water: A Strategic Resource Global Solutions to Regional Deficits. *Ground Water, 36*(4), 545–546. doi:10.1111/j.1745-6584.1998.tb02825.x

Alliance for a Green Revolution in Africa. (2014). *Africa Agriculture Status Report 2014: Climate Change and Smallholder Agriculture in Sub-Saharan Africa*. Nairobi: Alliance for a Green Revolution in Africa.

All-Russian Institute of Agrarian Problems and Informatics named after A. Nikonov. (2003). *Agricultural Policy and Russia's Accession to the WTO*. Moscow: All-Russian Institute of Agrarian Problems and Informatics.

Alonso, E. B., Cockx, L., & Swinnen, J. (2017). *Culture and Food Security*. Leuven: Centre for Institutions and Economic Performance.

Altukhov, A. (2014). Russia Needs a New Agricultural Policy. *The Economist*, *8*, 28-39.

Ambroszczyk, A. M., Jedrszczyk, E., & Nowicka-Polec, A. (2016). The Influence of Nano-Gro® Stimulator on Growth, Yield and Quality of Tomato Fruit (*Lycopersicon Esculentum Mill.*) in Plastic Tunnel Cultivation. *Acta Horticulturae*, (1123): 185–192. doi:10.17660/ActaHortic.2016.1123.26

Anderson, K., Jha, S., Nelgen, S., & Strutt, A. (2013). Re-Examining Policies for Food Security in Asia. *Food Security*, *5*(2), 195–215. doi:10.100712571-012-0237-5

Anderson, K., Martin, W., & Van Der Mensbrugghe, D. (2005). *Would Multilateral Trade Reform Benefit Sub-Saharan Africans?* Adelaide: Centre for International Economic Studies. doi:10.1596/1813-9450-3616

Anisimov, E., Gapov, M., Rodionova, E., & Saurenko, T. (2019). The Model for Determining Rational Inventory in Occasional Demand Supply Chains. *International Journal of Supply Chain Management*, *8*(1), 86–89.

Anisimov, V. (2009). *Optimization: An Adaptive Approach to Investment Management under Uncertainty*. Moscow: Publishing House of the Russian Customs Academy.

Antipov, S., Zhuravlev, A., Vinichenko, S., & Kazartsev, D. (2015). *Patent #2548209 "Vacuum Dryer of Continuous Operation with Ultra-High Frequency Power Supply."* Retrieved from https://findpatent.ru/patent/254/2548209.html

Antipova, L., Grebenshchikov, A., Mishchenko, A., Osipova, N., & Tychinin, N. (2017). Histochemical and Physiological-Biochemical Properties of Germinated Seeds Lentils as a Source of Nutrients. *Technologies of Food and Processing Industry of AIC – Healthy Food*, *4*, 69-79.

Antipova, L., & Presnyakova, O. (2008). Opportunities of Chickpea and Lupin Utilization in Condensed Milk Analogues Technology. *Storage and Processing of Farm Products*, *4*, 50–52.

Antonakakis, N., & Tondl, G. (2014). Does Integration and Economic Policy Coordination Promote Business Cycle Synchronization in the EU? *Empirica*, *41*(3), 541–575. doi:10.100710663-014-9254-2

Antonić, M. (1998). Razvoj pčelarstva u srednjevekovnoj Srbiji. *Pčelarstvo*, *1*(1), 3–6.

Antonova, N. (2012). Collaboration of APEC Member Countries in the Sphere of Food Security. *Spatial Economics*, *2*, 146–151. doi:10.14530e.2012.2.146-151

Arakelyan, M. (2013). *CIS Frontier Countries: Economic and Political Prospects*. Frankfurt am Main: Deutsche Bank.

Arayaa, A., Keesstra, S. D., & Stroosnijder, L. (2010). A New Agro-Climatic Classification for Crop Suitability Zoning in Northern Semi-Arid Ethiopia. *Agricultural and Forest Meteorology*, *150*(7-8), 1057–1064. doi:10.1016/j.agrformet.2010.04.003

Arskiy, A. (2016). Peculiarities of Calculation of Logistics Costs. Motor Fuel. *World of Modern Science*, *35*(1), 34–37.

Arskiy, A. (2018a). Management of Logistics Costs of Enterprises of Agro-Industrial Complex. *Bulletin of the Moscow University of Finance and Law*, *1*, 98–102.

Compilation of References

Arskiy, A. (2018b). Assessment of Efficiency of Management Decisions in Crisis Management of Agricultural Enterprise. *Marketing and Logistics, 16*(2), 6–11.

Arthur, W. B. (1994). *Increasing Returns and Path Dependence in the Economy*. Ann Arbor, MI: University of Michigan Press. doi:10.3998/mpub.10029

Arthur, W. B. (2013). *Complexity Economics: A Different Framework for Economic Thought*. Santa Fe, NM: Santa Fe Institute.

Asian Development Bank. (2001). *ADB Annual Report 2001*. Manila: Asian Development Bank.

Atanazevich, V. (2000). *Drying Food. Reference Manual*. Moscow: DeLi.

Atlas of Economic Complexity. (n.d.). *Global Rankings and Projections*. Retrieved from http://atlas.cid.harvard.edu/rankings/product/

Auat Cheein, F. A., & Carelli, R. (2013). Agricultural Robotics: Unmanned Robotic Service Units in Agricultural Tasks. *IEEE Industrial Electronics Magazine, 7*(3), 48–58. doi:10.1109/MIE.2013.2252957

Awulachew, S. B., Erkossa, T., & Namara, R. (2010). *Irrigation Potential in Ethiopia: Constraints and Opportunities for Enhancing the System*. Colombo: International Water Management Institute.

Axtell, R. L. (2001). Zipf Distribution of U.S. Firm Sizes. *Science, 5536*(293), 1818–1820. doi:10.1126cience.1062081 PMID:11546870

Ayatskov, D. (2002). *Land Ownership in Russia: History and Present Time*. Moscow: Russian Politic Encyclopedia.

Ayyappan, S. (2015). *India-Africa Cooperation in Agricultural Sector for Food Security*. Retrieved from https://www.mea.gov.in/in-focus-article.htm?25950/IndiaAfrica+Cooperation+in+Agricultural+Sector+for+Food+Security

Bailey, A. (1950). *Melting and Solidification of Fats*. New York, NY: Interscience Publishers.

Baker, W. E. (1984). The Social Structure of a National Securities Market. *American Journal of Sociology, 89*(4), 775–811. doi:10.1086/227944

Bakhir, V., Liakumovich, A., & Kirpichnikov, P. (1983). The Physical Nature of the Phenomena of Activation Substances. *News of the Academy of Sciences of Uzbekistan Soviet Socialist Republic, 1*, 60–64.

Bakhir, V., Prilutsky, V., & Shomovskaya, N. (2010). Electrochemically Activated Aqueous Media-Anolyte and Catholyte as a Means of Suppressing Infectious Processes. *Medical Alphabet, 13*(3), 40–42.

Bakst, D. (2018). *Agricultural Trade with China: What's at Stake for American Farmers, Ranchers, and Families*. Retrieved from https://www.heritage.org/agriculture/report/agricultural-trade-china-whats-stake-american-farmers-ranchers-and-families

Baldos, U. L. C., & Hertel, T. W. (2015). The Role of International Trade in Managing Food Security Risks from Climate Change. *Food Security, 7*(2), 275–290. doi:10.100712571-015-0435-z

Balfour, E. (n.d.). *Towards a Sustainable Agriculture. The Living Soil*. Journey to Forever. Retrieved from http://www.journeytoforever.org/farm_library/balfour_sustag.html

Bali Swain, R., & Varghese, A. (2013). Delivery Mechanisms and Impact of Microfinance Training in Indian Self-Help Groups. *Journal of International Development, 25*(1), 11–21. doi:10.1002/jid.1817

Baranovskiy, A. (2008). *Dietology: Guidance*. Saint Petersburg: Piter.

Barilla Center for Food and Nutrition. (2012). *Lo sprecoalimentare: cause, impatti e proposte*. Parma: Barilla Center for Food and Nutrition.

Barkhatov, V., Lisitsky, V., & Kozachenko, Z. (1993). Aseptic Preservation of Fruit and Vegetable Puree of Semi-Finished Products from Low-Acid Raw Materials. *News of Institutes of Higher Education. Food Technology, 3-4*, 51–52.

Barrett, C. B. (2010). Measuring Food Insecurity. *Science, 327*(5967), 825–828. doi:10.1126cience.1182768 PMID:20150491

Bates, B. C., Kundzewicz, Z. W., Wu, S., & Palutikof, J. (2008). *Climate Change and Water. Technical Paper of the Intergovernmental Panel on Climate Change*. Geneva: Intergovernmental Panel on Climate Change.

Baumane, V., & Pasko, O. (2014). Comparison of Land Reform of Latvia and Russia in Conditions of Transition Period. *International Scientific Journal, 1*, 40–44.

Bazeley, P. (2005). *Politics, Policies and Agriculture: The Art of the Possible in Agricultural Development*. ODI. Retrieved from https://www.odi.org/events/2390-politics-policies-and-agriculture-art-possible-agricultural-development

Bazga, B. (2015). Food Security Component of Sustainable Development – Prospects and Challenges in the Next Decade. *Procedia Economics and Finance, 32*, 1075–1082. doi:10.1016/S2212-5671(15)01570-1

Beatriz, M., Oliveira, P., & Ferreira, M. (1994). Evolution of the Quality of the Oil and the Product in Semi-Industrial Frying. *Grasas y Aceites, 45*(3), 113–118. doi:10.3989/gya.1994.v45.i3.982

Beck, J. M., & Jacobson, M. D. (2017). 3D Printing: What Could Happen to Product Liability When Users (and Everyone Else in Between) Become Manufactures. *Minnesota Journal of Law, Science & Technology, 18*, 143–150.

Bekić, B., Jeločnik, M., & Subić, J. (2013). Honey and Honey Products. *Works with XXVI Consulting Agronomists, Veterinarians. Technologists and Economists, 18*, 3–4.

Belousov, R. (1972). *History of National-State Construction in the USSR*. Moscow: Politizdat.

Belugin, A. (2017). *Food Security of the Russian Federation and Its Measurement in Modern Conditions*. Moscow: Moscow State University.

Belyanin, D. (2014). Factors of Peasant Migration to Siberia in the Second Half of XIX – Beginning of XX Century. *Bulletin of Tomsk State University, 378*, 100–108.

Ben-Iwo, J., Manovic, V., & Longhurst, P. (2016). Biomass Resources and Biofuels Potential for the Production of Transportation Fuels in Nigeria. *Renewable & Sustainable Energy Reviews, 63*, 172–192. doi:10.1016/j.rser.2016.05.050

Berg, A., & Krueger, A. (2003). *Trade, Growth, and Poverty: A Selective Survey*. Washington, DC: International Monetary Fund.

Berthet, E. T., Hickey, G. M., & Klerkx, L. (2018). Opening Design and Innovation Processes in Agriculture: Insights from Design and Management Sciences and Future Directions. *Agricultural Systems, 165*, 111–115. doi:10.1016/j.agsy.2018.06.004

Bessonov, V., Knyaginin, V., & Lipetskaya, M. (Eds.). (2017). *Nutritiology-2040. The Horizons of Science through the Eyes of Scientists*. Saint Petersburg: Center for Strategic Research.

Bestuzheva-Lada, S. (2011). The Reformer of All Russia. *Smena, 6*, 16–27.

Bezabih, M., Di Falco, S., & Mekonnen, A. (2014). *Is It the Climate or the Weather? Differential Economic Impacts of Climatic Factors in Ethiopia*. Leeds: Centre for Climate Change Economics and Policy.

Compilation of References

Bezuglov, V., & Konovalov, S. (Eds.). (2009). *Lipids and Cancer. Essays on Lipidology of Oncological Process.* Saint Petersburg: Prime-Evroznak.

Biomdv. (n.d.). *What Is Organic Food?* Retrieved from https://biomdv.ru/page/chto-takoe-organicheskie-produkty-pitaniya

Bishnoi, S., & Khetarpaul, N. (1993). Effect of Domestic Processing and Cooking Methods on In-Vitro Starch Digestibility of Different Pea Cultivars (*Pisum Sativum*). *Food Chemistry, 47*(2), 177–182. doi:10.1016/0308-8146(93)90240-G

Bluashvili, A., & Sukhanskaya, N. (2015). *AGRICISTRADE Country Report: Georgia.* Retrieved from http://www.agricistrade.eu/document-library

Bobrova, L. (2013). Food Security of China as a Reflection of the Economic Growth. *Approbation, 6*(9), 41–43.

Bodoshov, A. (2015). Amino-Acid Composition of Haricot Grains Cultivated in Kyrgyzstan. *Young Scientist, 104*(24), 94–96.

Boehringer, C., & Rutherford, T. (1999). *Decomposing General Equilibrium Effects of Policy Intervention in Multi-Regional Trade Models: Method and Sample Application.* Mannheim: Center for European Economic Research.

Bogdanov, D. (2016). *Evolution of Civil Liability from a Position of Justice: Comparative Legal Aspect.* Moscow: Prospect.

Bogdanov, V., Posternak, T., Pasko, O., & Kovyazin, V. (2016). The Issues of Weed Infestation with Environmentally Hazardous Plants and Methods of Their Control. *IOP Conference Series: Earth and Environmental Science, 43.* 10.1088/1755-1315/43/1/012036

Bogolubov, S. (1998). *Ecological Law: Manual.* Moscow: Norma-Infra.

Bohari, A. M., Hin, C. W., & Fuad, N. (2013). The Competitiveness of Halal Food Industry in Malaysia: A SWOT-ICT Analysis. *Geografia Online. Malaysian Journal of Society and Space, 9*(1), 1–9.

Boiteau, J. M. (2016). *Food Loss and Waste in the United States and Worldwide.* Retrieved August 20, 2019, from https://www.worldhunger.org/food-loss-and-waste-in-the-united-states-and-worldwide/

Bojang, F., & Ndeso-Atanga, A. (2013). *African Youth in Agriculture, Natural Resources and Rural Development.* Accra: FAO Regional Office for Africa.

Bokarev, D. (2016). *Russian-Chinese Agricultural Cooperation.* Retrieved from https://journal-neo.org/2016/04/08/russian-chinese-agricultural-cooperation/

Boltavin, A. (2006). Market for Frozen Convenience Foods from Vegetables. *Ice Cream and Frozen Products, 5,* 16–18.

Boltrik, O. (1999). *Parameters and Modes of Operation of the Electric Activator for Presowing Treatment of Seeds of Grain Crops. Zernograd*: Black Sea State Agroengineering Academy.

Bortolini, M., Cascini, A., Gamberi, M., Mora, C., & Regattieri, A. (2014). Sustainable Design and Life Cycle Assessment of an Innovative Multi-Functional Haymaking Agricultural Machinery. *Journal of Cleaner Production, 82,* 23–36. doi:10.1016/j.jclepro.2014.06.054

Bortolini, M., Mora, C., Cascini, A., & Gamberi, M. (2014). Environmental Assessment of an Innovative Agricultural Machinery. *International Journal of Operations and Quantitative Management, 20*(3), 243–258.

Botkin, O., Sutygina, A., & Sutygin, P. (2016). National Issues of Food Security Assessment. *Bulletin of Udmurt University, 26*(4), 20–27.

Bouet, A., & Laborde, D. (2017). *Building Food Security through International Trade Agreements.* IFPRI. Retrieved from http://www.ifpri.org/blog/building-food-security-through-international-trade-agreements

Bou, R., Guardiola, F., Barroeta, A. C., & Codony, R. (2005). Effect of Dietary Fat Sources and Zinc and Selenium Supplements on the Composition and Consumer Acceptability of Chicken Meat. *Poultry Science*, *84*(7), 1129–1140. doi:10.1093/ps/84.7.1129 PMID:16050130

Bourdieu, P. (1985). The Social Space and the Genesis of Groups. *Theory and Society*, *14*(6), 723–744. doi:10.1007/BF00174048

Boussard, J. M., Daviron, B., Gerard, F., & Voituriez, T. (2006). *Food Security and Agricultural Development in Sub-Saharan Africa*. Rome: Food and Agriculture Organization of the United Nations.

Bowles, S. (2006). *Microeconomics: Behavior, Institutions, and Evolution*. Princeton, NJ: Princeton University Press.

Bowles, S., Choi, J. K., & Hopfensitz, A. (2003). The Co-Evolution of Individual Behaviors and Social Institutions. *Journal of Theoretical Biology*, *223*(2), 135–147. doi:10.1016/S0022-5193(03)00060-2 PMID:12814597

Bozic, D., Bogdanov, N., & Sevarlic, M. (2011). *Agricultural Economics*. Belgrade: University of Belgrade.

Bräutigam, K. R., Jörissen, J., & Priefer, C. (2014). The Extent of Food Waste Generation across EU-27: Different Calculation Methods and the Reliability of their Results. *Waste Management & Research*, *32*(8), 683–694. doi:10.1177/0734242X14545374 PMID:25161274

Breene, K. (2016). *Food Security and Why It Matters*. WE Forum. Retrieved from https://www.weforum.org/agenda/2016/01/food-security-and-why-it-matters/

Brenton, P. (2012). *Africa Can Help Feed Africa: Removing Barriers to Regional Trade in Food Staples*. Washington, DC: World Bank.

Briggs, L. (1926). *USDA Departmental Bulletin #1379*. Washington, DC: United States Department of Agriculture.

Brock, W., & Durlauf, S. (2001). Interactions-Based Models. In J. Heckman & E. Leamer (Eds.), *Handbook of Econometrics* (pp. 3297–3380). Amsterdam: Elsevier. doi:10.1016/S1573-4412(01)05007-3

Brooks, J. (2012). *A Strategic Framework for Strengthening Rural Incomes in Developing Countries*. Paris: Organization for Economic Cooperation and Development. doi:10.1079/9781780641058.0023

Brooks, J., & Matthews, A. (2015). *Trade Dimensions of Food Security*. Paris: OECD Publishing.

Bruinsma, J. (2009). *The Resource Outlook to 2050: By How Much Do Land, Water, and Crop Yields Need to Increase by 2050?* Retrieved from http://www.fao.org/fileadmin/templates/esa/Global_persepctives/Presentations/Bruinsma_pres.pdf

Bruinsma, J. (Ed.). (2003). *World Agriculture: Towards 2015/2030*. London: Earthscan Publications Ltd.

Brunner, P. H., & Rechberger, H. (2017). *Handbook of Material Flow Analysis. For Environmental, Resource and Waste Engineers*. London: CRC Press.

Brussaard, L., de Ruiter, P. C., & Brown, G. G. (2007). Soil Biodiversity for Agricultural Sustainability. *Agriculture, Ecosystems & Environment*, *121*(3), 233–244. doi:10.1016/j.agee.2006.12.013

Buchenrieder, G., Hanf, J. H., & Pieniadz, A. (2009). 20 Years of Transition in the Agri-Food Sector. *German Journal of Agricultural Economics*, *58*(7), 285–293.

Buhaug, H., Benjaminsen, T. A., Sjaastad, E., & Theisen, O. M. (2015). Climate Variability, Food Production Shocks, and Violent Conflict in Sub-Saharan Africa. *Environmental Research Letters*, *10*(12), 125015. doi:10.1088/1748-9326/10/12/125015

Compilation of References

Bull, B. (2014). The Development of Business Associations in Central America: The Role of International Actors and Economic Integration. *Journal of Public Affairs*, *14*(3-4), 331–345. doi:10.1002/pa.1420

Burt, R. S. (1995). *Structural Holes*. Cambridge, MA: Harvard University Press.

Busnelli, F. D., Comande, G., Cousy, H., Dobbs, D. B., Dufwa, B. W., & Faure, M. G. … Widmer, P. (2005). Principles of European Tort Law. Vienna: Springer.

Buyakova, A., Da Costa, R. M., Veber, A., De Castro, J. C. M., & Kazydub, N. (2015). *Patent #2616864 "Process of Production of Fermented Milk Drink."* Retrieved from https://findpatent.ru/patent/261/2616864.html

Buyakova, A., Veber, A., De Castro, J. C. M., Da Costa, R. M., Kazydub, N., & Staurskaya, N. (2015). *Patent #2599569 "Functional Alimentary Product from Sprouted Kernels."* Retrieved from https://xn--90ax2c.xn--p1ai/catalog/000224_000128_0002599569_20161010_C1_RU/viewer/

Byerlee, D., de Janvry, A., & Sadoulet, E. (2009). Agriculture for Development: Toward a New Paradigm. *Annual Review of Resource Economics*, *1*(1), 15–35. doi:10.1146/annurev.resource.050708.144239

Bystrov, G. (2000). Land Reform in Russia: Legal Doctrine and Practice. *State and Law*, *4*, 49–50.

Cai, Z. (2005). *Characterisation of Electrochemically Activated Solutions for Use in Environmental Remediation*. Bristol: University of the West of England.

Caldeira, C., Corrado, S., & Sala, S. (2017). *Food Waste Accounting – Methodologies, Challenges and Opportunities*. Brussels: Publication Office of the European Union.

Calzadilla, A., Rehdanz, K., & Tol, R. S. (2011). Trade Liberalization and Climate Change: A Computable General Equilibrium Analysis of the Impacts on Global Agriculture. *Water (Basel)*, *3*(2), 526–550. doi:10.3390/w3020526

Campbell, J. (2019). *Unpacking Africa's 2019 GDP Growth Prospects*. Retrieved from https://www.cfr.org/blog/unpacking-africas-2019-gdp-growth-prospects

Central Statistical Authority. (2018). Agricultural Sample Survey.: Vol. II. *Report on Livestock and Livestock Characteristics (Private Peasant Holdings)*. Addis Ababa: Central Statistical Authority.

Cervantes-Godoy, D., & Dewbre, J. (2010). *Economic Importance of Agriculture for Poverty Reduction*. Paris: Organization for Economic Cooperation and Development.

Chambers, R. (1994). The Origins and Practice of Participatory Rural Appraisal. *World Development*, *22*(7), 953–969. doi:10.1016/0305-750X(94)90141-4

Chambers, R. (1996). *Relaxed and Participatory Appraisal: Notes of Practical Approaches and Methods*. Brighton: Institute of Development Studies.

Chambers, R. (1997). Editorial: Responsible Well-Being – A Personal Agenda for Development. *World Development*, *25*(11), 1743–1754. doi:10.1016/S0305-750X(97)10001-8

Chan, C., Sipes, B., Ayman, A., Zhang, X., LaPorte, P., Fernandes, F., ... Roul, P. (2017). Efficiency of Conservation Agriculture Production Systems for Smallholders in Rain-Fed Uplands of India: A Transformative Approach to Food Security. *Land (Basel)*, *6*(3), 58. doi:10.3390/land6030058

Charykova, O., & Nesterov, M. (2014). Theoretical Aspects of Substantiation of Rational Level of Self-Sufficiency of the Regions with Food. In Encyclopedia of Russian Villages (pp. 168-171). Academic Press.

Chaynikova, L. (2007). *Competitiveness of an Enterprise: Scientific and Practical Study*. Tambov: Tambov State Technical University.

Chekrygina, I., Bukreev, V., & Eremin, A. (2002). *Patent #2194228 "Method of Drying and Disinfection of Fruits and Berries."* Retrieved from http://www.freepatent.ru/patents/2194228

Chekurov, V., Sergeeva, S., & Zhalieva, L. (2003). New Growth Regulators. *Plant Protection Quarterly*, 9, 20–22.

Cheng, G. (2007). China's Agriculture within the World Trading System. In I. Sheldon (Ed.), *China's Agricultural Trade: Issues and Prospects* (pp. 81–104). Beijing: International Agricultural Trade Research Consortium.

Chen, J. (2007). Rapid Urbanization in China: A Real Challenge to Soil Protection and Food Security. *Catena*, 69(1), 1–15. doi:10.1016/j.catena.2006.04.019

Chernolutskaya, E. (2017). English-Language Historiography about Chinese Participation in the Agricultural Sector of the South of the Russian Far East during 1990-2010. *Regional Problems*, 20(3), 50–57.

Chikhuri, K. (2013). Impact of Alternative Agricultural Trade Liberalization Strategies on Food Security in the Sub-Saharan Africa Region. *International Journal of Social Economics*, 40(3), 188–206. doi:10.1108/03068291311291491

Chramkov, A. (1994). *Land Reform in Siberia (1896-1916) and Its Impact on the Situation of the Peasants. Allowance*. Barnaul: Altai State University.

Christiaensen, L., Demery, L., & Kuhl, J. (2010). *The (Evolving) Role of Agriculture in Poverty Reduction: An Empirical Perspective*. Helsinki: World Institute for Development Economics Research.

Chun, Y. (2015). Security at the Tip of the Tongue. *China Journal (Canberra, A.C.T.)*, 10, 25–26.

Cimino-Isaacs, C., & Schott, J. (2016). *Trans-Pacific Partnership: An Assessment*. Washington, DC: Peterson Institute for International Economics.

Ćirić, M., Ignjatijević, S., & Cvijanović, D. (2015). Research of Honey Consumers' Behavior in Province of Vojvodina. *Economics of Agriculture*, 62(3), 627–644.

CIS Council of Heads of Government. (2010). *Decision from November 19, 2010, "On the Concept of Improvement of Food Security of CIS Member States."* Retrieved from https://www.fsvps.ru/fsvps-docs/ru/news/files/3143/concept.pdf

Clapp, J. (2015). *Food Self-Sufficiency and International Trade: A False Dichotomy?* Rome: Food and Agriculture Organization of the United Nations.

Clapp, J., & Helleiner, E. (2012). Troubled Futures? The Global Food Crisis and the Politics of Agricultural Derivatives Regulation. *Review of International Political Economy*, 19(2), 181–207. doi:10.1080/09692290.2010.514528

Cling, J.-P. (2014). The Future of Global Trade and the WTO. *Foresight*, 16(2), 109–125. doi:10.1108/FS-06-2012-0044

Coalition of European Lobbies for Eastern African Pastoralism. (2017). *Recognising the Role and Value of Pastoralism and Pastoralists*. Retrieved from http://www.celep.info/wp-content/uploads/2017/05/Policybrief-CELEP-May-2017-Value-of-pastoralism.pdf

Colla, G., Rouphael, Y., Canaguier, R., Svecova, E., & Cardarelli, M. (2014). Biostimulant Action of a Plant-Derived Protein Hydrolysate Produced through Enzymatic Hydrolysis. *Frontiers in Plant Science*, 5, 448. doi:10.3389/fpls.2014.00448 PMID:25250039

Coltura & Cultura. (n.d.). *Un Volume, una Coltura*. Retrieved from https://www.colturaecultura.it/download

Compilation of References

Cornwall, A. (2000). *Making A Difference? Gender and Participatory Development*. Brighton: Institute of Development Studies.

Corrado, S., Caldeira, C., Eriksson, M., Hanssen, O. J., Hauser, H.-E., Van Holsteijn, F., ... Sala, S. (2019). Food Waste Accounting Methodologies: Challenges, Opportunities, and Further Advancements. *Global Food Security, 20*, 93–100. doi:10.1016/j.gfs.2019.01.002 PMID:31008044

Cotula, L., Dyer, N., & Vermeulen, S. (2008). *Fuelling Exclusion? The Biofuels Boom and Poor People's Access to Land*. Rome: Food and Agricultural Organization of the United Nations and International Fund for Agricultural Development. London: International Institute for Environment and Development.

Council of the European Union. (2007). *Council Regulation #834/2007 of June 28, 2007 On Organic Production and Labelling of Organic Products and Repealing Regulation (EEC) #2092/91*. Retrieved from https://eur-lex.europa.eu/legal-content/EN/TXT/?uri=celex%3A32007R0834

Creamer, R., Brennan, F. P., Fenton, O., Healy, M. G., Lalor, S., Lanigan, G. J., ... Griffiths, B. S. (2010). Implications of the Proposed Soil Framework Directive on Agricultural Systems in Atlantic Europe – A Review. *Soil Use and Management, 26*(3), 198–211. doi:10.1111/j.1475-2743.2010.00288.x

Csaki, C. (2000). Agricultural Reforms in Central and Eastern Europe and the Former Soviet Union: Status and Perspectives. *Agricultural Economics, 22*(1), 37–54. doi:10.1111/j.1574-0862.2000.tb00004.x

Csaki, C., & Jambor, A. (2009). *The Diversity of Effects of EU Membership on Agriculture in New Member States*. Budapest: FAO Regional Office for Europe and Central Asia.

Cui, K., & Shoemaker, S. (2018). A Look at Food Security in China. *NPJ Science of Food, 2*. doi:10.103841538-018-0012-x

Cvijanović, D., Mihajlović, B., & Cvijanović, G. (2012). General Trends on Agricultural-Food Products Market in Serbia. *Proceedings of the Third International Scientific Symposium Agrosym 2012*. Jahorina: University of East Sarajevo.

Dabbert, S., Haring, A. M., & Zanoli, R. (2004). *Organic Farming: Policies and Prospects*. London: Zed Books.

Dalin, C., Wada, Y., Kastner, T., & Puma, M. J. (2017). Groundwater Depletion Embedded in International Food Trade. *Nature, 543*(7647), 700–704. doi:10.1038/nature21403 PMID:28358074

Daman, P. (2003). *Rural Women, Food Security and Agricultural Cooperatives*. New Delhi: Rural Development and Management Centre.

Danilov, A. (2017). Influence of Growth Promoters on Yield and Quality of Grain Crops Products. *Vestnik of Mari State University Agricultural Economics, 9*(3), 28–32.

Danilov-Danilyan, V., & Losev, K. (2000). *Environmental Challenge and Sustainable Development*. Moscow: Progress-Tradition.

Deere, C. D., Oduro, A. D., Swaminathan, H., & Doss, C. (2013). Property Rights and the Gender Distribution of Wealth in Ecuador, Ghana and India. *The Journal of Economic Inequality, 11*(2), 249–265. doi:10.100710888-013-9241-z

Della Fave, L. R., & Hillery, G. A. Jr. (1980). Status Inequality in a Religious Community: The Case of a Trappist Monastery. *Social Forces, 59*(1), 62–84. doi:10.1093f/59.1.62

Demchenko, O., Shevchuk, V., & Yuzvenko, L. (2016). Investigation of the Resistance of Different Varieties of Buckwheat to Infectious Diseases after the Presowing Treatment of Seeds and Vegetating Plants with Biological Preparations. *Agrobiology, 1*, 57–66.

Demeke, M., Di Marcantonio, F., & Morales-Opazo, C. (2013). Understanding the Performance of Food Production in Sub-Saharan Africa and Its Implications for Food Security. *Journal of Development and Agricultural Economics*, *5*(11), 425–443. doi:10.5897/JDAE2013.0457

Demidov, S., Voronenko, B., & Bazhanova, I. (2015). Kinetics of Infrared Drying Shredded Carrots. *Scientific Journal NRU ITMO. Series. Processes and Food Production Equipment*, *3*, 158–163.

Desai, D. R., & Magliocca, G. N. (2014). Patents Meet Napster: 3D Printing and the Digitization of Things. *The Georgetown Law Journal*, *2*, 1691–1715.

Devaux, A., Torero, M., Donovan, J., & Horton, D. (2018). Agricultural Innovation and Inclusive Value-Chain Development: A Review. *Journal of Agribusiness in Developing and Emerging Economies*, *8*(1), 99–123. doi:10.1108/JADEE-06-2017-0065

Devereux, M. B. (1999). Growth and the Dynamics of Trade Liberalization. *Journal of Economic Dynamics & Control*, *23*(5-6), 773–795. doi:10.1016/S0165-1889(98)00043-8

Devilliers, D., & Mahe, E. (2007). Modified Titanium Electrodes. In M. Nunez (Ed.), *New Trends in Electrochemistry Research* (pp. 1–60). Hauppauge, NY: Nova Science Publishers.

Diouf, J. (2003). *Agriculture, Food Security and Water: Towards a Blue Revolution*. Retrieved from http://oecdobserver.org/news/fullstory.php/aid/942/Agriculture,_food_security_and_water_:_Towards_a_blue_revolution.html

Discovery Research Group. (2019). *Analysis of Healthy Food Market in Russia*. Moscow: Discovery Research Group.

Dobrinsky, V. (2017). Analysis of Placing Customs and Logistics Infrastructure in the Russian Federation and the Cargo Volumes Passing through Its Facilities. *Transport Business in Russia*, *1*, 10–13.

Donovan, G., & Casey, F. (1998). *Improving Soil Fertility in Sub-Saharan Africa*. Washington, DC: World Bank. doi:10.1596/0-8213-4236-3

Dorofeev, M. (2007). Land Reform in Siberia 1896-1916. Aspects of the Relations between the Authorities and the Peasantry. *Tomsk State University Journal*, *3*, 69–73.

Dorofeev, M. (2008). Reverse Migration from Siberia (End of XIX – Beginning of XX Century): The Climatic Aspect of the Problem. *Tomsk State University Journal*, *309*, 75–79.

Dorosh, P. A. (2004). Trade, Food Aid and Food Security: Evolving Rice and Wheat Markets. *Economic and Political Weekly*, *36*(39), 4033–4042.

Dorosh, P. A., Dradri, S., & Haggblade, S. (2009). Regional Trade, Government Policy and Food Security: Recent Evidence from Zambia. *Food Policy*, *34*(4), 350–366. doi:10.1016/j.foodpol.2009.02.001

Dotsenko, S., Bibik, I., Lyubimova, O., & Guzhel, Y. (2016). Kinetics of Biochemical Processes of Germination of Soybean Seeds. *The Bulletin of KrasGAU*, *1*, 66–74.

Doymaz, I. (2007). Air-Drying Characteristics of Tomatoes. *Journal of Food Engineering*, *78*(4), 1291–1297. doi:10.1016/j.jfoodeng.2005.12.047

Dragović, R. (2013). *Uspešne ekonomije srpskih manastira*. Novosti. Retrieved from http://www.novosti.rs/vesti/naslovna/reportaze/aktuelno.293.html:449678-Uspesne-ekonomije-srpskih-manastira

Drozdz, J., & Jurkenaite, N. (2017). Agri-Food Sector Potential in the Chosen CIS Countries. *Zagadnienia Ekonomiki Rolnej*, *352*(3), 103–115. doi:10.30858/zer/83035

Compilation of References

Dubuisson, T. (2014). *3D Printing and the Future of Complex Legal Challenges: The Next Great Disruptive Technology Opportunity or Threat?* Retrieved from https://papers.ssrn.com/sol3/papers.cfm?abstract_id=2718113

Dudwick, N., Fock, K. M., & Sedik, D. (2007). *Land Reform and Farm Restructuring in Transition Countries. The Experience of Bulgaria, Republic of Moldova, Azerbaijan, and Kazakhstan*. Washington, DC: The World Bank. doi:10.1596/978-0-8213-7088-9

Dugalić-Vrndić, N., Kečkeš, J., & Mladenović, M. (2011). The Authenticity of Honey in Relation to Quality Parameters. *Biotechnology in Animal Husbandry, 27*(4), 1771–1778. doi:10.2298/BAH1104771D

Duncan, J., & Ruetschle, M. (2002). Agrarian Reform and Agricultural Productivity in the Russian Far East. In J. Thornton & C. Ziegler (Eds.), *Russia's Far East: A Region at Risk* (pp. 193–220). Seattle, WA: University of Washington Press.

Ehnts, D., & Trautwein, H.-M. (2012). From New Trade Theory to New Economic Geography: A Space Odyssey. *Oeconomia, 2*(2-1), 35–66. doi:10.4000/oeconomia.1616

Elbehri, A., Elliott, J., & Wheeler, T. (2015). Climate Change, Food Security and Trade: An Overview of Global Assessments and Policy Insights. In A. Elbehri (Ed.), *Climate Change and Food Systems: Global Assessments and Implications for Food Security and Trade* (pp. 1–27). Rome: Food Agricultural Organisation of the United Nations.

Elfimova, Y. (2006a). Land and Will. Establishment and Development of Farming in Russia: Issues of Land Use. *Russian Entrepreneurship, 9*, 172–174.

Elfimova, Y. (2006b). Methodical Fundamentals of Economic Valuation of Agricultural Land. *Achievements of Modern Life Sciences, 10*, 90–92.

Elfimova, Y. (2009). Organization of Economic Assessment of Agricultural Land. *Vestnik Universiteta, 28*, 174–175.

Elfimova, Y., & Miroshnichenko, R. (2011). Innovative Aspect in Land Management. *Social Policies and Sociology, 70*(4), 247–264.

Ellen MacArthur Foundation. (2019). *What Is the Circular Economy?* Retrieved from https://www.ellenmacarthurfoundation.org/circular-economy/what-is-the-circular-economy

Endres, B. A. (2000). "GMO": Genetically Modified Organism or Gigantic Monetary Obligation? The Liability Scheme for GMO Damage in the United States and the European Union. *Loyola of Los Angeles International and Comparative Law Review, 22*, 453–462.

Engel, P. (2018). *Aligning Agricultural and Rural Development and Trade Policies to Improve Sustainable Development Impact*. Maastricht: European Centre for Development Policy Management.

Ereport.ru. (n.d.). *World Market of Agricultural Machinery and Equipment*. Retrieved from http://www.ereport.ru/articles/commod/mirovoj-rynok-selskohozjajstvennoj-tehniki.htm

Eriksson, D., Carlson-Nilsson, U., Ortíz, R., & Andreasson, E. (2016). Overview and Breeding Strategies of Table Potato Production in Sweden and the Fennoscandian Region. *Potato Research, 59*(3), 279–294. doi:10.100711540-016-9328-6

Erjavec, E., Volk, T., Rac, I., Kozar, M., Pintar, M., & Rednak, M. (2017). Agricultural Support in Selected Eastern European and Eurasian Countries. *Post-Communist Economies, 29*(2), 216–231. doi:10.1080/14631377.2016.1267968

Ermolaev, V., Fedorov, D., Sosnina, O., & Lifentseva, L. (2015). *Patent #2541395 "Method for Vacuum Drying of Fruit and Berries."* Retrieved from http://www.freepatent.ru/patents/2541395

Erokhin, V. (2015b). Structural Changes in International Trade in Food: Competitive Growth Models for Economies in Transition. *Proceedings of the 3rd International Conference "Economic Scientific Research – Theoretical, Empirical and Practical Approaches"*. Bucharest: Academia Romana.

Erokhin, V., Ivolga, A., & Heijman, W. (2014). Trade Liberalization and State Support of Agriculture: Effects for Developing Countries. *Agricultural Economics – Czech, 60*(11), 524-537.

Erokhin, V. (2015a). Russia's Foreign Trade in Agricultural Commodities in Its Transition to Liberalization: A Path to Go Green. In A. Jean-Vasile, I. Andreea, & T. Adrian (Eds.), *Green Economic Structures in Modern Business and Society* (pp. 253–273). Hershey, PA: IGI Global. doi:10.4018/978-1-4666-8219-1.ch014

Erokhin, V. (2017b). Factors Influencing Food Markets in Developing Countries: An Approach to Assess Sustainability of the Food Supply in Russia. *Sustainability, 9*(8), 1313. doi:10.3390u9081313

Erokhin, V. (2017c). Self-Sufficiency versus Security: How Trade Protectionism Challenges the Sustainability of the Food Supply in Russia. *Sustainability, 9*(11), 1939. doi:10.3390u9111939

Erokhin, V. (2018). Contemporary Foreign Trade Policy of China in the Region of Central and Northeast Asia. In A. C. Ozer (Ed.), *Globalization and Trade Integration in Developing Countries* (pp. 27–54). Hershey, PA: IGI Global. doi:10.4018/978-1-5225-4032-8.ch002

Erokhin, V. (Ed.). (2017a). *Establishing Food Security and Alternatives to International Trade in Emerging Economies*. Hershey, PA: IGI Global.

Erokhin, V., & Gao, T. (2018). Competitive Advantages of China's Agricultural Exports in the Outward-Looking Belt and Road Initiative. In W. Zhang, I. Alon, & C. Lattemann (Eds.), *China's Belt and Road Initiative: Changing the Rules of Globalization* (pp. 265–285). London: Palgrave Macmillan. doi:10.1007/978-3-319-75435-2_14

Erokhin, V., & Ivolga, A. (2012). How to Ensure Sustainable Development of Agribusiness in the Conditions of Trade Integration: Russian Approach. *International Journal of Sustainable Economies Management, 2*(1), 12–23. doi:10.4018/ijsem.2012040102

Erokhin, V., Ivolga, A., & Heijman, W. (2014). Trade Liberalization and State Support of Agriculture: Effects for Developing Countries. *Agricultural Economics, 60*(11), 524–537.

Erşan, Ş., & Çiftçi, A. (2016). Preservation Proposals for the Cultural Landscapes in Context of St. Ioannis Theologos Monastery and the Surrounding Vineyards. *Proceedings of TCL 2016 Conference "Tourism and Cultural Landscapes: Towards a Sustainable Approach"*. Budapest: Foundation for Information Society.

Estevadeordal, A., Freund, C., & Ornelas, E. (2008). Does Regionalism Affect Trade Liberalization Toward Nonmembers? *The Quarterly Journal of Economics, 123*(4), 1531–1575. doi:10.1162/qjec.2008.123.4.1531

Eurasian Economic Commission. (2010). *Agreement on Common Principles and Rules of Technical Regulation in the Republic of Belarus, the Republic of Kazakhstan and the Russian Federation*. Retrieved from http://www.eurasiancommission.org/en/act/texnreg/Pages/acts.aspx

Eurasian Economic Commission. (2011). *Technical Regulations of the Customs Union TR CU 021/2011 "On Food Safety."* Retrieved from http://www.eurexcert.com/TRCUpdf/TRCU-0021-On-food-safety.pdf

Eurasian Economic Commission. (2011). *Technical Requirements of the Customs Union #015/2011 "On Safety of Grain."* Retrieved from http://docs.cntd.ru/document/902320395

Eurasian Economic Commission. (2014). *Treaty of the Eurasian Economic Union*. Retrieved August 14, 2019, from https://docs.eaeunion.org/en-us

Compilation of References

Eurasian Economic Commission. (2015). *Analysis of the Barriers for Access of Agricultural Goods to the Market of the Countries of Southeast Asia.* Retrieved from http://www.eurasiancommission.org/ru/act/prom_i_agroprom/dep_agroprom/monitoring/Documents/Барьеры%20в%20Юго-Восточной%20Азии.pdf

Eurasian Economic Commission. (2019). *Statistics of the EAEU.* Retrieved from http://www.eurasiancommission.org/ru/act/integr_i_makroec/dep_stat/union_stat/Pages/default.aspx

European Commission. (2015). *Communication from the Commission to the European Parliament, the Council, the European Economic and Social Committee and the Committee of the Regions. Closing the Loop – An EU Action Plan for the Circular Economy.* Retrieved from https://eur-lex.europa.eu/legal-content/EN/TXT/?uri=CELEX:52015DC0614

European Commission. (2015). *Food and Framing – Focus on Jobs and Growth.* Retrieved from https://ec.europa.eu/agriculture/sites/agriculture/files/events/2015/outlook-conference/brochure-jobs-growth_en.pdf

European Commission. (2018). *Agri-Food Trade in 2017: Another Record Year for EU Agri-Food Trade.* Retrieved from https://ec.europa.eu/info/sites/info/files/food-farming-fisheries/news/documents/agricultural-trade-report_map2018-1_en.pdf

European Commission. (2019a). *Agri-Food Trade Statistical Factsheet.* Retrieved from https://ec.europa.eu/agriculture/sites/agriculture/files/trade-analysis/statistics/outside-eu/regions/agrifood-extra-eu-28_en.pdf

European Commission. (2019b). *Extra-EU Trade in Agricultural Goods.* Retrieved from https://ec.europa.eu/eurostat/statistics-explained/index.php/Extra-EU_trade_in_agricultural_goods

European Commission. (2019c). *Monitoring EU Agri-Food Trade: Development until January 2019.* Retrieved from https://ec.europa.eu/info/sites/info/files/food-farming-fisheries/trade/documents/monitoring-agri-food-trade_jan2019_en.pdf

European Commission. (2019d). *Record-Breaking Export Performance for EU Agri-Food Products.* Retrieved from https://ec.europa.eu/info/news/record-breaking-export-performance-eu-agri-food-products-2019-jan-10_en

European Commission. (2019e). *The Common Agricultural Policy at a Glance.* Retrieved from https://ec.europa.eu/info/food-farming-fisheries/key-policies/common-agricultural-policy/cap-glance_en

European Commission. (n.d.). *Agriculture and Rural Development.* Retrieved from https://ec.europa.eu/agriculture/trade-analysis/monitoring-agri-food-trade_en

European Geosciences Union. (n.d.). *General Assembly 2017-2018.* Retrieved from http://www.egu.eu

European Parliament. (2002). *Regulation (EC) #178/2002.* Retrieved from https://eur-lex.europa.eu/legal-content/EN/ALL/?uri=celex%3A32002R0178

European Parliament. (2007). *Regulation (EC) #864/2007.* Retrieved from https://eur-lex.europa.eu/legal-content/en/ALL/?uri=CELEX%3A32007R0864

European Parliament. (2008). *Directive 2008/98/EC of the European Parliament and of the Council of 19 November 2008 on Waste and Repealing Certain Directives.* Retrieved from https://eur-lex.europa.eu/legal-content/EN/TXT/?uri=celex%3A32008L0098

Eurostat. (2019). *Crop Production in EU Standard Humidity.* Retrieved from https://data.europa.eu/euodp/en/data/dataset/u33K8Gi1MFYGN7HyHUNhg

Eurostat. (n.d.a). *Comext.* Retrieved from https://ec.europa.eu/eurostat/web/international-trade-in-goods/data/focus-on-comext

Eurostat. (n.d.b). *International Trade in Goods – Overview*. Retrieved from https://ec.europa.eu/eurostat/web/international-trade-in-goods

Evsegneev, V. (2013). Unresolved Issues of Privatization of State-Owned Land in Russia. *Agrarian and Land Law*, *103*(7), 100–108.

Expert and Analytical Center for Agribusiness. (n.d.). *Agriculture of Russia*. Retrieved from https://ab-centre.ru/page/selskoe-hozyaystvo-rossii

Famine Early Warning Systems Network. (2010). *Cross-Border Livestock Trade Assessment Report: Impacts of Lifting the Livestock Import Ban on Food Security in Somalia, Ethiopia, and the Djibouti Borderland*. Washington, DC: Famine Early Warning Systems Network.

Fan, S. (1997). Production and Productivity Growth in Chinese Agriculture: New Measurement and Evidence. *Food Policy*, *22*(3), 213–228. doi:10.1016/S0306-9192(97)00010-9

Fatkhutdinov, R. (2005). *Strategic Competitiveness*. Moscow: Economy.

Federal Agency on Technical Regulating and Metrology. (2011). *Order #1575 from December 22, 2011, "On the Introduction of Interstate Standard."* Retrieved from http://docs.cntd.ru/document/902387886

Federal Customs Service of the Russian Federation. (2018). *Foreign Trade Statistics of the Russian Federation*. Retrieved from http://www.customs.ru/index.php?%20option=com_content&view=article&id=26274:20

Federal Service for Intellectual Property. (2009). *Patent #2349071. Russian Federation*. Retrieved from http://www.freepatent.ru/images/patents/116/2349071/patent-2349071.pdf

Federal Service for State Registration, Cadastre and Cartography. (n.d.). *State (National) Reports on Land Use in the Russian Federation in 1992-2015*. Moscow: Federal Service for State Registration, Cadastre and Cartography.

Federal Service for State Statistics of the Russian Federation. (2018a). *Statistics Yearbook of Russia. Statistics Digest*. Moscow: Federal Service for State Statistics of the Russian Federation.

Federal Service for State Statistics of the Russian Federation. (2018b). *Russia and Countries of the World. Statistics Digest*. Moscow: Federal Service for State Statistics of the Russian Federation.

Federal Service for State Statistics of the Russian Federation. (2019). *Russia's Foreign Trade in 2018*. Retrieved from http://www.gks.ru/

Federal Service for State Statistics of the Russian Federation. (2019). *Statistics*. Retrieved from http://www.gks.ru/

Fenner, D. C., Burge, B., Kayser, H. P., & Wittenbrink, M. M. (2006). The Anti-Microbial Activity of Electrolysed Oxidizing Water against Microorganisms Relevant in Veterinary Medicine. *Journal of Veterinary Medicine*, *53*(3), 133–137. doi:10.1111/j.1439-0450.2006.00921.x PMID:16629725

Ferapontov, A. (1994). One of Variants of Mathematical Model of Indicators of Competitiveness of Technical Production. *Standards and Quality*, *4*, 44–45.

FiBL. (n.d.). *Research Institute of Organic Agriculture*. Retrieved from https://www.fibl.org/en/homepage.html

Filippov, R. (2014). International Experience of Subsidizing of Agriculture as Basis of Food Security of Russia. *Naukovedenie*, *20*(1).

Filloon, W. (2018). *These Are All the Foods Being Affected by Trump's Trade War*. Eater. Retrieved from https://www.eater.com/2018/7/18/17527968/food-tariffs-trump-canada-china-mexico-eu

Compilation of References

Fingar, C., Loewendahl, H., Ewing, G., McMillan, C., Reynolds, J., & Rodriguez, E. ... Whitten, J. (2017). *The FDI Report 2017: Global Greenfield Investment Trends*. London: The Financial Times.

Fisinin, V., Yegorov, I., Yegorova, T., Rozanov, B., & Yudin, S. (2011). Enrichment of Eggs with Iodine. *Poultry and Poultry Products*, *4*, 37–40.

Follmer, H., Horst, U., & Kirman, A. (2005). Equilibria in Financial Markets with Heterogeneous Agents: A Probabilistic Perspective. *Journal of Mathematical Economics*, *41*(1-2), 123–155. doi:10.1016/j.jmateco.2004.08.001

Food and Agriculture Organization of the United Nations, & World Health Organization. (2019). *CODEX. Protecting Health, Facilitating Trade*. Rome: Food and Agriculture Organization of the United Nations.

Food and Agriculture Organization of the United Nations. (1951). *International Plant Protection Convention*. Retrieved from https://www.ippc.int/en/publications/1997-international-plant-protection-convention-new-revised-text/

Food and Agriculture Organization of the United Nations. (1981). *FAO: Its Origins, Formation and Evolution, 1945-1981*. Retrieved from http://www.fao.org/docrep/009/p4228e/P4228E04.htm

Food and Agriculture Organization of the United Nations. (1983). *Director-General's Report on World Food Security: A Reappraisal of the Concepts and Approaches: Item IV of the Provisional Agenda*. Rome: Food and Agriculture Organization of the United Nations.

Food and Agriculture Organization of the United Nations. (1991). *Protein Quality Evaluation*. Rome: Food and Agriculture Organization of the United Nations.

Food and Agriculture Organization of the United Nations. (1996). *Rome Declaration on World Food Security*. Retrieved from http://www.fao.org/3/w3613e/w3613e00.htm

Food and Agriculture Organization of the United Nations. (2003). *Trade Reforms and Food Security: Conceptualizing the Linkages*. Rome: Food and Agriculture Organization of the United Nations.

Food and Agriculture Organization of the United Nations. (2003). *WTO Agreement on Agriculture: The Implementation Experience. Developing Country Case Studies*. Rome: Food and Agriculture Organization of the United Nations.

Food and Agriculture Organization of the United Nations. (2006). *Food Security. Policy Brief*. Retrieved from http://www.fao.org/forestry/13128-0e6f36f27e0091055bec28ebe830f46b3.pdf

Food and Agriculture Organization of the United Nations. (2006). *International Standards for Phytosanitary Measures*. Retrieved from http://www.fao.org/3/a0450e/a0450e.pdf

Food and Agriculture Organization of the United Nations. (2009). *Declaration of the World Summit on Food Security*. Retrieved from http://www.fao.org/tempref/docrep/fao/Meeting/018/k6050e.pdf

Food and Agriculture Organization of the United Nations. (2009). *How to Feed the World in 2050*. Rome: Food and Agriculture Organization of the United Nations.

Food and Agriculture Organization of the United Nations. (2012). *Initiative on Soaring Food Prices: Ethiopia*. Rome: Food and Agriculture Organization of the United Nations.

Food and Agriculture Organization of the United Nations. (2012). *The State of Food and Agriculture: Investing in Agriculture for a Better Future*. Rome: FAO.

Food and Agriculture Organization of the United Nations. (2013). *Dietary Protein Quality Evaluation in Human Nutrition*. Rome: Food and Agriculture Organization of the United Nations.

Food and Agriculture Organization of the United Nations. (2013). *FAO Statistical Yearbook 2013. World Food and Agriculture*. Rome: Food and Agriculture Organization of the United Nations.

Food and Agriculture Organization of the United Nations. (2013). *Food Systems for Better Nutrition*. Rome: Food and Agriculture Organization of the United Nations.

Food and Agriculture Organization of the United Nations. (2013). *Food Wastage Footprint. Impacts on Natural Resources*. Rome: Food and Agriculture Organization of the United Nations.

Food and Agriculture Organization of the United Nations. (2014). *Biodiversity for Food and Agriculture: Contributing to Food Security and Sustainability in a Changing World*. Rome: Food and Agriculture Organization of the United Nations.

Food and Agriculture Organization of the United Nations. (2014). *Second International Conference on Nutrition. Rome Declaration on Nutrition*. Rome: Food and Agriculture Organization of the United Nations.

Food and Agriculture Organization of the United Nations. (2014). *Water Withdrawal and Pressure on Water Resources*. Retrieved from http://www.fao.org/nr/water/aquastat/didyouknow/index2.stm

Food and Agriculture Organization of the United Nations. (2015). *Global Initiative on Food Loss and Waste Reduction*. Rome: Food and Agriculture Organization of the United Nations.

Food and Agriculture Organization of the United Nations. (2015a). *The State of Agricultural Commodity Markets. Trade and Food Security: Achieving a Better Balance Between National Priorities and the Collective Good*. Rome: Food and Agriculture Organization of the United Nations.

Food and Agriculture Organization of the United Nations. (2015a). *Thirty-Ninth Session of the European Commission on Agriculture*. Retrieved from http://www.fao.org/europe/commissions/eca/eca-39/en/

Food and Agriculture Organization of the United Nations. (2015b). *Regional Overview of Food Security: Europe and Central Asia. Focus on Healthy and Balanced Nutrition*. Rome: Food and Agriculture Organization of the United Nations.

Food and Agriculture Organization of the United Nations. (2015b). *Towards a Water and Food Secure Future. Critical Perspectives for Policy*. Rome: Food and Agriculture Organization of the United Nations.

Food and Agriculture Organization of the United Nations. (2016). *Regional Food Security Review: Europe and Central Asia. Change the Status of Food Security*. Rome: Food and Agriculture Organization of the United Nations.

Food and Agriculture Organization of the United Nations. (2016a). *FAOSTAT*. Retrieved from http://www.fao.org/faostat/en/#data/QV

Food and Agriculture Organization of the United Nations. (2016b). *The State of Food and Agriculture. Climate Change, Agriculture and Food Security*. Rome: Food and Agriculture Organization of the United Nations.

Food and Agriculture Organization of the United Nations. (2017). *Africa. Regional Overview of Food Security and Nutrition. The Food Security and Nutrition-Conflict Nexus: Building Resilience for Food Security, Nutrition and Peace*. Accra: Food and Agriculture Organization of the United Nations.

Food and Agriculture Organization of the United Nations. (2017). *The Future of Food and Agriculture: Trends and Challenges*. Rome: Food and Agriculture Organization of the United Nations.

Food and Agriculture Organization of the United Nations. (2018). *The State of Agricultural Commodity Markets. Agricultural Trade, Climate Change and Food Security*. Rome: Food and Agriculture Organization of the United Nations.

Food and Agriculture Organization of the United Nations. (2018). *The State of Agricultural Commodity Markets: Agricultural Trade, Climate Change and Food Security*. Rome: Food and Agriculture Organization of the United Nations.

Food and Agriculture Organization of the United Nations. (2018). The State of Food Security and Nutrition in the World 2018. Building Climate Resilience for Food Security and Nutrition. Rome: Food and Agriculture Organization of the United Nations.

Food and Agriculture Organization of the United Nations. (2018). *World Food and Agriculture. Statistical Pocketbook 2018*. Rome: Food and Agriculture Organization of the United Nations.

Food and Agriculture Organization of the United Nations. (2018a). *Africa. Regional Overview of Food Security and Nutrition. Addressing the Threat from Climate Variability and Extremes for Food Security and Nutrition*. Accra: Food and Agriculture Organization of the United Nations.

Food and Agriculture Organization of the United Nations. (2018a). OECD-FAO Agricultural Outlook 2018-2027. Rome: Food and Agriculture Organization of the United Nations; Organization for Economic Cooperation and Development.

Food and Agriculture Organization of the United Nations. (2018a). *The Situation of Food Security and Nutrition in Europe and Central Asia*. Rome: Food and Agriculture Organization of the United Nations.

Food and Agriculture Organization of the United Nations. (2018b). *The State of Food Security and Nutrition in the World 2018. Building Climate Resilience for Food Security and Nutrition*. Rome: Food and Agriculture Organization of the United Nations.

Food and Agriculture Organization of the United Nations. (2019). *FAO's Work on Food Safety*. Retrieved from http://www.fao.org/publications/highlights-detail/en/c/1180272/

Food and Agriculture Organization of the United Nations. (2019). *Food Outlook – Biannual Report on Global Food Markets*. Rome: Food and Agriculture Organization of the United Nations.

Food and Agriculture Organization of the United Nations. (2019). *Save Food: Global Initiative on Food Loss and Waste Reduction*. Rome: Food and Agriculture Organization of the United Nations.

Food and Agriculture Organization of the United Nations. (n.d.). *Crops*. Retrieved from http://www.fao.org/faostat/en/#data/QC

Food and Agriculture Organization of the United Nations. (n.d.). *Russian Federation*. Retrieved from http://www.fao.org/countryprofiles/index/en/?iso3=RUS

Food and Agriculture Organization of the United Nations. International Fund for Agricultural Development, & World Food Programme. (2012). *Agricultural Cooperatives: Paving the Way for Food Security and Rural Development*. Retrieved from http://www.fao.org/3/ap088e/ap088e00.pdf

Food and Agriculture Organization of the United Nations. International Fund for Agricultural Development, United Nations International Children's Emergency Fund, World Food Programme, & World Health Organization. (2018). The State of Food Security and Nutrition in the World 2018. Building Climate Resilience for Food Security and Nutrition. Rome: Food and Agriculture Organization of the United Nations.

Food Loss + Waste Protocol. (2016a). *Food Loss and Waste Accounting and Reporting Standard*. Retrieved from https://flwprotocol.org/

Food Loss + Waste Protocol. (2016b). *Guidance on FLW Quantification Methods*. Retrieved from https://flwprotocol.org/wp-content/uploads/2016/05/FLW_Protocol_Guidance_on_FLW_Quantification_Methods.pdf

Food Security and Nutrition Working Group. (2018). *East Africa Crossborder Trade Bulletin*. Nairobi: Food Security and Nutrition Working Group.

Francois, J., Van Meijl, H., & Van Tongeren, F. (2005). Trade Liberalization in the Doha Development Round. *Economic Policy, 42*(20), 350–391. doi:10.1111/j.1468-0327.2005.00141.x

Frolova, N., & Boiko, T. (2015). Agriculture in the Context of Russia's Membership in WTO: Present-Day State, Problems, Perspectives. *Problems of Modern Economics, 56*(4), 312–315.

Frumin, G. (2006). *Geo-Ecology: Reality, Pseudoscientific Myths, Mistakes, and Delusions*. Saint Petersburg: Russian State Hydrometeorological University.

Fruska Gora National Park. (2018). *Fruškogorski manastiri*. Retrieved from http://www.strazilovo.org/page/np/manastiri.html

Gale, F., Hansen, J., & Jewison, M. (2015). *China's Growing Demand for Agricultural Imports*. Washington, DC: U.S. Department of Agriculture, Economic Research Service.

Galinovskaya, E. (2009). Land Legislation: Features of Formation and Development. *Journal of Russian Law, 155*(11), 14–25.

Gamel, T. H., Kiritsakis, A., & Petrakis, C. (1999). Effect of Phenolic Extracts on Trans-Fatty Acid Formation during Frying. *Grasas y Aceites, 50*(6), 421–425. doi:10.3989/gya.1999.v50.i6.689

Gans, K. M., & Lapane, K. (1995). Trans-Fatty Acid and Coronary Disease: The Debate Continues. 3. What Should We Tell Consumers? *American Journal of Public Health, 85*(3), 411–412. doi:10.2105/AJPH.85.3.411 PMID:7892934

Ganushchak-Yefimenko, L. (2013a). Economic Integration as a Basis for Small and Medium Enterprises Business. *Actual Problems of Economics, 141*(3), 70–77.

Ganushchak-Yefimenko, L. (2013b). Management of Innovation Potential Development of Small and Medium Business Based on Economic Integration. *Actual Problems of Economics, 144*(6), 72–79.

Gao, T. (2017). Food Security and Rural Development on Emerging Markets of Northeast Asia: Cases of Chinese North and Russian Far East. In V. Erokhin (Ed.), *Establishing Food Security and Alternatives to International Trade in Emerging Economies* (pp. 155–176). Hershey, PA: IGI Global.

Gao, T., Ivolga, A., & Erokhin, V. (2018). Sustainable Rural Development in Northern China: Caught in a Vice between Poverty, Urban Attractions, and Migration. *Sustainability, 10*(5), 1467. doi:10.3390u10051467

Gao, Y. (2017). Policy of Ensuring Food Security of China in the 1990s. *Nauchnyy Dialog, 10*(10), 201–207. doi:10.24224/2227-1295-2017-10-201-207

Garankin, N., & Komov, N. (2005). Rent as a Key Issue of Civil Turnover of Land. *Land Management. Land Monitoring and Cadaster, 12*, 54–56.

Garcia-Garcia, G., Stone, J., & Rahimifard, S. (2019). Opportunities for Waste Valorisation in the Food Industry – A Case Study with Four UK Food Manufacturers. *Journal of Cleaner Production, 211*, 1339–1356. doi:10.1016/j.jclepro.2018.11.269

Gassenmeier, K., & Schieberle, P. (1994). Formation of the Intense Flavor Compound Trans-4,5-Epoxy-(E)-2-Decenal in Thermally Treated Fats. *Journal of the American Oil Chemists' Society, 71*(12), 1315–1319. doi:10.1007/BF02541347

Gavrilyuk, O., Gaidaenko-Sher, I., & Merkulova, T. (2017). *Agrarian Legislation of Foreign Countries and Russia*. Moscow: Institute of Legislation and Comparative Law under the Government of the Russian Federation.

Gebbers, R., & Adamchuk, V. (2010). Precision Agriculture and Food Security. *Science, 327*(5967), 828–831. doi:10.1126cience.1183899 PMID:20150492

Compilation of References

Gebreegziabher, Z., Stage, J., Mekonnen, A., & Alemu, A. (2011). Climate Change and the Ethiopian Economy: A Computable General Equilibrium Analysis. *Environment and Development Economics, 21*(2), 1–21.

Geistfeld, M. A. (2006). The Doctrinal Unity of Alternative Liability and Market- Share Liability. *University of Pennsylvania Law Review, 155*(2), 447–501. doi:10.2307/40041311

General Directorate for Land Management and Agriculture. (1914). *Atlas of Asian Russia*. Saint Petersburg: A.F. Marks.

Ghironi, F., & Melitz, M. (2005). International Trade and Macroeconomic Dynamics with Heterogeneous Firms. *The Quarterly Journal of Economics, 120*(3), 865–915.

Ghose, B. (2014). Promoting Agricultural Research and Development to Strengthen Food Security in South Asia. *International Journal of Agronomy*. doi:10.1155/2014/589809

Ghosh, I., & Ghoshal, I. (2017). Implications of Trade Liberalization for Food Security under the ASEAN-India Strategic Partnership: A Gravity Model Approach. In V. Erokhin (Ed.), *Establishing Food Security and Alternatives to International Trade in Emerging Economies* (pp. 98–118). Hershey, PA: IGI Global.

Gibbs, D., & Muirhead, I. (1998). *The Economic Value and Environmental Impact of the Australian Beekeeping Industry*. Canberra: The Australian Honeybee Industry Council.

Gifford, D. G. (2005). The Challenge to the Individual Causation Requirement in Mass Products Torts. *Washington and Lee Law Review, 62*(3), 873–935.

Gintis, H. (2006). *The Economy as a Complex Adaptive System*. Retrieved from https://www.semanticscholar.org/paper/The-Economy-as-a-Complex-Adaptive-System-Gintis/dd497297c0745c744f07abb9da632439fe748bb5

Gintis, H. (2009). *Game Theory Evolving*. Princeton, NJ: Princeton University Press. doi:10.2307/j.ctvcm4gjh

Giro, T., & Chirkova, O. (2007). Meat Foodstuffs with Plant Ingredients for Functional Nutrition. *Meat Industry, 1*, 43–46.

Gitelman, C. (1994). *Methodical Recommendations of Institutions of Uzbekistan, the Russian Federation, and Ukraine on the Use of Electroactivated Aqueous Solutions for the Prevention and Treatment of the Most Common Human Diseases*. Moscow: Espero.

Glaeser, E. L., Sacerdote, B., & Scheinkman, J. (1996). Crime and Social Interactions. *The Quarterly Journal of Economics, 111*(2), 507–548. doi:10.2307/2946686

Global Water Partnership. (2017). *Virtual Water (C8.04)*. Retrieved from https://www.gwp.org/en/learn/iwrm-toolbox/Management-Instruments/Promoting_Social_Change/Virtual_water/

Gnedenko, E., & Kazmin, M. (2015). Agricultural Land and Regulation in the Transition Economy of Russia. *International Advances in Economic Research, 21*(3), 347–348. doi:10.100711294-015-9535-y

Golam, M., & Monowar, M. (2018). Eurasian Economic Union: Evolution, Challenges and Possible Future Directions. *Journal of Eurasian Studies, 9*(2), 163–172. doi:10.1016/j.euras.2018.05.001

Goldman Sachs. (2013). *The Search for Creative Destruction*. New York, NY: Goldman Sachs.

Golohvast, K., Ryzhakov, D., Chajka, V., & Gulikov, A. (2011). Application Potential of Solution Electrochemical Activation. *Water: Chemistry and Ecology, 2*, 23–30.

Gomez, M. I., & Ricketts, K. D. (2013). Food Value Chain Transformations in Developing Countries: Selected Hypotheses on Nutritional Implications. *Food Policy, 42*, 139–150. doi:10.1016/j.foodpol.2013.06.010

Goncalves, J. R. B., & Madi, M. A. C. (2013). Global Economic Integration, Business Expansion and Consumer Credit in Brazil, 1994-2010. *International Journal of Green Economics*, *7*(3), 213–225. doi:10.1504/IJGE.2013.058164

Gong, C., & Kim, S. (2013). Economic Integration and Business Cycle Synchronization in Asia. *Asian Economic Papers*, *12*(1), 76–99. doi:10.1162/ASEP_a_00188

Gorbatov, A. (2016). The Development of the Market for Organic Products in Russia. *Fundamental Research*, *11*(1), 154–158.

Gordeeva, E., Shoeva, O., Yudina, R., Kukoeva, T., & Khlestkina, E. (2016). Effect of Seed Pre-Sowing Gamma-Irradiation Treatment in Bread Wheat Lines Differing by Anthocyanin Pigmentation. *Cereal Research Communications*, *46*(1), 41–53. doi:10.1556/0806.45.2017.059

Gorlov, I., Slozhenkina, M., Danilov, Y., Semenova, I., & Miroshnik, A. (2018). Use of a New Food Ingredient in the Production of Meat Products of Functional Purposes. *Bulletin of Lower Volga Agrouniversity Complex*, *52*(4), 219–229.

Goryushkin, L., Bocharova, G., & Nozdrin, G. (1993). *Experience of Traditional Agriculture in Siberia (Second Half of XIX – Beginning of XX Century)*. Novosibirsk: Novosibirsk State University.

Gotz, C. (2017). *Upswing in Agricultural Machinery Industry*. Retrieved from https://lt.vdma.org/en/viewer/-/v2article/render/19870108

Government of France. (2016). *French Civil Code 2016*. Retrieved from https://www.trans-lex.org/601101/_/french-civil-code-2016/

Government of the People's Republic of China. (1993). *Product Quality Law of the People's Republic of China*. Retrieved from http://english.mofcom.gov.cn/article/policyrelease/Businessregulations/201303/20130300046024.shtml

Government of the People's Republic of China. (2006). *Law "On Quality and Safety of Agricultural Products."* Retrieved from https://www.fsvps.ru/fsvps-docs/ru/importExport/china/files/china_law_quality.pdf

Government of the People's Republic of China. (2009). *Law "On Food Safety."* Retrieved from https://www.fsvps.ru/fsvps-docs/ru/importExport/china/files/zakon1.pdf

Government of the People's Republic of China. (2010). *Tort Law of the People's Republic of China*. Retrieved from https://www.wipo.int/edocs/lexdocs/laws/en/cn/cn136en.pdf

Government of the People's Republic of China. (2013). *Law "On Protection of Consumers' Rights."* Retrieved from https://chinalaw.center/civil_law/china_consumer_rights_protection_law_revised_2013_russian/

Government of the Russian Federation. (1968). *Interstate Standard "Polished Pea. Specifications."* Retrieved from http://docs.cntd.ru/document/1200022305

Government of the Russian Federation. (1976). *Interstate Standard "Food Beans. Specifications."* Retrieved from http://docs.cntd.ru/document/1200023726

Government of the Russian Federation. (1977). *Order #755 from August 12, 1977, "On the Measures on Further Development of Organizational Forms of Work with the Use of Experimental Animals."* Retrieved from https://docplayer.ru/31723947-Ministerstvo-zdravoohraneniya-sssr-prikaz-12-avgusta-1977-g-n-755.html

Government of the Russian Federation. (1990). *Law #374-1 from November 23, 1990, "On Land Reform."* Retrieved from https://base.garant.ru/10107009/

Government of the Russian Federation. (1991). *Decree #86 from December 29, 1991, "On the Reorganization of Collective and State Farms."* Retrieved from http://base.garant.ru/10104699/

Compilation of References

Government of the Russian Federation. (1991). *Interstate Standard "Pea. Requirements for State Purchases and Deliveries."* Retrieved from http://docs.cntd.ru/document/gost-28674-90

Government of the Russian Federation. (1992). *Decree #708 from September 4, 1992, "On the Privatization and Reorganization of Agro-Industrial Enterprises and Organizations."* Retrieved from http://pravo.gov.ru/proxy/ips/?docbody=&nd=102018291&rdk=&backlink=1

Government of the Russian Federation. (1992). *Law #2300-I from February 7, 1992, "On Protection of Consumers' Rights."* Retrieved from http://base.garant.ru/10106035/

Government of the Russian Federation. (1993). *The Constitution of the Russian Federation.* Retrieved from http://www.constitution.ru/en/10003000-01.htm

Government of the Russian Federation. (1993a). *Constitution of the Russian Federation.* Retrieved from http://www.constitution.ru/en/10003000-01.htm

Government of the Russian Federation. (1993b). *Law #4979 from May 14, 1993, "On Veterinary Medicine."* Retrieved from http://base.garant.ru/10108225/

Government of the Russian Federation. (1994). *Decree # 874 from July 27, 1994, "On the Reforming of Agricultural Enterprises with Account of the Experience of Nizhny Novgorod Oblast."* Retrieved from http://www.consultant.ru/document/cons_doc_LAW_4234/

Government of the Russian Federation. (1995). *Decree #96 from February 1, 1995, "On the Procedure of Execution of Rights of the Owners of Land Shares and Property Shares."* Retrieved from http://www.consultant.ru/document/cons_doc_LAW_5724/

Government of the Russian Federation. (1996). *Law #14 from January 26, 1996, "Civil Code of the Russian Federation."* Retrieved from https://legalacts.ru/kodeks/GK-RF-chast-2/

Government of the Russian Federation. (1997). *Decree #1263 from September 29, 1997, "On the Regulations on the Examination of Low-Quality and Dangerous Food Raw Materials and Food Products, Their Use or Destruction."* Retrieved from https://legalacts.ru/doc/postanovlenie-pravitelstva-rf-ot-29091997-n-1263/

Government of the Russian Federation. (1999). *Law #52 from March 30, 1999, "On Sanitary and Epidemiological Wellbeing of Population."* Retrieved from http://base.garant.ru/12115118/

Government of the Russian Federation. (1999a). *National Standard "Vegetable Oils and Animal Fats. Determination by Gas Chromatography of Constituent Contents of Methyl Esters of Total Fatty Acid Content."* Retrieved from http://docs.cntd.ru/document/gost-r-51483-99

Government of the Russian Federation. (1999b). *National Standard "Vegetable Oils and Animal Fats. Preparation of Methyl Esters of Fatty Acids."* Retrieved from http://docs.cntd.ru/document/gost-r-51486-99

Government of the Russian Federation. (2000a). *Decree #883 from November 22, 2000, "On the Organization and Conduct of Quality, Food Safety and Public Health Monitoring."* Retrieved from https://legalacts.ru/doc/postanovlenie-pravitelstva-rf-ot-22112000-n-883/

Government of the Russian Federation. (2000b). *Law #29 from January 2, 2000, "On Quality and Safety of Food Products."* Retrieved from http://base.garant.ru/12117866/

Government of the Russian Federation. (2002). *Federal Law of the Russian Federation #101 from July 24, 2002, "On the Turnover of Agricultural Land."* Retrieved from http://base.garant.ru/12127542/

Government of the Russian Federation. (2002). *Law #184 from December 27, 2002, "On Technical Regulation."* Retrieved from http://base.garant.ru/5139626/

Government of the Russian Federation. (2003a). *National Standard "Margarines, Cooking Fats, Fats for Confectionery, Baking and Dairy Industry. Sampling Rules and Methods of Control."* Retrieved from http://docs.cntd.ru/document/1200036186

Government of the Russian Federation. (2003b). *National Standard "Spreads and Melted Blends. General Specifications."* Retrieved from http://docs.cntd.ru/document/1200032516

Government of the Russian Federation. (2005). *Decree #50 from February 2, 2005, "On the Regulations on the Use of Means and Methods of Control in the Implementation of the Passage of Persons, Vehicles, Cargoes, Goods, and Animals Across the State Border of the Russian Federation."* Retrieved from http://pravo.gov.ru/proxy/ips/?docbody=&nd=102090891

Government of the Russian Federation. (2008). *Norms of Physiological Requirements in Energy and Feedstuffs for Various Groups of Population in the Russian Federation.* Retrieved from http://docs.cntd.ru/document/1200076084

Government of the Russian Federation. (2008a). *National Standard "Animal and Vegetable Fats and Oils and Their Derivatives. Determination of Solid Fat Content. Pulsed Nuclear Magnetic Resonance Method."* Retrieved from http://docs.cntd.ru/document/1200074555

Government of the Russian Federation. (2008b). *Norms of Physiological Requirements in Energy and Feedstuffs for Various Groups of Population in the Russian Federation.* Retrieved from http://docs.cntd.ru/document/1200076084

Government of the Russian Federation. (2010). *Decree #1806 from October 18, 2010, "On the Approval of the Comprehensive Program for the Participation of the Russian Federation in International Cooperation in the Field of Agriculture, Fisheries and Food Security."* Retrieved from https://rulaws.ru/goverment/Rasporyazhenie-Pravitelstva-RF-ot-18.10.2010-N-1806-r/

Government of the Russian Federation. (2010). *Decree #1873 from October 25, 2010, "Foundations of State Policy of the Russian Federation in the Sphere of Healthy Nutrition of the Population till 2020."* Retrieved from https://rmapo.ru/medical/58-osnovy-gosudarstvennoy-politiki-rossiyskoy-federacii-v-oblasti-zdorovogo-pitaniya-naseleniya-na-period-do-2020-goda.html

Government of the Russian Federation. (2010). *Decree #2136-r from November 30, 2010, On Approval of the Concept of Sustainable Development of Rural Areas in the Russian Federation till 2020.* Retrieved from https://rg.ru/2010/12/14/sx-territorii-site-dok.html

Government of the Russian Federation. (2011). Decree *#501 from June 29, 2011, "On the Approval of the Rules for the Implementation of State Veterinary Supervision at Border Crossing Points of the Russian Federation."* Retrieved from http://pravo.gov.ru/proxy/ips/?docbody=&nd=102148718

Government of the Russian Federation. (2013). *Decree #839 from September 23, 2013, "On State Registration of Genetically Modified Organisms Intended for Release into the Environment, as well as Products Obtained Using Such Organisms or Containing Such Organisms, Including Specified Products Imported into the Territory of the Russian Federation."* Retrieved August 14, 2019, from http://www.consultant.ru/document/cons_doc_LAW_152217/

Government of the Russian Federation. (2014). *Law #206 from July 21, 2014, "On Plant Quarantine."* Retrieved from http://base.garant.ru/70699630/

Government of the Russian Federation. (2015). *Federal Law #11-FZ from February 12, 2015, "On the Amendments to Article 14 of the Federal Law "On Development of Agriculture.""* Retrieved from http://base.garant.ru/70866588/

Government of the Russian Federation. (2017). *Order #1455-r from July 7, 2017, "Strategy of Development of Agricultural Engineering in Russia".* Retrieved from http://government.ru/docs/28393/

Government of the Russian Federation. (2019). *National Project "International Cooperation and Export."* Retrieved from http://government.ru/rugovclassifier/866/events/

Grain. (2014). *Growing Corporate Hold on Farmland Risky for World Food Security.* Retrieved from https://ourworld.unu.edu/en/growing-corporate-hold-on-farmland-risky-for-world-food-security

Grand View Research. (2018). *Agricultural Machinery Market Analysis, Market Size, Application Analysis, Regional Outlook, Competitive Strategies, and Segment Forecasts, 2016 to 2024.* Retrieved from https://www.grandviewresearch.com/industry-analysis/agricultural-machinery-market/methodology

Granovetter, M. (1985). Economic Action and Social Structure: The Problem of Embeddedness. *American Journal of Sociology, 91*(3), 481–510. doi:10.1086/228311

Green, R. E., Cornell, S. J., Scharlemann, J. P. W., & Balmford, A. (2005). Farming and the Fate of Wild Nature. *Science, 307*(5709), 550–555. doi:10.1126cience.1106049 PMID:15618485

Greenville, J., Kawasaki, K., & Jouanjean, M. (2019). *Dynamic Changes and Effects of Agro-Food GVCS.* Paris: Organization for Economic Cooperation and Development.

Grishkova, Y., & Poluhin, I. (2014). Interaction Problems of Customs and Logistics and Transportation. *Reshetnikov's Reading, 18*, 433–434.

Gro Intelligence. (2018). *China's Road Map to Food Security.* Retrieved from https://gro-intelligence.com/insights/chinas-roadmap-to-food-security

Grubić, R. (2008). Pčelarstvo u Zrenjaninu. *Rad muzeja Vojvodine, 50*, 273-283.

Guseva, N. (2015). *Interview with Helena Bollesen.* Retrieved from https://lookbio.ru/obtshestvo/bio-portret/v-organike-rossiya-sejchas-kak-daniya-dvadcat-let-nazad-intervyu-s-xelenoj-bollesen/

Gusev, H., & Shpagina, O. (1981). *On the Importance of Studies of the State of Water. Issues of Water Exchange and Water Status in Plants.* Moscow: USSR Academy of Sciences.

Hadgu, G., Fantaye, K.T., Mamo, G., & Kassa, B. (2013). Trend and Variability of Rainfall in Tigray, Northern Ethiopia: Analysis of Meteorological Data and Farmers' Perception. *Academia Journal of Agricultural Research, 6*(1), 88-100.

Haggblade, S., Hazell, P., & Reardon, T. (2010). The Rural Non-Farm Economy: Prospects for Growth and Poverty Reduction. *World Development, 38*(10), 1429–1441. doi:10.1016/j.worlddev.2009.06.008

Haggblade, S., Me-Nsope, N. M., & Staatz, J. M. (2017). Food Security Implications of Staple Food Substitution in Sahelian West Africa. *Food Policy, 71*, 27–38. doi:10.1016/j.foodpol.2017.06.003

Hailu, T., Sala, E., & Seyoum, W. (2016). Challenges and Prospects of Agricultural Marketing in Konta Special District, Southern Ethiopia. *Journal of Marketing and Consumer Research, 28*, 1–7.

Halimi, R. A., Barkla, B. J., Mayes, S., & King, G. J. (2019). The Potential of the Underutilized Pulse Bambara Groundnut (*Vigna Subterranean (L.) Verdc.*) for Nutritional Food Security. *Journal of Food Composition and Analysis, 77*, 47–59. doi:10.1016/j.jfca.2018.12.008

Hallaert, J.-J., Cavazos-Cepeda, R. H., & Kang, G. (2011). *Estimating the Constraints to Trade of Developing Countries.* Paris: Organization for Economic Cooperation and Development.

Hausmann, R., Hidalgo, C., Bustos, S., Coscia, M., Simoes, A., & Yildirim, M. (2014). *The Atlas of Economic Complexity: Mapping Paths to Prosperity*. Cambridge, MA: MIT Press. doi:10.7551/mitpress/9647.001.0001

Havlík, P., Valin, H., Gusti, M., Schmid, E., Forsell, N., & Herrero, M. ... Obersteiner, M. (2015). Climate Change Impacts and Mitigation in the Developing World: An Integrated Assessment of the Agriculture and Forestry Sectors. Washington, DC: World Bank Group.

HeinOnline. (n.d.). *Restatement, Third, Torts: Liability for Physical and Emotional Harm*. Retrieved from https://home.heinonline.org/titles/American-Law-Institute-Library/Restatement-Third-Torts-Liability-for-Physical-and-Emotional-Harm/?letter=R

Heredia, A., Barrera, C., & Andrés, A. (2007). Drying of Cherry Tomato by a Combination of Different Dehydration Techniques. Comparison of Kinetics and Other Related Properties. *Journal of Food Engineering*, *80*(1), 111–118. doi:10.1016/j.jfoodeng.2006.04.056

Herrington, E. (1965). *Zone Melting of Organic Compounds*. Moscow: Peace.

History of Russia. (n.d.). *Economic Reform of 1965*. Retrieved from https://istoriarusi.ru/cccp/ekonomicheskie-reformi-1965-goda.html

Hoboetc. (n.d.). *Degassing Is... How Degassing of Water Is Done. Methods of Degassing*. Retrieved from https://hoboetc.com/biznes/1955-degazaciya-eto-kak-provoditsya-degazaciya-vody-sposoby-degazacii.html

Hodgson, G. M. (1998). The Approach of Institutional Economics. *Journal of Economic Literature*, *36*(1), 166–192.

Hoekstra, A. Y. (2010). *The Relation between International Trade and Freshwater Scarcity*. Geneva: World Trade Organization.

Hoekstra, A. Y., & Hung, P. (2002). *Virtual Water Trade: A Quantification of Virtual Water Flows between Nations in Relation to International Crop Trade*. Delft: IHE Delft Institute for Water Education.

Howard, A. (n.d.). *An Agricultural Testament*. Retrieved from http://www.journeytoforever.org/farm_library/howardAT/AT1.html

Howling Pixel. (n.d.). *Mokichi Okada*. Retrieved from https://howlingpixel.com/i-en/Mokichi_Okada

Hozayn, M., & Qados, A. M. S. A. (2010). Irrigation with Magnetized Water Enhances Growth, Chemical Constituent and Yield of Chickpea (*Cicer Arietinum L.*). *Agriculture and Biology Journal of North America*, *1*(4), 671–676.

Huang, H., Yun, Z., You, L., & Wu, J. (2011). Forecast of Subsidy for Purchasing Agricultural Machinery Based on Life Cycle Theory in China. *Paper presented at the International Conference on Management and Service Science*, Wuhan. Academic Press. 10.1109/ICMSS.2011.5998516

Huang, J., Wei, W., Cui, Q., & Xie, W. (2017). The Prospects for China's Food Security and Imports: Will China Starve the World via Imports? *Journal of Integrative Agriculture*, *16*(12), 2933–2944. doi:10.1016/S2095-3119(17)61756-8

Huang, J., & Yang, G. (2017). Understanding Recent Challenges and New Food Policy in China. *Global Food Security*, *12*, 119–126. doi:10.1016/j.gfs.2016.10.002

Huang, Y.-R., Hung, Y.-C., Hsu, S.-Y., Huang, Y.-W., & Hwang, D.-F. (2008). Application of Electrolyzed Water in the Food Industry. *Food Control*, *19*(4), 329–345. doi:10.1016/j.foodcont.2007.08.012

Hufbauer, G. C., & Cimino-Isaacs, C. (2015). How Will TPP and TTIP Change the WTO System? *Journal of International Economic Law*, *18*(3), 679–696. doi:10.1093/jiel/jgv036

Compilation of References

Hutchinson, R. (1984). *The Carp Strikes Back*. Henlow Camp. Beekay Publishers.

Ibragimov, M.-T., & Dokholyan, S. (2010). Methodological Approaches to the Assessment of Food Security of the Region. *Regional Problems of Economic Transformation, 26*(4), 172–193.

Ignjatijević, S., & Milojević, I. (2011). Kvalitet poljoprivredno-prehrambenih proizvoda kao faktor konkurentnosti na međunarodnom tržištu. *Proceedings of the 14 ICDQM International Conference*. Belgrade: Istraživački centar za upravljanje kvalitetom i pouzdanošću.

Ignjatijević, S., Ćirić, M., & Carić, M. (2013). International Trade Structure of Countries from the Danube Region: Comparative Advantage Analysis of Export. *Ekonomicky Casopis, 61*(3), 251–269.

Ignjatijević, S., Ćirić, M., & Ĉavlin, M. (2015). Analysis of Honey Production in Serbia Aimed at Improving the International Competitiveness. *Custos e Agronegocio, 11*(2), 194–213.

Ignjatijević, S., & Cvijanović, D. (2018). *Exploring the Global Competitiveness of Agri-Food Sectors and Serbia's Dominant Presence: Emerging Research and Opportunities*. Hershey, PA: IGI Global. doi:10.4018/978-1-5225-2762-6

Ignjatijević, S., Milojević, I., & Andžić, R. (2018). Economic Analysis of Exporting Serbian Honey. *The International Food and Agribusiness Management Review, 21*(7), 929–944. doi:10.22434/IFAMR2017.0050

Ignjatijević, S., Prodanović, R., Bošković, J., Puvača, N., Tomaš Simin, M., Peulić, T., & Đuragić, O. (2019). Comparative Analysis of Honey Consumption in Romania, Italy and Serbia. *Food & Feed Research, 46*(1), 125–136. doi:10.5937/FFR1901125I

India Brand Equity Foundation. (2019). *Indian Economy*. Retrieved from https://www.ibef.org/economy.aspx

India Times. (2018). *India Imports 50.8 Lakh Ton Pulses for Rs 17,280 Crore in April-December*. Retrieved from https://economictimes.indiatimes.com/news/economy/agriculture/india-imports-50-8-lakh-ton-pulses-for-rs-17280-crore-in-april-december/articleshow/62808888.cms

Interfax. (2015). *Putin Called Russia the World's Largest Potential Supplier of Ecologically Clean Products*. Retrieved from https://www.interfax.ru/business/482981

Intergovernmental Panel on Climate Change. (2007). *AR4 Climate Change 2007: The Physical Science Basis*. Cambridge: Cambridge University Press.

International Co-operative Alliance. (n.d.). *Statistical Information on the Cooperative Movement*. Retrieved from https://www.ica.coop/en/about-us/international-cooperative-alliance

International Food Policy Research Institute. (2018). *2018 Global Food Policy Report*. Washington, DC: International Food Policy Research Institute.

International Fund for Agricultural Development. (2010). *Rural Poverty Report: New Realities, New Challenges, New Opportunities for Tomorrow's Generation*. Rome: International Fund for Agricultural Development.

International Fund for Agricultural Development. (2017). *South-South and Triangular Cooperation (SSTC). Highlights from IFAD's Portfolio*. Rome: International Fund for Agricultural Development.

International Independent Institute of Agrarian Policy. (2016). *Structure of the World Food Market*. Retrieved from http://xn--80aplem.xn--p1ai/analytics/Struktura-mirovogo-prodovolstvennogo-rynka/

International Independent Institute of Agrarian Policy. (2017). *Global Market of Legume Crops*. Retrieved from http://xn--80aplem.xn--p1ai/analytics/Mirovoj-rynok-bobovyh-kultur/

International Labor Association. (2007). *COOP Fact Sheet #1. Cooperatives and Rural Employment*. Geneva: International Labor Association.

International Monetary Fund. (2018). *World Economic Outlook Update*. Washington, DC: International Monetary Fund.

International Monetary Fund. (2019). *World Economic Outlook (April 2019)*. Washington, DC: International Monetary Fund.

International Organization for Standardization. (2005). *ISO 9000:2005 "Quality Management Systems – Fundamentals and Vocabulary."* Retrieved from https://www.iso.org/standard/42180.html

International Organization for Standardization. (2008). *ISO 8292-1:2008 "Animal and Vegetable Fats and Oils – Determination of Solid Fat Content by Pulsed NMR – Part 1: Direct Method."* Retrieved from https://www.iso.org/standard/41256.html

Istituto di Servizi per il Mercato Agricolo Alimentare. (2014). *Patate: nel 2014 la produzione italiana cresce del 20%*.

Istituto di Servizi per il Mercato Agricolo Alimentare. (2019). *Osservatorio patate. Prezzi all'origine. Trend annui*. Retrieved from http://www.ismeamercati.it/flex/cm/pages/ServeBLOB.php/L/IT/IDPagina/4845#MenuV

Ivanov, V., & Sapunov, G. (2003). *Patent #2195824 "Method of Drying of Fruits and Vegetables."* Retrieved August 11, 2019, from http://allpatents.ru/patent/2195824.html

Ivanter, A., Khazbiev, A., & Yakovenko, D. (2014). *Time to Change Stereotypes*. Retrieved from https://expert.ru/expert/2014/49/pora-menyat-stereotipyi/

Ivolga, A. (2016). Sustainable Rural Development in the Conditions of Trade Integration: From Challenges to Opportunities. In V. Erokhin (Ed.), *Global Perspectives on Trade Integration and Economies in Transition* (pp. 262–280). Hershey, PA: IGI Global. doi:10.4018/978-1-5225-0451-1.ch013

Iwuchukwu, J. C., & Igbokwe, E. M. (2012). Lessons from Agricultural Policies and Programmes in Nigeria. *Journal of Law. Policy and Globalisation, 5*, 11–21.

Jespersen, L. M., Baggesen, D. L., Fog, E., Halsnaes, K., Hermansen, J. E., Andreasen, L., ... Halberg, N. (2017). Contribution of Organic Farming to Public Goods in Denmark. *Organic Agriculture, 7*(3), 243–266. doi:10.100713165-017-0193-7

Johnson, D. S. (2013). *Contributions of Co-operative to Agricultural Development: A Study of Agricultural Co-operatives in Awka North Local Government Area*. Nnamdi Azikiwe University.

Johnson, N., Njuki, J., Waithanji, E., Nhambeto, M., Rogers, M., & Kruger, E. H. (2015). The Gendered Impacts of Agricultural Asset Transfer Projects: Lessons from the Manica Smallholder Dairy Development Program. *Gender, Technology and Development, 19*(2), 145–180. doi:10.1177/0971852415578041

Jones, C. (2017). *Agricultural Opportunities in Africa: Crop Farming in Ethiopia, Nigeria and Tanzania*. London: Deloitte.

Jones, M. A. (2007). *Textbook on Torts*. Oxford: Oxford University Press.

Joob, B., & Wiwanitkit, V. (2017). Estimation of Cancer Risk Due to Exposure to Airborne Particle Emissions of a Commercial Three-Dimensional Printer. *Indian Journal of Medical and Paediatric Oncology: Official Journal of Indian Society of Medical & Paediatric Oncology, 38*(3), 409. doi:10.4103/ijmpo.ijmpo_118_17

Jorissen, J., Priefer, C., & Brautigam, K.-R. (2015). Food Waste Generation at Household Level: Results of a Survey among Employees of Two European Research Centers in Italy and Germany. *Sustainability, 7*(3), 2695–2715. doi:10.3390u7032695

Compilation of References

Josling, T., Anderson, K., Schmitz, A., & Tangerman, S. (2010). Understanding International Trade in Agricultural Products: One Hundred Years of Contributions by Agricultural Economists. *American Journal of Agricultural Economics*, 92(2), 424–446. doi:10.1093/ajae/aaq011

Just, F. (1990). Butter, Bacon and Organisational Power in Danish Agriculture. In F. Just (Ed.), *Co-operatives and Farmers' Unions in Western Europe – Collaboration and Tension* (pp. 137–156). Esbjerg: South Jutland University Press.

Kapitsa, S. (2012). Model of Earth Population Dynamics and Future of Humankind. *Svobodnaya Mysl*, 7-8, 141–152.

Karmas, E., & Harris, R. S. (1988). *Nutritional Evaluation of Food Processing*. Amsterdam: Springer Netherlands. doi:10.1007/978-94-011-7030-7

Kazantseva, I. (2016). *Scientific Foundation and Elaboration of Technological Solutions of Complex Processing of Chickpea Seeds for the Development of Healthy Food Products for People in Russia*. Saratov: Yuri Gagarin State Technical University of Saratov.

Kazydub, N., & Marakaeva, T. (2015). *Comparative Assessment of Economically Valuable Qualities of Haricot Samples (Phaseolus Vulgaris L.) and Development of New Selection Material on Their Basis in the Territory of the Southern Forest-Steppe of Western Siberia*. Omsk: Kant.

Kazymov, S., & Prudnikova, T. (2012). Germination Influence on Amino Acids Composition of Mash Beans. *News of Institutes of Higher Education. Food Technology*, 5-6, 25–26.

Ke La Xin, B. N. (1982). *Magnetization of Water*. Beijing: Measurement Press.

Keenan, R. J., Reams, G. A., Achard, F., de Freitas, J. V., Grainger, A., & Lindquist, E. (2015). Dynamics of Global Forest Area: Results from the FAO Global Forest Resources Assessment. *Forest Ecology and Management*, 352, 9–20. doi:10.1016/j.foreco.2015.06.014

Kennedy, C., Zhong, M., & Corfee-Morlot, J. (2016). Infrastructure for China's Ecologically Balanced Civilization. *Engineering*, 2(4), 414–425. doi:10.1016/J.ENG.2016.04.014

Khafizova, A., Galimardanova, Y., & Salmina, S. (2014). Tax Regulation of Activity of Agricultural Commodity Producers. *Mediterranean Journal of Social Sciences*, 24(5), 421–425.

Khalilov, H., Shalbuzov, N., & Huseyn, R. (2015). *AGRICISTRADE Country Report: Azerbaijan*. Retrieved from http://www.agricistrade.eu/document-library

Kharitonov, S. (2013). *Organization and Economic Aspects of Development of Organic Agriculture in Russia*. Moscow: Lomonosov Moscow State University.

Khlystun, V. (2005). Controversial Issues of Land Relation Development in Russia. In S. Volkov & A. Varlamov (Eds.), *Science and Education of Land Management in Russia at the Beginning of the Third Millennium* (pp. 82–91). Moscow: State University of Land Use Planning.

Khlystun, V. (2005). Structural Transformation and Development of Land Relations. *Farming*, 3, 20–21.

Khlystun, V. (2011). On State Land Policy. *Economy of Agricultural and Processing Enterprises*, 10, 1–4.

Khlystun, V. (2012a). Agrarian Transformations in Post-Soviet Russia (20th Anniversary of Agricultural Reform). *Economy of Agricultural and Processing Enterprises*, 6, 17–21.

Khlystun, V. (2012b). Land Relations in Agricultural Industry of Russia. *Domestic Notes*, 6, 78–84.

Khlystun, V. (2015). Quarter of a Century of Land Transformations: Intentions and Results. *Economy of Agricultural and Processing Enterprises, 10*, 13–17.

Khlystun, V., & Alakoz, V. (2016). Tools of Circulation of Unused Agricultural Land. *Economy of Agricultural and Processing Enterprises, 11*, 38–42.

Khudzhatov, M. (2017). The Study of Differentiation of Foreign Trade Prices by Using of Dispersion Analysis. *RUDN Journal of Economics, 25*(1), 91–101. doi:10.22363/2313-2329-2017-25-1-91-101

Khudzhatov, M. (2018a). *Enhancement of Customs Instruments of Promotion of Foreign Investments in Agricultural Mechanical Engineering in Russia*. Moscow: DPK Press.

Khudzhatov, M. (2018b). The Use of Customs Instruments for Stimulation of Foreign Investment in Agricultural Machinery in Russia. *Marketing and Logistics, 15*(1), 58–70.

Kidane, W., Maetz, M., & Dardel, P. (2006). *Food Security and Agricultural Development in Sub-Saharan Africa. Building a Case for More Public Support*. Rome: Food and Agriculture Organization of the United Nations.

Kireev, V. (1975). *Course of Physical Chemistry*. Moscow: Chemistry.

Klassen, V. (1982). *Magnetization of Water Systems*. Moscow: Nauka.

Kleinau, C., & Lin-Hi, N. (2014). Does Agricultural Commodity Speculation Contribute to Sustainable Development? *Corporate Governance, 14*(5), 685–698. doi:10.1108/CG-07-2014-0083

Klimova, J. (2011). Pilgrimages of Russian Orthodox Christians to the Greek Orthodox Monastery in Arizona. *Tourism: An International Interdisciplinary Journal, 59*(3), 305–318.

Kney, A. D., & Parsons, S. A. (2006). A Spectrophotometer-Based Study of Magnetic Water Treatment: Assessment of Ionic vs. Surface Mechanisms. *Water Research, 40*(3), 517–524. doi:10.1016/j.watres.2005.11.019 PMID:16386285

Kocira, A., Kornas, N., & Kocira, S. (2013). Effect Assessment of Kelpak SL on the Bean Yield (*Phaseolus Vulgaris L.*). *Journal of Central European Agriculture, 14*(2), 545–554. doi:10.5513/JCEA01/14.2.1234

Kofod, A. (1914). *Land Management in Russia*. Saint Petersburg: Selsky Vestnik.

Kokin, A., & Kokin, A. (2008). *Modern Environmental Myths and Utopias*. Saint Petersburg.

Kolesnikova, N. (2006). *The Development of Technology and Assessment of Consumers' Qualities of Foodstuffs on the Basis of Haricot for Pupils*. Krasnodar: Kuban State Technological University.

Kolodko, G. (1998). *Equity Issues in Policymaking in Transition Economies*. Washington, DC: International Monetary Fund.

Komarova, Z., Ivanov, S., & Nozhnik, D. (2012). Production of Table Eggs with a Predominantly Functional Properties. *Scientific Journal of Kuban State Agrarian University, 81*(7), 476–485.

Komov, N. (1995). *Management of Land Resources in Russia: Russian Model of Land Management and Land Tenure*. Moscow: RUSSLIT.

Komov, N., & Aratskiy, D. (2000). *Methodology of Land Management at Regional Level*. Nizhny Novgorod: Nizhny Novgorod Branch of the Russian Presidential Academy of National Economy and Public Administration.

Konar, M., & Caylor, K. K. (2013). Virtual Water Trade and Development in Africa. *Hydrology and Earth System Sciences, 17*(10), 3969–3982. doi:10.5194/hess-17-3969-2013

Compilation of References

Kondratyev, K., Krapivin, V., & Savinykh, V. (2003). *Development Prospects of the Civilization. Multidimensional Analysis*. Moscow: LOGOS.

Konikow, L. F., & Kendy, E. (2005). Groundwater Depletion: A Global Problem. *Hydrogeology Journal*, *13*(1), 317–320. doi:10.100710040-004-0411-8

Koptseva, N., & Kirko, V. (2017). Development of the Russian Economy's Agricultural Sector under the Conditions of Food Sanctions (2015-2016). *Journal of Environmental Management and Tourism*, *8*(1), 123–131.

Korsheva, I. (2017). Production of Iodine-Rich Edible Eggs. In *Proceedings of the Conference Biotechnology: Current State and Future Development*. Moscow: BioTech World.

Korsheva, I., & Trotsenko, I. (2015). The Effectiveness of Sel-Plex Usage for Production of Selenium Enriched Eggs. *Omsk Scientific Bulletin*, *144*(2), 199–201.

Kotler, P. (2003). *Marketing Management*. Upper Saddle River, NJ: Pearson Education International.

Kovach, R., & McGuire, B. (2003). *Philip's Guide to Global Hazards*. London: Philip's.

Kovalenko, E., Polushkina, T., & Yakimova, O. (2017). State Regulations for the Development of Organic Culture by Adapting European Practices to the Russian Living Style. Academy of Strategic Management Journal, 16(2).

Kovalev, A. (2009). Theoretical and Practical Issues of Economic Turnover of Unclaimed Land Shares. *Bulletin of Notarial Practice*, *5*, 44–47.

Kovarda, V., & Bezuglaya, Y. (2013). Influence of Infrastructure on the Development of the Agro-Industrial Complex in Russia. *Young Scientist*, *55*(8), 195–198.

Kozhevnikova, I. (n.d.). *Overview of the Russian Dairy Market*. Food Market. Retrieved from http://www.foodmarket.spb.ru/current.php?article=2279

Koziol, H. (2008). Punitive Damages – A European Perspective. *Louisiana Law Review*, *68*(3), 741–764.

Kozir, M. (1998). On Conceptual Development of Land Reform in Russia at Modern Stage. *Economy of Agricultural and Processing Enterprises*, *6*, 14.

Kozlobaev, A. (2016). *The Effectiveness of Growth Promoters and Micronutrients in Buckwheat*. Voronezh: Voronezh State University.

Kraft, A. (2008). Electrochemical Water Disinfection: A Short Review. *Platinum Metals Review*, *52*(3), 177–185. doi:10.1595/147106708X329273

Krasnov, N. (1993). Land Reform and Land Law in Modern Russia. *State and Law*, *12*, 3–13.

Krassov, O. (2000). *Land Law: Manual*. Moscow: Jurist.

Krassov, O. (2012). Permitted Use and the Specific Purpose of the Land. *Environmental Law (Northwestern School of Law)*, *2*, 16–20.

Kravchenko, A., & Sergeeva, O. (2014). China Policy in the Area of Food Security: Modernization of Agriculture. *Pacific Rim: Economics, Politics. Law*, *32*(4), 57–65.

Kravchenko, T., Volchyonkova, A., & Esina, Y. (2014). Small Business Development as a Factor of Higher Competitiveness of Agricultural Production. *Vestnik OrelGAU*, *49*(4), 44–50.

Kresnikova, N. (2008). Tools of Agricultural Land Turnover. *Land Management. Land Monitoring and Cadaster*, *4*, 52–54.

Kristjanson, P., Waters-Bayer, A., Johnson, N., Tipilda, A., Njuki, J., & Baltenweck, I. ... MacMillan, S. (2014). Livestock and Women's Livelihoods. In A. Quisumbing, R. Meinzen-Dick, T. Raney, A. Croppenstedt, J. Behrman, & A. Peterman (Eds.), Gender in Agriculture: Closing the Knowledge Gap (pp. 209-233). Amsterdam: Springer Netherlands.

Kruijssen, F., Keizer, M., & Giuliani, A. (2007). *Collective Action for Small-Scale Producers of Agricultural Biodiversity Products*. Washington, DC: Systemwide Program on Collective Action and Property Rights.

Krylatykh, E. (1997). Formation and Development of Economic Regulation of Land Relations. *Forecasting Issues, 1*, 31–39.

Krylatykh, E. (1998). Development of Land Relations in Agricultural Sector and Rural Areas. *Scientific and Technological Development in Agricultural Industry, 4*, 30–35.

Kulakova, S., Viktorova, E., & Levachev, M. (2008). Trans-Isomers of Fatty Acids in Food Products. *Oils and Fats, 85*(3).

Kulikov, S. (2018). *Dispatching to the East. China Intends to Increase Imports of Russian Grain*. Retrieved from https://rg.ru/2018/07/23/kitaj-gotov-uvelichit-import-zerna-iz-rossii.html

Kulomzin, A. (1903). *Siberian Railway: Past and Present. Historical Essay*. Saint Petersburg: State Printing House.

Kumar, A., & Mazumdar, R. (2017). *Feed Africa: Achieving Progress through Partnership*. Mumbai: Export-Import Bank of India.

Kurochkin, S., & Smolnyakova, V. (2012). Organic Farming. *Gardener's Bulletin, 1*, 46–49.

Kuzminykh, A., & Pashkova, G. (2016). Grain Yield and Quality of Winter Rye Depending on the Use of Growth Stimulants. *Vestnik of Mari State University. Chapter. Agricultural Economics, 5*(1), 26–29.

Kytzia, S., Faist, M., & Baccini, P. (2004). Economically Extended-MFA: A Material Flow Approach for a Better Understanding of Food Production Chain. *Journal of Cleaner Production, 12*(8-10), 877–889. doi:10.1016/j.jclepro.2004.02.004

Laboratory of Trends. (2018). *Study of the Russian Market of Healthy Food – 2018*. Retrieved from https://t-laboratory.ru/2018/05/01/issledovanie-rossijskogo-rynka-produktov-zdorovogo-pitanija-2018/

Lagioia, G., & Camaggio, G. (2002). *La trasformazione industrial della patata. Dal tubero al fast food*. Bari: Progedit.

Lambin, E. F. (2012). Global Land Availability: Malthus versus Ricardo. *Global Food Security, 1*(2), 83–87. doi:10.1016/j.gfs.2012.11.002

Lambin, E. F., & Meyfroidt, P. (2011). Global Land Use Change, Economic Globalization, and the Looming Land Scarcity. *Proceedings of the National Academy of Sciences of the United States of America, 108*(9), 3465–3472. doi:10.1073/pnas.1100480108 PMID:21321211

Lambrecht, E., Kuhne, B., & Gellynck, X. (2015). Asymmetric Relationships in Networked Agricultural Innovation Processes. *British Food Journal, 117*(7), 1810–1825. doi:10.1108/BFJ-05-2014-0183

Lawrence, R. Z., Hanouz, M. D., Doherty, S., & Moavenzadeh, J. (Eds.). (2010). *The Global Enabling Trade Report 2010*. Geneva: World Economic Forum.

Lawrence, G., & McMichael, P. (2012). The Question of Food Security. *International Journal of Sociology of Agriculture and Food, 2*(19), 135–142.

Le Goff, M., & Singh, R. J. (2014). *Does Trade Reduce Poverty? A View from Africa*. Washington, DC: The World Bank.

Le Pere, G. (2005). Emerging Markets – Emerging Powers: Changing Parameters for Global Economic Governance. *Internationale Politik und Gesellschaft, 2*, 36–51.

Compilation of References

Lee, R. (2007). *Food Security and Food Sovereignty*. Newcastle upon Tyne: Centre for Rural Economy, University of Newcastle upon Tyne.

Lee, R. (2013). *The Russian Far East and China: Thoughts on Cross-Border Integration*. Philadelphia, PA: Foreign Policy Research Institute.

Leke, A., & Barton, D. (2016). *3 Reasons Things Are Looking up for African Economies*. Retrieved from https://www.weforum.org/agenda/2016/05/what-s-the-future-of-economic-growth-in-africa/

Leke, A., Lund, S., Roxburgh, C., & van Wamelen, A. (2010). *What's Driving Africa's Growth*. Retrieved from https://www.mckinsey.com/featured-insights/middle-east-and-africa/whats-driving-africas-growth

Leppke, O. (1998). Economic and Legal Aspects of State Policy Related to Rational Use and Preservation of Land Resources (Analysis and Solutions). *International Agricultural Journal, 3*, 3–6.

Leppke, O. (2000). *Scientific and Organizational-Economic Regulation of Land Relations in Agro-Industrial Industry in a Transition Economy (Methodology and Regional Level)*. Moscow: Agripress.

Lerman, Z. (2008). Agricultural Development in Central Asia: A Survey of Uzbekistan, 2007-2008. *Eurasian Geography and Economics, 49*(4), 481–505. doi:10.2747/1539-7216.49.4.481

Lerman, Z. (2009). Land Reform, Farm Structure, and Agricultural Performance in CIS Countries. *China Economic Review, 20*(2), 316–326. doi:10.1016/j.chieco.2008.10.007

Lerman, Z., & Cimpoies, D. (2006). Land Consolidation as a Factor for Rural Development in the Republic of Moldova. *Europe-Asia Studies, 58*(3), 439–455. doi:10.1080/09668130600601933

Lerman, Z., Csaki, C., & Feder, G. (2004). *Agriculture in Transition: Land Policies and Evolving Farm Structures in Post-Soviet Countries*. Lanham, MD: Lexington Books.

Lerman, Z., Csaki, C., & Moroz, V. (1998). *Land Reform and Farm Restructuring in Moldova – Progress, and Prospects*. Washington, DC: The World Bank. doi:10.1596/0-8213-4317-3

Lerman, Z., & Sedik, D. (2008). *The Economic Effects of Land Reform in Tajikistan*. Budapest: FAO Regional Office for Europe and Central Asia.

Lerman, Z., Sedik, D., Pugachov, N., & Goncharuk, A. (2007). *Rethinking Agricultural Reform in Ukraine*. Halle: Leibniz Institute of Agricultural Development in Transition Economies.

Levushkina, S., & Elfimova, Y. (2012). Land Resources as One of the Main Factors of a Rural Entrepreneurship. *Polythematic Online Scientific Journal of Kuban State Agrarian University, 83*(9), 606–616.

Liefert, W. (2004). *Food Security in Russia: Economic Growth and Rising Incomes are Reducing Insecurity*. Washington, DC: Economic Research Service, United States Department of Agriculture.

Liefert, W. M., & Liefert, O. (2012). Russian Agriculture during Transition: Performance, Global Impact, and Outlook. *Applied Economic Perspectives and Policy, 34*(1), 37–75. doi:10.1093/aepp/ppr046

Liefert, W., & Swinnen, J. (2002). *Changes in Agricultural Markets in Transition Economies*. Washington, DC: Economic Research Service, United States Department of Agriculture.

Li, P., Mellor, S., Griffin, J., Waelde, C., Hao, L., & Everson, R. (2014). Intellectual Property and 3D Printing: A Case Study on 3D Chocolate Printing. *Journal of Intellectual Property Law and Practice, 9*(4), 322–332. doi:10.1093/jiplp/jpt217

Lipski, S. (2002). Specifics of Land Reform at Modern Stage. *Economist, 10*, 77–87.

Lipski, S. (2005). Land Shares: Actual Issues and Perspectives. *Real Estate and Investments. Legal Regulation, 4*, 86–90.

Lipski, S. (2014). *Land Relations and State Land Policy in Modern Russia (Theory, Methodology, Practice)*. Moscow: State University of Land Use Planning.

Lipski, S. (2015). Private Ownership for Agricultural Lands: Advantages and Disadvantages (Experience of Two Decades). *Studies on Russian Economic Development, 26*(1), 63–66. doi:10.1134/S1075700715010074

List, G. (2006). Reformulation of Food Products for Trans-Fatty Acid Reduction: A U.S. Perspective. *Proceedings of World Conference and Exhibition on Oilseed and Vegetable Oil Utilization*. Istanbul: Exponet.

Liu, J., & Diamond, J. (2005). China's Environment in a Globalizing World. *Nature, 435*(7046), 1179–1186. doi:10.1038/4351179a PMID:15988514

Liu, Y., Hu, W., Jette-Nantel, S., & Tian, Z. (2014). The Influence of Labor Price Change on Agricultural Machinery Usage in Chinese Agriculture. *Canadian Journal of Agricultural Economics, 62*(2), 219–243. doi:10.1111/cjag.12024

Livemint. (2017). *How Does Indian Investment in Africa Compare with China's?* Retrieved from https://www.livemint.com/Politics/Xd5t4vKx2dRCoPwnScENjO/How-does-Indian-investment-in-Africa-compare-with-Chinas.html

Livemint. (2019). *India's Forex Reserves Up by $1.3 Billion to over $420 Billion*. Retrieved from https://www.livemint.com/market/stock-market-news/india-s-forex-reserves-up-by-1-3-billion-to-over-420-billion-1558155104850.html

Li, X., & Jin, J. (2014). *Concise Chinese Torts Laws*. Heidelberg: Springer-Verlag. doi:10.1007/978-3-642-41024-6

Loayza, N., & Raddatz, C. (2016). *The Composition of Growth Matters for Poverty Alleviation*. Washington, DC: The World Bank.

Lobyshev, V., & Kirkina, A. (2012). *Influence of Isotopic Content Variation of Water on Its Biological Activity*. Retrieved from www.biophys.ru/archive/congress2012/proc-p21-d.pdf

Lok-Dessallien, R. (1998). *Review of Poverty Concepts and Indicators*. New York, NY: Social Development and Poverty Elimination Division, United Nations Development Program.

Lomakin, P. (2017). *Ensuring Food Security in Russia: Domestic and International Aspects*. Moscow: Moscow State Institute of International Relations.

Lomborg, B. (2001). *The Skeptical Environmentalist: Measuring the Real State of the World*. Cambridge: Cambridge University Press. doi:10.1017/CBO9781139626378

Lomborg, B. (2004). *Global Crises, Global Solutions*. Cambridge: Cambridge University Press. doi:10.1017/CBO9780511492624

Loskutova, Z. (1980). *Vivarium*. Moscow: Medicine.

Loyko, P. (2001). *Agricultural Reform: Issues and Practice*. Moscow: AS Plus.

Loyko, P. (2009). *Land Management: Russia and the World (Future Outlook)*. Moscow: State University of Land Use Planning.

Lukianchikova, S. (2016). «Exceptional Privatization»: How to Avoid Legal Violations upon Acquisition of Unclaimed Land Shares on a Preferential Basis. *Legal Issues of Real Estate, 2*, 37–40.

Lupton, D., & Turner, B. (2018). Both Fascinating and Disturbing': Consumer Responses to 3D Food Printing and Implications for Food Activism. In T. Schneider, K. Eli, C. Dolan, & S. Ulijaszek (Eds.), *Digital Food Activism* (pp. 151–167). London: Routledge.

Mahapatra, R., Rattani, V., Sengupta, R., Pandey, K., & Goswami, S. (2017). *How to Make Africa Food Self-Sufficient, Again?* Retrieved from https://www.downtoearth.org.in/coverage/agriculture/why-farmers-now-dread-a-normal-monsoon-58206

Mahendra Dev, S., & Zhong, F. (2015). Trade and Stock Management to Achieve National Food Security in India and China? *China Agricultural Economic Review*, *7*(4), 641–654. doi:10.1108/CAER-01-2015-0009

Ma, L. (2011). Sustainable Development of Rural Household Energy in Northern China. *Journal of Sustainable Development*, *5*(4), 115–124.

Malevannaya, N. (2001). Plant Growth Regulators in Agricultural Production. *Fertility*, *1*, 29.

Malozemov, S. (2014). Experience of State Support of Agriculture in Foreign Countries. *Science and Modernity*, *32*(1), 136–141.

Malthus, T. (1798). *An Essay of the Principle of Population*. London: J. Johnson.

Manyika, J., Bughin, J., Lund, S., Nottebohm, O., Poulter, D., Jauch, S., & Ramaswamy, S. (2014). *Global Flows in a Digital Age: How Trade, Finance, People, and Data Connect the World Economy*. New York, NY: McKinsey Global Institute.

Manyika, J., Lund, S., Bughin, J., Woetzel, J., Stamenov, K., & Dhingra, D. (2016). *Digital Globalization: The New Era of Global Flows*. New York, NY: McKinsey Global Institute.

Mariani, B. (2013). *China's Role and Interests in Central Asia*. London: Saferworld.

Marinković, S., & Nedić, N. (2010). *Analysis of Production and Competitiveness on Small Beekeeping Farms in Selected Districts of Serbia*. Budapest: Agroinform Publishing House. doi:10.19041/Apstract/2010/3-4/10

Markel, D. (2009). Retributive Damages: A Theory of Punitive Damages as Intermediate Sanction. *Cornell Law Review*, *94*, 239–340.

Markelova, H., & Mwangi, E. (2010). Collective Action for Smallholder Market Access: Evidence and Implications for Africa. *The Review of Policy Research*, *27*(5), 621–640. doi:10.1111/j.1541-1338.2010.00462.x

Markov, A. (Ed.). (1995). *History of the National Economy*. Moscow: Law and Right, Unity.

Martinelli, L. A., Batistella, M., Silva, R. F. B. D., & Moran, E. (2017). Soy Expansion and Socioeconomic Development in Municipalities of Brazil. *Land (Basel)*, *6*(3), 62. doi:10.3390/land6030062

Martin, W., & Laborde Debucquet, D. (2018). Trade: The Free Flow of Goods and Food Security and Nutrition. In *International Food Policy Research Institute, 2018 Global Food Policy Report* (pp. 20–29). Washington, DC: International Food Policy Research Institute.

Matejčeková, Z., Liptáková, D., & Valík, L. (2017). Functional Probiotic Products Based on Fermented Buckwheat with Lactobacillus Rhamnosus. *Lebensmittel-Wissenschaft + Technologie*, *81*, 35–41.

Matusevich, Y., Kolbatsky, P., Zelepukhin, I., Ostryakov, I., & Zelepukhin, V. (1983). *Copyright Certificate of the USSR #1001965 "Thermal Degasser"*. Retrieved from http://patents.su/3-1001965-termicheskijj-degazator.html

Matz, J. A., Kalkuhl, M., & Abegaz, G. A. (2015). The Short-Term Impact of Price Shocks on Food Security – Evidence from Urban and Rural Ethiopia. *Food Security*, *7*(3), 657–679. doi:10.100712571-015-0467-4

Mazilkina, E., & Panichkina, G. (2009). *Competitiveness Management*. Moscow: Omega-L.

McCarthy, U., Uysal, I., Badia-Melis, R., Mercier, S., Donnell, C. O., & Ktenioudaki, A. (2018). Global Food Security – Issues, Challenges and Technological Solutions. *Trends in Food Science & Technology*, *77*, 11–20. doi:10.1016/j.tifs.2018.05.002

McConnell, W. J., & Vina, A. (2018). Interactions between Food Security and Land Use in the Context of Global Change. *Land (Basel)*, *7*(2), 53. doi:10.3390/land7020053

McMinimy, M. (2016). *TPP: American Agriculture and the Trans-Pacific Partnership (TPP) Agreement*. Washington, DC: Congressional Research Service.

Meadows, D. H., Meadows, D. L., Randers, J., & Behrens, W. W. (1972). *The Limits to Growth*. New York, NY: Universe Books.

Means, E., & McMahon, R. (1978). *Patent USA 4,073,712 Electrostatic Water Treatment*. Retrieved from http://patents.com/us-4073712.html

Mechanical Engineering Portal. (n.d.). *Analytics*. Retrieved from http://www.mashportal.ru/analytics.aspx

Medin, H. (2014). *New Trade Theory: Implications for Industrial Policy*. Oslo: Norwegian Institute of International Affairs.

Medprom.ru. (n.d.). *Overhead Magnetron*. Retrieved from http://medprom.ru/medprom/1394898?TESTROBOT=YES

Medvedeva, A. (2018). *World Market of Agricultural Machinery – Stability and Need for Innovations*. Retrieved from https://www.agroxxi.ru/selhoztehnika/novosti/mirovoi-rynok-selskohozjaistvennoi-tehniki-stabilnost-i-potrebnost-v-innovacijah.html

Megakhim. (n.d.). *Electrolyzer*. Retrieved from http://www.megahim.ru/equipment/electrolysis

Melitz, M. J. (2003). The Impact of Trade on Intra-Industry Reallocations and Aggregate Industry Productivity. *Econometrica*, *71*(6), 1695–1725. doi:10.1111/1468-0262.00467

Meredith, S., Lampkin, N., & Schmid, O. (Eds.). (2018). *Organic Action Plans: Development, Implementation and Evaluation*. Brussels: IFOAM EU.

Merkulov, G. (1969). *Course of Pathology Techniques*. Saint Petersburg: Medicine.

Migina, E. (2016). Prospective Usage of Soybean Seeds and Products of Their Processing in Development of New Feed Additives. *Young Scientist*, *125*(21), 284–288.

Mihailović, B., Cvijanović, D., Milojević, I., & Filipović, M. (2014). The Role of Irrigation in Development of Agriculture in Srem District. *Economics of Agriculture*, *61*(4), 989–1004.

Ministry of Agriculture of the Russian Federation. (2018). *National Report "On the Progress and Results of the Implementation of the State Program for the Development of Agriculture and Regulation of Markets of Agricultural Products, Raw Materials, and Food for 2013-2020."* Retrieved from http://mcx.ru/upload/iblock/ec8/ec8f3b2c7fa3b4642f76d3fbda07804b.pdf

Ministry of Agriculture of the Russian Federation. (n.d.). *Reports on the Conditions and Use of Agricultural Lands in 2010-2016*. Moscow: Ministry of Agriculture of the Russian Federation.

Ministry of Finance and Economic Development of Ethiopia. (2012). *Ethiopia's Progress towards Eradicating Poverty: An Interim Report on Poverty Analysis Study (2010/11)*. Addis Ababa: Development Planning and Research Directorate, Ministry of Finance and Economic Development of Ethiopia.

Ministry of Finance of India. (2018). *Economic Survey 2018*. Retrieved from http://mofapp.nic.in:8080/Economicsurvey/#

Ministry of Finance of India. (2019). *Key Highlights of Economic Survey 2018-19*. Retrieved from http://pib.nic.in/newsite/PrintRelease.aspx?relid=191213

Ministry of Health and Social Development of the Russian Federation. (2008). *Order #194 from April 24, 2008, "On the Approval of Medical Criteria for Determining the Severity of Harm Caused to Human Health."* Retrieved from http://base.garant.ru/12162210/

Mintusov, V. (2016). *Directions of Improvement of Regulation of Food Imports in the Russian Federation*. Moscow: State University of Management.

Mironenko, O. (2016). Organic Market of Russia. Results of 2016. *Prospects*. Retrieved from http://rosorganic.ru/files/statia%20org%20rinok%20rossii.pdf

Mitin, A., & Rozhdestvensky, V. (n.d.). *The Analysis of State of Domestic Agriculture in Conditions of Fierce and International Competition*. Retrieved from http://bmpravo.ru/show_stat.php?stat=323

Mitrovic, I. (2011). *Beekeepers Association "Jovan Zivanovic". Study on Protection Geographical Indications Product Fruškogorski Med*. Retrieved from http://www.pcelarins.org.rs/ns/fruskogorski-lipov-med/

Modi, R., Desai, D. D., & Venkatachalam, M. (2019). *South-South Cooperation: India-Africa Partnerships in Food Security and Capacity-Building*. Mumbai: Observer Research Foundation.

Molchanova, E. (2013). *Physiology of Nutrition*. Saint Petersburg: Piter.

Molchanova, E., & Suslyanok, G. (2013). Quality Assessment and Food Proteins Value. *Storage and Processing of Farm Products*, *1*, 16–22.

Molden, D. (2007). *Water for Food, Water for Life: A Comprehensive Assessment of Water Management in Agriculture*. London: Earthscan.

Moller, H., Hanssen, J., Gustavsson, J., Ostergren, K., Stenmarck, A., & Dekhtyar, P. (2014). *Report on Review of (Food) Waste Reporting Methodology and Practice*. Krakeroy: Ostfold Research.

Molotoks, A., Kuhnert, M., Dawson, T. P., & Smith, P. (2017). Global Hotspots of Conflict Risk between Food Security and Biodiversity Conservation. *Land (Basel)*, *6*(4), 67. doi:10.3390/land6040067

Moltz, J. (2002). Russo-Chinese Normalization from an International Perspective. Coping with the Pressures of Change. In T. Akaha (Ed.), *Politics and Economics in the Russian Far East* (pp. 187–197). New York, NY: Routledge.

Mordor Intelligence. (2017). *Agricultural Machinery Market – Segmented by Type (Tractors, Plowing and Cultivating Machinery, Planting Machinery, Harvesting Machinery, Haying and Forage Machinery, Irrigation Machinery), by Geography – Analysis of Growth, Trends and Progress (2019-2024)*. Retrieved from https://www.mordorintelligence.com/industry-reports/agricultural-machinery-market?gclid=CjwKCAiAiJPkBRAuEiwAEDXZZW41MxYmd5jvxtxaw-6prmgjEvU3F_UY1hMAd9lfiZtwtE8rLKLcMxoCqwwQAvD_BwE

Moroz, V., Stratan, A., Ignat, A., & Lucasenco, E. (2015). *AGRICISTRADE Country Report: Moldova*. Retrieved from http://www.agricistrade.eu/document-library

Morozova, I., & Litvinova, T. (2014). Russian Market of Agricultural Equipment: Challenges and Opportunities. *Asian Social Science*, *23*(10), 68–77.

Morozova, I., Litvinova, T., Rodina, E., & Prosvirkin, N. (2015). Marketing Mix in the Market of Agricultural Machinery: Problems and Prospects. *Mediterranean Journal of Social Sciences*, *36*(6), 19–26.

Morris, M. C., Evans, D. A., Bienias, J. L., Tangney, C. C., Bennett, D. A., Aggarwal, N., ... Wilson, R. S. (2003). Dietary Fats and the Risk of Incident Alzheimer Disease. *Archives of Neurology, 60*(2), 194–200. doi:10.1001/archneur.60.2.194 PMID:12580703

Mozaffarian, D., Katan, M. B., Ascherio, A., Stampfer, M. J., & Willett, W. C. (2006). Trans-Fatty Acids and Cardiovascular Disease. *The New England Journal of Medicine, 354*(15), 1601–1613. doi:10.1056/NEJMra054035 PMID:16611951

Muimba-Kankolongo, A. (2018). Leguminous Crops. In A. Muimba-Kankolongo (Ed.), *Food Crop Production by Smallholder Farmers in Southern Africa* (pp. 173–203). Amsterdam: Elsevier. doi:10.1016/B978-0-12-814383-4.00010-4

Mukasa, A. N., Woldemichael, A. D., Salami, A. O., & Simpasa, A. M. (2017). Africa's Agricultural Transformation: Identifying Priority Areas and Overcoming Challenges. *Africa Economic Brief, 8*(3), 1–16.

Mukhachev, V. (1975). *Living Water*. Moscow: Nauka.

Mulatu, A., & Wossink, A. (2014). Environmental Regulation and Location of Industrialized Agricultural Production in Europe. *Land Economics, 90*(3), 509–537. doi:10.3368/le.90.3.509

Muller, M., & Bellmann, C. (2016). *Trade and Water: How Might Trade Policy Contribute to Sustainable Water Management?* Geneva: International Centre for Trade and Sustainable Development.

Munang, R. (2015). *Winning Africa's Future: Food Security for All*. Retrieved from https://intpolicydigest.org/2015/07/23/winning-africa-s-future-food-security-for-all/

Museum of Local History in Chulym. (n.d.). *The History of Settlement of Chulym Land*. Retrieved from https://vmuseum.ucoz.ru/index/istorija_zaselenija_zemli_chulymskoj/0-9

Muzlera, J. (2014). Capitalization Strategies and Labor in Agricultural Machinery Contractors in Argentina. *Research in Rural Sociology and Development, 20*, 57–74. doi:10.1108/S1057-192220140000020002

Naidenova, L. (2013). Life Within Monastery Walls (A Case Study of the Monastery at Solovki). *Russian Studies in History, 52*(1), 66–88. doi:10.2753/RSH1061-1983520103

Narayan, D., Chambers, R., Shah, M. K., & Petesch, P. (2000). *Voices of the Poor. Crying Out for Change*. New York, NY: Oxford University Press. doi:10.1596/0-1952-1602-4

Nath, R., Luan, Y., Yang, W., Yang, C., Chen, W., Li, Q., & Cui, X. (2015). Changes in Arable Land Demand for Food in India and China: A Potential Threat to Food Security. *Sustainability, 7*(5), 5371–5397. doi:10.3390u7055371

National Academy of Sciences. (2005). *Dietary Reference Intakes for Energy, Carbohydrate, Fiber, Fat, Fatty Acids, Cholesterol, Protein, and Amino Acids*. Washington, DC: National Academies Press.

National Bank of Ethiopia. (2018). *National Bank of Ethiopia Annual Report*. Addis Ababa: National Bank of Ethiopia.

National Rating Agency. (2016). *Agriculture of Russia in 2015 and First Half of 2016*. Retrieved from http://www.ra-national.ru/sites/default/files/analitic_article/Сельское%20хозяйство%20и%20семена%202016%202_0.pdf.

Ncube, B., French, A., & Mupangwa, W. (2018). Precision Agriculture and Food Security in Africa. In P. Mensah, D. Katerere, S. Hachigonta, & A. Roodt (Eds.), *Systems Analysis Approach for Complex Global Challenges* (pp. 159–178). Heidelberg: Springer. doi:10.1007/978-3-319-71486-8_9

Ndamani, F., & Watanabe, T. (2014). Rainfall Variability and Crop Production in Northern Ghana: The Case of Lawra District. *Proceedings of the 9th International Symposium on Social Management Systems SSMS 2013*. Sydney: Society for Social Management Systems.

Compilation of References

Nebbia, G. (1995). *Lezioni di Merceologia*. Rome: Laterza & Figli Spa.

Negassa, A., Rashid, S., & Gebremedhin, B. (2011). *Livestock Production and Marketing*. Addis Ababa: International Food Policy Research Institute.

Nekrasov, R. (2019). *The Results of Plant Breeding Branch and Engineering and Technical Services in 2018, Realization of Tasks of the State Programme for 2013-2020*. Moscow: Ministry of Agriculture of the Russian Federation.

Nelson, R. A. (2007). *Electro-Culture (The Electrical Tickle)*. Retrieved from https://pdfs.semanticscholar.org/f1a3/7c7a653e7cd6c546205989650c5c22b3864a.pdf?_ga=2.126908106.613024882.1549709931-55171684.1549709931

Nelson, R., & Winter, S. (1982). *An Evolutionary Theory of Economic Change*. Cambridge, MA: Harvard University Press.

Neschadin, A. (2008). Experience of Governmental Regulation and Support of Agriculture Abroad. *Society and Economy*, *8*, 132–150.

New Partnership for Africa's Development. (2002). *Comprehensive Africa Agriculture Development Programme*. Midrand: New Partnership for Africa's Development.

New Zealand Foreign Affairs and Trade. (2018). *Comprehensive and Progressive Agreement for Trans-Pacific Partnership Text and Resources*. Retrieved from https://www.mfat.govt.nz/en/trade/free-trade-agreements/free-trade-agreements-in-force/cptpp/comprehensive-and-progressive-agreement-for-trans-pacific-partnership-text

Newman, M. (2010). *Networks: An Introduction*. Oxford: Oxford University Press. doi:10.1093/acprof:oso/9780199206650.001.0001

Nikolsky, S. (2016). *Agrarian Reform of the 1990s: Regional Models and Ideological Foundations*. Retrieved April 27, 2019, from http://lawinrussia.ru/content/agrarnaya-reforma-90-h-godov-regionalnye-modeli-i-ih-ideologicheskie-osnovaniya

Nivievskyi, O., Stepaniuk, O., Movchan, V., Ryzhenkov, M., & Ogarenko, Y. (2015). *AGRICISTRADE Country Report: Ukraine*. Retrieved from http://www.agricistrade.eu/document-library

Njuki, J., Kaaria, S., Chamunorwa, A., & Chiuri, W. (2011). Linking Smallholder Farmers to Markets, Gender and Intra-Household Dynamics: Does the Choice of Commodity Matter? *European Journal of Development Research*, *23*(3), 426–443. doi:10.1057/ejdr.2011.8

Nolasco, C. L., Soler, L. S., Freitas, M. W., Lahsen, M., & Ometto, J. P. (2017). Scenarios of Vegetable Demand vs. Production in Brazil: The Links between Nutritional Security and Small Farming. *Land (Basel)*, *6*(3), 49. doi:10.3390/land6030049

Norton, A., Bird, B., Brock, K., Kakande, M., & Turk, C. (2001). *A Rough Guide to PPAs Participatory Poverty Assessment: An Introduction to Theory and Practice*. London: Overseas Development Institute.

Nosworthy, M. G., Medina, G., Franczyk, A. J., Neufeld, J., Appah, P., Utioh, A., ... House, J. D. (2018). Effect of Processing on the In Vitro and In Vivo Protein Quality of Red and Green Lentils (*Lens Culinaris*). *Food Chemistry*, *240*, 588–593. doi:10.1016/j.foodchem.2017.07.129 PMID:28946315

Novikova, V. (2015). Problems of Development of Agriculture in Russia. In *Proceedings of the International Conference "Relevant Issues of Economics, Management, and Finance in Modern Conditions."* Saint Petersburg: Innovation Center for the Development of Education and Science.

Novitsky, I. (2017). *Organic Agriculture: Profitability and Basic Principles*. Retrieved from https://xn--80ajgpcpbhkds4a4g.xn--p1ai/articles/organicheskoe-selskoe-hozyajstvo-rentabelnost-i-osnovnye-printsipy/

O'Brien, R. (2007). *Fats and Oils: Formulating and Processing for Applications*. Saint Petersburg: Profession.

Odum, J. (1986). *Ecology*. Moscow: Mir.

Office of the U.S. Trade Representative. (2016). *How TPP Benefits U.S. Agriculture*. Retrieved from https://ustr.gov/sites/default/files/TPP-Benefits-for-US-Agriculture-Fact-Sheet.pdf

Ogarkov, A. (2000). Agriculture and Its Resource and Production Potential. *Economy of Agricultural and Processing Enterprises*, *5*, 7–9.

Oh, P., & Monge, P. (2016). *Network Theory and Models*. Retrieved from https://onlinelibrary.wiley.com/doi/full/10.1002/9781118766804.wbiect246

Okrenilov, V. (1998). *Quality Management: Scientific and Practical Study*. Moscow: Economics.

Olson, D. R. (2003). *Towards Food Sovereignty: Constructing an Alternative to the World Trade Organization's Agreement on Agriculture*. Minneapolis, MN: Institute for Agriculture and Trade Policy.

One-Straw Revolution. (n.d.). *About "The One-Straw Revolution."* Retrieved from https://onestrawrevolution.net/

Onyima, J.K.C., & Okoro, C.N. (2009). *Co-operatives: Elements, Principles and Practices*. Awka: Maxiprint.

Organisation for Economic Cooperation and Development. (1997). *Agricultural Outlook*. Paris: OECD.

Organisation for Economic Cooperation and Development. (2013). Better Policies for Development. In *Focus: Policy Coherence for Development and Global Food Security*. Paris: OECD.

Organisation for Economic Cooperation and Development. (2018). *Monitoring and Evaluation: Reference Tables*. Retrieved from https://stats.oecd.org/viewhtml.aspx?datasetcode=MON2018_REFERENCE_TABLE&lang=en#

Organization for Economic Cooperation and Development. (2015). *OECD-FAO Agricultural Outlook 2015*. Paris: OECD Publishing.

Organization for Economic Cooperation and Development. (2016). *Agriculture in Sub-Saharan Africa: Prospects and Challenges for the Next Decade*. Paris: OECD Publishing.

Organization for Economic Cooperation and Development. (2017). *OECD-FAO Agricultural Outlook 2017-2026*. Paris: OECD Publishing.

Organization for Economic Cooperation and Development. (2018). *Economic Outlook for Southeast Asia, China and India 2018*. Paris: OECD Publishing.

Osipova, A. (2016). *China's Policy in the Sphere of Food Security: Environmental Aspect*. Retrieved from https://scienceforum.ru/2016/article/2016026522

Oulhaj, L. (2013). *Evaluation of the Agricultural Strategy of Morocco (Green Morocco Plan) with a Dynamic General Equilibrium Model*. Marseille: Femise Research Programme.

Ovcharenko, M. (2001). Humates Are the Activators of Productivity of Agricultural Crops. *Agrochemical Bulletin*, *2*, 13–14.

Paganopoulos, M. (2009). The Concept of "Athonian Economy" in the Monastery of Vatopaidi. *Journal of Cultural Economics*, *2*(3), 363–378. doi:10.1080/17530350903345595

Palm Oil Research Institute of Malaysia. (1993). *Selected Readings on Palm Oil and Its Uses. Bandar Baru Bangi*. Palm Oil Research Institute of Malaysia.

Panitchpakdi, S., & Clifford, M. L. (2002). *China and the WTO: Changing China, Changing World Trade*. Singapore: J. Wiley & Sons.

Compilation of References

Pankova, K. (2000). Land Shares and Collective Land Use. *Agrarian Science*, *1*, 3–5.

Papargyropoulou, E., Lozano, R., Steinberger, J. K., Wright, N., & Bin Ujang, Z. (2014). The Food Waste Hierarchy as a Framework for the Management of Food Surplus and Food Waste. *Journal of Cleaner Production*, *76*, 106–115. doi:10.1016/j.jclepro.2014.04.020

Paradikovic, N., Vinkovic, T., Vinkovic Vrcek, I., & Tkalec, M. (2013). Natural Biostimulants Reduce the Incidence of BER in Sweet Yellow Pepper Plants (*Capsicum Annum L.*). *Agricultural and Food Science*, *22*(2), 307–317. doi:10.23986/afsci.7354

Paraušić, V., & Cvijanović, D. (2007). Serbian Agriculture: Programmes of Credit Support by the State and Commercial Banks between 2004-2007. *Economic Annals*, *52*(174-175), 186–207. doi:10.2298/EKA0775186P

Park, A., Jin, H., Rozelle, S., & Huang, J. (2002). Market Emergence and Transition: Arbitrage, Transaction Costs, and Autarky in China's Grain Markets. *American Journal of Agricultural Economics*, *84*(1), 67–82. doi:10.1111/1467-8276.00243

Park, H., Hung, Y.-C., & Brackett, R. E. (2002). Antimicrobial Effect of Electrolyzed Water for Inactivating *Campylobacter Jejuni* during Poultry Washing. *International Journal of Food Microbiology*, *72*(1-2), 77–83. doi:10.1016/S0168-1605(01)00622-5 PMID:11843416

Parry, M., Canziani, O., Palutikof, J., van der Linden, P., & Hanson, C. (Eds.). (2007). *Climate Change 2007: Impacts, Adaptation and Vulnerability*. Cambridge: Cambridge University Press.

Pasko, O., Semenov, A., & Dirin, V. (2000). *Patent #2147446 Water Activation Device*. Russian Federation. Retrieved from http://www.freepatent.ru/patents/2144506

Pasko, O., Semenov, A., Smirnov, G., & Smirnov, D. (2009). *Patent #2350568 Non-Replaceable Electrolyzer*. Russian Federation. Retrieved from http://www.freepatent.ru/patents/2350568

Pasko, O. (2000). *Activated Water and Its Use in Agriculture*. Tomsk: Tomsk State University.

Pasko, O. (2010). Property Changes of Piped Water Treated with Different Methods. *Water: Chemistry and Ecology*, *7*, 40–45.

Pasko, O. (2013a). Economic Development and Perspectives of Cooperation Between the USA, Europe, Russia, and CIS States. *Cibunet Publishing*, *1*, 45–60.

Pasko, O. (2013b). The Use of Agricultural Land in Tomsk Region. *Agrarian Science*, *6*, 9–10.

Pasko, O. (2014). Dynamics of Changes in Agricultural Land in Tomsk Region. *Lucrari Stintifice*, *33*, 28–33.

Pasko, O., & Gomboev, D. (2011). *Activated Water and Possibilities of Its Application in Plant Growing and Animal Husbandry*. Tomsk: Tomsk Polytechnic University.

Pasko, O., Semenov, A., Smirnov, G., & Smirnov, D. (2007). *Activated Fluids, Electromagnetic Fluids, and Flicker Noise. Their Application in Medicine and Agriculture*. Tomsk: Tomsk State University of Operation Systems and Radioelectronics.

Paull, J. (2011). Attending the First Organic Agriculture Course: Rudolf Steiner's Agriculture Course at Koberwitz, 1924. *European Journal of Soil Science*, *21*(1), 64–70.

Paull, J. (2014). Lord Northbourne, the Man Who Invented Organic Farming, a Biography. *Journal of Organic Systems*, *9*(1), 31–53.

Pauw, K., Thurlow, J., & Van Seventer, D. (2010). *Droughts and Floods in Malawi: Assessing the Economywide Effects*. Washington, DC: International Food Policy Research Institute.

Pechatnova, A. (2014). Innovative Development of Agriculture: Problems and Prospects. *Young Scientist, 4*, 427–429.

Pellegrini, G., Sillani, S., Gregori, M., & Spada, A. (2019). Household Food Waste Reduction: Italian Consumers' Analysis for Improving Food Management. *British Food Journal, 121*(6), 1382–1397. doi:10.1108/BFJ-07-2018-0425

Pelley, J. (2018). Safety Standards Aim to Rein in 3-D Printer Emissions. *ACS Central Science, 4*(2), 134–136. doi:10.1021/acscentsci.8b00090 PMID:29532010

Pena, K. (2008). Opening the Door to Food Sovereignty in Ecuador. *Food First News & Views, 111*(30), 1–4.

Perkel, R. (1990). *Research, Technology Development and Organization of Production of Interesterified Edible Fats*. Saint Petersburg: All-Russian Research Institute of Fats.

Petibskaya, V. (1999). Inhibitors of Proteolytic Enzymes. *News of Institutes of Higher Education. Food Technology, 5-6*, 6–10.

Petrichenko, V., & Loginov, S. (2010). Influence of Plant Growth Regulators and Microelements on Productivity of Sunflower and Oil Content of Seeds. *Agrarian Russia, 4*, 24–26.

Petrikov, A. (2000). Agricultural Reform in Russia and Agricultural Policy Issues. *Economy of Agricultural and Processing Enterprises, 12*, 7–8.

Petrikov, A. (2012). It Is Necessary to Increase Adaptation of Russian Agrarian Sector to WTO Conditions. *Economy of Agricultural and Processing Enterprises, 6*, 6–8.

Petrikov, A. (2016). Major Directions of Implementation of Modern Agrifood and Rural Policy. *International Agricultural Journal, 1*, 3–9.

Petrikov, A. (2016). The Main Directions of the Modern Agri-Food and Rural Policy. *International Agricultural Journal, 1*, 3–9.

Petruláková, M., & Valík, L. (2015). Legumes as Potential Plants for Probiotic Strain *Lactobacillus Rhamnosus GG. Acta Universitatis Agriculturae et Silviculturae Mendelianae Brunensis, 63*(5), 1505–1511. doi:10.11118/actaun201563051505

Pettinger, T. (2017). *New Trade Theory*. Retrieved from https://www.economicshelp.org/blog/6957/trade/new-trade-theory/

Philippidis, G., Sartori, M., Ferrari, E., & M'Barek, R. (2019). Waste not, Want not: A Bio-Economic Impact Assessment of Household Food Waste Reductions in the EU. *Resources, Conservation and Recycling, 146*, 514–522. doi:10.1016/j.resconrec.2019.04.016 PMID:31274960

Pigato, M. (2009). *Strengthening China's and India's Trade and Investment Ties to the Middle East and North Africa*. Washington, DC: The World Bank.

Pigorev, I., Zasorina, E., Rodionov, K., & Katunin, K. (2011). Use of Growth Regulators in the Agricultural Complex in Potato Cultivation in the Central Black Earth Region. *Agricultural Science, 2*, 15–18.

Pikabu. (n.d.). *How the Standard of Living Has Changed in Russia over the Past 130 Years*. Retrieved from https://pikabu.ru/story/kak_menyalsya_uroven_zhizni_v_rossii_za_poslednie_130_let_5695251

Pinstrup-Andersen, P. (2009). Food Security: Definition and Measurement. *Food Security, 1*(1), 5–7. doi:10.100712571-008-0002-y

Pitersky, V. (1999). *Strategical Potential of Russia. Natural Resources*. Moscow: Geoinformmark.

Compilation of References

Plotnikov, V. (2013). If Tens of Billions Rubles Are Not Sent Now – Hundreds Will Not Save Tomorrow. *Krestyanskij Dvor, 21*(6), 8-9.

Pocol, C.B., Bârsan, A., & Popa, A. (2012). A Model of Social Entrepreneurship Developed in Barcău Valley, Sălaj County. *Analele Universității din Oradea, Fascicula: Ecotoxicologie, Zootehnie și Tehnologii de Industrie Alimentară*, 183-190.

Pocol, C. B. (2012). Consumer Preferences for Deferent Honey Varieties in the North West Region of Romania. *Agronomy Series of Scientific Research, 55*(2), 263–266.

Pocol, C. B., Ignjatijević, S., & Cavicchioli, D. (2017). Production and Trade of Honey in Selected European Countries: Serbia, Romania and Italy. In V. A. A. De Toledo (Ed.), *Honey Analysis* (pp. 1–20). London: IntechOpen Limited. doi:10.5772/66590

Pocol, C. B., & Popa, A. A. (2012). Types of Beekeeping Practiced in the North West Region of Romania-Advantages and Disadvantages. *Bulletin of University of Agricultural Sciences and Veterinary Medicine Cluj-Napoca. Horticulture, 69*(2), 239–243.

Podberezkina, O. (2015). Transport Corridors in Russian Integration Projects, the Case of the Eurasian Economic Union. *MGIMO Review of International Relations, 40*(1), 57–65.

Podkopaev, O. (2013). State Support of the Agricultural Sector of the Economy in the Conditions of Russia's Membership in the WTO: The Issue of Food Security of the Country. *Successes of Modern Natural Science, 3*, 156–157.

Podolny, J. M. (1993). A Status-Based Model of Market Competition. *American Journal of Sociology, 98*(4), 829–872. doi:10.1086/230091

Pomelov, A., Pasko, O., & Baranova, A. (2015). Comparative Analysis of Land Management in the World. *IOP Conference Series: Earth and Environmental Science, 27*. 10.1088/1755-1315/27/1/012040

Ponomareva, Y., & Zaharova, O. (2015). The Effect of Mineral Fertilizers and Growth Regulator on Yield and Quality of Malt Barley When Drought. *Bulletin of Kostychev Ryazan State Agrotechnological University, 27*(3), 36–42.

Ponomarev, S. (2011). *The Development of Resource Saving Technology of Byproducts of Pea Processing*. Orenburg: Orenburg State University.

Popa, A.A., Mărghitaş, L.A., & Pocol, C.B. (2011). Economic and Socio-Demographic Factors that Influence Beekeepers' Entrepreneurial Behavior. *Agronomy Series of Scientific Research, 54*(2).

Popa, A. A., & Pocol, C. B. (2011). A Complex Model of Factors that Influence Entrepreneurship in the Beekeeping Sector. *Bulletin of University of Agricultural Sciences and Veterinary Medicine Cluj-Napoca. Horticulture, 68*(2), 188–195.

Popescu, A. (2010). Study on the Economic Efficiency of Romania's Honey Foreign Trade. *Scientific Papers Series D, Zootechnics. Faculty of Animal Science Bucharest, 53*, 176–182.

Popova, I., & Klimenko, E. (2018). Prospects for the BRICS Countries Cooperation in the Field of Food Security in the Grain Production Sector. *Technico-Tehnologicheskie Problemy Servisa, 43*(1), 43–48.

Porshnev, B. (1963). *Modern Status of Question or Relict Hominids*. Moscow: All-Russian Institute of Scientific and Technical Information of the Russian Academy of Science.

Portes, A., & Sensenbrenner, J. (1993). Embeddedness and Immigration: Notes on the Social Determinants of Economic Action. *American Journal of Sociology, 98*(6), 1320–1350. doi:10.1086/230191

PotatoPro.com. (n.d.). *The Potato Sector*. Retrieved from https://www.potatopro.com/world/potato-statistics

Potenko, T., & Emelyanov, A. (2018). Export Potential of Agriculture in the Far East of Russia. *Agrarian Bulletin of the Far East*, *45*(1), 125–133.

Potrebitely. (n.d.). *What Are Food Products and What Applies to Them*. Retrieved from https://potrebitely.com/tovary/prodovolstvennye-tovary-eto.html

Pouch, T. (2012). *Crisis, Market Instability and Agricultural Policy: Analysis and Prospective Elements*. Retrieved from http://www.augurproject.eu/IMG/pdf/Pouch_Traduc_rf1_WP6-2.pdf

Powell, W. (1990). Neither Market nor Hierarchy: Network Forms of Organization. *Research in Organizational Behavior*, *12*, 295–336.

Pravda.ru. (2017). *Top 10 Companies That Control the World's Food and Drinks*. Retrieved from https://www.pravda.ru/news/economics/1329909-monopoly/

President of the Russian Federation. (1991). *Decree #323 from December 27, 1991 "On Urgent Measures for the Implementation of Land Reform in Russian Socialist Federative Soviet Republic."* Retrieved from http://www.consultant.ru/document/cons_doc_LAW_206/

President of the Russian Federation. (1992). *Decree #213 from March 2, 1992, "On Establishment of the Norm of Free Land Allotment to the Individuals."* Retrieved from http://www.consultant.ru/document/cons_doc_LAW_365/

President of the Russian Federation. (1993). *Decree #1767 from October 27, 1993, "On the Regulation of Land Relations and Agrarian Reform in Russia."* Retrieved from http://www.consultant.ru/document/cons_doc_LAW_2601/

President of the Russian Federation. (1993). *Decree of the President of the Russian Federation #1767 from October 27, 1993, "On the Regulation of Land Relations and Development of Agrarian Reform in Russia*. Moscow: Kremlin.

President of the Russian Federation. (1996). *Decree #337 from March 7, 1996, "On the Implementation of Citizens' Constitutional Rights for Land."* Retrieved from https://base.garant.ru/10105753/

President of the Russian Federation. (2009). *Decree #537 from May 12, 2009, "On the National Security Strategy of the Russian Federation till 2020."* Retrieved from https://www.garant.ru/products/ipo/prime/doc/95521/

President of the Russian Federation. (2010). *Decree #120 from January 30, 2010, "Food Security Doctrine of the Russian Federation"*. Retrieved from http://www.garant.ru/hotlaw/federal/228793/

President of the Russian Federation. (2010). *Decree #120 from January 30, 2010, "On the Approval of Food Security Doctrine of the Russian Federation."* Retrieved from http://base.garant.ru/12172719/

President of the Russian Federation. (2010). *Decree #120 from January 30, 2010, "On the Approval of Food Security Doctrine of the Russian Federation."* Retrieved from http://www.gks.ru/free_doc/new_site/import-zam/ukaz120-2010.pdf

President of the Russian Federation. (2013). *Decree #2573 from November 1, 2013, "Fundamentals of the State Policy in the Field of Ensuring Chemical and Biological Security of the Russian Federation for the Period up to 2025 and Beyond."* Retrieved from https://www.garant.ru/products/ipo/prime/doc/70423098/

President of the Russian Federation. (2013). *Presidential Address to the Federal Assembly*. Retrieved from http://en.kremlin.ru/events/president/news/19825

President of the Russian Federation. (2015). *Decree #683 from December 31, 2015, "On National Security Strategy of the Russian Federation."* Retrieved from http://www.consultant.ru/document/cons_doc_LAW_191669/

President of the Russian Federation. (2015). *Decree #683 from December 31, 2015, "On the National Security Strategy of the Russian Federation."* Retrieved from http://base.garant.ru/71296054/

President of the Russian Federation. (2018). *The President Signed Executive Order on National Goals and Strategic Objectives of the Russian Federation through to 2024*. Retrieved from http://en.kremlin.ru/events/president/news/57425

PricewaterhouseCoopers. (2015). *Food Security in Africa. Water on Oil*. Retrieved from https://www.pwc.com/gx/en/issues/high-growth-markets/assets/food-security-in-africa.pdf

PricewaterhouseCoopers. (2016). *India-Africa Partnership in Agriculture: Current and Future Prospects*. Retrieved from https://www.pwc.in/assets/pdfs/publications/2016/india-africa-partnership-in-agriculture-current-and-future-prospects.pdf

Priefer, C., Jörissen, J., & Bräutigam, K.-R. (2016). Food Waste Prevention in Europe – A Cause-Driven Approach to Identify the Most Relevant Leverage Points for Action. *Resources, Conservation and Recycling, 109*, 155–165. doi:10.1016/j.resconrec.2016.03.004

Principato, L., Ruini, L., Guidi, M., & Secondi, L. (2019). Adopting the Circular Economy Approach on Food Loss and Waste: The Case of Italian Pasta Production. *Resources, Conservation and Recycling, 144*, 82–89. doi:10.1016/j.resconrec.2019.01.025

Prodanović, R., Bošković, J., & Ignjatijević, S. (2016). Organic Honey Production in Function of Enviromental Protection. *Ecologica, 82*(23), 315–321.

Prodanov, M., Sierra, I., & Vidal-Valverde, C. (2004). Influence of Soaking and Cooking on the Thiamin, Riboflavin and Niacin Contents of Legumes. *Food Chemistry, 84*(2), 271–277. doi:10.1016/S0308-8146(03)00211-5

Ptukha, A. (n.d.). *Overview of the Russian Market of Healthy Food*. Step by step. Retrieved from http://www.step-by-step.ru/articles/our/RFDM%204%202017%20kartofel.pdf

Putnam, R. (2000). *Bowling Alone – The Collapse and Revival of American Community*. New York, NY: Simon & Schuster.

Qi, J., Qian, B., Shang, J., Huffman, T., Liu, J., Pattey, E., ... Wang, J. (2017). Assessing the Options to Improve Regional Wheat Yield in Eastern Canada Using the CSM-CERES-Wheat Model. *Agronomy Journal, 109*(2). doi:10.2134/agronj2016.06.0364

Rabinovich, B., Fedoseev, I., & Ignatiev, A. (1995). *Land Reform in Russia: Principles and Implementation*. Moscow: Stars and Co.

Rajalahti, R., Janssen, W., & Pehu, E. (2008). *Agricultural Innovation Systems: From Diagnostics Toward Operational Practices*. Washington, DC: The World Bank.

Razvedskaya, L. (2001). *The Development of the Technology of Production of Plant Preserves with Addition of Milk and Products of Leguminous Crops Processing*. Krasnodar: Kuban State Agrarian University.

Reinert, S. (2015). *How Rich Countries Have Become Rich and Why Poor Countries Remain Poor*. Moscow: Higher School of Economics.

Relf, D. (2015). *Sprouting Seeds for Food*. Retrieved from http://pubs.ext.vt.edu/content/dam/pubs_ext_vt_edu/426/426-419/426-419_pdf.pdf

Research and Markets. (2018). *Global Agricultural Machinery Market – Industry Trends, Opportunities and Forecasts to 2023*. Retrieved from https://www.researchandmarkets.com/research/b6lzkj/global?w=5

Reserve Bank of India. (n.d.). *Weekly Statistical Supplement*. Retrieved from https://www.rbi.org.in/scripts/WSSViewDetail.aspx?TYPE=Section&PARAM1=1

Resettlement Directorate. (1911). Resettlement beyond the Urals in 1911: A Reference Book for Walkers and Settlers with a Road Map of Asian Russia. Saint Petersburg: Printing House "Village of the Messenger".

Revenko, L. (2003). *World Food Market in the Era of the "Gene" Revolution.* Moscow: Economy.

Revenko, L. (2015). Parameters and Risks of Food Security. *International Processes, 41*(13), 6–20.

Rezaei, M., & Liu, B. (2017). Food Loss and Waste in the Food Supply Chain. *Nutfruit,* 26-27.

RIA Information Agency. (2018). *Russian Agricultural Imports to China Grew by 48% over Three Quarters.* Retrieved from https://ria.ru/20181123/1533386548.html

Robinson, G. O. (1982). Multiple Causation in Tort Law: Reflections on the Des Cases. *Virginia Law Review, 68*(4), 713–769. doi:10.2307/1072725

Robinson, S., Strzepek, K., & Cervigni, R. (2013). *The Cost of Adapting to Climate Change in Ethiopia: Sector-Wise and Macro-Economic Estimates.* Addis Ababa: Ethiopia Strategy Support Program.

Rodimov, B., Marshunina, A., Yafarov, I., Sadovnikova, V., & Labina, I. (1975). Biological Role of Heavy Water in Living Organisms. In *Questions of Radiobiology and Hematology* (pp. 118-126). Publishing House of Tomsk University.

Rodimov, B. (1961). Snow Water-Stimulator of Growth and Productivity of Animals and Plants. *Agriculture of Siberia and the Far East, 7,* 66–69.

Rogachevskaya, M. (n.d.). *P.A. Stolypin: Agrarian Reform and Siberia.* Retrieved from http://econom.nsc.ru/eco/arhiv/ReadStatiy/2002_09/Rogachevska.htm

Rogatnev, Y. (2001). *Fundamentals of Planning in Land Management.* Omsk: Omsk State Agrarian University.

Rogatnev, Y. (2003). *Theoretical and Methodological Fundamentals of Land Management under Conditions of Forming of Market Relations in Western Siberia.* Omsk: Omsk State Agrarian University.

Romanov, M., & Stepanko, A. (2018). Dynamics of Territorial Structures of Agriculture of the Far East of Russia. *Agrarian Bulletin of the Far East, 45*(1), 133–143.

Romero, A., Cuesta, C., & Sanchez-Muniz, F. J. (2000). Trans-Fatty Acid Production in Deep Fat Frying of Frozen Foods with Different Oils and Frying Modalities. *Nutrition Research (New York, N.Y.), 20*(4), 599–608. doi:10.1016/S0271-5317(00)00150-0

Roos, E., Mie, A., Wivstad, M., Salomon, E., Johansson, B., Gunnarsson, S., ... Watson, C. A. (2018). Risks and Opportunities of Increasing Yields in Organic Farming. A Review. *Agronomy for Sustainable Development, 38*(2), 14. doi:10.100713593-018-0489-3

Rose, D. C., & Chilvers, J. (2018). Agriculture 4.0: Broadening Responsible Innovation in an Era of Smart Farming. *Frontiers in Sustainable Food Systems, 2,* 87. doi:10.3389/fsufs.2018.00087

Rosenberg, D. (1984). The Causal Connection in Mass Exposure Cases: A "Public Law" Vision of the Tort System. *Harvard Law Review, 97*(4), 849–929. doi:10.2307/1341021

Roser, M., & Ritchie, H. (2013). *Hunger and Undernourishment.* Retrieved from https://ourworldindata.org/hunger-and-undernourishment#depth-of-the-food-deficit

Rosser, J. B. Jr. (1999). On the Complexities of Complex Economic Dynamics. *The Journal of Economic Perspectives, 13*(4), 169–192. doi:10.1257/jep.13.4.169

Ross, U. (1844). The Patent Commissioner of the United States Report. *Journal Scientific American, 27,* 370.

Roumeliotis, E., Kloosterman, B., Oortwijn, M., Kohlen, W., Bouwmeester, H., Visser, R., & Bachem, C. (2012). The Effects of Auxin and Strigolactones on Tuber Initiation and Stolon Architecture in Potato. *Journal of Experimental Botany*, *63*(12), 4539–4547. doi:10.1093/jxb/ers132 PMID:22689826

Rounsevell, M., Pedroli, B., Erb, K.-H., Gramberger, M., Busck, A. G., Haberl, H., ... Wolfslehner, B. (2012). Challenges for Land System Science. *Land Use Policy*, *29*(4), 899–910. doi:10.1016/j.landusepol.2012.01.007

Ruggeri Laderchi, C. (2001). *Participatory Methods in the Analysis of Poverty: A Critical Review*. Oxford: University of Oxford.

Rumyantsev, F. (2012). On Unclaimed Land Shares. *Economy and Law*, *5*, 111–115.

Russian Century. (2011). *Figures: Who Lost and Who Won from the Collapse of the USSR*. Retrieved from http://www.ruvek.ru/?module=articles&action=view&id=6031

Russian Research Institute of Information and Technical and Economic Studies on Engineering and Technical Support of Agriculture. (2018). *Report on the Conditions of Rural Areas in the Russian Federation in 2016*. Moscow: Russian Research Institute of Information and Technical and Economic Studies on Engineering and Technical Support of Agriculture.

Ryazanova, K. (2014). Convenience Meat Product with Fillings. In *Proceedings of the Conference "Youth. Science. Future – 2014."* Magnitogorsk: Magnitogorsk State Technical University.

Rybakov, I. (1995). Quality and Competitiveness in Market Relations. *Standards and Quality*, *12*, 43–47.

Rybas, S., & Tarakanova, L. (1991). *The Reformer: The Life and Death of Peter Stolypin*. Moscow: Nedra.

Rylko, D., Khotko, D., Abuzarova, S., Yunosheva, N., & Glazunova, I. (2015). *AGRICISTRADE Country Report: Russia*. Retrieved from http://www.agricistrade.eu/document-library

Ryumkin, S., & Malykhina, I. (2018). Development of Agriculture as a Solution to the Food Problem of Subarctic North of Russia. *Proceedings of the Conference Arctic 2018: International Collaboration, Environment and Security, Innovation Technologies and Logistics, Legislation, History and Modern State*. Krasnoyarsk: Krasnoyarsk State Agrarian University.

Ryumkin, S., Ryumkina, I., Nakonechnaya, O., & Essaulenko, D. (2018). Prospects for the Introduction and Use of Nature-Like Technologies in Agriculture. *Proceedings of the Conference Green Economy in Agriculture: Challenges and Perspectives*. Krasnodar: Kuban State Agrarian University.

Ryumkin, S., & Malykhina, I. (2017). On the Issue of "Smart" Agriculture: State, Problems and Prospects of Development. *Proceedings of the XX International Conference Agrarian Science to Agricultural Production of Siberia, Mongolia, Kazakhstan, Belarus and Bulgaria*. Novosibirsk: Novosibirsk State Agrarian University.

Sadler, P., & Cossich, J. (1997). *Patent UK 1487052 "Electrolytic Treatment of Drinking Water"*. Retrieved from https://patents.google.com/patent/US5807473

Saidi, A., & Diouri, M. (2017). Food Self-Sufficiency Under the Green-Morocco Plan. *Journal of Experimental Biology and Agricultural Sciences*, *5*(Spl-1- SAFSAW), 33–40. doi:10.18006/2017.5(Spl-1-SAFSAW).S33.S40

Samedova, E. (2013). *Meat Embargo: Help or Bad Turn for Russian Farmers?* Retrieved from http://dw.de/p/17emB

Samofalova, L. (2010). *The Scientific Foundation of Application of Germinated Seeds of Biobated Plants in the Production of Plant Basis and Substitutes of Dairy Products for Special Purpose*. Orel State Technical University.

Sanctuary, M., Tropp, H., & Haller, L. (2005). *Making Water a Part of Economic Development: The Economic Benefits of Improved Water Management and Services*. Stockholm: Stockholm International Water Institute.

Sapunov, V. (2012). *Environmental Challenges of the Mankind and Their Solutions*. Saarbrucken: Palmarium Academic Publishing.

Sasson, A. (2012). Food Security for Africa: An Urgent Global Challenge. *Agriculture & Food Security, 1*(2). doi:10.1186/2048-7010-1-2

Savchenko, E. (2001). Contemporary Issues of Land Relations Regulation. *International Agricultural Journal, 1*, 3–6.

Sayouti, N. S., & El Mekki, A. A. (2015). Le Plan Maroc Vert et l'autosuffisance alimentaire en produits de base à l'horizon 2020. *Alternatives Rurales, 3*(1), 78–90.

Schaafsma, G. (2012). Advantages and Limitations of the Protein Digestibility-Corrected Amino Acid Score (PDCAAS) as a Method for Evaluating Protein Quality in Human Diets. *British Journal of Nutrition, 108*(S2), 333–336. doi:10.1017/S0007114512002541 PMID:23107546

Schmidt, L., Ryumkin, S., & Ryumkina, I. (2018). NBICS-Evolution: Factors of Production in Digital Economy for Sustainable Agriculture and Rural Development. *Proceedings of the Conference Role of Agrarian Science in Sustainable Development of Rural Territories*. Novosibirsk: Novosibirsk State Agrarian University.

Schmidt, S. (2000). Formation of Trans-Unsaturation during Partial Catalytic Hydrogenation. *European Journal of Lipid Science and Technology, 102*(10), 646–648. doi:10.1002/1438-9312(200010)102:10<646::AID-EJLT646>3.0.CO;2-2

Schmitz, A., Moss, C., Schmitz, T., Furtan, W., & Schmitz, H. (2010). *Agricultural Policy, Agribusiness, and Rent-Seeking Behaviour*. Toronto: University of Toronto Press.

School of Active Longevity. (n.d.). *Activator*. Retrieved from http://www.mzk.ru/

Schott, J., Kotschwar, B., & Muir, J. (2013). *Understanding the Trans-Pacific Partnership*. Washington, DC: Peterson Institute for International Economics.

Schulte, R. P. O., Creamer, R. E., Donnellan, T., Farrelly, N., Fealy, R., O'Donoghue, C., & O'hUallachain, D. (2014). Functional Land Management: A Framework for Managing Soil-Based Ecosystem Services for the Sustainable Intensification of Agriculture. *Environmental Science & Policy, 38*, 45–58. doi:10.1016/j.envsci.2013.10.002

Schwartzstein, P. (2016). *African Farmers Say They Can Feed the World, and We Might Soon Need Them to*. Retrieved from https://qz.com/africa/736626/african-farmers-say-they-can-feed-the-world-and-we-might-soon-need-them-to/

Schwarz, W. (2000). Formation of Trans-Polyalkenoic Fatty Acids during Vegetable Oil Refining. *European Journal of Lipid Science and Technology, 102*(10), 648–649. doi:10.1002/1438-9312(200010)102:10<648::AID-EJLT648>3.0.CO;2-V

Sebedio, J. L., Dobarganes, M. C., Marquez-Ruiz, G., Wester, I., Christie, W. W., Dobson, F., ... Lahtinen, R. (1996). Industrial Production of Crisps and Prefried French Fries Using Sunflower Oils. *Grasas y Aceites, 47*(1-2), 5–13. doi:10.3989/gya.1996.v47.i1-2.836

Secretariat of the Convention on Biological Diversity. (2000). *Cartagena Protocol on Biosafety to the Convention on Biological Diversity*. Montreal: Secretariat of the Convention on Biological Diversity.

Secretariat of the Convention on Biological Diversity. (2005). *The Impact of Trade Liberalization on Agricultural Biological Diversity, Domestic Support Measures and their Effects on Agricultural Biological Diversity*. Montreal: Secretariat of the Convention on Biological Diversity.

Segneanu, A.E., Grozescu, I., Cepan, C., Cziple, F., Lazar, V., & Velciov, S. (2018). Food Security into a Circular Economy. *HSOA Journal of Food Science and Nutrition, 4*, 38.

Segrè, A., Falasconi, L., & Politano, A. (2016). *Crisi dei prezzi agricoli, sostenibilità e sprechi alimentari.* Retrieved from https://www.researchgate.net/publication/228837174_Crisi_dei_prezzi_agricoli_sostenibilita_e_sprechi_alimentari

Segrè, A., & Azzurro, P. (2016). *Spreco alimentare: dal recupero alla prevenzione. Indirizzi applicativi della legge per la limitazione degli sprechi.* Milan: Fondazione Giangiacomo Feltrinelli.

Segrè, A., & Falasconi, L. (2011). *Il libro nero dello spreco in Italia: il cibo.* Milan: Edizioni Ambiente.

Sen, A. (1998). *The Possibility of Social Choice.* Nobel Prize Organization. Retrieved from https://www.nobelprize.org/prizes/economic-sciences/1998/sen/lecture/

Serdyukova, M., & Nikolaeva, E. (2017). Development of Agricultural Cooperation and Small Forms of Economic Activities in Foreign Countries. *Bulletin of Chelyabinsk State University. Economic Sciences, 406*(10), 147–155.

Šerić, N. (2003). Importance of Remodeling of Marketing Strategies for the Market in the Countries in Transition. *Proceedings of the 5th International Conference "Enterprise in Transition"*. Split: University of Split.

Šerić, N., & Uglešić, D. (2014). The Marketing Strategies for Market Niches during Recession. *Proceedings of the 3rd REDETE 2014 Conference "Economic Development and Entrepreneurship in Transition Economies: Challenges in the Business Environment, Barriers and Challenges for Economic and Business Development"*. Banja Luka: University of Banja Luka.

Serova, E. (2000). Agricultural Reforms in Transition Economies: Mutual Objectives and Diverse Tools. Moscow: Encyclopedia of Russian Villages.

Shabin, S., Tyshkevich, E., & Ershova, T. (2017). Influence on the Yield of Spring Wheat of Presowing Treatment of Seeds by Ozone Air Agent. *Modern Science-Intensive Technologies, 49*(1), 130–136.

Shagaida, N., & Alakoz, V. (2017). *Land for People.* Moscow: Center for Strategic Research. doi:10.2139srn.3090400

Shagaida, N., & Fomin, A. (2017). Improvement of Land Policy in the Russian Federation. *Moscow Economic Journal, 3*, 1–26.

Shapiro, S. (1995). Trans-Fatty Acid and Coronary Disease: The Debate Continues. 2. Confounding and Selection Bias in the Data. *American Journal of Public Health, 85*(3), 410–413. doi:10.2105/AJPH.85.3.410-a PMID:7892932

Shapkina, L. (2012). Regional Aspects of Management of Food Security. *Terra Economicus, 1-2*, 128–131.

Shaskolskiy, V., & Shaskolskaya, N. (2007). Antioxidant Activity of Germinated Seeds. *Bread Products, 8*, 58–59.

Shcherbakova (Ponomareva), A. (2017). Organic Agriculture in Russia. *Siberian Journal of Life Sciences and Agriculture, 9*(4), 151–173.

Shelepina, N. (2010). Main Trends in Pea Grains Cultivation. Security and Quality of Goods. *Proceedings of IV International Scientific and Practical Conference*. Saratov: KUBiK.

Shelepina, N., & Parshutina, I. (2013). Comparative Characteristics of Amino-Acid Composition of Wheat Flour and Embryonic Foodstuffs from Pea. *Proceedings of the III International Conference "New in Technologies and Techniques of Food for Special Purpose with Medical and Biological Approach."* Voronezh: Voronezh State University of Engineering Technologies.

Sheynin, L. (2011). *Real Estate: Legislative Gaps.* Moscow: Delovoy Dvor.

Shields, D. (2015). *Farm Safety Net Programs: Background and Issues.* Washington, DC: Congressional Research Service.

Shiferaw, A., & Bedi, W. (2013). The Dynamics of Job Creation and Job Destruction in an African Economy: Evidence from Ethiopia. *Journal of African Economies*, *22*(5), 651–692. doi:10.1093/jae/ejt006

Shilovsky, M. (2003). *Siberian Relocation. Documents and Materials*. Novosibirsk: Novosibirsk State University.

Shilovsky, M. (2012). Branch Records as a Source on the History of Land Management in Tomsk Province in the Beginning of XX Century. *Tomsk State University Journal. History (London)*, *18*(2), 74–78.

Shramko, G., Alexandrova, E., & Knyazeva, T. (2011). Improving the Technology of Foliar Feeding of Winter Wheat Using Electrochemically Activated Water. *Proceedings of Kuban State Agrarian University*, *33*(6), 69–72.

Silva, R. F. B. D., Batistella, M., Dou, Y., Moran, E., Torres, S. M., & Liu, J. (2017). The Sino-Brazilian Telecoupled Soybean System and Cascading Effects for the Exporting Country. *Land (Basel)*, *6*(5), 53. doi:10.3390/land6030053

Simakova, I., Perkel, R., Lomnitsky, I., & Terentiev, A. (2015). Clinical Studies on the Safety of Frying Fats Containing Trans-Isomers of Oleic Acid. *Scientific Review (Singapore)*, *2*, 52–56.

Singer, J. (2006). Framing Brand Management for Marketing Ecosystems. *The Journal of Business Strategy*, *27*(5), 50–57. doi:10.1108/02756660610692716

Skiba, I. (1988). *Agro-Industrial Complex: Economic Reform and Democratization: On the Experience and Problems of Economic Reform and Strengthening of Democratization in Agri-Industrial Complex of a Country*. Moscow: Politizdat.

Skopichev, V., Eisymont, T., Alekseev, N., Bogolyubova, I., Enukashvili, A., & Karpenko, L. (2004). *Animal Physiology and Etiology*. Moscow: Kolos.

Skripko, O., Isaycheva, N., & Pokotilo, O. (2015). Preparation of Protein-Vitamin-Mineral Products with Use of Soy for a Healthy Nutrition. *Food Industries*, *5*, 34–37.

Skurikhin, I., & Tutelyan, V. (2002). *Chemical Composition of Foodstuffs in Russia: Guidance*. Moscow: DeLi Print.

Smith, P. (2013). Delivering Food Security without Increasing Pressure on Land. *Global Food Security*, *2*(1), 18–23. doi:10.1016/j.gfs.2012.11.008

Smith, P., Gregory, P. J., van Vuuren, D., Obersteiner, M., Havlik, P., Rounsevell, M., ... Bellarby, J. (2010). Competition for Land. *Philosophical Transactions of the Royal Society of London. Series B, Biological Sciences*, *365*(1554), 2941–2957. doi:10.1098/rstb.2010.0127 PMID:20713395

Snider, M. (2019). *Asbestos and Additive Manufacturing: Addressing Early Concerns Surrounding Manufacturing 3D-Printing Technology Using Asbestos Litigation as a Model*. Retrieved from https://papers.ssrn.com/sol3/papers.cfm?abstract_id=3343881

Solomintsev, M., & Mogilny, M. (2009). Determination of Proteinase Inhibitors Activity in Food Products. *News of Institutes of Higher Education. Food Technology*, *1*, 13–16.

Solovchuk, K. (2015). Regulation and Support for Innovations in the Agricultural Sector of the European Union. *Actual Problems of Economics*, *165*(3), 62–68.

Sozinova, S., & Kolpakova, T. (2015). Trends in the Development of Cooperation between Russia and China in the Sphere of Agriculture. *Russia and China: Problems of Strategic Cooperation. Proceedings of the Eastern Center*, *16*(2), 70–81.

Spielman, D. J., & Birner, R. (2008). *How Innovative Is Your Agriculture? Using Innovation Indicators and Benchmarks to Strengthen National Agricultural Innovation Systems*. Washington, DC: The World Bank.

Spring, J. (2018). *Brazil Record Soy Exports to China Could Expand Further – Official*. Retrieved from https://www.reuters.com/article/brazil-agriculture-trade/brazil-record-soy-exports-to-china-could-expand-further-official-idUSL4N1XP5YI

Spring, J., & Polansek, T. (2018). *Trump Trade War Delivers Farm Boom in Brazil, Gloom in Iowa*. Retrieved from https://www.reuters.com/article/us-usa-trade-china-brazil-insight/trump-trade-war-delivers-farm-boom-in-brazil-gloom-in-iowa-idUSKCN1ML0E7

Standing Committee for Economic and Commercial Cooperation of the Organization of Islamic Cooperation. (2017). *Improving Agricultural Market Performance : Creation and Development of Market Institutions*. Ankara: COMCEC Coordination Office.

Staus, A., & Becker, T. (2012). Attributes of Overall Satisfaction of Agricultural Machinery Dealers Using a Three-Factor Model. *Journal of Business and Industrial Marketing, 27*(8), 635–643. doi:10.1108/08858621211273583

Stenmarck, A., Jensen, C., Quested, T., & Moates, G. (2016). *Estimates of European Food Waste Levels*. Stockholm: IVL Swedish Environmental Research Institute.

Stolypin, P., & Krivoshein, A. (1911). *Trip to Siberia and Volga Region*. Retrieved from https://xn--90anbaj9ad0j.xn--80asehdb/documents/poezdka-v-sibir-i-povolzhe-zapiska-pa-stolypina-i-av-krivosheina-spb-1911/

Stroev, E. (Ed.). (2001). *Mixed Agrarian Economy and Russian Village (Middle of the 1980-1990s)*. Moscow: Kolos.

Stroev, E., & Volkov, S. (2001). *Land Issues in Russia in the Beginning of the XXI Century (Challenges and Solutions)*. Moscow: State University of Land Use Planning.

Stroh de Martinez, C., Feddersen, M., & Speicher, A. (2016). *Food Security in Sub-Saharan Africa: A Fresh Look on Agricultural Mechanisation. How Adapted Financial Solutions Can Make a Difference*. Bonn: German Development Institute.

Study.com. (2018). *New Trade Theory (NTT): Definition and Analysis*. Retrieved from https://study.com/academy/lesson/new-trade-theory-ntt-definition-analysis.html

Sugar.ru. (2016). *Ten Biggest Agricultural Producers in the World*. Retrieved from http://sugar.ru/node/14299

Sukhomirov, G. (2012). World Food Crisis and Food Security of the Far Eastern Federal District. *Spatial Economics, 2*(30), 158–160. doi:10.14530e.2012.2.158-160

Sunder, S. (2018). *India Economic Survey 2018: Farmers Gain as Agriculture Mechanisation Speeds up, but More R&D Needed*. Retrieved from https://www.financialexpress.com/Budget/India-Economic-Survey-2018-For-Farmers-Agriculture-Gdp-Msp/1034266/

Sun, P., & Heshmati, A. (2010). *International Trade and Its Effects on Economic Growth in China*. Bonn: Institute for the Study of Labor.

Supreme Eurasian Economic Council. (2018a). *Decision #3 from May 8, 2018, "On the Approval of the Agreement on Trade and Economic Cooperation between the Eurasian Economic Union and Its Member States, from One Side, and the People's Republic of China, from Another Side."* Retrieved from https://docs.eaeunion.org/docs/ru-ru/01418737/scd_10052018_3

Supreme Eurasian Economic Council. (2018b). *Decision #19 from December 6, 2018, "On Major Directions of International Activity of the Eurasian Economic Union in 2019."* Retrieved from https://docs.eaeunion.org/docs/ru-ru/01420197/scd_07122018_19

Syroedov, N. (1997). Notarization of Land Shares Transactions. *Bulletin of Legal Information. Land and Law, 3*, 6.

Syzdykov, R., Aitmambet, K., & Dautov, A. (2015). *AGRICISTRADE Country Report: Kazakhstan*. Retrieved April 27, 2019, from http://www.agricistrade.eu/document-library

Taheripour, F., & Tyner, W. E. (2018a). Impacts of Possible Chinese 25% Tariff on U.S. Soybeans and Other Agricultural Commodities. *Choices, 33*(2).

Taheripour, F., & Tyner, W. E. (2018b). *Impacts of Possible Chinese Protection on US Soybeans*. West Lafayette, IN: Purdue University.

Tarasov, N., & Volodin, V. (1998). Land Relation Issues in Agricultural Cooperatives – Collective Farms. In *Proceedings of the Conference "Land Relations in Agro-Industrial Industry of Russia."* Uglich.

Tavano, O. L., Neves, V. A., & Da Silva, S. I. J. (2016). In Vitro Versus in Vivo Protein Digestibility Techniques for Calculating PDCAAS (Protein Digestibility-Corrected Amino Acid Score) Applied to Chickpea Fractions. *Food Research International, 89*(1), 756–763. doi:10.1016/j.foodres.2016.10.005 PMID:28460976

Technopolis Group. (2016). *Regulatory Barriers for the Circular Economy. Lessons from Ten Case Studies*. Amsterdam: Technopolis Group.

Teri, E. (1914). *Russia in 2014. Economic Review*. Retrieved from http://www.mysteriouscountry.ru/wiki/index.php/%D0%AD%D0%B4%D0%BC%D0%BE%D0%BD_%D0%A2%D1%8D%D1%80%D0%B8_%D0%A0%D0%9E%D0%A1%D0%A1%D0%98%D0%AF_%D0%92_1914_%D0%B3._%D0%AD%D0%BA%D0%BE%D0%BD%D0%BE%D0%BC%D0%B8%D1%87%D0%B5%D1%81%D0%BA%D0%B8%D0%B9_%D0%BE%D0%B1%D0%B7%D0%BE%D1%80/%D0%9F%D1%80%D0%B5%D0%B4%D0%B8%D1%81%D0%BB%D0%BE%D0%B2%D0%B8%D0%B5

Tesfatsion, L., & Judd, K. L. (2006). *Handbook of Computational Economics II: Agent-Based Computational Economics*. London: North-Holland.

Tesser, F., & Cavicchioli, D. (2014). Economic Aspects of Beekeeping and Honey Productions in Italy and in Lombardy Region. *Proceedings of the Seminarion sugli insetti utili – sala dei Cavalieri, Castelo Visconteo di Sant`Angelo Lodigiano*. Lodi: Lombardo Museum of the History of Agriculture.

The American Law Institute. (1965). *Restatement of the Law*. Washington, DC: American Law Institute Publishers.

The Economist Intelligence Unit. (2018). *Global Food Security Index*. Retrieved from https://foodsecurityindex.eiu.com/Home/DownloadIndex

The Economist. (2018). *Africa Has Plenty of Land. Why Is It so Hard to Make a Living from It?* Retrieved from https://www.economist.com/middle-east-and-africa/2018/04/28/africa-has-plenty-of-land-why-is-it-so-hard-to-make-a-living-from-it

The Intergovernmental Panel on Climate Change. (2019). *Final Report. Summary for Politics*. Retrieved from http://www.ipcc.ch/pdf/assessment-report/ar5/syr/AR5_SYR_FINAL_SPM_ru.pdf

The International Bank for Reconstruction and Development. (1992). *Food and Agricultural Policy Reforms in the Former USSR. An Agenda for the Transition Country Department III Europe and Central Asia Region*. Washington, DC: The International Bank for Reconstruction and Development.

The White House. (2016). *Historical Tables: Budget of the U.S. Government. Fiscal Year 2016*. Retrieved from https://www.whitehouse.gov/omb/historical-tables/

The World Bank. (1992). *International Agriculture and Trade Reports*. Washington, DC: The World Bank.

The World Bank. (2018). *Central Asia. China (and Russia) 2030 – Implications for Agriculture in Central Asia*. Washington, DC: The World Bank.

The World Bank. (2019). *World Bank Open Data*. Retrieved from https://data.worldbank.org

The World Bank. (n.d.). *DataBank*. Retrieved from https://databank.worldbank.org/source/exporter-dynamics-database-%E2%80%93-indicators-at-country%5Eproduct-hs4%5Eyear-level/Type/TABLE/preview/on

The World Bank. (n.d.). *Liberalizing Trade in Agriculture: Africa and the New WTO Development Agenda*. Washington, DC: The World Bank.

Thirtle, C., Lin, L., & Piesse, J. (2003). The Impact of Research-Led Agricultural Productivity Growth on Poverty Reduction in Africa, Asia and Latin America. *World Development, 31*(12), 1959–1975. doi:10.1016/j.worlddev.2003.07.001

Thow, A. M., & Hawkes, C. (2009). The Implications of Trade Liberalization for Diet and Health: A Case Study from Central America. *Globalization and Health, 28*(1), 5. doi:10.1186/1744-8603-5-5 PMID:19638196

Tikhonov, P. (1985). *Competitiveness of Industrial Products*. Moscow: Publishing House of Standards.

Tilman, D., Balzer, C., Hill, J., & Befort, B. L. (2011). Global Food Demand and the Sustainable Intensification of Agriculture. *Proceedings of the National Academy of Sciences of the United States of America, 108*(50), 20260–20264. doi:10.1073/pnas.1116437108 PMID:22106295

Tirelli, D. (n.d.). *Richieste del consumatore*. Retrieved from https://www.colturaecultura.it/capitolo/richieste-del-consumatore

Tiwari, S., & Dhuria, S. (2018). Effect of Pre-Sowing Treatment on Seed Germination and Seedlings Growth Characteristics of *Albizia Procera*. *Asian Journal of Research in Agriculture and Forestry, 2*(1), 1–6. doi:10.9734/AJRAF/2018/42370

Toleubaeva, D., & Patlasov, O. (2018). State Regulation of the Market of Organic Products in Russia. *Human Science: Humanitarian Studies, 31*(1), 182–191.

Tran, J. (2016a). 3D-Printed Food. *Minnesota Journal of Law, Science & Technology, 17*, 855–880.

Tran, J. (2016b). Press Clause and 3D Printing. *Northwestern Journal of Technology and Intellectual Property, 14*(1), 75–80.

Tsafack Matsop, A. S., Muluh Achu, G., Kamajou, F., Ingram, V., & Vabi Boboh, M. (2011). Comparative Study of the Profitability of Two Types of Bee Farming in the North West Cameroon. *Tropicultura, 29*, 3–7.

Tsareva, N. (2007). *The Use of Foaming Qualities of Leguminous Crops in the Technology of Whipped Cottage Cheese Deserts. Orel*. Orel State Technical University.

Turk, C. (2001). *Linking Participatory Poverty Assessments to Policy and Policymaking: Experience from Vietnam*. Hanoi: The World Bank.

Tutelyan, V. (n.d.). *Priorities of Government Policy in Healthy Nutrition in Russia on the Federal and Regional Levels*. Retrieved from http://pfcop.opitanii.ru/articles/state_feed_prioritets.shtml

Tutelyan, V. (2005). Nutrition and Health. *Food Industries, 5*, 6–7.

Tyukavkin, V. (1986). Agrarian Resettlement in Russia in the Era of Imperialism. In *Socio-Demographic Processes in Rural Areas of Russia (XVI – beginning of XX century)* (pp. 214-225). Academic Press.

Uauy, R., Aro, A., Clarke, R., Ghafoorunissa, R., L'Abbe, M., Mozaffarian, D., & Tavella, M. (2009). WHO Scientific Update on Trans-Fatty Acids: Summary and Conclusions. *European Journal of Clinical Nutrition, 63*(S2), S68–S75. doi:10.1038/ejcn.2009.15

Unified Interdepartmental System of Information and Statistics. (2019). *The Share of Imported Food Products in the Commodity Resources of Retail Trade in Food Products.* Retrieved from https://fedstat.ru/indicator/37164

United Nations Comtrade Database. (n.d.). *International Trade Statistics Database.* Retrieved from http://comtrade.un.org

United Nations Conference on Environment and Development. (1992). *Rio Declaration on Environment and Development.* Retrieved from http://www.unesco.org/education/pdf/RIO_E.PDF

United Nations Conference on Trade and Development. (2009). *The Role of South-South and Triangular Cooperation for Sustainable Agriculture Development and Food Security in Developing Countries.* Geneva: United Nations.

United Nations Conference on Trade and Development. (2017). *World Investment Report 2017. Investment and the Digital Economy.* Geneva: United Nations.

United Nations Conference on Trade and Development. (2018a). *Investment and New Industrial Policies: Key Messages and Overview.* Geneva: United Nations.

United Nations Conference on Trade and Development. (2018b). *Economic Development in Africa. Report 2018. Migration for Structural Transformation.* New York, NY: United Nations.

United Nations Conference on Trade and Development. (2019). *Statistics Database* [Data file]. Retrieved from http://unctad.org/en/Pages/Statistics.aspx

United Nations Environment Programme. (2010). *Africa Water Atlas.* Nairobi: United Nations Environment Programme.

United Nations Office for South-South Cooperation. (2019). *India-UN Development Partnership Fund.* New York, NY: UNOSSC.

United Nations. (1974). *Communication from the Commission to the Council. SEC (74) 4955 Final.* New York, NY: United Nations.

United Nations. (1992). *Convention on Biological Diversity.* Retrieved from https://www.cbd.int/convention/text/

United Nations. (2000). *United Nations Millennium Declaration.* Retrieved from https://www.un.org/millennium/declaration/ares552e.pdf

United Nations. (2017). *Forecast of the World's Population by 2050.* Retrieved from https://www.mirprognozov.ru/prognosis/society/oon-prognoz-naseleniya-zemli-k-2050-godu/

United Nations. (2017). *World Population Prospects: The 2017 Revision.* New York, NY: United Nations.

United Nations. (2018). *World Economic Situation and Prospects.* Washington, DC: United Nations.

United Nations. (n.d.). *Agreement on Agriculture: Recognition of Interests in Negotiations (Article 20 & the Preamble).* Retrieved from https://www.un.org/ldcportal/agreement-on-agriculture-recognition-of-interests-in-negotiations-article-20-the-preamble/

United Nations. (n.d.). *UN Comtrade Database.* Retrieved from https://comtrade.un.org/

United States Department of Agriculture. (2010). *National Export Initiative: Importance of U.S. Agricultural Exports.* Retrieved from https://webarchive.library.unt.edu/web/20130216190207/http://@fas.usda.gov/info/NEI/NEInewrev.pdf

United States Department of Agriculture. (2018). *Top U.S. Agricultural Exports in 2017.* Retrieved from https://www.fas.usda.gov/data/top-us-agricultural-exports-2017

United States Government. (1985). *99-198 Food Security Act of 1985*. Retrieved from http://legcounsel.house.gov/Comps/99-198%20-%20Food%20Security%20Act%20Of%201985.pdf

Urquia Grande, E., Cano, E., Estebanez, R. P., & Chamizo-Gonzalez, J. (2018). Agriculture, Nutrition and Economics through Training: A Virtuous Cycle in Rural Ethiopia. *Land Use Policy, 79*, 707–716. doi:10.1016/j.landusepol.2018.09.005

Urquia Grande, E., & Rubio-Alcocer, A. (2015). Agricultural Infrastructure Donation Performance: Empirical Evidence in Rural Ethiopia. *Agricultural Water Management, 158*, 245–254. doi:10.1016/j.agwat.2015.04.020

Urutyan, V., Yeritsyan, A., & Mnatsakanyan, H. (2015). *AGRICISTRADE Country Report: Armenia*. Retrieved from http://www.agricistrade.eu/document-library

Usenko, L. (2014). Food Security of Russia: Challenges and Implementation Mechanisms. *Proceedings of VEO, 187*, 198–203.

Ushachev, I. (2014). *Scientific Support of the State Program of Development of Agriculture and Regulation of Markets of Agricultural Products, Raw Materials, and Food for 2013-2020 Report at the General Meeting of the Russian Agricultural Academy*. Retrieved from http://www.vniiesh.ru/news/9671.html

Ushachev, I. (2003). To Improve State Agricultural Policy. *Agro-Industrial Complex: Economy. Management, 11*, 3–9.

Ushachev, I. (2013). *Food Security of Russia in the Framework of the Global Partnership*. Moscow: All-Russian Research Institute of Agrarian Economics and Social Development of Rural Territories.

Ushachev, I. (2015). Strategic Approaches to the Development of Agriculture in Russia in the Context of Interstate Integration. *Proceedings of the International Scientific Conference "Agrarian Sector of Russia in the Conditions of International Sanctions: Challenges and Responses."* Moscow: Russian State Agrarian University.

Uskova, R. (2014). *Regional Food Security*. Vologda: Institute for Social and Economic Development of Territories of the Russian Academy of Science.

Ustyukova, V. (2007). Citizens Collective Ownership Rights for Land Plots of Agricultural Purpose: Myth or Reality? *Ecological Law, 2*, 19–25.

Uzun, V. (2012). Russian Policy of Agriculture Support and the Necessity of Its Modification after WTO Accession. *Issues of Economics, 10*, 132–149.

Uzun, V., & Lerman, Z. (2017). Outcomes of Agrarian Reform in Russia. In S. Gomez y Paloma, S. Mary, S. Langrell, & P. Ciaian (Eds.), *The Eurasian Wheat Belt and Food Security: Global and Regional Aspects* (pp. 81–101). Amsterdam: Springer. doi:10.1007/978-3-319-33239-0_6

Uzun, V., & Shagaida, N. (2015). *Agrarian Reform in Post-Soviet Russia: Tools and Results*. Moscow: Delo.

Vakhidov, V., Mamadzhanov, U., Kasymov, A., Bakhir, V., Alekhin, S., & Iskhakova, H. ... Goncharov, P. (1999). *Patent #1121906 The Method of Obtaining a Liquid with Biologically Active Properties*. Russian Federation. Retrieved from http://www.bakhir.ru/inventions/03/

Vakhitova, R. (2015). *The Formation of Pea Harvest Depending of the Elements of Cultivation Technology in the Cis-Ural region of the Republic of Bashkortostan*. Ufa: Bashkir State Agrarian University.

Valuyeva, T., & Mosolov, V. (1995). Proteins-Inhibitors of Proteolytic Enzymes of Plants. *Applied Biochemistry and Microbiology, 31*(6), 579–589.

Valuyeva, T., & Mosolov, V. (2011). Inhibitors of Proteolytic Enzymes while Plants Abiotic Stress. *Applied Biochemistry and Microbiology, 47*(5), 501–507.

Van der Molen, P. (2017). Food Security, Land Use and Land Surveyors. *Survey Review, 353*(49), 147–152.

Vavilov, N. (1987). *Five Continents*. Leningrad: Nauka.

Veber, A., Kazydub, N., Leonova, S., Zhiarno, M., Nadtochii, L., Vorobiev, D., & Fialkov, D. (2019). *Patent #2685911 "Process of Foodstuff from Haricot Production."* Retrieved from http://www1.fips.ru/registers-doc-view/fips_servlet?DB=RUPAT&DocNumber=2685911&TypeFile=html

Veber, A., Kazydub, N., Leonova, S., & Zhiarno, M. (2017). The Process of Biologically Active Component Derivation from Germinated Haricot Seed for Further Development. *Bread Products, 17*(6), 35–38.

Veber, A., Leonova, S., Kazydub, N., Simakova, I., & Nadtochii, L. (2019). Special Legume-Based Food as a Solution to Food and Nutrition Insecurity Problem in the Arctic. In V. Erokhin, T. Gao, & X. Zhang (Eds.), *Handbook of Research on International Collaboration, Economic Development, and Sustainability in the Arctic* (pp. 570–592). Hershey, PA: IGI Global. doi:10.4018/978-1-5225-6954-1.ch027

Veber, A., Maradudin, M., Strizhevskaya, V., Romanova, H., & Simakova, I. (2019). The Research of Functional and Technological Qualities of Composite Mixtures from Wheat and Haricot of Domestic Selection. *Food Industries, 3*, 45–49.

Venchiarutti, A. (2018). The Recognition of Punitive Damages in Italy: A Commentary on Cass Sez Un 5 July 2017, 16601, AXO Sport, SpA v NOSA Inc. *Journal of European Tort Law, 9*(1), 104–122. doi:10.1515/jetl-2018-0105

Vernadsky, V. (2002). *Biosphere and Noosphere*. Moscow: Rolf.

Vershinin, V., & Petrov, V. (2015). Development of Tools Ensuring Agricultural Circulation of Unused Agricultural Land. *International Agricultural Journal, 5*, 9–11.

Vinogradov, P. (1914). *The Memorial Book of Tomsk Province in 1914*. Tomsk: Printing House of Provincial Administration.

Visser, O., Mamonova, N., Spoor, M., & Nikulin, A. (2015). "Quiet Food Sovereignty" as Food Sovereignty without a Movement? Insights from Post-socialist Russia. *Globalizations, 12*(4), 1–16. doi:10.1080/14747731.2015.1005968

Vlahovich, S. (2008). Frozen Food Market Today and Forecasts of Its Development for the Future. *Ice-Cream and Frozen Products, 12*, 28–31.

Volgareva, G., & Pasko, O. (2014). *Land Reforms at the Beginning and the End of XX Century in Russia*. Tomsk: Demos.

Volkov, S. (2001). *Land Management. Land Use Planning. Intra-Farm Land Tenure*. Moscow: Kolos.

Volkov, S., & Khlystun, V. (2014). Land Policy: How to Improve Effectiveness? *International Agricultural Journal, 1-2*, 3–6.

Volkov, S., & Lipski, S. (2017). Legal and Land Use Planning Measures for Involvement of Unused Agricultural Land into Economic Circulation and for Ensuring Their Effective Use. *Land Management. Land Monitoring and Cadaster, 145*(2), 5–10.

Von Braun, J., Serova, E., Seeth, H., & Melyukhina, O. (1996). *Russia's Food Economy in Transition: What Do Reforms Mean for the Long-term Outlook?* Washington, DC: International Food Policy Research Institute.

Von Lampe, M., Willenbockel, D., Ahammad, H., Blanc, E., Cai, Y., Calvin, K., ... van Meijl, H. (2014). Why Do Global Long-Term Scenarios for Agriculture Differ? An Overview of the AgMIP Global Economic Model Intercomparison. *Agricultural Economics, 45*(1), 3–20. doi:10.1111/agec.12086

Vorobiev, D., Zhiarno, M., Leonova, S., Veber, A., Kazydub, N., & Zabudsky, A. ... Ponomareva, E. (2017). *Patent #2676166 "Process of Ice-Cream Production."* Retrieved from https://findpatent.ru/patent/267/2676166.html

Compilation of References

Vosta, M. (2014). The Foodstuffs Market in the CR and Its Regulation within the Framework of the EU Agricultural Policy. *Agricultural Economics – Czech, 60*, 279-286.

Vozhyan, V., Taran, M., Yakobutsa, M., & Avadrniy, L. (2013). The Nutrition Value of Pea, Soybean, Haricot Cultivars and Antinutritional Stuffs in Them. *Scientific and Industrial Journal. Leguminous and Cereal Foodstuffs, 5*(1), 26–29.

Vujičić, M. D., Vasiljević, D. A., Marković, S. B., Hose, T. A., Lukić, T., Hadžić, O., & Janićević, S. (2011). Preliminary Geosite Assessment Model (GAM) and Its Application on Fruška Gora Mountain, Potential Geotourism Destination of Serbia. *Acta Geographica Slovenica, 51*(2), 361–376. doi:10.3986/AGS51303

Wagesho, N., Goel, N., & Jain, M. (2013). Temporal and Spatial Variability of Annual and Seasonal Rainfall over Ethiopia. *Hydrological Sciences Journal, 58*(2), 354–373. doi:10.1080/02626667.2012.754543

Walker, M. (2007). Globalisation 3.0. Prospects for a New World Order. *The Wilson Quarterly*, 16–24.

Walugembe, M., Tebug, S., Tapio, M., Missohou, A., Juga, J., Marshall, K., & Rothschild, M. F. (2016). *Gendered Intra-Household Contributions to Low-Input Dairy in Senegal*. Ames, IA: Iowa State University. doi:10.31274/ans_air-180814-208

Wang, W., Jiang, D., Chen, D., Chen, Z., Zhou, W., & Zhu, B. (2016). A Material Flow Analysis (MFA)-Based Potential Analysis of Eco-Efficiency Indicators of China's Cement and Cement-Based Materials Industry. *Journal of Cleaner Production, 112*(1), 787–796. doi:10.1016/j.jclepro.2015.06.103

Wang, X., Cai, D., Grant, C., Hoogmoed, W. B., & Oenema, O. (2015). Factors Controlling Regional Grain Yield in China over the Last 20 Years. *Agronomy for Sustainable Development, 35*(3), 1127–1138. doi:10.100713593-015-0288-z

Warr, P. (2011). *Food Security vs. Food Self-Sufficiency: The Indonesian Case*. Canberra: The Australian National University.

Wasserman, S., & Faust, K. (1994). *Social Network Analysis: Methods and Applications*. Cambridge: Cambridge University Press. doi:10.1017/CBO9780511815478

Wegren, S. (2012). Institutional Impact and Agricultural Change in Russia. *Journal of Eurasian Studies, 3*(2), 193–202. doi:10.1016/j.euras.2012.03.010

White, C. (2012). *Understanding Water Scarcity: Definitions and Measurements*. Retrieved from http://www.globalwaterforum.org/2012/05/07/understanding-water-scarcity-definitions-and-measurements/

White, D., Burton, M., & Dow, M. (1981). Sexual Division of Labor in African Agriculture: A Network Autocorrelation Analysis. *American Anthropologist, 83*(4), 824–849. doi:10.1525/aa.1981.83.4.02a00040

Wiebe, K., Lotze-Campen, H., Sands, R., Tabeau, A., van der Mensbrugghe, D., Biewald, A., ... Willenbockel, D. (2015). Climate Change Impacts on Agriculture in 2050 under a Range of Plausible Socioeconomic and Emissions Scenarios. *Environmental Research Letters, 10*(8), 085010. doi:10.1088/1748-9326/10/8/085010

Wierzbowska, J., Cwalina-Ambroziak, B., Glosek, M., & Sienkiewicz, S. (2015). Effect of Biostimulators on Yield and Selected Chemical Properties of Potato Tubers. *Journal of Elementology, 20*(3), 757–768.

Wirtz, J., Tuzovic, S., & Ehret, M. (2015). Global Business Services: Increasing Specialization and Integration of the World Economy as Drivers of Economic Growth. *Journal of Service Management, 26*(4), 565–587. doi:10.1108/JOSM-01-2015-0024

Wishnick, E. (2005). Migration and Economic Security: Chinese Labour Migrants in the Russian Far East. *Crossing National Borders: Human Migration Issues in North Asia*, 68-92.

Wittman, H., Desmarais, A., & Wiebe, N. (2010). *Food Sovereignty: Reconnecting Food, Nature and Community.* Halifax, Oakland: Fernwood Publishing and Food First Books.

Wittman, H., Desmarais, A., & Wiebe, N. (2010). *Food Sovereignty: Reconnecting Food, Nature and Community. Halifax: Fernwood Publishing.* Oakland: FoodFirst Books.

World Bank. (2008). *Sustainable Land Management Sourcebook.* Washington, DC: World Bank.

World Bank. (2014). *Renewable Internal Freshwater Resources.* Retrieved from https://data.worldbank.org/indicator/ER.H2O.INTR.K3

World Bank. (2015). *Improving Food Security in West Africa: Removing Obstacles to Regional Trade Markets.*

World Bank. (n.d.). *World Development Indicators.* Retrieved from https://databank.worldbank.org/reports.aspx?source=2&type=metadata&series=SP.POP.TOTL

World Health Organization. (2017). *Progress on Drinking Water, Sanitation and Hygiene.* Geneva: World Health Organization.

World Health Organization. (n.d.). *Population Nutrient Intake Goals for Preventing Diet-Related Chronic Diseases.* Retrieved from https://www.who.int/nutrition/topics/5_population_nutrient/en/

World Trade Organization. (1995). *Agreement on Agriculture.* Retrieved from https://www.wto.org/english/docs_e/legal_e/14-ag_01_e.htm

World Trade Organization. (2012). *Trade Policy Review: United States of America.* Retrieved from https://www.wto.org/english/tratop_e/tpr_e/tp375_e.htm

World Trade Organization. (2013a). *Trade Policy Review: Brunei Darussalam.* Retrieved from https://www.wto.org/english/tratop_e/tpr_e/s309_e.pdf

World Trade Organization. (2013b). *Trade Policy Review: Mexico.* Retrieved from https://www.wto.org/english/tratop_e/tpr_e/s279_e.pdf

World Trade Organization. (2013c). *Trade Policy Review: Peru.* Retrieved from https://www.wto.org/english/tratop_e/tpr_e/s289_e.pdf

World Trade Organization. (2013d). *Trade Policy Review: Vietnam.* Retrieved from https://www.wto.org/english/tratop_e/tpr_e/s287_e.pdf

World Trade Organization. (2014). *Trade Policy Review: Malaysia.* Retrieved from https://www.wto.org/english/tratop_e/tpr_e/s292_e.pdf

World Trade Organization. (2015a). *Trade Policy Review: Australia.* Retrieved from https://www.wto.org/english/tratop_e/tpr_e/s312_e.pdf

World Trade Organization. (2015b). *Trade Policy Review: Canada.* Retrieved from https://www.wto.org/english/tratop_e/tpr_e/s314_e.pdf

World Trade Organization. (2015c). *Trade Policy Review: Chile.* Retrieved from https://www.wto.org/english/tratop_e/tpr_e/s315_e.pdf

World Trade Organization. (2015d). *Trade Policy Review: Japan.* Retrieved from https://www.wto.org/english/tratop_e/tpr_e/s310_e.pdf

Compilation of References

World Trade Organization. (2015e). *Trade Policy Review: New Zealand*. Retrieved from https://www.wto.org/english/tratop_e/tpr_e/s316_e.pdf

World Trade Organization. (2018). *World Tariff Profiles 2018*. Retrieved from https://www.wto.org/ENGLISH/res_e/publications_e/world_tariff_profiles18_e.htm

World Trade Organization. (2019). *World Trade Statistical Review 2018*. Geneva: World Trade Organization.

World Trade Organization. (n.d.). *Agreement on Agriculture*. Retrieved from https://www.wto.org/english/docs_e/legal_e/14-ag_01_e.htm

World Trade Organization. (n.d.). *Statistics Database*. Retrieved from http://stat.wto.org/Home/WSDBHome.aspx?Language=E

Wright, R. W. (1985). Causation in Tort Law. *California Law Review*, *73*(6), 1735–1828. doi:10.2307/3480373

Xie, W. (1983). *Magnetized Water and Its Application*. Beijing: Science Press.

Yakutin, Y. (2014). Food Security – A Strategic Component of National Security of Russia. *Proceedings of VEO*, *187*, 166–167.

Yakymenko, O. (2013). Peculiarities in Strategic Management of Enterprise Development in Agricultural Machinery Sector. *Actual Problems of Economics*, *147*(9), 138–144.

Yamaguti, S., Misawa, S., & Asanuma, G. (1999). *Patent US 5 051161 Apparatus Producing Continuously Electrolyzed Water*. Russian Federation. Retrieved from https://patents.google.com/patent/US5051161A/en

Yamaguti, S., Ukon, M., Misawa, S., & Arisaka, M. (1994). *Patent #0612694A1 Method and Device for Producing Electrolytic Water*. Russian Federation. Retrieved from https://patents.google.com/patent/EP0612694A1/da

Yawson, D. O., Mulholland, B. J., Ball, T., Adu, M. O., Mohan, S., & White, P. J. (2017). Effect of Climate and Agricultural Land Use Changes on UK Feed Barley Production and Food Security to the 2050s. *Land (Basel)*, *6*(4), 74. doi:10.3390/land6040074

You, L., Ringler, C., Wood-Sichra, U., Robertson, R., Wood, S., Zhu, T., ... Sun, Y. (2011). What Is the Irrigation Potential for Africa? A Combined Biophysical and Socioeconomic Approach. *Food Policy*, *36*(6), 770–782. doi:10.1016/j.foodpol.2011.09.001

Yu, X., Leng, Z., & Zhang, H. (2012). Optimal Models for Impact of Agricultural Machinery System on Agricultural Production in Heilongjiang Agricultural Reclamation Area. *Paper presented at the 24th Chinese Control and Decision Conference*, Taiyuan. Academic Press. 10.1109/CCDC.2012.6244128

Yushkevych, O. (2013). Regulation Mechanisms in the Development of Agricultural Enterprises. *Actual Problems of Economics*, *147*(9), 132–137.

Yussefi, M., & Willer, H. (2003). *The World of Organic Agriculture: Statistics and Future Prospects*. Bonn: International Federation of Organic Agriculture Movements.

Yu, W., Elleby, C., & Zobbe, H. (2015). Food Security Policies in India and China: Implications for National and Global Food Security. *Food Security*, *7*(2), 405–414. doi:10.100712571-015-0432-2

Yu, Y., Feng, K., Hubacek, K., & Sun, L. (2016). Global Implications of China's Future Food Consumption. *Journal of Industrial Ecology*, *20*(3), 593–602. doi:10.1111/jiec.12392

Zaghdaoui, H., Jaegler, A., Gondran, N., & Montoya-Torres, J. (2017). Material Flow Analysis to Evaluate Sustainability in Supply Chains. *Proceedings of the 20th IFAC World Congress*. Toulouse: The International Federation of Automatic Control.

Zelepukhin, V., & Zelepukhin, I. (1987). *The Key to Living Water*. Alma-Ata: Kainar.

Zelepukhin, V., & Zelepukhin, I. (1973). Stimulation of Physiological Processes with Biologically Active Water. *Works of Works of Kazakh Agricultural Institute, 18*(14), 143–148.

Zhang, H. (2016). Food in Sino-U.S. Relations. From Blessing to Curse? In F. Wu & H. Zhang (Eds.), *China's Global Quest for Resources. Energy, Food and Water* (pp. 100–118). London: Routledge.

Zhang, H., & Chen, G. (2016). China's Food Security Strategy Reform: An Emerging Global Agricultural Policy. In F. Wu & H. Zhang (Eds.), *China's Global Quest for Resources. Energy, Food and Water* (pp. 23–41). London: Routledge.

Zhang, M. (2011). Tort Liabilities and Torts Law: The New Frontier of Chinese Legal Horizon. *Richmond Journal of Global Law and Business, 10*(4), 415–495.

Zhiarno, M., Zabodalova, L., Veber, A., Petushkova, Y., & Kazydub, N. (2017). *Patent #266119 "Process of Production of Fermented Product."* Retrieved from https://findpatent.ru/patent/266/2661119.html

Zhou, J. (2015). *Chinese Agrarian Capitalism in the Russian Far East*. Retrieved from https://www.tni.org/files/download/bicas_working_paper_13_zhou.pdf

Zhou, Z. (2010). Achieving Food Security in China: Past Three Decades and Beyond. *China Agricultural Economic Review, 2*(3), 251–275. doi:10.1108/17561371011078417

Zhovnovach, R. (2014). Satisfaction of Consumers' Demand as the Basis for Planning Competitiveness of Agricultural Machinery Enterprises. *Actual Problems of Economics, 155*(5), 171–180.

Zhudro, M. (2009). *Development of Economic Instruments of Increase of Competitiveness of the Use of Agricultural Technology*. Gorki: Belsha.

Zhu, Y. (2016). International Trade and Food Security: Conceptual Discussion, WTO and the Case of China. *China Agricultural Economic Review, 8*(3), 399–411. doi:10.1108/CAER-09-2015-0127

Zinchuk, G. (2006). Types of Food Market and Development Problems in Russia. *Russian Economic Internet Journal, 4*, 1–15.

Zotikov, V., Naumkina, T., & Sidorenko, V. (2014). Role of Leguminous Crops in Economy of Russia. *Farming, 4*, 6–8.

Zuckerman, E. W. (1999). The Categorical Imperative: Securities Analysis and the Legitimacy Discount. *American Journal of Sociology, 104*(5), 1398–1438. doi:10.1086/210178

About the Contributors

Vasilii Erokhin is an Associate Professor, School of Economics and Management, Harbin Engineering University, China. Since 2017, Dr. Erokhin is a researcher at the Center for Russian and Ukrainian Studies (CRUS) and Arctic Blue Economy Research Center (ABERC) Harbin Engineering University, China. In 2018, Dr. Erokhin also joined Key West University (Florida, USA) as an Adjunct Professor in Macroeconomics, Microeconomics, and History of Economic Thought. Dr. Erokhin is an author of over 170 scientific works in the areas of international trade, globalization, sustainable development, food security issues with a focus on emerging markets, developing countries, and economies in transition. His major book titles include Global Perspectives on Trade Integration and Economies in Transition (2016), Establishing Food Security and Alternatives to International Trade in Emerging Economies (2017), and Handbook of Research on International Collaboration, Economic Development, and Sustainability in the Arctic (2018). Dr. Erokhin is a Guest Editor at MDPI Sustainability and MDPI Economies; an Associate Editor at Springer's Journal of Knowledge Economy; a member of the Editorial Review Board at IGI International Journal of Sustainable Economies Management (IJSEM); an editorial board member at a number of international journals and publications. Dr. Erokhin is a holder of the honorary awards from the Ministry of Agriculture of the Russian Federation and the Ministry of Education and Science of the Russian Federation. He is a holder of Publons Peer Review Award 2018 "Top 1% of Reviewers in the Spheres of Environment and Ecology" for elite contribution to scholarly peer review and editorial pursuits internationally and outstanding commitment to protecting the integrity and accuracy of published research.

Gao Tianming is an Associate Professor, School of Economics and Management, Harbin Engineering University, China. He obtained his Bachelor degree in Industrial External Trade from Harbin Engineering University, Master degree in Management from the Russian Presidential Academy of National Economy and Public Administration under the President of the Russian Federation, and Ph.D. degree in International Economics from Moscow State Institute of International Relations under the Ministry of Foreign Affairs of the Russian Federation. He has over 15 years of professional experience as a general representative of Chinese transnational corporation in Russia and the CIS. He is a Director and Chief Expert of the Center for Russian and Ukrainian Studies (CRUS) and Arctic Blue Economy Research Center (ABERC) at Harbin Engineering University, Deputy Head of the Heilongjiang International Economic and Trade Association, leading consultant of governmental bodies and commercial organizations in the sphere of economic collaboration between China, Russia, and the Republic of Korea. Dr. Gao is an author of many publications related to China's Arctic policy and developments, China-Russia and China-Nordic collaboration in the spheres of economic development, industrial policy, and investment. He is a member of the Scientific Board of the Research Network on Resources Economics and Bioeconomy (RebResNet) and a regular member of the editorial boards in Marketing and Logistics journal, Public Administration Aspects journal, Grani almanac, and Agricultural Bulletin of Stavropol Region.

* * *

Ahmed Abduletif Abdulkadr is a Ph.D. student at the Doctoral School of Regional Sciences, Szent Istvan University, Hungary. He received BSc in Statistics from Hawassa University, Ethiopia in 2009; MSc in Economics (focus on Statistical and Mathematical Methods for Economic Analysis and Forecast) from Moscow State University of Economics, Statistics and Informatics, Russia (2014). He is a full academic staff of Samara University, Ethiopia. Main research interests are related to the spheres of agricultural and rural development.

Sinmi Abosede is an Associate Professor at the School of Management and Social Sciences, Pan Atlantic University, Nigeria. She teaches Introduction to Quantitative Reasoning and Introduction to Mathematical Methods to first-year undergraduates. Her current research activities are centered on the water and waste water sector in Nigeria, specifically on various issues such as water scarcity, water regulation, water resources management, and the relationship between water and other sectors such as energy and agriculture. Prior to joining Pan Atlantic University, she worked as a Chemical Engineer in the UK water industry for companies such as Thames Water and Atkins Limited and as a technical specialist in water regulation, audit, engineering design, and engineering project management. Currently, she provides consultancy services to various clients providing various engineering solutions for homes, offices, estates, or entire districts. Dr Abosede holds a BEng degree in Chemical Engineering from the University of Leeds, UK and a Ph.D. in Chemical Engineering from Imperial College London, UK.

Aleksey Aleshkov is an Associate Professor at the Department of Commodity Research, Khabarovsk State University of Economics and Law, Russia. Dr. Aleshkov graduated Khabarovsk State Academy of Economics and Law, Russia in 2003. He teaches the disciplines in the field of commodity research and examination of goods. His research interests are focused on the development and study of meat products enriched with functional ingredients.

Vera Amicarelli is an Associate Professor at Department of Economics, Management, and Business Law, Commodity Science Section at the University Aldo Moro of Bari, Italy. She has a Ph.D. in Commodity Science. Dr. Amicarelli teaches Industrial Ecology, Quality Theory and Technique, and Resource and Waste Management. She is an author of approximately 70 papers published in scientific journals and academic volumes. She has been involved in several national and international projects related to her principal research interests. Her studies refer to the following topics: material flow analysis, environmental indicators such as water and carbon footprint and circular economy. Her main academic activities are related to Erasmus + exchange program. Dr. Amicarelli is an Erasmus Department Delegate and coordinator of several agreements with foreign universities. She is included into Ph.D. Academic Board in Economics and Management. Dr. Amicarelli is a member of the Italian Commodity Science Academy (AISME).

About the Contributors

Alexander Arskiy is an Associate Professor in the Department of Agrarian Relations and Staffing of in Agriculture, Russian Academy of Personnel Support for the Agroindustrial Complex, Russia. Dr. Arskiy is an author of over 60 publications on logistics, marketing, and customs, including textbooks, monographs, and research papers. He is an Executive Editor of the Marketing and Logistics journal and a regular member of the editorial boards in a number of Russian journals and international publications.

Dmitry Bogdanov is a Professor in the Department of Civil Law, Kutafin Moscow State Law University, Russia. Prof. Bogdanov is an author of over 100 publications, including 7 monographs. Research interests include civil liability, contract law, corporate law, comparative law, torts, and legal regulation of genomic research, and bioprinting.

Svetlana Bogdanova is an Associate Professor at the Department of Law, Intellectual Property and Forensic Expertise, Bauman Moscow State Technical University, Russia. Dr. Bogdanova is an author of over 30 publications, including one monograph. Her research interests include civil liability, contract law, food safety, business law, and comparative law.

Christian Bux is a Ph.D. student in Economics and Management at the Department of Economics, Management, and Business Law, Commodity Science Section, University of Bari Aldo Moro, Italy. His main field of interest is the relationship between natural resources, commodity production, and consumption and environmental management systems, as well as food loss and waste management, and circular economy.

Sukalpa Chakrabarti is an Associate Professor (International Relations and Public Policy) and Deputy Director at the Symbiosis School of International Studies, Symbiosis International (Deemed University), India. She has been teaching across the specializations in politics, international relations, and international business and economy at undergraduate to postgraduate levels for over twelve years now. Her areas of research interest include Asian regionalism, security and diplomacy, global political economy, and public policy. She has many publications to her credit.

Elena Chaunina is the Head of Department of Animal Science, Omsk State Agrarian University, Russia. Dr. Chaunina is Ph.D. in Agriculture, Associate Professor. Her research is focused on the improvement of the feeding of animals.

Drago Cvijanović is a Full Professor at the Faculty of Hotel Management and Tourism in Vrnjacka Banja, University of Kragujevac, Republic of Serbia. In 1981, he graduated from the Agro-Economic Department of the Faculty of Agriculture in Belgrade, in 1986, obtained a master's degree at the same school, in 1993, and defended a Ph.D. thesis "Former Development and Perspectives of the Industrial Plants Production in Yugoslavia" at the Faculty of Economics in Belgrade. In 2014, Prof. Cvijanovic was elected as Honorary Professor at Stavropol State Agrarian University, Russia. Prof. Cvijanovic has over 30 years' length of service, during which he has worked in numerous scientific-research institutions, enterprises, and governmental administrative bodies. His master's thesis, doctoral thesis, monographs, and papers have been published in foreign and domestic expert journals, and presented at scientific meetings of the national and international significance, he was engaged in the market and marketing of agro-food products, economics of agriculture, multifunctional agriculture and rural development, management in agriculture, investments in agriculture, tourist market, marketing in tourism, as on the macro, as well as on the micro level. He was also engaged in the subjects regarding the agro-economic and economic profession. As the author/co-author, he has published over 350 biographical units, which have been cited in domestic and foreign professional literature. He has been engaged in scientific-research work at numerous projects (44) as the project manager, head of the research team, professional coordinator, and the member of the research teams. He was in charge of the international project in Brcko District "Competitiveness and Comparative Sustainable Development of Istra and Kolubara District" and joint bilateral Croatian-Serbian project financed by the Ministry of Science, Education, and Sport of the Republic of Croatia and the Ministry of Science and Technological Development of the Republic of Serbia (2008).

Yulia Elfimova is an Associate Professor, Department of Tourism and Service, Stavropol State Agrarian University, Russia. Dr. Elfimova is an author of 3 monographs, with 15 papers in peer-reviewed journals, including those indexed in Scopus and Web of Science. Research interests are focused on the organization of innovative activity of agricultural enterprises, economics and management, and tourism and service.

Zhanbota Esmurzaeva is an Associate Professor at the Department of Foreign Languages, Omsk State Agrarian University, Russia. Dr. Esmurzaeva is an author and co-author of over 70 scientific papers and theses. In 2018, she participated in the study of technology of functional foodstuffs of complex raw composition with the use of haricot suspension. The results were published in the proceedings of the international conference. In 2017, Dr. Esmurzaeva participated in the study of the problems of development of technical and technological solutions on domestic selection haricot use and their application in the production of specialized products of various functional orientations. The results were published in the proceedings of International Congress "Biotechnology: State of the Art and Perspectives". In 2011, Dr. Esmurzaeva was a scholar of ERASMUS Mundus Action 2 IAMONET – RU (staff mobility) at the Czech University of Life Science Prague, Czech Republic.

Maria Fedorova is a Teacher and Russian-English translator at Omsk State Technical University, Russia, as well as a manager and translator for International forums INNOSIB organized annually by the Ministry of Economy of Omsk region and the Center for Social Innovation. Scientific interests include social innovation, international communication, and terminology.

Teodoro Gallucci is a researcher with national scientific qualifications as an Associate Professor in the Department of Economics, Management and Business Law, DEMDI Commodity Science Section, University of Bari Aldo Moro, Italy. He has a Ph.D. in Commodity Science at the University of Bari-Italy and he teaches at the University of Bari Aldo Moro, Italy and Catholic University Our Lady of Good Counsel of Tirana, Albania. He is responsible for APRE PUGLIA Regional Office of the Agency for European Research for the University in 2010-2012. Member of the working group SUA RD (University Departmental Research Unit) and Member of the working group of the CVR (Research Evaluation Committee) for the qualification of the scientific research in the University. He participated in different national and international project. Since 2018, Dr. Gallucci is a member of the national platform ICESP (Italian Circular Economy Stakeholder Platform).

About the Contributors

Ishita Ghosh is an Assistant Professor at the Symbiosis School of Economics, Symbiosis International (Deemed University), India. Dr. Ghosh is specialized in International Economics and Trade. Her thesis dealt with trade in services and the Gravity Model, pertaining to trade costs arising from non-tariff barriers. She has over 15 years of experience in the corporate and academia, combined. Dr. Ghosh has had a stint in urban policy and planning for Maharashtra (one of the largest states of India) with special emphasis on Energy, Economic Policy, and Governance. Some of the courses she has taught at the undergraduate and graduate programs are Trade and related, Research Methodology, Macroeconomics, Economic Journalism, Econometrics, etc. Dr. Ghosh has a bunch of publications under her belt and her research interests lie in new-new trade theories, trade and aid, services trade, trade and gender, the behavioral angle in trade and trade-related development.

Ishita Ghoshal is an Assistant Professor at Fergusson College (Autonomous), India. Dr. Ghoshal holds a Ph.D. from Gokhale Institute of Politics and Economics, India. Her thesis focused on causality analysis of inflation in India with frequency domain analysis and structural VAR models. Her areas of interest are macroeconomics, applied econometrics, and international trade. She is B.Sc. in Economics (Hons) with Mathematics and Statistics from Lady Brabourne College, University of Calcutta, India, and M.A in Economics from Gokhale Institute of Politics and Economics, India. Currently, she teaches macroeconomics, mathematical economics, and econometrics to graduate and post-graduate students. Dr. Ghoshal has published in reputed journals like Journal of Quantitative Economics, Global Economy Journal, and Procedia – Economics and Finance.

Alexey Gorodilov is a student at the School of Earth Sciences and Engineering, Tomsk Polytechnic University, Russia.

Sergei Greizik is a Research Fellow at the Research Centre of Shanghai Cooperation Organization and the Asia-Pacific Region, Khabarovsk State University of Economics and Law, Russia. In 2011-2015, he served as a consultant at the Interregional Association for Economic Interactions Far East and Trans-Baikal Region.

Wim Heijman is a Professor of Regional Economics, Wageningen University and Research, the Netherlands. He received MSc degrees respectively in Economics and Human Geography from Tilburg University and the University of Utrecht in the Netherlands. He received his Ph.D. degree from Wageningen University. In 2000, he was appointed Professor of Regional Economics at the latter university. His major research topics are rural development, landscape economics, and the assessment of the regional economic impact of tourism. Prof. Heijman was awarded honorary doctor's degree by the Universities of Debrecen in 1999 and Nitra in 2015. In 2004, 2014, 2015, and 2016, he received honorary professor titles from Czech University of Life Sciences in Prague, Mongolian University of Life Sciences in Ulaanbaatar, Stavropol State Agrarian University (Russia), and Belgorod State Agricultural University (Russia), respectively. He has been affiliated with AGRIMBA (the network for the MBA in Agribusiness and Commerce) since the start of the organization in 1995. In 2015, he was appointed its Secretary General.

Svetlana Ignjatijević is an Associate Professor at the Faculty of Economics and Engineering Management in Novi Sad, University Business Academy in Novi Sad, Republic of Serbia. In 1993, she graduated from the Department of Marketing of the Faculty of Economics in Subotica, in 2006, obtained a master's degree at the same faculty, in 2011, defended the Ph.D. thesis "Comparative Advantages of the Agriculture in Serbia in Foreign Trade" at the Faculty of Economics and Engineering Management in Novi Sad. Prof. Ignjatijevic has over twenty years' length of service, during which she was working in the economy. In her master's thesis, doctoral thesis, and papers published in foreign and domestic expert journals and presented at scientific meetings of the national and international significance, she was engaged in the comparative advantages, competitiveness, business terms, foreign trade business, market and marketing of agro-food products, economics of agriculture, tourist market, and marketing in tourism. As the author/co-author, she has numerous scientific papers in domestic and foreign journals. She was engaged in scientific-research work at numerous projects, as the project manager and the member of the research teams. She was in charge of the project "Competitiveness Improvement of Fruska Gora Linden Honey on Domestic and International Market in Terms of the Sustainable Development of AP Vojvodina", Provincial Secretariat of AP Vojvodina. As a member of the research team, she participated in the project IPA "Service for Creation of Marketing Plan for the Project: Lime Trees & Honey Bees for Sustainable Development of the Danube Microregion".

Anna Ivolga is an Associate Professor, Head of the Department of Tourism and Service, Deputy Dean for Research, Faculty of Social and Cultural Service and Tourism, Stavropol State Agrarian University, Russia. She holds a Ph.D. degree in Economics and Management (2006) and a Specialist degree in International Economics (2003) from Stavropol State Agrarian University, Russia. Dr. Ivolga published over 120 papers in the spheres of tourism, innovations, sustainable development, rural development, and land management. She is a holder of the honorary awards from the Ministry of Education of Stavropol Region and the Ministry of Tourism and Recreation of Stavropol Region. Since 2017, Dr. Ivolga is a certified World Skills Expert in Hotel Administration.

Azaharaini Bin Hj. Mohd. Jamil is a retired civil servant currently holding a position of Senior Lecturer at the Institute of Policy Studies, University Brunei Darussalam. Prior to retirement, the position held was the Director of the Brunei Institute of Civil Service. Immediately after retirement, the position held was the Executive Director of the Brunei Darussalam Centre for Strategic and Policy Studies. Research interests include educational management, public sector management, and policy development. Dr. Jamil has been involved in consultancy work with ILIA and UBD (Executive Development Program), Royal Brunei Armed Forces (Junior and Senior programs), and Ministry of Finance (Junior staff program).

Mikail Khudzhatov is an Associate Professor in the Customs Department, Peoples' Friendship University of Russia, Russia. He is a specialist in the field of customs affairs and a permanent author of the materials for the information base "Financier" in the JSC "Consultant Plus", Russia. Dr. Khudzhatov is a co-editor at the scientific and practical journal "Marketing and Logistics". Areas of research interests include international trade, customs and tariff regulation, economic security, and customs business. Dr. Khudzhatov is a Ph.D. in Economics and an author of many publications in the spheres of international trade and customs business.

About the Contributors

Inna Korsheva is an Associate Professor at the Department of Animal Science, Omsk State Agrarian University named after P.A. Stolypin, Russia. The spheres of research interests include scientific support of the development of agricultural production and organic farming

Giovanni Lagioia is a Full Professor, Head of Department of Economics, Management, and Business Law – DEMDI, University of Bari Aldo Moro, Italy and Vice-Rector concerning business cooperation and labor market of the University Our Lady of Good Counsel – Tirana (Albania). He has a Ph.D. in Commodity Science at the University of Bari, Italy and he teaches the following courses: commodity science; technology and environmental certification, resources, and waste management. Prof. Lagioia is an academic coordinator of different Erasmus agreements and Member of the Board of the Doctoral School in Economics and Management. He is affiliated to International Society of Industrial Ecology, IGWT (International Society of Commodity Science and Technology), and the Circular Economy and LCA (Life Cycle Assessment) Italian Networks. The main interests of research are ecological economics, circular economy, material flow analysis (MFA), life cycle analysis (LCA), environmental management, and biomass and bioenergy.

Svetlana Leonova is a Professor in the Department of Catering Technology and Processing of Plant Raw Materials, Bashkir State Agrarian University, Russia. Prof. Leonova is a Head of the scientific direction devoted to the development of the system of estimating and forming the quality of wheat. Another focus of her research is the improvement of a phytochemical potential of food products. Prof. Leonova is an author of over 100 publications.

Stanislav Lipski is a Professor and Head of Department of Land Law, State University of Land Use Planning, Russia. In 1989, upon graduation from Moscow Engineering Institute of Land Management, he started serving as a junior research fellow. In 1990, he entered the State Land Reform Committee as a leading specialist, chairperson assistant, and consultant. In 1992-2011, he worked at the Government Office of the Russian Federation as the leading specialist, expert, consultant, advisor, deputy head of the department, head of department, and referent. Prof. Lipski is a Public Counselor of the Russian Federation (third rank). Prof. Lipski has over 350 academic and educational publications in the areas of regulation of land turnover, rational and efficient use of land, and land management.

Elena Meleshkina is a Director of All-Russian Research Institute of Grain and Grain Products, Russia. Prof. Meleshkina is a Doctor in Technical Sciences. The spheres of her research include grain processing, nanotechnologies, and cascade production of food from raw materials of plant origin."

Henrietta Nagy is a Habilitated Associate Professor, Szent Istvan University, Hungary. She received a Ph.D. degree in 2016. Since then, she has had experience in teaching at bachelor, master, and Ph.D. levels, as well as in the research projects at the Szent Istvan University. In addition to academic activities, Dr. Nagy has been responsible for the development of international relations at the Faculty of Economics and Social Sciences. Her major research fields are regional/cohesion policy, local economic development, and enterprise development. Two of her Ph.D. students have defended their dissertations successfully. Dr. Nagy is the member of the Hungarian Academy of Science.

György Iván Neszmélyi is an Associate Professor, Head of International Relations at the Faculty of Commerce, Hospitality and Tourism, Budapest Business School, Hungary. He has been an Honorary Professor of Szent Istvan University for twenty years. Dr. Neszmélyi is Academic Director of Advenio eAcademy, a Malta-based e-learning higher education institution. He received a Ph.D. degree in Economics in 1997, and then he worked as the Head of Department in the Research and Information Institute of Agricultural Economics (AKII). Between 1998 and 2012, he served as a diplomat in several countries in Asia and Africa nearly a decade, while as part-time lecturer and researcher he had continuously been involved in various research and academic activities of Szent Istvan University and Budapest Business School. In 2004, he earned his MBA degree at Sejong University in Seoul, the Republic of Korea accomplishing the joint Global Management course of Sejong and Syracuse (NY, USA) universities. While he teaches in Hungarian and English-taught BA, MA and Ph.D. programs he is a supervisor of five Ph.D. students published more than 130 scientific papers, including 50 journal articles. He is an author and editor of six books and wrote twenty book chapters. Many of his papers were published in English, and some of them in the USA, Canada, Republic of Korea, Slovakia, Czech Republic, Romania, Russia, and Taiwan. His research field is the examination of economic, political and social factors in the development of Asian and African countries. He earned the post-doctoral title "Dr. habil" in regional sciences in 2016 at Szent Istvan University. He is a member of the Public Body of the Hungarian Academy of Science and several other scientific and civilian organizations, like Regional Science Association International, Hungarian Regional Science Association, Hungarian Economic Association, Hungarian, Hungarian Foreign Policy Association, and the Hungary-Korea Society.

Tamara Nikiforova is a Professor at the Department of Food Production Technology, Orenburg State University, Russia. Prof. Nikiforova is a Doctor in Technical Sciences. Her research interests are focused on the bioconversion of secondary raw materials of grain processing enterprises and the creation of functional and specialized foods.

Olga Pasko is a Professor at the Department of Geology and Land Management, Institute of Natural Resources, Tomsk Polytechnic University, Russia; a Professor at the Department of Hydrogeology, Engineering Geology and Hydrogeoecology, Tomsk Polytechnic University, Russia. She is a leading teacher in Land Management, Cadaster, Land Monitoring, and Earth Studies. Prof. Pasko is an author of over 250 publications, an editor of over ten scientific proceedings, including materials of the all-Russian conference "The Arctic and Its Development" (Tomsk, Russia, 2017). The areas of her research interests include monitoring and rational use of natural resources. Prof. Pasko supervises the research projects of postgraduate students related to the analysis of specially protected natural areas in the Arctic and control of anthropogenic pollution of the Arctic zone of Russia.

Roman Perkel is a Professor at the Higher School of Biotechnology and Food Science, Peter the Great Saint Petersburg Polytechnic University, Russia. Prof. Perkel is a Doctor in Technical Sciences.

Fedor Pertsevyi is an Associate Professor at the Department of Food Technology, Sumy National Agrarian University, Ukraine.

About the Contributors

Muhamad Rusliyadi is a lecturer at Polytechnic of Agricultural Development Yogyakarta-Magelang, Agricultural Extension and Human Resource Development Agency, Ministry of Agriculture, Indonesia. Educational background: Bachelor of Agriculture (SP), Faculty of Agriculture, Jenderal Soedirman University Purwokerto, Indonesia (2003); Master of Science in Rural Development, China Agricultural University, China (2011); Ph.D. in Policy Studies, Faculty of Business, Economics and Policy Studies, Universiti Brunei Darussalam (2017). Dr. Rusliyadi is a former researcher at the Assessment Institute for Agricultural Technology Gorontalo, Agency of Agricultural Research and Development. Part of the Ministry of Agriculture, Indonesia.

Ivan Ryazantsev is an Associate Professor, Department of Tourism and Service, Stavropol State Agrarian University, Russia. In 2001, Dr. Ryazantsev obtained a Specialist degree in Finance and Management; in 2003, he defended a Ph.D. thesis in Economics and Management. Dr. Ryazantsev is an author of over 80 publications in peer-reviewed journals and conference proceedings, including those indexed in Scopus and Web of Science. Research interests are focused on economic and organizational issues of land management, agricultural production, and tourism and service.

Sergei Ryumkin is an Associate Professor at the Department of Economics, Novosibirsk State Agrarian University, Russia. In 2000, he obtained a Specialist degree in Accounting and Audit at Irkutsk State Agricultural Academy, Russia. In 2003, Dr. Ryumkin defended a Ph.D. thesis in Economics and Management of National Economy at Irkutsk State Agricultural Academy, Russia. Since 2017, at Novosibirsk State Agrarian University, Dr. Ryumkin has been pursuing doctorate studies in Economics, since 2018 – Master program in Biology.

Inga Ryumkina is an Associate Professor at the Department of Economics of Agro-Industrial Complex, Irkutsk State Agrarian University named after A.A. Ezhevsky, Russia. In 2001, she obtained a Specialist degree in International Economics at the International Academy of Entrepreneurship, Russia. In 1998-1999, Dr. Ryumkina studied Chinese at Beijing International Studies University, China. In 2006, she obtained a Ph.D. degree in Economics at the Buryat Scientific Center of the Siberian branch of the Russian Academy of Sciences. Since 2016, at Novosibirsk State Agrarian University, Dr. Ryumkina has been pursuing doctorate studies in interdisciplinary sciences (Economic and Ecology), Russia; since 2018 – Master program in Biology and Ecology.

Valentin Sapunov is a Professor in the Department of Management of Social and Economic Processes and History, Saint Petersburg State Agrarian University, Russia; an Honorary Member of the European Geosciences Union; a Fellow of the Peter the Great Academy of Sciences and Art; a Fellow of the New York Academy of Sciences. His research areas include fundamental issues of environmental processes in the Northwestern territories of Russia. Prof. Sapunov elaborated novel research area, namely, physiological genetics of microevolution and developed the theories of environmental mates and latent reserve of the biosphere. He is an author of over 800 academic and scientific popular publications, including papers in peer-reviewed journals and conference proceedings, monographs, and textbooks. Prof. Sapunov participated in many international forums in Russia, Poland, Czech Republic, Germany, Singapore, the USA, and other countries, as well as in the dozens of expeditions in Russia and Central Asia.

Inna Simakova is a Professor in the Department of Food Technology, Saratov State Vavilov Agrarian University, Russia. She is an author of over 150 publications in the areas of food technology and food products for mass and special types of food, including biotechnology of food products, food safety, nutrition, nutrigenomics, resource saving and environmentally-friendly food production, equipment and high-tech food production, nanostructured absorption materials for the food industry.

Natalia Staurskaya is the Head of International Relations Office at Omsk State Technical University, Russia. In 2007, she graduated from Omsk State Pedagogical University with honors as a specialist in Foreign Languages and Educational Science. In 2015, she graduated from Omsk Agrarian University College with a degree in Law. In 2017, Dr. Staurskaya obtained a Ph.D. degree in Philology. Dr. Staurskaya took over twenty vocational training courses in project management, sustainable development, and bio-based economy. In 2007-2015, she worked as a translator and a project manager and subsequently as the Head of International Relations Office at Omsk State Agrarian University, Russia. In 2015-2017, Dr. Staurskaya worked as the Head of International Research Projects Office at Omsk State Institute of Service. She worked as a manager for over fifteen international research and academic projects. Dr. Staurskaya is an author of over twenty publications and an Erasmus alumna.

Olga Storozhenko is an Associate Professor at the Department of Law, Intellectual Property and Forensic Expertise, Bauman Moscow State Technical University, Russia. Dr. Storozhenko has 20 years of experience in teaching law, managing international educational programs, and providing private businesses with legal consulting services in higher educational institutions (Bauman Moscow State Technical University, Russia; Russian Open Transport Academy, Russia; Moscow University of Finance and Law, Russia; Kyiv National University of Trade and Economics, Ukraine; Wisconsin International University, Ukraine; Borys Grinchenko Kyiv University, Ukraine). Dr. Storozhenko is an author of over 45 scientific works, including five monographs, three manuals prepared in the framework of the international project "Revival", methodical recommendations, teaching course curriculums, reference lecture notes, and papers. In 2008, Dr. Storozhenko earned a Ph.D. degree in Law. In 2010, she completed internships at the University of Antwerp, Belgium, and the Academy of the World Intellectual Property Organization, Switzerland. In the framework of the Erasmus Mundus Partnership Programme, Dr. Storozhenko passed the training at Mykolas Romeris University, Lithuania. In 2016, Dr. Storozhenko obtained a Certificate in Law course from the Institute of Commercial Management, the UK.

Victoria Strizhevskaya is an Associate Professor in the Department of Food Technology, Saratov State Vavilov Agrarian University, Russia. She is Ph.D. in Technical Sciences and an author of over 40 publications in the sphere of food technology.

Ismail Taaricht is a Professor at Cadi Ayyad University, Morocco. Experience in agriculture, producing oranges and olives, teaching at primary school: Regional Academy of Education and Training of Beni Mellal-Khénifra. Education: high school diploma (2002), Cadi Ayyad University, Faculty of Arts and Humanities, Beni Mellal; high school diploma (2017), Faculté des Lettres et des Sciences Humaines Sultan Moulay Slimane Beni-Mellal, Tadla-Azilal, Morocco. Certifications: Professional Proficiency Certificate at Tourism Communication, Religious, Cultural, and Historical.

About the Contributors

Mikhail Tomilov is a Research Fellow at the Economic Research Institute, Far Eastern Branch of the Russian Academy of Sciences, Russia. In 2012, Mr. Tomilov graduated from Khabarovsk State University of Economics and Law, Russia. In 2013-2015, he served as a research assistant in the same university.

Alexander Trukhachev is the Minister of Tourism and Recreation of Stavropol Region, Russia; a Professor at the Department of Tourism and Service, Stavropol State Agrarian University, Russia. Research interests include tourism and hospitality, tourist market, tourist clusters, establishment and operation of recreational zones, sustainable development of tourism sector, innovations in tourism, and rural tourism. Prof. Trukhachev has over 80 publications, including books, monographs and chapters, and papers in peer-reviewed journals and conference proceedings. He is a holder of many awards, including the National Distinguished Medal for Development of Tourism in Russia. In 2017, Prof. Trukhachev received the Presidential Grant for the development of rural tourism entitled "Conceptual Foundations of State Policy for the Development of Rural Tourism in the Russian Federation."

Bistra Vassileva is an Associate Professor of Marketing, University of Economics–Varna, Bulgaria. She graduated as Master of Science of Commodities. Since 1992 she is lecturing and consulting in the field of Marketing Research, International Marketing, Marketing Communications, TQM, Marketing Management. Dr. Vassileva was a visiting professor in Portugal, France, Germany, Spain, UK and a guest lecturer in Belgium. She implemented more than fifteen international and national projects of different donors. During the last few years, she took part in the EU funded projects for various research issues and problems as an expert. She was a Marie Currie fellowship holder as a Senior Researcher in 2007-2008 in Lodz, Poland. As a Director of the Centre for Innovation and Development at the University of Economics-Varna, she is responsible for the organization of research and ICT projects with different scope. Dr. Vassileva is a member of CIM, ESOMAR, and EMAC. Her scientific interests are focused on nonlinear dynamics and network theory in the field of marketing, marketing analytics, and digital marketing.

Anna Veber is an Associate Professor at the Department of Food Biotechnology, Omsk State Agrarian University named after P.A. Stolypin, Russia. She has been the head of research groups in several international scientific projects in the field of sustainable development, biotechnology, and bio-based economy. Dr. Veber is an author of over 50 publications. Among other topics, she is also focused on the study of the culture and traditions of indigenous peoples from the point of view of the bio-based economy.

Alexander Voronenko is an Associate Professor and Director at the Research Center for Shanghai Cooperation Organization and Asia Pacific Region, Khabarovsk State University of Economics and Law, Russia. In 2007, he graduated Khabarovsk State University of Economics and Law, Russia, specialization "Economics and Management". In 2007-2009, Dr. Voronenko was the scholar at the Economic Research Institute, Far Eastern Branch of the Russian Academy of Sciences. In 2009-2016, he worked as a consultant at the Interregional Association for Economic Interactions "Far East and Trans-Baikal Region". In 2015, Dr. Voronenko was a scholar at the Far Eastern Institute of State Service under the President of the Russian Federation. In 2016, Dr. Voronenko obtained a Ph.D. degree in Economics from Khabarovsk State University of Economics and Law, Russia. In 2016-2018, he served as a senior expert at Research Center for Shanghai Cooperation Organization and Asia-Pacific Region, Khabarovsk State University of Economics and Law. His research focuses on interregional cooperation in the Asia-Pacific Region and the Arctic.

Igor Vorotnikov is the Head of the Department of Organization of Production and Business Management in Agriculture and Vice-Rector for Research and Innovations, Saratov State Vavilov Agrarian University, Russia. Prof. Vorotnikov is Doctor in Economics and Ph.D. in Technical Sciences. The major areas of his research are actual issues of the innovation economy and environmental management.

Alexander Zakharchenko is a Senior Research Fellow at the Institute of the Problems of Northern Development, Siberian Branch of the Russian Academy of Sciences, Russia. Dr. Zakharchenko is a Doctor of Biological Sciences. In 1980, he graduated from Tomsk State University with a degree in Agrochemistry-Soil Science. He worked in the Research Institute of Biology and Biophysics, Tomsk State University and passed all stages from laboratory assistant to senior researcher at Soil Science Laboratory. In 1998-2006, he worked as an Associate Professor at Tomsk State University of Control Systems and Radio Electronics. In 2002, he entered doctoral studies and in 2006, he defended his doctoral thesis "Spatial Organization and Morphogenesis of Forest and Anthropogenically Modified Soils". Since 2006, he is an Associate Professor at the Department of Ecology at Yugorsk State University of Khanty-Mansiysk, Russia. In 2007, he received the title of Associate Professor in Ecology and moved to the position of Professor at Ugra State University, Russia.

Kirill Zemliak is an Associate Professor at the Department of Commodity Research, Khabarovsk State University of Economics and Law, Russia. Dr. Zemliak obtained degrees at Pacific Ocean State Economic University, Russia, in 2010 and Khabarovsk State Academy of Economics and Law, Russia, in 2007. He teaches the disciplines related to the commodity research and examination of goods. His research interests are focused on the development and research of foods from wild-growing plants.

Anna Zhebo is a Vice-Rector for Research, Khabarovsk State University of Economics and Law, Russia. Dr. Zhebo is Ph.D. and Associate Professor. Her research interests are focused on the development of foods for rational nutrition and non-waste technologies for processing food raw materials.

Index

3D printing 176-177, 187

A

AG 173, 495, 499
agrarian reform 141, 189-190, 192, 199-202, 205-206, 213-214, 473
agribusiness 66, 99, 201-202, 298, 403, 405
Agricultural Cooperative 373, 468-469, 473, 475-476, 479
agricultural engineering 105, 107-108, 110, 112, 121-122, 127, 173
agricultural investment 374, 423, 427
agricultural lands 3, 44, 215, 217, 219, 222, 226-228, 231, 234, 239, 321, 330, 371
agricultural productivity 22, 27, 112, 329, 368, 384, 388-389, 392-393, 425-427, 436-438, 442
agricultural trade 13-14, 17-18, 20, 22-23, 27-28, 32, 34-35, 91, 217, 242, 252, 319, 323, 325, 328-329, 362-363, 365, 367-370, 384-385, 388-389, 392, 400, 405, 407, 417, 423, 428, 430
agro-industrial complex 62, 86, 105-107, 120-122, 127, 201, 206-207, 213, 348-349, 428
amino acid 255-258, 263, 276
arable land 146, 200, 205, 216, 321-322, 349, 354, 371, 376, 382, 388, 403-405, 418, 473
Association of Southeast Asian Nations 367
Autotrophicy 12

B

balanced diet 67, 86, 243, 248, 250, 255
Belt and Road Initiative 328, 335
Bioactive Substances 317
Biological Studies 294
biologically active substances 243, 245, 250, 339
Biotechnology 1, 3, 10, 12, 368, 370, 406
Breakthrough Development 79, 85

C

causation 178-179
chromatography 281
Circular Economy 163, 169
climate change 18, 20, 32, 34, 242, 338, 354, 371, 384-385, 387-389, 392, 400, 492
Codex Alimentarius (Food Code) 173, 187
collective farm 3, 6, 225, 382
Comprehensive and Progressive Agreement for Trans-Pacific Partnership (CPTPP) 382
Comprehensive Regional Economic Partnership (CREP) 382
Cooperative Payments 467, 479
credit facilities 402, 467, 480, 492
crop 6, 8, 43-45, 51, 56, 65, 68-69, 71, 97, 110, 130, 150, 196, 202, 214, 242, 244-245, 328, 372, 376-377, 384-388, 404-405, 427-432, 436-437, 442, 477, 488
Customer Profile 463

D

damage 4, 9, 44, 62, 97, 178, 180, 182, 188, 190, 207, 297, 307, 315, 365, 473
dehydration 296, 298, 300-302, 307, 310, 314, 317
Dehydration of Vegetables or Fruits 317
demarcation of state-owned lands 231, 235
developing countries 14, 23, 28, 34, 89-90, 93, 99, 104, 109, 153, 157, 298, 319-320, 329-330, 340, 365, 368, 370, 387, 392, 404, 424, 426, 428, 436, 456, 476
DMP Program 481, 485-486, 488-490, 493-495, 497, 499
Domestic Trade 430, 442

E

Ecological Niche 6, 12

ecological pyramid 4, 6, 12
ecological safety 62, 370
economic crisis 111, 206, 213, 338, 386
economic efficiency 119, 121, 206-207, 248, 310, 313, 315, 317
economic reforms 189, 213, 321
economy 8, 14, 17, 28, 62, 66, 75, 81, 94-95, 100, 104, 106-107, 109-110, 127, 129, 131, 133, 136, 141, 149, 163, 169, 189, 199-203, 205-207, 214, 244, 321, 325, 343, 345, 347, 360, 371, 373, 376-377, 401, 405, 419, 425-429, 431-433, 436, 445, 468, 472-473, 497
efficiency of agricultural production 109, 207, 214, 349, 443
Energy Value 275
Ensurance of Food Safety 187
entropy 116-117, 127-128
environmental sustainability 76, 85-86, 97, 99
Eurasian Economic Union 120, 127, 173-174, 259, 348
Eutrophication 12
export 6, 14, 16, 20, 23-24, 28-29, 31, 70, 72-73, 80-81, 89, 92-93, 96, 99-100, 202, 255, 270, 324-326, 328, 336, 340, 342-343, 350-351, 353-354, 363, 365-370, 372, 374, 378, 389-392, 405-407, 409, 413-416, 425-428, 430, 432-437, 444, 446-447, 459, 470, 472, 474
Extension Policy 481

F

Falkenmark Water Stress Indicator 387, 398
fatty acid composition 281, 288-290, 294
Feed Ingredients 250
Fishburne's Rule 127
food access 149, 360, 386, 392, 398
Food and Agriculture Organization 14, 18, 148, 171, 173, 205, 243, 253, 297, 338-339, 361-362, 384, 386, 401, 403, 424, 426, 468
food availability 149, 345, 348, 360, 386-392, 398-399
food independence 345-346, 351, 354, 360
food loss 147, 149-151, 153-154, 169, 297, 306, 353
food market 63, 65-68, 70, 72, 75, 77-81, 86, 88-90, 104, 206, 297, 300, 320-322, 329, 335, 342-343, 346, 348-349, 354, 370, 374
food quality 148, 174-176, 181, 187, 245
food safety 55, 62, 162, 170-176, 181-182, 187-188, 252, 270, 352-353, 360, 402, 474
food sector 147, 149, 163, 169, 344, 428
food self-reliance 389, 391, 399
food self-sufficiency 105, 201, 320-321, 329, 336, 339-341, 345, 351, 360, 370, 384, 389, 391, 399, 488
food sovereignty 90, 93, 344
food supply chain 147, 149-150, 169
food utilization 149, 386, 388, 392, 399
food waste 147, 149, 163-164, 169, 354
foodstuffs 67-68, 70, 79, 106, 256-258, 260, 262-266, 269-270, 275-276, 474
foreign trade 89, 92-94, 100, 201-202, 320-321, 323-324, 329-330, 336, 342, 346-347, 425-426
Fortified Food (Functional Foods) 67, 247, 250, 252
fractionation 277, 280, 291, 294
Free Trade Zone of Asia-Pacific (FTZAP) 382
Fruska Gora 443-445, 447-449, 452, 455-459
FVT 495, 499

G

GDP 14, 22-23, 29, 91, 94, 152, 205, 255, 340, 353, 363, 385, 387, 392, 401-402, 404-405, 418, 426-427, 429-430, 436-438, 472-473, 489, 499
Geographical Indication (GI) 463
global climate 12, 338
Global Food Security Index 340, 351
globalization 13-14, 16-17, 28, 66, 88, 91, 107, 109, 244, 297, 320
Group Farming 467, 480

H

haricot 252-255, 257-270
harm to health 174, 177, 179, 181, 188
healthy nutrition 67, 96, 245, 265, 303, 345
hidden hunger 242, 244, 249-250, 297-298, 315
Holder 442
honey production 443-444, 447-448, 452, 454-458
hunger 4, 7-8, 18, 148, 163, 242-244, 248-250, 253, 297-298, 315, 338-339, 355, 368, 370, 384-386, 400-401, 424, 427, 486, 492
hydrogenated fat 277-278, 282-284, 294
hydrogenated fats 277-278, 280-282, 286-288, 291
hydrogenation 278, 280, 283, 291, 294

I

import 3, 14, 20, 23-25, 28-29, 68, 70, 89, 91-93, 99-100, 106, 112, 201-202, 207, 255, 319-321, 325, 329-330, 336, 342-344, 346-349, 353-354, 362-363, 367-370, 372, 390-391, 401, 404-407, 409-413, 416, 418, 420, 433, 436, 474

Index

India-Africa Institute of Agriculture and Rural Development 402, 424
Indo-Japan Asia-Africa Growth Corridor (AAGC) 402, 424
Innosib Forum (Omsk) 252, 258, 260, 269, 478
Innovation System 90, 104
Innovative approach 79, 85, 89
interesterification 277, 280, 289-291, 294
international trade 13-14, 16-17, 19, 32, 89-90, 92, 99-100, 107-109, 128, 243, 245, 249, 320, 329, 336, 338, 340, 368, 385, 388, 391, 393, 401, 425, 432, 436
intervention policies 13-14
iodine 242, 244, 246-248, 266, 340

L

land cadaster 138, 140-141, 146
land management 129-141, 146, 173, 190-192, 194, 198, 214, 217-219, 388-389
land plot 225-226, 239, 472
land relations 130-133, 135, 138, 195, 201, 207, 218
land share 139, 222, 225-226, 239
Land Sharing 239
land tenure 140-141, 146, 322
land use 130-134, 138-139, 141, 146, 189, 191, 219, 222, 362
liability 164, 170-171, 176-181, 467
liberalization 28, 35, 89-91, 99-100, 104, 201, 319-320, 329, 336, 362, 367-370, 372, 392, 436, 438
livestock 65, 68-69, 71, 97, 110, 150, 192, 196, 201, 207, 242, 244, 247-248, 351, 353-354, 365-366, 371-372, 385, 388, 405, 428, 430-437, 470, 475, 487-489, 492-493, 495-497
LKD 493, 495, 497, 499
logistics costs 105, 119-121, 127

M

Macronutrients 70, 275
market niche 444, 446, 448-449, 455, 457-459, 463
Material and Technical Resources 310, 317
Material Flow Analysis 147, 149, 159, 169
Meher (Main) Season 442
member of a cooperative 466-467, 480
micronutrients 245, 248, 275
minor components 298, 302, 306-307, 314-315
Minor Food Components 297, 317
mobile shop 311, 313
mobile workshop 310, 315, 317
monasteries 443-445, 447-449, 452, 454-459, 463

N

National Bank for Agriculture and Rural Development Consultancy Service (NABCONS) 402, 424
national security 62, 88, 106, 122, 171, 325, 337, 344-346
nature-likeness technologies 76, 78, 86
NGOs 483, 490, 499
NTFS 488, 499
NTPFPs 489, 499
nutrients 65, 206, 242-245, 249-251, 253, 266, 269, 275-276, 300, 303, 314, 317, 339
Nutrition Structure 251
nutrition value 257-258, 260, 265, 267, 269, 276, 305

O

OECD 18, 20, 22, 24-25, 32, 88, 91-92, 94-97, 354, 387, 401-402
organic agriculture 64-66, 76, 255, 276
organic products 63, 65-66, 75-76, 78, 81, 86
Orthodox Monastery 463

P

palm oil 277-278, 280-283, 286-288, 290-291, 294, 321, 325, 369, 401
Participatory Poverty Assessment (PPA) 482
PDCAAS 263-264, 276
pesticides 44, 62, 64-65, 76, 81, 86, 140, 259, 322, 447, 475
phytochemical potential 256, 259, 261-262, 265
PNPM 490, 499
post-Soviet space 205-206, 214
PRA 482, 499
Principle of Maximum Entropy 128
privatization 201, 203, 215-222, 225-226, 228, 232, 234, 239-240
Product Differentiation 464
Productivity Stimulant 62
Proper nutrition 67, 86, 242-243, 253
protein digestibility-corrected amino acid score 263
Protein Quality 263, 276
proteins 243, 251, 253, 255-256, 258, 261, 263-265, 275-276, 305

Q

quercetin 307, 310, 317
quintal 442

R

rainfall variability 386-390, 392-393
rational management decisions 79, 85-86
RMU 495, 499
rural area 444-445, 448, 464
rural development 31, 81, 112, 402, 424-425, 433-434, 436, 443-445, 448, 457-459, 465, 490, 495, 497
rural territories 95, 136, 205, 438, 444, 465-466, 477
Russia's Far East 371-373, 376-377, 383

S

Safe Product 175, 188
Segmentation 446, 464
selenium 242, 244, 246-248
Share Contribution 467, 480
smallholder 90, 387, 426, 431, 436-437, 442
Smart Agriculture 77-78, 86
South-South and Triangular Cooperation 402, 406, 424
soybeans 23, 27, 253, 280, 321-322, 325-326, 328, 330, 371-373, 375
Specialized Fortifiers 251
Spot Privatization 240
stability 50, 91, 97, 106, 149, 179, 243, 278, 286-288, 298, 320, 339-340, 343-344, 351, 360, 386, 388, 390, 399, 474, 492
State of Food Insecurity (SOFI) 424
state policy 64, 214, 218, 245, 253
state support 66, 68, 80, 88-91, 94-97, 99-100, 104, 146, 201-202, 205, 298, 341, 343, 348-349, 354, 356, 369, 372, 377, 446, 449, 452
Stolypin 189-191, 193-194, 196, 205-206, 214
Stolypin's Agrarian Reform 214
Sub-Saharan Africa 23, 32, 153, 384-393, 400-401, 405-406, 408-409, 413, 416-418, 424
support measures 20, 91, 365-369
Sustainable Agricultural Intensification 426, 442
sustainable development 18, 55, 66, 99, 136, 163, 173, 215-216, 298, 315, 355, 368, 384, 401, 424-428, 434, 436, 438, 442, 465-466, 475-477, 486
Sustainable Development Goals (SDG) 401, 424, 427

T

Tariff Escalation 328, 336
Taymate cooperative 475
tort 170, 176-181, 188
Toudarte cooperative 476
trade in food 88, 90, 93, 99-100, 172-173, 243, 320, 330, 346-347, 368, 375

trade liberalization 28, 89-91, 100, 104, 319-320, 329, 336, 362, 367, 369-370, 392, 436
Trade Protectionism 336
trade war 336, 344
trans-isomers of fatty acids 278
trans-isomers of unsaturated fatty acids 279-280, 295
Trans-Pacific Partnership 367, 382

U

unclaimed land shares 215, 227-229, 231

V

vegetable raw materials 296, 298, 303, 307, 315
Vernadsky Law 12
virtual water 391-392, 399
vitamins 243, 245, 248-249, 251, 253, 255, 258, 266, 275, 283, 297-299, 303, 307, 310

W

water resources 296, 353-354, 384-385, 387-388, 391, 399, 447, 472
water scarcity 373, 387, 391-392, 398-399
World Food Market 63, 66, 72, 78, 80, 86, 348
World Trade Organization 18, 90, 112, 128, 173, 201, 320, 348, 361-363

Purchase Print, E-Book, or Print + E-Book

IGI Global's reference books are available in three unique pricing formats:
Print Only, E-Book Only, or Print + E-Book.
Shipping fees may apply.
www.igi-global.com

Recommended Reference Books

ISBN: 978-1-5225-7727-0
© 2019; 335 pp.
List Price: $195

ISBN: 978-1-5225-5577-3
© 2019; 291 pp.
List Price: $195

ISBN: 978-1-5225-5291-8
© 2019; 252 pp.
List Price: $175

ISBN: 978-1-5225-5766-1
© 2019; 501 pp.
List Price: $195

ISBN: 978-1-5225-5909-2
© 2019; 308 pp.
List Price: $195

ISBN: 978-1-5225-8027-0
© 2019; 337 pp.
List Price: $205

Do you want to stay current on the latest research trends, product announcements, news and special offers?
Join IGI Global's mailing list today and start enjoying exclusive perks sent only to IGI Global members.
Add your name to the list at **www.igi-global.com/newsletters.**

Publisher of Peer-Reviewed, Timely, and Innovative Academic Research

www.igi-global.com   Sign up at www.igi-global.com/newsletters   facebook.com/igiglobal   twitter.com/igiglobal   linkedin.com/igiglobal

Ensure Quality Research is Introduced to the Academic Community

Become an IGI Global Reviewer for Authored Book Projects

The overall success of an authored book project is dependent on quality and timely reviews.

In this competitive age of scholarly publishing, constructive and timely feedback significantly expedites the turnaround time of manuscripts from submission to acceptance, allowing the publication and discovery of forward-thinking research at a much more expeditious rate. Several IGI Global authored book projects are currently seeking highly-qualified experts in the field to fill vacancies on their respective editorial review boards:

Applications and Inquiries may be sent to:
development@igi-global.com

Applicants must have a doctorate (or an equivalent degree) as well as publishing and reviewing experience. Reviewers are asked to complete the open-ended evaluation questions with as much detail as possible in a timely, collegial, and constructive manner. All reviewers' tenures run for one-year terms on the editorial review boards and are expected to complete at least three reviews per term. Upon successful completion of this term, reviewers can be considered for an additional term.

If you have a colleague that may be interested in this opportunity, we encourage you to share this information with them.

IGI Global Proudly Partners With eContent Pro International

Receive a 25% Discount on all Editorial Services

Editorial Services

IGI Global expects all final manuscripts submitted for publication to be in their final form. This means they must be reviewed, revised, and professionally copy edited prior to their final submission. Not only does this support with accelerating the publication process, but it also ensures that the highest quality scholarly work can be disseminated.

English Language Copy Editing

Let eContent Pro International's expert copy editors perform edits on your manuscript to resolve spelling, punctuaion, grammar, syntax, flow, formatting issues and more.

Scientific and Scholarly Editing

Allow colleagues in your research area to examine the content of your manuscript and provide you with valuable feedback and suggestions before submission.

Figure, Table, Chart & Equation Conversions

Do you have poor quality figures? Do you need visual elements in your manuscript created or converted? A design expert can help!

Translation

Need your documjent translated into English? eContent Pro International's expert translators are fluent in English and more than 40 different languages.

Hear What Your Colleagues are Saying About Editorial Services Supported by IGI Global

"The service was very fast, very thorough, and very helpful in ensuring our chapter meets the criteria and requirements of the book's editors. I was quite impressed and happy with your service."

– Prof. Tom Brinthaupt,
Middle Tennessee State University, USA

"I found the work actually spectacular. The editing, formatting, and other checks were very thorough. The turnaround time was great as well. I will definitely use eContent Pro in the future."

– Nickanor Amwata, Lecturer,
University of Kurdistan Hawler, Iraq

"I was impressed that it was done timely, and wherever the content was not clear for the reader, the paper was improved with better readability for the audience."

– Prof. James Chilembwe,
Mzuzu University, Malawi

Email: customerservice@econtentpro.com   www.igi-global.com/editorial-service-partners

 **Celebrating Over 30 Years** of Scholarly Knowledge Creation & Dissemination

www.igi-global.com

InfoSci®-Books

A Database of Over 5,300+ Reference Books Containing Over 100,000+ Chapters Focusing on Emerging Research

GAIN ACCESS TO **THOUSANDS** OF REFERENCE BOOKS AT **A FRACTION** OF THEIR INDIVIDUAL LIST **PRICE**.

InfoSci®-Books Database

The **InfoSci®Books** database is a collection of over 5,300+ IGI Global single and multi-volume reference books, handbooks of research, and encyclopedias, encompassing groundbreaking research from prominent experts worldwide that span over 350+ topics in 11 core subject areas including business, computer science, education, science and engineering, social sciences and more.

Open Access Fee Waiver (Offset Model) Initiative

For any library that invests in IGI Global's InfoSci-Journals and/or InfoSci-Books databases, IGI Global will match the library's investment with a fund of equal value to go toward **subsidizing the OA article processing charges (APCs) for their students, faculty, and staff** at that institution when their work is submitted and accepted under OA into an IGI Global journal.*

INFOSCI® PLATFORM FEATURES

- No DRM
- No Set-Up or Maintenance Fees
- A Guarantee of No More Than a 5% Annual Increase
- Full-Text HTML and PDF Viewing Options
- Downloadable MARC Records
- Unlimited Simultaneous Access
- COUNTER 5 Compliant Reports
- Formatted Citations With Ability to Export to RefWorks and EasyBib
- No Embargo of Content (Research is Available Months in Advance of the Print Release)

*The fund will be offered on an annual basis and expire at the end of the subscription period. The fund would renew as the subscription is renewed for each year thereafter. The open access fees will be waived after the student, faculty, or staff's paper has been vetted and accepted into an IGI Global journal and the fund can only be used toward publishing OA in an IGI Global journal. Libraries in developing countries will have the match on their investment doubled.

 To Learn More or To Purchase This Database:
www.igi-global.com/infosci-books

eresources@igi-global.com • Toll Free: 1-866-342-6657 ext. 100 • Phone: 717-533-8845 x100

Publisher of Peer-Reviewed, Timely, and Innovative Academic Research Since 1988

IGI Global's Transformative Open Access (OA) Model:
How to Turn Your University Library's Database Acquisitions Into a Source of OA Funding

In response to the OA movement and well in advance of Plan S, IGI Global, early last year, unveiled their OA Fee Waiver (Offset Model) Initiative.

Under this initiative, librarians who invest in IGI Global's InfoSci-Books (5,300+ reference books) and/or InfoSci-Journals (185+ scholarly journals) databases will be able to subsidize their patron's OA article processing charges (APC) when their work is submitted and accepted (after the peer review process) into an IGI Global journal.*

How Does it Work?

1. When a library subscribes or perpetually purchases IGI Global's InfoSci-Databases including InfoSci-Books (5,300+ e-books), InfoSci-Journals (185+ e-journals), and/or their discipline/subject-focused subsets, IGI Global will match the library's investment with a fund of equal value to go toward subsidizing the OA article processing charges (APCs) for their patrons.

   Researchers: Be sure to recommend the InfoSci-Books and InfoSci-Journals to take advantage of this initiative.

2. When a student, faculty, or staff member submits a paper and it is accepted (following the peer review) into one of IGI Global's 185+ scholarly journals, the author will have the option to have their paper published under a traditional publishing model or as OA.

3. When the author chooses to have their paper published under OA, IGI Global will notify them of the OA Fee Waiver (Offset Model) Initiative. If the author decides they would like to take advantage of this initiative, IGI Global will deduct the US$ 1,500 APC from the created fund.

4. This fund will be offered on an annual basis and will renew as the subscription is renewed for each year thereafter. IGI Global will manage the fund and award the APC waivers unless the librarian has a preference as to how the funds should be managed.

Hear From the Experts on This Initiative:

"I'm very happy to have been able to make one of my recent research contributions, 'Visualizing the Social Media Conversations of a National Information Technology Professional Association' featured in the *International Journal of Human Capital and Information Technology Professionals*, freely available along with having access to the valuable resources found within IGI Global's InfoSci-Journals database."

– **Prof. Stuart Palmer**, Deakin University, Australia

For More Information, Visit: www.igi-global.com/publish/contributor-resources/open-access or contact IGI Global's Database Team at eresources@igi-global.com.

Printed in the United States
By Bookmasters